POSTHARVEST BIOLOGY

POSTHARVEST BIOLOGY

Stanley J. Kays

The University of Georgia
Athens

Robert E. Paull

University of Hawaii
Honolulu

Published by Exon Press
Athens, GA

Exon Press
P.O. Box 80803
Athens, Georgia 30608-0803
United States

PREFACE

Postharvest Biology represents the second edition of *Postharvest Physiology of Perishable Plant Products* with the new title more accurately reflecting the subject matter dealt with in the text. *Postharvest Biology* has the same organizational structure of the first edition; however, a new chapter (Molecular Genetics, Signal Transduction and Recombinant DNA) has been added, reflecting the extensive changes that have occurred in plant biology over the past decade. In addition, several appendices have been included to provide detailed reference information that would otherwise be too cumbersome to be included within the body of the text. The size of the task of revising the diverse topics dealt with in the text and the increasing level of sophistication of the technology lead me to solicit the participation of Dr. Robert E. Paull in the revision. Dr. Paull is well know for his research in postharvest biology and has been an invaluable addition, facilitating the evaluation and interpretation of recently published results.

There are two primary reasons for offering a broad overview of postharvest biology. First, many critical functions that are operative during plant growth shift after harvest when the input of energy, water and other essential requisites ceases in most products. Consequently, postharvest biology differs substantially from what is covered in a typical plant physiology/biology text. Second, the value of the majority of live agricultural plant products approximately doubles between harvest and retail sales. The total cost of losses occurring late in the production–harvest–marketing sequence are substantially greater than those incurred during the production phase. Likewise, inputs essential to prevent or minimize these losses are often only a minute fraction of the overall costs for the product. Therefore, a better understanding of the functional processes after harvest makes both biological and economic sense.

This book focuses on the environmental factors modulating and functional processes controlling physical and chemical changes in live plants and plant products after harvest. Our objective is to provide a concise overview of the theoretical principles and processes governing these changes. Examples include agronomic crops, fruits, nuts, vegetables, flowers, woody ornamentals, seeds and other forms of plant propagules and turf grasses, with examples ranging from intact plants to pollen. Emphasis is placed on the basic principles operative rather than on detailing optimum storage and handling conditions for individual crops. The latter information can be found in several excellent reference books cited in the text. A solid understanding of the basic processes operating should provide the reader with an understanding of the rationale for specific commodity recommendations and the potential to anticipate appropriate conditions for lesser known or new products for which handling and storage recommendations are not yet available.

This book is intended to be a useful overview reference for professionals in botany, crop science, forestry, food science, horticulture, pest management and agricultural and biological engineering. It can also serve as a suitable textbook for junior/senior-level undergraduates and for first year graduate students in these academic disciplines. Due to the diverse backgrounds of undergraduate students taking a course in postharvest biology, the text is written so that only introductory botany and chemistry are essential prerequisites. The text can complement a plant physiology course in the student's normal curriculum.

There is generally more material provided than can be covered in a typical semester course. Based on the varied background of the students, instructors may choose to select certain portions of the text and substitute material form their own field of interest. For example, chapter 2 describes the structure of individual products from distinct organs to the subcellular level. If this information is covered in prerequisite courses, other information may be emphasized. Sections from the text not covered in a course provide a valuable reference source for the students.

References are presented at the end of each chapter and are separated into two groups: research papers cited in the text, and books and reviews for additional reading. The research papers cited (indicated in the text as superscript numerals to minimize disrupting the narrative) represent only a small selection of the published information available. It would be impossible to include all references and still provide a readable text. Rather, specific references are included to provide examples of a particular concept. It is hoped that handling the references in this manner will enhance the reader's access to the literature and interest in pursuing the subject further.

The development of this text required the effort and expertise of a number of individuals. We are grateful to Drs. Rob M. Alba, Randolph M. Beaudry, Virginia L. Butler, John Fellman, Glenn A. Galau, Judy Jernstedt, Edward S. Law, Yuriy Posudin, Albert C. Purvis, Mikal E. Saltveit, Jörg Schönherr and Ronald W. Walcott for reviewing individual chapters of the manuscript, Charles C. Doyle for editing the entire text, and to David H. Simons for the many contributions he made to the original edition. We are also indebted to the many colleagues that have provided photographs and have consented to having their work included in the text. Special thanks goes to Asha R. Kays for revising and coordinating the illustrations and to Betty Schroeder for assistance with many aspects of the finalization of the text.

CONTENTS

1

THE SCIENCE AND PRACTICE OF POSTHARVEST PLANT BIOLOGY

1. NATURE OF THE DISCIPLINE

The handling, storage and marketing of plants and plant parts is one of the major preoccupations of human societies. It is an integral component of the human food supply chain and is, therefore, as diverse as are the cultures and the foods used. This diversity is increased by the use of perishable plants and their parts for decorative purposes and for modification of the environment. A major concern with the handling of perishable plant material is the maintenance of quality; therefore, it is necessary to consider not only the nature of the plant material but also the technological and economic aspects associated with getting the product to the consumer.

Postharvest biology is a division of plant biology dealing with the study of plant material after it has been harvested. Postharvest biology is concerned with plants or plant parts that are handled and marketed in a living state. This includes grains, fruits, vegetables, cut flowers and foliage, nursery products, turf, vegetative propagules, seeds and edible fungi. Some of these products are intact plants, but the majority are isolated plant parts. Included in the range of products are all the various plant organs, and there are examples of all stages of development from germinating seed and juvenile shoots to mature plants, storage organs and dormant seeds.

Postharvest biology deals with the time period from harvest or removal of the plant from its normal growing environment until the time of ultimate utilization, deterioration or death. When preharvest and harvesting factors have a direct influence on postharvest quality, these become vital components of the complete postharvest picture.

Just what is meant by utilization varies with the product. For seeds and cuttings used for propagation and transplants, utilization is planting; for cut flowers and foliage, it is displaying and maintaining them for decorative purposes. Continued development may be desired during utilization, while senescence and death typically occur towards the end of their useful life. However, for the majority of products, utilization is the time of death due to consumption or processing. Unfortunately, much of the harvested plant material never reaches this point of utilization but is discarded because of deterioration due to senescence, stress responses, pathogen activity, insect attack or mechanical damage.

2. FUNDAMENTAL NATURE OF PERISHABLE PLANT PRODUCTS

The most important characteristic of perishable plant products is that they are alive and, therefore, continue to function metabolically. However, their metabolism is not identical with that of the parent plant growing in its original environment, since the harvested product is under varying degrees of stress.

In the case of severed plant parts, harvesting, packaging and handling interfere with or eliminate entirely some of the essential requisites for plant growth. Water and mineral nutrient supply from the soil are eliminated as is commonly the flow of carbon and energy from photosynthesis. Exposure to light is changed and in many cases virtually eliminated, and the availability of oxygen and the concentration of carbon dioxide are altered. The process of harvesting often results in substantial wounding, while packaging and transport can cause further mechanical damage. The gravitational orientation of the harvested produce is commonly altered. It is frequently subjected to physical pressure, to a substantially altered temperature regime, and to an undesirable gaseous environment. For intact plants such as containerized nursery products and transplants, the stresses incurred may be less extreme; however, the same factors are involved and the plants respond in a similar manner. All in all, the plant material is typically subjected to very harsh treatment during its postharvest life.

Living organisms respond in a multitude of interacting ways to counter the effects of stresses to which they are exposed to maintain as nearly as possible a homeostatic condition. If the stress is so severe that it exceeds the physical or physiological tolerance of the organism, then death occurs. In this respect plants and humans react similarly. It is useful to look at the handling of a typical product such as lettuce* to recognize what is being done at each stage.

In harvesting lettuce, a substantial wound is inflicted by severing the stem. The supply of water from the roots is eliminated, and the detached leafy portions of the plant continue to lose water. Unless this water loss from the leaves and wounded tissue is inhibited, the inevitable consequence is loss of turgor pressure and wilting. Harvest also eliminates the supply of mineral nutrients essential for metabolic activity. Thus the harvested product is now dependent on recycling of those nutrients already present.

With the handling of lettuce, whether during harvesting and marketing or in a retail store, leaf breakage is common. The removal of outer leaves results in more wounds; and additional breakage of the leaves occurs when the heads are packed tightly into containers. Some heads are placed upright whereas others are turned upside down, a distinctly abnormal gravitational orientation. Packing into the container effectively eliminates photosynthetic light and, therefore, the plants' external source of energy. In the confined environment within the container, gas exchange is restricted.

While the lettuce head is injured and its circumstance radically altered in contrast to the preharvest state, it is still alive and continues to respond metabolically, adjusting to its new set of conditions. In fact, its rate of respiration is increased due to the treatment it has received. This situation in turn may more rapidly deplete the oxygen concentration within the container and increase that of carbon dioxide. Both of these changes will precipitate further changes in the metabolism of the lettuce. Injury also causes the tissue to produce ethylene, which, if allowed to accumulate in the confined space of the container, will accelerate the rate of senescence of the lettuce itself.

Respiration results in the release of energy as heat, causing the temperature of the lettuce

---

*The Latin binomials for plants cited in the text by their common name are listed in the Species Index. *Hortus Third* and *Cultivated Vegetables of the World*[31] are used as the authority for the nomenclature when applicable.

to rise, accelerating deteriorative processes such as water loss, senescence, and rate of growth of pathogens. The latter may find an ideal environment with the rising temperature, high humidity and ready access to the host tissue through wounds. In addition, during transport to the consumer the harvested produce is subject to physical pressure, vibration, and bruising and frequently to temperatures and humidities which accelerate senescence. After all this, the product is presented in retail markets as "farm fresh" produce.

Nevertheless, perishable produce must be harvested and moved through some handling and transportation system to its site of utilization while maintaining its live status and condition. There is an inherent conflict between the requirements of human societies and the biological nature of the harvested perishable produce. Harvesting for many plant products irreversibly initiates senescence and the eventual death of the plant part, yet living plant material must be moved from sites of production to sites of consumption that are commonly distant from each other. Even in less extreme cases, containerized plants cannot be moved from their growing environment to the suboptimal environments of transport vehicles and of most marketing situations without some deterioration occurring.

Given this inherent conflict between our need to harvest plant parts, which precipitates their death, and our need to keep them alive, we must accept compromises. Compromises are an essential element at each level of postharvest handling of perishable plant products. They may take the form of compromises in temperature to minimize metabolic activity while avoiding chilling injury, or in oxygen concentration to minimize aerobic respiration yet avoid anaerobic respiration, or in the tightness of packing to minimize pressure damage while avoiding vibration damage, and so on.

Understanding the nature of the harvested product and the effects of our handling practices are essential for arriving at the most appropriate compromises to maintain optimum condition of the produce. There is unfortunately no fixed recipe or solution for each product; rather the most appropriate practices must be worked out by the individual operator for each particular situation, taking into account physiological, physical, personnel, and economic factors.

3. EVOLUTION AND HISTORY OF POSTHARVEST STORAGE

An evolutionary analysis of storage can be approached from several different biological positions. In some cases animals store food internally. For example, prior to their winter hibernation, bears indulge in luxury consumption of food that is stored as fat for subsequently utilization as an energy source. In many cases, however, food is stored external to the animal and typically, at least from a historical context, in a non-processed and readily perishable form.

A surprisingly large number of animal and insect species utilize some form of food storage.[67] For some, the act of storage represents an instinctive behavioral process; however, for many others, storage is a learned response. Squirrels and chipmunks store food in underground caches for use during the winter months. Leopards often place for later use unconsumed portions of animals in crotches or branches of trees as protection from scavengers. Ants in the genus *Pogonomyrmex* collect and store seeds in their underground nests. Leaf cutter ants of the neotropics have evolved an even more complex system in which harvested leaves are used to culture fungi in specialized underground storage areas of the nest. The symbiotic relationship between the ants and microorganism is not new; most fungal growing ants appear to have clonally propagated the same fungal line for at least 23 million years.[17]

Only with man, however, has storage evolved to a highly complex level, and even this development is very recent on the evolutionary time scale. Storage was no doubt utilized by early hominids other than *Homo sapiens;* however, this has been little explored. As one analyzes our current knowledge of the evolution of man, it becomes apparent that development of the ability to store food represented an extremely important step in the evolutionary process. However, questions such as when did man or the predecessor of man first start storing food and what impact did storage have on survival potential, mobility, delineation of roles between the sexes, changes in diet, population sedentation, and the evolution of agriculture have not been adequately addressed. This deceptively simple concept of food storage has been intimately associated with the evolutionary development of mankind.

3.1. Prehistoric Period

Man evolved slowly over many millennia; however, during the past 5 million years, major changes leading to modern man occurred (Figure 1.1). Early hominids progressed through a series of evolutionary stages, our understanding of which is based upon a relatively small number of fossil remains.[34] The sequence for the emergence of modern man remains in a state of transition, with alterations being proposed virtually every few months. Current thinking is that modern man evolved in a sequence from *Ardipithecus ramidus* (syn. *Australopithecus ramidus*) to the genus *Australopithecus* and subsequently *Homo* with species-level stages of *H. habilis* → *H. erectus (H. ergaster)* → *H. heidelbergensis* → *H. sapiens*. The genus *Paranthropus* along with *H. rudolfensis, H. neanderthalensis* and *H. erectus* (in Southeast Asia and China) appear to have been side-branches in the evolutionary tree that subsequently became extinct. During the progression of these early hominids, there were periods when several species were present concurrently. In some instances species were separated by location (e.g., *H. neanderthalensis* in Europe and *H. erectus* in Asia); in others they appeared to cohabit the same geographical areas (e.g., *H. rudolfensis, H. habilis, P. boisei* and *P. robustus* are thought to have co-existed about 1.8 to 2 million years ago).

Based on mitochondrial DNA evidence, modern man evolved in Africa[15,16] and is of surprisingly recent origin (i.e., only 100,000 to 200,000 year ago).[63] *Homo sapiens* spread outward from Africa and diversified throughout the world, displacing other *Homo* populations from previous migrations.[1] What happened to these more archaic forms? A possible scenario is that modern man brought with him diseases, for which he had immunity, that decimated the existing populations.

The ability of early hominids to function on two feet (bipedalism) rather than on all fours evolved approximately 5 million years ago and was an extremely important evolutionary step. Subsequently these early predecessors of man migrated from the tropical rainforest to the savannah, a move that is thought to have occurred after *Ardipithecus ramidus*. The reason for the move is not known; however, the consequences were monumental. Not only the availability of food but the plant species that could be utilized changed radically. Seasonal variability in food was the result of alterations between wet and dry seasons, an essential element in the maintenance of a grassland. Food acquisition became more critical, and "feast and famine" cycles in food availability were common. Selection conditions for developing the ability to store food, even in the most primitive sense (i.e., gathering more than was immediately needed with the intent of using it later), were operative. Many anthropologists believe that during this period important changes in diet occurred, such as the introduction of seeds to supplement fruits, roots, insects and meat. Dried seeds and nuts store well, whereas fruits and meat rot quickly. Thus the ability to gather more durable foods and ration them as needed helped to even out fluctuations in food supply.

More distinct differences in the roles between males and females probably began to emerge 2 to 2.5 million years ago with the advent of hunting (Figure 1.1), a technological advance that probably evolved gradually from the scavenging of animal carcasses killed by other

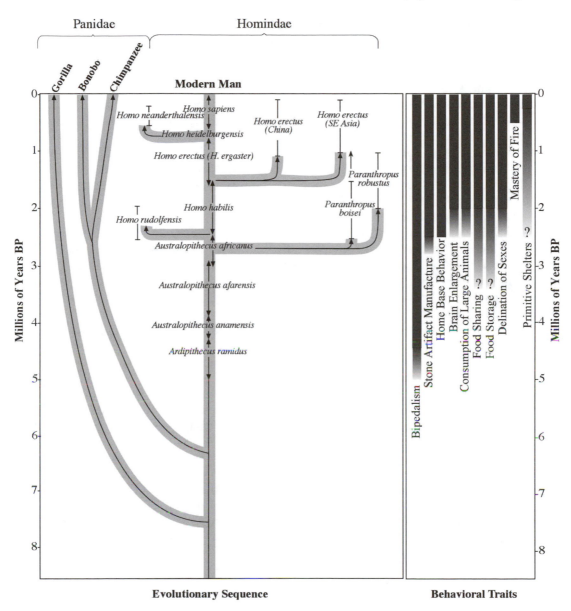

Figure 1.1. Timeline for the emergence of man with key prehistoric behavioral, anatomical and technological advances.[8]

animals or natural causes. During this period, males increasingly concentrated on hunting while females gathered fruit, nuts, seeds, roots and other plant material. While the predecessor of modern man is often depicted as a hunter, in reality plant products typically represented substantially more than 50% of the diet. Since plant remains rarely survive, an indication of the importance of vegetable materials in the diet comes from recent hunter-gatherers.[6] Estimates of up to 80% of the diet being of plant origin have been made. Therefore, women in hunter-gather societies were the first to be involved in postharvest handling of plant material as its acquisition and storage was within their sphere of responsibility.

The impact of storage on early hunter-gatherer societies depended in part upon the amount of product stored and its relative importance to survival. Trestart[65] separates hunter-gatherers into two groups, storing and non-storing, the former utilizing extensive storage of seasonal food resources and the latter, only short-term storage as the food was used within a very short time interval. The development of large-scale storage required the following conditions to be met: an abundance of food, seasonality in its availability, an efficient means of food collecting, and knowledge of food storage techniques. Storing hunter-gatherers, therefore, were absent from desert areas where food resources are not abundant and from tropical areas where there is no marked seasonality in food availability. Storing, therefore, tended to be found in medium and high latitudes.

A diverse range of foods were stored for short intervals by both storing and non-storing groups; however, long-term storage involved predominantly plant material (e.g., nuts, seeds) with the exception of dried fish and, in more arctic areas, frozen meat. Most animal material either spoils rapidly or requires such an extensive investment in processing that only small quantities were available (e.g., pemmican produced by some North American Indian tribes).

The ability to successfully store substantial food resources had a pronounced impact on groups practicing storage.[65] Large reserves allowed a sedentary life style in which the group stayed in one location near their stored reserves. In contrast, non-storing hunter-gatherers were nomadic, following the changing availability in resources. Extensive food storage also enabled the population to stabilize at a higher density. Liebig's Law stipulates that the smallest quantity of resources available during the year, rather than the yearly total, will dictate population size. Finally, storage led to socioeconomic inequities in that some individuals/groups were better able to acquire resources than others. Surpluses allowed the presence of nonproductive individuals (e.g., priest, bureaucrats) and the emergence of class distinctions, a condition that evolved rapidly after the advent of agriculture. Sedentarism was also a prerequisite for the accumulation of material goods (e.g., baskets, pots, storage pits, granaries) that do not lend themselves to the mobility required by nomadic societies. Thus, storage led to and was essential for the development of agriculture, which in turn paved the way for the development of civilization, class society and the state.

The first technological advances that have withstood time are stone tools which date to approximately 2.4 million years ago.[64] Initially these were very primitive; basically stone flakes, called choppers, that were chipped from hand sized stones and used for butchering animal carcasses (Figure 1.1). By 1.4 to 1.7 million years ago, these had progressed to bifacial tools, sharpened on two sides, yielding axes and cleaves (called Acheulean tools) which could have been used for hunting and other activities.*

The mastery of fire was also an extremely significant technological advance. While there is circumstantial evidence for the use of fire by *Homo erectus* 1.4 to 1.6 million years ago,[12] the first hearths date from only 400,000 to 500,000 year ago.[55,72] Fire provided light, heat and protection from carnivores, and eventually led to altering the state of foods, increasing their palatability and nutritional value.

The use of large animals as part of the diet solidified behavioral changes such as food sharing and gathering around a home base. In addition to caves, primitive shelters built of branches and grass probably began to be used, although the first definitive evidence dates to only 300,000 to 400,000 years ago (i.e., postholes which apparently held supports for relatively large shelters[38]).

When did the first primitive forms of food storage begin? Based upon behavioral changes that evolved with the consumption of large animals (e.g., food sharing and the use of a home base), storage of vegetable and other food material must have also been prevalent at least 2 million years ago (Figure 1.1). How much earlier hominids gathered more than they intended to

---

*Recent evidence from China dates Acheulean tools there to 800,000 years ago.[71]

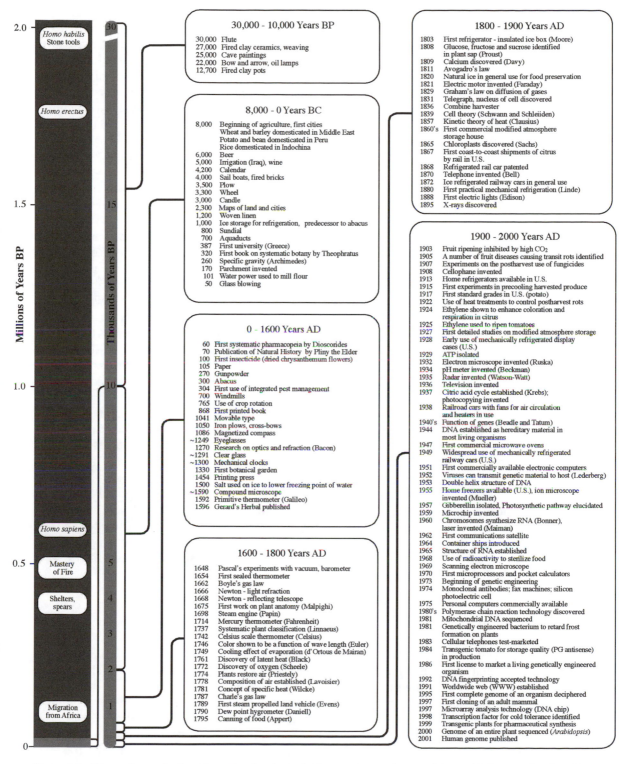

Figure 1.2. Chronology of scientific and technological advances impacting postharvest biology and technology.

use at a particular moment for later use is not known. The migration from the tropical rain forest to the savannah must have set the stage for food storage in its most primitive sense. There is no evidence of food gathering containers being used by early hominids; however, materials that might have been used, such as rawhide, animal organs, and plant materials like hollow, dried fruits (similar to the gourd vessels that are used to this day), are themselves highly perishable. Materials utilized to facilitate food gathering/storage would have progressed in complexity from animal skins to woven plant material and eventually pots. The first fired clay pots date to only 12,000 to 13,000 year ago,[62] although fired clay figurines from 27,000 to 28,000 years ago have been found.[68]

Around 15,000 years ago the last major glacial period was drawing to a close[56] marking the end of a long series of extreme climatic changes. Diverse changes in the flora began to occur at the close of the ice age, and these changes were largely completed by 8,000 years before the present. These changes set the stage for the first steps toward agriculture, which occurred in an area of the Near East called the Fertile Crescent.

The impetus resulting in the development of agriculture is not known; however, the impact of agriculture on the development of civilization was monumental.[10] Man was no longer forced to adapt to his environment; instead he began to alter and shape it. In a remarkably short period of time, agriculture became the dominant mode of subsistence, opening the door for the development of all of the complex societies and civilizations that have followed.

Storage must have played a critical role in the development of agriculture. Seeds of wheat and other early annual domesticates had to be stored in a manner that would maintain their viability until the next growing season. In addition, grain to be utilized for food had to be stored in such a way as to minimize losses. Rotting due to excess moisture and losses due to insects and rodents, typical problems confronted by early farmers, were well documented during the Roman era.

While storage must have evolved very early in the development of hunter-gatherer societies, with the onset of agriculture its importance became much more critical. Technological advances such as pottery which could be used as storage vessels began to occur, and this progress in turn affected the subsequent development of agriculture. The sequence of plant domestication in the Near East underscores the importance of storage and its subsequent effect on agricultural development. Wheat was domesticated at least 9,000 years ago, followed by chickpeas, lentils, and fava beans during the next millennium (Figure 1.2). All were crops that were relatively stable in storage if handled properly. Later species to be domesticated in the centuries following (around 6,000 years ago) were almonds, pistachios, walnuts, figs, dates, apricots, grapes, and olives. These too were crops which could be stored successfully for extended periods in the natural or processed forms.

3.2. Early Advances in Storage Technology

Advances in technology occurred very slowly over the past 2.5 million years. It was really not until the 1500's that there was a marked acceleration in the rate of technological developments (Figure 1.2) and the beginning of an exponential growth in our understanding of the world. Prehistoric advances in technology enhanced the ability of humans to successfully store harvested plant products. Containers made of plant (gourds, baskets) and animal (bladders, skins) material were probably in widespread use during the last ice age; however, due to their perishable nature none are known to have survived. The ability to manipulate reeds, grasses, bark and other plant material into mats, baskets, nets, cords, braids and eventually cloth greatly enhanced the capacity of modern man to adapt,[27] facilitating food gathering and storage. Basket-making, for example, was a highly developed art by 9,000 years ago with baskets

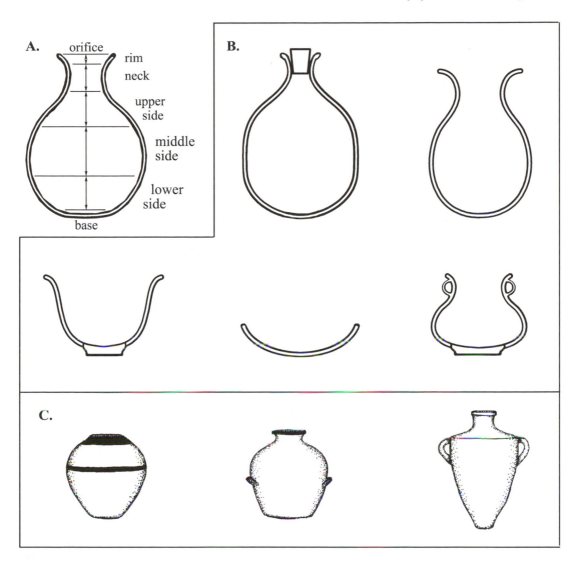

Figure 1.3. Pottery was an extremely important technological innovation. The size and shape of pots reflect their diverse functions [e.g., dry storage (grain), wet storage (water, alcoholic beverages), transport, cooking, fermentation, processing, serving, individual consumption and washing]. (A) By making very detailed measurements of the pot's morphology (e.g., volume, height, shape, orifice diameter), archeologists can categorize pots, even fragments of pots. Examples of several shape types are represented (B). Large wide-mouthed vessels were routinely used for cooking, washing and processing of foodstuffs while those with narrow necks and orifices held liquids and grains that were small enough to pour. Small orifices allowed the use of tight fitting lids or stoppers that sealed the contents, creating a modified atmosphere within the container and eliminated access by herbivores. Examples of Palestinian pottery (C), from left to right, are from the Chalcolithic period, the early Bronze age, and late Bronze age (*after Gonen*[26]).

being used both for gathering seeds, nuts, fruits, and other vegetable material and for storage until their subsequent use.

The ability to make pots of fired clay was an extremely significant advance in storage capability that appears to have been invented at different times in a number of societies around the world.[29] The earliest pots are from the end of the Pleistocene and earliest Holocene periods; e.g., the earliest Japanese pottery dates to about 12,700 years ago.[3] Pottery does not appear to have emerged with the onset of agriculture as once thought. Rather, the use of pots by foraging societies predates agricultural societies, although pottery was far more prevalent in the latter.[7] Pottery allowed stored products to be sealed in containers, significantly decreasing the chances of loss (Figure 1.3). By 6,500 years ago, pottery had become common, and specialized craftsmen made a wide range of pots for various uses (e.g., storage, cooking, processing, transportation). Without successful storage of surplus food, labor specialization could not have occurred and the subsequent development of arts and crafts would have stagnated.

Another early technique for the storage of grain was the underground pit or silo (Figure 1.4). These are known to have been used by pre-neolithic societies in the Middle East, 9,000 to 11,000 years ago. Early pits or silos were small and shallow, but with time the use of larger pits developed.[13,51] Around 2,500 years ago in northern China, fresh vegetables were stored stacked in trenches and covered with earth and straw.[57] By the Roman era, underground silos were one of the major means of long-term storage of grain, and they continued to be one of the main storage facilities used until the early 19th century. Even today grain and other products may be stored in pits in the major grain producing and exporting countries of the world (Figure 1.4). Pits and other simple structures are also currently used in traditional societies for storage of a wide range of vegetable and even fruit products (Figure 1.5).

The earliest record of modified atmosphere storage of plant products comes from the Roman era. Varro gave a detailed account of the construction of underground grain storage pits and sealing them after they were filled. He also stressed the dangers of entering them too quickly after opening. Because entering the silos without adequate aeration could be fatal, the Romans devised the technique of lowering a burning lamp into the silo to detect "foul air." During storage, respiration of the grain decreased the oxygen and elevated the carbon dioxide concentrations, greatly decreasing the chances of insect and rodent losses. As today, techniques and variations of greater or lesser utility were advocated for storage of grain in the Roman era. Pliny recognized the importance of harvesting a crop for storage at the proper stage of ripeness and discussed various techniques used.[52] He relates that "some people tell us to hang up a toad by one of its longer legs at the threshold of the barn before carrying the corn into it." Even today, misconceptions and superstitions sometimes still influence how harvested products are stored.

Processing of plant products was also developed to facilitate storage. Domesticated fruits such as figs, grapes, apricots and many of those gathered from the wild were dried (Figure 1.6). In some instances, grains and nuts were parched, and grains and grapes were often fermented. The net effect was to enhance the storage potential of the products.

3.3. Developments Leading Up to Modern Storage Technology

The importance of the management of storage temperature, gas atmosphere, product moisture content, transportation, storage facilities, and insects and pathogens was appreciated to varying degrees by the time of the Roman Empire. While few of the basic principles governing these factors were understood, many were utilized to enhance the reliability and quality of stored products.

3.3.1. Storage Temperature

By the beginning of the Roman Empire, the importance of refrigeration was known but not widely utilized for harvested products. Varro discusses the undesirable effect caused by respi-

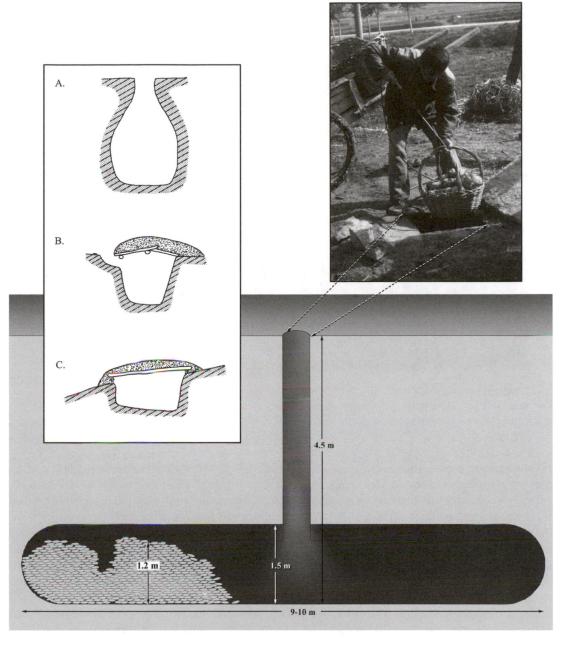

Figure 1.4. A wide range of relatively simple storage structures have been utilized by man over the centuries. Storage pits have been described from the later stages of the Paleolithic period (Shanidar cave).[58,59] Storage pits are found with increasing frequency at sites dating from 9000–7000 BC and some display distinct technological improvements (e.g., plastered walls).[49] Taylor[60] suggests that the "Neolithic Revolution"—the time period when man began to make the transition from a hunter-gatherer subsistence—could with equal justification be called the "storage revolution." Storage facilities can be separated into above and below ground structures. Below ground structures were often preferred due to the simplicity of construction and greater security, especially during more turbulent periods, since the store can be more easily hidden. Shown are examples of grain (A) and potato (B and C) storage pits[6] and present day sweetpotato storage in the Pingyin area of China.

Figure 1.5. Structures, some of which have undergone little technological improvement since their use 8,000 to 10,000 years ago, remain widely used by farmers in many areas of the world today. For example, a traditional potato store, pictured here, is currently used by farmers in the Puma of Southern Peru.[54]

ratory heating during prolonged storage of olives and in grain stored before being sufficiently mature. The importance of cool temperatures for the storage of apples was also well known. Initially caves and pits provided storage areas that were cooler than outside temperatures and were used. Because ice provided a means of keeping products much colder, wealthy citizens had cellars packed with ice or snow to help preserve luxury goods. Obtaining ice and storing it became a significant form of commerce. The earliest record of specialized structures constructed for the storage of ice is from Mesopotamia approximately 4000 years ago.[8] Ice gathering and use became a common practice in many countries, for example, in China approximately 3000 years ago. In Great Britain in the 18th and 19th centuries, large estates had underground ice houses, and various texts of the day gave directions on their placement[39] and construction.[53] By the early 20th century, ice cellars in shops and homes were common in Russia, where there were an estimated 10,000 in St. Petersburg alone.[35]

While the basic principles governing the effect of salt on the freezing point of water were established in 1550,* the application of salt on ice was used much earlier (i.e., Roman era) to cool wine and drinking water. Salt was to play an important role in the refrigerated transport of fruits and vegetables well into the 20th century.

Improvements in utilization of temperature management for harvested plant products progressed slowly until the Industrial Revolution.[28] In 1803, Thomas Moore invented the refrigerator, essentially an insulated ice-box.[40] By 1820, ice collected during the winter months in northern areas of the temperate zone came into general use for the preservation of foods

---

*Reportedly discovered by the Spanish physician Blasius Villafrance.

(A)

(B)

Figure 1.6. The mural above depicts: (A) storing the harvest; and (B) a scribe checking the storage of raisins. These were recorded in hieroglyphics found in the Egyptian tomb of Beni Hasan dating from around 2500 years ago (*from Newberry*[42]).

(Figure 1.7). A thriving ice business developed in the northeastern United States; shipments were made to Havana, the West Indies, Rio de Janeiro, Ceylon, Bombay, Madras, Batavia, Manila, Singapore, Mauritius and Australia.[19] A key figure in the ice trade was Frederick Tudor, who made a fortune shipping ice from New England to the tropics,[24] and returning with cotton, jade and other goods. The business was particularly lucrative during the American Civil War (1861–65) when the South's cotton production was severely curtailed. The ships' holds were insulated with sawdust to minimize the loss of ice during transit. Ice shippers also exported luxury products such as apples from North America that in India sold for as much as 50 to 75 cents each—a remarkable price in the mid-1800's.[48]

Ice was used for refrigeration in specially constructed railway cars as early as 1868. The cars were insulated, and galvanized-iron tanks held a salt and ice mixture for cooling.[60] The tanks were subsequently replaced with V-shaped boxes each containing around 3,000 pounds of ice that transversed the length of the car. By 1872, the refrigerated railroad car came into general use in the United States.

The availability and price of natural ice varied considerably from year to year. Cold winters meant cheap ice while mild winters, such as in 1890 in the United States, resulted in a much reduced supply and higher prices. The need for a mechanical means of refrigeration was apparent. A machine for refrigerating air by the evaporation of water in a vacuum was patented in 1755 by a Scotsman named Cullen. However, it was not until 1859 that a machine, using ether, was commercially marketed by a firm in Australia. The ammonia compression machine developed by Professor Carl Linde of the University of Munich opened the way for the large scale production of ice. By 1889, there were more than 200 commercial ice plants in the United States and by 1909 around 2,000.

Early use of refrigeration machinery was primarily for the production of ice rather than

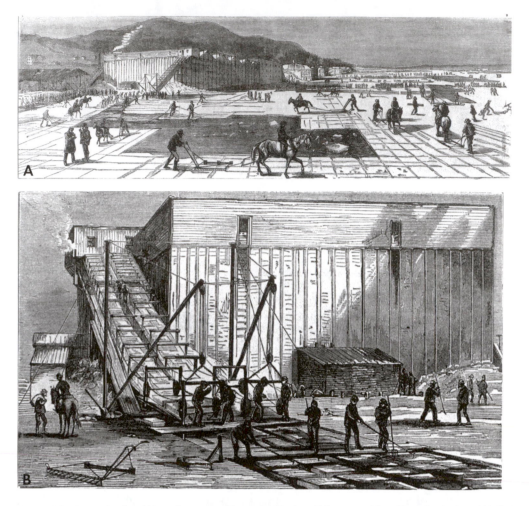

Figure 1.7. Ice gathering (A) and storage (B) on the Hudson River near New York in the late 1800's.[4]

the direct cooling of storage houses. By the late 1800's interest in the use of refrigerated warehouses for the commercial storage of fruit developed.[60] In 1876, a storage house cooled with barrels of ice was constructed by the Western Cold Storage Company in Chicago for the preservation of fruit. An estimate made in 1901 indicated that 600 commercial establishments were in use for refrigerated storage of fruit and other produce.

Use of mechanical refrigeration spread rapidly, and by 1928 the use of cold temperature display cases in retail food stores was common. Refrigeration had become an indispensable part of the food distribution system in many countries.

Low temperature stress, both pre- and postharvest, has also long been a matter of concern. During Roman times, Columellia recommended the burning of piles of chaff around vineyards and fields to prevent injury to the crop when frost threatened. Fruit rooms in the 1700's and onward typically had a small stove that when absolutely essential was used to elevate the temperature. Much of the postharvest research around the beginning of this century was directed toward determining the optimum temperature for storage of various plant produce. It was commonly believed that temperatures as low as 0°C were undesirable for fruits such as apples and pears and higher temperatures (i.e., ≥2°C) were essential. This was shown, however, not to be the case; quality of many cultivars could be maintained substantially longer at the lower temperatures.

The inability to control storage room temperature precisely created problems that are seldom encountered with the controlled mechanical refrigeration systems used today. From the Roman times until well into the 20th century, cool dry conditions were recommended for fruit storage, which caused serious losses in fruit quality due to desiccation. Dry conditions were recommended because the inability to precisely control the temperature of the storage room resulted in significant temperature fluctuations. When a moist, warm period followed cold weather, the fruit temperature was lower than the air temperature, and water condensed on the surface of the fruit. If these conditions continued, they had disastrous consequences due to stimulated pathogen activity. It was, therefore, recommended that fruit rooms be kept dry and aired only when the internal and external temperatures were the same. Techniques that dry the air (e.g., fire, $CaCl_2$) were also often employed. Accurate control of storage temperature in modern storage rooms has to a large extent eliminated this problem. However, with traditional storage systems and occasionally with mechanical failure of modern refrigeration systems, the problems encountered with imprecise control of temperature are dramatically illustrated.

3.3.2. *Alteration of the Storage Gas Atmosphere*

The effect of sealing grain in underground silos on the storage gas atmosphere was documented during Roman times and probably known prior to that period. The primary benefit, the control of insects and rodents, as well as the dangers to humans, was recognized by the Romans. Cato recommended storing grapes in clay pots, a practice which would have also resulted in a modified atmosphere. Various elaborations of this basic technique resulting in a modified storage gas atmosphere have been described since that period. For example, R. Brookes[11] gave the following description in 1763 of a fruit storage technique "communicated to the public by the Chevalier *Southwell,* and which has been used in France with success."

> As many expedients have beeen tried among us, for preserving fruit fresh all the year, I shall beg leave to give one communicated to the public by the Chevalier *Southwell,* and which has been used in *France* with success. Take of Salt Petre one pound, of Bole Armenic two pounds, of common Sand well freed from its earthy parts, four pounds, and mix all together ; after this let the fruit be gathered with the hand before it be thorough ripe, each fruit being handled only by the stalk ; lay them regularly, and in order, in a large wide mouthed glafs veffel ; then cover the top of the glafs with an oiled paper, and carrying it into a dry place, fet it in a box filled all round to about four inches thicknefs, with the aforefaid preparations, fo that no part of the glafs veffel shall appear, being buried in a manner in the prepared Nitre ; and at the end of the year fuch fruits may be taken out as beautiful as they were when firft put in.

The first scientific studies on modified storage atmospheres were conducted in 1819 by Jacques Berard in France.[9] Berard demonstrated that harvested fruits utilize oxygen and liberate carbon dioxide during storage and that fruits placed in containers devoid of oxygen did not ripen. Benhamin Nyce in the 1860's built and operated a fruit storage warehouse in Cleveland, Ohio, which utilized the basic principles of modified atmosphere storage.[20] Unfortunately, little interest was generated among fruit storage workers of that time, and the idea did not take hold.

During the early part of the 20th century, the effect of high CO_2 on inhibiting the ripening of apples and softening of peaches was reported. In 1927, Franklin Kidd and Cyril West published their studies on the utilization of storage gas atmosphere to control ripening and longevity of apples.[32] Their studies pointed to an essential element of gas atmosphere storage: the necessity to accurately control the gas composition of the storage atmosphere. Although the term "controlled atmosphere storage" was not used until 1941, the need for control was apparent. Research during this period concentrated on optimum storage gas atmospheres for various products and their cultivars and methods for constructing and operating controlled atmosphere storage rooms.

Commercial utilization of controlled atmosphere storage began in England in 1929 when a grower stored around 30 tons of apples in 10% CO_2. As the news of the success of this storage technique spread, it stimulated research and utilization in many countries. By 1999, in the United States alone, 97 million bushels of apples were stored under controlled atmosphere conditions.[5]

3.4. The Future of Storage Technology

Storage technology and our understanding of the basic genetic and biochemical processes governing postharvest alterations have made tremendous advances since the beginning of storage by prehistoric man. There is every reason to believe that our ability to maintain the quality of harvested products will continue to progress at an increasing rate; however, in some instances technological advances pose new problems. For example, the volume of information available has increased in a logarithmic manner (Figure 1.2) to the point where we are approaching an information overload. It is estimated that 1 billion pages of new information are created per day in the U.S. alone, along with 736,000 new books each year.[21] Most postharvest decisions are relatively easy to make if the essential information is available; however, we have entered an age in which finding the appropriate information represents the Achilles' heel of the decision making process. The sheer volume of information now makes traditional approaches (e.g., scholarly books, reviews, research articles) to accessing information more difficult, and the emergence of the World Wide Web has additionally complicated the picture. While the Web has facilitated our access to information, it has been at the cost of credibility. In the past, most published information was rigorously reviewed by competent third parties. Publishing houses and professional societies built their reputations on the quality and accuracy of the information disseminated. Traditional methods of credibility control are not operative on the Web, where any information, including intentionally incorrect information, can be posted. This act has placed the burden of determining the quality of the information on the eventual user, a responsibility for which we are often not trained.

4. THE SIGNIFICANCE OF POSTHARVEST PLANT BIOLOGY AND HANDLING OF PERISHABLE PLANT PRODUCTS

When plants or plant parts are used by humans, whether it is for food, for aesthetic purposes or for environmental modification, there is always a postharvest component. The postharvest component encompasses everything that happens to a plant product from the time it is harvested until utilization or death. Just how important the postharvest component is in the overall business of satisfying our requirements for plant products varies widely. It is influenced by the nature of the product itself, particularly its perishability, the intended use of the product, the environmental and handling conditions to which it is exposed, the relative abundance of the product at the time, the culture of the society, and socioeconomic factors. The greater the time lapse for a product between harvest and use, the more important is the postharvest component.

4.1. Factors Influencing the Postharvest Component

There is a very wide range in the degree of perishability of plants and plant products. Dry seeds such as cereal grains and pulses are perhaps the least perishable of the major food items. They are comparatively low in moisture content (i.e., 3.4–15%) and are protected from excessive moisture loss, microbial infection, and mechanical damage by specialized tissues such as the seed coat. Under appropriate conditions many species of seed can be maintained in good condition for years with relatively simple storage facilities and treatments. However, as the storage duration is extended, the importance of the postharvest component increases. In contrast, immature products such as okra or lettuce have little protection, are harvested when they are actively growing, and often contain 85% or more water. In addition, they have a high metabolic rate, little protection against water loss or microbial infection, and high susceptibility to mechanical damage. Under the best of conditions, these plant parts can be maintained for a few weeks; under ambient conditions, they last only a few days.

Differences in postharvest perishability probably delineate the separation of agronomic and horticultural crops better than any other single characteristic. Most agronomic products tend to be relatively stable in contrast to most horticultural products, which are highly perishable. This difference unfortunately has tended to deemphasize the importance given to the postharvest period of agronomic crops even though extensive losses do occur.

The way in which harvested products are to be utilized also affects how they can be handled during the postharvest period. For example, produce that is to be used for fresh market is handled quite differently from that being processed. Produce for the fresh market commonly passes through several stages or operations, each requiring time and handling, and is often transported over long distances. Therefore, much care and effort is needed to maintain its condition. In contrast, fresh produce for processing is usually grown close to the processing plant where it is harvested and moved quickly with a minimum of handling. The produce has similar physiological and deteriorative characteristics in both cases, but the time required between harvesting and processing limits the extent of deterioration and, therefore, the benefits to be gained from intensive postharvest care.

In many instances, particularly in subsistence agriculture situations, perishable crops are used by the grower or sold and exchanged locally over short distances in a short time span. In such cases, the postharvest requirements are minimized but not eliminated. An understanding of the physiology of the produce and of the cause of deterioration can lead to simple and inexpensive changes in postharvest handling practices that can greatly reduce losses.

In more mechanized societies, the trend is for the production of perishable products to shift further and further away from the major markets. The spread of cities over nearby production centers and the subsequent increase in land costs are common problems that force production to more distant sites. Mechanization of production is an important factor in minimizing costs in many countries. For it to be economical, large tracts of land not usually available close to the markets are required. In addition, the demand from consumers for year-round supplies of major products encourages production to shift location with seasonal changes of climate. Hence, winter production of summer crops can be achieved in subtropical regions;

Figure 1.8. Occasionally decisions in a seemingly unrelated sector of the economy unfavorably impact postharvest handling practices. For example, in Tanzania shipping costs are based upon a per container rate. To reduce costs, the container (small white bag at the base) has been artificially enlarged using netting to absurd proportions, greatly increasing the potential for mechanical damage to the cassava roots (*photograph courtesy of A. Graffham*).

however, the harvested produce must then be transported back to the markets in the temperate regions. In such cases, careful attention to the postharvest requirements of the crops is essential, and a substantial cost factor is added to these products.

The importance attached to the postharvest handing of produce is also influenced by current supply/demand situations. When supply is high in relation to demand, the price paid is lower. Low prices can result in less effort and expense during marketing and the acceptance of greater losses of product.

Finally, cultural and socioeconomic factors have a major influence on the importance attached to the postharvest component. Traditional postharvest practices have evolved over long periods of time and are intimately associated with the local culture and the structure of the society. Although these practices have withstood the test of time, our current understanding of postharvest behavior may offer opportunities for improvement. In some instances, inappropriate practices have stemmed from decisions made in other sectors of the economy that inadvertently impact postharvest quality (Figure 1.8). Acceptance of changes in postharvest practices by traditional societies, as well as more developed ones, may be difficult to achieve. Clear demonstration of the economic or social value to those involved in production and marketing may be required before they are prepared to place a greater emphasis on the postharvest care of their products.

Governments may also exert a substantial influence on the emphasis placed on postharvest care of produce through their policies relating to national food supply goals. Policies on transportation and communication systems, support programs, and taxation greatly affect how products are handled after harvest. Government policy may not necessarily improve postharvest practices or reduce postharvest losses. Changes in postharvest practices may have undesirable secondary effects in other areas such as employment and distribution of population.

4.2. Postharvest Losses

A postharvest loss is any change in the quantity or quality of a product after harvest that prevents or alters its intended use or decreases its value. Postharvest losses vary greatly in kind (e.g., ranging from losses in volume to subtle losses in quality), magnitude, and where in the postharvest handling system they occur.[50] Likewise, individual crops differ greatly in their susceptibility to loss.

How important are postharvest losses in the overall production, marketing and utilization scheme of agricultural plant products? Accurate estimates of net losses simply are not available. While a detailed cost analysis can be made for an individual crop at a specific time and place (e.g., within an individual village or production brigade), these measurements do not extrapolate accurately to an all-encompassing province, country or worldwide estimate for the crop. Losses in one village may be radically different from another a few kilometers away. Likewise, losses one year within an individual village may differ tremendously with the next. Ideally a summation of losses for each crop at the local, state, national and finally international level is needed for an accurate assessment. A very general estimate of the potential importance of postharvest losses can be seen in the increase in value of a product after harvest, i.e., the distribution of costs between production and marketing segments of the overall system. While it is often difficult to obtain precise data on the actual costs of production and marketing of plant products from the grower right through to consumer, general estimates can be made from production and marketing statistics together with estimates of the general profit margins (Table 1.1). Based upon these estimates it is apparent that approximately 50% of the retail value of fresh produce is accrued after harvest.

Another way to assess the significance of the postharvest component is to look at the extent of losses that occur during this phase. Once again a meaningful determination of these losses is extremely difficult to obtain since they are specific to location, time, storage duration, and even what is considered a postharvest loss. As a consequence, losses vary widely between studies (e.g., average yam losses of 3.2,[45] 5,[30] 25,[30] 40,[69] 30–50,[23] and up to 50%[18] and average papaya losses of 20–26,[46] 24,[37] 75,[47] and 40–100%[43] have been reported). Quantification of losses typically represents only a general estimate with but a few studies[46] providing a precise assessment of the situation. As a consequence, at present it is impossible to substantiate statistically these losses on a national or international basis. However, authoritative groups such

Table 1.1. Estimated Distribution of Total Costs Between Production and Postharvest Costs as Percentage of Cost to Consumer.*

Crop	Production (%)	Postharvest (%)
Snap bean	53.7	46.3
Lima bean	61.5	38.5
Okra	50.6	49.4
Sweetpotato	50.5	49.5
Sweet corn	49.0	51.0
Cabbage	46.7	53.3
Apple	44.4	55.6
Peach	53.1	46.9
Karume azalea	45.5	54.5
Pin oak	33.3	66.7
Juniper	46.0	54.0

*Source: Data derived in part from Cooperative Extensive Service, University of Georgia publications, 1981–1982.

Table 1.2. Reported Production and Loss Figures in Less Developed Countries.

Commodity	Production (1,000 tonnes)	Estimated Losses (% of Total Crop)
Roots/Tubers		
Carrots	557	44
Potatoes	26,909	5–40
Sweetpotatoes	17,630	35–95
Yams	20,000	10–60
Cassava	103,486	10–25
Vegetables		
Onions	6,474	16–35
Tomatoes	12,755	5–50
Plantain	18,301	35–100
Cabbage	3,036	37
Cauliflower	916	49
Lettuce	—	62
Fruits		
Banana	36,898	20–80
Papaya	931	40–100
Avocado	1,020	43
Peaches, apricots, nectarines	1,831	28
Citrus	22,040	20–95
Grapes	12,720	27
Raisins	475	20–95
Apples	3,677	14

Source: NSF. 1978.[44]

Table 1.3. Postharvest Losses of Wheat, Rice and Maize.

Crop	Country	Storage Method	Cause of Loss	Storage Period (Months)	% Loss	Reference
Wheat	India	gunny sacks	insects	6–12	28–50	33
	India	bulk	insects	—	3.5	70
	Egypt	stack	—	12	36–48	2
Rice	Sri Lanka, Nepal, Malaysia, Vietnam	—	insects	30	30	36
	West Africa	farm storage	—	12	2–10	44
Maize	USA	—	insects	12	0.5	29
	Tanzania	—	—	—	20–100	52
	India	farm storage	—	9	2–11	33

as the National Research Council of the United States[43] and the Food and Agriculture Organization of the United Nations[22] have made estimates for planning purposes in developing countries. Their conservative estimates are that the minimum overall loss of the more durable cereal grains and legumes is around 10% while that for more perishable food products is approximately 20%. Reported losses for individual crops are often much higher than this, ranging from 0.5 to 100% (Tables 1.2 and 1.3). Postharvest losses in developed countries are generally considered to be less than in the developing regions of the world, but reliable estimates

PRODUCT IN FIELD

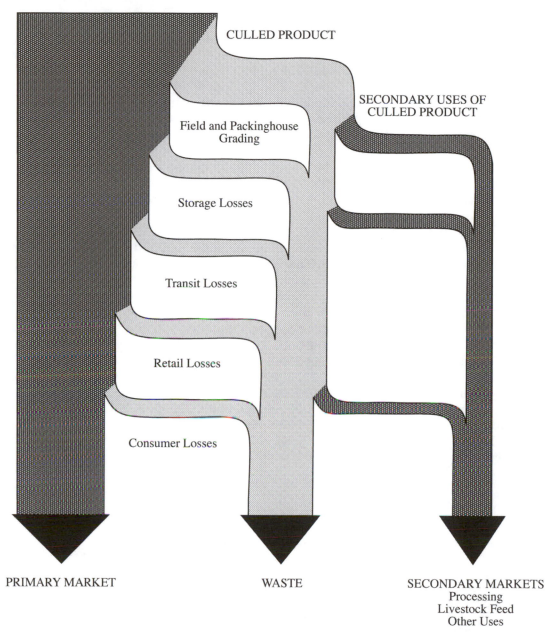

CULLED PRODUCT

SECONDARY USES OF
CULLED PRODUCT

Field and Packinghouse
Grading

Storage Losses

Transit Losses

Retail Losses

Consumer Losses

PRIMARY MARKET

WASTE

SECONDARY MARKETS
Processing
Livestock Feed
Other Uses

Figure 1.9. A flow chart of postharvest losses for tomatoes as they move toward final utilization (data from Campbell[14]). Inferior grades that would be discarded in some countries often remain in the food chain in developing countries. These find alternate markets that have different quality tolerances.[41]

of these are not readily available. A USDA study[66] that dealt with isolated segments of the total postharvest component and with a variable range of products, reported losses in durable products such as grain, dried fruits, and nuts due to insects and other storage problems to be 3.62%. Losses of 5.42% of fruits and 10.3% of vegetables during transit, unloading, and retail marketing were also reported. The figures do not take into account postharvest losses on the farm, during sorting, grading, packaging, storage and wholesale marketing or at the consumer end of the chain.

The importance of the large number of stages, each commonly under the control of a different organization, in the marketing of perishable plant products is illustrated in Figure 1.9. While each operator along the chain may be prepared to accept some loss, the cumulative effect of all these losses can be substantial. The rubbish bin is too frequently a major consumer of produce.

While complete physical loss of produce is the most dramatic example, much of the total loss is due to reduced quality and nutritional value. Postharvest care, therefore, is as much concerned with maintenance of quality as with quantity. It should be applied to all produce, not only to the top quality material. The aim is to maintain as much as possible of the existing quality, whatever that may be. It can be argued that even greater care is needed for produce that has already started to deteriorate than for top quality material since the established deterioration sequence must be inhibited (e.g., prevent water loss from wounded tissue and infection of the wound as well as maintain physiological condition) and the existing condition maintained.

The full significance of quality loss may not be immediately obvious as in the case of storage of seed or other propagules. Seed that has been allowed to deteriorate during storage may still be viable, but if planted, its rate of establishment and subsequent yield will be impaired. Thus what may superficially appear to be a small postharvest loss in quality of the seed is in fact greatly amplified into a serious loss in yield of the crop.

Although it is not possible to eliminate losses, the extent of these losses can be reduced through better understanding of the nature of harvested produce and use of appropriate technology. There is, however, a cost associated with the use of many postharvest techniques, but the cost of increasing the useable supply of a particular product by reducing losses is considerably less than the cost of producing more to compensate for postharvest losses. One example often quoted is that the total energy cost of good grain storage practices is about 1% of the energy cost of producing that grain. Other advantages gained through using improved postharvest practices to increase useable supply, in contrast to increased production, are fewer adverse effects on health, safety or the environment.

ADDITIONAL READING

Bellingham, C. 1604. *The Fruiterers Secrets.* R. Bradock, London.

Braun, D. 1983. Pots as tools. Pp. 107–134. In: *Archaeological Hammers and Theories.* J.A. Moore and A.S. Keene (eds.). Academic Press, New York.

Bynum, W.F., E.J. Browne and R. Porter (eds.). 1981. *Dictionary of the History of Science.* Princeton University Press, Princeton, NJ.

Campbell, B.G. 2000. *Humankind Emerging.* HarperCollins, New York.

Cohen, M.N. 1977. *The Food Crisis in Prehistory: Overpopulation and the Origins of Agriculture.* Yale Univ. Press, New Haven, CT.

Cooper, E. 1972. *A History of Pottery.* Longman, London.

FAO. 1977. *Analysis of FAO Survey of Postharvest Crop Losses in Developing Countries.* Food an Agric. Org. United Nations (AGPP/MISC/27), Rome.

FAO. 1981. *Food Loss Prevention in Perishable Crops.* FAO Agricultural Services Bull. 43, 72p.

Gonen, R. 1973. *Ancient Pottery.* Cassell, London.

Green, A.W. 1977. *An Analysis of an FAO Survey of Post-harvest Food Losses in Developing Countries.* Food and Agric. Org. United Nations, Rome.

Harlan, J.R. 1976. *Origins of African Plant Domestication.* Aldine, Chicago, IL.

Hellemans, A., and B. Bunch. 1988. *The Timetables of Science.* Simon and Schuster, New York.

Ho, P.-T. 1975. *The Cradle of the East: An Inquiry into the Indigenous Origins of Technique and Ideas of Neolithic and Early Historic China, 5000–1000 BC.* University of Chicago Press, Chicago, IL.

Kaplan, H., and K. Hill. 1992. The evolutionary ecology of food acquisition. Pp. 167–201. In: *Evolutionary Ecology and Human Behavior.* E.A. Smith and B. Winterhalder (eds.). Aldine, New York.

Kent, S. (ed.). 1989. *Farmers as Hunters. The Implications of Sedentism.* Cambridge University Press, Cambridge.

NSF. 1978. *Report of the Steering Committee for Study on Postharvest Food Losses in Developing Countries.* National Research Council, National Academy of Sciences. Washington, D.C..

Odell, G.H. (ed.). 1996. *Stone Tools. Theoretical Insights into Human Prehistory.* Plenum Press, New York.

Partridge, M. 1974. *Farm Tools through the Ages.* New York Graphic Soc., Boston, MA.

Plog, S. 1980. *Stylistic Variation in Prehistoric Ceramics.* Cambridge University Press, Cambridge.

Reed, C.A. 1977. *Origins of Agriculture.* Aldine, Chicago, IL.

Rice, P.M. 1987. *Pottery Analysis: A Sourcebook.* University of Chicago Press, Chicago, IL.

Sankalia, H.D. 1964. *Stone Age Tools; Their Techniques, Names and Probable Functions.* Deccan College Postgrad. and Res. Inst., Poona, India.

Spier, R.F.G. 1970. *From the Hand of Man, Primitive and Preindustrial Technologies.* Houghton Mifflin Co., Boston, MA.

Vander Wall, S.B. 1990. *Food Hoarding in Animals.* University of Chicago Press, Chicago, IL.

Varro, M. Terentius. 1800. *The Three Books of M. Torentina Varro Concerning Agriculture,* translated by T. Owen, Oxford University Press, Oxford, England.

Walters, H.B. 1905. *History of Ancient Pottery.* 2 vols. John Murray, London.

White, K.D. 1967. *Agricultural Implements of the Roman World.* Cambridge University Press, Cambridge.

REFERENCES

1. Abbate, E., A. Albianelli, A. Azzaroli, M. Benvenuti, B. Tesfamariam, P. Bruni, N. Cipriani, R.J. Clarke, G. Piccarelli, R. Macchiarelli, G. Napoleone, M. Papini, L. Rook, M. Sagri, T.M. Tecle, D. Toree and I. Villa. 1998. A one-million-year old *Homo* cranium from the Danakil (Afar). *Nature* 393:458–460.

2. Abou-Nasr, S., H.S. Salama, I.I. Ismail and S.A. Salem. 1973. Ecological studies on insects infesting wheat grains in Egypt. *Zeit. Fuer. Ango. Ento.* 73:203–212.

3. Aikens, C.M. 1995. First in the World. The Jomon pottery of early Japan. Pp. 11–21, In: *The Emergence of Pottery.* W.K. Barnet and J.W. Hoopes (eds.). Smithsonian Inst. Press, Washington, DC.

4. Anon. 1875. Gathering and storing ice. *Scient. Amer.* XXXII, March 27, p.195.

5. Anon. 1999. *U.S. Apple Market News* 5(2):1–9.

6. Ardrey, R. 1976. *The Hunting Hypothesis.* Atheneum, New York.

7. Arnold, D. 1985. *Ceramic Theory and Cultural Process.* Cambridge University Press, Cambridge.

8. Beamon, S.P., and S. Rolf. 1990. *The Ice-houses of Britain.* Routledge, London.

9. Berard, J.E. 1821. Memoire sur la maturation des fruits. *Annales de chimie et de Physique* XVI:152–183, 225–251.

10. Braidwood, R.J., and C.A. Reed. 1957. The achievement and early consequences of food production: A consideration of the archeological and natural-historical evidence. *Cold Springs Harbor Symp. on Quant. Biol.* 22:19–31.

11. Brookes, R. 1763. *The Natural History of Vegetables.* London.

12. Brian, C.K., and A. Sillen. 1988. Evidence from the Swartkrans Cave for the earliest use of fire. *Nature* 336:464–466.

13. Buttler, W. 1936. Pits and pit-dwellings in Southeast Europe. *Antiquity* 10:25–36.

14. Campbell, D.T. 1988. A systems analysis of postharvest injury of fresh fruits and vegetables. M.S. Thesis, Univ. Georgia, Athens, GA.

15. Cann, R.L., M. Stoneking and A.C. Wilson. 1987. Mitochondrial DNA and human evolution. *Nature* 329:111–112.

16. Cann, R.L., O. Rickards and J.K. Lum. 1994. Mitochondrial DNA and human evolution: Our one Lucky Mother. Pp. 135–148, In: *Origins of Anatomically Modern Humans.* M.H. Nitecki and D.V. Nitecki (eds.). Plenum Press, New York.

17. Chapela, I.H., S.A. Rehner, T.R. Schultz and U.G. Mueller. 1994. Evolutionary history of the symbiosis between fungus-growing ants and their fungi. *Science* 266:1691–1694.
18. Coursey, D.G. 1967. *Yams.* Longman, London.
19. Cummings, R.O. 1949. *The American Ice Harvests.* University of California Press, Berkley, CA.
20. Debrymple, D.G. 1969. The development of an agricultural technology: Controlled atmosphere storage of fruit. *Tech. and Cult.* 10:35–48.
21. Dobert, R.C. 1997. Technology transfer: Biotechnology information and the federal government. Pp. 1–11. In: *Technology Transfer of Plant Biotechnology.* P.M. Gresshoff (ed.). CRC Press, Boca Raton, FL.
22. FAO. 1981. Food loss prevention in perishable crops. *FAO Agri. Ser. Bull.* 43.
23. FAO. 1977. *Analysis of FAO Survey of Postharvest Crop Losses in Developing Countries.* Food and Agric. Org. United Nations (AGPP/MISC/27), Rome.
24. Forbes, F.H. 1876. *Scribner's Monthly.* July.
25. Greeley, M. 1991. Postharvest technologies. Implications for food policy analysis. Pp. 1–19. In: *Analytical Case Studies.* Economic Development Inst., World Bank, New York.
26. Gonen, R. 1973. *Ancient Pottery.* Cassell, London.
27. Good, I. 2001. Archaeological textiles: A review of current research. *Annu. Rev. Anthropol.* 30:209–226.
28. Goosman, J.C. 1924–1927. History of Refrigeration. *Ice and Refrigeration* 66:297, 446, 541; 67:33, 110, 181, 227, 428; 68:70, 135, 335, 413, 478; 69:99, 149, 203, 267, 372; 70:123, 197, 312, 503, 612; 71:81.
29. Hoopes, J.W., and W.K. Barnett. 1995. The shape of early pottery studies. Pp. 1–7, In: *The Emergence of Pottery.* W.K. Barnet and J.W. Hoopes (eds.). Smithsonian Inst. Press, Washington, DC.
30. Ihekoronye, A.I., and P.O. Nyoddy. 1985. *Integrated Food Science and Technology for the Tropics.* Macmillan, London.
31. Kays, S.J., and J.C. Silva Dias. 1996. *Cultivated Vegetables of the World.* Exon Press, Athens, GA.
32. Kidd, F., and C. West. 1927. A relation between the concentration of oxygen and carbon dioxide in the atmosphere, rate of respiration, and length of storage of apples. *Rept. Fd. Invest. Bd.,* London for 1925,1926. Pp. 41–42.
33. Khare, B.P. 1973. *Insect Pests of Stored Grain and Their Control in Uttar Pradesh.* G.B. Pant University Agric. Tech., Pantnager, U.P., India.
34. Klein, R.G. 1999. *The Human Career. Human Biological and Cultural Origins.* University of Chicago Press, Chicago, IL.
35. Kohl, G.J. 1842. *Russia and the Russians.* 2 vol. Colburn, London.
36. Lever, R.J.A.W. 1971. Losses in rice and coconut due to insect pests. *World Crops* 23:66–67.
37. Liu, M.S., and P.C. Ma. 1984. Postharvest problems of vegetables and fruit in the tropics and subtropics. Pp. 26–35. In: *Workshop on Postharvest Technology of Food Industry Research and Development.* Inst. Agric. Produce, Taipei, Taiwan.
38. Lumley, H. de, and Y. Boone. 1976. Les structures d'habitat au Paléolithique inférieur. Vol.1:625–643. In: *Le Préhistorie Française.* H. de Lumley (ed.). Centre National de la Recherche Scientifique, Paris.
39. Miller, P. 1768. *The Gardener's Dictionary.* London.
40. Moore, Thomas. 1803. *Essay on the Most Eligible Construction of Ice-Houses; Also, a Description of the Newly Invented Machine Called the Refrigerator.* Bonsal and Niles, Baltimore, MD.
41. Mukai, M.K. 1987. Postharvest research in a developing country: A view from Brazil. *HortScience* 22:7–9.
42. Newberry, P.E., F.L. Griffith and G.W. Fraser. 1893. *Beni Hasan.* Part I. K. Paul, Trench,Trübner and Co., London.
43. NAS. 1978. *Postharvest Food Losses in Developing Countries.* Board on Science and Technology for International Development, National Acad. Sci., Washington, DC.
44. NSF. 1978. *Report of the Steering Committee for Study on Postharvest Food Losses in Developing Countries.* Nat. Res. Coun., National Science Foundation, Washington, D.C.
45. Okoh, R.N. 1997. Economic evaluation of losses in yams stored in traditional barns. *Trop. Sci.* 38:125–127.
46. Pantastico, E.B. 1979. *Postharvest Losses of Fruits and Vegetables in Developing Countries – An Action Program.* SEARCA Professional Chair Lecture, Postharvest Teaching and Research Center, Los Banos, Philippines.
47. Paull, R.E., W. Nishijima, M. Reyes and C. Cavaletto. 1997. Postharvest handling and losses during marketing of papayas (*Carica papaya* L.). *Postharv. Biol. Tech.* 11:165–179.

48. Procter, D.V. 1981. *Ice Carrying Trade at Sea.* National Maritime Museum, London.

49. Perrot, J. 1966. Le gisement natufiende Mallaha (Èynan), Israël. *L'Anthropologie* 70:437–484.

50. Pierson, T.R., J.W. Allen and E.W. McLaughlin. 1982. Produce losses. *Mich. State Univ., Agric. Econ. Rept.* 422, 48p.

51. Puleston, D.E. 1971. An experimental approach to the function of classic Maya chultuns. *Amer. Antiq.* 36:322–335.

52. Rackham, H. 1938. *Pliny Natural History.* Vol. 5, Harvard Univ. Press, Cambridge, MA.

53. Rees, A. 1819. *The Cyclopedia or Universal Dictionary of Arts, Science and Literature.* 47 vols. Bradford and Firman, Philadelphia, PA.

54. Rhoades, R., M. Benavides, J. Recharte, E. Schmidt and R. Booth. 1988. Traditional potato storage in Peru: farmer's knowledge and practices. *Potatoes in Food Syst. Res. Ser. Rept.* 4, International Potato Center, Lima, Peru.

55. Roebroeks, W., and T. van Kolfschoten. 1994. The earliest occupation of Europe: A short chronology. *Antiquity* 68:489–503.

56. Severinghaus, J.P., and E.J. Brook. 1999. Abrupt climate change at the end of the last glacial period inferred from trapped air in polar ice. *Science* 286:930–934.

57. Shih, Sheng-Han. 1982. *A Preliminary Survey of the Book – Ch'I Min Yao Shu, an Agricultural Encyclopedia of the 6th Century.* Science Press, Beijing.

58. Solecki, R.S. 1964. "Shanidar" Cave, a late pleistocene site in Northern Iraq. *VIth Intern. Cong. on the Quaternary Rept.* 4:413–423.

59. Solecki, R.S. 1964. "Zawi Chemi Shanidar," a post-pleistocene village site in Northern Iraq. *VIth Intern. Cong. on the Quaternary Rept.* 4:405–412.

60. Taylor, W.A. 1900. The influence of refrigeration on the fruit industry. Pp. 561–580. In: *USDA Yearbook Agric.* 1900, Washington, DC.

61. Taylor, W.W. 1973. Storage and the neolithic revolution. In: *Estudios Dedicados al Prof. Dr. Luis Pericot.* University of Barcelona, Barcelona, Spain.

62. Teruya, E. 1986. The origins and characteristics of Jomon Ceramic Culture: A brief introduction. Pp. 223–228. In: *Windows on the Japanese Past: Studies in Archaeology and Prehistory.* R.J. Pearson, G.L. Barnes, and K.L. Jutterer (eds.). Center for Japanese Studies, University of Michigan, Ann Arbor, MI.

63. Tishkoff, S.A., E. Dietzsch, W. Speed, A.J. Pakstis, J.R. Kidd, K. Cheung, B. Bonné-Tamir, A.S. Santachiara-Benerecetti, P. Moral, M. Krings, S. Pääbo, E. Watson, N. Risch, T. Jenkins and K.K. Kidd. 1996. Global patterns of linkage disequilibrium at the CD4 locus and modern human origins. *Science* 271:1380–1387.

64. Toth, N. 1985. The Oldowan reassessed: A close look at early stone artifacts. *J. Arch. Sci.* 12:101–120.

65. Testart, A. 1982. The significance of food storage among hunter-gatherers: Residence patterns, population densities, and social inequalities. *Curr. Anthro.* 23:523–530.

66. USDA. 1965. Loss in Agriculture. *USDA Agri. Res. Ser., Agri. Hbk.* No. 291, Washington, DC.

67. Van der Wall, S.B. 1990. *Food Hoarding in Animals.* University of Chicago Press, Chicago, IL.

68. Vandiver, P.B., O.Soffer, B. Klima and J. Svoboda. 1989. The origins of ceramic technology at Dolni Vestonice, Czechosloviakia. *Science* 246:1001–1008.

69. Waitt, A.W. 1961. Review of yam research in Nigeria 1920–1961. *Federal Agric. Res. and Teaching Sta.,* Memo 31, Umudike, Nigeria.

70. Wilson, H.R., A. Singh, O.S. Bindra and T.R. Evertt. 1970. *Rural Wheat Storage in Ludhiana District, Punjab.* Ford Foundation, New Delhi.

71. Yamei, H., R. Potts, Y. Baoyin, G. Zhengtang, A. Deino, W. Wei, J. Clark, X. Guangmao and H. Weiwen. 2000. Mid-Pleistocene Acheulean-like stone technology of the Bose Basin, South China. *Science* 287:1622–162.

72. Zune, L. 1985. Reply to Binford and Ho. *Curr. Anthro.* 26:432–433.

2

NATURE AND STRUCTURE
OF HARVESTED PRODUCTS

How plants and plant products are handled after harvest and the changes that occur within them during the postharvest period are strongly influenced by their basic structure. For example, plant parts that function in nature as storage organs behave after harvest in many ways that are distinctively different than structurally dissimilar parts such as leaves or flowers. Consequently, it is desirable to be familiar with the structure not only of the general product but with the tissues and cells that comprise it. This chapter describes general morphological groupings of harvested products, the tissues that are aggregated to form these products, and the structure of the cells that make up these tissues.

One extremely important concept of plant morphology is that structure, whether at the organ, tissue or the cellular or subcellular level, is not fixed, but is in a state of transition. Changes in structure are especially important during the postharvest period. A structural unit once formed is eventually destined to be degraded and recycled back to carbon dioxide. Marked changes in structure occur with eventual use of the product (e.g., consumption of a food product) or loss due to pathogen invasion. Although less dramatic, distinct and important structural alterations occur in plant products during storage and marketing. Formation of fiber cells, loss of epicuticular waxes during handling, and chloroplast degradation or transformation are but a few of many changes that can occur. At the cellular and subcellular level, very pronounced changes occur and these intensify as the product approaches senescence, or as in the case with seeds and intact plants, during the beginning of renewed growth. It is of paramount importance, therefore, that we view organs, tissues, cells and subcellular bodies of harvested products as structural units in a state of change. Decreasing this rate of change is in most cases an essential requisite for successful extended storage of the product.

1. GROUPING OF HARVESTED PRODUCTS BASED ON MORPHOLOGY

The range of plant products that are harvested and used by mankind is vast and the characteristic responses after harvest are so varied that some system of grouping or classification is necessary. Our interest in postharvest biology stems from the practical requirement of getting perishable plant products from producers to consumers and maintaining the desired supply and quality throughout the year. The classification most useful for this purpose is one which brings together products with similar environmental requirements after harvest for maintenance of quality or those that are susceptible to chilling or other types of injury. Although this system is extremely useful to the practical operator, it does not help us to understand the nature of the harvested produce. Likewise, classification according to the product's use, that is, fruits, vegetables, florist items, nursery products, agricultural seed products, does not aid in understanding those products and there is often considerable overlap of the categories. For example, foliage is not only used as a florist item but also as vegetative cuttings and edible vegetables. Their characteristic postharvest behavior is very similar, irrespective of the end use. Fruits such as tomato are used as vegetables but still ripen like other fruits. Understanding the nature of the product, rather than for what it is to be used, is more useful and allows prediction of the likely response of harvested products.

Botanical classification into family, genus, and species is of only limited value with regard to postharvest handling.* Whole plants or any of the separate organs from that plant may have vastly different characteristics and behavior, yet all are from the same botanical specimen.

A classification according to plant part and stage of development is a satisfactory alternative. It allows an understanding of the nature of the harvested product, and therefore a means of predicting behavior. Both physical and physiological characteristics are indicated. It should be recognized that classifications are systems devised by scientists and applied to the subject of their choice to serve a particular purpose. In this case, the intent is to classify harvested plant products in order to reduce their number for easy management and to be able to predict the behavior and handling requirements of products for which there is no specific information available. Because of the very nature of biological systems and the inherent variation both within and among species, as well as the wide range of uses of plant products, no classification system can be perfect. For example, the storage temperature requirements for sweetpotato and beetroot, both root crops, differ substantially. As a consequence, scientists must be prepared to allow for the exceptions and to refine or alter the classification for specific purposes.

Classifying according to plant part and knowing the characteristics of those parts makes it easier to understand the current and potential processes that may be operative during the postharvest period.

1.1. Intact Plants

Whole plants are considered to be harvested when they are removed from the production environment. They retain both the shoot and root system which may or may not still be associated with soil or growing media. Whole plants should have little or no harvest injury and maximum capacity to continue or to recommence growth and development. They commonly have access to all the normal requirements for plant growth, that is, water, mineral nutrients, oxygen, carbon dioxide and light (energy) and are susceptible to all the influences on plant growth: physical, chemical and biological.

Germinated seed such as bean or alfalfa sprouts have the endosperm as an energy source and do not require light. They have a very high metabolic rate and release a substantial amount of heat. In the juvenile shoot, there has been little development of the cuticle and the roots are exposed. The seedlings are therefore very susceptible to water loss and to mechanical damage.

Bare root seedlings and rooted cuttings are physiologically more developed than germinated seeds, but are also very subject to water loss and mechanical damage. These young plants do not have stored reserves, as do germinated seeds, so are dependent on either continued photosynthesis or minimized metabolic rate for maintenance of quality. More mature whole plants can tolerate harsher conditions as they are in general less brittle, have a better-developed cuticle and have accumulated some stored reserves. The extreme is seen in dormant

---

*Botanical classification can, however, be useful in tracing the evolution of certain traits.

woody plants where metabolic rate is suppressed and leaves and active feeder roots are absent, so susceptibility to water loss, mechanical damage and microbial infection is greatly reduced.

It is easier to maintain the quality of intact plants marketed in containers. The root systems are not exposed and as a consequence, are less subject to stress and mechanical damage. The medium around the plant roots provides a reservoir of mineral nutrients and water, increasing the amount of water available to the plant. Water, however, will be depleted unless it is periodically replenished or the plant is kept in an environment where water loss does not occur. Very often during retail sales, containerized plants are subjected to water stresses of sufficient magnitude to result in quality losses.

Tissue cultured plants are a special type of containerized intact plants. Due to the special environment under which they are produced, they are particularly delicate, but at the same time, they are relatively well protected within the microenvironment of their containers. Their compact size and controlled environment often enables them to be shipped through normal mail and parcel delivery systems that are not accustomed to handling live material. Special precautions, however, are necessary to maintain orientation, to protect from temperature extremes and to avoid prolonged delays.

1.2. Detached Plant Parts

Plant organs utilized by humans come from virtually every portion of the plant. This fact is perhaps best illustrated by vegetables (Figure 2.1), which range from immature flowers to storage roots.

1.2.1. Aboveground Structures

a. Leaves

Leaves, widely consumed as foods, are also utilized for ornamental purposes. In some cases, they may also represent propagules for asexually reproducing plants. Morphologically the leaves of most dicotyledonous plants are comprised of a leaf blade, the thin flattened portion of the leaf, and a petiole which attaches the blade to the stem. Many of our leafy crops have intermediate-to-large petioles (e.g., spinach, collards, Ceylon spinach), while others have only sessile (having the blade attached directly to the stem) or very much reduced petioles (e.g., Chinese cabbage, lettuce, cabbage). Several species have been developed in which the leaf petiole is the product of interest (e.g., celery, rhubarb). However, these tend to have significantly different postharvest responses and as a consequence are covered separately.

While attached to the plant, the primary function of a leaf blade is the acquisition of carbon through photosynthesis. Leaves also control, to a large extent, transpiration by the plant, which helps to regulate their temperature. After harvest, these functions are seldom operative. The leaf loses its potential to acquire additional carbon (energy) and is cut off from its supply of transpirational water. Hence the energy required by the cells of the leaf for the maintenance of life processes must now come from recycled carbon found within. Leaves of most species do not act as long-term carbon storage sites. This lack of energy reserves decreases their potential postharvest life expectancy. As a consequence, leaves are stored under conditions that will minimize the rate of utilization of these limited energy reserves.

Water loss from harvested leaves is an additional limitation to long-term storage. With detachment from the plant, the leaves can no longer replenish water lost through transpiration. The leaf responds with closure of the stomata, the primary avenue of water loss. Closure greatly decreases the rate of water loss, but does not eliminate it.

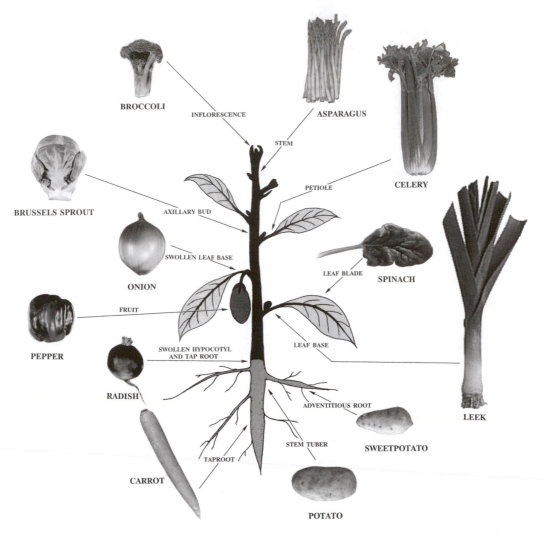

Figure 2.1. Plant organs utilized by man are derived from virtually every portion of the plant. The diagram illustrates examples of various plant organs used as vegetables.

The exterior surface of leaves of some species (e.g., cabbage and collards) is covered with a relatively thick waxy cuticle that helps to decrease water loss. The rate of water loss, therefore, is modulated by both the nature of the product and the environmental conditions in which it is stored. Thus leafy products are usually stored under conditions of high relative humidity and a reduced temperature to minimize the loss of water.

Another method that is used to help increase the longevity of leafy products is harvesting the leaves and stem intact. Although the stem may not be consumed, it acts as a reservoir for both water and carbon, helping to increase the life expectancy of the leaves.

b. Petioles

With intact plants, petioles of dicotyledonous leaves function as conduits for the transport of photosynthates from the leaf to sites of utilization and for the transport of water and nutrients from the root system to the leaf. The petiole also provides support and positioning for the leaf within the aerial canopy of the plant. In a few cases, the petiole acts as a site for the storage of photosynthetic carbon.

Several plants have well-developed fleshy petioles which are the primary morphological component. Celery, rhubarb and pak-choi are utilized either exclusively or largely for their edible petioles.[27] These petioles tend to have more stored energy reserves than the leaf blades. They also lose water less readily due to a smaller surface-to-volume ratio and fewer disruptions in the continuity of the surface. Thus products such as celery and rhubarb have a considerably greater potential storage duration than whole leaves. Nevertheless, losses of water and carbon remain of critical importance and storage conditions very similar to those for whole leaves are utilized for petioles.

c. Stems, shoots and spikes

Postharvest products in which the stem is a primary or essential component can be separated into two general classes. (1) Products such as asparagus, coba, sugarcane, celtuce, kohlrabi, tsatsai and bamboo shoots are utilized almost exclusively for their stem tissue even though rudimentary leaves may be present. (2) With many ornamental foliages and floral spikes, the stem represents an essential part of the product although it is generally considered as secondary to the leaves and/or flowers present. Common examples would be gladiolus spikes or any one of many species used for cut foliage. The former class tends to be largely meristematic or composed of very young tissues and is metabolically highly active. They typically can continue to take up water and elongate if placed on moist pads or in shallow pans of water. Exceptions to this would be kohlrabi and tsatsai which are formed at the base of the stem and are more mature and metabolically less active. Products such as ornamental asparagus and floral spikes have developed leaves and/or flowers. These also have relatively high metabolic rates and can continue to take up water if handled properly.

Several stem products (e.g., gladiolus and flowering ginger spikes or asparagus spears) exhibit strong gravitropic responses after harvest. If stored horizontally, the apical portion of the stem will elongate upward, producing a bent product of diminished quality.

Optimum storage conditions vary in this relatively diverse group; however, moist cool conditions predominate. Holding the cut base of the stem in water is either desirable or essential for many stem crops.

d. Flowers

Flowers are compressed shoots made up of specialized foliar parts that are adapted for reproduction. A significant number of harvested plant products are, in fact, flowers. Flowers represent a diverse group varying in size, structure, longevity and use. Uses range from aesthetic appeal to articles of food (e.g., cauliflower, broccoli, lily blossoms).

Flowers are made up of young, diverse, metabolically active tissues, typically with little stored carbon. There are distinct limits on potential longevity when detached from the parent plant, even when held under optimal conditions. In fact, while attached to the parent plant, many of the flowers structural components display only a brief functional existence. Anthesis or flower opening represents a very short period in the overall sequence from flower initiation to seed maturation. Almost invariably, floral products are highly perishable, seldom lasting more than a few weeks after harvest.

From a handling and storage perspective, flowers can be separated into two primary groups: those that are detached from the parent plant at harvest and those that remain attached. Most floral products fall into the former classification. As with other detached plant parts, after harvest they are unable to fix additional carbon through photosynthesis* nor are they able to import photosynthate from adjacent leaves on the parent plant. With many floral products, attached stems represent a reserve of carbon and water that can in part be utilized by the flower. In many cases, this reserve greatly extends the potential longevity of the individual flowers. Thus flower crops are typically made up of much more than just the floral tissues. Attached stems and leaves represent important components and these structures often strongly influence the postharvest behavior of the flower.

Some flowers are marketed attached to the parent plant (e.g., potted chrysanthemum or azalea). The flower or flowers may represent the sole reason for purchase or their presence may simply enhance the attractiveness of the plant, which is the article of primary interest. Both advantages and disadvantages are realized during the postharvest period by having the flowers attached to the parent plant. While the flowers benefit from a continued supply of photosynthate, water and minerals, storage conditions are generally dictated by what is best for the entire plant and not for the flowers alone. Optimal storage temperature for the plant may not necessarily coincide with the optimal storage temperature of the flower.

Individual flowers are borne on a stem or flower stalk, the structure and arrangement of which varies widely between species. They may be found as a solitary flower or on spikes, umbels, panicles and other variations. Structurally a complete flower is made up of sepals, petals, stamens, and pistil borne on a receptacle. Incomplete flowers lack one or more of these floral parts.

The sepals, from the Greek word for covering, are often leaf-like scales that make up the outermost part of the flower. Although leaf-like, the sepals are structurally not as complex as leaves. Usually they are green, although on some species they may have the same coloration as the petals. Collectively, the sepals of a single flower are called the calyx.

The petals (the Greek word for flower leaves) are also modified leaves, and as with the sepals, they are of greater structural simplicity than an actual leaf. The petals of the flower, especially those utilized for their ornamental appeal, are generally brightly colored and showy. Collectively the petals are called the corolla and the petals and sepals the perianth.

Interior to the petals are the pollen bearing parts of the flower, the stamens. The upper portion of a stamen, which produces the pollen, is the anther. This is supported on a slender stalk called the filament. While this male floral part is occasionally found singly, usually there are multiple stamen within a flower. Together, the stamens within a single flower make up the androecium.

The female portion of the flower is the pistil. Structurally it is comprised of three major parts: the stigma, which is the apical tip of the pistil that acts as the receptive surface for the pollen; the style, which is an elongated column of tissue connecting the stigma with the ovary; and the ovary, which is at the base and is the reproductive organ of the flower. Within the ovary are individual ovules attached to the placenta that develop into seeds when fertilized. The pistil may be composed of a single carpel or several carpels fused together. Collectively the carpels make up the gynoecium.

e. Fruits

A fruit is a matured ovary plus associated parts; this includes both the fleshy fruits of commerce such as banana and apple—products normally associated with the term "fruit"—and dry fruits such as nuts, legumes, and siliques. Many of the vegetables we utilize are, in fact, botanically fruits (e.g., tomato, squash, melons, snapbean). It is evident, therefore, that the

---

*Photosynthesis can occur under the appropriate conditions; however, the rate is exceedingly low.

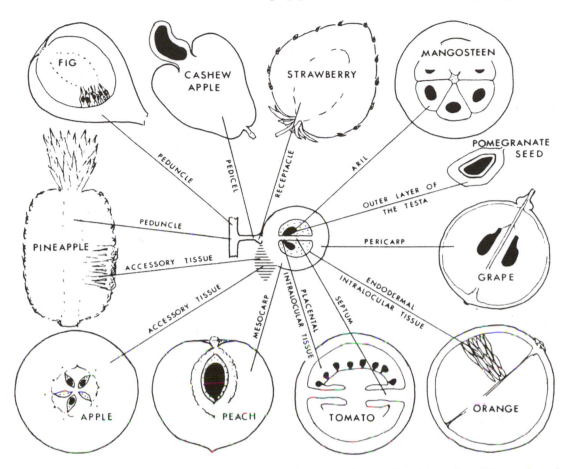

Figure 2.2. A wide range of inflorescence structures can become fleshy and make-up the edible portion of fruits. The diagram illustrates the origin of the fleshy portion of a number of common fruits (*from Coombe*[13]).

term "fruit" encompasses a very broad range in morphological, biochemical and physiological variation.

i. Fleshy Fruits

The actual morphological part of fleshy fruits that is consumed varies widely (Figure 2.2). The fleshy portion can be derived from either the pistil or from parts other than the pistil (accessory parts). The ovary wall, which matures into the pericarp, can be substantial in some fruits. It is subdivided into three regions: the exocarp, mesocarp, and endocarp (from the outside inward). Practically all parts of the floral structure can develop, in various species, into the fleshy part of the fruit. In the peach, the edible portion is principally mesocarp tissue. In many species the accessory tissues of the fruit dominate over the carpellary tissue, making up a majority of the fruit's volume. For example, with the apple the edible portion is perianth tissue, whereas in the strawberry it is largely receptacle tissue. Because of this diversity in morphological makeup of various fruits, an equally wide range in chemical composition between various types of fruits and in biochemical changes within these tissues after harvest should be anticipated.

Fleshy fruits may be divided into several subclasses based on their morphology.

Berry—A pulpy fruit from a single pistil with one or more carpels with several-to-many seeds (e.g., tomato, papaya).

Hesperidium—Fruits of several carpels with a leathery rind and inner pulp juice sacs or vesicles [e.g., orange, lime, grapefruit, lemon).

Pepo—Fruits derived from an inferior ovary that develops from multiple carpels bearing many seeds (e.g., squash, cucumber, melons).

Drupe—A simple fruit where the mesocarp tissue becomes thick and fleshy and the endocarp stony (peach, plum, cherry, olive).

Pome—A simple fruit comprised of several carpels, the edible portion of which is made up of accessory tissue (e.g., apple, pear).

Thus mesocarp, aril, peduncle, pedicel, receptacle, testa, placental, endodermal and other tissues may, in respective fruits, represent all or a major portion of the fleshy edible part of the crop (Figure 2.2). As a consequence, classification of fruits on a strict morphological basis is not overly useful after harvest in that handling and storage requirements do not follow the same classification.

From a postharvest viewpoint, it is more useful to separate fleshy fruits into those that have the potential to ripen after harvest (climacteric fruits) and those that must be ripe when gathered (nonclimacteric fruits). Climacteric fruits such as apples, tomatoes, pears and bananas typically have much more flexibility in the rate at which they may be marketed. For example, it is possible to store mature, unripe apples for as long as 9 months before eventual ripening. This flexibility greatly alters how we approach the postharvest handling, storage and marketing of these products. Most nonclimacteric fruits are ripe at harvest and, although there are notable exceptions to their potential storage duration (e.g., dates and citrus), many have relatively short storage lives. As a consequence, these fruits tend to be marketed relatively quickly.

ii. Dried Fruits

With dried fruits, the fruit wall is sclerenchymatous and dry at maturity. Seldom is the entire fruit utilized; in food crops, typically only the seed represents the article of commerce. Crops such as wheat, rice and soybeans or seeds of many other species are examples of species in which a portion of the dried fruit is used. In contrast to the fleshy parenchymatous tissue of succulent fleshy fruits, the fruit wall of these species is dry at maturity. In a number of dried fruit species, the integuments are completely fused to the ovary wall such that the fruit and seed are one entity (grains, grasses). Likewise, they may also have accessory parts such as bracts that remain attached. Many dried fruits are dehiscent with the fruit wall splitting open at maturity. Most dehiscent fruits contain multiple ovules. Thus dehiscence represents a means for dispersal of the individual seeds. Dried fruits, both dehiscent and indehiscent, can be divided into several subclasses based on fruit morphology. The following are subclasses of dehiscent fruits.

Capsule—The fruit is formed from two or more united carpels each of which contains one-to-many seeds. Common examples of capsules would be the fruits of the poppy, iris and Brazil nut.

Follicle—The fruit is formed from a single carpel that splits open only along the front suture at maturity. Follicles are found widely in floral crops such as the peony, delphinium and columbine.

Silique—The fruit is formed from two united carpels splitting two long halves that are separated longitudinally with the seeds attached. Examples would be fruits in the Brassicaceae family such as those of the radish or mustard.

Legume—The fruit is formed from a single carpel which splits along two sides when mature. Examples would be soybean and the common bean.

Indehiscent fruits often contain a single seed and do not split open upon reaching maturity. The following are subclasses of dehiscent fruits.

Achene—It is a thin-walled fruit containing one seed and in which the seed coat is free, attached to the pericarp at only one point. Examples of species having achenes are the strawberry, sunflower and *Clematis* spp. Although the fruit of the strawberry is botanically a dried fruit, the crop is treated after harvest as a fleshy fruit because the fleshy receptacle is the edible portion of interest.

Caryopsis—It is a one seeded fruit in which the thin pericarp and the seed coat are adherent in the fruit. Rice, wheat and barley are plants with a caryopsis fruit.

Nut—With nuts, the single seed is enclosed within a thick, hardened pericarp. Examples of species having nuts are the filbert, pecan and acorn.

Samara—They are one- or two-seeded fruits which possess a wing-like appendage formed from the ovary wall. The winged propagules of dispersal of the ash, maple and elm are examples of plants with this type of fruit.

Schizocarp—The fruit is formed from two or more carpels that split upon reaching maturity, yielding usually one seeded carpels. The fruits of many of the Apiaceae (e.g., carrot, parsley) are schizocarps.

Mature ovules or seeds contained within the ripened dried fruit represent the primary article of interest for agriculturists. In most instances the seeds are removed from the fruit during harvest or before storage. Seeds are comprised of an embryo, endosperm and testa.

The embryo consists of an axis that at one end is the root meristem and at the other the cotyledon(s) and shoot meristem. The level of structural complexity varies widely; for example, lateral shoot or adventitious root primordia may be present. The endosperm represents the storage tissue of the seed. Not all seeds have endosperms, although those utilized in agriculture almost exclusively have well-developed endosperms. Endosperm structure varies widely, for example, cell wall thickness and amount and nature of stored materials. The principal storage component in most seeds is starch with lower amounts of protein and lipids. In oil seed crops and some nuts, lipids are the primary storage component. In many seeds (e.g., Poaceae), the outer layer of the endosperm (the aleurone layer) is an important site for the synthesis of the enzymes required for the remobilization of the stored material during germination. The testa or seed coat represents a protective barrier for the seed. Its structure (e.g., thickness), composition and physical properties vary considerably.

Like fleshy fruits, dried fruits and seeds begin a progressive, irreversible series of deteriorative changes after maturation and harvest. While it is not possible to stop the process of deterioration, we can, through proper storage conditions, greatly decrease the rate. One important postharvest consideration influencing the way in which the product is handled is the eventual use of the dried fruit or individual seed. Seeds that are to be used for reproduction typically have more stringent storage requirements than seeds that are to be processed into foods or other products. Deteriorative processes in seed stocks result in losses of seedling vigor and subsequently in the ability of the seed to germinate. In most tests, seeds that will no longer germinate due to deterioration are considered dead. In many cases, however, the majority of the cells may be alive. While no longer useful for reproduction, these seeds remain, in many cases, useful for processing. It is also possible for seed to lose its functional utility as a raw product for processing (due to postharvest deterioration) without the complete loss in the ability to germinate. An example would be when low levels of lipid-peroxidation cause sufficient rancidity in an oil seed crop destined for human consumption causing the quality to be impaired (e.g., pecans). Postharvest peroxidation of seed lipids, however, normally is thought to be a primary factor in storage deterioration of seed germination potential.[40,52]

In general there are four critical factors that affect the rate of postharvest losses of dried fruits and seeds. These are the nature of the fruit or seed, its moisture content and the temperature and oxygen concentration of the storage environment. Therefore, proper handling and storage can greatly extend the functional utility of these products.

f. Other structures

Mushrooms, members of the Ascomycetes and Basidiomycetes, represent another form of detached aboveground structures. Some, such as truffles, may be formed belowground. Unlike seed-producing Angiosperms, mushrooms reproduce from spores. The actual mushroom of commerce is the fruiting body of the organism. It is comprised of three distinct parts: the pileus or umbrella-like cap; the lamellae or delicate spore forming gills at the base of the pileus; and the stipe, the stalk on which the pileus is held. The fruiting body may also be thin, ear-shaped and gelatinous or various other structural variations.

Mushrooms sold undried as a fresh product have a relatively short storage and marketing potential. After harvest, the fruiting body continues to develop with the pileus, initially in the form of a tight button or closed cap, opening exposing the gills. With spore shed, the fruiting body begins to deteriorate with development of off-odors. Mushrooms are also subject to quality losses after harvest due to breakage, bruising and discoloration. As the pileus opens, the mushroom becomes much more susceptible to mechanical damage.

1.2.2. Belowground Structures

Subterranean storage organs are specialized structures in which products of photosynthesis accumulate and serve to maintain the plant during periods of environmental stress such as winter or drought. These underground structures do not normally contain chlorophyll but derive the energy necessary for a continued low rate of metabolism from their stored reserves. Commonly buds in these storage organs or in parent tissue associated with them are dormant when harvested and therefore have a suppressed rate of metabolism. While it is desirable for this dormancy to be maintained, it is not always possible. If suitable environmental conditions prevail, active growth commences at the expense of the storage reserves.

a. Roots

Roots are modified to form storage organs in several crops. They are characteristically swollen structures which may contain reserves, primarily of starch and sugars.

The edible radish is partly root and partly hypocotyl tissue. The secondary xylem parenchyma continues to grow and divide and forms the bulk of the radish. There is relatively little stored reserve in the radish; it is mainly cellulose and water. Small amounts of glucose, fructose and starch may accumulate under different cultural conditions. Similarly, the carrot storage root is actually formed from hypocotyl and taproot tissue. The cortex sloughs off and secondary growth results in a central xylem core surrounded by phloem and pericycle tissue. In all these tissues, there is extensive development of parenchyma in which starch is stored. The fleshy root of beet is also of hypocotyl-root origin, but in this case, secondary development is different. A series of cambia develop outside the primary vascular core, each producing strands of xylem and phloem embedded in parenchyma. This structure results in the concentric ring pattern characteristic of cross-sections of beet roots. Starch and sugars accumulate in the parenchyma cells.

The sweetpotato storage organ is an adventitious root in which an anomalous secondary growth occurs. There is a large proportion of parenchyma in the primary xylem and cambia develop in this tissue. These produce xylem, phloem and many storage parenchyma cells in the tissue that was originally xylem. The normal phloem surrounds this unusual development of the xylem, as does the periderm which originates in the pericycle. During secondary growth of these storage roots the cortex and epidermis do not continue to grow but are split by the expanding secondary vascular tissue and eventually slough off. Periderm arises in the pericycle during secondary growth and forms a protective tissue to replace the epidermis. Its protective capacity is due to the accumulation of layers of suberin, a fatty acid substance, and wax in the cell walls.

The capacity to form a protective periderm is not only important in the normal development of storage roots but also in wound healing of roots, particularly sweetpotato, and of modified stems such as the potato tuber. The exposed cells are first sealed with suberin and other fatty materials. High humidity, proper temperature, and adequate aeration are necessary for sealing of the wound, providing an environment that stimulates cell division and formation of a periderm. If excessive moisture is present, suberization may be inhibited and callus tissue formed instead.

b. Rhizomes and tubers

While rhizomes and tubers are underground structures with the superficial appearance of roots, they are anatomically stems. They arise from lateral buds near the base of the main stem and grow predominately horizontally through the soil. They have nodes and internodes with leaves, sometimes reduced to scale leaves, at the nodes. Buds, often referred to as eyes, form in the axil of leaves. These buds may elongate into shoots and adventitious roots may develop from the stem tissue. Rhizomes may be somewhat enlarged and function as storage organs. In some species, they are used for vegetative propagation of the plant, for example many of the members of the *Iris* genus. For others, the rhizomes may be utilized for both propagation and consumption as in ginger and lotus.

Potato tubers form as swelling at the end of short rhizomes. In potatoes, the swelling begins by division of the parenchyma cells in the pith followed by division in the cortex and vascular regions. Vascular elements become separated by parenchyma tissue. Starch begins to accumulate in the cortex and later in the deeper tissues of the vascular region and the pith. The epidermis is replaced by a suberized periderm derived from the subepidermal layer. Numerous lenticels are formed by the production of loose masses of cells under the stomates of the original epidermis. Under favorable conditions the lenticels proliferate and rupture the epidermis and are evident as small white dots on the surface of the tuber. Tissues of the tuber retain their capacity to form periderm even after harvest, provided satisfactory humidity and temperature conditions are maintained. This capacity to heal periderm that is damaged during harvest and handling or even to heal cuts across the tuber in preparation of sets (seed pieces) for propagation is a very important characteristic. Without the formation of this protective layer, the damaged tuber would be subject to excessive water loss and microbial infection.

Yams are another very important food crop usually regarded as tubers. However, Onwueme[38] points out some important differences, particularly the lack of scale leaves, nodes or buds. These so-called tubers do not arise from stem tissue but rather as outgrowths of the hypocotyl. While the tubers do not have preformed buds, they are able to form buds from a layer of meristematic tissue just below the surface. Once dormancy is broken, these buds grow and can be the cause of major postharvest losses. The tubers are covered by several layers of cork (phellem) which arise from successive cork cambia (phellogen). Periderm formation may continue after harvest and is essential for wound healing.

c. Bulbs

Bulbs, as occur in onion, tulip and hyacinth, are underground buds in which the stem is reduced to a plate with very short internodes and the sheathing leaf bases are swollen to form a storage organ. At horticultural maturity, the aboveground parts of leaves shrivel and die but the swollen leaf bases remain alive. There is no anatomical distinction between these two parts and therefore no formation of a protective layer such as occurs in abscission zones or periderms. Therefore, adequate curing or drying of the neck and outer leaf bases is needed if quality is to be maintained during the postharvest period.

Bulbs, as storage organs, have a dormant period during which their metabolic rate is low and keeping quality is good. However, they contain intact shoots and are therefore capable of growth if dormancy is broken and suitable environmental conditions exist.

d. Corms

Corms are short, thickened, underground stems. When dormancy is broken the terminal bud or lateral buds grow into new plants with adventitious roots arising from the base of parent corms and from the new shoots. The dry leaf bases and hollow flower stem of *Gladiolus* spp. tend to remain more or less attached to the corm and afford protection from water loss and mechanical damage.

Taro, on the other hand, does not retain its leaf bases but has a well developed periderm on the corm. Secondary corms or cormels arise from lateral buds on the corm. The Chinese water chestnut forms corms at the end of slender rhizomes. These corms have a scaly, brown periderm.

e. Non-storage organs

Products included in this category are primarily used as propagation material. To be effective as such, they must contain some stored reserves to supply energy and nutrients for the early development of the new plant.

i. Root Cuttings

While a number of species can be propagated from root cuttings (e.g. peony and bramble fruits are occasionally marketed commercially), there is relatively little use of this technique, since other propagation systems are generally less labor intensive and are quicker.

Root cuttings are commonly taken during the dormant stage so they do not display a high level of metabolic activity. Typically, root cuttings are thickened secondary roots with a well-developed periderm. There are often no obvious anatomical features that can be used to determine the orientation of root cuttings, but polarity is maintained, with shoots being produced from the proximal end and roots from the distal. It is, therefore, important to use a handling system that identifies the ends. One such system is to cut the proximal end straight and the distal end slanted.

ii. Crowns

Crown is a general term referring to the junction of shoot and root system of a plant. Those marketed in the horticultural trade are most commonly herbaceous perennials and consist of numerous branches from which adventitious roots arise. Crowns are divided by cutting, preferably during the dormant period, and are marketed before or soon after the onset of active growth. Some plants like asparagus develop thick storage roots as well.

Crowns with young shoots and active roots are susceptible to mechanical damage, and the cut surfaces are subject to microbial infection and water loss. Stored reserves are present in the mature stems and roots and the metabolic rate is comparatively low.

2. TISSUE TYPES

The structure of harvested products can be further subdivided into four general tissue types of which the products are composed. These include the dermal, ground, vascular and meristematic tissues. Dermal tissues form the interface between the harvested product and its external environment and as a consequence, are extremely important after harvest. They often have a dominant influence on gas exchange (e.g., water vapor, oxygen, carbon dioxide) and the resistance of the product to physical and pathological damage during handling and storage. With some postharvest products, the dermal tissues are also important components of the products visual appeal. The luster imparted by the epicuticular waxes on an apple or the presence of desirable pigmentation in the surface cells are examples of this.

Ground tissues make up the bulk of many edible products, especially fleshy products such as roots, tubers, seeds and many fruits. These often act as storage sites for carbon; however, they have many other functions. Maintenance of their characteristic texture, composition, flavor and other properties after harvest is especially important. Vascular tissues are responsible for movement of water, minerals and organic compounds throughout the plant. Their function is especially critical during growth; however, in detached plant parts their primary role is diminished in importance. With some products, for example, carrots, the vascular tissues make up a significant portion of the edible product. Within the vascular and ground tissue systems are found support tissues such as collenchyma and sclerenchyma that give structural support to the plant. Collenchyma tissues are found widely within plants, whereas sclerenchyma tissues, being much more lignified and rigid, are less common in most edible products. Lastly, meristematic tissues are comprised of cells that have the capacity for active cell division. As a consequence, they are of particular importance during growth.

2.1. Dermal Tissue

Dermal tissue covers the outer surface of the plant or plant part and constitute its interface with the surrounding environment. The two principal types of dermal tissues are the epidermis and the periderm.

2.1.1. Epidermis

Epidermal cells covering the surface of the plant are quite variable in size, shape and function. Most are relatively unspecialized, although scattered throughout these unspecialized cells are often highly specialized cells or groups of cells such as stomates, trichomes, nectaries, hydathodes and various other glands.

The epidermis is typically only one layer of cells in thickness, although with some species, parts of the plant may have a multilayered epidermis. Epidermal cells are typically tabular in shape and vary in thickness with species and location on the plant. The cells are alive, metabolically active and may contain specialized organelles and pigments within the vacuoles. Cu-

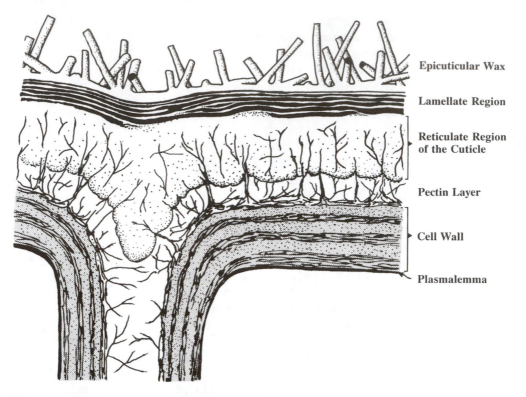

Epicuticular Wax

Lamellate Region

Reticulate Region
of the Cuticle

Pectin Layer

Cell Wall

Plasmalemma

Figure 2.3. Diagrammatic representation of a plant cuticle. Successive layers from the plasma-lemma of the surface cell outward are the cell wall, pectin layer and the pectin layer of the middle lamella, the reticulate region of the cuticle in which cutin and waxes are traversed by cellulose fibrils, the lamellate region where there are separate layers of cutin and wax, and the exterior epicuticular wax (*after Juniper and Jeffree*[26]).

ticle is found on the outer surface of epidermal cells of all aboveground plant parts (e.g., leaves, flowers, stems, fruits, seeds) and often on significant portions of the root system. The cuticle functions as a barrier to water loss and protection against pathogens and minor mechanical damage. It also affects the wetability of the surface, requiring the use of wetting agents (surfactants) for many agricultural chemicals, and the surface optical properties of the product. Cutin, comprised of complex fatty substances found intermeshed within the outer cell wall and on the outer surface of the epidermis, is a primary component of the cuticle, as are waxes (Figure 2.3). Surface cuticular waxes may be seen as flat plates or as rods or filaments protruding outward from the surface (Figure 2.4).

a. Stomata

Stomata are specialized openings in the epidermis which facilitate the bidirectional exchange of gases (water vapor, carbon dioxide, oxygen, etc.).[50] Although varying widely in appearance (Figure 2.5), stomata are comprised of a pair of specialized cells, called the guard cells, which through changes in their internal pressure alter the size of the stomatal opening (Figure 2.6).[20] Below the guard cells is the substomatal chamber (Figure 2.7).

Stomata are found on most aerial portions of the plant including fruit; however, they are most abundant on leaves. The number of stomata per unit surface area of leaf varies widely among species, cultivar and the environmental conditions under which the plants are grown.

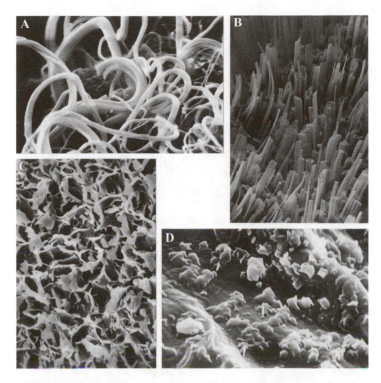

Figure 2.4. The surface of many aerial plant parts is covered with a layer of epicuticular wax. The deposition and continuity of these waxes varies widely from long coiled cylindrical rods (A), long tubules (B), fringed plates (C), to irregularly shaped plates (D) (*from Baker*[3]).

Typically the lower surface of the leaf has the greatest number of stomata; in some cases, they are found only on the lower surfaces, in others (e.g., waterlily) only on the upper surface, and in submerged aquatics generally absent. A density of approximately 100 stomata per mm^2 is common; however, greater than 2,200 per mm^2 are found on *Veronica cookiana* Colenso.[44] In young apple fruit (cv. Golden Delicious) stomatal function is similar to that in leaves, the density of which (25 mm^{-2}) decreases markedly as the fruit increases to its final size (<1 mm^{-2}).[7]

Stomatal opening is altered by light, carbon dioxide concentration and plant water status. The mechanics of opening and closing of the stomatal aperture appears to be largely controlled by the movement of potassium ions between the guard cells and their adjacent neighboring cells. When the potassium ion concentration in the guard cells is high, the stomate is open, when it is low, closure occurs.

Stomatal closure represents a means of conserving water within the tissue. When plant parts are severed at harvest from the parent plant, the supply of water from the root system is eliminated and the stomata typically decrease their aperture markedly. During the postharvest period, stomata also represent potential sites for entry of pathogenic fungi. This allows the fungi to bypass the plants' surface defense mechanisms (chiefly the cuticle and epicuticular waxes) and gain rapid entry into the interior.

b. Trichomes

Trichomes represent another type of specialized epidermal cells that may be found on virtually every part of the plant.[21] These extend outward from the surface, greatly expanding the

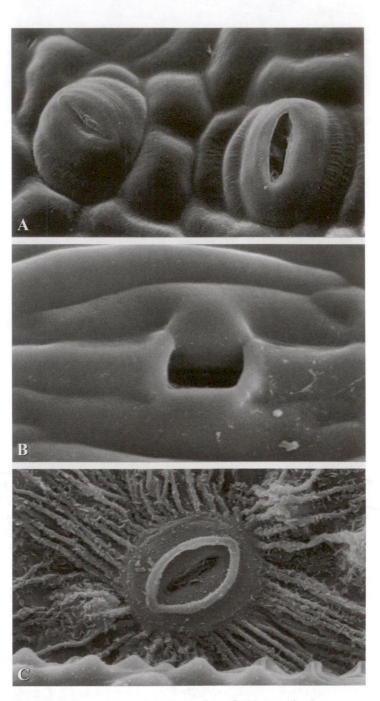

Figure 2.5. Stomata seen from the surface of: (A) *Liquidambar styraci-flua* L.; (B) a *Pinus caribaea* Morelet. × *P. palustris* Mill. hybrid; and (C) *Cornus florida* L. illustrate variations in morphology (*Photographs courtesy of H. Y. Weitzstein*).

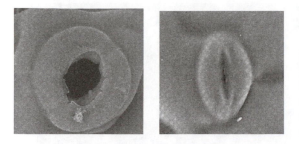

Figure 2.6. Open and closed stomata of *Arabidopsis thaliana* (L.) Heynh., the opening of which is caused by the transport of potassium into the guard cells mediating an increasing turgor pressure and their outward swelling (*Scanning electron micrographs courtesy of E. Grill and H. Ziegler*[20]).

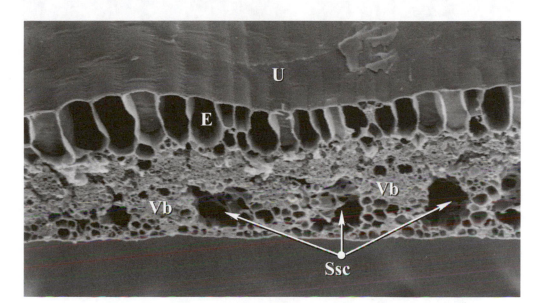

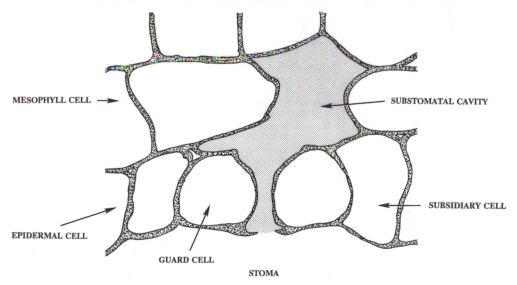

Figure 2.7. (A). An electron micrograph of a *Cyperus* spp. leaf displaying stomata with large substomatal cavities (Ssc) on the lower surface (U—upper surface of the leaf; E—epidermal cell; Vb—vascular bundle). (B). Diagrammatic representation of a lemon leaf stomata in cross-section. (*Scanning electron micrograph courtesy of H. Y. Wetzstein*).

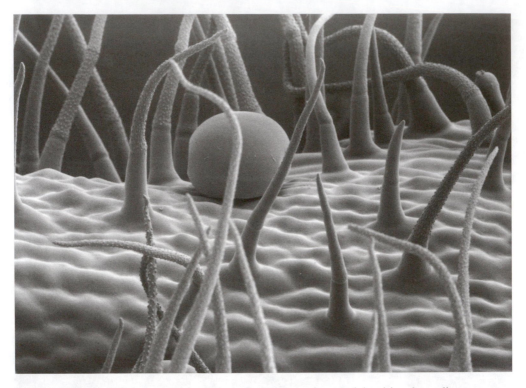

Figure 2.8. Surface trichomes are found in various shapes, sizes, and densities, depending upon species and position on the plant. The cyro scanning electron micrograph displays simple multicellular hairs and a peltate gland on the surface of a *Nepeta racemosa* Lam. leaf (*Micrograph courtesy of R.J. Howard, DuPont Co.*).

surface area (Figure 2.8). As a class, trichomes are highly variable in structure, ranging from glandular to nonglandular, single celled to multicelled. The tremendous diversity makes classification into precise groups difficult,[39] as a consequence, they are often separated into three very general groups: glandular, nonglanular and root hairs. The surface fuzz on peach and okra fruit is an example of trichomes.

Root hairs greatly expand the water and nutrient absorbing surface of the root system as well as anchoring the elongating root from which they are borne. With aerial trichomes, the role is less well defined. In some species, they represent one means by which the plant combats insect predation. The specialized hooked trichomes of some species trap insects, thus providing a physical barrier (Figure 2.9). Trichomes may contain a number of chemicals which attract or repel specific insects. In addition, some trichomes function as sites for the sequestering and/or secreting of certain chemicals. Excess salts taken up by some species are removed *via* the trichomes. Trichomes are also known to exert a pronounced effect on the boundary layer of air around the organ, affecting the exchange of gases.

During the postharvest period, trichomes which are broken during harvesting or handling provide primary entry sites for pathogens. Operations such as defuzing peaches (trichome removal), as a consequence, can greatly decrease the life expectancy of the product if appropriate treatments are not utilized.

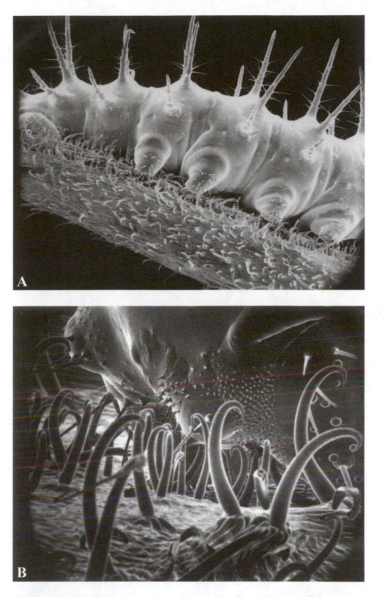

Figure 2.9. Trichomes may in some species function as defense structures discouraging insect herbivores. The photographs show the trapping of second (top) and third (bottom) instar larvae of *Heliconius melpomene* by the hooked trichomes on the leaves and petioles of *Passiflora adenopod* (*from Gilbert*[18]). Damage to trichomes during or after harvest, however, provides entry sites for postharvest pathogens.

c. Nectaries

Nectaries are multicellular surface glands found on flowers and other aerial plant parts that secrete sugars and certain other organic compounds (Figure 2.10). Their position on flowers, the most common site, varies with species; they may be found on the petals, sepals, stamens, ovaries or receptacle. Nectaries may be flush with adjacent surface cells or be found as out-

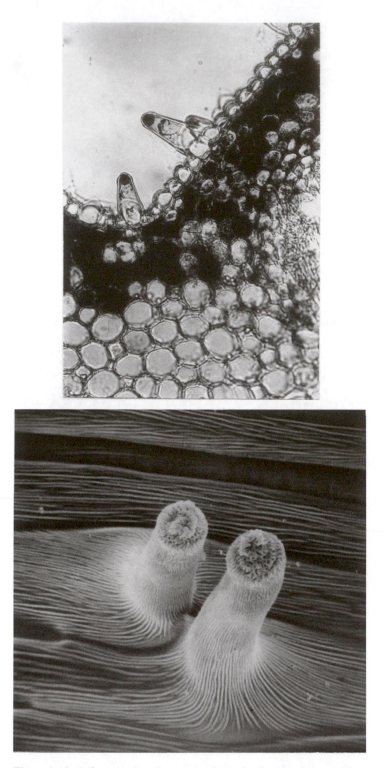

Figure 2.10. Micrographs of nectaries from the flower of *Tropaeolum majus* L. (*left*—cross-section; *right*—scanning micrograph). On some species nectaries may be found on other organs (e.g., leaves, bracts, sepals). Their postharvest importance lies largely in their potential role as sites for the entry of pathogens (*Micrographs courtesy of Rachmile-vitz and Fahn*[41]).

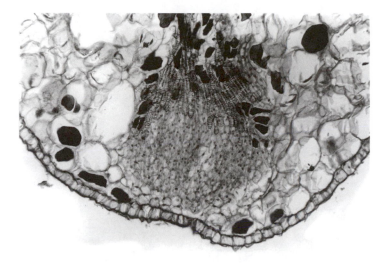

Figure 2.11. Light micrograph of a hydrothode. Hydrathodes represent one of several types of specialized secretory structures. They secrete water which is brought to the leaf periphery by tracheids. The water then moves through an area of loosely packed parenchyma cells called the epithem, exiting through modified stomata (pores), no longer capable of closing. Secretion of water, called guttation, results in the formation of droplets along the border of the leaf (*Micrograph courtesy A. Fahn[17]*).

growths or sunken. They consist of an epidermis, with or without trichomes, and specialized parenchyma.[15] Sugars are the primary group of organic compounds secreted, which exude from either modified epidermal cells or trichomes of the nectary. The primary function of nectaries is the facilitation of flower pollination by insects.[47]

d. Hydathodes

Hydathodes are much more complex than just modified epidermal cells; however, their function is in many ways similar. They represent a modification of both the vascular and ground tissues along the margins of leaves, that permits the passive release of water through surface pores that remain permanently open[45] (Figure 2.11). Guttation, the discharge of water in the liquid state, results in small droplets forming on the leaf margins or tip and occurs through the hydathodes. Guttation is most noticeable when transpiration is suppressed, such as when the relative humidity is high during the night. The surface pore or opening is of stomatal origin; however, unlike stomates, a hydathode is not capable of altering its aperture. Hydathodes represent potential sites for water loss and pathogen entry in harvested leafy products.

2.1.2. Periderm

On plant parts such as roots and stems that increase in thickness due to secondary growth, the epidermis is replaced by a protective tissue called the periderm. Periderm is also formed in response to wounding in most species. When roots are damaged during harvesting, wound periderm forms over the wound surfaces, decreasing the risk of pathogen entry.

The periderm is composed of three tissue types: (1) the phellogen or cork cambium from

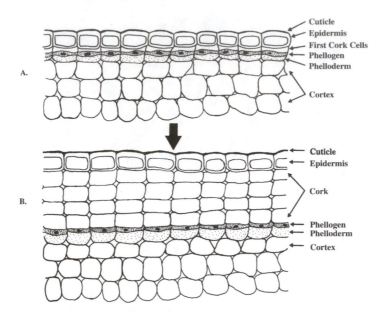

Figure 2.12. When plant parts increase in thickness due to secondary growth, the epidermis is replaced by periderm (A) which is composed of the phellogen, the phellem or cork and the phelloderm. The phellogen divides (B) forming successive layers of cork cells which act as a protective barrier (*redrawn from Kramer and Kozlawski*[28]).

which the other cells of the periderm arise; (2) the phellem or cork, which is the protective tissue or the exterior surface; and (3) the phelloderm, which is tissue found interior to the cambium (Figure 2.l2). The phellem or cork cells are not alive when mature and typically have suberized cell walls. They are tightly arranged, having virtually no intercellular space between adjacent cells, and are found in layers of varying cell number in thickness.

Some harvested products such as sweetpotato roots or white potato tubers must be held under conditions favorable for wound periderm formation prior to storage. Warm temperature and high relative humidity (i.e., 5–7 days at 29°C, 90–95% RH for sweetpotato roots) favor rapid wound periderm formation, thus decreasing water loss and pathogen invasion during storage.

Interspersed on the periderm of many species are lenticels comprised of groups of loosely packed cells having substantial intercellular space (Figure 2.l3). These appear to be present to facilitate the diffusion of gases into and out of the plant. Unlike stomata, lenticels are not capable of closure and as a consequence, their presence enhances the potential for water loss from the product after harvest.

Lenticels range in size from extremely small to up to 1 cm in diameter depending on the species and location on the plant. They are often seen as groups of cells with a more vertical orientation, protruding above the surface of the periderm (Figure 2.l3).

2.2. Ground Tissue

2.2.1. *Parenchyma*

Parenchyma cells form the ground tissue of most postharvest products. In fleshy fruits and roots and in seeds they act as storage sites for carbohydrates, lipids or proteins and make up

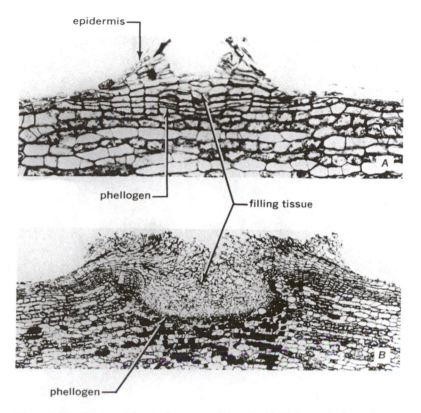

Figure 2.13. Cross-sections of a young (A) and old (B) lenticels from *Persea americana* Mill. stems. Note the phellogen at the base of the lenticels and the extensive filling of the older lenticel (B) (*from Esau*[15]).

the bulk of the edible portion. In leaf tissue, parenchyma cells have numerous chloroplasts and have a photosynthetic function. Parenchyma cells may also act as secretory cells and can resume meristematic activity in response to wounding.

Parenchyma cells have numerous sides or facets and are highly variable in shape. The number of sides ranges from approximately 9 to 20 or more. Within a single mass of relatively homogenous parenchyma cells, both the number of sides and the actual size of the individual cells often vary. In fruit, roots and tubers there is considerable intercellular space, whereas in seeds the parenchyma cells are much more compacted. In some aquatic species, the parenchyma cells are very loosely packed forming a tissue called aerenchyma, facilitating diffusion of gases.

The characteristics of individual parenchyma cells are to a large extent dependent on the function of the tissue and its composition. Photosynthetic parenchyma in the mesophyll of leaves forms chlorenchyma due to the abundance of chlorophyll. Storage parenchyma cells exhibit characteristics that are in part dependent on the organic compounds sequestered within their specialized plastids. Parenchyma cells of tubers, roots and some fruits contain amyloplasts that store starch. The parenchyma cells in flowers contain chromoplasts or vacuoles with various colorful pigments.

Parenchyma cells typically have relatively thin primary cell walls, although in some seeds these walls may be rather thick. This general lack of structural rigidity of the wall makes the parenchyma cells rapidly loose their shape and textural properties when water is lost. In many

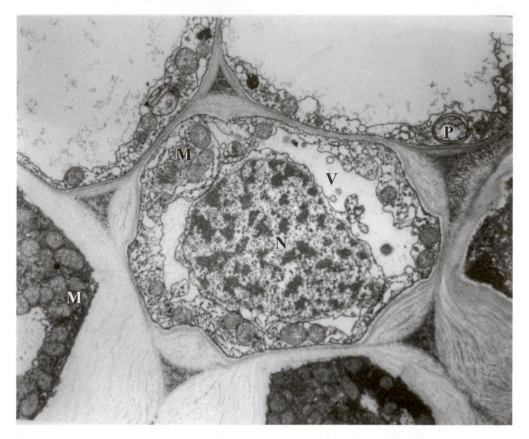

Figure 2.14. Electron micrograph of collenchyma. Particularly noticeable are the much thickened primary cell walls which lend structural support to the tissue and the presence of protoplasm containing a central nucleus (N), vacuole (V) and a number of mitochondria (M). The cell wall is non-lignified, containing large amounts of pectin and water (*Electron micrograph ×16,500 from Ledbetter and Porter*[30]).

postharvest products, the actual percentage of the total water present that must be lost before significant alterations occur in the tissues' physical properties is often relatively small.

2.2.2. Collenchyma

Collenchyma is in many ways similar to parenchyma; however, collenchyma cells have thickened cell walls that provide structural support for the plant (Figure 2.14). The cells are strong and flexible with nonlignified cell walls. The walls are primary in nature but considerably thicker than those found surrounding parenchyma cells. In addition, collenchyma cells tend to be more elongated in shape than parenchyma cells. The walls of collenchyma are much more pliable than their structural counterparts, sclerenchyma (Figure 2.15). In addition, they remain metabolically active and have the ability to degrade much of the wall if induced to resume meristematic activity.

The walls are composed primarily of cellulose, pectins and hemicelluloses but are not lignified. Thickening of the wall occurs as the cell grows. Mechanical stress caused by wind in-

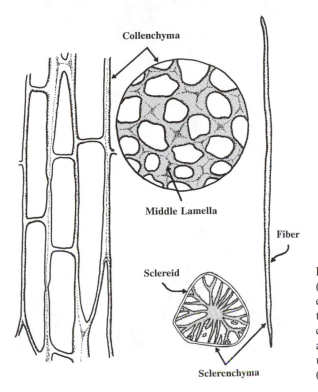

Collenchyma

Middle Lamella

Fiber

Sclereid

Sclerenchyma

Figure 2.15. Supporting cells of collenchyma (left) and sclerenchyma (right) tissues are often elongated. Sclereids, one form of sclerenchyma, tend to be shorter and compact. Collenchyma cells have irregularly thickened primary cell walls and differ from sclerenchyma in that the latter usually have well developed secondary cell walls (*after Esau*[15]).

creases the extent to which the walls thicken. In older parts of the plant, collenchyma cells may form secondary cell walls, becoming sclerenchyma.

As a support tissue, collenchyma is found in the aerial parts of the plant and not in the roots. The cells generally are located just below the surface of leaves, petioles, and herbaceous stems. The strands of cells found in the edible petioles of celery are collenchyma cells and vascular tissue.

2.2.3. Sclerenchyma

Sclerenchyma lends hardness and structural rigidity to plants and plant parts. Cells of this tissue have lignified secondary cell walls which are formed after the completion of expansion (Figure 2.15). At maturity most are nonliving and no longer contain protoplasts. Sclerenchyma is found throughout the plant but seldom in the extensive homogenous masses in which parenchyma and collenchyma are found. Rather, sclerenchyma is usually found as individual cells or in small clusters interspersed among other cell types. Two general cell types may be distinguished based primarily on shape. Sclereids tend to be shorter and more compact than fibers which typically are quite elongated. Sclereids are highly variable in shape; some are compact and more-or-less regular, while others may be highly branched. They are often found in layers or clusters in epidermal, ground and vascular tissues of stems, leaves, seeds and some fruits. The stone cells in pears are an example of sclereids (Figure 2.16).

Fibers are elongated cells (Figure 2.17) varying from quite short to as long as 250 mm in the ramie fibers of commerce. They function as support elements in non-elongating plant parts, especially stems and the leaves and fruit of some species. The formation of fiber can oc-

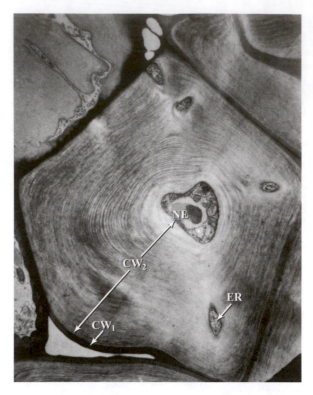

Figure 2.16. Stone cells such as those found in pear fruit represent an example of a sclereid. Extensive secondary cell wall (CW_2) surrounds the cytoplasm which contains mitochondria, plastids and nuclear envelope (NE). The primary cell wall (CW_1), is found external to the outer layers of the secondary wall (ER—endoplasmic reticulum), (*Electron micrograph ×12,000 courtesy of Ledbetter and Porter*[30]).

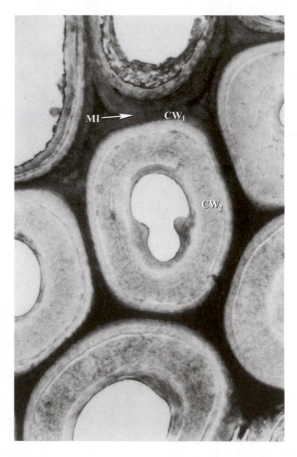

Figure 2.17. Fibers seen in cross section. The central fiber has an empty central cavity which is surrounded by a thick secondary cell wall (CW_2) composed of three layers. Adjacent to the exterior layer of the secondary cell wall is the primary cell wall (CW_1), followed by the middle lamella (Ml), (*Electron micrograph ×17,000 courtesy of Ledbetter and Porter*[30]).

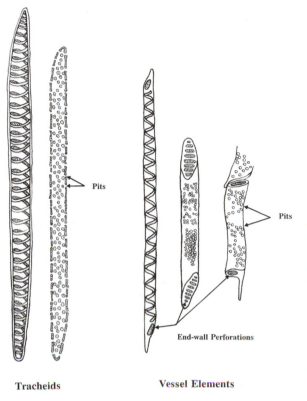

Figure 2.18. Water and minerals are moved throughout the plant in specialized xylem cells, tracheids and vessel elements. Tracheids have tapered end walls and represent a more primitive form of xylem than vessel elements. The geometry of the secondary cell wall varies from spiral to extensive. When there is an extensive secondary wall, numerous pits are generally present through which water moves from tracheid to tracheid. Vessel elements are found end to end, with water moving through their perforated end walls (*redrawn from I.P. Ting*[46]).

cur after harvest in some products, decreasing their acceptability. This is often a problem, for example, in asparagus and okra.

2.3. Vascular Tissue

Vascular tissues provide the conduits for the movement of water and nutrients throughout the plant. Of the tissues found in the plant, vascular tissue is the most complex, being composed of several types of cells. The xylem and phloem are the two types of vascular tissues. Within the xylem, water, minerals and some organic compounds from the root system move upward throughout the plant. Carbohydrates (chiefly sucrose) and to a much lesser extent other organic compounds formed in the leaves or apical meristem are transported both acropetally and basipetally in the phloem.

2.3.1. Xylem

The xylem functions primarily as a tissue for the conduction of water, but it also has storage and support functions. This diversity in roles is in part due to the occurrence of several types of cells making up the xylem. Tracheids and vessel members are the water-conducting cells (Figure 2.18). Tracheids are elongated cells, tapering toward the ends, with secondary cell walls that impart structural support to the plant. At maturity they are nonliving. Water moves through openings, called pits, in the sides of the cells into adjacent pits of neighboring tracheids.

Vessel elements, also elongated cells but often with flattened porous ends, are joined end-

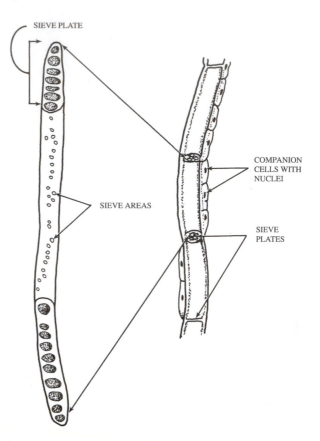

Figure 2.19. Photosynthates and other organic constituents move through the cells of the phloem. Illustrated at the left is an elongated sieve-tube element with sieve plates on the end walls formed from groups of sieve areas. Sieve areas are also found on the side walls through which some lateral transport takes place. At the right are three sieve-tube cells attached end to end forming an elongated conduit (*adapted from Wilson and Loomis*[51]).

to-end, forming vessels. These may be as much as a meter or greater in length in some species. Vessels are a more complex form of water-conducting cells found in angiosperms. They appear to have evolved from tracheids during the evolutionary development of the angiosperms.

Associated with tracheids and vessel elements are parenchyma cells which act as storage sites. These cells may contain starch, lipids, tannins or other material and are particularly important in the secondary xylem of woody perennials. Fibers are also part of the xylem, providing further structural support.

The xylem may be of either primary or secondary origin. Primary xylem is formed during the initial development of the plant, arising from the procambium. Secondary xylem develops during the secondary thickening of stems and roots, and is derived from the vascular cambium.

2.3.2. *Phloem*

The phloem is the photosynthate-conducting tissue of vascular plants. Carbohydrates formed in the leaves are transported both acropetally to the growing tip of the plant and basipetally toward the root system. The directional allocation depends on the position and strength of competing sinks within the plant for photosynthate, the position of the individual leaf on the plant, its stage of development, time of day and other factors. In plant parts that are detached from the parent plant at harvest, phloem transport is essentially eliminated.

Phloem, like the xylem, is composed of several types of cells; however, phloem tends to be less sclerified. Photosynthate and other organic compounds are transported through specialized elongated cells called sieve elements (Figure 2.19). While sieve elements are much like

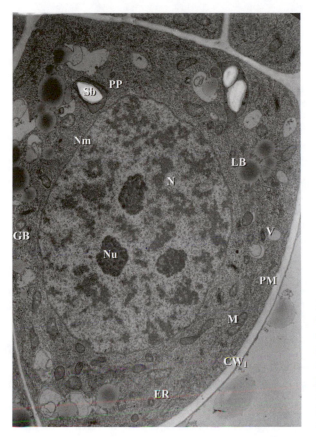

Figure 2.20. Meristematic cells are characteristically small, have only a few small vacuoles (V) and a relatively thin primary cell wall (CW_1), (N—nucleus; Nu—nucleoli; NM—nuclear membrane; PM—plasma membrane; PP—protoplastid; ER—endoplasmic reticulum; LB—lipid body; M—mitochondrion; Sb—starch body; GB—Golgi bodies) (*Electron micrograph courtesy of H. Ammerson*).

the tracheids of the xylem, they differ in being alive at maturity, although they do not contain a nucleus or vacuole. Individual cells are joined end to end, forming sieve tubes. At the ends of each cell is a porous plate called the sieve area or sieve plate which allows, and may in part control, the movement of material from one cell to the next. The remainder of the cell wall is variable in thickness and may also have perforated regions.

Associated with a sieve element are one or more parenchymatous companion cells. These appear to partially control the enucleate sieve element and are joined by an interconnecting membrane system. Companion cells also function during the loading of photosynthate into the phloem and the subsequently unloading upon arrival at the sink site. These parenchyma cells may also act as storage sites for an array of organic compounds.

Fibers are also associated with the phloem. These provide structural support and rigidity to the system.

Phloem, like the xylem, can be divided into two general classes based on origin. The primary phloem is derived from the procambium while the secondary phloem arises from the vascular cambium during secondary thickening.

2.4. Meristems

Meristems are comprised of groups of cells that retain the ability for cell division. Their primary function is in protoplasmic synthesis and the formation of new cells. Meristematic cells are typically small with a thin primary wall and few vacuoles (Figure 2.20). Certain of these cells undergo division, forming new cells. Using *Arabidopsis* as a model, the mechanisms controlling turning on the cell division process in meristematic cells and switching them from vegetative to reproductive growth[48] are beginning to be elucidated.

An apical meristem is found in the growing tip of shoots and roots. It gives rise to the primary growth and structure of the plant. Lateral meristems give rise to the secondary growth of tubers, storage roots and woody stems. A third type of meristematic tissue is the intercalary meristem found in the growing stems of many monocots, such as grasses.

When plant parts are decapitated at harvest, there is generally little meristematic activity. Some tissues, however, can and do recycle nutrients and water into these cells resulting in growth. After extended cold storage, the apical portion of the stem of cabbage will resume growth. With intact plants during storage, conditions are generally selected to minimize growth, and as a consequence, meristematic activity is also repressed. Maintenance of meristems in a healthy condition in intact plants, however, is essential. With improper storage, the apical meristems can readily die, decreasing both the quality of the product and the rate of which it recovers upon subsequent planting.

3. CELLULAR STRUCTURE

Cells are the structural units of living organisms, aggregations of which form tissues. Plant cells vary widely in size, organization, function and response after harvest. They differ from

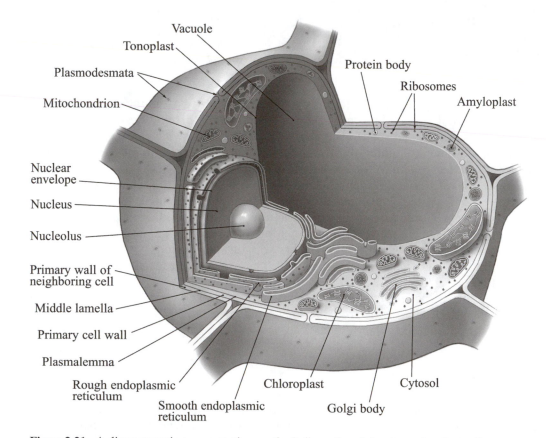

Figure 2.21. A diagrammatic representation on the 3-dimentional structure of a plant cell.

Table 2.1. Comparison of the Size and Number of Subcellular Structures.

Constituent	Size (diameter)	Number per cell*
Vacuole	up to 95% of cell volume	1–several
Nucleus	5–15 μm	1
Chloroplasts	5–10 × 2–4 μm	0–200[†]
Chromoplasts	3–10 μm × 2–4 μ	0–200
Amyloplasts	7–25 μm × 7–45 μm	0–30
Nucleolus	3–5 μm (dia)	1
Mitochondria	1–4 × 0.5–1.0 μm	500–2,000
Primary cell wall	1–3 μm (thick)	1
Microbodies	0.5–1.5 μm (dia)	few–>1000
Peroxisomes	0.5–1.0 μm (dia)	highly variable
Glyoxysomes	1.0–1.5 μm (dia)	highly variable
Golgi apparatus	0.5–2.0 μm (dia)	few–>100
Ribosomes	0.015–0.025 μm (dia)	$5–50 \times 10^5$

*The specific number of many organelles can be highly variable depending upon cell type, age, condition and other factors.
[†]Beet leaf parenchyma cells contain 40–50 chloroplasts.[6]

animal cells in the presence of a rigid cell wall and a large central vacuole. Interior to the cell wall is the plasma membrane or plasmalemma, which separates the interior of the cell and its contents, the protoplasm, from the cell's external environment. Much of the cell's energy transfer and synthetic and catabolic reactions occur within the cytoplasm, as does information storage, processing and transfer systems. Within the cytoplasm of eukaryotic cells (organisms other than bacteria and blue-green algae) are numerous organelles and cytoplasmic structures. These provide a means of compartmentalization of areas within the cell that have specific functions. Organelles such as the nucleus, mitochondria, plastids, microbodies and Golgi bodies and cytoplasmic structures such as microtubules, ribosomes and the endoplasmic reticulum are found within the cytoplasm (Figure 2.21). While the number of various organelles varies with cell type, age, and location within the plant, a general example of the relative number per cell is presented in Table 2.1.

The response of plant products after harvest is a function of the collective responses of these subcellular organelles. An understanding of the structure and function of cells and their individual components provides the basis for a more thorough understanding of postharvest alterations occurring in the product.

3.1. Cell Wall

The cell wall has long been known to act as a cytoskeleton, providing mechanical support for plants and harvested products, but only recently has it been established that the wall plays a much more dynamic role.[10] For example the wall appears to be involved in instigating certain pathogen defense mechanisms. Likewise, postharvest structural changes in the cell walls of some products are of tremendous importance in that they can result in alterations in texture and quality. There are two types of walls, primary (Figure 2.22) and secondary, the latter being much more rigid and formed interior to the primary wall (Figure 2.16). While all cells have primary wall, not all have secondary. Typically very few of the cells in edible products contain secondary walls.

The cell wall varies widely in composition and appearance among cell types, species, location and other factors. The wall can vary in thickness, number of plasmodesmata (small,

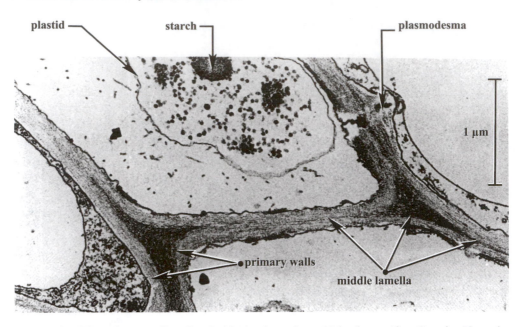

Figure 2.22. The primary cell walls of phloem tissue from *Nelumbo nucifera* Gaertin. Note the middle lamella, found externally to the primary cell wall, and the absence of a secondary cell wall (*from Esau*[15]).

membrane-lined channels transversing the wall and connecting neighboring cells through which constituents are transported and communication occurs), and other factors depending upon the location of the cell. Epidermal cells, for example, typically have substantially thinner interior than the exterior walls, the latter being embedded with cutin and waxes. The primary cell walls of onion are only about 0.1 μm thick; however, the wall can range up to 3 μm in thickness in the cells of some species.

The primary wall is formed from the cytoplasm during cell division and contains cellulose, hemicellulose, pectin, both structural and non-structural proteins,[12] water, and other organic and inorganic substances. A substantial amount of metabolic energy is invested in the synthesis of these components at precise times and in appropriate amounts during development when the actual size of the cell increases from 10 to 1000 fold. The wall also has a diverse mosaic of specialized carbohydrates and proteins that interact with molecules both inside and outside of the cell. This interaction is especially important during development, when the fate of a cell depends on the neighboring wall(s) it is touching. The wall's role in turning undifferentiated cells into specific organs has established it as far more than simply a rigid enclosure for the protoplasm.

In the primary cell wall, cellulose molecules, composed of long chains of glucose subunits, occur in orderly strands arranged into microfibrils. Microfibrils make up the cellulose framework of the wall, which is embedded with a hydrated complex of polysaccharides. The microfibrils are aligned and cross-linked with other constituents to give a very stable, rigid structure (Figure 2.23). The cellulose component is quite stable, and as a consequence, there is little alteration after harvest in the cellulosic structure of the cell wall, exceptions being in abscission zone cells and during senescence. As a consequence, alterations in the wall leading to softening do not involve alters in microfibril composition.

Hemicelluloses are flexible polysaccharides that are attached to the surface of the mi-

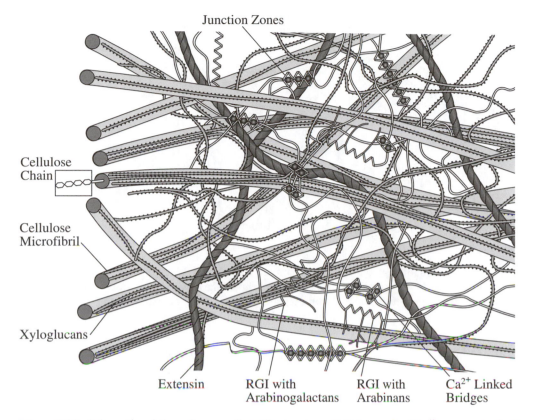

Junction Zones

Cellulose Chain

Cellulose Microfibril

Xyloglucans

Extensin RGI with Arabinogalactans RGI with Arabinans Ca^{2+} Linked Bridges

Figure 2.23. Schematic of the primary cell wall's structure. Cellulose microfibrils, comprised of straight chains of glucose molecules, are coated and cross-linked by xyloglucans. In addition, rhamnogalacturonans (RGI) with arabinogalactans or arabinans attached, protein strands (extensin), cross-linkages *via* Ca^{2+} bridges and junction zones collectively form a three-dimensional framework. Interspersed within the structure are enzymes that facilitate structural alterations in the rigidity of the wall, for example, during fruit ripening (*after Buchanan, et al., Carpita and Gibeart,*[11] *and Rose and Bennett*[42]).

crofibrils, tethering them together. Embedded within this structure is a watery gel of pectin molecules and structural proteins. Pectins are found in the middle lamella, the area between neighboring cell walls, and act as a cementing gel holding adjacent walls together. They are also interspersed in the cellulose microfibrils along with hemicellulose and structural proteins. The ratio of these components on a dry weight basis in cell walls varies from approximately 25% cellulose, 25% hemicellulose, 35% pectin and >8% protein in many cells to 60–70% hemicellulose, 20–25% cellulose, and 10% pectin in grass coleoptiles. Tomato fruit cell walls, for example, have relatively high levels of hemicellulose. Water also represents a major component of the walls, comprising 75–80% of the total fresh weight.

Secondary cell walls, formed interior to the primary cell wall, provide much greater rigidity to the cell due to the presence of lignin. The secondary cell walls also contain cellulose and hemicelluloses, but very little pectin. Most edible plant products do not have a large number of cells with secondary cell walls.

After harvest or during ripening pectin and hemicellulose molecules in the primary cell walls of many fruit are enzymatically altered, leading to significant changes in cell-to-cell bonding and the rigidity of the structural framework of the tissue. Therefore postharvest al-

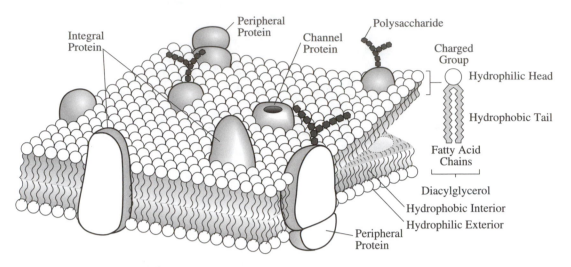

Figure 2.24. Diagrammatic representation of a plant membrane bilayer separated in one corner as would be visualized using freeze fracture preparation. Present is the lipid bilayer with hydrophobic head groups and hydrophobic tails, transport proteins and glycoproteins. The structure of a common membrane phospholipid, illustrates the hydrophilic head group and the hydrophobic tail (*after Wolfe*[53]).

terations in the wall play an important role in textural changes in products, such as softening in apple and pear fruit.

3.2. Cellular Membranes

Diverse membranes are integral components of the cell. Cellular membranes are no longer viewed as separate, autonomous entities, but rather as part of a dynamic and interaction system which plays a major role in controlling the movement of molecules within and secretion outside of the cell, as a site for the biosynthesis of compounds, and in the production of certain cytoplasmic organelles. Postharvest disruption of the membrane system due to low temperature, mechanical or other stresses can have a disastrous impact on the cell.

Cellular membranes are comprised of a lipid bilayer where two layers of polar lipids (predominately diacylglycerols) are oriented with their hydrophobic tails toward the interior of the membrane (Figure 2.24) and their hydrophilic end facing the aqueous exterior. Phospholipids or glycolipids, depending upon the particular membrane, make up the bulk of the membrane structure; however, proteins, both lipoproteins and non-conjugated proteins, are critical components that have transport, catalytic and other functions. The lipid composition of the membrane imparts a flexible, fluid-like structure under ambient conditions that is essential for its proper function. At low temperatures fluidity can be greatly reduced, especially in certain areas of the membrane. This reduction in fluidity can impair membrane function, leading to physiological disorders (chilling injury) in certain crops.

The cell's membrane system is often separated into several very general segments—the **plasmalemma** (or plasma membrane) which surrounds the cell, the **endomembrane system** (endoplasmic reticulum, nuclear envelope, tonoplast, Golgi apparatus, microtubules, etc.), and the mitochondrial and plastid membranes, the latter being self-reproducing and as a consequence, more independent of the endomembrane system.

The cell is separated from its surrounding environment by the plasmalemma, a thin membrane (approximately 7.5 nm wide) found just interior to the cell wall. The plasmalemma is composed of a viscous lipid bilayer (Figure 2.24). Interspersed on the surface and within the

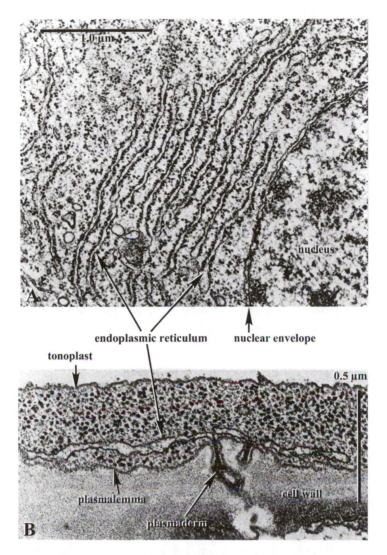

Figure 2.25. Transmission electron micrograph of leaf cells of *Nicotiana tabacum* L. (A) and *Beta vulgaris* L. (B) displaying the endoplasmic reticulum (ER), plasmalemma, nuclear envelope and tonoplast membrane (*from Esau*[15]).

membrane are a number of other molecules (e.g., proteins) which carry out an array of critical functions. For example, some sites on the membrane control the specificity of transport of molecules across the membrane into and outside of the cell, while others catalyze specific reactions. Interspersed along the plasmalemma are plasmodesmata[34], narrow cylindrical strands of cytoplasm bound by the plasmalemma that penetrate through the cell walls and into the neighboring cell, transforming the plant from a collection of individual isolated cells into a interconnection network of protoplasts. Molecules that due to their size or charge can not readily cross the plasmalemma can move between cells by way of the plasmodesmata. For example, transcription factors that instigate gene expression can move from cell to cell *via* plasmodesmata.[37]

A number of other membranes are found within the cell (Figure 2.25). Organelles such as the nucleus, mitochondria, and plastids are bound by two membranes. The vacuole and microbodies, on the other hand, are enclosed by a single membrane. Also found throughout the cytoplasm is the endoplasmic reticulum. It forms a continuous membrane system that functions as a reactive surface for many biochemical reactions and in compartmentalization of certain compounds. Ribosomes, which carry out the assembly of new proteins, are often attached to the endoplasmic reticulum (called the rough endoplasmic reticulum). The smooth endoplasmic reticulum (areas without ribosomes) is believed to be involved in transporting and secreting sugars and lipids.

3.3. Cytoplasm

The cytoplasm is the viscous matrix which envelops all of the more differentiated parts and organelles of the protoplasm. It is the cellular mass interior to the plasma membrane and exterior to the vacuolar membrane, the tonoplast. Individual organelles (e.g., nucleus, mitochondria, plastids) while found in the cytoplasm, are not considered part of it. The cytoplasm contains proteins, carbohydrates, amino acids, lipids, nucleic acids, and other substances that are water soluble. Generally, nonorganelle structures such as microtubules, ribosomes, and the endoplasmic reticulum are also considered as part of the cytoplasm. Although under certain conditions the cytoplasm can assume a gel-like structure, typically it is viscous and can move within the cell. This movement, called protoplasmic streaming, is quite substantial in some types of cells.

3.4. Nucleus

Nearly all living cells of agricultural plant products are uninucleate. Exceptions would be mature sieve tube cells which no longer contain a nucleus and certain specialized cells which are multinucleate. The nucleus represents the primary repository of genetic information within the cell and thus functions in the replication of this information during cell division, and more importantly from a postharvest context, as the cell's command center, controlling protein synthesis. Specific enzymes essential for ripening and other postharvest changes are assembled in the cytoplasm. The nucleus is delimited by two porous membranes separated by a perinuclear space (Figure 2.26).

Interior to the nuclear membrane is the nuclear matrix or nucleoplasm which contains deoxyribonucleic acid (DNA), ribonucleic acid (RNA), nucleic acids, proteins, lipids, and other substances. Also found in the nuclear matrix is a dark spherical body, the **nucleolus.** It functions in the storage of RNA and the synthesis of ribosomal RNA. The chromosomes are also found here; however, with the exception of during cell division, they are not in a tightly coiled state, but are found as a matrix of DNA and protein called **chromatin.** These associated proteins, a significant number of which are histones, act as repressors and activators of transcription, impart a three dimensional structure, aid in protection and have various other functions.

As the primary control center, a critical question in plant and animal biology has been how does the nucleus communicate with the rest of the cell? What chemical signals move into the nucleus to activate the expression of specific genes* and since enclosed in a double membrane that excludes molecules of any size, how do these chemical signals enter and RNA exit to instigate protein synthesis in the cytoplasm? Recent work has begun to identify the molecular structure and control of a series of pores that transverse the membrane, facilitating the

---

*The control of gene expression is discussed in Chapter 5.

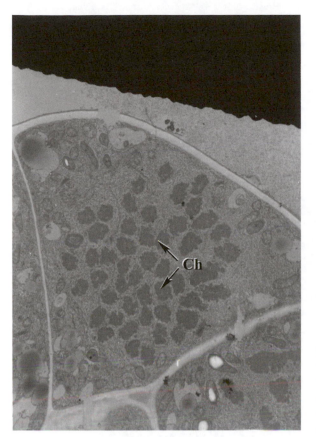

Figure 2.26. As meristematic cells begin to divide, the nuclear membrane disappears and the chromatin becomes organized into chromosomes (Ch) (dark masses in the center of the above *Pinus taeda* L. cell) (*Electron micrograph courtesy of H. Ammerman*).

movement of large molecules across. With the assistance of transport proteins and proteins that control loading and unloading of the cargo from the transport proteins,[35] the flow of large molecules is very precisely regulated.

3.5. Mitochondria

Mitochondria are small, elongate, occasionally spherical structures, 1 µm to 5 µm long. The number of mitochondria per cell varies with cell type, age and species, ranging from a few hundred to several thousand. Their primary function is in cellular respiration. Within the mitochondria occur the energy conversion processes of the tricarboxylic acid cycle and the electron transport system. As a consequence, mitochondria are of critical importance in the recycling of stored energy after harvest.

Mitochondria, like the nucleus, are enclosed within a double membrane. The outer membrane is relatively porous while the inner membrane contains numerous tubular folds or extensions called cristae (Figure 2.27). The energy transfer proteins of the electron transport system are situated on the surface of cristae. Found free within the mitochondrial matrix are many of the enzymes of the tricarboxylic cycle (some, e.g., succinic dehydrogenase, are located in the inner membrane). Also located in this aqueous portion of the mitochondria are RNA and DNA that control the synthesis of certain mitochondrial enzymes.[32] The presence of genetic information within the mitochondria has led to the belief that they were originally au-

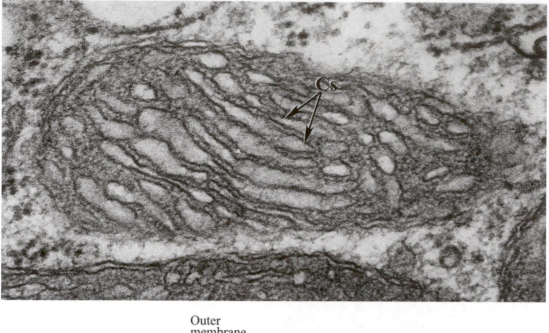

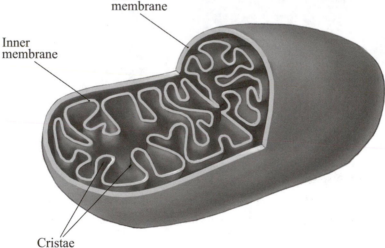

Figure 2.27. Electron micrograph of a mitochondrion showing the outer membrane and cristae (Cs) and diagrammatic illustration of what is thought to be the arrangement of the double membrane forming the outer membrane, inner membrane and cristae (*Electron micrograph ×250,000 courtesy of W.W. Thomson; drawing after Ting*[46]).

tonomous organisms that became associated with eukaryotic cells early in the evolution of life.[19]

Increases in the number of mitochondria within a cell appear to begin initially with an increase in their size, which is followed by division into two separate organelles. An interesting facet of the mitochondria story is that mitochondrial DNA (mtDNA) comes entirely from the female parent, hence the mtDNA in the progeny of crosses between two plants does not change through the input of genes from the male parent, as with nuclear DNA. Changes in the mtDNA arise from mutations, seen as nucleotide substitutions in the mtDNA. As a consequence, mtDNA has be used as a "molecular clock": the fewer the number of substitutions, the older the organism. Utilizing this approach, researchers have estimated that our common female ancestor or "mitochondrial Eve" lived about 200,000 years ago in Africa. The critical assumption of the mitochondrial clock is that the rate of insertion of mutations is relatively constant over time; however, recent evidence suggests that this is not always the case.

An important additional role for mitochondria that is beginning to be elucidated is in the regulation of programmed cell death or **apoptosis.** In animals, and to a much lesser extent in plants, there is a continual turnover in certain cells; old cells die and new cells are formed. A central question has been, "What controls programmed cellular death?" Since mitochondria are essential for energy metabolism, shutting down the mitochondria would lead to rapid death.[9] It now appears there are several means by which mitochondria can affect apoptosis (e.g., disruption of energy metabolism; release of proteins that mediate death). Deterioration in the mitochondria also appears to be closely associated with the aging of organs and organisms.

3.6. Plastids

Plant cells contain a distinct group of organelles called plastids. As with the nucleus and mitochondria, these organelles are enclosed within a double membrane. Plastids are found with differing form, size, and function. The three principal types are **chloroplasts** (chlorophyll containing plastids in which photosynthesis occurs), **chromoplasts** (plastids containing other pigments such as carotene or lycopene in red tomato fruit), and the non-pigmented **leucoplasts.** One type of leucoplast (i.e., amyloplasts) acts as a storage site for starch and is prevalent in a wide range of harvested products. Much of the research on plastids has focused on chloroplasts, however, the structure, physiology and biochemistry of leucoplasts is of considerable interest in postharvest biology in that they represent sites of stored energy, e.g., amyloplasts (Figure 2.28).

Plastids arise from proplastids inherited with the cytoplasm in newly formed cells. These proplastids appear to divide and differentiate into the various types of plastids (Figure 2.29), depending upon the nature of the cell in which they exist. Their final form is not static, however, in that there is often considerable interconversion between types of plastids (Figure 2.30). In many cases, pronounced changes are associated with major physiological events such as fruit ripening or senescence.

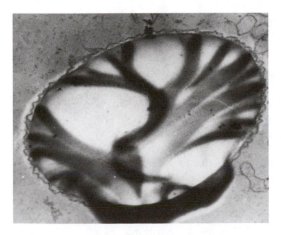

Figure 2.28. Electron micrograph of a potato tuber amyloplast containing a single starch grain (*Electron micrograph ×74,000 courtesy of H.Y. Wetzstein*).[49]

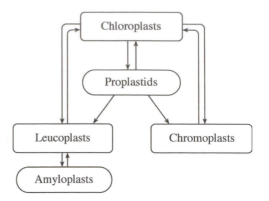

Figure 2.29. Schematic of the formation of various types of plastids.

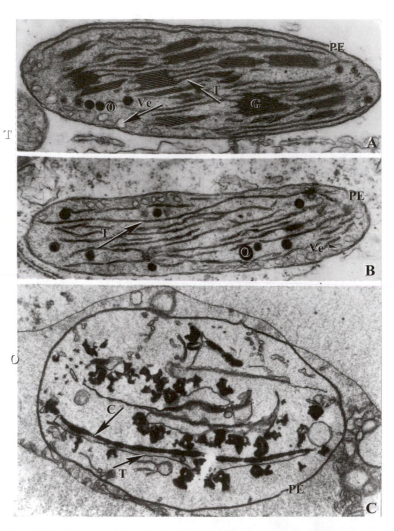

Figure 2.30. The transition during ripening of an elongated chloroplast (A) from tomato fruit with distinct granal and stromal thylakoids, to a metamorphosing chloroplast (B) with only a few vestiges of granal thylakoids remaining and finally a chromoplast (C) containing lycopene crystals (C—crystal; G—granum; O—osmiophilic globule; PE—plastid envelope; T—thylakoid; Ve—vesicle) (*Electron micrographs: (a) ×46,800, (b) ×43,700 and (c) ×23,900 courtesy of Rosso*[43]).

Contained within a plastid such as a chloroplast are proteins, lipids, starch grains, DNA, RNA, and various organic compounds. Chloroplasts have a complex inner lamellar membrane system of varying levels of complexity and an embedding matrix called the stroma. The membrane system is composed of grana, seen as flattened thylakoids stacked in the shape of a cylinder, and frets, an inner-connecting membrane system transversing the stroma between individual grana (Figure 2.31). In the stroma the photosynthetic CO_2 fixation reactions occur, while the photosynthetic photosystems are found on the thylakoid membranes.

Like mitochondria, chloroplasts contain their own DNA and synthesize a portion of the proteins required there. It is estimated, however, that greater than 80% of their protein is encoded by the nucleus and assembled outside of the chloroplast. An area of research interest has been how proteins formed in the cytosol are targeted to and transported across the chloroplast's outer membrane. Transmembrane trafficking of proteins is now known to be controlled by an import apparatus in the membrane called a translocon.[25] Targeting is accomplished by amino acid sequences at the end of the protein, which are removed once the molecule has arrived at its destination.

3.7. Microbodies

Microbodies are small (0.5–1.5 μm) spherical organelles (Figure 2.32) found in a variety of plant species and types of cells. There may be from 500 to 2000 microbodies in a single cell. They are bound by a single membrane and contain specific enzymes. The enzymes present and the function of the microbody vary depending upon its biochemical type. Microbodies that are associated with chloroplasts and function in glycolate oxidation during photorespiration are called **peroxisomes.** Those present in oil rich seeds and that function in the conversion of lipids to sugars during germination (glyoxylate cycle) are called **glyoxysomes.**

Microbodies appear to be formed from invaginations on the smooth endoplasmic reticulum. Their internal matrix is generally amorphous although crystalline substances may be present.

3.8. Vacuole

In mature cells, the vacuole is seen as the central feature within the cytoplasm (Figure 2.33). Young meristematic cells have numerous small vacuoles, which may with time enlarge and coalesce into the large central vacuole of the mature cell. Structurally, the vacuole is bound by a single membrane, called the tonoplast, which exhibits a differential permeability to various molecules. Thus the movement of many molecules into and out of the vacuole is closely controlled. Contained within the vacuole is a diverse array of possible compounds. For example, the vacuole may contain sugars, organic acids, amino acids, proteins, tannins, calcium oxalate, anthocyanins, phenolics, alkaloids, gums and other compounds. These may be dissolved in the aqueous medium, found as crystals (Figure 2.33), or congealed into distinct bodies.

The vacuole has several critical functions within the cell.[31] It acts as a storage site for a wide range of compounds, including pigments and chemicals that would be harmful to the cytosol (e.g., alkaloids, crystals) if allowed to reside there. Thus the vacuole acts as a disposal site for certain "waste" material from the cytosol. This function is essential since most plant cells can not readily excrete unwanted substances outside of the cell. As a consequence the vacuole acts as repository for these substances.

The cytosolic concentration of hydrogen ions (pH) is modulated by the vacuole, with excess levels being transported there. The concentrating of hydrogen ions accounts for the more

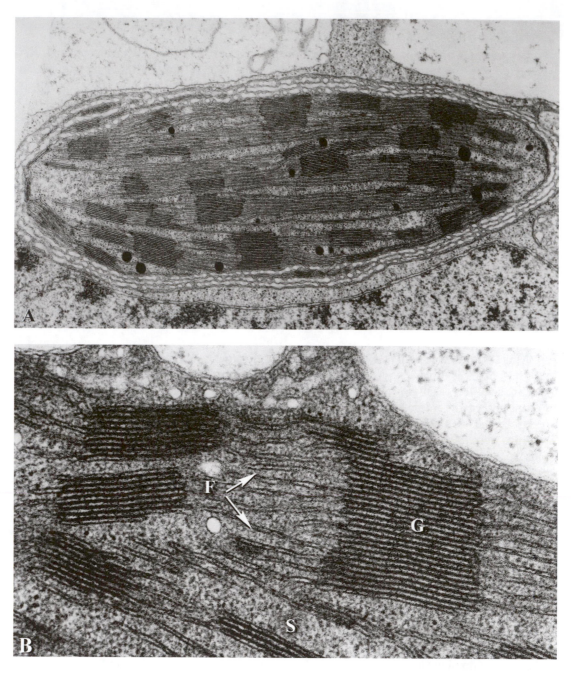

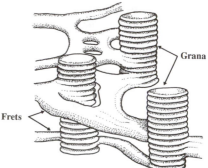

Figure 2.31. Electron micrograph of a chloroplast (A) with enlarged view of a granal region (B) (G—grana; S—stroma; F—fret). Diagramed is the arrangement of the grana and fret structure within the granal region (*Electron micrograph courtesy of W.W. Thompson; drawing after Bidwell[5]*).

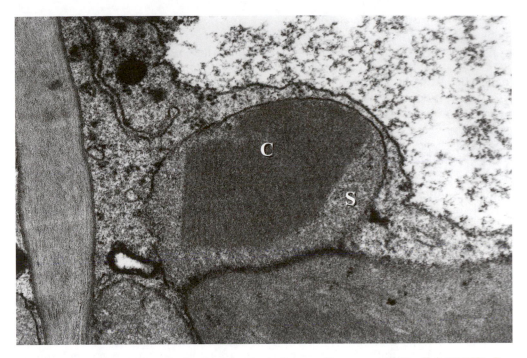

Figure 2.32. Electron micrograph of a microbody from a *Citrus* mesophyll cell (×81,700). S = stroma, C = crystalline inclusion of protein, possibly the enzyme catalase (*Electron micrograph courtesy of W.W. Thompson*).

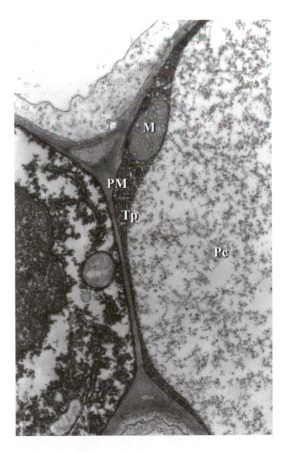

Figure 2.33. Electron micrograph of the central vacuole in a mature *Liquidambar styraciflua* L. cell. Within the vacuole are precipitated components (Pc), (PM—plasma membrane; Tp—tonoplast; M—mitochondria) (*Electron micrograph courtesy of H.Y. Wetzstein*).

acidic pH of the vacuole (as low as 3.0 versus around 7.0 in the cytosol). In certain fruit (lemon, orange) the accumulation of citric or other organic acids in the vacuole contributes to their characteristic tart taste.

The Ca^{2+} and phosphate ion concentrations in the cytosol are also maintained at appropriate levels through their transport across the tonoplast. Due to the lower pH of the vacuole, Ca often forms crystals with oxalate, phosphate or sulfate. Nitrogen is likewise stored in the vacuole, and when its concentration (or phosphate) becomes too low in the cytosol, it can move outward.

The vacuole also functions in the maintenance of the turgor pressure of the cell. The concentrating of solutes within the vacuole alters the osmotic gradient, causing water to move into the vacuole until reaching equilibrium with the cytosol. Absorption of water by the vacuole provides the outward force that contributes to the shape, texture and volume of the cell. Loss of turgor pressure after harvest readily diminishes product quality.

Finally the vacuole contains a large number of hydrolytic enzymes, e.g., proteases, lipases and phosphatases. Under normal conditions these enzymes act in part in the recycling of compounds from the cytosol. However, when the tonoplast is ruptured due to injury or senescence, these hydrolytic enzymes are released into the cytosol. Here they attack a wide range of cellular constituents, accelerating the rate of disorganization and death of the cell. Conversely, certain enzymes are present in the cytosol and their substrates are released with disruption of the vacuole. The discoloration reactions occurring after cells sustain mechanical injury (e.g., bruising) are the result of the action of phenol oxidase on phenols released from the vacuole into the cytosol.

3.9. Oleosomes

Oleosomes, also called oil bodies and spherosomes, are 0.6–2.5 μm diameter bodies (Figure 2.34) bound by a single membrane; they contain an amorphous mass of lipids.[24] Within the cell they may be compressed into irregular shapes; however, when isolated they are spherical. In maize kernels, oleosomes contain 97% neutral lipids and small amounts of protein, phospholipids and free fatty acids. They are formed by vesiculation of the rough endoplasmic reticulum where the lipids are synthesized.[13] Oleosomes in seeds are degraded during germination and growth of the seedling with the constituent lipids converted to carbohydrates for growth.

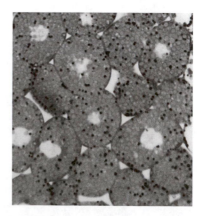

Figure 2.34. Photomicrograph of oleosomes, small dark, oil containing spherical bodies present in cortical nodule parenchyma cells of *Medicago sativa* L. (*courtesy of A.K. Bal*).

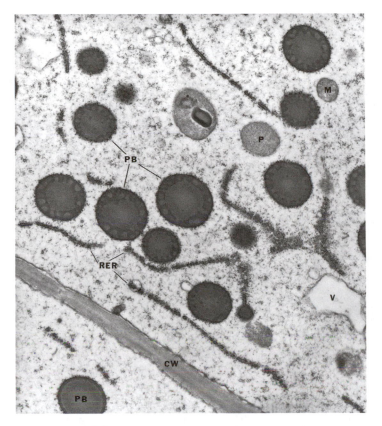

Figure 2.35. Maize endosperm rough endoplasmic reticulum (ER) connected to protein bodies (PB); cell wall (CW) (*courtesy of B.A. Larkins*).[22]

3.10. Protein Bodies

Plants sequester proteins within the cells of certain organs to provide carbon, nitrogen and sulfur for subsequent growth and development (e.g., seeds, certain vegetative tissues). Typically these are storage proteins and accumulate within storage protein vacuoles and protein bodies (Figure 2.35).[22] Protein bodies are 0.1–22 µm diameter spherical, single-membrane organelles. Protein deposits may also be found free within the cytoplasm.[33] Protein bodies differ from protein storage vacuoles in that they arise from the endoplasmic reticulum, the site of protein synthesis. In some seeds, protein bodies are found in aleurone cells located near the seed surface; in legumes they tend to be within cotyledon parenchyma cells. Protein bodies are also found in non-seed tissues (roots, leaves, flowers) but seldom numerically as concentrated as in seeds.

3.11. Golgi Bodies

The interior of the cell is comprised of a maze of compartments, each bound within a lipid membrane. When proteins and polysaccharides are synthesized in the cytoplasm and are des-

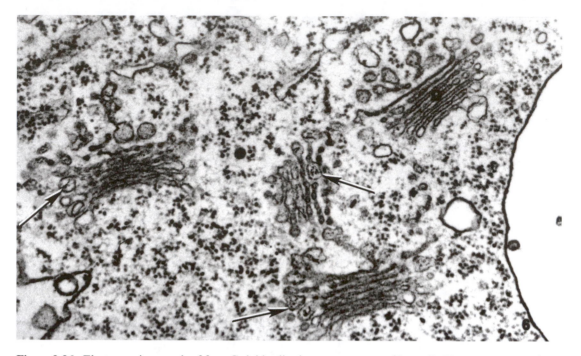

Figure 2.36. Electron micrograph of four Golgi bodies in a young cotton fiber cell. Also present are a large number of ribosomes (*Electron micrograph ×121,000 courtesy of J.D. Berlin*[15]).

tined for a specific organelle or secretion outside of the cell, there must be some orderly way to direct them to the appropriate place. Golgi bodies (named after Camillo Golgi who first identified them in 1898), working in tandem with the endoplasmic reticulum, accomplish this.

Golgi bodies (also known as dictyosomes) are small (1–3 µm in diameter by 0.5 µm thick) subcellular organelles which are composed of stacks of flat circular cisternae, each enclosed by a single membrane (Figure 2.36) and are surrounded by smaller membranous tubules and vesicles. The number of cisternae per Golgi body and number of Golgi bodies per cell varies. There are typically 3 to 10 cisternae in a single Golgi body and from a small number to over 100 Golgi bodies per cell, depending upon cell type, which are dispersed throughout the cytoplasm and are a critical part of the cell wall assembly system. Collectively, the Golgi bodies within a cell are called the Golgi apparatus. Cells that are actively secreting compounds from the cytoplasm, for example, during cell wall development or mucilage from the tips of roots, tend to have an abundance of Golgi bodies. As a consequence, a primary function of Golgi bodies in plant cells is believed to be in the secretion of cellular compounds, chiefly polysaccharides and glycoproteins, exterior to the plasma membrane.

Secretion is accomplished through the formation of small spherical vesicles of polysaccharides or glycoproteins by the cisternae (in animal cells it is primarily proteins). The endoplasmic reticulum participates in the process that now appears to be more complex than originally thought (i.e., protein molecules are enzymatically altered in the Golgi, for example, carbohydrate groups may be attached).[4] Proteins assembled on the endoplasmic reticulum that are destined for secretion are sequestered within membrane-enclosed vesicles that separate from the endoplasmic reticulum and migrate to the Golgi complex. The vesicles fuse with the cisternae and the material progresses, *via* a yet-to-be-ascertained mechanism, from the back cisternae to the most forward. While within the Golgi body, proteins are often structurally modified. Once reaching the most forward cisternae, a second vesicle forms containing

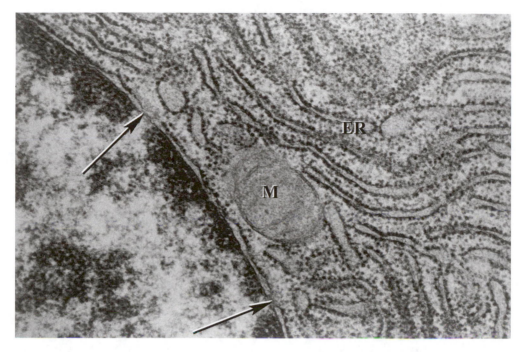

Figure 2.37. Arrangement of ribosomes on the rough endoplasmic reticulum (*Electron micrograph ×59,000 courtesy of Wolfe*[53]). Arrows denote nuclear pores; mitochondria (M).

the molecules for secretion. The vesicle then migrates to the plasma membrane where the vesicle membrane and the plasma membrane fuse, emptying its contents to the exterior for incorporation into the cell wall. Due to the transitory nature of the cisternae, Golgi bodies are constantly changing, with some cisternae growing in size and others disappearing altogether.

3.12. Ribosomes

Ribosomes are small bodies (0.017–0.025 µm in diameter) which are the site of protein synthesis within the cell (Figure 2.37). They are found in the cytoplasm, both associated with the endoplasmic reticulum and free, dispersed singly or in small groups. Ribosomes are also found in nuclei, plastids and mitochondria. These, however, appear to be distinct from those found in the cytoplasm. Because of the relatively short life expectancy of many proteins and the large number required by the cell, many ribosomes are needed. Estimates of the number of ribosomes per cell range from 500,000 to 5,000,000.

When a messenger RNA carrying the code for a specific protein from the nucleus unites with multiple ribosomes, the resulting complex (similar to a string with beads) is called a polysome or polyribosome. Transfer RNAs within the cytoplasm bind to amino acids, bringing these basic building blocks of proteins to the polysome for incorporation into the new protein molecule. By having multiple ribosomes attached to a single messenger RNA, a number of identical proteins can be assembled from the single template. Smaller ribosomes (0.015 µm) are found in the chloroplasts and mitochondria and synthesize a portion of the proteins required in these organelles. The remaining proteins are assembled in the cytoplasm and transported into the chloroplast or mitochondria.

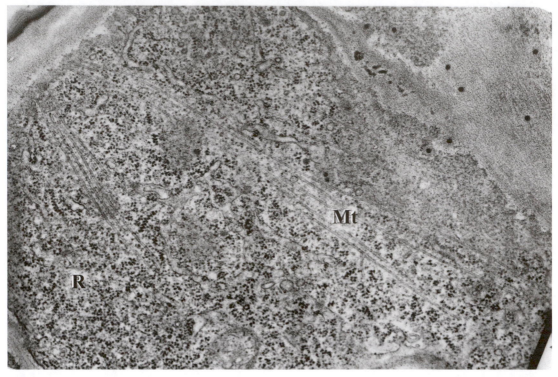

Figure 2.38. Electron micrograph of microtubules (Mt) in a young plant cell. The inset illustrates the arrangement of the fibril making up the elongated tubular nature of the microtubule (*Electron micrograph courtesy of W.W. Thomson; drawing after Ting*[46]).

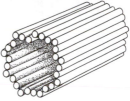

Microtubules

3.13. Cytoskeleton

While the wall provides the cell a 3-dimensional structure, the interior of the cell also has a structural framework of filamentous proteins called the **cytoskeleton.** The cytoskeleton provides spatial organization for the organelles, scaffolding for movement of organelles and materials within the cell, and is important in cell wall synthesis. It is comprised of three components: microtubules, microfilaments, and intermediate filaments.

3.13.1. Microtubules

Microtubules appear as hollow cylinders approximately 25 nm in diameter and varying from a few nm to several μm in length (Figure 2.38). They are composed of 13 protein fibrils (columns) comprised of hundreds of thousands of subunits of a globular protein called tubulin.

Microtubules appear to have several functions within the cell. During cell division, microtubules form the spindle fibers which control chromosome migration. Colchicine, a chemical used by plant breeders to double the number of chromosomes per cell, binds to tubulin, preventing chromosome migration during anaphase. Thus the cell retains both sets of chromosomes.

Microtubules also appear to be involved in coordinating the development of the cell wall. They are found in greatest numbers in the peripheral cytoplasm, adjacent to regions with growing cell walls. Disruption of the microtubules with colchicine also appears to alter the orderly arrangement of the new wall. Likewise, exposure to the phytohormone ethylene causes the microtubules to become more elongated, which alters the microfibril structure of the wall. This causes the cell's growth in length to be inhibited and the radial diameter to increase, a classical symptom of ethylene exposure to plants.[2,29] Microtubules are also involved in the transport of proteins and lipids to their appropriate destination within the cell,[23] serving as a rail on which motor proteins convey these components.

3.13.2. Microfilaments

Microfilaments are smaller than microtubules, typically 5 to 7 nm in diameter, and found as two protein chains intertwined in a helical fashion around each other, rather than a tubular structure. They are composed of the protein actin and appear to be responsible for cellular movements such as protoplasmic streaming. Both microtubules and microfilaments can assemble and disassemble, depending upon conditions within the cell; the former being generally more stable.

3.13.3. Intermediate filaments

Intermediate filaments are linear keratin or keratin-like polypeptides aligned in pairs and helically coiled around each other. The double coil then aligns with a similar structure producing a tetramer (4 proteins) which in turn pack together to form a 10 nm diameter filament.[1] Their function in plant cells is not known, but they may act similar to their counterparts in animal cells by providing structure to the interior of the cell. Intermediate filaments have been observed connecting the nucleus surface to the cell periphery[36] and are more stable than microtubules and microfilaments.

ADDITIONAL READING

Alberts, B., D. Bray, J. Lewis, M. Raff, K. Roberts and J.D. Watson. 1994. *Molecular Biology of the Cell.* Garland, New York.

Bahadur, B. 1998. *Nectary Biology: Structure, Function, and Utilization.* Dattsons, Nagpur, India.

Becker, W.M., J.B. Reece and M.F. Poenie. 1996. *The World of the Cell.* Addison-Wesley, Menlo Park, CA.

Berger, E.G., and J. Roth (eds.). 1997. *The Golgi Apparatus.* Birkhäuser Verlag, Basel, Switzerland.

Blatt, M.R., R.A. Leigh and D. Sanders (eds.). 1994. *Membrane Transport in Plants and Fungi: Molecular Mechanisms and Control.* Soc. Exp. Biol., Cambridge, England.

Brett, C., and K. Waldron. 1996. *Physiology and Biochemistry of Plant Cell Walls.* Chapman and Hall, London.

Bryant, J.A., and V.L. Dunham (eds.). 1988. *DNA Replication in Plants.* CRC Press, Boca Raton, FL.

Callow, J.A. (ed.). 1997. *The Plant Vacuole.* Academic Press, New York.

Cassab, G.I. 1998. Plant cell wall proteins. *Annu. Rev. Plant Physiol. Plant Mol. Biol.* 49:281–309.

Cheville, N.F. 1994. *Ultrastructural Pathology: An Introduction to Interpretation.* Iowa State Univ. Press, Ames, IA.

Dickison, W.C. 2000. *Integrative Plant Anatomy.* Academic Press, New York.

Esau, K. 1977. *Anatomy of Seed Plants.* Wiley, New York.

Fahn, A. 1990. *Plant Anatomy.* Pergamon Press, Oxford, England.

Fink, S. 1999. *Pathological and Regenerative Plant Anatomy.* Gebruder Borntraeger Verlagsbuchhandlung, Berlin.

Fischer, R.L., and A.B. Bennett. 1991. Role of cell wall hydrolases in fruit ripening. *Annu. Rev. Plant Physiol. Plant Mol. Biol.* 42:675–703.

Fry, S.C. 1995. Polysaccharide-modifying enzymes in the plant cell wall. *Annu. Rev. Plant Physiol. Plant Mol. Biol.* 46:497–520.

Fukui, K., and S. Nakayama. 1996. *Plant Chromosomes: Laboratory Methods.* CRC Press, Boca Raton, FL.

Ghoshroy, S., R. Lartey, J. Sheng and V. Citovsky. 1997. Transport of proteins and nucleic acids through plasmodesmata. *Annu. Rev. Plant Physiol. Mol. Biol.* 48:27–50.

Gunning, B.E.S., and M.W. Steer. 1996. *Plant Cell Biology: Structure and Function of Plant Cells.* Jones & Bartlett, Boston, MA.

Hallahan, D.L., and J.C. Gray (eds.). 2000. *Plant Trichomes.* Academic Press, New York.

Herman, E.M., and B.A. Larkins. 1999. Protein storage bodies and vacuoles. *Plant Cell* 11:601–613.

Herrmann, R.G. (ed.). 1992. *Cell Organelles.* Springer Wien, New York.

Huang, A.H.C. 1992. Oil bodies and oleosins in seeds. *Annu. Rev. Plant Physiol. Plant Mol. Biol.* 43:177–200.

Karp, G. 1996. *Cell and Molecular Biology. Concepts and Experiments.* Wiley, New York.

Kerstiens, G. (ed.). 1996. *Plant Cuticles: An Integrated Functional Approach.* Bios Scientific Publ., Oxford, England.

Kigel, J., and G. Galili (eds.). 1995. *Seed Development and Germination.* Marcel Dekker, New York.

Kindl, H., and P.B. Lazarow (eds.). 1982. *Peroxisomes and Glyoxysomes.* New York Acad. Sci. Vol. 386.

Krishnamurthy, K.V. 1999. *Methods in Cell Wall Cytochemistry.* CRC Press, Boca Raton, FL.

Larkins, B.A., and I.K. Vasil (eds.). 1997. *Cellular and Molecular Biology of Plant Seed Development.* Kluwer Academic, London.

Leigh, R.A., and D. Sanders (eds.). 1997. *The Plant Vacuole.* Academic Press, New York.

Levings, C.S., and I.K. Vasil. 1995. *The Molecular Biology of Plant Mitochondria.* Kluwer Academic, Norwell, MA.

Linskens, H.F., and J.F. Jackson (eds.). 1996. *Plant Cell Wall Analysis.* Springer-Verlag, New York.

Lodish, H., D. Baltimore, A. Berk, S. Zipursky, P. Matsidaira and J. Darnell. 1995. *Molecular Cell Biology.* Scientific American Books, New York.

Lucus, W.J. 1995. Plasmodesmata: intercellular channels for macromolecular transport in plants. *Curr. Opin. Cell Biol.* 7:673–680.

Lucas, W.J., B. Ding and C. Van Der Schoot. 1993. Plasmodesmata and the supracellular nature of plants. Tansley Reviews No. 58, *New Phytol.* 125:435–476.

Marks, M.D. 1997. Molecular genetic analysis of trichome development in Arabidopsis. *Annu. Rev. Plant Physiol. Plant Mol. Biol.* 48:137–163.

Miller, I.M., and P. Brodelius (eds.). 1996. *Plant Membrane Biology.* Oxford Univ. Press, Oxford, England.

Møller, I.M. (ed.). 1996. *Plant Membrane Biology.* Clarendon Press, Oxford, England.

Møller, I.M. (ed.). 1998. *Plant Mitochondria: From Gene to Function.* Proc. Intern. Cong. Plt. Mitochondria, Backhuys, Aronsborg, Sweden.

Moore, A.L., C.K. Wood and F.Z. Watts. 1994. Protein import into plant mitochondria. *Annu. Rev. Plant Physiol. Plant Mol. Biol.* 45:545–575.

Nicols, M., and V. Gianinazzi-Pearson (eds.). 1996. *Histology, Ultrastructure & Molecular Cytology of Plant-Microorganism Interactions.* Kluwer Academic, Norwell, MA.

Olsen, L.J., and J.J Harada. 1995. Peroxisomes and their assembly in higher plants. *Annu. Rev. Plant Physiol. Plant Mol. Biol.* 46:123–146.

Percy, K.E. 1994. *Air Pollutants and the Leaf Cuticle.* NATO Adv. Res. Workshop, Springer-Verlag, New York.

Petrini, O., and G.B. Ouellette. 1994. *Host Wall Alterations by Parasitic Fungi.* Amer. Phytopath. Soc., St. Paul, MN.

Post-Beittenmiller, D. 1996. Biochemistry and molecular biology of wax production in plants. *Annu. Rev. Plant Physiol. Plant Mol. Biol.* 47:405–430.

Raven, P.H. 1999. *Biology of Plants.* Freeman, New York.

Reddy, J.K., T. Suga, G.P. Mannaerts, P.B. Lazarow and S. Subramani (eds.). 1996. *Peroxisomes: Biology and Role in Toxicity and Disease.* Ann. New York Acad. Sci., vol. 804.

Romberger, J.A., Z. Hejnowicz and J.F. Hill. 1993. *Plant Structure: Function and Development.* Springer-Verlag, Berlin.

Roth, I. 1995. *Leaf Structure.* Gebruder Borntraeger Verlagshuchhandlung, Berlin.

Shewry, P.R., and K. Stobart (eds.). 1993. *Seed Storage Compounds: Biosynthesis, Interactions, and Manipulation.* Oxford Science, Oxford, England.

Staehelin, L.A., and I. Moore. 1995. The plant Golgi apparatus: Structure, functional organization, and trafficking mechanisms. *Annu. Rev. Plant Physiol. Plant Mol. Biol.* 46:261–288.

Tanner, W., and T. Caspari. 1996. Membrane transport carriers. *Annu. Rev. Plant Physiol. Plant Mol. Biol.* 47:595–626.

Tobin, A.J., and R.E. Morel. 1997. *Asking about Cells.* Harcourt Brace, Fort Worth, TX.

Tobin, A.K. (ed.). 1992. *Plant Organelles.* Cambridge Univ. Press, Cambridge.

Tzagoloff, A. 1982. *Mitochondria.* Plenum, New York.

Verma, D.P.S. (ed.). 1996. *Signal Transduction in Plant Growth and Development.* Springer Wien, New York.

Vogel, S. 1990. *The Role of Scent Glands in Pollination.* Smithsonian, Washington, DC.

Werker, E. 1997. *Seed Anatomy.* Gebruder Borntraeger Verlagshuchhandlung, Berlin.

Williamson, R.E. 1993. Organelle movements. *Annu. Rev. Plant Physiol. Plant Mol. Biol.* 44:181–202.

Willmer, C.M., and M. Fricker. 1995. *Stomata.* Chapman & Hall, New York.

REFERENCES

1. Alberts, B., D. Bray, J. Lewis, M. Raff, K. Roberts and J. Watson. 1994. *Molecular Biology of the Cell.* Garland, New York.

2. Apelbaum, A., and S. Burg. 1971. Altered cell microfibrillar orientation in ethylene-treated *Pisum sativum* stems. *Plant Physiol.* 48:648–652.

3. Baker, E.A. 1982. Chemistry and morphology of plant epicuticular waxes. Pp. 139–165, In: *The Plant Cuticle.* D.F. Cutler, K.A. Alvin and C.E. Price, (eds.). Academic Press, London.

4. Berger, E.G., and J. Roth (eds.). 1997. *The Golgi Apparatus.* Birkhäuser Verlag, Basel, Switerland.

5. Bidwell, R.G.S. 1979. *Plant Physiology.* Macmillan, New York.

6. Birky, C.W., Jr. 1978. Transmission genetics of mitochondria and chloroplasts. *Annu. Rev. Genet.* 12:471–512.

7. Blanke, M., and F.L. Bonn. 1985. Spaltöffnungen, Fruchtoberflache und Transpiration wachsender Apelfrüchte der Sorte 'Golden Delicious'. *Erwebsobstbau* 27:139–143.

8. Bonner, J., and R.W. Galston. 1958. *Principles of Plant Physiology.* Freeman, San Francisco, CA.

9. Brenner, C., and G. Kroemer. 2000. Mitochondria—the death signal integrators. *Science* 289:1150–1151.

10. Brett, C., and K. Waldron. 1996. *Physiology and Biochemistry of Plant Cell Walls.* Chapman and Hall, London.

11. Buchanan, B.B., W. Gruissem and R.L. Jones. 2000. *Biochemistry and Molecular Biology of Plants.* Amer. Society of Plant Physiologist, Rockville, MD.

12. Carpita, N.C., and D.M. Gibeaut. 1993. Structural models of primary cell walls in flowering plants: consistency of molecular structure with the physical properties of the walls during growth. *Plant J.* 3:1–30.

13. Cassab, G.I. 1999. Plant cell wall proteins. *Annu. Rev. Plant Physiol. Plant Mol. Biol.* 49:281–309.

14. Coombe, B.G. 1976. The development of fleshy fruits. *Annu. Rev. Plant Physiol.* 27:207–228.

15. Cummins, I., and D.J. Murphy. 1990. Mechanism of oil body synthesis and maturation in developing seeds. Pp. 231–233. In: *Plant Lipid Biochemistry, Structure, and Utilization.* P.J. Quinn and J.L. Harwood (eds.). Portland Press, London.

16. Esau, K. 1977. *Anatomy of Seed Plants.* John Wiley, New York.

17. Fahn, A. 2000. Structure and function of secretory cells. *Adv. Bot. Res.* 31:37–75.

18. Fahn, A. 1990. *Plant Anatomy.* Pergamon Press, Oxford, England.

19. Gilbert, L.E. 1971. Butterfly-plant coevolution: Has *Passiflora adenopoda* won the selectional race with Heliconine butterflies? *Science* 172:585–586.

20. Gray, W.M., G. Burger and B.F. Lang. 1999. Mitochondrial evolution. *Science* 283:1476–1481.
21. Grill, E., and H. Ziegler. 1998. A plant's dilemma. *Science* 282:252–253.
22. Hallahan, D.L., and J.C. Gray (eds.). 2000. *Plant Trichomes.* Academic Press, New York.
23. Herman, E.M., and B.A. Larkins. 1999. Protein storage bodies and vacuoles. *Plant Cell* 11:601–613.
24. Hirokawa, N. 1998. Kinesin and dynein superfamily proteins and the mechanism of organelle transport. *Science* 279:519–526.
25. Huang, A.H.C. 1992. Oil bodies and oleosins in seeds. *Annu. Rev. Plant Physiol. Plant Mol. Biol.* 43:177–200.
26. Jarvis, P., L.-J. Chen, H. Li, C.A. Peto, C. Fankhauser and J. Chory. 1998. An *Arabidopsis* mutant defective in the plastid general protein import apparatus. *Science* 282:100–103.
27. Juniper, B.E., and C.E. Jeffrie. 1983. *Plant Surfaces.* Edward Arnold, London.
28. Kays, S.J., and J.O. Silva Dias. 1996. *Cultivated Vegetables of the World.* Exon Press, Athens, GA.
29. Kramer, P.J., and T.T. Kozlowski. 1979. *Physiology of Woody Plants.* Academic Press, New York.
30. Lang, J.M., W.R. Eisinger and P.B. Green. 1982. Effects of ethylene on the orientation of microtubules and cellulose microfibrils on pea epicotyl cells with polylamellate cell walls. *Protoplasma* 110:5–14.
31. Ledbetter, M.C., and K.R. Porter. 1970. *Introduction to the Fine Structure of Plant Cells.* Springer-Verlag, Berlin.
32. Leigh, R.A., and D. Sanders. 1997. *The Plant Vacuole.* Academic Press. San Diego.
33. Levings, C.S., and I.K. Vasil. 1995. *The Molecular Biology of Plant Mitochondria.* Kluwer Academic, Norwell, MA.
34. Lott, J.N.A. 1980. Protein bodies. Pp. 589–623. In: *The Biochemistry of Plants.* Vol.1. N.E. Tolbert (ed.), Academic Press, New York.
35. Lucus, W.J. 1995. Plasmodesmata: intercellular channels for macromolecular transport in plants. *Curr. Opin. Cell Biol.* 7:673–680.
36. Melchior, F., and L. Gerace. 1998. Two-way trafficking with RNA. *Trends in Cell Biology* 8:175–179.
37. Mizuno, K. 1995. A cytoskeletal 50 kDa protein in higher plants that forms intermediate-sized filaments and stabilizes microtubules. *Protoplasma* 186:99–112.
38. Nakajima, K., G. Sena, T. Nawy and P.N. Benfey. 2001. Intercellular movement of the putative transcription factor SHR in root patterning. *Nature* 413:307–311.
39. Onwueme, I.C. 1978. *The Tropical Tuber Crops.* Wiley, New York.
40. Payne, W. 1978. Glossary of plant hair terminology. *Brittonia* 30:239–255.
41. Priestley, D.A., and A.C. Leopold. 1983. Lipid changes during natural aging of soybean seed. *Plant Physiol.* 59:467–470.
42. Rachmilevitz, T., and A. Fahn. 1975. The floral nectary of *Tropaeolum majus* L.—The nature of the secretory cells and the manner of nectar secretion. *Ann. Bot.* 39:721–728.
43. Rose, J.K.C., and A.B. Bennett. 1999. Cooperative disassembly of the cellulose-xyloglucan network of plant cell walls: parallels between cell expansion and fruit ripening. *Trends Plant Sci.* 4:176–183.
44. Rosso, S.W. 1968. The ultrastructure of chromoplast development in red tomatoes. *J. Ultrastructure Res.* 25:307–322.
45. Salisbury, E.J. 1927. On the causes and ecological significance of stomatal frequency with special reference to woodland flora. *Phil. Trans. Roy. Soc. Lond.,* Ser. B, 216:1–65.
46. Stevens, A.B.P. 1956. The structure and development of hydathodes of *Caltha palustris* L. *New Phytol.* 55:339–345.
47. Ting, I.P. 1982. *Plant Physiology.* Addison-Wesley, Reading, MA.
48. Vogel, S. 1990. *The Role of Scent Glands in Pollination.* Smithsonian, Washington, DC.
49. Wagner, D., R.W.M. Sablowski and E.M. Meyerowitz. 1999. Transcriptional activation of APETALA 1 by LEAFY. *Science* 285:582–584.
50. Wetzstein, H.Y., and C. Sterling. 1978. Integrity of amyloplasts membranes in stored potato tubers. *Z. Pflanzenphysiol.* 90:373–378.
51. Willmer, C., and M. Fricker. 1996. *Stomata.* Chapman & Hall, London.
52. Wilson, C.L., and W.E. Loomis. 1967. *Botany.* Holt, Rinehart and Winston, New York.
53. Wilson, D.O., Jr., and M.B. McDonald. 1986. The lipid peroxidation model of seed aging. *Seed Sci. Technol.* 14:269–300.
54. Wolfe, S.L. 1985. *Cell Ultrastructure.* Wadsworth Pub. Co., Belmont, CA.

3

METABOLIC PROCESSES IN HARVESTED PRODUCTS

Metabolism is the entirety of biochemical reactions occurring within cells. Many components of metabolism, especially those which are beneficial or detrimental to the quality of postharvest products, are of major interest to postharvest biologists. The acquisition and storage of energy and the utilization of stored energy are central processes in the control of the overall metabolism of a plant. The acquisition of energy through photosynthesis and its recycling *via* the respiratory pathways are compared in Table 3.1. Respiration occurs in all living products, while photosynthesis does not occur in products devoid of the green pigment chlorophyll.

The various organs of intact plants have a high degree of specialization as to carbon acquisition, allocation and storage. Leaves, for example, photosynthesize but seldom act as long-term storage sites for photosynthates. Petioles and stems transport fixed carbon, but typically have only a limited photosynthetic potential and when utilized for storage, often only act as temporary sinks (e.g., the stems of the Jerusalem artichoke). Flowers, roots, tubers, and other organs or tissues likewise have relatively specific roles with regard to the overall acquisition of carbon. While attached to the plant, these plant parts (organ or tissue) derive the energy required to carry out their specific functions from photosynthesizing leaves. There is, therefore, in intact plants an interdependence among these different parts with divergent primary functions. Severing these parts from the plant at harvest disrupts this interdependence and can, therefore, influence postharvest behavior. For example, the detaching of leaves, whose primary function is to fix carbon dioxide rather than the storage of carbon, markedly restricts or terminates photosynthesis, leaving them with extremely low reserves that can be used for maintenance. Storage organs, on the other hand, if sufficiently mature, have substantial stored carbon that can be recycled for utilization in maintenance and synthetic reactions.

In contrast to this high degree of specialization among parts in the acquisition of energy, respiration occurs in all living cells and is essential for the maintenance of life in products after harvest. The factors affecting these two general processes, energy acquisition (photosynthesis) and energy utilization (respiration), are reviewed in this chapter. These processes are affected by both internal (commodity) and external (environmental) factors that often interact. Important commodity factors include species, cultivar, type of plant part, stage of development, surface to volume ratio, surface coating, previous cultural and handling conditions, and chemical composition. Among the major external factors influencing respiratory rate are temperature, gas composition, moisture conditions, light and other factors that induce stress conditions within the harvested product.

Table 3.1. General Comparison of Photosynthesis and Respiration in Plants

	Photosynthesis	*Respiration*
Function	Energy acquisition	Energy utilization and Formation of carbon skeletons
Location	Chloroplasts	Mitochondria and cytoplasm
Role of light	Essential	Not required
Substrates	CO_2, H_2O, light	Stored carbon, O_2
End products	O_2, stored carbon	CO_2, H_2O, energy
Overall effect	Increase in weight energy	Decrease in weight
General reaction	$6CO_2 + 6 H_2O \xrightarrow{\text{chloroplast}}$ $C_6H_{12}O_6 + 6O_2$	$C_6H_{12}O_6 + 6O_2 \xrightarrow{\text{mitochondria}}$ $6CO_2 + 6H_2O + \text{energy}$

1. RESPIRATION

Respiration is a central process in all living cells that mediates the release of energy through the breakdown of carbon compounds and the formation of carbon skeletons necessary for maintenance and synthetic reactions after harvest. The rate of respiration is important because of these main effects but also because it gives an indication of the overall rate of metabolism of the plant or plant part. All metabolic changes occurring after harvest are important, especially those that have a direct bearing on product quality. The central position of respiration in the overall metabolism of a plant or plant part and its relative ease of measurement allow us to use respiration as a measure of metabolic rate. The relationship between respiration and metabolism, however, is very general since specific metabolic changes may occur without measurable changes in net respiration. This is illustrated by comparing changes in a number of the physical and chemical properties of pineapple fruit during development, maturation and senescence (Figure 3.1). Neither changes in the concentration of chlorophyll, reducing sugars, acidity, carotenoids, nor esters correlates well with changes in respiratory rate. Therefore, it is important to view respiration as it fits into the overall process of harvested product metabolism rather than as an end in itself.

There are two general types of respiratory processes in plants—those that occur at all times regardless of the presence or absence of light (dark respiration) and those occur only in the light (photorespiration).

1.1. Dark Respiration

The living cells of all plant products respire continuously, utilizing stored reserves and oxygen (O_2) from the surrounding environment and releasing carbon dioxide (CO_2). The ability to respire is an essential component of the metabolic processes that occur in live harvested products. The absence of respiration is the major distinction between processed plant products and living products. Respiration is the term used to represent a series of oxidation-reduction reactions where a variety of substrates found within the cells are oxidized to carbon dioxide. At the same time, oxygen absorbed from the atmosphere is reduced to form water. In its simplest form, the complete oxidation of glucose can be written as:

$$C_6H_{12}O_6 \;+\; 6\,O_2 \;+\; 6\,H_2O \;\rightarrow\; 6\,CO_2 \;+\; 12\,H_2O \;+\; \text{energy}$$

glucose　　　oxygen　　　water　　　carbon　　　water
　　　　　　　　　　　　　　　　　dioxide

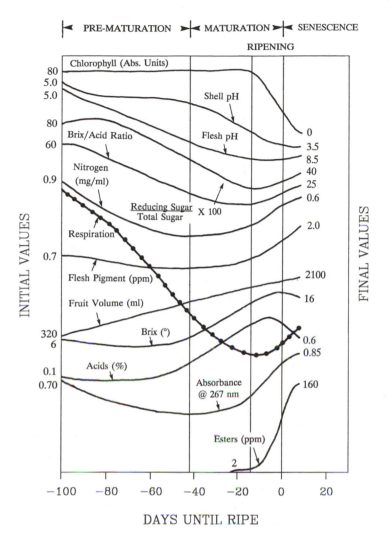

Figure 3.1. Chemical and physical changes in the pineapple fruit during prematuration, maturation, ripening and senescence (*after Gortner et al.*).[40] While respiratory rate can give a general indication of the overall rate of metabolism, it often does not correlate with specific changes occurring, e.g., changes in the % acids.

The products of this reaction are carbon dioxide, water and the energy that is required for essential cellular processes. Some of the energy generated in respiration of harvested produce is lost as heat (~46%); however, significant amounts are retained by the cells in chemical forms that may be used for these essential processes. The rate of the reaction is determined by substrate (e.g., glucose) and oxygen availability, and most importantly, temperature.

Respiration is much more complex than the generalized reaction just presented. The glycolytic, tricarboxylic acid, pentose phosphate and electron transport system pathways are involved in the breakdown of many of the common substrates utilized by the cell. Often during the oxidation of a substrate, the conversion to carbon dioxide is not complete and intermediates are utilized by the cells for synthetic reactions such as the formation of amino acids, nucleotides, pigments, lipids and flavor compounds. Hence, the respiratory pathways provide precursors, often called carbon skeletons, required for the formation of a large number of plant products.

During the preharvest growth of a plant or plant product, a major portion of the carbon trapped during photosynthesis is diverted into synthetic reactions. It is through the respiratory pathways that the carbon from photosynthesis begins its transformation into the majority of the other compounds in the plant. Since the synthesis of these compounds also requires energy, derived from the respiratory pathways, a portion of the photosynthetic carbon fixed is utilized for this purpose. Therefore, a balance is reached between respiratory substrate availability and the demand for energy production and carbon skeletons. Since neither availability nor demand is static, the system is continually changing this balance during the day and over the developmental cycle of the plant or plant part.

At harvest, the relationship between carbon acquisition and utilization is radically changed when the plant product is severed from its readily replenishible supply of carbon provided by photosynthesis. Hence, a new balance must be reached; energy and carbon skeletons must now come from already existing sources within the severed product. Mature plant parts which function as carbon storage organs (e.g., seeds, roots, bulbs, tubers) have substantial stores of carbon that can be utilized *via* the respiratory pathways for an extended period. Leaves and flowers do not function as carbon storage sites and hence have very little reserves. As a consequence, the balance shifts to a situation where demand can readily deplete the supply.

After harvest, the objective is to maintain the product as close to its harvested condition (quality) as possible, thus in most products growth is considered undesirable. Postharvest conditions for these products often result in an extensive reduction to a total elimination of photosynthesis, necessitating the reliance upon existing reserves. The respiratory pathways that are operative after harvest in both intact plants and severed plant parts are the same as those prior to harvest. The major changes are the now finite supply of respiratory substrate available to the various pathways and the new equilibrium established between supply and tissue demand for it.

There are a series of steps in the respiratory oxidation of sugar or starch that involve three interacting pathways. The initial pathway is glycolysis, where sugar is broken down into pyruvic acid, a three carbon compound. The pathway occurs in the cytoplasm and can operate in the absence of oxygen. The second pathway is the tricarboxylic acid (TCA) or Krebs cycle that occurs in the mitochondria, where pyruvic acid is oxidized to carbon dioxide. Oxygen, although not reacting directly in these steps, is required for the TCA pathway to proceed, as are several organic acids. The third pathway, the electron transport system, occurs in the mitochondrial inner membrane and transfers hydrogen atoms (reducing power), removed from organic acids in the tricarboxylic acid cycle and from 3-phosphoglyceraldehyde during glycolysis, to oxygen. The electrons are moved through a series of oxidation-reduction steps that terminate upon uniting with oxygen, forming water. The energy is used to pump protons that are then allowed to flow back through a proton channel that converts the gradient to chemical energy in the form of adenosine triphosphate (ATP). ATP is then utilized to drive various energy requiring reactions within the cell. A fourth respiratory pathway, the pentose phosphate system, while not essential for the complete oxidation of sugars, functions by providing carbon skeletons, reduced NADP required for certain synthetic reactions and ribose-5-phosphate for nucleic acid synthesis. The pentose phosphate pathway appears to be operative to varying degrees in all respiring cells.

While oxygen is not required for the operation of the glycolytic pathway, it is essential for the tricarboxylic acid cycle, the pentose phosphate pathway and the electron transport system. Glycolysis can proceed therefore under anaerobic conditions, i.e., in the absence of oxygen.

The occurrence of anaerobic conditions poses a serious problem in the postharvest handling of plant products. When the oxygen concentration within the tissue falls below a threshold level (around 2%), pyruvic acid can no longer proceed through the tricarboxylic acid cycle. Pyruvic acid instead is converted to lactate and/or ethanol that can accumulate to toxic levels. Prolonged exposure to anaerobic conditions, therefore, results in cellular death and loss of the harvested product. Exposure for short periods often results in the formation of off-flavors in edible products. Depending on the tissue and length of exposure to low oxygen, the off-flavors may be eliminated upon returning to aerobic conditions.

1.1.1. Glycolysis

Glucose, derived from sucrose or starch, is broken down by the glycolytic pathway in a sequence of steps to form pyruvic acid. In the initial step, glucose has a phosphate added (i.e., is phosphorylated) (Figure 3.2). If the starting compound is free glucose, the reaction is catalyzed by the enzyme hexokinase to form glucose-6-phosphate. If, as found in many postharvest products, the glucose occurs as part of a starch molecule, phosphate is added by the enzyme starch phosphorylase, forming glucose-1-phosphate, which is subsequently converted to glucose-6-phosphate. The phosphorylation of free glucose requires energy, in the form of 1 ATP, while the phosphorylation of glucose when it is part of a starch molecule does not. The six-carbon glucose molecule progresses through fructose-1-phosphate to fructose-1,6-bisphosphate before being split by the enzyme aldolase into two 3-carbon compounds, dihydroxyacetone phosphate and 3-phosphoglyceraldehyde. The 3-phosphoglyceraldehyde molecule is the first compound to lose electrons in the respiratory pathway, forming 1,3-bisphosphoglycerate, when 2 hydrogen atoms are removed and accepted by NAD (nicotinamide adenine dinucleotide). 1,3-Bisphosphoglycerate undergoes four additional enzymatic steps, resulting in the formation of pyruvic acid. Two of the four steps from 1,3-bisphosphoglycerate to pyruvic acid yield chemical energy in the form of ATP. None of the reactions from glucose or starch to pyruvic acid require oxygen, so the glycolytic pathway can proceed normally under anaerobic conditions.

If anaerobic conditions occur in the harvested tissue due to restricted entry of oxygen or an insufficient supply in the atmosphere surrounding the commodity, pyruvic acid cannot enter the tricarboxylic acid cycle and be oxidized. The inability to enter the cycle is due to an absence of oxidized flavin adenine dinucleotide (FAD) and NAD required for the cycle to proceed. When this occurs, pyruvic acid accumulates and is usually decarboxylated to form CO_2 and acetaldehyde, which is subsequently reduced to ethanol. Pyruvate may also be reduced to form lactic acid. Alcohol (ethanol) and to a lesser extent lactic acid accumulate within the tissue. Both reactions require energy, which is provided by NADH formed during the oxidation of 3-phosphoglyceraldehyde previously in the pathway. The overall reaction in simplified form is:

$$\text{glucose} + 2\,\text{ATP} + 2\,\text{Pi} + 2\,\text{ADP} \rightarrow 2\,\text{ethanol} + 2\,CO_2 + 4\,\text{ATP}$$

When ethanol is produced from glucose, two ATP molecules are required but four are formed from each free glucose molecule giving a net yield of two ATPs. This represents one fourth of the energy yield that would be derived from the glycolytic pathway when sufficient oxygen is present and is only 1/16 that derived when glucose is fully oxidized (glycolysis and the tricarboxylic acid cycle). As cells switch their carbon flow toward lactate and alcohol formation, the production of CO_2 increases (Figure 3.2). The increase is due to the reduced energy yield under anaerobic conditions, as much more glucose must be oxidized to meet the cell's energy requirements. The complete oxidation of one glucose molecule under aerobic conditions yields

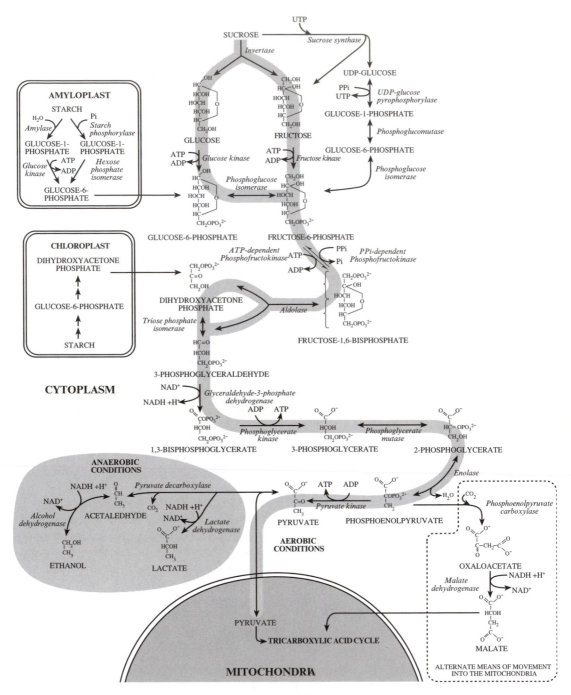

Figure 3.2. The glycolytic pathway for the aerobic oxidation of glucose or glucose-1-phosphate to pyruvate. The pathway can flow in either the gluconeogenic (forming sugars) or the glycolytic direction, the latter being the predominant direction. During ripening of fruit such as banana, the gluconeogenic direction may predominate.[7] Under anaerobic conditions the movement of pyruvate into the tricarboxylic acid cycle is inhibited, and NADH + H formed in the oxidation of 3-phosphoglyceraldehyde is utilized to reduce acetaldehyde to ethanol or pyruvate to lactate by way of the shaded portion of the pathway.

~32 ATP equivalents, while under anaerobic conditions a net of only 2 ATPs are formed from a glucose molecule. While anaerobiosis has disastrous consequences for living tissue in terms of loss of stored reserves and accumulation of undesirable compounds, it is also the basis of a very important processing technique, fermentation. Potential energy remains stored in the form of alcohol that can be recaptured if oxygen is supplied.

1.1.2. Tricarboxylic Acid Cycle

Pyruvic acid produced by the glycolytic pathway is further broken down in the tricarboxylic acid (TCA) cycle, also known as the Krebs cycle, and the citric acid cycle. Tricarboxylic acid refers to the three carboxyl groups that are present on some of the acids in the cycle, while citric acid is an important early intermediate in the sequence of reactions. The reactions of the tricarboxylic acid cycle occur in the matrix of the mitochondria and on the surface of the inner membrane. Pyruvic acid, therefore, must move from the cytoplasm, where glycolysis occurs, into the mitochondria for further oxidation to proceed.

   In the initial step, pyruvic acid is decarboxylated as it combines with Coenzyme A forming the 2 carbon compound acetyl CoA (Figure 3.3). Acetyl CoA then combines with the 4 carbon molecule oxaloacetic acid, yielding citric acid that undergoes a series of oxidative and decarboxylation reactions ending with the formation of oxaloacetic acid, allowing the cycle to begin again. Energy is captured as reduced NAD (i.e., NADH) at the conversion of isocitric acid to α-ketoglutaric, α-ketoglutaric acid to succinyl CoA, and malic acid to oxaloacetic acid. A single ATP is produced on the conversion of succinyl CoA to succinic acid, and FAD is reduced at the conversion of succinic acid to fumaric acid. Carbon dioxide is liberated from pyruvic, isocitric and α-ketoglutaric acids.

   Each revolution of the tricarboxylic acid cycle, a three carbon pyruvate molecule releases three carbon dioxide molecules and produces reducing power in the form of four NADH molecules and one $FADH_2$ molecule. Combined with the two NADH molecules from the glycolytic pathway, a total of 10 reduced NADs are formed with the complete oxidation of a single glucose molecule. Only 12 of the 24 protons (H) are from glucose; the remaining 12 are from water that is added at various steps in the cycle.

1.1.3. Electron Transport or Cytochrome System

NAD reduced to NADH in the TCA cycle, in glycolysis and by other reactions in the cell is recycled by the removal of the electrons. NADH cannot, however, directly reduce oxygen to form water. The electrons are removed through a series of reactions forming a positive potential gradient, from compounds of low reduction potentials to higher reduction potentials (i.e., from lower to greater tendency to accept electrons), culminating in a reaction with oxygen that has the greatest tendency to accept electrons (Figure 3.4). During the process, protons are pumped across the inner mitochondrial membrane, forming a proton gradient. The proton gradient is released through a protein complex (ATP synthase), and energy is conserved in a biologically usable form as ATP. ATP is used to drive reactions, especially synthetic, that require energy inputs. In actively metabolizing cells, the efficiency of energy trapping in the electron transport system is only about 54%. A mole of glucose has a calorie potential of approximately 686 kcal·mole^{-1}. Only a small amount of energy is lost in the initial transfer of energy as electron pairs to NAD and FAD in glycolysis and the tricarboxylic acid cycle. However, during the transfer of the energy to ATP in the electron transport system, the energy potential drops to approximately 263 kcal·mole^{-1}. The remaining energy escapes as respiratory (vital)

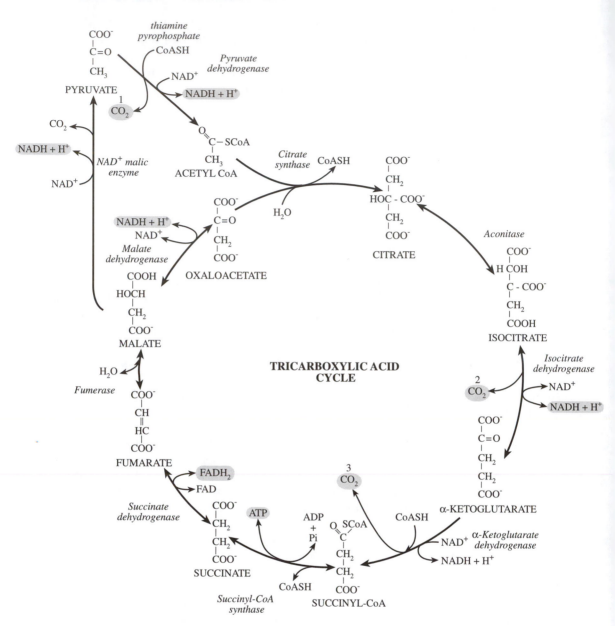

Figure 3.3. The tricarboxylic acid cycle results in the complete oxidation of 1 pyruvate molecule with each complete sequence through the cycle forming CO_2, ATP, NADH + H, and $FADH_2$. NADH + H and $FADH_2$ are then oxidized in the electron transport system (Figure 3.4).

heat, a normally detrimental factor that must be dealt with during the postharvest handling and storage period. Therefore, the overall function of the electron transport system is to trap energy in a biologically usable form (ATP) and recycle NAD and FAD required for certain reactions in the various metabolic pathways. The major components of the electron transport system have been elucidated. Figure 3.4 shows the sequence of steps involved. Each component enzyme is specific and can only accept electrons from the previous component in the

ELECTRON TRANSPORT   **OXIDATIVE PHOSPHORYLATION**

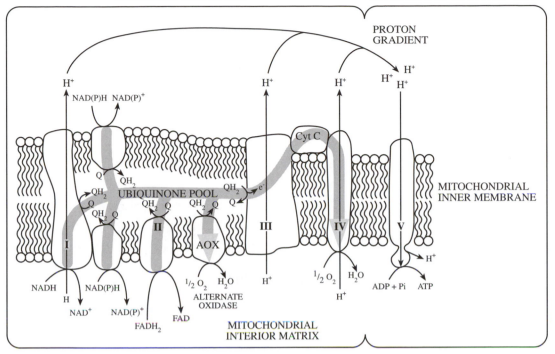

Figure 3.4. This model illustrates the organization of the electron transport system, located within the inner membrane of the mitochondria. Electrons released during oxidative steps in glycolytic pathway and the tricarboxylic acid cycle (trapped as NADH and $FADH_2$) move (shaded arrow) through a series of complexes to the terminal acceptor, oxygen. Four protein complexes participate in the process with complex I accepting energy from NADH and complex II from $FADH_2$. The free energy released during electron transfer is coupled to the translocation of protons (H^+) across the membrane, creating an electrochemical proton gradient. Protons on the exterior flow back through complex V, an ATP synthase complex that is coupled to the conversion of ADP + P_i to ATP, retrapping the free energy. When the alternative oxidase (AOX) is operative, only 1 proton is transferred from NADH. FAD reduction is associated with succinate dehydrogenase activity (i.e., succinate → fumerate), and when the AOX is operative, energy in QH_2 is transferred directly to oxygen, bypassing the formation of ATP. Thus, when the alternative pathway is operative, only 1 ATP equivalent is produced from NADH with the remaining energy being lost as heat.

chain. NADH and $FADH_2$, being different in energy potential, enter the chain at different points.

The total energy balance from the oxidation of one molecule of glucose remains a subject of debate. When ADP:O ratios (the number of ATPs synthesized per 2 electrons transferred to oxygen) are calculated in isolated mitochondria, consensus values are 2.5 ATPs per NADH and 1.5 per $FADH_2$. In the glycolytic pathway, the energy balance for a single glucose molecule under aerobic conditions is -2 ATP + 4 ATP + 2 NADH. Since glycolytic NADH is in the cytosol and cannot diffuse into the mitochondria for conversion to ATP *via* the ETS, it must go through a shuttle where the energy from one NADH is transferred to the mitochondria. Two options are postulated: the glycerol phosphate shuttle (Figure 3.5) which yields only 1.5 ATP's per glycolytic NADH; and the malate-asparate shuttle which yields 2.5. Therefore, either 5 or 7 ATPs are derived during glycolysis, depending upon the shuttle method (Table

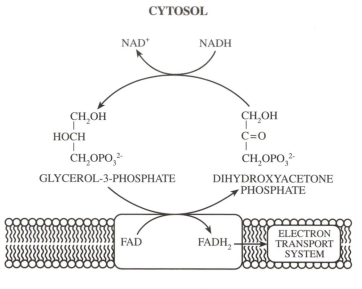

CYTOSOL

MITOCHONDRIA

Figure 3.5. The energy from reduced NAD produced in the glycolytic pathway can be transported from the cytosol into the mitochondria (site of electron transport pathway) *via* the glycerol phosphate shuttle. NADH reduces dihydroxyacetone phosphate to glycerol phosphate, and then at the mitochondrial membrane it is oxidized back to dihydroxyacetone phosphate yielding one reduced FAD within the membrane that enters the electron transport system. Thus, the energy potential drops from 2.5 ATP equivalents to 1.5 for each reduced glycolytic NAD. An alternative malate-aspartate shuttle (not shown) is more complex, and involving the transport of glutamate and malate across the mitochondrial membrane. The malate-aspartate shuttle, however, does not result in an energy drop.

3.2). When added with the ATPs from the TCA cycle (8 NADH + 2 $FADH_2$ + 2 ATP) per molecule of glucose, the net yield is 30 or 32 ATPs.[51] The exact number (32 or 30) remains in question, each version being found in various textbooks and articles.

Carbon monoxide (CO), azide (N_3) and hydrogen cyanide (CN) are potent inhibitors of electron transport, in that they combine with the metals in the terminal cytochrome oxidase, the final enzyme in the electron transport chain. In plants, CN does inhibit the terminal cytochrome oxidase, but it also stimulates the rate of respiration. That is because plant tissues have a second terminal oxidase called the alternative oxidase which is insensitive to inhibitors of cytochrome oxidase. Electrons follow the normal electron path to ubiquinone, at which point they are transferred to oxygen to form water by the alternative oxidase. It is important to note that less than one ATP is generated per NADH in the alternative pathway; in contrast to the 2.5 ATPs generated by the normal pathway, the remainder of the energy is lost as heat. Therefore, the alternative pathway represents an inefficient energy conserving system that bypasses the normal pathway, which can substantially increase the respiratory heat load of the product.

The alternate electron transport pathway has been found in all plant tissues studied to

Table 3.2. Yield of ATPs from the Oxidation of One Glucose Molecule *via* Two Shuttle Pathways.*

Pathway	ATP Yield per Glucose	
	Glycerol-Phosphate Shuttle	Malate-Asparate Shuttle
Glycolysis (cytosol)		
Glucose phosphorylation (1 molecule)	-1	-1
Fructose-6-P phosphorylation (1 molecule)	-1	-1
1,3-bisphosphate glyceric acid dephosphorylation (2 molecules)	+2	+2
Phosphoenol pyruvic acid dephosphorylation (2 molecules)	+2	+2
Glyceraldehyde-3-phosphate oxidation (2 molecules) yields 2 NADH		
Pyruvate to acetyl-CoA (mitochondria) yields 2 NADH		
Tricarboxylic acid cycle (mitochondria)		
Succinyl-CoA → succinate	+2	+2
Succinate oxidation (2 molecules) yields 2 $FADH_2$		
Isocitrate, α-ketoglutarate and malate oxidation (2 molecules each) yields 6 NADH		
Oxidative phosphorylation (mitochondria)		
2 NADH from glycolysis (yielding 1.5 or 2.5 ATPs depending upon shuttle method)	+3	+5
2 NADH from oxidation of pyruvate at 2.5 ATPs	+5	+5
2 $FADH_2$ from succinate at 1.5 ATPs	+3	+3
6 NADH from tricarboxylic acid cycle at 2.5 ATPs	+15	+15
Net ATP Yield	+30	+32

*ATP yields are based upon consensus P/O ratios which give ATP equivalents for mitochondrial oxidation of NADH and $FADH_2$ of 2.5 and 1.5, respectively. Two shuttle pathways are given for the oxidative phosphorylation of glycolytic NADH in the mitochondria.

date. The abundance of the alternative oxidative protein increases in many plant tissues exposed to any one of several environmental and biotic stresses, as well as during the ripening and/or senescence of some fruit tissues (e.g., cold stored potato tubers, parsnip and carrot roots; ripening avocado and banana fruits). In nature, its only well documented role appears in the thermogenesis associated with the flowering of some species belonging to the families Annonaceae, Araceae, Aristolochiaceae, Cyclanthaceae, and Nymphaeaceae.[79] The elevated temperature (e.g., up to 15°C above the ambient air)[78] associated with the alternative path in certain flower parts results in the volatilization of odoriferous compounds that attract insects, thereby facilitating pollination. The alternative pathway is activated on the day of flowering and remains active for only a few hours. Although the existence of the alternative pathway has been known for over 70 years,[37] its physiological function in most tissues remains speculative. While not fully documented, several roles have been proposed, based on the principle that electron transport through the alternative path supports a high rate of respiration that is not constrained by respiratory control (i.e., when the level of ADP is low and the level of ATP is high). The nonphosphorylating alternative pathway may support higher respiration rates that would ensure a stable supply of metabolites, such as organic acids, required for biosynthetic reactions in the cells. Related to this is the "energy overflow hypothesis" by Lambers,[69] which considers the alternative path as a coarse control of carbohydrate metabolism operative when carbohydrates accumulate in greater quantities that required for growth, storage and ATP synthesis. A more recent hypothesis postulates that the alternative pathway alleviates the over-reduction of the electron transport chain, which could lead to the formation of superoxide anions and other deleterious reactive oxygen species.[92] Reactive oxygen species react with phospholipids, proteins, DNA and other cellular components, ultimately resulting in cell death.

1.1.4. Pentose Phosphate Pathway

In addition to glycolysis and the tricarboxylic acid cycle, the pentose phosphate pathway can be used to oxidize sugars to carbon dioxide. The name is derived from the fact that many of the intermediates in the pathway are five carbon (penta)phosphorylated sugars. The pentose phosphate system is found in the cytoplasm, and its main function does not appear to be energy production *via* the formation of ATP in the electron transport system, but rather as a source of ribose-5-phosphate for nucleic acid production, as reduced NADP for synthetic reactions and as a means of interconversion of sugars to provide 3, 4, 5, 6, and 7 carbon skeletons for biosynthetic reactions. One example is the formation of erythrose-4-phosphate used as a backbone for shikimic acid and aromatic amino acids. In addition, NADPH is required for the synthesis of fatty acids and sterols from acetyl CoA. A major difference between the pentose phosphate pathway and the tricarboxylic acid-glycolysis systems is that NADP rather than NAD accepts electrons from the sugar molecule. NADPH is specifically required in some metabolic reactions, and it can enter into the mitochondrial electron transport system *via* an NADPH dehydrogenase.

Initial reactions in the pentose phosphate pathway include the irreversible oxidation of glucose-6-phosphate from glycolysis to 6-phosphogluconic acid, yielding a reduced NADP (Figure 3.6). Subsequently, 6-phosphogluconic acid is converted through the removal of carbon dioxide and hydrogen to a 5 carbon sugar, ribulose-5-phosphate which upon isomerization forms a ribose-5-phosphate that is essential for nucleic acid synthesis. The conversion of phosphogluconic acid to ribulose-5-phosphate is also not reversible, and reduced NADP is formed. The two initial reactions are the only oxidative (i.e., removal of electrons) steps in the pathway, and the second is the only point in the entire pathway at which carbon dioxide is removed. Subsequent steps are reversible and can recycle back to glucose-6-phosphate, the initial substrate.

Since the pentose phosphate pathway is an alternative means of oxidizing sugars, it is of interest to know which system is operative in harvested tissue. Existing evidence indicates that the glycolysis, tricarboxylic acid and pentose phosphate pathways are operative to some extent in all tissue; however, it is difficult to accurately measure the precise contribution of each pathway. In tomato fruit, the pentose phosphate pathway is thought to account for only about 16% of the total carbohydrates oxidized, a level probably common in many tissues. However, in some tissues such as storage roots, the pentose pathway appears to be responsible for 25 to 50% of the oxidation of sugars.

1.2. Photorespiration

The acquisition of carbon *via* photosynthesis and the loss of carbon through respiration can be seen as opposing processes in chlorophyll containing plant tissues. Growth is achieved when the gain in carbon exceeds losses, i.e., is above the carbon dioxide compensation point. In most species, it is known that the respiratory rate of chlorophyll containing tissue, as measured by the loss of CO_2 from the tissue, proceeds at a higher rate in the light than in the dark. This light-stimulated loss of carbon, termed photorespiration, is a process that occurs in ad-

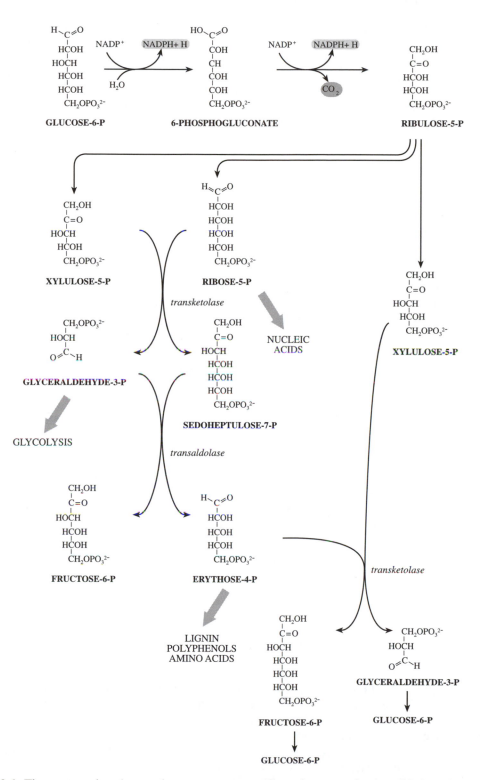

Figure 3.6. The pentose phosphate pathway, represents an alternative means for the oxidation of sugars and provides a ready mechanism for the formation of 3, 4, 5, 6 and 7 carbon skeletons for synthetic reactions. The pathway also provides NADPH + H$^+$ and ribose-5-phosphate, needed for nucleic acid production.

dition to or superimposed upon the normal dark respiratory processes in the plant, discussed previously.

If one uses a strict interpretation, photorespiration should not be considered a respiratory process since there is no transfer of energy between molecules, a classical requirement for respiration. Rather, it represents a form of oxidative photosynthesis. A significant portion of the carbon that is fixed into sugars in many species actually moves through this pathway. Since it has, however, generally been viewed as a respiratory process, for continuity we utilized this conventional approach.

In contrast to photorespiration, dark respiration (glycolysis, tricarboxylic acid cycle, pentose phosphate pathway and electron transport system) proceeds at essentially the same rate whether in the dark or in the light. The rate is determined by both metabolic demand and temperature. It has been estimated that 30 to 50% of the photosynthetically assimilated carbon in the leaves of some C_3 plants may be lost through photorespiration.[109]

The relative importance of photorespiration, and for that matter photobiology in general during the postharvest period, has not been studied to any appreciable extent since most products are stored in the dark or at low light levels. As a consequence, the degree to which we need to be concerned with detrimental effects of light and the potential usefulness of light during this time frame remains to be ascertained. Since photorespiration occurs in chlorophyll containing tissues that are actively photosynthesizing, it is assumed to be of greater importance in intact plants (e.g., bedding plants, woody ornamentals, transplants) than in detached plant parts. Since photorespiration decreases with both decreasing light intensity and oxygen concentration, both conditions common in postharvest handling, its rate could be readily altered.

The primary objective during the postharvest period is to maintain the product as close to the preharvest condition as possible (i.e., no significant growth in intact plants). Consequently, the balance between photosynthesis and respiratory losses may be more critical than the actual rates of each process.

Of the three primary photosynthetic carbon fixation pathways operative in higher plants, approximately 500 species possess the C_4 pathway, 250 species the CAM pathway; the remaining 300,000 are generally thought to utilize the C_3 pathway (for additional details of the pathways, see 2.2. Dark Reactions). In comparing the two primary groups, C_3 and C_4, there are a number of important characteristics that distinguish them. For example, plants having the C_3 photosynthetic pathway for carbon fixation have distinctly higher levels of photorespiration and carbon dioxide compensation points than do C_4 species (Table 3.3). The C_3 species, which comprise the majority of the woody and herbaceous ornamentals and transplants in postharvest handling and marketing, also differ in a number of other important characteristics. Photosynthesis in C_3 species is significantly inhibited by ambient oxygen levels (21%), and as a consequence, net photosynthesis is elevated and photorespiration depressed with low oxygen conditions. In addition, photosynthesis in many C_3 species also tends to saturate at lower light intensities than in C_4 species, and the optimum temperature for photosynthesis is significantly lower (Table 3.3).

During photosynthesis in C_3 species, a relatively large amount of glycolic acid is synthesized; however, the molecule cannot be metabolized in the chloroplasts. Upon movement out of the chloroplast and into peroxisomes, glycolic acid is oxidized to glyoxylic acid, which is subsequently converted to glycine (Figure 3.7). Glycine then moves into adjacent mitochondria where two molecules of glycine react to produce one molecule of serine and carbon dioxide. Since the oxidation step is not linked to ATP formation, photorespiration results in a loss of both energy and photosynthetic carbon from the plant.

The inhibition of photosynthesis by oxygen was first observed by Otto Warburg in 1929

Table 3.3. Several Characteristics Which Distinguish C_3 and C_4 Species.*

Characteristics	C_3 *Plants*	C_4 *Plants*
Leaf anatomy	No significant differentiation between mesophyll and bundle sheath cells	Bundle sheath cells containing large numbers of chloroplasts and other organelles
Major pathway of CO_2 fixation in light	Reductive pentose-phosphate cycle (i.e., Calvin-Benson cycle)	C_4 pathway plus reductive pentose cycle
Photorespiration	High	Low
Inhibitory effect of O_2 on photosynthesis and growth	Yes	No
CO_2 compensation point in photosynthesis (ppm CO_2)	30–70	0–10
Net photosynthesis vs. light intensity	Saturation at ca. 1000–4000 foot candles	No saturation
Maximum net photosynthetic rate (mg CO_2/dm² leaf area/hr)	15–35	40–80
Optimum temperature for net photosynthesis (°C)	15–25	30–45
Transpiration rate (g H_2O/g dry wt)	450–950	250–350

*After: Kanai and Black[56]

and has subsequently been known as the **Warburg effect,** in the same manner as the inhibition of sugar breakdown by oxygen was named the Pasteur effect after Louis Pasteur. The inhibition of photosynthesis by oxygen involves the competition between molecules of carbon dioxide and oxygen for the same binding site on ribulose bisphosphate carboxylase, the primary photosynthetic carboxylation enzyme. The higher the oxygen level, the more favored the oxygenation reaction and the greater the production of glycolic acid, the substrate for photorespiration.

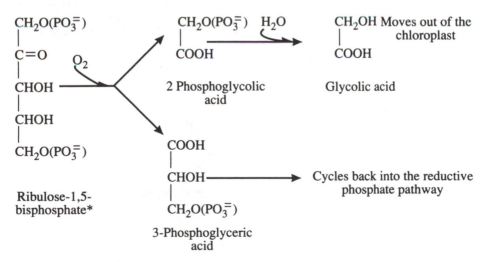

When the oxygen concentration is lowered, the carboxylation reaction is increasingly favored.

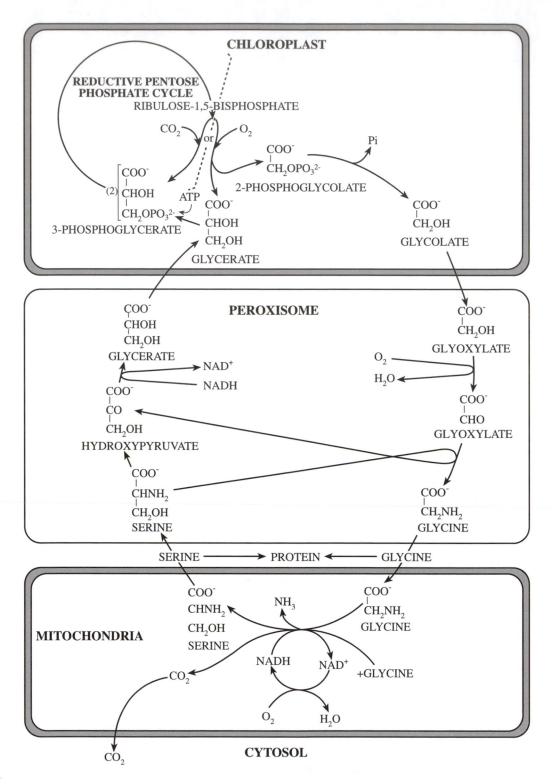

Figure 3.7. The metabolic pathway for carbon metabolism during photorespiration. Glycolate produced in the chloroplasts is transported to peroxisomes, where it is oxidized to glycine. Glycine is then converted in the mitochondria, forming serine and liberating carbon dioxide. Serine can then be cycled back through the peroxisomes and converted to glycerate, which re-enters the C_3 cycle giving a net loss of 1 molecule of carbon dioxide per molecule of glycolate formed.

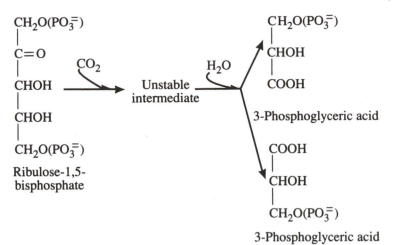

3-Phosphoglyceric acid

The rate of photorespiration is difficult to measure precisely in an illuminated leaf since a portion of the carbon dioxide respired is photosynthetically refixed before it escapes from the leaf. The **carbon dioxide compensation point,** the concentration of carbon dioxide in the atmosphere where carbon dioxide fixed equals that respired, is often used as an index of photorespiration. Species which have high compensation points (30–70 ppm carbon dioxide) have high rates of photorespiration, and conversely, those with low compensation points (0–10 ppm carbon dioxide) have low photorespiration rates. For C_3 species, the difference in the rate of photosynthesis at 21% oxygen and 2% oxygen is also used as a measure of photorespiration, since photorespiration is almost totally blocked (the oxidation of ribulose-1,5-bisphosphate) by low oxygen.

2. PHOTOSYNTHESIS

Photosynthesis is the process by which green plants capture light energy and convert it into chemical energy that is allocated between growth and maintenance reactions.[104] Photosynthesis is not commonly considered a significant postharvest metabolic process, since many harvested products contain few chloroplasts and/or are usually stored in the dark. However, a number of products have the potential to photosynthesize, and many, although not all, may derive a benefit from this process upon removal from the production area. These products fall into two major groups: 1) intact plants such as ornamentals, leafy cuttings and tissue cultures; and 2) chlorophyll containing detached plant parts such as green apples or pepper fruits, petioles, shoots, leaves and others. Therefore, a distinct group of postharvest products are, at least theoretically, not totally severed from an external source of energy that may be used for maintenance. In some cases, even small inputs of free energy after harvest may substantially reduce or eliminate the products' dependence upon stored reserves.[68]

With intact plants, there are two general options for handling the product. Conditions can be created or selected that will maintain the plants' photosynthetic environment. This requires light, an appropriate carbon dioxide concentration and temperature, and sufficient water to maintain an adequate moisture balance within the plant. In contrast to the site of production, the postharvest environment is maintained at a lower level of these parameters, a level that will ensure maintenance of the product rather than enhanced growth and development. In many postharvest environments for intact plants, appropriate plant moisture status is the parameter that is most commonly handled improperly.

A postharvest environment may also be selected for intact plants that will minimize the metabolic rate of the product. Therefore, in contrast to an environment conducive for photosynthesis, an environment can be selected to minimize the utilization of stored energy reserves. This is the primary option selected for the handling of both intact plants and detached plant parts and is accomplished largely by product temperature management.

Products that were photosynthetic organs prior to being severed from the plant at harvest (e.g., lettuce, amaranths, spinach) are logical candidates to derive a benefit from light during storage. This, however, is rarely the case. One reason is that the light energy trapping efficiency of plants, even under optimum conditions, is low (usually under 5%), the remaining energy being dissipated primarily as heat. This elevates the leaf temperature and leads to counterproductive increases in the use of stored energy reserves *via* the respiratory pathways. In intact plants, leaf temperature is decreased through the cooling effect of evapotranspiration. One gram of water removes 540 calories of heat upon being transformed from a liquid to a gas. Severed plant parts, however, do not have a readily replenishable source of water that can be used for cooling *via* evapotranspiration. As a consequence, product temperature increases.

An additional problem with utilizing photosynthesis to help maintain harvested chlorophyll containing plant parts is that the temperatures at which the products are normally stored are substantially below those required for optimum photosynthesis. The lower temperatures are essential, however, for successful storage since they decrease the metabolic rate of the product and the utilization of stored energy reserves.

In products that benefit from photosynthesis after harvest, the amount of external energy needed prior to harvest differs from that required after harvest. This difference is based on a distinction between the primary goals of the product before and after harvest. Prior to harvest, growth is a primary goal; therefore, carbon and energy acquisition must be greater than respiratory utilization. After harvest, during the postharvest handling period, growth is seldom desirable. Rather, the objective is to maintain the product as close to its harvested condition as possible (i.e., minimize change). Therefore, photosynthesis after harvest is seen as way of maintaining the energy balance within the plant, rather than as a means of providing excess energy for the purpose of carbon accumulation.

Photosynthesis occurs within specialized plastids, the chloroplasts, found primarily in leaves. The most important pigment in these plastids is chlorophyll, but other pigments such as carotenoids and phycobilins also participate in photosynthesis. The simplified overall reaction occurring in photosynthesis can be written as:

$$6\,CO_2 + 6\,H_2O + \text{light (h}\nu) \xrightarrow{\text{green plant}} C_6H_{12}O_6 + 6\,O_2$$

where carbon dioxide is fixed and oxygen from water is released. Photosynthesis can be divided into two interconnected processes: the light reactions that trap energy from light and release oxygen from water, and the dark reactions that use the energy to fix carbon dioxide.

2.1. Light Reactions

The light reactions involve the splitting of water with the release of oxygen:

$$\text{light (h}\nu) + H_2O + NADP \rightarrow \tfrac{1}{2}O_2 + NADPH + H$$

and the light driven formation of ATP from ADP and Pi (photophosphorylation). The reactions trap light energy (photons) and transport it in the form of electrons from water through a series of intermediates to NADP, where it can be stored as NADPH (Figure 3.8). Two separate light reactions act cooperatively in elevating the electrons to the energy level required for

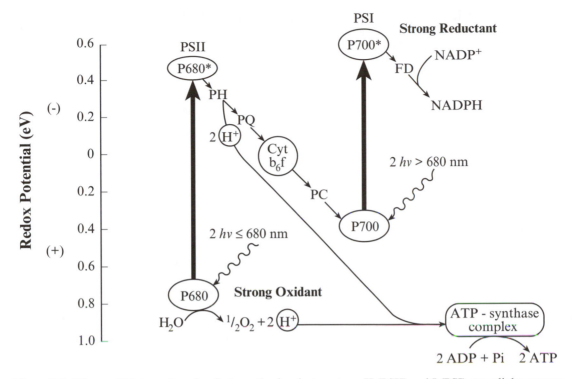

Figure 3.8. The two light reactions in photosynthesis, photosystems II (PSII) and I (PSI), trap light energy and convert it to ATP and NADPH + H$^+$; oxygen is also liberated in the process. Energy is absorbed by photosystem II which results in the splitting of water (photolysis), releasing O$_2$ and H$^+$ and the excitation of electrons to a high energy level where they can be accepted by the first carrier in a series which transfers the electrons to the chlorophyll of photosystem I. Additional light energy absorbed by the chlorophyll molecule increases the energy level of the electrons which are trapped by an electron acceptor and subsequently transferred to ferredoxin (FD). NADP is reduced, utilizing the H$^+$ formed in the photolysis of water, yielding NADPH + H$^+$.

their transfer to NADP. In this process, the electrons are transported *via* an electron transport chain that operates on the same alternating oxidation-reduction principle as the respiratory electron transport system, though it is distinctly different.

2.2. Dark Reactions

The energy trapped in the light reactions as NADPH and ATP can be used in a number of reactions within the plant; however, its primary role is in the fixation of carbon from atmospheric CO$_2$ (dark reactions). In plants commonly encountered after harvest, there are three primary means of fixation of CO$_2$: the C$_3$, C$_4$, and CAM pathways.

2.2.1. C$_3$–Reductive Pentose Phosphate Pathway

The C$_3$ or reductive pentose phosphate pathway (PPP) is operative within the majority of plant species. The name reductive PPP is to distinguish it from the oxidative PPP which shares some of the same enzymes. The pathway is also referred to as the Calvin cycle after Melvin Calvin, who with his colleagues elucidated the cyclic pathway in the 1950's. In the pathway, CO_2 from the air is fixed by reacting with ribulose-1,5-bisphosphate (5 carbon sugar) to form two 3-carbon phosphoglycerate (PGA) molecules.

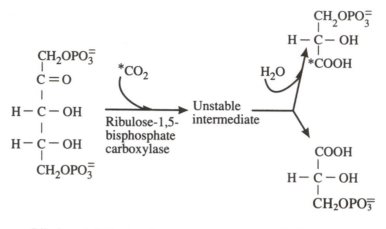

Ribulose-1,5-bisphosphate   (2) Phosphoglyceraldehydes

The energy captured in the light reactions is used to convert PGA back to ribulose-1,5-bisphosphate (Figure 3.9) for the continuation of the process. Each cycle fixes a single carbon

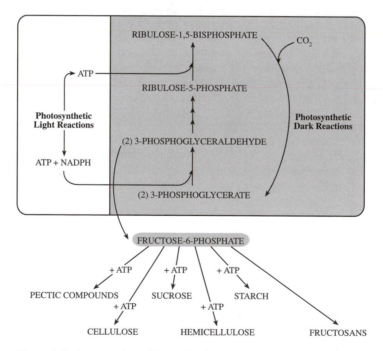

Figure 3.9. An overview of the reductive pentose phosphate or Calvin-Benson photosynthetic cycle. NADPH and ATP formed in the light reactions (Figure 3.8) are used to convert 3-phosphoglyceraldehyde back to ribulose-1,5-diphosphate to complete the cycle. As the number of 3-phosphoglyceraldehyde molecules increases, they are converted to hexose sugars and subsequently into the diverse array of carbon compounds found within the plant.

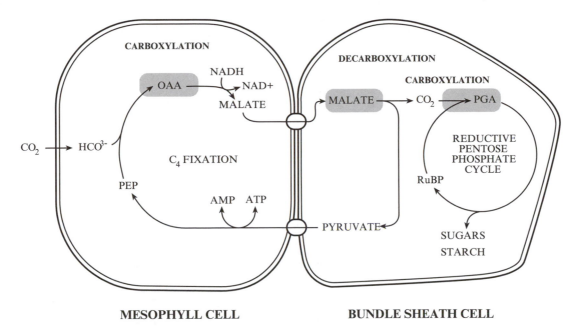

Figure 3.10. In C_4 plants, carbon dioxide is initially fixed (carboxylated) in the mesophyll cells of the leaf by reacting with phosphoenolpyruvate, forming oxaloacetate and subsequently malate. Malate is transported into bundle sheath cells, where it is decarboxylated, liberating CO_2 that is refixed *via* the reductive pentose phosphate pathway. Decarboxylation in the bundle sheath cells greatly increases the CO_2 concentration, increasing the efficiency of the C_3 pathway.

dioxide molecule. The chemical energy captured (nine ATP equivalents) is required for the fixation of one molecule of carbon dioxide, three as ATP and six equivalents in the reducing power of two NADPH molecules.

2.2.2. C_4 Pathway

In some species of plants, carbon dioxide reacts with phosphoenolpyruvic acid in the mesophyll cells, forming the four carbon compound oxaloacetate (hence the name C_4 pathway). Oxaloacetate is then converted to malate (Figure 3.10) that diffuses into the bundle sheath cells of the vascular bundles, where a carbon dioxide molecule is removed (decarboxylated) from the malate, yielding pyruvate that subsequently recycles back to phosphoenolpyruvate. The carbon dioxide removed is not lost but is refixed *via* the reductive pentose phosphate pathway in the bundle sheath cells. Here the oxygen concentration is low, and because of the release of carbon dioxide, its concentration is higher, greatly increasing the efficiency of the carboxylation reaction of the C_3 pathway (i.e., very low photorespiration).[19] Two variations of the pathway have also been found: 1) oxaloacetate → asparate → oxalacetate → malate → pyruvate; and 2) oxaloacetate → asparate → oxalacetate → phosphoenolpyruvate. In each case, the product formed in the bundle sheath cells with the removal of carbon dioxide (e.g., pyruvate or phosphoenolpyruvate) cycles through a series of reactions back to phosphoenolpyruvate in the mesophyll cells. Therefore in C_4 plants, the enzymes required for both the C_4 and C_3 pathways are present but in different cells. The C_4 pathway, however, is a more efficient means of carbon fixation than the reductive pentose phosphate pathway. Many species with the C_4 pathway (e.g., corn) evolved in geographical regions with hot, dry climates, enhancing their resistance to high temperatures and water use efficiency.

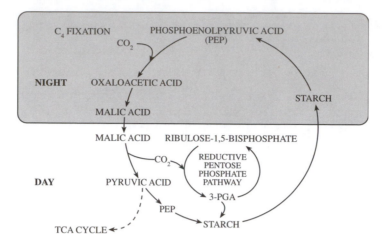

Figure 3.11. The CAM or crassulacean acid cycle found in some plants fixes CO_2 at night *via* the C_4 pathway when the stomata are open, with the formation of malic acid that is stored in the vacuole. During the day (stomata closed), malic acid moves out of the vacuole and is decarboxylated. The CO_2 is refixed in the chloroplasts, using the reductive pentose phosphate pathway. Some of the carbon is converted to starch that is recycled in a series of reactions, eventually forming phosphoenolpyruvic acid, starting the CAM cycle over again.

2.2.3. *Crassulacean Acid Metabolism*

A third means of fixing carbon is found in crassulacean acid metabolism (CAM) plants (e.g., pineapple). These plants trap carbon dioxide at night when their stomates are open rather than during the day as in C_3 and C_4 species whose stomates open in the light. Carbon dioxide is fixed through the action of the enzyme phosphoenolpyruvate carboxylase, forming oxaloacetate from phosphoenolpyruvate (Figure 3.11). During the day, when the stomates are closed, malate formed from oxaloacetate has carbon dioxide removed (decarboxylated) and refixed *via* the reductive pentose phosphate cycle. In CAM plants, both the C_3 and C_4 cycles are operative and found within the same cells as the CAM cycle. The pathway has evolved in plants that grow in very hot, arid regions where opening their stomates at night, rather than during the day, minimizes water loss.

3. METABOLIC CONSIDERATIONS IN HARVESTED PRODUCTS

3.1. Dark Respiration

3.1.1. *Effects of Respiration*

Photosynthesis provides the carbohydrates that plants use for growth and storage, while respiration is a mechanism by which the energy stored in the form of carbon compounds is released. In the general equation for the oxidation of a hexose sugar, the substrate and oxygen are converted into carbon dioxide, water and energy. The rate of conversion is modulated by temperature and the concentration of oxygen and carbon dioxide. The conversion is therefore significant for both the stored product and the environment surrounding the product. The two major functions of dark respiration are the release of energy stored in chemical form as starch, sugars, lipids, and other substrates, and the formation of carbon skeletons to be used in various synthetic and maintenance reactions. The effects of respiration substantially alter the methods employed in handling and storing many products and, hence, are of considerable commercial importance.

The loss of substrate from stored plant products results in a decrease in energy reserves within the tissue. This loss decreases the length of time the product can effectively maintain its condition. Loss of energy reserves eventually results in tissue starvation and accelerated senescence, and is especially critical in products such as leaves, flowers and other structures that are not carbon storage sites. Likewise, in a marketing system based on weight, respiratory losses of carbon represent weight losses in the product, hence a decreased value. The rate of respiration can in fact be used to predict the loss of dry weight from stored products (Figure 3.12). Respiratory losses also decrease the total food value (i.e., energy content) of edible products.

Respiration removes oxygen from the storage environment. If the oxygen concentration in the environment is severely depleted, anaerobic conditions occur that can rapidly spoil most plant products. As a consequence, the rate of respiration is important for determining the amount of ventilation required in the storage area. It is also critical in determining the type and design of packaging materials to be used, as well as the use of artificial surface coating on the product (e.g., waxes on citrus or cucumbers). The respiratory reduction in oxygen concentration in the storage environment can also be used as a tool to extend the storage life of a product. Since oxygen concentration has a pronounced effect on the rate at which respiration proceeds, a respiration mediated decrease in the ambient oxygen concentration can create a modified environment that may be used to slow respiration. This principle, used since Roman times, represents the basis for present day storage practices for several highly perishable products.

Elevated ambient levels of carbon dioxide generated by respiration can also be used to decrease respiration since its accumulation impedes the rate at which the process proceeds. The degree of inhibition of respiration by carbon dioxide and the sensitivity of the tissue to high carbon dioxide concentration varies widely among products. Carbon dioxide produced during the respiratory process, if allowed to accumulate, can be harmful to many stored products. For example, lettuce,[72] mature green tomatoes, bell peppers[82] and other products are damaged by high carbon dioxide. So it is essential that the carbon dioxide concentration be maintained at a safe level through adequate ventilation or absorption.

Water is produced during the respiratory process (termed metabolic water) and becomes part of the water present within the tissue. Metabolic water, however, represents only a very minor addition to the total volume of water within the tissue and hence is of minimal significance (Figure 3.12).

Energy, the final product in the respiratory equation, has a significant influence both upon the maintenance of the product and the preferred storage environment. The complete oxidation of one mole of a six carbon sugar such as glucose results in the formation of 686 kcal ($2,868 \text{ kJ} \cdot \text{mol}^{-1}$) of energy. In actively growing tissues, a significant portion of this energy is utilized in chemical forms by the cell for synthetic and maintenance reactions. A substantial amount of energy is lost, however, as heat, also referred to as "vital" heat, since the energy conservation during the transfer among molecules is not 100% efficient. In actively metabolizing tissues, around 46% of the total respiratory energy is lost as heat. This amount, however, varies among different types of plants and organs, and the general condition of the tissue. The amount of heat produced by the product can be calculated (i.e., within ~10%)[41] directly from the respiratory rate of the product (Figure 3.12). Knowledge of the amount of heat produced is important in determining the cooling requirements for a product and therefore the size of

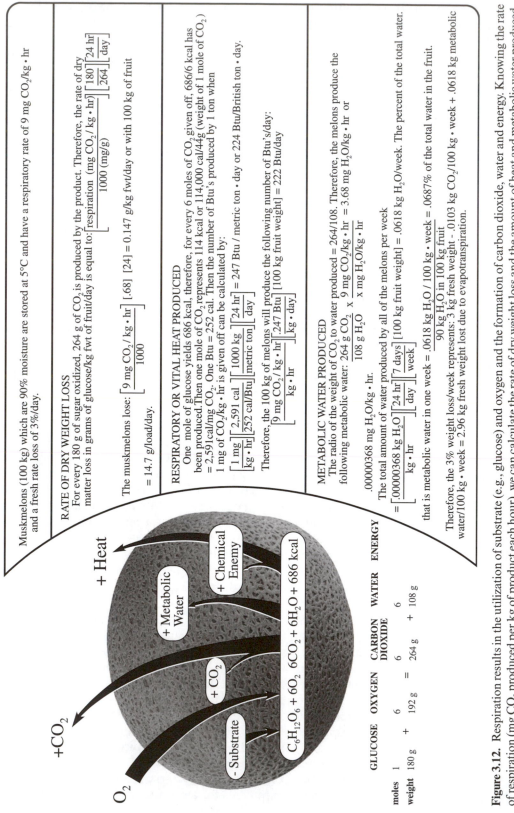

Muskmelons (100 kg) which are 90% moisture are stored at 5°C and have a respiratory rate of 9 mg CO_2/kg · hr and a fresh rate loss of 3%/day.

RATE OF DRY WEIGHT LOSS

For every 180 g of sugar oxidized, 264 g of CO_2 is produced by the product. Therefore, the rate of dry matter loss in grams of glucose/kg fwt of fruit/day is equal to:

$$\left[\frac{respiration\ (mg\ CO_2/kg \cdot hr)}{1000\ (mg/g)}\right][.68]\left[\frac{180}{264}\right]\left[\frac{24\ hr}{day}\right]$$

The muskmelons lose: $\left[\dfrac{9\ mg\ CO_2/kg \cdot hr}{1000}\right][.68][24] = 0.147$ g/kg fwt/day or with 100 kg of fruit

= 14.7 g/load/day.

RESPIRATORY OR VITAL HEAT PRODUCED

One mole of glucose yields 686 kcal, therefore, for every 6 moles of CO_2 given off, 686/6 kcal has been produced. Then one mole of CO_2 represents 114 kcal or 114,000 cal/44g (weight of 1 mole of CO_2) = 2.591 cal/mg CO_2. One Btu = 252 cal. Then the number of Btu's produced by 1 ton when 1 mg of CO_2/kg · hr is given off can be calculated by:

$$\left[\frac{1\ mg}{kg \cdot hr}\right]\left[\frac{2,591\ cal}{mg}\right]\left[\frac{1000\ kg}{metric\ ton}\right]\left[\frac{24\ hr}{day}\right]\left[\frac{1}{252\ cal/Btu}\right] = 247\ Btu\ /\ metric\ ton \cdot day\ or\ 224\ Btu/British\ ton \cdot day.$$

Therefore, the 100 kg of melons will produce the following number of Btu's/day:
$$\left[\frac{9\ mg\ CO_2/kg \cdot hr}{kg \cdot hr}\right]\left[\frac{247\ Btu}{kg \cdot day}\right][100\ kg\ fruit\ weight] = 222\ Btu/day$$

METABOLIC WATER PRODUCED

The radio of the weight of CO_2 to water produced = 264/108. Therefore, the melons produce the following metabolic water: $\dfrac{264\ g\ CO_2}{108\ g\ H_2O} \times \dfrac{9\ mg\ CO_2/kg \cdot hr}{x\ mg\ H_2O/kg \cdot hr} = 3.68$ mg H_2O/kg · hr or

.00000368 mg H_2O/kg · hr.

The total amount of water produced by all of the melons per week

$$= \left[\frac{.00000368\ kg\ H_2O}{kg \cdot hr}\right]\left[\frac{24\ hr}{day}\right]\left[\frac{7\ days}{week}\right][100\ kg\ fruit\ weight] = .0618\ kg\ H_2O/week.$$ The percent of the total water.

that is metabolic water in one week = .0618 kg H_2O / 100 kg · week = .0687% of the total water in the fruit.
$$\frac{.0618\ kg\ H_2O\ in\ 100\ kg\ fruit}{90\ kg\ H_2O\ in\ 100\ kg\ fruit}$$

Therefore, the 3% weight loss/week represents: 3 kg fresh weight - .0103 kg CO_2/100 kg · week + .0618 kg metabolic water/100 kg · week = 2.96 kg fresh weight lost due to evapotranspiration.

O_2
+CO_2
+ Heat
+ Metabolic Water
+ Chemical Enemy
+ CO_2
- Substrate

$$C_6H_{12}O_6 + 6O_2 \quad 6CO_2 + 6H_2O + 686\ kcal$$

	GLUCOSE		OXYGEN		CARBON DIOXIDE		WATER		ENERGY
moles	1	+	6	=	6	+	6		6
weight	180 g	+	192 g	=	264 g	+	108 g		

Figure 3.12. Respiration results in the utilization of substrate (e.g., glucose) and oxygen and the formation of carbon dioxide, water and energy. Knowing the rate of respiration (mg CO_2 produced per kg of product each hour), we can calculate the rate of dry weight loss and the amount of heat and metabolic water produced. Examples of these calculations for muskmelons are illustrated in the figure.

the refrigeration system needed to maintain the desired temperature of the storage room. The amount of heat produced also influences the size of the fans required to move air around the product in storage, package design and stacking method.

3.1.2. *Respiratory Substrates*

Respiration depends on the presence of a substrate. In many tissues, this substrate is a storage form of carbohydrate such as starch in the sweetpotato root or inulin in Jerusalem artichoke tubers. These more complex molecules are broken down into simple sugars that enter the respiratory pathways to provide energy for the plant. In some species, carbon may be stored as lipids. The avocado fruit contains approximately 25% lipid on a fresh weight basis and pecan kernels approximately 74%. Organic acids, proteins and other molecules may also be utilized as respiratory substrates in plants, although in most cases these secondary substrates are not produced for this purpose. Under conditions where the tissue is depleted or "starved" of carbohydrate or lipid reserves, these secondary respiratory substrates are utilized. This situation is more likely to occur in postharvest products such as leaves or flowers that do not represent storage organs and therefore have relatively little reserve substrate. Proteins may also be hydrolyzed into their component amino acids and catabolized in the glycolytic pathway and tricarboxylic acid cycle.

When these various substrates are utilized and completely oxidized, different amounts of oxygen are consumed in relation to the amount of carbon dioxide evolved. The ratio of the two is called the **respiratory quotient** (RQ). The RQ provides a general indication of the particular substrate being used as the primary source of respiratory energy. For example, the oxidation of a common carbohydrate, lipid and organic acid give the following respiratory quotients.

Type of Substrate	Substrate	Reaction	Respiratory Quotient (CO_2/O_2)
Carbohydrate	glucose	$C_6H_{12}O_6 + 6\,O_2 \rightarrow 6\,CO_2 + 6\,H_2O$	1.00
Lipid fatty acid	palmitic acid	$C_{16}H_{32}O_2 + 11\,O_2 \rightarrow C_{12}H_{22}O_{11} + 4\,CO_2 + 5\,H_2O$	0.36
Organic acid	malic acid	$C_4H_6O_5 + 3\,O_2 \rightarrow 4\,CO_2 + 3\,H_2O$	1.33

The RQ was of greater interest early last century when analytical techniques were limited. There is still a diversity of opinions as to the actual value of the RQ. Along with tissue type, temperature and tissue age, a number of other factors significantly alter the RQ. In addition, substrates are not always completely oxidized, and several types of substrates may be used simultaneously by the cells, each greatly complicating interpretation of the RQ obtained. Other factors that affect the apparent RQ are the differential permeabilities of the tissues to oxygen and carbon dioxide, as well as respiration at oxygen levels approaching anaerobic conditions.

3.1.3. *Control Points in the Respiratory Pathway*

Changes in the cell's internal environment by external (e.g., temperature) or internal (e.g., substrate availability) factors often result in significant alterations in the respiration rate. The alteration may be due to shifts in the activity at the regulatory sites in the pathways or due to changing priorities among different pathways. Control of respiration rate in plant cells can be regulated at various points in the respiratory pathways and by a number of means. Substrate supply can control the rate of respiration by regulating substrate available for a particular reaction. For example, if glucose-6-phosphate levels are high, the reaction catalyzed by phosphoglucoisomerase shifts toward the formation of fructose-6-phosphate in order to maintain an equilibrium. Substrate control is probably more important when demand for intermediates generated by the tricarboxylic acid cycle is high. Control can occur through the activity of an enzyme and to a lesser extent the enzymes concentration. Enzyme activities are modulated by substrate and product concentration, cofactors such as metal ions, compounds that activate or inhibit the enzyme, and the rate of enzyme synthesis and degradation. For rate limiting reactions, the concentration of an enzyme is thought to represent a coarse control. In contrast, enzyme activation is considered a means of fine control. The availability of phosphate acceptors (ADP) represents an extremely important means of respiratory control. Restricting the rate of flow of electrons through the electron transport chain and, hence, the rate of oxidation of NADH limits the rate of a number of reactions. However, if NADH is reoxidized by an alternative reaction, oxidative phosphorylation is diminished in its regulatory role. High levels of ATP also directly inhibit certain enzyme reactions (e.g., phosphofructokinase and pyruvate kinase). Therefore, the levels of ADP, NAD and NADP and their reduced products represent important modulators of respiration.

The tricarboxylic acid cycle appears to be largely regulated by mitochondrial energy status (ADP, ATP). However, low oxygen and high carbon dioxide are also known to have a pronounced effect on the rates of specific enzymes in the cycle. High carbon dioxide inhibits the conversion of succinate to malate, and malate to pyruvate, in apple fruit tissue.[63] Key enzymes controlling the rate of the glycolytic pathway are phosphofructokinase and pyruvate kinase, while in the pentose phosphate pathway the activity of glucose-6-phosphate dehydrogenase is controlled by the NADPH/NADP ratio.

3.1.4. *Factors Affecting the Rate of Dark Respiration*

The control of postharvest respiratory responses is strongly influenced by a number of commodity and environmental factors. For many products, high respiratory rates are closely correlated with reduced storage life. Proper management of these factors is imperative for maintaining quality and maximizing storage life.

a. *Temperature*

Temperature has a pronounced effect on the respiratory rate of harvested products. As product temperature increases, reaction rates increase;[90] however, the degree of increase is not the same for all the reactions within a tissue (e.g., the optimum temperature for photosynthesis is usually lower than the optimum temperature for respiration). The rate of change in reactions due to temperature is commonly characterized using a measure called the Q_{10}, which is the ratio of the rate of a reaction at one temperature (T_1) versus the rate at that temperature plus 10°C [(rate at $T_{1+10°C}$)/rate at T_1]. The Q_{10} is often quoted for respiration, in that it gives a very general estimate of the effect of temperature on the overall tissue metabolic rate. There are, however, many exceptions in different metabolic pathways; for example the respiratory rate of potato tubers decreases with decreasing temperature while the formation of sugars from starch increases below 10°C (Figure 3.13). For many products the Q_{10} for respiration is between 2.0 and 2.5 for every 10°C increase in temperature within the 5°C to 25°C range. If we are interested in maintaining a product as close to its condition at harvest, the use of low temperatures to reduce changes due to metabolism is essential. By decreasing the temperature from 25°C to 15°C when the Q_{10} is 2, the respiration rate will be half that at 25°C and halved again if the temperature is reduced from 15°C to 5°C. As the temperature increases from 25°C

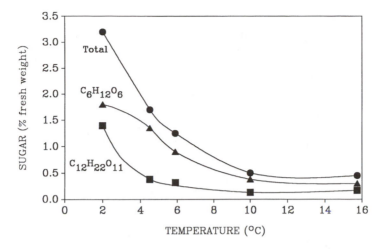

Figure 3.13. The effect of temperature on the formation of sugars from starch in potato tubers cv. Majestic (*redrawn from Burton*).[24]

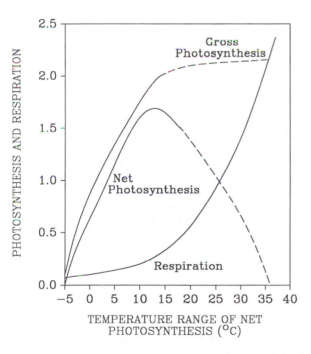

Figure 3.14. The effect of temperature on photosynthesis, respiration and net or apparent photosynthesis of Swiss stone pine seedlings (broken lines represent estimates, *redrawn from Tranquillini*).[101]

into the 30°C to 35°C range, the Q_{10} declines for most products, and at very high temperatures reaction rates are actually depressed, probably due to the loss of enzyme activity. The actual temperature range over which there is a linear increase in Q_{10} and the maximum and minimum temperature for a particular metabolic process vary widely among species and the type of tissue monitored (Figure 3.14). For example, respiration in *Populus tremuloides* stems can be measured at -11°C,[35] a temperature at which an apple fruit would be frozen solid, terminating respiration.

It is important to note that while the ambient temperature at which the product is stored is of critical importance in determining the metabolic rate, the product temperature is typically

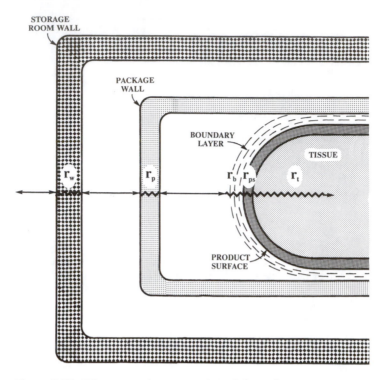

STORAGE
ROOM WALL

PACKAGE
WALL

BOUNDARY
LAYER

TISSUE

r_w

r_p

r_b / r_{ps}

r_t

PRODUCT
SURFACE

Figure 3.15. Diagrammatic presentation of the resistances to gas exchange in a harvested product. Total resistance is r_w (storage walls) + r_p (package wall) + r_b (boundry layer) + r_{ps} (product surface) + r_t (tissue). The resistances for r_w, r_p and r_{ps} are often manipulated for certain postharvest products to extend storage life.

slightly higher than the ambient temperature due to the heat liberated from the respiratory process. This often slight difference in temperature is quite important due to its effect on maintaining the moisture balance of the harvested product (see chapter 9).

b. Gas Composition

The gas composition of the atmosphere that surrounds the product can influence both its respiratory and general metabolic rate. Oxygen, carbon dioxide and ethylene are the most important gases influencing respiration. Pollutant gases such as sulfur dioxide, ozone, propylene, and others can also have a significant effect if their concentration becomes sufficiently high.

During normal plant growth and development in the field, there are seldom large or long term alterations in gas atmosphere composition. After harvest, however, plant products are normally bulked tightly together and placed in containers and storage areas that have restricted air flow (Figure 3.15). Restricted air flow creates additional resistances for gas movement into and out of the product and hence alters the concentration of gases within the tissue. Typically, reduced gas exchange leads to a decrease in the internal oxygen and an increase in carbon dioxide. However in some crops, such as submerged aquatics (e.g., Chinese water chestnut, lotus root) and some root and tuber crops that grow normally under conditions of high gas diffusion resistance, the opposite may be true. Therefore, postharvest conditions commonly result in significant alterations in the gas environment to which the product experienced prior to harvest, and these changes can influence metabolic activity.

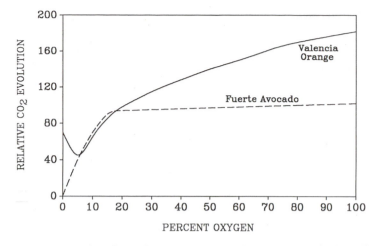

Figure 3.16. The effect of oxygen concentration on the respiration of Fuerte avocadoes and Valencia oranges. Note the "Pasteur effect" (stimulated respiration) on oranges at very low oxygen levels (*redrawn from Baile*).[10,12]

The effect of oxygen concentration on harvested plant products has been known since around the beginning of the 19th century.[50] Berard in 1821 noted that fruits held in an environment devoid of oxygen did not ripen, and if not kept too long under the low oxygen conditions, ripened normally upon return to air.[9] The rate of respiration is closely tied to the oxygen concentration in harvested products. As the internal oxygen concentration decreases, respiration decreases (Figure 3.16) until the oxygen concentration reaches the **extinction point** or critical concentration at about 1 to 3% oxygen. This concentration represents the point at which aerobic respiration *via* the tricarboxylic acid cycle is blocked and anaerobic fermentation begins. Below the extinction point, the rate of respiration increases. The increase in respiration at very low oxygen concentrations is known as the **Pasteur effect,** after Louis Pasteur, who first studied the phenomenon in microorganisms. Although the Pasteur effect is found widely in plants, it does not occur in all harvested products. For example, there is no increase in respiration of the avocado fruit even at 0% oxygen (Figure 3.16). The Pasteur effect is due to a cessation of the tricarboxylic acid cycle, as NAD and NADP are not available, having been converted to the reduced forms (NADH and NADPH). The reduced forms cannot transfer their energy to the electron transport chain in the mitochondria to produce ATP as the final step in the chain requires oxygen. For the cell to maintain itself, ATP is required; hence the rate of glycolysis has to be increased, since it does not directly require oxygen. This change also leads to the production of ethanol and lactic acid that requires the NADH produced in the glycolytic pathway.

The precise reason why the respiration rate is reduced at oxygen concentrations above the critical concentration in most products is not known. A number of possible explanations have been suggested. The enzyme phosphofructokinase, which catalyzes the conversion of fructose-6-phosphate to fructose-1,6-bisphosphate in the glycolytic pathway, is inhibited by ATP and citric acid, both of which are formed in the oxygen dependent tricarboxylic acid cycle and may represent the modulating factors. It is also possible that oxygen concentration modulates one or more of the glycolytic enzymes. Respiration rate could also be decreased through an effect on several of the enzymes in the tricarboxylic acid cycle. In green banana fruit, low oxygen limits the rate of the enzymatic steps between either oxaloacetate or pyruvate and citrate and between 2-oxoglutarate and succinate.[78]

Table 3.4. Effect of Temperature and Oxygen Concentration on the Respiratory Rate of Various Commodities.*

	Carbon Dioxide Production (mg · kg⁻¹ · hr⁻¹)					
	In Air (°C)			*In 3% O_2 (°C)*		
	0	*10*	*20*	*0*	*10*	*20*
Asparagus	28	63	127	25	45	75
Beans, broad	35	87	145	40	55	80
Beans, runner	21	36	90	15	25	46
Beetroot, storing	4	11	19	6	7	10
Beetroot, bunching with leaves	11	22	40	7	14	32
Blackberries 'Bedford Giant'	22	62	155	15	50	125
Blackcurrants 'Baldwin'	16	39	130	12	30	74
Brussels sprouts	17	50	90	14	35	70
Cabbage 'Primo'	11	30	40	8	15	30
Cabbage 'January King'	6	26	57	6	18	28
Cabbage 'Deccma'	3	8	20	2	6	12
Carrots, storing	13	19	33	7	11	25
Carrots, bunching with leaves	35	74	121	28	54	85
Calabrese	42	105	240	—	70	120
Cauliflower 'April Glory'	20	45	126	14	45	60
Celery, white	7	12	33	5	9	22
Cucumber	6	13	15	5	8	10
Gooseberries 'Leveller'	10	23	58	7	16	26
Leeks 'Musseburgh'	20	50	110	10	30	57
Lettuce 'Unrivalled'	18	26	85	15	20	55
Lettuce 'Kordaat'	9	17	37	7	12	25
Lettuce 'Klock'	16	31	80	15	25	45
Onion 'Bedfordshire Champion'	3	7	8	2	4	4
Parsnip 'Hollow Crown'	7	26	49	6	12	30
Potato, main crop 'King Edward'	6	4	6	5	3	4
Potato, new (immature)	10	20	40	10	18	30
Peas (in pod) early, 'Kelvedon Wonder'	40	130	255	29	84	160
Peas main crop, 'Dark Green Perfection'	47	120	250	45	60	160
Peppers, green	8	20	35	9	14	17
Raspberries 'Malling Jewel'	24	92	200	22	56	130
Rhubarb (forced)	14	35	54	11	20	42
Spinach 'Prickly True'	50	80	150	51	87	137
Sprouting broccoli	77	170	425	65	115	215
Strawberries 'Cambridge Favorite'	15	52	127	12	45	86
Sweetcorn	31	90	210	27	60	120
Tomato 'Eurocross BB'	6	15	30	4	6	12
Turnip, bunching with leaves	15	30	52	10	19	39
Watercress	18	80	207	19	72	168

*Source: Robinson et al.[94]

The respiratory rate of most stored products can be decreased by lowering the oxygen to a concentration that is not below the extinction point for that product (Table 3.4). The actual critical concentration of oxygen appears to vary among products. In addition, the external concentration of oxygen that gives the appropriate internal concentration for minimizing respiration varies with the rate of oxygen utilization by the tissue, the tissue's diffusive resistance, and the differential in oxygen partial pressures between the interior and the exterior. Therefore, at higher temperatures, the external concentration of oxygen must be increased to maintain a given oxygen level within the tissue due to the increased rate of utilization of the oxygen by the tissue.

In general, a significant decrease in the respiratory rate for most stored products does not occur until the external oxygen concentration is below 10%. The optimum external concentration for a number of products held in cold storage is in the 1 to 3% range; however, there are exceptions. For example, sweetpotatoes shift to anaerobic metabolism at external oxygen concentrations below 5 to 7 percent.[25] Much of the variation in optimum external oxygen concentration among types of products and even among cultivars can be accounted for by factors other than the external oxygen concentration.

The use of low oxygen in the storage of plant products has the potential to decrease the overall metabolic rate and a diverse array of specific biochemical changes. At the product level, the net effect may be seen as delayed ripening, aging or the development of certain storage disorders. Low oxygen environments, however, are not commercially used for many commodities for several reasons. The very short time span between harvest and the retail sale of many products and the availability of the product year round from different production locations often makes its use unnecessary. Likewise, for most products the costs are substantially greater than the benefits obtained, though there are notable exceptions. Approximately a half a million metric tons of apples are stored each year, in the United States alone, utilizing low oxygen conditions. This storage method extends the availability of the crop for 4 to 10 months over conventional storage practices, greatly increasing the net worth of the industry.

Carbon dioxide impedes respiration, resulting in a net and often quite significant decrease in respiration in some products. The effect of carbon dioxide, although not universal, has been shown in seedlings, intact plants and detached plant parts, and occurs both under aerobic and anaerobic respiratory conditions. The degree to which respiration is impeded increases in relation to the concentration of carbon dioxide in the atmosphere. For example, in pea seedlings, the inhibitory action of carbon dioxide at concentrations up to 50% increases approximately with the square root of the concentration.[58] Carbon dioxide, therefore, appears to retard the rate of respiration but does not totally block it.[75]

Under aerobic conditions respiration is impeded by high carbon dioxide when sufficient respiratory reserves are present, a common condition with most postharvest products. However, under conditions where the tissue is depleted of a ready source of stored reserves, respiration is no longer decreased by high carbon dioxide. The precise mechanisms of action of high carbon dioxide that results in a decrease in respiration are not known. The inhibitory effect is not due to permanent injury to the tissue, since upon removal of the carbon dioxide, respiration returns to normal. High carbon dioxide concentrations under aerobic conditions affect the tricarboxylic acid cycle at the conversion of succinate to malate and malate to pyruvate in apple fruit.[63] Succinate dehydrogenase, which converts succinate to malate, is the enzyme most significantly impeded. The level of reduction in the presence of 15% carbon dioxide results in toxic levels of succinate accumulating in apples, causing damage to the tissue.[53] The influence of carbon dioxide on other tricarboxylic acid cycle enzymes appears negligible.

A high carbon dioxide concentration during storage does not depress respiration in all tissues. In some cases, respiration may be unaffected or even significantly increased by elevated carbon dioxide. The respiratory rate of potato tubers, onion and tulip bulbs, and beetroot has been shown to be substantially increased, in some instances up to 200%, upon exposure to extremely high levels of carbon dioxide (30 to 70%)[100] and in lemon fruits by 10% carbon dioxide. Carrot roots are not affected.

The mechanisms that lead to the stimulation of respiration by high carbon dioxide may be attributable in part to the fixation of carbon dioxide by malic enzyme and phosphoenolpyruvate carboxylase (Figure 3.2). The initial products formed in lemon fruit after brief

Table 3.5. Effect of Ethylene and Cyanide on the Respiratory Rate as Measured by Oxygen Uptake of Various Types of Plant Tissue.*

Tissue	$\mu L\ O_2 \cdot g^{-1}$ fresh wt $\cdot$ hour^{-1}		
	Control	Ethylene	Cyanide
Fruit			
Apple	6	16	18
Avocado	36	150	150
Cherimoya	35	160	152
Lemon	7	16	21
Grapefruit	11	30	40
Stem			
Irish Potato	3	14	14
Rutabaga	9	18	23
Root			
Beet	11	22	24
Carrot	12	20	30
Sweetpotato	18	22	24

Source: Solomos and Biale.[98]

exposure to $^{14}CO_2$ are malic, citric and aspartic acids.[108] High concentrations of carbon dioxide may, therefore, facilitate the formation of tricarboxylic acid cycle intermediates and thereby stimulate respiration. The stimulation may also be related to secondary effects of the carbon dioxide molecule on the pH of the cytoplasm. The effect of elevated carbon dioxide on cellular pH is complex and varies with tissue. Carbon dioxide is readily soluble in the cytoplasm and the vacuole, existing as bicarbonate and hydrogen ions with the dissociation of carbonic acid. When lettuce was exposed to 15% carbon dioxide[96] and avocado to 20%,[71] the pH of the cytoplasm declined 0.4 units while the lettuce vacuole pH declined 0.1 units. When the lettuce and avocado were moved back into air, the pH returned to near the pretreatment level. The change in pH of freshly harvested green peas exposed to elevated carbon dioxide was compensated for by a decrease in malic acid concentration giving essentially no net change.[105] Short transitory changes in pH caused by returning the tissue to ambient carbon dioxide conditions may activate the carboxylation of phosphoenolpyruvate to oxaloacetate and subsequently malate.[106] At present the effect of even small changes in cellular pH on respiration is not known. It is known that along with a change in cellular pH in avocado, there is a decline in ATP levels and respiratory enzymes, which are also transitory.[70,71]

Ethylene is a phytohormone that can significantly stimulate the respiratory rate of a number of harvested products. This was first illustrated by the work of Denny[30,31] on citrus fruit and later with bananas.[48] The effect of ethylene is of considerable interest to postharvest biologists in that harvested products synthesize ethylene. In most cases, however, an increase in respiratory rate *per se* represents only a minor concern in relation to major biochemical changes in quality that may also be induced by ethylene (e.g., accelerated floral senescence, loss of chlorophyll, abscission).

A relatively wide range of vegetative and reproductive tissues respond to ethylene by increasing their respiratory rate (Table 3.5), and the increase is dependent upon the continued presence of the gas.[93,98] Respiration is not stimulated by ethylene in all tissue. For example, the respiration of strawberry fruit, pea[39] and wheat seedlings,[73] and peanut leaves[57] is not stimulated. Flower respiration typically declines after harvest, followed by an increase as they begin to senesce. Enhanced respiration induced by ethylene in flowers may be an indirect effect, through an acceleration of the senescence process. Many of the tissues in which respiration is

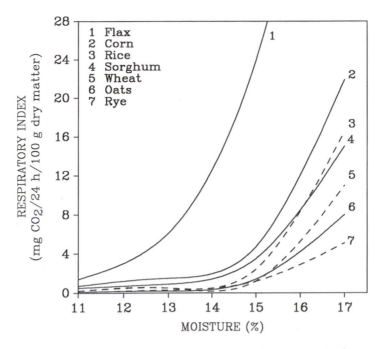

Figure 3.17. Effect of seed moisture level on respiration rates (*redrawn from Bailey*).[4]

accelerated by ethylene have significant storage energy reserves. It has been suggested that the ability of a tissue to respond to ethylene with an increase in respiration is closely correlated with the presence of the alternative electron transport pathway (Table 3.5). In these tissues, both cyanide and ethylene stimulate respiration, although through different mechanisms. The respiratory increase in either case does not necessarily involve the induction of ripening or stimulated ethylene synthesis, since it can be found in potato tuber tissue that neither ripens nor has autocatalytic ethylene synthesis.

c. Moisture content of the tissue

In general, respiration and metabolic processes decrease with decreasing tissue moisture content. There are, however, many interacting factors such as species, tissue type and physiological condition that significantly alter the plant or plant part's response to a particular moisture status. The change in respiratory rate can be influenced by moisture changes after harvest or preharvest differences in moisture content. For example, the respiratory rate of the storage roots of sweetpotato cultivars is closely related to the cultivar's percent moisture.

Seeds show the most dramatic effect of postharvest changes in the product's moisture status on respiration (Figure 3.17). The respiratory rate closely parallels the seed's moisture content as the seed imbibes or loses water. The respiration rate of leaves on an intact plant also declines with decreasing leaf water content; however, under severe dehydration it may temporarily increase.[21] Leaf water status is determined by soil moisture deficits or excesses.[26] In fleshy fruit, postharvest respiratory changes normally occur only after a substantial change in the internal water concentration. Hence, with most fleshy or succulent postharvest products, decreasing the internal moisture content is not a viable method for controlling the rate of respiration of the tissue. In these tissues, moisture content is often closely tied to product quality,[90] and a decrease in moisture is counterproductive. In many other products, however, longevity can be greatly extended through the repression of respiratory and metabolic activity with reduced product moisture. For example, in the storage of grains, seeds, dates, and most nuts, moisture status alteration represents an excellent means of extending a product's useful life.

d. Wounding

Wounding of plant tissue stimulates the respiratory rate of the affected cells, a response that has been known for nearly a century. Boehn[18] demonstrated that cutting potato tubers resulted in an abnormal rate of carbon dioxide production. Cutting the roots of trees[46,47] and the handling of leaves[3,38] also results in a stimulation of respiration.

Respiratory increases from wounds to plant tissue are often grouped into two general classes: a) those caused by mechanical damage—**wound respiration,** and b) those caused by infection by another organism such as fungi or viruses—**infection-induced respiration.**[103] This classification is not, however, definitive in that wounds induced by other means, although normally less frequent, are found (e.g., chemical sprays, light, pollutants). In addition, several types of wounds may occur simultaneously in the same tissue.

Mechanically induced wounds include those caused by harvesting, handling, wind, rain, hail, insects and animals. The wounds can be separated into subclasses based upon the presence or absence of surface punctures, cuts or lesions. Injuries which facilitate the diffusion of gasses to or from the underlying tissue often result in a substantial, but transient, increase in apparent respiration due to the escape of carbon dioxide that has accumulated in the intercellular spaces of the tissue. As a consequence, it is often difficult to make a clear distinction between altered gas exchange and wound effects on respiration when carbon dioxide production is used as sole means of measuring respiration.

Uritani and Asahi[103] characterized the differences in respiratory response between mechanically wounded and infected tissue (Figure 3.18), illustrating two distinct patterns. In both cases, increases in respiration coincided with increases in storage carbohydrate catabolism and an increase in soluble sugars in some tissues. Both the glycolytic[55] and pentose phosphate pathways are stimulated in response to increased demand for both primary and secondary plant products needed for wound healing. Healing includes the formation of lignin, suberin and in some cases, callus. Wound respiration, therefore, facilitates the supply of precursors and cofactors required for the biosynthesis of these wound healing layers (see Bloch[16,17]).

Infection-induced-respiratory increases are related to primary and secondary defense reactions by the cells. Plants have evolved multiple and varied techniques to combat the invasion of microorganisms. For example, rapid cell death resulting in necrotic areas confines the mycelia, limiting the number of cells infected (hypersensitive response). In addition, secondary products such as phytoalexins may be formed to minimize invasion. These infection induced processes, like those of mechanical wounding, require respiratory derived energy and secondary metabolic products, resulting in the observed increases in tissue respiration.

e. Species and plant part

Extremely large differences in respiratory rates can be found among different plant parts. The respiration rate of harvested vegetables, representing a range of plant parts, could be ranked on a dry weight basis in the following order: asparagus, lettuce, green bean, okra, green onion, carrot, tomato, beet root, green mango, red pimento[8] and other crops.[88] Differences include variation due to species and cultivar. Likewise, the respiratory rate of different plant parts, even within the same species and cultivar, can vary significantly, and there is often a wide range among parts of the same organ (e.g., the respiratory activity of the wheat seed embryo was twenty times that of the adjoining endosperm).[23]

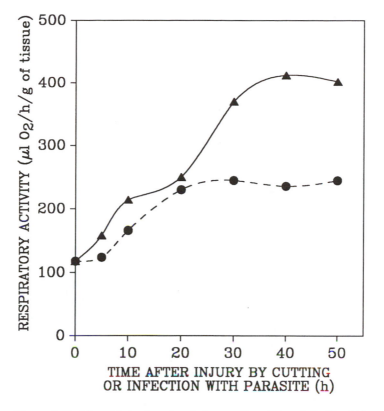

Figure 3.18. The effect of mechanical wounding (●) or infection (▲) by the fungus *Ceratocystis fimbriata* on the respiratory rate of sweet-potato storage root tissue (*redrawn from Uritani and Asahi*).[103]

The intensity of respiration of harvested products can vary widely among similar plant parts from different species. For example, flax seed have a respiratory rate that is 14 times that of barley seed at the same temperature and moisture content (11%).[4] Avocado fruit have a maximum rate of respiration at their climacteric peak nearly 8 times that of apple.[11]

f. Cultivar

While one could anticipate a significant range in respiratory rates among species, differences at the cultivar level may also be substantial. For example, cut flowers of the chrysanthemum cultivar Indianapolis White had a respiratory rate of 1.6 times that of the cultivar Indianapolis Pink, based on flower fresh weight, and 4.3 times greater expressed on a per flower basis.[67] Likewise 'McIntosh' apple fruit have been shown to have preharvest respiratory rates double that of 'Delicious'.[42,43] While not all cultivars exhibit this degree of variation, it is common and, depending on the postharvest conditions, may pose a factor for consideration during storage.

g. Stage of development

The stage of development of a plant or plant part can have a pronounced effect on the respiratory and metabolic rate of the tissue after harvest. In general, young actively growing cells tend to have higher respiratory rates than older, more mature cells. A number of factors, how-

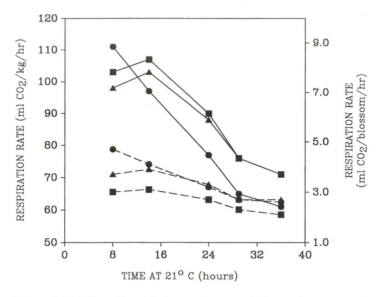

Figure 3.19. The effect of chrysanthemum (*Chrysanthemum morifolium* Ramat. cv. Indianapolis White) flower maturity (■—intermediate; ▲—mature; ●—fully mature) on flower respiration (--- = ml CO_2/kg/hr; — = ml CO_2/blossom/hr) at harvest (0 time) and during storage (*redrawn from Kuc and Workman*).[67]

ever, affect this relationship between maturity and respiratory rate, for example, species, plant part and stage of maturity. Therefore, generalizations on the effect of maturity on respiratory rate should be kept within fairly strict commodity-species bounds, and in some cases exceptions become more numerous than the rule.

The effect of maturity within a general commodity type is well illustrated in cut flowers. Carnations ('White Sims') harvested at varying stages of maturity between the tight bud stage and fully opened displayed significant differences in their respiratory rate expressed on either a weight or per blossom basis.[67] When flowers were harvested at a more mature stage of development (Figure 3.19), the respiratory rate rose. In addition, these differences tended to be maintained during the postharvest period. In contrast, however, chrysanthemum (cv. Indianapolis White) displayed the opposite trend, with the respiratory rate decreasing with maturity. Even more pronounced effects of maturation on respiratory rate can be seen in climacteric fruits when the range in maturities tested include preclimacteric, climacteric and postclimacteric stages.[33] Young apple leaves and stems have from three to seven times the rate of respiration of corresponding fully developed organs from the same plant.[85]

h. Surface area to volume ratio

The surface to volume ratio may influence the respiratory rate of some products due to its effect on gas exchange. As an object increases in size, assuming the shape is not altered, there is a progressive decrease in surface area relative to its volume. This is because volume increases as the cube of length (length × length × length) while surface area increases only as the square (length × length), as illustrated in Figure 3.20 with the comparison of two sizes of spherical fruit placed in a cubic box (20 × 20 × 20 cm). One sphere 20 cm in diameter will fit into the container or eight 10 cm diameter spheres. The composite surface area of the smaller spheres is double that of the larger sphere while the total volumes are equal. This difference provides in a

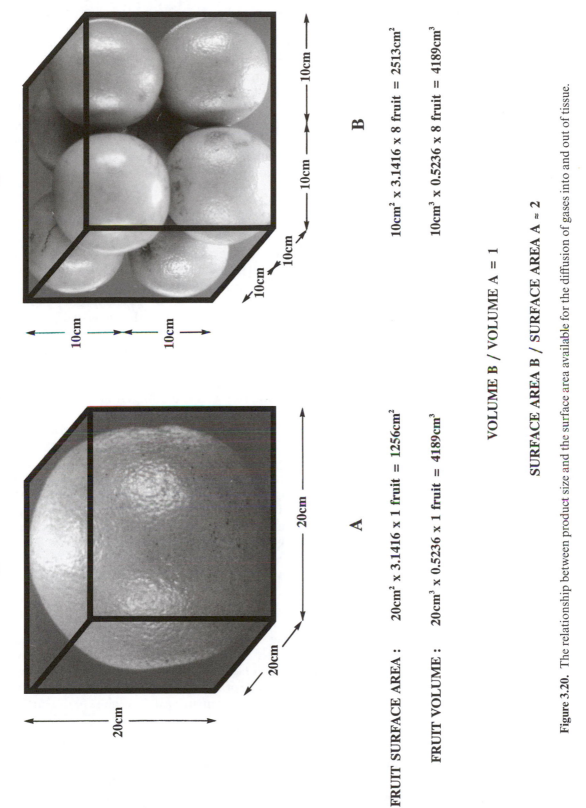

A

B

FRUIT SURFACE AREA : $20cm^2$ x 3.1416 x 1 fruit = $1256cm^2$

FRUIT VOLUME : $20cm^3$ x 0.5236 x 1 fruit = $4189cm^3$

B

$10cm^2$ x 3.1416 x 8 fruit = $2513cm^2$

$10cm^3$ x 0.5236 x 8 fruit = $4189cm^3$

VOLUME B / VOLUME A = 1

SURFACE AREA B / SURFACE AREA A ≈ 2

Figure 3.20. The relationship between product size and the surface area available for the diffusion of gases into and out of tissue.

small diameter product a larger surface area for gas exchange for the underlying cells, shifting the uptake-utilization (oxygen) and production-emanation (carbon dioxide) equilibrium.

The shape of the majority of harvested products (leaves, flowers, nuts, etc.) deviates substantially from spherical and in some cases, they also have rough or uneven surfaces. This increases the surface area to volume ratio, facilitating diffusion. In products where the surface represents a significant barrier to diffusion due to the presence of the cuticle or periderm, increased surface area may be important. Although the actual surface to volume ratio can be substantially altered by environmental conditions during growth (e.g., the effect of thinning on fruit size), little control can be exerted over it during the postharvest period. The postharvest environment, however, can be adjusted to compensate for surface/volume conditions of a specific product to prevent undesirable internal conditions from developing.

i. Nature of the harvested product's surface

The composition of the gas atmosphere within most harvested products has an effect on their respiratory rate. Both high carbon dioxide and low oxygen concentrations have been shown to decrease the respiratory rate of cells. The internal gas composition is controlled by the rate of oxygen use and carbon dioxide production by the tissue, differences in the partial pressures of these gases between the interior and exterior environment, and the gas permeability of the tissue and any applied surface coatings. The nature of the surface of the harvested product, therefore, can have an impact upon gas diffusion resistance. High diffusion resistances result in a greater difference between the internal and external gas atmosphere. If the differential between internal and external oxygen and carbon dioxide concentration is sufficiently large, the respiratory rate of the internal cells can be altered.

Surface resistances are generally much larger than internal diffusion resistances, since there is a significant volume of intercellular air space. Surface cells are arranged much more tightly (little intercellular space), and compounds that resist gas movement (e.g., cutin, waxes) are present on the surface. Therefore, the nature of the surface of harvested products and postharvest practices that alter these surface characteristics (e.g., application of waxes) can exert a considerable influence over respiratory and metabolic rates. Lenticels, stomates, surface cuts or abrasions, fruit stem or peduncle scars, and other openings provide localized areas that have lower diffusion resistances than the majority of the product surface. Natural surface coatings of epicuticular waxes and cutin tend to increase the diffusive resistance to oxygen, carbon dioxide and water movement into and out of the tissue.[65] When gas diffusion is sufficiently restricted, the internal concentration of carbon dioxide increases and oxygen decreases. For example, apples of the 'Granny Smith' cultivar held in air had an internal oxygen concentration of 17% at 7°C and 2% at 29°C, while the respective internal carbon dioxide concentrations were 2% and 17%.[102] The surface of the tomato fruit restricts all but around 5% of the total gas exchange between the interior and the exterior, the primary path of exchange being *via* the stem scar.

In some postharvest products, it is advantageous to apply an artificial coating of waxes or similar material on the surface. Citrus fruit, apples, cucumbers, pineapples, rutabagas, cassava and dormant rose plants are commonly waxed. This not only alters the internal gas concentration of the product[32] but has the additional advantage of decreasing water loss. In many cases, the waxes also enhance the appearance of the product by imparting a shiny gloss to the surface

j. Preharvest cultural and postharvest handling conditions

Preharvest factors can significantly influence the respiratory behavior of a harvested product. The nutrient composition of the harvested product is strongly affected by the nutrition of the parent plant. Preharvest nutrition alters not only the elemental composition of the product but also the relative amount of many organic compounds.[83] Plant tissues low in potassium and

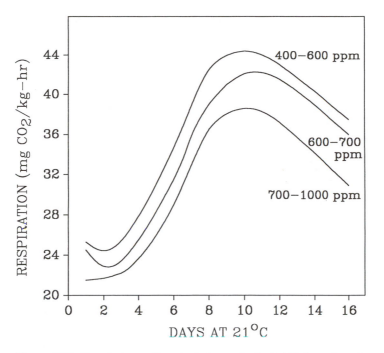

Figure 3.21. Respiratory climacteric in apple fruit (*Malus sylvestris* Mill. cv. Baldwin) with varying peel calcium contents (*redrawn from Bramlage et al.*).[20]

calcium often have substantially higher respiratory rates.[1,2,34,44] This is illustrated by the correlation between the peel calcium content of apples and their subsequent postharvest respiratory rate (Figure 3.21). Apples with a low calcium concentration have higher respiratory rates at the preclimacteric stage, the climacteric peak and during the postclimacteric period than apples with a higher calcium content.[20] High tissue nitrogen concentration (apples, strawberries) is also correlated with elevated respiration; in the case of apple fruit the effect of high nitrogen is pronounced only when the fruit calcium concentration is low.[34,87]

Other factors such as preharvest sprays,[97] rough handling,[74] orchard temperature,[80] production year[20] and acclimatization can also significantly influence postharvest respiratory responses.

3.1.5. Methods of Measuring Respiration

The respiratory rate of a stored product can be used as an indicator for adjusting the storage conditions to maximize the longevity of the commodity. As a consequence, it is often desirable to measure respiration in commercial storage houses. These measures can also be used in many cases as a general index of the potential storage life of the product. In addition, the rate of respiration can be used to calculate the loss of dry matter from the product during storage, the rate of oxygen removal from storage room air, and the heat generated during storage.

As discussed in section 3.1.1. of this chapter, respiration consumes oxygen from the surrounding environment and substrate from the commodity. Carbon dioxide, water and energy (both chemical and heat) are produced.

$$\text{Substrate} + O_2 \rightarrow CO_2 + H_2O + \text{energy (chemical and heat)}$$

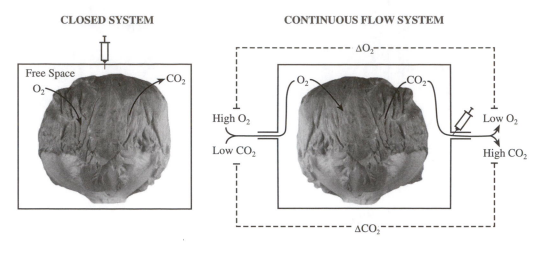

CLOSED SYSTEM

Free Space
O_2
CO_2

CONTINUOUS FLOW SYSTEM

ΔO_2
O_2
CO_2
High O_2
Low CO_2
Low O_2
High CO_2
ΔCO_2

ΔCO_2= concentration time 2 - concentration time 1

ΔO_2= concentration time 1 - concentration time 2

ΔCO_2= concentration out - concentration in

ΔO_2= concentration in - concentration out

$$\frac{(\Delta \% \times 10)(\text{free space volume of container in liters})}{(\text{product fwt in kg})(\text{time container is closed in hours})} = ml\ kg^{-1}\ hr^{-1}$$

$$\frac{(\Delta \% \times 10)(\text{air flow rate in ml/min} \times 60)/1000}{(\text{product fwt in kg})} = ml\ kg^{-1}\ hr^{-1}$$

* Milliliters of gas are normally converted to milligrams to remove the effect of temperature on the volume of gas so that direct comparisons can be made. To do this, a temperature correction must be used.

One mole of gas is equal to 22.4 L at O°C at 1 atmosphere, therefore, its volume (V_1) at the temperature of the product can be calculated with the following equation:

$$V_1 = 22.4\ (1 + \frac{\text{Temperature of product in °C}}{273°\ \text{Kelvin}})$$

For example, the volume of 1 mole of CO_2 at 25°C = 24.45 L. The volume of gas per gram is calculated by dividing the correct volume by the molecular weight of the gas (CO_2 = 44, O_2 = 32), i.e. 24.45 L/44 = .556 L/g or 556 ml/1000 mg (the volume of CO_2 at 25°C divided by its molecular weight). Then the weight of gas from the respiration sample can be calculated by:

$$\frac{556\ ml}{1000\ mg} = \frac{\text{measured ml from sample}}{x}$$

Corrections for common temperatures are:
0°C = 509 ml CO_2/1000 mg or 700 ml O_2/1000 mg
5°C = 518 ml CO_2/1000 mg or 712 ml O_2/1000 mg
10°C = 528 ml CO_2/1000 mg or 726 ml O_2/1000 mg
15°C = 537 ml CO_2/1000 mg or 738 ml O_2/1000 mg
20°C = 546 ml CO_2/1000 mg or 751 ml O_2/1000 mg
25°C = 556 ml CO_2/1000 mg or 764 ml O_2/1000 mg
30°C = 565 ml CO_2/1000 mg or 777 ml O_2/1000 mg

Figure 3.22. Techniques for collecting respiratory gas samples and calculating respiratory rates of harvested products.

Theoretically, changes in any of these reactants or products could be used as a measure of respiration. In general practice, however, measurement of carbon dioxide production is used due to its relative ease of measurement and accuracy. Oxygen is more difficult to accurately detect in that relatively small changes in concentration are against a high background oxygen concentration (21%) in air, while changes in carbon dioxide are large compared to the background (0.033%). Since the respiratory reactions take place in an aqueous medium, the small quantity of water produced in relation to the total volume of water present in the tissue cannot be accurately measured. Similarly, relatively large rates of respiration over a short measurement period result in only small total substrate or dry matter changes. Energy production, whether chemically trapped or liberated as heat, is also difficult to measure precisely. As a consequence, either the production of carbon dioxide or the utilization of oxygen is almost invariably used to monitor respiration.

Several techniques may be used for collecting gas samples from a respiring product. The product may be placed in a closed (gas tight) container and the decrease in oxygen or increase in carbon dioxide measured over a known period of time. Small samples are withdrawn from the enclosed atmosphere and either or both gases are measured (Figure 3.22). By measuring the change (Δ) in concentration as a function of time (i.e., concentration of oxygen at time$_1$ minus the concentration of oxygen at time$_2$ divided by time$_2$ minus time$_1$ gives Δoxygen/unit of time), the volume of free space in the container and the weight of the enclosed product, the respiratory rate can be expressed as weight of gas/weight of product unit of time (see Figure 3.22). In closed systems, care must be taken not to leave the product enclosed for too long, since the decreasing oxygen and increasing carbon dioxide concentrations will begin to affect the rate of respiration of the product. For many products, it is not desirable to let the carbon dioxide concentration increase to much above 0.5%. A second technique employs a continuous flow of air or gas of known composition through the container holding the product (Figure 3.22). The difference (Δ) between the concentration of oxygen and/or carbon dioxide going into the container and that leaving the container is used to calculate the respiratory rate. In addition to the difference in gas concentration, the rate of air flow through the container and the weight of the product must be known. Care must be taken to adjust the air flow rate through the container to an appropriate level. An excessively high flow rate results in extremely small differences between incoming and exiting gases, making measurement with an acceptable level of accuracy difficult. Air flow rates that are too slow result in the same problem that can be encountered with a closed system, the buildup of carbon dioxide or depletion of oxygen altering the rate of respiration. Care should also be taken to allow the system sufficient time to develop a steady-state equilibrium before measurements are made.

The oxygen and/or carbon dioxide concentrations in gas samples from either system can be measured by utilizing any of a number of different methods (Table 3.6).

Table 3.6. Comparison of Techniques Available for Measuring Respiration of Plants.

Technique	Gas measured	Sample Type	Sample size	Initial Expense	Recurring Expense	Requirement for Electricity
Gas chromatography	CO_2 and/or O_2	Discrete	0.2–5.0 ml	Very high	Medium	Yes
Infrared	CO_2	Continuous flow		Very high	Low	Yes
Paramagnetic	O_2	Continuous flow		High	Very low	Yes
Titration/colorimetry	CO_2	Continuous flow		Low	Low	No
Kitagawa	CO_2 or O_2	Discrete		Very low	Very low	No
Pressure/volume changes	CO_2 or O_2	Discrete	100 ml	Medium	Low	No
Polarography	O_2	Continuous flow		Medium	Low	Yes

a. Gas chromatography

Carbon dioxide and oxygen can be measured with gas chromatographs equipped with thermal conductivity detectors and dual columns. Gas chromatographic analysis is used widely because of its accuracy and the small gas samples (i.e., 0.1–5 mL) needed.

b. Infrared gas analyzer for carbon dioxide

This instrument is used to measure carbon dioxide in a continuous flow of air and has the advantage over many techniques of being extremely accurate at very low carbon dioxide concentrations. The molecules of carbon dioxide in the sample absorb infrared radiation at a specific wavelength, and this absorption is used as a measure of the carbon dioxide concentration in the air stream.

c. Paramagnetic oxygen analyzers

Oxygen is strongly paramagnetic, and since no other gases commonly present in the air exert a magnetic influence, this characteristic can be monitored and used as a measure of the oxygen concentration in a continuous stream of air.

d. Titration/colorimetry for carbon dioxide

When a stream of air is passed through a sodium hydroxide or calcium hydroxide solution, the carbon dioxide is absorbed, and sodium or calcium carbonate is formed, decreasing the solution's pH. The change in alkalinity (decrease in pH) is used to determine the quantity of carbon dioxide absorbed,[27] a technique that can measure carbon dioxide concentrations up to 1.0%. The pH change is measured either by titration or colormetrically with the addition of bromthymol blue[91] and monitored with a spectrophotometer. As with other continuous flow systems, the flow rate of air needs to be known to calculate the final respiratory rate. This technique has the advantage of being relatively inexpensive and requiring only a limited amount of equipment.

e. Kitagawa detectors

Carbon dioxide, oxygen and a number of other gases can be measured quickly, relatively accurately and without significant expense using Kitagawa detectors. Gas is pulled into a reaction tube (specific for each gas monitored) where it is absorbed and reacts with a chemical reagent. The color change produced is used as a measure of the concentration of the gas. Carbon dioxide can be accurately measured between 0.01 and 2.6% and oxygen from 2 to 30%.

f. Pressure/volume changes

Samples of gas are placed in sealed containers of known volume and either carbon dioxide or oxygen is absorbed by a suitable reactant. The change in the internal pressure of the container or in the volume of gas within the container is used as a measure of the concentration of the respective gas.[29] The absorption principle is similar to that used in the titration/colorimetry method; however, instead of measuring changes in the absorbing material (e.g., pH), pressure or volume changes within the chamber are determined. Relatively large gas samples (e.g., 100 mL) are required; however, analyses are accurate to approximately 0.5%.

g. Polarography

The oxygen concentration in a gas sample can be measured with a polarographic oxygen electrode. The differential in electrical potential across a pair of electrodes is measured.

3.1.6. Respiratory Patterns

In the 1920's, Kidd and West studied the respiratory patterns of sunflower plants and their component parts during an entire growing season (Figure 3.23). In general they found that respiration closely paralleled the rate of plant growth (i.e., young, rapidly metabolizing cells have the highest respiratory rates). The high demand for energy and carbon compounds in actively growing cells results in a stimulation of respiration. As the age of the plant or individual organ (stem, leaves, flowers) increases, the respiratory rate decreases. This decline in respiration of the whole plant could not be attributed simply to an increased percentage of non-respiring structural material in the plant, since the initial respiratory rate of successive new leaves also decreased with the age of the plant. Hence, internal factors have a pronounced influence on respiration.

A number of external (environmental) and internal (commodity) factors have a pronounced influence on the rate of respiration of plant tissues (section 3.1.4). While environmental factors such as temperature are routinely studied with each postharvest product, considerable effort has been directed toward understanding the more elusive commodity factors that influence respiration. Changes in respiratory rate during growth, development and senescence of a plant or plant part under standard conditions display distinctive patterns, and these can often be related to other functional processes that occur concurrently.

Fruits are typically classed into one of two groups based on their pattern in respiratory

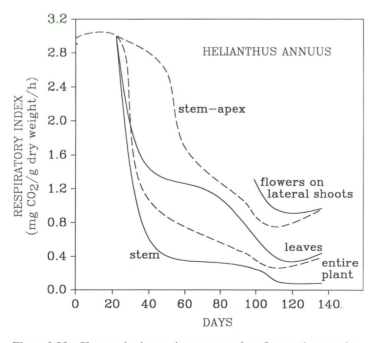

Figure 3.23. Changes in the respiratory rate of sunflower plants and selected plant parts during development (*data from Kidd et al.*).[61]

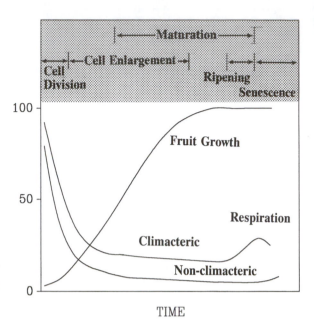

Figure 3.24. General respiratory pattern of climacteric and non-climacteric fruits during development, maturation, ripening and senescence.

behavior during the final stages of ontogeny of the organ (e.g., fruit ripening, Figure 3.24). Fruits classified as having a **respiratory climacteric** exhibit a marked upsurge in respiratory activity at the end of the maturation phase. These marked changes in the rate of respiration of climacteric fruit have long fascinated postharvest biologists. Although climacteric fruit represent an extremely small percentage of the plant products handled in agriculture, the dramatic shift in respiration during ripening has stimulated research into the function and control of respiratory patterns. The respiratory climacteric represents a transition between maturation and senescence. Nonclimacteric fruit (Figure 3.24) do not exhibit an upsurge in respiration, but rather a progressive, slow decline during senescence until microbial or fungal invasion.

The climacteric rise in respiration was described as early as 1908[84] in apple and pear fruit. Later, Kidd and West[60] detailed the relationship between changes in respiratory rate and changes in quality attributes occurring during the climacteric period. The dark respiration of an unripe fruit declines to what is termed the **preclimacteric minimum** just before the climacteric rise in respiration (Figure 3.24). Subsequently, respiration increases dramatically, often to levels 2 to 4 times that of the preclimacteric minimum. A similar trend occurs if the fruit is allowed to ripen on the tree,[64] although the respiratory pattern is modified somewhat [e.g., the rate at which it proceeds (slower) and the peak value (higher)]. Interpretation of the overall response is complicated by the fact that photosynthesis and photorespiration are occurring concurrently. An exception is found in some avocado cultivars, where the respiratory upsurge is inhibited while the fruit remains attached to the tree.

The respiratory climacteric is substantially altered by temperature. At both low and high temperatures, the climacteric can be suppressed. As storage temperature decreases from around 25°C, the duration of the climacteric rise is prolonged and the rate of respiration at the climacteric peak depressed. In addition, the ambient oxygen and carbon dioxide concentration can markedly alter the respiratory climacteric. Low oxygen and high carbon dioxide (up to approximately 10%) can prolong the length of time to the climacteric peak in a number of fruit, thus extending storage life. In many nonclimacteric fruit, respiration can also be depressed by low oxygen and high carbon dioxide concentrations. There are exceptions, however. For example, high carbon dioxide tends to stimulate the respiration of lemon fruit,[14]

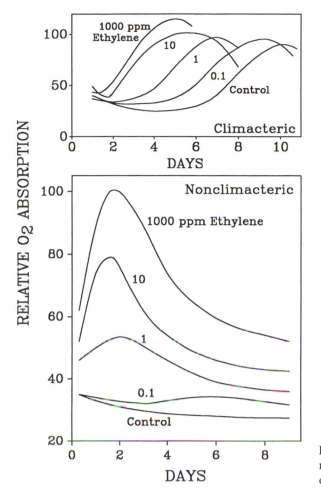

Figure 3.25. Oxygen uptake by climacteric and non-climacteric fruits exposed to varying levels of ethylene (*redrawn from Biale*).[13]

probably through the fixation and incorporation of the molecule into organic acids utilized in the tricarboxylic acid cycle.

The unsaturated hydrocarbon gas ethylene stimulates the respiration of a wide range of plant tissues. The response differs for climacteric and nonclimacteric fruit (Figure 3.25). Exposure of climacteric fruits to relatively low levels of ethylene decreases the preclimacteric period without a substantial effect on the rate of respiration at the climacteric peak. The concentration required for maximum acceleration varies with different fruits (e.g., 1 µL · L⁻¹ for banana, 10 µL · L⁻¹ for avocado). The shortening of the preclimacteric period is approximately proportional to the logarithm of the concentration applied.[22] Once ripening is initiated, removal of the external ethylene has no effect on the subsequent respiratory rate or pattern. Respiration in nonclimacteric fruit is likewise stimulated by ethylene; however, upon removal of ethylene the respiratory rate returns to near the value found prior to treatment (Figure 3.25).

Ethylene is known to be a natural product of fruit ripening.[36] The synthesis and the internal concentration of ethylene increases in most climacteric fruits around the same time the upsurge in respiration. Ethylene is thought to accelerate the outset of ripening and to coordinate ripening in the whole fruit. Exposure to an external source of ethylene results in an increased synthesis of ethylene. This stimulated synthesis is due to the effect of ethylene on the *de novo* synthesis of ACC synthase, a critical enzyme in the ethylene synthesis pathway.[6] There are at least five ACC synthase genes involved in tomato ripening[6] that are expressed in a specific order and at different stages of fruit ripening. Exogenous ethylene, therefore, can be substituted for the endogenous ethylene normally produced by the plant to initiate the respiratory climacteric and ripening response, allowing earlier ripening of the fruit. Ethylene application (exogenous) is most widely used to commercially ripen banana fruit.

Theories as to the precise cause of the respiratory climacteric have been numerous. In 1928, Blackman and Parija[15] proposed that the increase in respiration was due to the loss of organizational resistance between enzymes and substrates. Several other theories have enjoyed popularity, including: 1) the presence of "active" substrate,[59] 2) availability of phosphate acceptors, 3) availability of cofactors, 4) uncouplers of oxidation and phosphorylation,[81] 5) shifts in metabolic pathways[54] and 6) an increase in mitochondria content and/or activity.[49] After 90 years of research, the precise cause of the respiratory climacteric in fruit has yet to be elucidated, although our understanding of the physiological, chemical and enzymatic changes occurring has increased tremendously.

Leaves undergo distinct changes in their respiratory behavior at certain stages of their development.[99] Generally, there is an increase in respiration during the early stages of senescence (the period of chlorophyll degradation) followed by a steady decline in the later stages. The respiratory increase, although not universal, occurs in a wide range of species in both attached[45] and detached leaves. Severing the leaf from the plant enhances the rate of senescence;[95] however, the timing of the respiratory increase relative to other biochemical and physical changes occurs at essentially the same stage in the senescence process. Low light (100–200 lux) delays senescence in detached oat leaves and thus can influence the respiratory strategy utilized.[97] When leaves are held in the dark, approximately 25% of the respiratory increase could be accounted for by increases in free amino acids and sugars from catabolic events. The remaining respiration (~75%) appears to be due to a partial uncoupling of respiration from phosphorylation.

Many flowers also undergo marked changes in respiration rate, the pattern and control of which has many parallels with the respiratory changes in climacteric fruits. In fact, the term respiratory climacteric is often used in studies on flower storage and senescence. Respiration in many species of cut flowers declines after harvest and then increases as the flowers begin to senesce.[86] This trend, however, is not universal. For example, cut roses progressively decline in respiration following harvest (Figure 3.26). In flowers that exhibit a postharvest respiratory rise, the increase in respiration, like that in many climacteric fruits, appears to be closely tied to the endogenous synthesis of ethylene by the flower (Figure 3.27). There is an autocatalytic synthesis of ethylene in flowers like the carnation, and the increase in ethylene precedes changes in membrane permeability and other senescence-related phenomena.[77] Chemicals such as aminovinylglycine, aminooxyacetic acid, Ag^{2+} ions, and 1-methylcyclopropene (MCP) that inhibit ethylene synthesis or action, can delay senescence in a wide range of ethylene sensitive flowers.

As the flower proceeds toward the final stages of senescence, there is a gradual reduction in respiration that may reflect a decline in respiratory substrate availability. Carbohydrates are known to be transported from the petals into the ovary during this period, with the reallocation stimulated by ethylene.

In summary, many plant products undergo substantial changes in their respiratory pattern after harvest. These changes often reflect significant alterations in metabolism and concurrent physical and chemical alterations within the tissues. Changes in respiration are of interest from an applied point of view, in that specific handling strategies may be required. Changes in respiration in part reflect the physiological state of the commodity, which can help in predicting the product's storage potential and life expectancy.

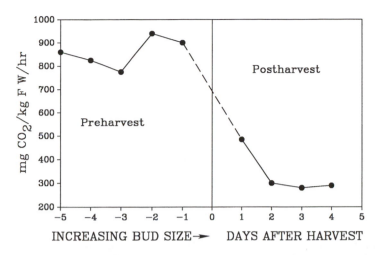

Figure 3.26. The effect of time of harvest on the respiratory rate of 'Velvet Times' roses held at 21°C (*redrawn from Coorts et al.*).[28] Zero time corresponds to the stage of development that the cultivar would be commercially harvested.

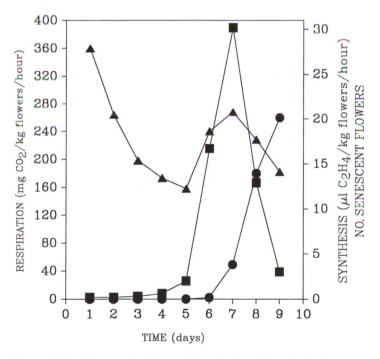

Figure 3.27. Changes in the rate of respiration (▲) and ethylene synthesis (■) with the development of senescence (●) after harvest of 'Improved White Sims' carnation flowers at 20°C (*redrawn from Maxie et al.*).[76]

3.3. Photosynthesis

The potential for photosynthesis and the rate at which it proceeds in harvested products varies widely, influenced by both internal and external factors. Many plant products are devoid of chlorophyll, or they are held after harvest under environmental conditions (low light and temperature) that are not conducive to photosynthesis. The importance of photosynthesis in detached chlorophyll containing tissues is probably minimal, but it should not be ignored.

A significant number of postharvest products not only have the potential for photosynthesis but need to photosynthesize to maintain the product's existing level of quality. Intact plants such as actively growing herbaceous and woody ornamentals, vegetable transplants and rooted cuttings are typical examples of postharvest products that normally photosynthesize during the handling and marketing period. For some protea flowers, photosynthesis can minimize postharvest leaf blackening, and illumination of broccoli plantlets during storage, even at the light compensation point, maintains their dry weight and subsequent growth potential.[68] The rate at which these plants photosynthesize after removal from the production area is governed by a number of internal and external factors. While the internal factors, such as stomatal number, photosynthetic pathway and others, influence the rate of photosynthesis, it is not possible to alter them. However, a number of external factors such as light, temperature, moisture, carbon dioxide and exogenous chemicals that influence photosynthesis can be altered. Manipulation of these external factors, therefore, provides the potential to exert a significant level of control over the rate of photosynthesis in many harvested products.

3.3.1. Tissue Type and Condition

The ability to photosynthesize and the rate of photosynthesis vary considerably among plant species as well as types of tissue. Chloroplasts, which carry out photosynthesis in the cell, are found primarily in leaf tissue; however, they also occur in petioles, stems, specialized floral parts of many plants and the epidermis of certain fruit. Roots, tubers and other structures, normally devoid of chloroplasts and hence chlorophyll, are capable of synthesizing chlorophyll when exposed to sufficient light. The pre- or postharvest formation of chlorophyll in some products (e.g., potato, Jerusalem artichoke) is detrimental to quality and needs to be avoided. The contribution of chloroplasts in organs other than leaves to the total assimilation of carbon is generally small due in part to the low number or absence of stomates, reducing carbon dioxide availability. In some instances, however, such as the cortical tissue of dormant dogwoods, photosynthesis by non-leaf structures may offset a significant portion of respiratory loss of carbon. The capacity of a fruit's epidermal cells to carry out photosynthesis declines with ripening, with the majority of photosynthate being supplied from the plant leaves.[89]

Photosynthetic rate is affected by leaf age (Figure 3.28). Rates are commonly highest when leaves first mature and shortly thereafter but tend to decline gradually with age.[66] The effect of leaf age on photosynthesis is a general phenomenon, found in annuals and perennials, including evergreen species.

3.3.2. Light

Whether from the sun or from an artificial source, light provides the energy needed by plants to fix carbon from carbon dioxide, allowing them to offset respiratory losses incurred. Light is one of the most important postharvest external variables affecting photosynthesis. During

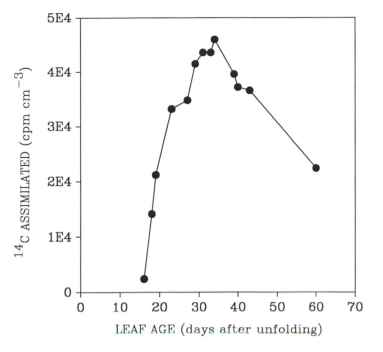

Figure 3.28. The effect of leaf age on the rate of photosynthesis ($^{14}CO_2$ assimilated) of grape leaves (*redrawn from Kriedemann et al.*).[66]

the postharvest handling of plants, especially intact plants, light intensity, quality and duration are crucial in determining storage potential.

Individual leaves, when exposed to increasing light intensity, exhibit a typical light response pattern (Figure 3.29). As the intensity of the light is increased, the **light compensation point** is reached. At this point, the amount of carbon dioxide trapped is equal to the amount of carbon dioxide lost from the tissue due to respiratory processes. Additional increases in intensity result in a proportional increase in photosynthetic rate of carbon fixed, eventually reaching a point at which photosynthesis becomes light saturated. At this point, additional increases in light intensity have only a slight effect on increasing the carbon fixation rate. Light saturation of individual leaves of full sun plants is often only 1/4 to 1/2 that of full sunlight; however, with entire plants, saturation is seldom reached because of mutual shading of the leaves within the canopy. With further increases in light, the point of maximum photosynthesis is reached, and additional increases result in a decrease in carbon fixation and sometimes damage (Figure 3.29).

Light is necessary for the formation of chlorophyll in plants, and there is a continuous turnover (synthesis and degradation) of chlorophyll molecules under normal conditions. Insufficient light, therefore, can result in the net loss of leaf chlorophyll. With prolonged exposure, an indirect loss occurs through abscission of leaves, decreasing the plant's surface area of photosynthetic tissue. Normally, abscission progresses from the oldest leaves to the youngest, with leaves that are shaded by their position at the bottom or interior of the canopy being more susceptible to abscission.

Plant species vary widely in their tolerance to light, and excess light may present a serious postharvest problem. For example, prolonged exposure of full shade plants such as the African violet to full sunlight may result in chlorophyll degradation, leaf burning and a decrease in net photosynthesis. The effect of excess light can also be seen in heliotropic responses,

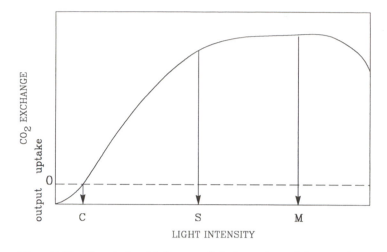

Figure 3.29. The relationship between light intensity and the rate of carbon uptake by photosynthesizing leaves (C—light compensation point; S—light saturation intensity; M—maximum photosynthetic rate) (*redrawn from Rabinowitch*).[93] At the light compensation point (C), the amount of carbon fixed by the leaf is equal to the amount lost through respiration. At very high light intensities, photosynthesis is inhibited.

where the plant's growth pattern is altered in the direction of the light, which may potentially be in an undesirable orientation.

It is normally desirable to maintain the product in the same qualitative condition as at harvest (removal from the production zone for intact plants), and hence the net carbon balance (acquisition vs. utilization) is important. The precise postharvest photosynthetic requirements for individual species at various stages in their life cycle are not currently known. It is probable, however, that the requirements for photosynthetic carbon input needed to maintain the existing condition of the plant are slightly above the plant's gross respiratory utilization, due to non-respiratory uses of carbon in maintenance reactions. Photosynthetic acquisition of carbon can be maintained at or above this critical maintenance point with a range of light intensity and duration combinations. When some species are exposed to prolonged periods at higher light intensities, the chloroplasts are unable to store the additional starch formed, and photosynthesis is inhibited.

Of the total spectrum of radiant energy, plants utilize light only from a region between 400 and 700 nm for photosynthesis (Figure 3.30). Peak photosynthesis and absorption of light by chlorophyll and other pigments come from the red and blue portions of the spectrum. In natural canopies, light entering the upper leaves is selectively absorbed from the red and blue regions, with less photosynthetically active light being transmitted to the lower and inner leaves. As the spectral distribution of the light shifts toward a greater percentage of green light, the rate of photosynthesis per photon of photosynthetically active light declines.

Changes in the amount of energy at particular wavelengths (light quality) a plant receives after harvest are often also dramatically altered. These changes occur primarily from the use of artificial light that does not have the same spectral quality as sunlight and in the use of shading material that selectively absorbs light from certain regions of the spectrum. While photomorphogenic changes in the plant due to light quality are normally of greater postharvest importance than changes in the net photosynthetic rate, prolonged exposure of plants to light that is not spectrally suited for photosynthesis will compromise product quality maintenance.

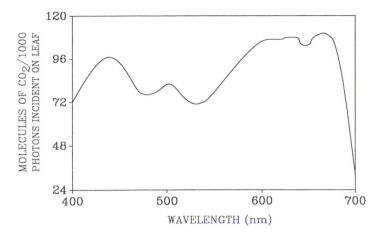

Figure 3.30. The change in photosynthetic activity with changes in wavelength of the light striking the leaf (*redrawn from Balegh and Biddulph*).[5]

3.3.3. Temperature

The rate of photosynthesis is highly dependent upon temperature. This dependence varies widely, with some species capable of photosynthesizing at temperatures near 0°C, while others require substantially higher temperatures. Generally, net photosynthesis (total photosynthesis minus respiration) increases with temperature until reaching a maximum and then declines (Figure 3.14). The decline is probably mediated by several factors, one of which is an elevation in respiration that occurs with increasing temperature. As temperature is progressively increased, respiratory losses will eventually be greater than the carbon fixed through photosynthesis, giving a net loss of carbon. During the postharvest handling of photosynthetically active products, both excessively low and high temperatures present potential problems in maintaining an adequate carbon input-output balance. At very low temperatures, photosynthesis is insufficient, while at high temperatures, respiratory losses are greater than the carbon dioxide fixed.

3.3.4. Moisture Stress

The availability of water is an important factor governing the rate of photosynthesis (Figure 3.31). Upon removal of plants from the production zone, the potential for moisture stress increases substantially due in part to the altered environmental conditions to which the plants are exposed. In addition, water management often becomes less organized and structured. Lack of precise control over the moisture balance in postharvest products can result in either water deficit or excess; each can substantially impede photosynthesis. Leaf water content has a direct effect on water's chemical role in photosynthesis and an indirect effect on hydration of the protoplasm and closure of stomates. Since the total amount of water that participates directly in the biochemical reactions of photosynthesis is extremely small, indirect effects on stomatal aperture appear to present the greater hindrance to photosynthesis under conditions of low moisture. The degree of inhibition of photosynthesis due to water deficit depends to a large extent upon both the level of stress imposed and the species involved. Wilted sunflower leaves are photosynthetically as much as 10 times less efficient than turgid leaves. As a conse-

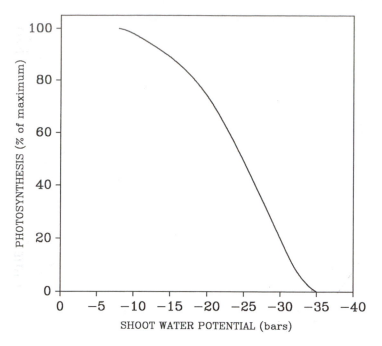

Figure 3.31. The effect of shoot water potential on the photosynthetic rate of Douglas fir trees (*redrawn from Brix*).[21]

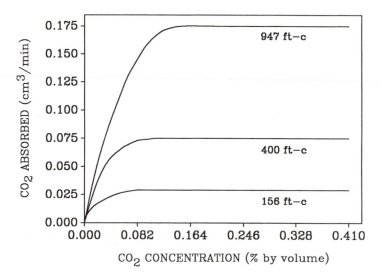

Figure 3.32. The effect of carbon dioxide concentration on photosynthetic rate (CO_2 absorbed) at varying light intensities (*redrawn from Hoover et al.*).[52]

quence, prolonged closure of stomates can result in a serious deficit in carbohydrates within the plant.

3.3.5. Carbon Dioxide

Photosynthesis in plants exposed to sufficient light is limited primarily by the low carbon dioxide concentration in the ambient atmosphere ($\sim$330 $\mu L \cdot L^{-1}$). As the carbon dioxide concentration increases up to 1000 to 1500 $\mu L \cdot L^{-1}$, photosynthesis also increases in many species. At high light intensities, the effect of additional carbon dioxide is even greater (Figure 3.32).

Elevated carbon dioxide concentration has been used successfully to increase the growth rate of a number of agricultural crops grown in controlled environment conditions (e.g., greenhouses).[107] To date, the beneficial effect of elevated carbon dioxide concentration on the photosynthetic rate of plants has been used only in the production zone for these crops and not during the postharvest handling and sales period.

3.3.6. Chemicals

A number of chemical compounds applied directly to plants (e.g., pesticides) or found in the ambient atmosphere (air pollutants), can depress photosynthesis. The response may be due to a direct effect on the biochemical reactions of photosynthesis or indirect effects such as increases in the diffusive resistance of carbon dioxide into the leaf, changes in the optical properties of the leaf, changes in the leaf's thermal balance, or the loss of photosynthetic surface area. The common air pollutants that can reduce photosynthesis are sulfur dioxide, ozone, fluorides, ethylene and particulate matter such as dusts. Brief exposure to sulfur dioxide (4.1 mg $\cdot m^{-3}$ for 2 hours) has been shown to inhibit photosynthesis by as much as 80%. While the effect of chemicals on postharvest photosynthesis may be pronounced, more direct effects such as the formation of lesions, chlorosis, discoloration, and leaf and flower drop are often economically much more significant than impaired photosynthesis in compromising product quality.

ADDITIONAL READING

Barber, J. (ed.). 1992. *The Photosystems: Structure, Function and Molecular Biology.* Elsevier, Amsterdam.

Borochov, A., and R. Woodson. 1990. Physiology and biochemistry of flower petal senescence. *Hort. Rev.* 11:15–43.

Boyer, P.D. 1997. The ATP synthase: A splendid molecular machine. *Annu. Rev. Biochem.* 66:717–749.

Buchanan, B.B., W. Gruissem and R.L. Jones. 2000. *Biochemistry & Molecular Biology of Plants.* American Soc. Plant Physiologists, Rockville, MD.

Buchanan-Wollaston, V. 1997. The molecular biology of leaf senescence. *J. Exp. Bot.* 48:181–199.

Burton, W.G. 1982. *Postharvest Physiology of Food Crops.* Longman, New York.

Chitnis, P.R. 1996. Photosystem I. *Plant Physiol.* 111:661–669.

Cohen, Y., S. Yalovsky and R. Nechushtai. 1995. Integration and assembly of photosynthetic protein complexes in chloroplast thylakoid membranes. *Biochem. Biophys. Acta* 1241:1–30.

Davies, D.D. (ed.). 1988. *Biochemistry of Metabolism.* Vol. 11. *The Biochemistry of Plants.* P.K. Stumpf and E.E. Conn (eds.). Academic Press, London.

Davies, D.D. (ed.). 1988. *Physiology of Metabolism.* Vol. 12. *The Biochemistry of Plants.* P.K. Stumpf and E.E. Conn (eds.). Academic Press, London.

Day, D.A., J. Whelan, A.H. Millar, J.N. Siedow and J.T. Wiskich. 1995. Regulation of the alternative oxidase in plants and fungi. *Aust. J. Plant Physiol.* 22:497–509.

Dennis, D.T., Y. Huang and F.B. Negm. 1996. Glycolysis, the pentose phosphate pathway and anaerobic respiration. Pp.105–123. In: *Plant Metabolism*. D.T. Dennis, D.H. Turpin, D.D. Lefebvre and D.B. Layzell (eds.), Longman Press, Harlow, UK.

Emes, M.J., and D.T. Dennis. 1997. Regulation by compartmentation. Pp.69–80. In: *Plant Metabolism*. D.T. Dennis, D.H. Turpin, D.D. Lefebvre and D.B. Layzell (eds.), Longman Press, Harlow, UK.

Fluhr, R., and A.K. Mattoo. 1996. Ethylene-biosynthesis and reception. *Critical Rev. Plant Sci.* 15:479–523.

Gan, S., and R.M. Amasino. 1997. Making sense of senescence. *Plant Physiol.* 113:313–319.

Gutteridge, S., and A.A. Gatensby. 1995. Rubisco synthesis, assembly, mechanism and regulation. *Plant Cell* 7:809–819.

Halevy, A.H., and S. Mayak. 1979. Senescence and postharvest physiology of cut flowers. Part I. *Hort. Rev.* 1:204–236.

Hartman, F.C., and M.R. Harpel. 1994. Structure, function, regulation and assembly of D-ribulose-1, 5-bisphosphate carboxylase/oxygenase. *Annu. Rev. Biochem.* 63:197–234.

Horton, T., A.V. Ruban and R.G. Walters. 1996. Regulation of light harvesting in green plants. *Annu. Rev. Plant Physiol. Plant Mol. Biol.* 47:655–684.

James, W.O. 1953. *Plant Respiration.* Clarendon Press, Oxford, England.

Kostychev, S. 1927. *Kostychev's Plant Respiration.* Translated by C.J. Lyon, Blakiston's Son & Co., Philadelphia, PA.

Kozaki, A., and G. Takeba. 1996. Photorespiration protects C_3 plants from photooxidation. *Nature* 384:557–560.

Kromer, S. 1995. Respiration during photosynthesis. *Annu. Rev. Plant Physiol. Plant Mol. Biol.* 46:45–70.

Lambers, H. 1985. Respiration in intact plants and tissues. Its regulation and dependence on environmental factors, metabolism and invaded organisms. Pp.418–473. In: *Higher Plant Cell Respiration.* Vol. 18. *Encyclopedia of Plant Physiology.* (new series), R. Douce and D.A. Day (eds.), Springer, Berlin.

Long, S.P., S. Humphries and P.G. Falkowski.1994. Photoinhibition of photosynthesis in nature. *Annu. Rev. Plant Physiol. Plant Mol. Biol.* 45:633–662.

Mackenzie, S., and L. McIntosh. 1999. Higher plant mitochondria. *Plant Cell* 11:571–586.

Mir, N., and R. Beaudry. 2002. Atmospheric control using oxygen and carbon dioxide. Pp. 122–156. In: *Fruit Quality and its Biological Basis.* M. Knee (ed.), CRC Press, Boca Raton, FL.

Moller, I.M., P. Gardestom, K. Glimelius and E. Glaser. 1998. *Plant Mitochondria: From Gene to Function.* Backhuys Pub., Leiden, The Netherlands.

Nugent, J.H.A. 1996. Oxygenic photosynthesis: Electron transfer in photosystem I and photosystem II. *Eur. J. Biochem. Sci.* 237:519–531.

Ort, D.R., and C.F. Yocum (eds.). 1996. *Oxygenic Photosynthesis: The Light Reactions.* Vol. 4. *Advances in Photosynthesis.* Kluwer, Dordrecht, The Netherlands.

Pakrasi, H.B. 1995. Genetic analysis of the form and function of Photosystem I and Photosystem II. *Annu. Rev. Genet.* 29:755–776.

Plaxton, W.C. 1996. The organization and regulation of plant glycolysis. *Annu. Rev. Plant Physiol. Plant Mol. Biol.* 47:185–214.

Rogner, M., E.J. Boekema and J. Barber. 1996. How does photosystem 2 split water? The structural basis of efficient energy conversion. *Trends Biochem. Sci.* 21:44–49.

Seymour, G.B., J.E. Taylor and G.A. Tucker (eds.). 1993. *Biochemistry of Fruit Ripening.* Chapman & Hall, London.

Siedow, J.N. 1995. Bioenergetics: The plant mitochondrial electron transfer chain. Pp.281–312. In: *The Molecular Biology of Plant Mitochondria.* C.S. Levings III and I. Vasil (eds.), Kluwer, Dordrecht, The Netherlands.

Siedow, J.N., and A.L. Umbach. 1995. Plant mitochondrial electron transfer and molecular biology. *Plant Cell* 7:821–831.

Taiz, L., and E. Zeiger. 1998. *Plant Physiology.* Benjamin/Cummins Pub., Redwood City, CA.

Vanlerberghe, G.C., and L. McIntosh. 1997. Alternative oxidase: From gene to function. *Annu. Rev. Plant Physiol. Plant Mol. Biol.* 48:703–734.

Wagner, A.M., and K. Krab. 1995. The alternative respiration pathway in plants: Role and regulation. *Physiol. Plantarum* 95:318–325.

Whitehouse, D.G., and A.L. Moore. 1995. Regulation of oxidative phosphorylation in plant mitochondria.

Pp.313–344. In: *The Molecular Biology of Plant Mitochondria,* C.S. Levings III and I. Vasil (eds.), Kluwer, Dordrecht, The Netherlands.

REFERENCES

1. Alban, E.K., H.W. Ford and F.S. Nowlett. 1940. A preliminary report on the effect of various cultural practices with greenhouse tomatoes on the respiration rate of the harvested fruit. *Proc. Amer. Soc. Hort. Sci.* 52:385–390.
2. Amberger, A. 1953. Zur rolle des kaliums bei atmung svorgängen. *Biochem. Z.* 323:437–438.
3. Audis, L.J. 1935. Mechanical stimulation and respiration rate in the cherry laurel. *New Phytol.* 34:386–402.
4. Bailey, C.H. 1940. Respiration of cereal grains and flaxseed. *Plant Physiol.* 15:257–274.
5. Balegh, S.E., and O. Biddulph. 1970. The photosynthetic action spectrum of the bean plant. *Plant Physiol.* 46:1–5.
6. Barry, C.S., M.I. Llop-Tous and D. Grierson. 2000. The regulation of 1-aminocyclopropane-1-carboxylic acid synthase gene expression during the transition from system-1 to system-2 ethylene synthesis in tomato. *Plant Physiol.* 123:979–9.
7. Beaudry, R.M., R.F. Severson, C.C. Black and S.J. Kays. 1989. Banana ripening: Implications of changes in glycolytic intermediate concentrations, glycolytic and gluconeogenic carbon flux, and fructose 2,6-bisphosphate concentration. *Plant Physiol.* 91:1436–1444.
8. Benoy, M.P. 1929. The respiration factor in the deterioration of fresh vegetables at room temperature. *J. Agr. Res.* 39:75–80.
9. Berard, J.E. 1821. Memoire sur la maturation des fruits. *Annales de Chimie et de Physique* XVI:152–183, 225–251.
10. Biale, J.B. 1946. Effect of oxygen concentration on respiration of the Fuerte avocado fruit. *Amer. J. Bot.* 33:363–373.
11. Biale, J.B. 1950. Postharvest physiology and biochemistry of fruits. *Annu. Rev. Plant Physiol.* 1:183–206.
12. Biale, J.B. 1954. Physiological requirements of citrus fruits. *Citrus Leaves* 34:6–7, 31–33.
13. Biale, J.B. 1960. Respiration of fruits. pp. 536–592. In: *Handbuch der Pflanzenphysiologie,* Vol. XII/2, W. Rukland (ed.). Springer-Verlag, Berlin.
14. Biale, J.B., and R.E. Young. 1962. The biochemistry of fruit maturation. *Endeavor* 21:164–174.
15. Blackman, F.F., and P. Parija. 1928. Analytic studies in plant respiration. II. The respiration of a population of senescent ripening apples. *Proc. Roy. Soc. Lond.,* Ser. B. 103:422–445.
16. Bloch, R. 1941. Wound healing in higher plants. *Bot. Rev.* 7:110–146.
17. Bloch, R. 1964. Wound healing in higher plants. *Bot. Rev.* 18:655–679.
18. Boehm, J.A. 1887. Üeber die respiration der kartoffel. *Bot Ztg.* 45:671–675, 680–691.
19. Bowyer, J.R., and R.C. Leegood. 1997. Photosynthesis. In: *Plant Biochemistry.* P.M. Dey and J.B. Harbone (eds.). Academic Press, New York.
20. Bramlage, W.J., M. Drake and J.H. Baker. 1974. Relationships of calcium content to respiration and seedlings. *Physiol. Plant.* 15:10–20.
21. Brix, H. 1962. The effect of water stress on the rates of photosynthesis and respiration on tomato plants and loblolly pine seedlings. *Plant Physiol.* 15:10–20.
22. Burg, S.P., and E.A. Burg. 1962. Role of ethylene in fruit ripening. *Plant Physiol.* 37:179–189.
23. Burlakow, G. 1898. Üeber athmung des keimes des weizens, *Triticum vulgare. Bot. Cent.* 74:323–324.
24. Burton, W.G. 1965. The sugar balance in some British potato varieties during storage. I. Preliminary observations. *European Potato J.* 8:80–91.
25. Chang, L.A., and S.J. Kays. 1981. Effect of low oxygen on sweet potato roots during storage. *J. Amer. Soc. Hort. Sci.* 106:481–483.
26. Childers, N.F., and D.G. White. 1942. Influence of submersion of the roots on transpiration, apparent photosynthesis, and respiration of young apple trees. *Plant Physiol.* 17:603–618.
27. Claypool, L.L., and R.M. Keefer. 1942. A colorimetric method for CO_2 determination in respiration studies. *Proc. Amer. Soc. Hort. Sci.* 40:177–186.

28. Coorts, G.D., J.B. Gartner and J.P. McCollum. 1965. Effect of senescence and preservative on respiration in cut flowers of *Rosa Hybrida,* 'Velvet Times'. *Proc. Amer. Soc. Hort. Sci.* 86:779–790.

29. deWild, H.P.J., and H.W. Peppelenbos. 2001. Improving the measurement of gas exchange in closed systems. *Postharv. Biol. Tech.* 22:111–119.

30. Denny, F.E. 1924. Hastening the coloration of lemons. *J. Agr. Res.* 27:757–771.

31. Denny, F.E. 1924. Effect of ethylene upon respiration of lemons. *Bot. Gaz.* (Chicago) 7:327–329.

32. Eaks, I.L., and W.A. Ludi. 1960. Effects of temperature, washing, and waxing on the composition of the internal atmosphere of orange fruits. *Proc. Amer. Soc. Hort. Sci.* 76:220–228.

33. Emmert, F.H., and F.W. Southwick. 1954. The effect of maturity, apple emanations, waxing and growth regulators on the respiration and red color development of tomato fruit. *Proc. Amer. Soc. Hort. Sci.* 63:393–401.

34. Faust, M., and C.B. Shear. 1972. The effects of calcium on respiration of apples. *J. Amer. Soc. Hort. Sci.* 97:437–439.

35. Foote, K.C., and M. Schaedle. 1976. Diurnal and seasonal patterns of photosynthesis and respiration by stems of *Populus tremuloides* Michx. *Plant Physiol.* 58:651–655.

36. Gane, R.1934. Production of ethylene by some ripening fruits. *Nature* (Lond.) 7:1465–1470.

37. Genevois, M.L. 1929. Sur la fermentation et sur la respiration chez les végétaux chlorophylliens. *Rev. Gen. Bot.* 41:252–271.

38. Godwin, H. 1935. The effect of handling on the respiration of cherry laurel leaves. *New Phytol.* 34:403–406.

39. Goeschl, J.D., L. Rappaport and H.K. Pratt. 1966. Ethylene as a factor regulating the growth of pea epicotyls subjected to physical stress. *Plant Physiol.* 42:877–884.

40. Gortner, W.A., G.G. Dull and B. Krauss. 1967. Fruit development, maturation, ripening, and senescence: A biochemical basis for horticultural terminology. *HortScience* 2:141.

41. Green, W.P., W.V. Hukill and D.H. Rose. 1941. Calorimetric measurements of the heat of respiration of fruits and vegetables. *USDA Tech. Bull.* 771:1–21.

42. Greene, D.W., W.J. Lord and W.J. Bramlage. 1977. Mid-summer applications of ethephon and daminozide on apples. I. Effect on 'McIntosh'. *J. Amer. Soc. Hort. Sci.* 102:491–49.

43. Greene, D.W., W.J. Lord and W.J. Bramlage. 1977. Mid-summer applications of ethephon and daminozide on apples. II. Effect on 'Delicious'. *J. Amer. Soc. Hort. Sci.* 102:494–497.

44. Gregory, F.G., and F.J. Richards. 1929. Physiological studies in plant nutrition. I. The effect of manurial deficiency on the respiration and assimilation rate of barley. *Ann. Bot.* 43:119–161.

45. Hardwick, K., M. Wood and H.W. Woolhouse. 1968. Photosynthesis and respiration in relation to leaf age in *Perilla frutescens* (L.) Britt. *New Phytol.* 67:79–86.

46. Harris, G.H. 1929. Studies on tree root activities. I. An apparatus for studying root respiration and factors which influence it. *Sci. Agr.* 9:553–565.

47. Harris, G.H. 1930. Studies on tree root activities. II. Some factors which influence tree root respiration. *Sci. Agr.* 10:564–585.

48. Harvey, R.B. 1928. Artificial ripening of fruits and vegetables. *Minn. Agr. Exp. Sta. Bull.* 247.

49. Hatch, M.D., J.A. Pearson, A. Millerd and R.N. Robertson. 1959. Oxidation of Krebs cycle acids by tissue slices and cytoplasmic particles from apple fruit. *Aust. J. Biol. Sci.* 12:167–174.

50. Hill, G.R. 1913. Respiration of fruits and growing plant tissues in certain gases with reference to ventilation and fruit storage. *Cornell Agr. Exp. Sta. Bull.* 330:374–408.

51. Hinkle, M.A., A. Kumar, A. Resetar and D.L. Harris. 1991. Mechanistic stoichiometry of mitochondrial oxidative phosphorylation. *Biochemistry* 30:3576–3582.

52. Hoover, W.H., E.S. Johnston and F.S. Brackett. 1933. Carbon dioxide assimilation in a higher plant. *Smithsonian Inst. Misc. Coll.* 87:1–19.

53. Hulme, A.C. 1956. Carbon dioxide injury and the presence of succinic acid in apples. *Nature* (Lond.) 178:218–219.

54. Hulme, A.C., J.D. Jones and L.S.C. Wooltorton. 1963. The respiratory climacteric in apple fruits. *Proc. Roy. Soc. London,* Ser. B. 158:514–535.

55. Kahl, G. 1974. Metabolism in plant storage tissue slices. *Bot. Rev.* 40:263–314.

56. Kanai, R., and C.C. Black, Jr. 1972. Biochemical basis for net CO_2 assimilation in C_4-plants. Pp. 75–93. In: *Net Carbon Dioxide Assimilation in Higher Plants.* C.C. Black, Jr. (ed.). Symp. Southern Reg. Amer. Soc. Plant Physiol.

57. Kays, S.J., and J.E. Pallas, Jr. 1980. Inhibition of photosynthesis by ethylene. *Nature* 285:51–52.

58. Kidd, F. 1917. The controlling influence of carbon dioxide. Part III. The retarding effect of carbon dioxide on respiration. *Proc. Royal Soc. London* 89B:136–156.

59. Kidd, F. 1934. The respiration of fruits. *Royal Inst. of Great Britain* (as cited by Baile, 1960).

60. Kidd, F., and C. West. 1925. The course of respiratory activity throughout the life of an apple. *Great Britain Dept. Sci. Ind. Res., Food Invest. Bd. Rept.* 1924:27–33.

61. Kidd, F., C. West and G.E.A. Briggs. 1921. A quantitative analysis of the growth of *Helianthus annuus*. Part I. The respiration of the plant and of its parts throughout the life cycle. *Proc. Royal Soc. London,* B92:361–384.

62. Kirk, J.T., and R.A. Tilney-Bassett. 1967. *The Plastids; Their Chemistry, Structure, Growth, and Inheritance.* W.H. Freeman, San Francisco.

63. Knee, M. 1973. Effects of controlled atmosphere storage on respiratory metabolism of apple fruit tissue. *J. Sci. Food Agr.* 24:1289–1298.

64. Knee, M. 1995. Do tomatoes on the plant behave as climacteric fruits? *Physiol. Plant.* 95:211–216.

65. Kolattukudy, P.E. 1980. Cutin, suberin and waxes. Pp. 571–645. In: *Lipids: Structure and Function.* R.K. Stumpf (ed.). Vol. 4. *The Biochemistry of Plants.* Academic Press, New York.

66. Kriedemann, P.E., W.M. Kleiwer and J.M. Harris. 1970. Leaf age and photosynthesis in *Vitis vinifera* L. *Vitis* 9:97–104.

67. Kuc, R., and M. Workman. 1964. The relationship of maturity to the respiration and keeping quality of cut carnations and chrysanthemums. *Proc. Amer. Soc. Hort. Sci.* 84:575–581.

68. Kubota, C., and T. Kozai. 1995. Low-temperature storage of transplants at the light compensation point: Air temperature and light intensity for growth suppression and quality preservation. *Sci. Hort.* 61:193–204.

69. Lambers, H. 1980. The physiological significance of cyanide-resistant respiration in higher plants. *Plant Cell Environ.* 3:293–302.

70. Lange, D.L., and A.A. Kader. 1997. Effects of elevated carbon dioxide on key mitochondrial respiratory enzymes in 'Hass' avocado fruit and fruit disks. *J. Amer. Soc. Hort. Sci.* 122:238–244.

71. Lange, D.L., and A.A. Kader. 1997. Elevated carbon dioxide exposure alters intracellular pH and energy charge in avocado fruit tissue. *J. Amer. Soc. Hort. Sci.* 122:253–257.

72. Lipton, W.J. 1977. Toward an explanation of disorders of vegetables induced by high CO_2 and low O_2. *Proc. Second Nat. Controlled Atmos. Res. Conf.,* Mich. State Univ. Hort. Rept. 28:137–141.

73. Mack, W.B., and B.E. Livingstone. 1933. Relation of oxygen pressure and temperature to the influence of ethylene or carbon dioxide production and shoot elongation in very young wheat seedlings. *Bot. Gaz.* 94:625–687.

74. Massey, L.M., Jr., B.R. Chase and M.S. Starr. 1982. Effect of rough handling on CO_2 evolutions from 'Howes' cranberries. *HortScience* 17:57–58.

75. Mathooko, F.M. 1996. Regulation of respiratory metabolism in fruits and vegetables by carbon dioxide. *Postharv. Biol. Tech.* 9:247–264.

76. Maxie, E.C., D.S. Farnham, F.G. Mitchell, N.F. Sommer, R.A. Parsons, R.G. Snyder and H.L. Rae. 1973. Temperature and ethylene effects on cut flowers of carnations (*Dianthus caryophyllus* L.). *J. Amer. Soc. Hort. Sci.* 98:568–572.

77. Mayak, S., and A.H. Halevy. 1980. Flower senescence. Pp. 131–156. In: *Senescence in Plants.* K.V. Thimann (ed.). CRC Press, Boca Raton, FL.

78. McGlasson, W.B., and R.B.H. Wills. 1972. Effects of oxygen and carbon dioxide on respiration, storage life and organic acids of green bananas. *Aust. J. Biol. Sci.* 25:35–42.

79. Meeuse, B.J.D. 1975. Thermogenic respiration in Aroids. *Annu. Rev. Plant Physiol.* 26:117–126.

80. Mellenthin, W.M., and C.Y. Wang. 1976. Preharvest temperatures in relation to postharvest quality of d'Anjou pears. *J. Amer. Soc. Hort. Sci.* 101:302–305.

81. Millerd, A., J. Bonner and J.B. Biale. 1953. The climacteric rise in plant respiration as controlled by phosphorylative coupling. *Plant Physiol.* 28:521–531.

82. Morris, L.L., and A.A. Kader. 1977. Physiological disorders of certain vegetables in relation to modified atmospheres. *Proc. Second Nat. Controlled Atmos. Res. Conf.,* Mich. State Univ. Hort. Rept. 28:142–148.

83. Mulder, E.G. 1955. Effect of mineral nutrition of potato plants on respiration of the tubers. *Acta Bot. Neerlandica* 4:429–451.

84. Müller-Thurgan, H., and O. Schneider-Orelli. 1908. Reifevorgänge bei Kernobst früchten. *Landwirtsch. J. B. Schwei.* 22:760–774.

85. Nicholas, G. 1918. Contribution á l'etude des variatins de la respiration des végétaux avec l'age. *Rev. Gen. Bot.* 30:214–225.

86. Nichols, R. 1968. The response of carnations (*Dianthus caryophyllus*) to ethylene. *J. Hort. Soc.* 43:335–349.

87. Overholser, E.L., and L.L. Claypool. 1931. The relation of fertilizers to respiration and certain physical properties of strawberries. *Proc. Amer. Soc. Hort. Sci.* 28:220–224.

88. Peiris, K.H.S., J.L. Mallon and S.J. Kays. 1997. Respiration rate and vital heat of some specialty vegetables at various storage temperatures. *HortTechnology* 7:46–49.

89. Piechulla, B., R.E. Glick, H. Bahl, A. Melis and W. Gruissem. 1987. Changes in photosynthetic capacity and photosynthetic protein pattern during tomato fruit ripening. *Plant Physiol.* 84:911–917.

90. Paull, R.E. 1999. Effect of temperature and relative humidity on fresh commodity quality. *Postharv. Biol. Tech.* 15:263–277.

91. Pratt, H.K., and D.B. Merrdoza, Jr. 1979. Colorimetric determination of carbon dioxide for respiration studies. *HortScience* 14:175–176.

92. Purvis, A.C., and R.L. Shewfelt. 1993. Does the alternative pathway ameliorate chilling injury in sensitive plant tissues? *Plant Physiol.* 88:712–718.

93. Rabinowitch, E.I. 1951. Photosynthesis and related processes. Vol. II, Pt. I. *Spectroscopy and Fluorescence of Photosynthetic Pigments; Kinetics of Photosynthesis.* Interscience, New York.

94. Robinson, J.E., K.M. Browne and W.G. Burton. 1975. Storage characteristics of some vegetables and soft fruits. *Ann. Appl. Biol.* 81:399–408.

95. Simon, E.W. 1967. Types of leaf senescence. *Symp. Soc. Exp. Biol.* 21:215–230.

96. Siripanich, J., and A.A. Kader. 1986. Changes in cytoplasmic and vacuolar pH in harvested lettuce tissue as influenced by CO_2. *J. Amer. Soc. Hort. Sci.* 111:73–77.

97. Smock, R.M., L.J. Edgerton and M.B. Hoffman. 1954. Some effects of stop drop auxins and respiratory inhibitors on the maturity of apples. *Proc. Amer. Soc. Hort. Sci.* 63:211–219.

98. Solomos, T., and J.B. Biale. 1975. Respiration and fruit ripening. *Colloq. Int. C.N.R.S.* 238:221–228.

99. Thimann, K.V. 1980. The senescence of leaves. Pp. 85–115. In: *Senescence in Plants.* K.V. Thimann (ed.). CRC Press, Boca Raton, FL.

100. Thornton, N.C. 1933. Carbon dioxide storage. III. The influence of carbon dioxide on oxygen uptake by fruits and vegetables. *Contrib. Boyce Thompson Inst.* 5:371–402.

101. Tranquillini, W. 1955. Die bedeutung des lichtes und der temperatur fur die kohlensaureassimilation von *Pinus cembra* jungwuchs an einem hochalpinen standort. *Planta* 46:154–178.

102. Trout, S.A., E.G. Hall, R.N. Robertson, M.V. Hackney and S.M. Sykes. 1942. Studies in the metabolism of apples. *Aust. J. Exp. Biol. Med. Sci.* 20:219–231.

103. Uritani, I., and T. Asahi. 1980. Respiration and related metabolic activity in wounded and infected tissue. Pp. 463–487. In: *Metabolism and Respiration.* D.D. Davis (ed.). Vol. 2. *The Biochemistry of Plants.* Academic Press, New York.

104. van Iersel, M.W., and L. Seymour. 2000. Growth respiration, maintenance respiration, and carbon fixation of Vinca: A time series analysis. *J. Amer. Soc. Hort. Sci.* 125:702–706.

105. Wager, H.G. 1974. The effect of subjecting peas to air enriched with carbon dioxide. I. The path of gaseous diffusion, the content of CO_2 and the buffering of the tissue. *J. Exp. Bot.* 25:330–337.

106. Wager, H.G. 1974. The effect of subjecting peas to air enriched with carbon dioxide. II. Respiration and the metabolism of the major acids. *J. Exp. Bot.* 25:338–351.

107. Wittwer, S.H., and W. Robb. 1964. Carbon dioxide enrichment of greenhouse atmospheres for food crop production. *Econ. Bot.* 18:34–56.

108. Young, R.E., and J.B. Biale. 1956. Carbon dioxide fixation by lemons in a CO_2 enriched atmosphere. *Plant Physiol.* 32 (suppl.):23.

109. Zelitch, I. 1979. Photorespiration: Studies with whole tissues. Pp. 353–367. In: *Photosynthesis II.* M. Gibbs and E. Latzko (eds.). Vol. 6. *Encyclopedia of Plant Physiology.* Springer-Verlag, Berlin.

4

SECONDARY METABOLIC PROCESSES AND PRODUCTS

The synthesis and degradation of carbohydrates, organic acids, proteins, lipids, pigments, aromatic compounds, phenolics, vitamins and phytohormones are classified as secondary processes (i.e., secondary to respiration and photosynthesis), but the distinction is somewhat arbitrary. The metabolism of most of these products is absolutely essential in both the pre- and postharvest life of a product. During the postharvest period, there is continued synthesis of many compounds (for example, the volatile flavor components of apples and bananas) and a degradation of other compounds to provide energy and precursors for synthetic reactions. Many of these changes occurring after harvest, however, are not desirable. As a consequence, we strive to store products in a manner that minimizes undesirable changes.

Excluding the synthesis of carbohydrates and proteins, there are three primary pathways that lead to the diverse array of chemical compounds found in plants. These include: 1) the shikimic acid pathway that leads to the formation of lignin, coumarins, tannins, phenols and various aromatics; 2) the acetate-malonate pathway which forms the precursors of fatty acids, phospholipids, glycerides, waxes, and glycolipids; and 3) the acetate-mevalonate pathway which results in various terpenoids (gibberellins, carotenoids, abscisic acid) and steroids (Figure 4.1).

This chapter covers the major classes of plant constituents in postharvest products, critiques some of the quantitative and qualitative changes that occur after harvest and outlines the important environmental factors that accelerate these postharvest alterations.

1. CARBOHYDRATES

Carbohydrates are the most abundant biochemical constituent in plants, representing 50–80% of the total dry weight. They function as forms of stored energy reserves and make up much of the structural framework of the cells. In addition, simple carbohydrates such as the sugars sucrose and fructose, impart important quality attributes to many harvested products. The concentration of sugars alone can range from slight, as in lime fruit, to as much as 61% of the fresh weight in the date (Table 4.1).

Carbohydrates are molecules comprised of carbon, hydrogen and oxygen, however, many may also contain other elements such as nitrogen and phosphorous. As a group, they are defined as polyhydroxy aldehydes or ketones, or substances that yield either of these compounds

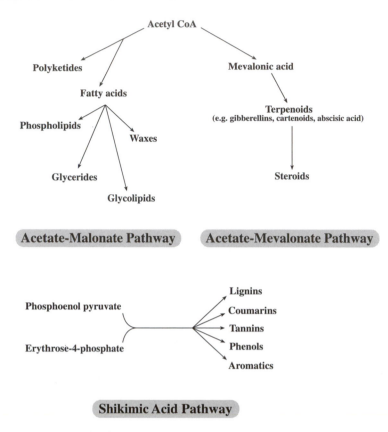

Figure 4.1. The three dominant pathways responsible for the synthesis of secondary plant products.

upon hydrolysis. Glucose and fructose, structural isomers of each other (both are $C_6H_{12}O_6$), illustrate the differences between the two types of sugars, aldoses and ketoses.

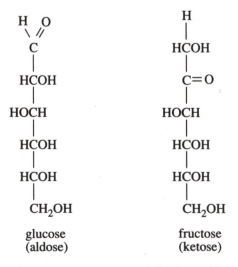

In addition, sugars that have a free or potentially free aldehyde group are classified as reducing sugars based on their ability to act as a reducing agent (accept electrons) in an alkaline solution. Most of the common sugars in plants are reducing sugars (e.g., glucose, fructose,

Table 4.1. General Range of Total Sugars and Relative Amounts of Specific Sugars Found in a Cross-Section of Fruits.

	Total Sugars[†] (% Fresh Weight)	Specific Sugars* (% Fresh Weight of Edible Portion)		
		Glucose	Fructose	Sucrose
Apple	11.6	1.7	6.1	3.6
Avocado	0.4			
Currant, red	5.1	2.3	1.9	0.2
Date	61.0	32.0	23.7	8.2
Grape	14.8	8.2	7.3	
Lime	0.7			
Pineapple	12.3	2.3	1.4	7.9
Pear	10.0	2.4	7.0	1.0
Tomato	2.8	1.6	1.2	

*Source: Widdowson and McCance.[246]
[†]Source: Biale;[16] Money;[167] Money and Christian;[168] Swisher and Swisher;[222] Widdowson and McCance.[246]

galactose, mannose, ribose and xylose). Sucrose and raffinose are the most common nonreducing sugars. The level of reducing sugars is important in several postharvest products such as white potato, to be used for chips (crisps). When the reducing sugar concentration in the potatoes is high, there is a greater incidence of undesirable browning reactions during frying. Improper handling and storage conditions prior to cooking can significantly increase the level of free reducing sugars, leading to a lower quality product.

Carbohydrates can be further classified, based on their degree of polymerization, into monosaccharides, oligosaccharides and polysaccharides. Simple sugars or monosaccharides represent the most fundamental group and cannot be further broken down into smaller sugar units. These basic units of carbohydrate chemistry are subclassed based upon the number of carbon atoms they contain (Figure 4.2). This number ranges from three in the triose sugars to seven in the heptose sugars, although occasionally octuloses are found. Both the glycolytic and pentose pathways are important in the synthesis of these molecules that are the building blocks for more complex carbohydrates. Monosaccharides may also be modified to form several types of compounds that are essential metabolic components of the cells. For example, amino and deoxy sugars, and sugar acids and alcohols, although seldom high in concentration in harvested products, are common (Figure 4.2). Less common are branched sugars such as apiose.

1.1. Monosaccharides

In many products the monosaccharides comprise a major portion of the total soluble sugars present (Table 4.1). Glucose and fructose are the predominant simple sugars found, especially in fruits, however, mannose, galactose, arabinose, xylose and various others are found in a number of harvest products.

1.2. Oligosaccharides

Oligosaccharides are more complex sugars that yield two to six molecules of simple sugars upon hydrolysis. For example, a disaccharide such as sucrose yields two monosaccharides

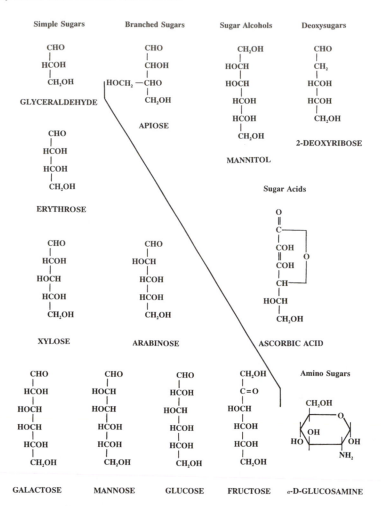

Figure 4.2. Common sugars and sugar derivatives.

(glucose and fructose) upon hydrolysis while a pentasaccharide yields five monosaccharide molecules. Both the monosaccharides and oligosaccharides are water-soluble and together they comprise the total soluble sugars in a product. The most abundant oligosaccharide is sucrose which is the primary transport form of carbohydrate in most plants. Sucrose phosphate synthetase appears to be the prevalent *in vivo* enzyme catalyzing the reversible reaction.[101]

$$\text{UDP-glucose} + \text{fructose-6-P} \rightarrow \text{sucrose-P} + \text{UDP}$$
$$\text{sucrose-P} \rightarrow \text{sucrose} + \text{Pi}$$

Sucrose can also be synthesized by sucrose synthase. This reaction is also reversible; however, the opposite direction (i.e., the hydrolysis of sucrose) is generally favored. In addition to sucrose synthase, sucrose may also be hydrolyzed by the enzyme invertase yielding glucose and fructose. Other common oligosaccharides are the disaccharide maltose found in germinating seeds and the trisaccharide raffinose and tetrasaccharide stachyose act as translocatable sugars in several species (Figure 4.3).

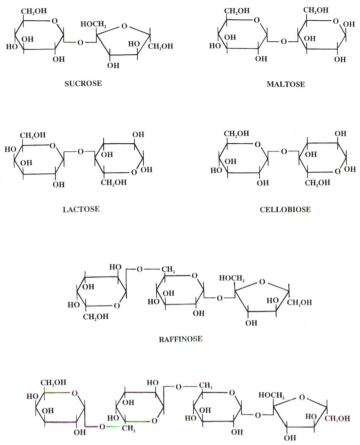

Figure 4.3. Molecular structure of some common oligosaccharides found in plants.

1.3. Polysaccharides

1.3.1. Cellulose

Cellulose is a straight-chain crystalline polymer of glucose and is one of the most abundant compounds in plants. Cellulose is not, however, the major component in many storage organs where storage carbohydrates or lipids often abound. For example, the cellulose content of dates in only 0.8%[35] whereas cotton fibers are 98% cellulose. Individual cellulose molecules can be extremely long, 1,000 to 10,000 glucose subunits, with molecular weights of 200,000 to 2,000,000 Da. The attachment between neighboring glucose molecules in the chain is a β-linkage between carbon 1 of one glucose molecule and carbon 4 of the next glucose and is referred to as a β-(1-4) linkage (Figure 4.4). Cellulose is found largely in the primary and secondary cell walls of the tissue.

The synthesis of cellulose takes place in a plasma membrane bound protein complex that is guided by the cellular cytoplasm cytoskeleton.[43] The basic subunit for insertion into the chain is cellobiose (two glucoses) rather than individual glucose molecules.[44] UDP-glucose is an essential participant in the synthesis scheme.[36] In most detached products where there is little, if any, growth, cellulose synthesis is usually limited.

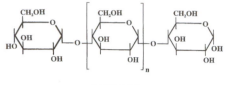

AMYLOSE

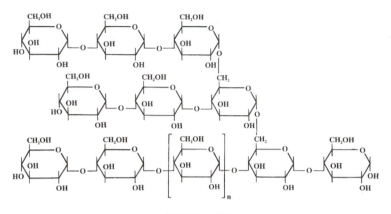

AMYLOPECTIN

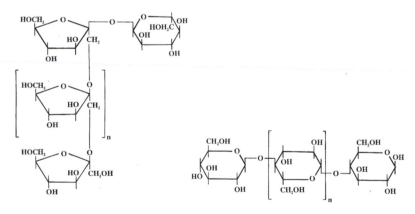

INULIN

CELLULOSE

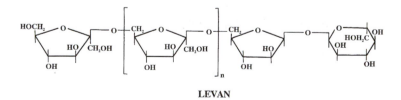

LEVAN

Figure 4.4. The structure of several common polysaccharides found in plants.

Cellulose molecules are extremely stable and can be broken down (hydrolyzed) only with strong acids or by enzymes such as cellulases, found in bacteria and fungi. There is often little change in the cellulose structure in fruits during ripening (e.g., peach[220]). Plants have a glucanase, often referred to as cellulase, that can degrade water-soluble molecules having the same glucose bonding but not crystalline cellulose. Glucanases are known to function during the abscission of leaves or other organs from the parent plant but do not act on cellulose.

1.3.2. Starch

Starch, composed of a mixture of branched and straight-chained glucose polymers, represents the major storage carbohydrate in most postharvest products. It is stored in the form of starch grains, found in specialized storage plastids (amyloplasts) and in leaf chloroplasts. Starch is comprised of two compounds: a straight-chained molecule, amylose that contains 200 to 1,000 glucose subunits and amylopectin, a branched chain molecule that is substantially larger, 2,000 to 200,000 subunits. Amylose has individual glucose molecules linked by α-(1-4) glucosidic bonds (Figure 4.4), the bonding angle of which imparts a helical structure to the molecule. Amylopectin has similar α-(1-4) bonds between glucose subunits, however, every 20 to 25 glucose molecules there is a branch formed through an α-(1-6) glucosidic linkage (Figure 4.4). The ratio of amylose to amylopectin is genetically controlled with amylopectin comprising the dominant form, e.g., 60 to >95%.

Biosynthetic or hydrolytic changes in the starch concentration are extremely important during the postharvest period for many commodities. For example, in banana and many other climacteric fruits, the conversion of starch to sugars in the fruit is an important component of the ripening process, giving the fruit its distinctive sweet flavor as well as precursors for many of the aromatic flavor compounds. On the other hand, in some products (e.g., sweet corn) free sugars can be readily converted to starch after harvest, decreasing the quality of the product. Because of these differing postharvest scenarios, an understanding of starch metabolism is advantageous.

a. Starch synthesis

Starch synthesis occurs in the chloroplasts of leaves and plastids in non-green tissue.[174] In plastids, synthesis probably begins with sucrose, the primary transport carbon source, which is converted to nucleotide diphosphate glucose. Subsequent polymerization is carried out by the enzyme starch synthetase, which requires an existing α-(1-4) glucan primer molecule of at least two glucose residues (e.g., maltose). There are several forms of starch synthetase (e.g., ADP-glucose and UDP-glucose transglucosylase) that catalyze the addition of glucose from either ADP or UDP-glucose nucleotides. Sucrose is converted to ADP-glucose either directly by sucrose synthase or indirectly through the action of invertase. Invertase first produces glucose and fructose with the glucose activated by ATP to ADP-glucose.

- ADP-glucose formation from sucrose

$$\text{sucrose} + H_2O \rightarrow \text{glucose} + \text{fructose}$$
$$\textit{invertase}$$

$$\text{glucose} + \text{ATP} \rightarrow \text{glucose-6-P} \rightarrow \text{glucose-1-P}$$

$$\text{glucose-1-P} + \text{ATP} \rightarrow \text{ADP-glucose} + \text{PPi}$$

• Addition of glucose to an existing glucan chain

$$n \text{ ADP-glucose} + \alpha\text{-}(1\text{–}4) \text{ glucan primer} \rightarrow \text{starch} + n\text{ADP}$$

The formation of amylopectin, requiring the addition of α-(1-6) branches, proceeds similarly; however, a branching enzyme (amylo-(1,4-1,6)-transglycosylase) catalyzes the branch addition every 20 to 25 glucose residues.

Several mutants of sweet corn have been found that give both higher sugar levels in the endosperm and decreased incorporation of the sugars into starch after harvest.[71,169] In the case of one mutant, both the branching enzyme and starch synthase levels are considerably reduced.[187]

b. Starch breakdown

Starch can be broken down to glucose by at least three different enzymes, α- and β-amylase and starch phosphorylase. The amylases hydrolyze starch into two glucose segments (maltose) that are then further hydrolyzed by the enzyme maltase to glucose.

$$\text{starch} + n\text{ H}_2\text{O} \rightarrow n \text{ maltose}$$
amylase

$$\text{maltose} + \text{H}_2\text{O} \rightarrow 2 \text{ glucose}$$
maltase

The α-amylases rapidly hydrolyze the α-(1-4) linkages of amylose at random points along the chain, forming fragments of approximately 10 glucose subunits called maltodextrins. These are more slowly hydrolyzed to maltose by the enzyme. Alpha-amylase also attacks the α-(1-4) linkages of amylopectin, however, in the regions of the α-(1-6) branch points it is inactive, leaving dextrins (>3 glucosyl units).[151] Beta-amylase removes maltose units starting from the nonreducing end of the starch chain and hydrolyses up to a α-(1-6) branching point. This yields maltose and dextrins.

Starch phosphorylase also attacks α-(1-4) linkages, but forms glucose-1-phosphate. Unlike the hydrolysis reactions of the amylases where a single water molecule is used in each bond cleavage, the phosphorylase enzyme incorporates phosphate.

$$\text{starch} + n\text{ Pi} \rightarrow n \text{ glucose-1-P}$$
starch phosphorylase

Neither starch phosphorylase nor either of the amylases will attack the α-(1-6) branch points of amylopectin, so complete breakdown by these enzymes is not possible. Several debranching enzymes have been isolated from plants; however, neither their importance nor action are well understood.

1.3.3. Pectic Substances

The bulk of the primary cell walls in plants is comprised of dense gel-like, noncellulosic polysaccharides called pectic substances.[29] Pectin is found extensively in the middle lamella where it functions as a binding agent between neighboring cell walls. Each molecule is composed largely of α-(1-4) linked D-galacturonic acid subunits, although a number of other monosaccharides may by present (i.e., xylose, glucose, rhamnose, mannose, galactose, arabinose) (Table 4.2). There are three general classes of pectins based upon their chemical composition: a) homogalacturonans, b) xylogalacturons and c) rhamnogalacturonan I (Figure 4.5A).

Table 4.2. Gross Composition of the Cell Wall for Apple and Strawberry: Monomers Yielded Upon Hydrolysis of Wall Polymers.*

Component Monomers	% Total Accounted for	
	Apple	*Strawberry*
Rhamnose	0.4	1.1
Fucose	0.7	N.D.
Arabinose	19.5	6.5
Xylose	5.9	1.9
Mannose	1.9	0.7
Galactose	5.8	7.6
Glucose	47.5	31.1
Galacturonic acid	16.6	40.3
α-Amino acid	1.7	10.7
Hydroxyproline	0.04	0.1

Source: Knee.[129]
ND = not detected.

The structure of pectic substances varies widely with source. For example, jackfruit pectin[8] is comprised of only galacturonic acid residues; however, most species have rhamnose interspersed between segments of galacturonic acid residues. A 3-dimensional structure is conferred to the pectin matrix *via* cross-linking between neighboring pectin molecules by a) glucomannans, b) galactomannans and c) mannans (Figure 4.5B). Calcium (or magnesium) bridges between neighboring pectin molecules are also important in conferring a more ridged structure [Figure 4.5B(d)].

Pectic acids, found in the middle lamella and the primary cell wall, are the smallest of three general size classes (a very heterogenous group based upon solubility) and are usually around 100 galacturonic acid subunits in size. Pectic acids are soluble in water but may become insoluble if many of the carboxyl groups combine with Ca^{2+} or Mg^{2+} to form salts (see Ca^{2+} bridge in Figure 4.5B). Pectins are usually larger than pectic acids (e.g., 200 subunits) and have many of their carboxyls esterified by the addition of methyl groups. They are also found in the middle lamella, primary cell wall and in some cells as constituents of the cytoplasm. Protopectins are larger in molecular weight than pectins and intermediate in the degree of methylation between pectic acids and pectins. They are insoluble in hot water and are found primarily in the cell wall.

a. Synthesis of pectic substances

Pectic substances are synthesized from UDP-galacturonic acid on other UDP-sugars, although several other nucleotides may, in some cases, function in place of uridine diphosphate.[27,73] Methylation occurs after the subunit is placed in the chain and is catalyzed by methyl transferase. The methyl donor is S-adenosylmethionine.

b. Breakdown of pectic substances

The activity of pectic enzymes correlates with increases in softening during the latter stages of ripening of many fruits and a concurrent increase in soluble pectins. For example, the soluble pectin content of apples increases more than 3-fold during a 1.4 kg decrease in fruit firmness.[10] The increase is due to hydrolytic cleavage of the long pectic chains increasing their solubility. The principal pectic enzymes are pectinesterase, endopolygalacturonase and exopolygalacturonase. The early stages of fruit softening, however, are not apparently associated with the solubilization of pectin, however, the enzymes are operative in the later stages.

Pectinesterase or pectinmethylesterase catalyzes the hydrolysis of methyl esters along the pectic chain producing free carboxyl groups (Figure 4.5C). The enzyme deesterifies in a linear manner, moving down the chain and producing segments with free carboxyl groups. Deesterification by pectinesterase must precede degradation by polygalacturonases that require at least four galacturonic acid groups in sequence without methylation.

The polygalacturonases represent a class of pectolytic enzymes that degrade deesterified pectin chains into smaller molecular weight polymers and component monosaccharides.[81] Often two polygalacturonases are found in fruit tissues. Exopolygalacturonase cleaves single galacturonic acid subunits from the non-reducing end of the protopectin molecule (Figure 4.5C), while endopolygalacturonase attacks the chain randomly. Cleavage within the chain by endopolygalacturonase has a much more pronounced effect on the degree of solubilization of the pectic molecule and the pectin viscosity. Both enzymes are found in a number of fruits and often the increases in their activity tend to parallel the formation of water-soluble pectins and are thought to be involved in the later stages of fruit softening during ripening.[188]

1.3.4. Hemicelluloses

The hemicelluloses represent a heterogeneous group of polysaccharide compounds that are closely associated with cellulose, hence the name hemi (*half*) cellulose. They are stable, a major component of cell walls and normally can only be extracted with a strong base. Hemicelluloses do not represent carbohydrate reserves that can be recycled as an energy source for cells, though an exception may be mannans. These carbohydrates are composed largely of glucose, galactose, mannose, xylose, and arabinose molecules linked in various combinations and with varying degrees of branching. The total number of subunits ranges from 40 to 200. In monocots, the major hemicellulosic components of the cell walls are arabinoxylans[28] while in dicots, they are xyloglucans.[29] Hemicelluloses are synthesized from nucleotide sugars (e.g., UDP-xylose, UDP-arabinose, UDP-glucuronic acid) in reactions catalyzed by transferase enzymes.[27,73]

1.3.5. Fructosans

Some plant species store polymers of fructose as carbohydrate reserves rather than glucose. The polymers include the inulins and levans and are common in the Compositae, Campanulaceae and Graminae families.[186] Inulin is composed of a chain of 25 to 35 fructose subunits joined by β-linkages through C-1 of one molecule and C-2 of the adjacent, i.e., β-(2-1), and is terminated with a sucrose molecule. Inulin represents a population of structurally similar molecules that differ in the number of fructose subunits. Inulin is a straight-chained polymer, however, some fructosans are branched. They are substantially smaller than starch molecules and more soluble in water. Inulins are more commonly found stored in roots and tubers rather than above-ground plant parts. The tubers of Jerusalem artichokes and dahlia, the bulbs of iris and the roots of dandelion and chicory are high in inulin.

The synthesis of inulin begins with the addition of fructose to a terminal sucrose molecule forming a trisaccharide.[49] Complete synthesis of the polymer requires several enzymes:

- sucrose-sucrose 1-fructosyltransferase (SST) catalyzes the formation of the trisaccharide from two sucrose molecules;

$$Glu–Fru + Glu–Fru \rightarrow Glu–Fru–Fru + Glu$$

- Glucose is subsequently converted *via* several steps into sucrose;
- β-(2-1) fructan 1-fructosyltransferase (FFT) catalyzes the transfer of a fructose sub-unit from a donor to an acceptor, both of which are trisaccharides or greater in size.

$$\text{Glu–Fru–Fru}_N + \text{Glu–Fru–Fru}_M \rightarrow \text{Glu–Fru–Fru}_{N-1} + \text{Glu–Fru–Fru}_{M+1}$$

donor   acceptor   donor   acceptor

This reaction is reversible and also functions during depolymerization.

The degradation or depolymerization of inulin can follow one of two possible pathways. During cold storage, inulin is broken down into shorter chain length oligomers. This involves the action of hydrolases, β-(2-1′) fructan 1-fructosyl-transferase (FFT) and the enzymes involved in sucrose synthesis. During sprouting, inulin is degraded completely to fructose by hydrolases and the fructose converted to sucrose for export to the growing apices. Several yeasts have the enzymes required to hydrolyze the inulin polymer and subsequently convert the sub-units to alcohol, thus enhancing the attractiveness of using inulin as a carbon substrate for alcohol production.[78]

Levans, another type of fructose polymer, are formed through a β-linkage between C-2 and C-6 of two adjacent fructose subunits, i.e., β-(2-6). As with inulin, levans are also terminated with a sucrose molecule. Levans are found primarily in the Gramineae family; for example, a levan called phlein is found stored in the roots of timothy.

1.3.6. Gums and Mucilages

Gums and mucilages are composed of a wide cross-section of sugar subunits and as a consequence, generalizations about their individual composition cannot easily be made. Hydrolysis of gum from plum fruits yields a mixture D-galactose, D-mannose, L-arabinose, D-xylose, L-rhamnose, D-glucuronic acid and traces of 4-0-methyl glucuronic acid.

These polymers may be found free in the cytoplasm or in some cases sequestered in specialized cells. As a group they are hydrophilic in nature and their function in the plant is not well understood. Gums are thought to be involved in sealing mechanical or pathogenic wounds to the plant while mucilages may function by modulating water uptake in seeds or water loss from some succulent species.

2. ORGANIC ACIDS

A number of harvested plant products contain significant concentrations of organic acids, many of which play a central role in metabolism. In addition, the levels of organic acids present often represent an important quality parameter; this is especially so in many fruits (Table 4.3).

Organic acids are small mono-, di- and tricarboxylic acids that exhibit acidic properties due to the presence of their carboxyl group(s) (COOH) that can give up a hydrogen atom. They exist as free acids or anions, or are combined as salts, esters, glycosides or other compounds. Organic acids are found in active pools that are utilized in the cytoplasm for metabolism and to a greater extent, as storage pools in the vacuole. For example, only about 30% of the malate is in the mitochondria, with the remainder thought to be in the vacuole. In some plant cells, certain organic acids may, to a large extent, be in the form of insoluble salts, e.g., calcium oxalates in rhubarb or potassium bitartrate in grapes. When in the ionized anion form (-COO⁻), the name of the acid ends in "ate" (e.g., malate); while in the protonated state (-COOH), the ending is "ic" (e.g., malic acid).

Organic acids can be classified or grouped in a number of ways. For example, they may be grouped based on the number of carbon atoms present (typically 2 to 6) or on their specific function(s) within the cell. Another means of separation is based on the number of carboxylic groups present. This nomenclature gives a greater indication of how the acid will act chemically. The organic acids found in postharvest products (Figure 4.6), include monocarboxylic acids, monocarboxylic acids with alcohol, ketone or aldehyde groups, monocarboxylic carbocyclic aromatic and alicyclic acids, dicarboxylic acids and tricarboxylic acids. The type of acid present and the absolute and relative concentration of each, vary widely among different postharvest products (Table 4.3). Most organic acids are found in only trace amounts, however, several, such as malic, citric and tartaric, tend to be found in abundance in some tissues. The concentrations of these abundant acids varies widely among products, for example, lemon fruit contain 70–75 meq of citric acid · 100 g^{-1} fresh weight while the banana fruit has only 4 meq of malic acid · 100 g^{-1} fresh weight. In addition, in the first case (lemon), high acidity is a desirable flavor attribute while in the second it is not.

Organic acids play a central role in the general metabolism of postharvest products. A number of organic acids are essential components of the respiratory tricarboxylic acid cycle and phosphoglyceric acid plays an essential role in photosynthesis. Many organic acids have multiple functions in the plant. In some tissues with high concentrations, organic acids repre-

Table 4.3. The Organic Acids Present in the Fruit of Apple, Pear, Grape, Banana and Strawberry.*

Acids	Apple	Pear	Grapes	Banana	Strawberry
Glycolic	+	+	+	+	tr
Lactic	+	+	+	+	
Glyceric	+	+	+	+	tr
Pyruvic	+	+	+		
Glyoxylic		+	+	+	
Oxalic	+		+	+	
Succinic	+	+	+	+	+
Fumaric	+		+		
Malic	++	++	++	++	+
Tartaric			++		
Citramalic	+	+		+	
Citric	+	+	+	+	+++
Isocitric	+		+		
Cis aconitic			+		
Oxaloacetic	+		+	+	
α-Oxoglutaric	+	+	+	+	
Galacturonic	+	+	+		
Glucuronic	+		+		
Caffeic	+		+		
Chlorogenic	+	+	+		
p-Coumarylquinic	+				
Quinic	+	+	+	+	+
Shikimic	+	+	+	+	tr

* *Source:* Ulrich[235]; data derived from Hane;[83] Hulme and Wooltorton;[103] Kliewer;[127] Kollas;[132] Steward, et al.;[221] and Wyman and Palmer.[251] Increasing number of + signs denotes increased concentration; tr = trace.

A.) PECTIC SUBSTANCES

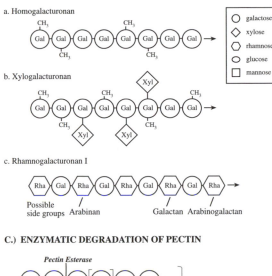

a. Homogalacturonan

b. Xylogalacturonan

c. Rhamnogalacturonan I

Possible side groups Arabinan Galactan Arabinogalactan

B.) GROSS-LINKING SUBSTANCES

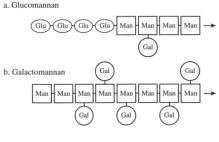

a. Glucomannan

b. Galactomannan

c. Mannan

d. Calcium bridges between homogalacturonan molecules

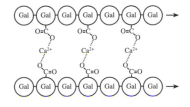

C.) ENZYMATIC DEGRADATION OF PECTIN

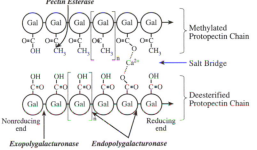

Pectin Esterase

Methylated Protopectin Chain

Salt Bridge

Deesterified Protopectin Chain

Nonreducing end Reducing end

Exopolygalacturonase *Endopolygalacturonase*

Figure 4.5. The cell wall region is comprised of cellulose, hemicellulose, protein, pectic substances (A) and a number of relatively complex cross-linking compounds (B), the latter of which confer rigidity to the three-dimensional wall structure. Calcium bridges (B-d) can form between carboxyl groups of neighboring pectin chains, tethering the molecules together and enhancing the structure. Degradation of pectic compounds during softening involves three enzymes (C). Pectinesterase removes methyl groups leaving free hydroxyls; exopolygalacturonase removes single galacturonic acid subunits starting at the non-reducing end of the molecule; and endopolygalacturonase cleaves the linkage between neighboring galacturonic acids randomly within the chain.

sent a readily available source of stored energy that can be utilized during the postharvest period. In food products, organic acids may impart a significant portion of the characteristic flavor, both taste and odor. Aromatic compounds, such as esters of organic acids, given off by various plant products are diverse in type (Table 4.4) and in some cases, such as isoamyl acetate in banana fruit, impart a major portion of the characteristic aroma.

Organic acids are synthesized primarily through oxidations, decarboxylations and, in some cases, carboxylations reactions in the respiratory tricarboxylic acid pathway. Some, however, are formed from sugars during the early stages of the photosynthetic dark reactions. Therefore, in most cases, the immediate precursors of organic acids are either other organic acids or sugars.

After harvest and during storage, the concentration of total organic acids tends to decline. Postharvest changes vary with the specific acid in question, the type of tissue, handling and storage conditions, cultivar, year and a number of other parameters. For example, the con-

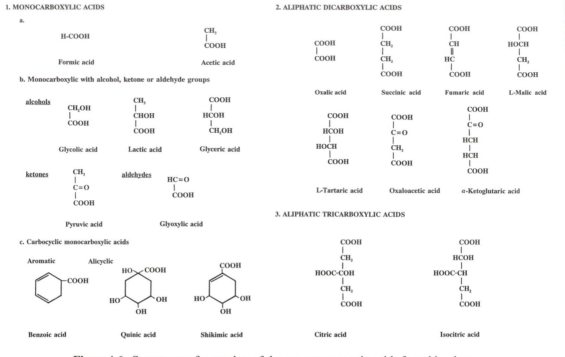

Figure 4.6. Structures of a number of the common organic acids found in plants.

Table 4.4. Organic Acids Emanating as Esters from Apples (cv. 'Cabville blanc').*

Methyl formate	tr	Methyl butyrate	tr
Propyl formate	tr	Ethyl butyrate	
Hexyl formate	tr	Propyl butyrate	tr
Isobutyl formate	tr	Butyl butyrate	
Methyl acetate	tr	Amyl butyrate	
Ethyl acetate		Hexyl butyrate	tr
Propyl acetate		Isopropyl butyrate	tr
Butyl acetate		Isobutyl butyrate	tr
Amyl acetate		Isoamyl butyrate	tr
Hexyl acetate		Ethyl isobutyrate	tr
Isobutyl acetate		Ethyl valerianate	tr
Butyl (second)	tr	Butyl valerianate	tr
Isoamyl acetate		Methyl caproate	tr
Methyl propionate	tr	Ethyl caproate	
Ethyl propionate	tr	Butyl caproate	
Propyl propionate	tr	Ethyl octanoate	tr
Butyl propionate			
Amyl propionate	tr		
Isoamyl propionate			

* *Source:* Paillard;[179] tr = trace.

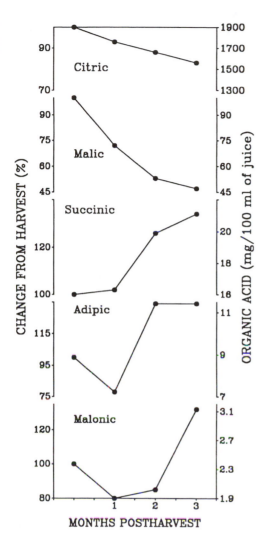

Figure 4.7. Changes in the organic acid composition of the juice of 'Shamouti' oranges during storage at 17°C. Changes are expressed as the % change from harvest and as mg/100 mL of juice (*redrawn from Sasson and Monselise*[201]).

centration of citric and malic acids in the juice of 'Shamouti' oranges declined with storage time while malonic, succinic and adipic acids increased (Figure 4.7). Changes in albedo (white interior portion of the peel) and flavedo (pigmented exterior portion of the peel) organic acids, however, did not parallel changes in the juice.[201] Likewise, both cultivar and year of production can have a pronounced effect on the concentration of specific acids and total titratable acidity (Table 4.5).

Controlled atmosphere storage has been shown to alter the changes in organic acids occurring after harvest. The juice of 'Valencia' oranges stored at 3% O_2 and 5% CO_2 (3.5°C) lost less acid than oranges held in air (0°C). Principal differences were a decreased rate of malic acid loss and an increase in quinic and shikimic acids in controlled atmosphere stored fruit.[132] These effects may in part be due to the dark fixation of CO_2 by the fruit.

Table 4.5. Differences in the Organic Acid Concentration and Titratable Acidity for 12 Muscadine Grape Cultivars (*Vitis rotundifolia* Michx).*

Cultivar	Malate (%)	Tartrate (%)	Citrate (%)	Titratable acidity (%)
Albemarle	0.49[†]	0.25	0.05	0.75
Carlos	0.54	0.19	0.04	0.85
Chowan	0.50	0.23	0.05	0.82
Dearing	0.45	0.23	0.04	0.71
Hunt	0.54	0.32	0.04	0.93
Magnolia	0.33	0.23	0.05	0.64
Magoon	0.43	0.41	0.02	0.94
Pamlico	0.62	0.24	0.04	1.05
Roanoke	0.70	0.29	0.04	0.10
Scuppernong	0.52	0.28	0.04	0.87
Thomas	0.51	0.27	0.05	0.84
Topsail	0.37	0.19	0.04	0.51
Roanoke 1965	0.26	0.23	0.05	0.52
1966	0.67	0.27	0.06	1.22
1967	1.06	0.38	0.02	1.55

*Source: Carroll et al.[30] and Kliewer.[127]
[†]Mean of three seasons.

3. PROTEINS AND AMINO ACIDS

Proteins are extremely important components of living cells in that they regulate metabolism, act as structural molecules and in some products, represent storage forms of carbon and nitrogen. Proteins are composed of chains of amino acids each joined together by a peptide bond (Figure 4.8). When there are fewer than 10 amino acids, they are referred to as peptides, 10 to 100 amino acids are polypeptides and more than 100 amino acids are proteins. Many of the properties of a polypeptide or protein are a function of which amino acids are in the chain and their particular sequence in the molecule. In plants, approximately 20 amino acids are commonly found, however, over 100 nonprotein amino acids are known.

Each of the amino acids that make up a protein has distinct properties that, in turn, influence the properties of the polypeptide in which they are found. Amino acids are small in size

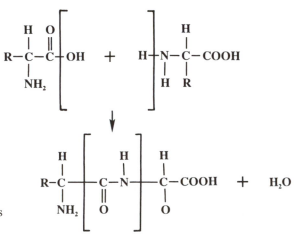

Figure 4.8. Proteins are composed of amino acids joined together in long chains by peptide bonds.

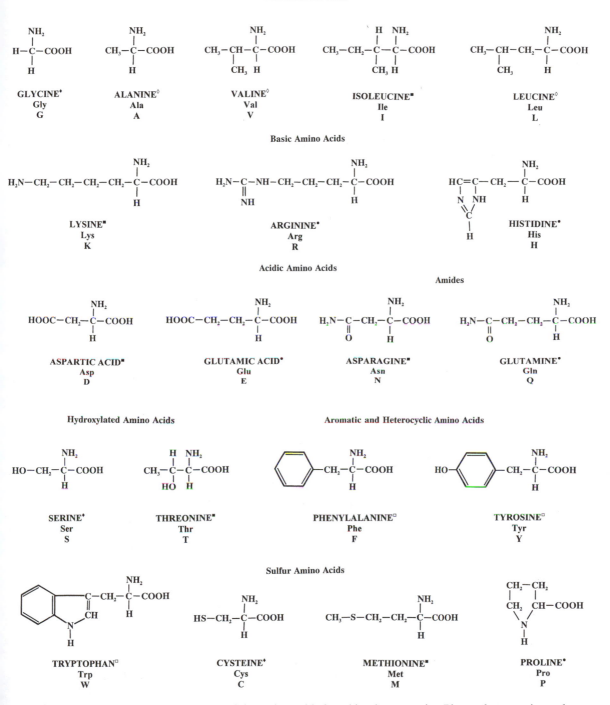

Figure 4.9. The classes and structures of the amino acids found in plant proteins. Plants also contain a relatively diverse cross-section of non-protein amino acids in addition to the above. Symbols indicate precursors: ■ oxaloacetate; ◆ 3-phosphoglycerate; □ phosphoenolpyruvate; ● α-ketoglutarate; and ◇ pyruvate.

Met-Ala-Thr-Lys-Ile-Leu-Ala-Leu-Leu-Ala-Leu-Leu-Ala-Leu-Leu-Val-Ser-Ala-

Thr-Asn-Ala-Phe-Ile-Ile-Pro-Gln-Cys-Ser-Leu-Ala-Pro-Ser-Ala-Ser-Ile-Pro-Gln-

Phe-Leu-Pro-Pro-Val-Thr-Ser-Met-Gly-Phe-Glu-His-Pro-Ala-Val-Gln-Ala-Tyr-

Arg-Leu-Gln-Leu-Ala-Leu-Ala-Ala-Ser-Ala-Leu-Gln-Gln-Pro-Ile-Ala-Gln-Leu-

Gln-Gln-Gln-Ser-Leu-Ala-His-Leu-Thr-Leu-Gln-Thr-Ile-Ala-Thr-Gln-Gln-Gln-

Gln-Gln-Gln-Phe-Leu-Pro-Ser-Leu-Ser-His-Leu-Ala-Met-Val-Asn-Pro-Val-Thr-

Tyr-Leu-Gln-Gln-Gln-Leu-Leu-Ala-Ser-Asn-Pro-Leu-Ala-Leu-Ala-Asn-Val-Ala-

Ala-Tyr-Gln-Gln-Gln-Gln-Gln-Leu-Gln-Gln-Phe-Met-Pro-Val-Leu-Ser-Gln-Leu-

Ala-Met-Val-Asn-Pro-Ala-Val-Tyr-Leu-Gln-Leu-Leu-Ser-Ser-Ser-Pro-Leu-Ala-

Val-Gly-Asn-Ala-Pro-Thr-Tyr-Leu-Gln-Gln-Gln-Leu-Leu-Gln-Gln-Ile-Val-Pro-

Ala-Leu-Thr-Gln-Leu-Ala-Val-Ala-Asn-Pro-Ala-Ala-Tyr-Leu-Gln-Gln-Leu-Leu-

Pro-Phe-Asn-Gln-Leu-Ala-Val-Ser-Asn-Ser-Ala-Ala-Tyr-Leu-Gln-Gln-Arg-Gln-

Gln-Leu-Leu-Asn-Pro-Leu-Ala-Val-Ala-Asn-Pro-Leu-Val-Ala-Thr-Phe-Leu-Gln-

Gln-Gln-Gln-Gln-Leu-Leu-Pro-Tyr-Asn-Gln-Phe-Ser-Leu-Met-Asn-Pro-Ala-Leu-

Gln-Gln-Pro-Ile-Val-Gly-Gly-Ala-Ile-Phe

Figure 4.10. The amino acid sequence for one of four distinct groups forming the storage protein zein (approximate molecular weight = 22,000), as determined by sequencing its mRNA (*after Marks and Larkins*[153]).

and soluble in water. Each contains both a carboxyl group (-COOH) and an amino group (-NH) and some may also have hydroxyl groups (-OH), sulfhydryl groups (-SH) or amide groups (-CONH) present.

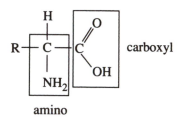

Typically, amino acids are classified into one of six types based upon the properties of their R group (Figure 4.9) and are designated by their common name, abbreviation or by a single letter. When denoting the amino acid sequence in a large protein, the single letter designation is commonly used.

a) Neutral amino acids—where R is a hydrogen, aliphatic or hydroxyl group (glycine, alanine, valine, leucine and isoleucine).

b) Basic amino acids (arginine and lysine).
c) Acidic amino acids (aspartic acid, glutamic acid), and their amides (asparagine and glutamine).
d) Hydroxylated amino acids (serine and threonine).
e) Aromatic and heterocyclic amino acids (phenylalanine, tyrosine, proline, tryptophan and histidine).
f) Sulfur containing amino acids (cysteine and methionine).

3.1. Protein Synthesis

During the postharvest period, the metabolic processes within plant cells continue, requiring specific proteins to be synthesized at precise times. Synthesis and degradation are the two primary means of modulating the level of a specific protein. As a consequence, protein synthesis, especially the synthesis of specific proteins after harvest, is of interest to postharvest biologists. The sequence for each protein, as illustrated by one of the component groups of the storage protein zein (Figure 4.10), is found coded in the cell's DNA (Figure 4.11). This coded sequence is transcribed by the formation of a special type of ribonucleic acid, the messenger RNAs (mRNA), thus transferring the required amino acid sequence to a molecule that can move from the nucleus into the cytoplasm where the actual synthesis of the protein molecule occurs. The messenger RNA subsequently has a ribosome attached that translates the code from the mRNA and assembles each amino acid in the appropriate sequence to form the polypeptide. Several ribosomes are often attached to an mRNA strand forming a polysome. One protein molecule is assembled per ribosome at a time. In some cases, there are modifications of the amino acids in the protein after assembly of the amino acids; this process is called post-translational modifications.

A very general overview of the steps in protein synthesis is given in Figure 4.12. As one would anticipate, the precise method in which DNA is transcribed to mRNA and mRNA is translated to form the polypeptide is much more complex than indicated by this brief overview. Several references listed in the back of this section give a more detailed account.[21,152]

3.2. Protein Structure

The proteins formed are folded into a three-dimensional structure that is a function of the kind, number and sequence of the amino acids present and the type of nonprotein (prosthetic) groups attached to the amino acids. Protein structure is typically broken down into 4 levels of organization: primary, secondary, tertiary and quaternary (Figure 4.13). The primary structure is the kind, number and sequence of the amino acids present in the chain, while the secondary structure is the conformation of the chain of amino acids due to hydrogen bonding between the oxygen of a carboxyl group and a nitrogen atom of a neighboring amino group. Tertiary structure, which leads to folding or bending of the chain, is due to interactions between side chains on certain amino acids and adjacent portions within the chain and is facilitated by proteins called chaperones. These interactions may be by hydrogen and ionic bonds, and allow the aggregation of polypeptides into a globular, planar, or fibrous forms. Quaternary structure is when two or more different proteins come together to form a complex to carry out a certain function.

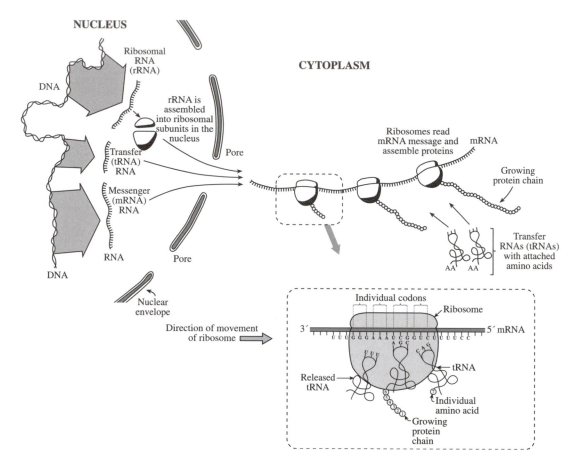

Figure 4.11. Diagrammatic representation of the synthesis of a protein. The amino acid sequence for an individual protein is transcribed with the synthesis of a specific messenger RNA (mRNA) for the protein. Ribosomal RNA (rRNA) and transfer RNA (tRNA), also derived from the DNA molecule, are likewise present. Once in the cytoplasm, several ribosomes attach to an mRNA molecule and moving down the molecule, translating the code. Insertion of the appropriate amino acid, carried to the ribosome by its tRNA, allows assembly of the protein in the appropriate sequence, one amino acid at a time. Many proteins undergo some post-translational modifications once the initial amino acid sequence is assembled.

3.3. Protein Classification

Proteins can be classified based on: a) their physical and/or chemical properties, b) the type of molecules that may be joined to the protein, or c) by the function of the protein within the cell. The first type of classification, based on chemical and/or physical properties, typically separates proteins into groups based on their size, structure, solubility or degree of basicity (kind and number of basic amino acids, e.g., lysine, arginine and histidine). The second means of classification, based on the prosthetic group attached, separates proteins into classes such as lipoproteins (lipid), nucleoproteins (nucleic acid), chromoproteins (pigment), metalloproteins (metal) and glycoproteins (carbohydrate). Proteins can also be classified based upon their function in the plant. They are typically grouped into three general classes; structural proteins (e.g., membrane-bound proteins, cell wall proteins), storage proteins and enzymes. In some instances, however, the classes overlap. For example, many en-

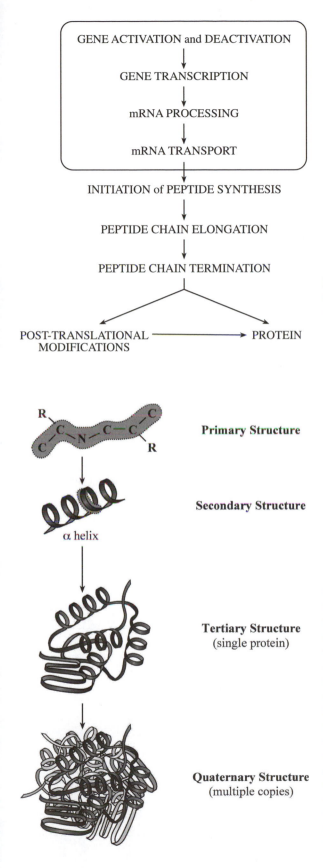

GENE ACTIVATION and DEACTIVATION

↓

GENE TRANSCRIPTION

↓

mRNA PROCESSING

↓

mRNA TRANSPORT

↓

INITIATION of PEPTIDE SYNTHESIS

↓

PEPTIDE CHAIN ELONGATION

↓

PEPTIDE CHAIN TERMINATION

POST-TRANSLATIONAL ⟶ PROTEIN
MODIFICATIONS

Figure 4.12. The major sequence of steps in protein synthesis and their location in the cell is presented in this flow chart.

Primary Structure

Secondary Structure

α helix

Tertiary Structure
(single protein)

Quaternary Structure
(multiple copies)

Figure 4.13. The three dimensional structure of proteins is a critical component of their functional specificity. The primary structure is conferred by the peptide bond of each amino acid with the secondary structure, forming an α-helix or pleated sheet (not shown), due to bonding between neighboring amino acids within the chain. The tertiary structure is created by the folding and cross-linking of various parts of the protein, a process facilitated by two classes of molecules called chaperones. The quaternary structure is due to the positioning of multiple copies of the protein forming stacks in a specific orientation. Not all proteins are grouped into a quaternary structure, however, storage proteins such as zein are commonly found in this manner, consolidating the protein and decreasing the space required.

zymes are also components of membranes and may, therefore, have a multiple role within the cell.

Storage proteins, found in abundance in seeds, serve as a source of nitrogen and amino acids during germination. Cereal grains contain on an average 10% protein, largely as storage protein, while legumes 20–30%. Together these make up the major portion of the protein consumed by man, e.g., ~70%. Most other edible crops are lower in protein (Table 4.6).

Enzymatic proteins are extremely important in that they regulate virtually all of the biochemical reactions within cells. Primarily through enzyme synthesis, activation and degradation, control is exerted over the rate of specific processes, thus allowing the plant product to adjust its metabolism to changes in the environment in which it is held and to genetically controlled metabolic shifts (e.g., ripening).

Enzymes are grouped based on their type of catalytic function:

a) Oxidoreductases—catalyze oxidation-reduction reactions, e.g., malate dehydrogenase.
b) Transferases—catalyze the transfer of a specific group from one molecule to another, e.g., methionine transferase in the ethylene synthesis pathway.
c) Hydrolases—catalyze hydrolysis by the addition of water, e.g., amylase.
d) Lyases—catalyze the addition or removal of groups without the involvement of water, e.g., phosphoenolpyruvate carboxylase.
e) Isomerases—catalyze isomerizations, e.g., the conversion of glucose-6-phosphate to fructose-6-phosphate by phosphoglucoisomerase.
f) Ligases—catalyze condensing reactions, e.g., pyruvate carboxylase.

3.4. Protein Degradation

Most of the proteins within cells are in a continuous state of synthesis and degradation. After synthesis, they begin to progress toward eventual degradation and recycling of their component

Table 4.6. Amino Acid Composition of Several Types of Postharvest Products.*

Food Group	Plant Part Used	Common Name	Protein Content $(g \cdot 100g^{-1})$	Isoleucine	Leucine	Lysine	Methionine	Cystine	Phenylalanine	Tyrosine	Threonine
				Amino Acids (mg/g of nitrogen)							
Cereal	Seed	Rice	7.1	306	563	219	225	100	350	300	288
	Seed	Wheat	12.2	204	417	179	94	159	282	187	183
Pulses	Seed	Chickpea	20.1	277	468	428	65	74	358	183	235
	Seed	Bean	22.1	262	476	450	66	53	326	158	248
Roots and Tubers	Tuber	White potato	2.0	236	377	299	81	37	251	171	235
	Root	Sweetpotato	1.3	230	340	214	106	69	241	146	236
Vegetables	Leaves	Lettuce	1.3	238	394	238	112	—	319	169	256
	Stems	Celery	1.1	244	425	150	138	—	281	—	213
	Flowers	Cauliflower	2.8	302	436	356	99	—	225	—	264
	Immature seed	Pea	6.6	260	435	456	58	60	275	194	235
	Immature seed	Green bean	2.4	234	432	344	81	53	266	209	241
Fruits	Fruit	Apple	0.4	220	390	370	49	84	160	94	230
	Fruit	Banana	1.2	181	294	256	125	169	244	163	213
Nuts	Nut	Coconut	6.6	244	419	220	120	76	283	167	212
	Nut	Brazil nut	14.8	175	431	175	363	131	244	169	163
Milk		Cow's milk	3.5	295	596	487	157	51	336	297	278
Eggs		Hen egg	12.4	393	551	436	210	152	358	260	320

* *Source:* Data from FAO.[55]

parts. The length of their life expectancy varies widely among different proteins; it may be from as short as a few minutes to as long as years. The pathway for degradation involves recognition of a protein targeted for degradation, attachment of a small protein (ubiquitin) to the targeted protein and subsequent degradation of the protein by various proteases.[237] The degradation of storage proteins in seeds during germination has been the focus of much research but differs from the process in cells. The turnover of proteins in organs such as leaves, fruits, roots and tubers is under developmental control, though at this time it is not well understood.

Enzymes that cleave the peptide bonds in protein chains (peptidases, proteinases or proteases) are classified as either endopeptidases or exopeptidases. Endopeptidases cleave peptide bonds within the protein chain while exopeptidases remove individual amino acids from either the carbon or nitrogen end of the molecule. These two classes are further broken down into subgroups based on the enzyme's mechanism of action, i.e., there are currently four types of endopeptidases and two types of exopeptidases. The site of degradation may be in the cytoplasm, the vacuole or exterior to the plasma membrane, and individual steps at a specific locale may be sequestered in separate compartments within the cell.

3.5. Changes in Amino Acids and Proteins after Harvest

The relative change in the protein and amino acid content and composition in harvested plant organs is greatest in those products that undergo significant changes during the postharvest period (e.g., leaf senescence). Products such as seeds that are relatively stable when properly stored, do not undergo substantial changes in their protein composition. Two general phenomena that precipitate large changes in protein and amino acid content and composition, are the onset of senescence and fruit ripening.

Leaf tissue has been used widely as a model system for studying senescence, since the changes are substantial and occur quite rapidly. Senescence in leaves is characterized by a decline in photosynthesis and the loss of protein and chlorophyll (Figure 4.14). Proteolysis, the

Table 4.6. (*continued*)

										Amino Acids (mg/g of nitrogen)	
Tryptophane	Valine	Arginine	Histidine	Alanine	Aspartic Acid	Glutamic Acid	Glycine	Proline	Serine	Total Essential Amino Acids	Total Amino Acids
73	463	644	188	—	275	731	381	313	363	2887	—
—	276	288	143	226	308	1866	245	621	287	2049	6033
—	284	588	165	271	726	991	251	263	318	2426	5998
—	287	355	177	262	748	924	237	223	347	2389	5662
—	292	311	94	278	775	639	237	235	259	2082	4910
—	283	307	84	298	825	541	234	219	255	1972	4735
—	338	281	100	269	719	638	256	325	206	—	—
81	300	289	120	—	—	—	—	—	—	—	—
86	347	250	94	—	—	—	—	—	—	—	—
—	296	548	133	281	620	910	246	240	281	2332	5591
—	306	266	147	275	750	669	238	238	334	2253	5170
58	250	170	120	280	1300	700	240	200	270	1905	5205
—	250	469	469	275	656	575	263	256	244	1968	5175
—	339	822	128	279	553	1171	281	233	303	2148	5918
119	269	831	144	219	463	1163	275	300	269	2239	5903
—	362	205	167	217	481	1390	123	571	362	2947	6463
—	428	381	152	370	601	796	207	260	478	3201	6446

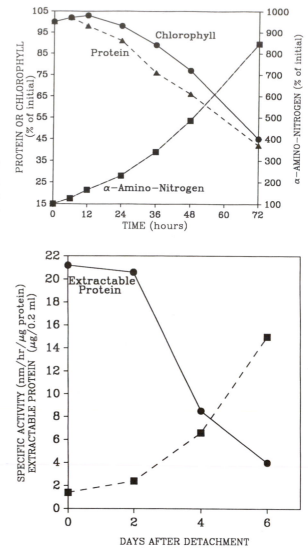

Figure 4.14. Changes in concentration of chlorophyll, protein, and α-amino nitrogen in detached *Avena* leaves held for varying intervals in the dark (*redrawn from Martin and Thimann*[155]).

Figure 4.15. The increase in peptidase (protease) specific activity and concurrent decrease in protein concentration with time in detached *Avena* leaves held in the dark (*redrawn from Martin and Thimann*[155]).

breakdown of protein, begins fairly rapidly after the individual leaf is detached from the parent plant.[155,156] Peptidases (proteinases) which cleave proteins are always present within the leaves, however, their concentrations increase substantially at the onset of senescence (Figure 4.15). While most enzymes are declining, certain specific enzymes increase in activity and/or concentration during senescence.[22] For example, glutamate dehydrogenase activity increases by as much as 400% in spinach leaves held in the dark[110] where it appears to stimulate the deamination of amino acids.

Therefore, while proteins are being broken down and the component amino acids recycled, a small but extremely important number of specific proteins are also being synthesized.[22] Their importance in the development of proteolysis and senescence can be inferred from the fact that inhibitors of protein synthesis strongly decrease the rate at which senescence proceeds. The amino acids formed are largely transported, often after conversion to glutamine, to other parts of the plant and this transport is greatest to areas that represent strong sinks (high demand) such as reproductive organs.[34] In leaves detached at harvest, transport out of the organ is not possible thus the protein degradation products tend to accumulate.

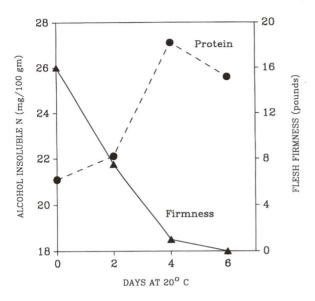

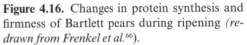

Figure 4.16. Changes in protein synthesis and firmness of Bartlett pears during ripening *(redrawn from Frenkel et al.[66])*.

During the onset of ripening of several climacteric fruits, it has been shown that the actual concentration of protein increases (Figure 4.16). In apples, avocados and several other climacteric fruits, enhanced synthesis of both RNA and protein occurs.[102,194,195] The net effect is an enhanced activity of certain enzymes during ripening (Table 4.7). As with protein synthesis during the onset of leaf senescence, these new proteins appear to be essential since ripening is inhibited if protein synthesis is inhibited. The increase in synthesis of specific enzymes has been monitored using labeled amino acids, then identifying the enzymes that are radioactive and by determining what genes are expressed during ripening.[81] Malic enzyme, which catalyzes the decarboxylation of malic acid, the primary organic acid in apples and certain pear cultivars, is an example of one enzyme that increases markedly during the climacteric. The increase in malic enzyme activity increases the concentration of pyruvic acid, the product of the reaction, which can then enter the respiratory tricarboxylic acid cycle.

4. LIPIDS

Plant lipids represent a very broad group of compounds that play diverse roles in the physiology and metabolism of harvested products. In addition, the absolute concentration of these compounds also varies widely among different species and plant parts. Most postharvest products, however, are relatively low in total lipids, exceptions include avocados, olives and seeds that are high in lipids (Table 4.8). A majority of lipids present are in the form of storage compounds that, in the case of seeds, can be used as an energy source during germination. Plant lipids, in addition to representing a storage form of carbon, also function as components of cellular membranes, as cuticular waxes forming a protective surface on many products and in some cases, as vitamins, pigments, sterols and secondary products.

Biochemically, lipids are normally grouped into neutral lipids, phospholipids, glycolipids, waxes, and terpenoids. Neutral lipids are comprised of fats and oils and represent primarily carbon storage compounds. Phospholipids and glycolipids are components of cellular membranes. Waxes are typically long-chain fatty acids or esters of fatty acids and long-chain alcohols, although numerous other compounds may be found.[130] These compounds form the thin waxy layer on the surface of leaves, fruits and other plant parts. Terpenoids are primarily water-insoluble acyclic and cyclic compounds such as steroids, essential oils and rubber.

Table 4.7. Changes in the Activity of Various Enzymes and Isoenzymes During the Development and Ripening of a Tomato Fruit.*

Enzyme	Stage with Peak Activity	State with Maximum Number of Bands	Activity Trend During Development	Number of Bands at		
				SG	MG	OR
Tyrosinase	LG†	MG	Decrease	1	4	2
Peroxidase	OR	MG-OR	Increase	1	4	4
Esterase	MG	MG	Peak at MG	11	13	8
Acid phosphatase	SG	SG	Decrease	8	6	3
Glycerophosphatase	SG	red	Decrease	1	3	2
ATPase	SG	SG	Decrease	6	5	5
NADH$_2$-diaphorase	MG	MG	Peak at MG	10	12	11
Fumarase	MG	MG	Peak at MG	1	4	2
Malate dehydrogenase	LG red	all	Maximum LG-OR	4	4	4
NADP$^+$ malic enzyme	MG	MG	Peak at MG	3	4	2
Iso-citrate dehydrogenase	SG	MG-OR	Decrease	1	2	2
Glutamate dehydrogenase	MG	SG-MG	Peak at MG	6	6	1
Phosphofructokinase	MG?	MG	Peak at MG?	2	4	3
6-Phosphogluconate dehydrogenase	MG	MG	Peak at MG	2	6	2
Phosphoglucomutase	MG	MG	Peak at MG	3	6	1
Phosphohexose isomerase	SG-MG	MG	SG-MG then a decrease	6	10	6
Glucose-6-phosphate dehydrogenase	SG-MG	MG	Peak at MG	3	4	3
Glutamate-oxaloacetate transaminase	SG-MG	all	Decrease	4	4	4
Leucine aminopeptidase	red?	all	Peak at red?	3	3	3

*Source: After Hobson.[97]

†Notes: SG, small green; LG, large green (preclimacteric); MG, mature green (close to the beginning of the climacteric rise); red, nearly fully ripe; and OR, overripe (postclimacteric).

4.1. Fatty Acids

A substantial portion of the physical and chemical properties of lipids are due to the long chains of fatty acids present. These fatty acids may be saturated (no double bonds) or unsaturated to varying degrees. The most common fatty acids in plants range from 4 to 26 carbons in size (Table 4.9), with oleic and linoleic being the most prevalent in nature. Their structure has a zigzag configuration (Figure 4.17) and double bonds, as seen in the example of oleic and linoleic acids, tend to result in curvature of the molecule.

Several methods are utilized to designate fatty acids. They may be referred to by their common name, systematic name or quite commonly by the use of an abbreviation. The short or abbreviated form simply denotes the number of carbon atoms in the molecule and the number of double bonds present. For example, linoleic acid (18:2) has 18 carbons with 2 double bonds. In some cases, the position of the double bond may be designated next to the common name of the unsaturated fatty acids (Table 4.9).

Most fatty acids have an even number of carbons although trace amounts of straight-chain, odd-numbered carbon compounds from C_7 to C_{35} have been detected. For example in pecan kernels, $C_{15:0}$, $C_{15:1}$, $C_{15:2}$, $C_{17:0}$, $C_{17:1}$, $C_{17:2}$ and $C_{21:0}$ fatty acids have been found (Table 4.10) but their concentration represents only approximately 2.1% of the total fatty acids.[209,210]

Table 4.8. Lipid Content of Several Types of Harvested Products.*

Product	Lipid Content of Edible Portion	
	% Dry weight	*% Fresh weight*
Fruits		
avocado	63.0	16.4
banana	0.8	0.2
olive	69.0	13.8
Seeds		
peanut	50.3	47.5
rice	0.5	0.4
walnut	61.2	59.3
Leaves		
amaranth	3.8	0.5
cabbage	2.6	0.2
lettuce	2.2	0.1
Roots and tubers		
parsnip	*2.4*	*0.5*
potato	0.4	0.1
radish	1.8	0.1

* *Source:* Data from Watt and Merril.[242]

Table 4.9. Common Fatty Acids in Harvested Plant Products.

Abbreviation	Systematic Name	Common Name	Formula
Saturated Fatty Acids			
4:0	Butanoic	Butyric acid	$CH_3(CH_2)_2COOH$
6:0	Hexanoic	Caproic acid	$CH_3(CH_2)_4COOH$
8:0	Octanoic	Caprylic acid	$CH_3(CH_2)_6COOH$
10:0	Decanoic	Capric acid	$CH_3(CH_2)_8COOH$
12:0	Dodecanoic	Lauric acid	$CH_3(CH_2)_{10}COOH$
14:0	Tetradecanoic	Myristic acid	$CH_3(CH_2)_{12}COOH$
16:0	Hexadecanoic	Palmitic acid	$CH_3(CH_2)_{14}COOH$
18:0	Octadecanoic	Stearic acid	$CH_3(CH_2)_{16}COOH$
20:0	Eicosanoic	Arachidic acid	$CH_3(CH_2)_{18}COOH$
22:0	Docosanoic	Behenic acid	$CH_3(CH_2)_{20}COOH$
24:0	Tetracosanoic	Lignoceric acid	$CH_3(CH_2)_{22}COOH$
Unsaturated Fatty Acids			
16:1	9-Hexadecenoic*	Palmitoleic acid	$CH_3(CH_2)_5CH=CH(CH_2)_7COOH$
18:1	9-Octadecenoic	Oleic acid	$CH_3(CH_2)_7CH=CH(CH_2)_7COOH$
18:2	9, 12-Octadecadienoic	Linoleic acid	$CH_3(CH_2)_3(CH_2CH=CH)_2(CH_2)_7COOH$
18:3	9, 12,15-Octadecatrienoic	Linolenic acid	$CH_3(CH_2CH=CH)_3(CH_2)_7COOH$

*The double bond is between carbons 9 and 10.

4.2. Triacylglycerols

Triacylglycerols (previously called triglycerides) are comprised of three fatty acids linked through ester bonds to a glycerol molecule. Mono- and diacylglycerols may also be present with the fatty acid-free position(s) on the glycerol molecule being bonded to other compounds, thus yielding a wide range of other types of lipids.

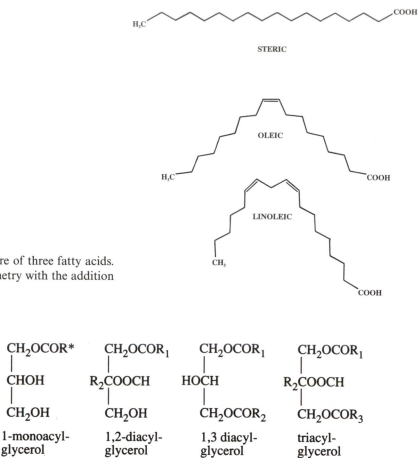

Figure 4.17. The structure of three fatty acids. Note the change in geometry with the addition of each double bond.

$$
\begin{array}{ccccc}
CH_2OH & CH_2OCOR^* & CH_2OCOR_1 & CH_2OCOR_1 & CH_2OCOR_1 \\
| & | & | & | & | \\
CHOH & CHOH & R_2COOCH & HOCH & R_2COOCH \\
| & | & | & | & | \\
CH_2OH & CH_2OH & CH_2OH & CH_2OCOR_2 & CH_2OCOR_3 \\
\text{glycerol} & \text{1-monoacyl-} & \text{1,2-diacyl-} & \text{1,3 diacyl-} & \text{triacyl-} \\
 & \text{glycerol} & \text{glycerol} & \text{glycerol} & \text{glycerol}
\end{array}
$$

*R = alkyl group

Triacylglycerols that are present in nature are normally mixtures and these mixtures are often very complex due to the different fatty acids that can be esterified to each of the three hydroxyl positions. For example, the combination of only two fatty acids at all possible positions on the molecule yields six potential triacylglycerol compounds.

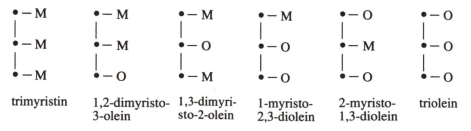

| trimyristin | 1,2-dimyristo-3-olein | 1,3-dimyristo-2-olein | 1-myristo-2,3-diolein | 2-myristo-1,3-diolein | triolein |

4.3. Neutral Lipids

The neutral lipids are largely triacylglycerols that make up the fats and oils found in plants. Waxes are also grouped here; however, due to their distinctly different physiological role in harvested products, they will be treated separately. As a group, neutral lipids do not have

Table 4.10. Fatty Acid Composition of Pecans.*

Abbreviation	Systematic Name	Common Name	Concentration (g/100 g nut meat)
Saturated fatty acids			
10:0	decanoic	capric	<0.20
12:0	dedecanoic	lauric	<0.20
14:0	tetracecanoic	myristic	1.20
15:0	pentadecanoic		<0.2
16:0	hexadecanoic	palmitic	4.09
17.0	heptadecanoic	margaric	0.27
18:0	octadecanoic	stearic	1.51
20:0	eicosanoic	arachidic	0.44
21:0	heneicosanoic		<0.20
Unsaturated fatty acids			
12:1	dedecenoic		<0.20
14:1	tetradecenoic		<0.20
14:2	tetradecadienoic		<0.20
15:1	pentadecenoic		<0.20
15:2	pentadecadienoic		<0.20
16:1	hexadecenoic	palmitolic	0.42
16:2	hexadecadienoic		<0.20
17:1	heptadecenoic		0.26
17:2	heptadecadienoic		<0.20
18:1	octadecenoic	oleic	37.90
18:2	octadecadienoic	linoleic	22.53
18:3	eicosenoic		0.54
20:2	eicosodienoic		<0.20

*Source: Data from Senter and Horvat.[209, 210] Data for the major fatty acids represent means of six cultivars.

charged functional groups attached. The distinction between fats and oils is based simply on their physical form at room temperature. Oils are liquid and tend to contain a larger percentage of unsaturated fatty acids (e.g., oleic, linoleic and linolenic) than fats that are solid at room temperature. Fats have saturated fatty acids as their major component.

The role of these storage lipids has been studied most extensively in oil containing seed crops (linseed, canola, castor bean). Here they represent a source of energy and carbon skeletons during the germination period. In the avocado, which has high levels of lipids in the mesocarp tissue, existing evidence suggests that it does not represent a source of respiratory substrate during the ripening process.

4.4. Phospholipids and Glycolipids

Phospholipids and glycolipids are important components of cellular membranes. Phospholipids are often diacylglycerols that yield inorganic phosphate upon hydrolysis. They are characteristically high in the fatty acid linoleic. The fatty acid portion forms the hydrophobic tail which strongly influences the orientation of the molecule. Common examples of phospholipids in plants are phosphatidyl serine, phosphatidyl ethanolamine, phosphatidyl choline, phosphatidyl glycerol and phosphatidyl inositol. They represent important components of both the cytoplasmic and mitochondrial membranes.

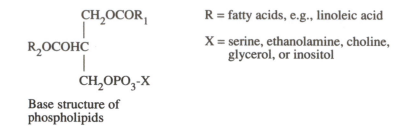

Base structure of
phospholipids

Glycolipids, in contrast, have carbohydrate substitutions without phosphate and are important components of the chloroplast membranes. Common examples would be mono-galactosyldiglyceride and digalactosyldiglyceride. Approximately 70% of the fatty acid component in these photosynthetic membranes is linolenic acid.

4.5. Waxes, Cutin and Suberin

The outer surface of plants is protected by three general types of lipid compounds.[111] Waxes, typically esters of a larger molecular weight fatty acid and a higher aliphatic alcohol, and cutin, hydroxy fatty acid polymers, act as a protective coating on much of the above ground parts of the plant. Similarly, suberin, a lipid-derived polymeric material, is found on underground plant parts and on healed surfaces of wounds. Like cutin, suberin is often imbedded with waxes.[238]

Plant waxes are extremely important during the postharvest storage and marketing of plant products in that they limit water loss from the tissue and impede pathogen invasion. As a group, waxes are chemically very heterogeneous. In addition to being esters of a higher fatty acid and a higher aliphatic alcohol, waxes contain alkanes, primary alcohols, long chain free fatty acids and other groups. Alkanes may represent over 90% of the hydrocarbons in waxes of apple fruit and *Brassica oleracea*.[131,162] Some plant products (e.g., rutabaga, orange, pineapple) may benefit from the application of supplemental wax after harvest due to insufficient natural waxes on the product, inadvertent removal of surface waxes during washing operations, or the desire to enhance surface glossiness.

The precise chemistry and structure of both cutin and suberin are not fully elucidated, although substantially more is known about the former. Both represent very complex heterogeneous compounds. Cutin, a polyester, is composed primarily of C_{16} and/or C_{18} monomers.[111] General differences in the composition of cutin and suberin are given in Table 4.11 with tentative models in Figure 4.18.

4.6. Fatty Acid and Lipid Synthesis

Fatty acids are synthesized within the cytosol of the cell and in many cases within certain plastids (e.g., chloroplasts, chromoplasts). There are two distinct pathways, one forming saturated

Table 4.11. Differences in the Monomer Composition of Cutin and Suberin.*

	Cutin	*Suberin*
Dicarboxylic acids	Minor	Major
In-chain-substituted acids	Major	Minor
Phenolics	Low	High
Very long-chain (C_{20}-C_{26}) acids	Rare and Minor	Common and Substantial
Very long-chain alcohols	Rare and Minor	Common and Substantial

*Source: Kolattukudy.[130]

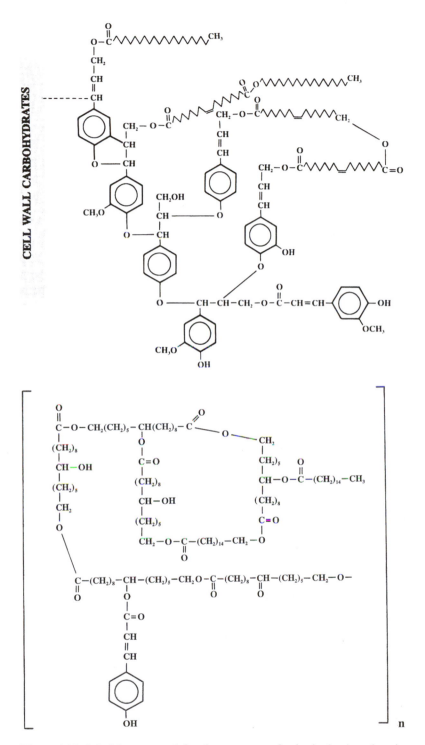

Figure 4.18. Models proposed for the structure of suberin (top) and cutin (bottom) (*redrawn from Kolattukudy*[130]).

fatty acids and the other unsaturated fatty acids.[86] To form saturated fatty acids, acetyl subunits are condensed *via* a series of enzymatically controlled steps forming fatty acids of up to 16 carbons in length.[203] Longer fatty acids require a separate chain lengthening sequence, usually building from a palmitic acid base unit. With the synthesis of unsaturated fatty acids, the introduction of the first double bond forming oleic acid is fairly well established. This can occur *via* one of two options, an anaerobic system or an aerobic one. Introduction of the second and third double bonds forming linoleic and α-linolenic acids, however, is not well understood.[86]

Triacylglycerols are synthesized in plants either directly from carbohydrates or through the modification of existing glycerides.[177,178] The glycerol backbone is derived from glycerol-3-phosphate from either the glycolytic or pentose phosphate pathway, or by direct phosphorylation of glycerol. Esterification of positions 1 and 2 (addition of a fatty acid) is by specific acyltransferases. In some cases the monoacyl (position one) is formed from dihydroxyacetone phosphate in a separate two-step sequence; however, the number 2 position is esterified by a 2-specific acyltransferase. The phosphate moiety is then cleaved from the molecule by phosphaitidate phosphohydrolase to yield diacylglycerol that then serves as a precursor for triacylglycerols, phosphoglycerides and glycosylglycerides. The final step in the synthesis of a triacylglycerol is the esterification of the 3-hydroxy position by a 3-specific transferase.[176]

4.7. Lipid Degradation

Each lipid class undergoes varying degrees of degradation during the postharvest period as the product approaches senescence or in the case of seeds, as they begin to germinate. Of the constituent lipids, the storage lipids are known to undergo marked changes in many products. Storage lipids are composed primarily of triacylglycerols that represent the most common lipids found in the plant kingdom, although they are not the predominant form in all plants. In that their degradation is fairly well understood, the discussion will focus on this class of lipids.

The first step in the recycling of carbon stored as triacylglycerols is the removal (hydrolysis) of the three acyl (fatty acid) units from the glycerol molecule.[89] This is accomplished through the action of the enzyme lipase (acyl hydrolase) or more specifically triacylglycerol acyl hydrolase.[69] The sequence involves the removal of the fatty acid from the number 3 position yielding a 1,2 diacylglycerol, followed by the removal of a second fatty acid yielding either 1- or 2-monoacylglycerol. The final acyl group is then hydrolyzed leaving glycerol. The free acids can then be converted to acetate, a starting point for a number of synthetic reactions and a respiratory substrate. In addition, acetate can also be converted to sucrose, the primary transport form of carbon in plants and an essential step during the germination of many seeds.

Free fatty acids can be metabolized by several possible mechanisms in the plant.[89] The most prevalent mechanism for the degradation of fatty acids is β-oxidation (Figure 4.19). It results in the formation of acetyl-CoA, in which a major portion of the stored energy remains trapped in the thioester bond. This trapped energy can either be converted to ATP by the movement of acetyl-CoA through the tricarboxylic acid cycle or acetyl-CoA can move through the glyoxylate cycle and provide carbon skeletons for synthetic reactions. The second option (i.e., *via* the glyoxylate cycle) does not appear to be operative in most harvested products, however, it is extremely important in germinating seed that are high in lipids.

The β-oxidation reactions occur both in the cytosol and in the glyoxysomes in many oil containing seeds. In this scheme the two terminal carbons of the fatty acid are cleaved sequentially, moving down the chain. With each acetyl-COA cleaved, 5 ATP equivalents (1 $FADH_2$ and 1 NADH) are produced. In the initial step of the reaction (Figure 4.19), catalyzed

β-OXIDATION

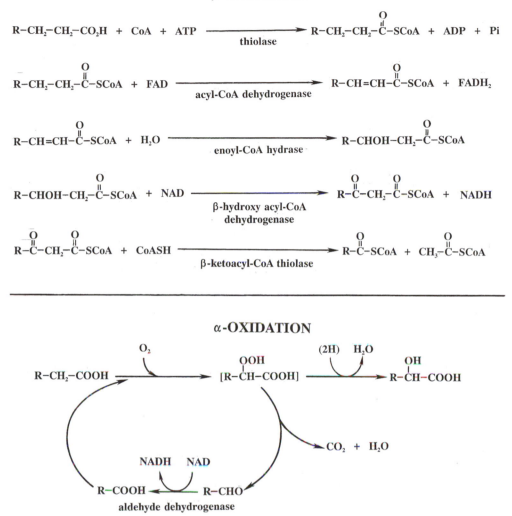

Figure 4.19. A comparison of β-oxidation and α-oxidation of free fatty acids (*redrawn from Shine and Stumpf*[216]).

by the enzyme thiolase, the free fatty acid combines with coenzyme A, a step requiring ATP to form acyl-CoA. Only one ATP is required to activate the fatty acid for complete degradation regardless of the number of carbon atoms in the chain. The next step is an oxidative reaction, catalyzed by acyl-CoA dehydrogenase that produces a double bond between the number 2 and 3 carbons and results in the formation of $FADH_2$. This is followed by the addition of water to the double bond (carbon 3) by enoyl-CoA hydrase and subsequent oxidation of the hydroxyl of the number 3 carbon, with the production of NADH. In the final step, acetyl-CoA (the two terminal carbons) is cleaved from the fatty acid molecule. This series of steps is then repeated, removing additional acetyl-CoA's.

Free fatty acids may also be degraded yielding CO_2, H_2O and energy by the α-oxidation pathway; however, its role appears to be only minor at least from the standpoint of energy production from stored lipids.[169] Unlike β-oxidation where the reactions involve an acyl thioester,

Table 4.12. Hydroperoxides and Aldehydes (with Single Oxygen Function) Possibly Formed in Autoxidation of Some Unsaturated Fatty Acids.*

Fatty acid	Methylene Group Involved	Isomeric Hydroperoxides Formed From Structures Contributing to Intermediate Free Radical Resonance Hydrid	Aldehydes Formed by Decomposition of Hydroperoxides
Oleic	11	11-Hydroperoxy-9-ene	Octanal
		9-Hydroperoxy-10-ene	2-Decenal
	8	8-Hydroperoxy-9-ene	2-Undecenal
		10-Hydroperoxy-8-ene	Nonanal
Linoleic	11	13-Hydroperoxy-9,11-diene	Hexanal
		11-Hydroperoxy-9,12-diene	2-Octenal
		9-Hydroperoxy-10,12-diene	2,4-Decadienal
Linolenic	14	16-Hydroperoxy-9,12,14-triene	Propanal
		14-Hydroperoxy-9,12,15-triene	2-Pentenal
		12-Hydroperoxy-9,13,15-triene	2,4-Heptadienal
	11	13-Hydroperoxy-9,11,15-triene	3-Hexenal
		11-Hydroperoxy-9,12,15-triene	2,5-Octadienal
		9-Hydroperoxy-10,12,15-triene	2,4,7-Decatrienal
Arachidonic	13	15-Hydroperoxy-5,8,11,13-tetraene	Hexanal
		13-Hydroperoxy-5,8,11,14-tetraene	2-Octenal
		11-Hydroperoxy-5,8,12,14-tetraene	2,4-Decadienal
	10	12-Hydroperoxy-5,8,10,14-tetraene	3-Nonenal
		10-Hydroperoxy-5,8,11,14-tetraene	2,5-Undecadienal
		8-Hydroperoxy-5,9,11,14-tetraene	2,4,7-Tridecatrienal
	7	9-Hydroperoxy-5,7,11,14-tetraene	3,6-Dodecadienal
		7-Hydroperoxy-5,8,11,14-tetraene	2,5,8-Tetradecatrienal
		5-Hydroperoxy-6,8,11,14-tetraene	2,4,7,10-Hexadecatetraenal

Source: Badings.[6]
Note: Only the most active methylene groups in each acid are considered.

α-oxidation acts directly on the free fatty acids. The proposed scheme for α-oxidation is presented in Figure 4.19.

In most cases, direct oxidation of fatty acids to CO_2 does not appear to be the primary physiological role of α-oxidation in plants. Under extreme conditions, for example, wounding of tissue slices, the initial rise in respiration appears to be largely due to α-oxidation. In normally metabolizing cells, α-oxidation is thought to function by creating odd-numbered fatty acids by the removal of a terminal carbon. It may also be used as an adjunct to β-oxidation for the removal of a carbon when the number 3 carbon of the fatty acid has a side group preventing the β-oxidation process.

4.7.1. Lipid Peroxidation

The oxidation of lipids in harvested plant products may occur either in biologically mediated reactions catalyzed by lipoxygenases or through direct chemical or photochemical reactions.[169] In enzymatically controlled oxygenation reactions, polyunsaturated fatty acids are attacked producing hydroperoxides that can be further degraded often forming characteristic tastes and odors, both desirable and undesirable (e.g., rancidity). For example, linolenic acid (18:3) can be oxidized forming a number of hydroperoxides that decompose into aldehydes

Lipid Peroxidation

LIPOXYGENASE REACTION

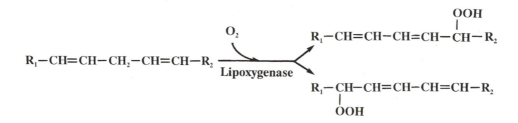

AUTOXIDATION

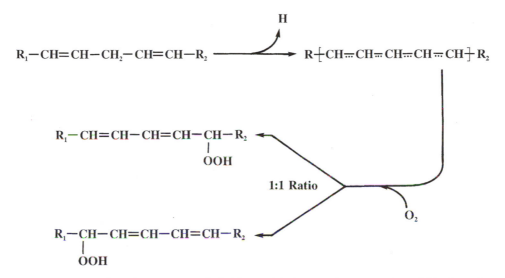

Figure 4.20. Enzymatic and non-enzymatic lipid peroxidation reactions occurring within the cell.

(Table 4.12). In cucumber fruit, linoleic acid is attacked by lipoxygenase forming both 9-hydroperoxide and 13-hydroperoxide which are cleaved to form the volatile flavor components *cis*-3-nonenal and hexenal, respectively.[70] While these enzymes occur widely in higher plants, the level of activity does not appear to follow any set botanical or morphological pattern. Very high activities have been recorded for bean seed, potato tubers, eggplant fruit and immature artichoke flowers.[185]

Peroxidation reactions also include autocatalytic oxidation, the direct reaction with oxygen, and a non-autocatalytic process mediated by light. As with enzymatic reactions, hydroperoxides are formed. The sequence of autoxidation is presented in Figure 4.20. In general, the rate of lipid oxidation is largely dependent upon the degree of unsaturation of the component fatty acids. Numerous other factors, both internal (e.g., antioxidants, pro-oxidants), and external (e.g., oxygen concentration, temperature and light intensity), exert a pronounced influence.[3]

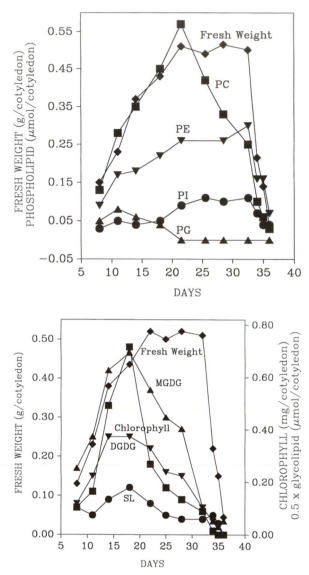

Figure 4.21. Changes in membrane phospholipids in cotyledons during the onset and development of senescence (PC = phosphatidyl choline; PE = phosphatidyl ethanolamine; PG = phosphatidyl glycerol; PI = phosphatidyl inositol) (*redrawn from Ferguson and Simon*[58]).

Figure 4.22. Changes in cotyledon chlorophyll and chloroplast glycolipids during the onset and development of senescence (MGDG = monogalactosyl diglyceride, DGDG = digalactosyl diglyceride; SL = sulpholipid) (*redrawn from Ferguson and Simon*[58]).

4.8. Galactolipases and Phospholipases

Plant cells contain enzymes capable of breaking down both glycolipids and phospholipids.[169] They function in the normal turnover of these molecules but also have additional roles. For example, galactolipases may be important in the breakdown of prolamellar bodies during the greening of etioplasts.[230] In addition, both types of enzyme appear to function during the onset of senescence of harvested products. During this period there is general breakdown of lipids with subsequent disorganization of the integrity of the cellular membranes.

Enzymes capable of attacking galactolipids, the predominant form of plant glycolipids, are found both in the cytosol and in the chloroplasts.[169] Lipolytic acyl hydrolase catalyzes the hydrolysis of the ester bonds yielding free fatty acids and the corresponding galactosylglycerols. The galactosylglycerols are subsequently attached by α- and β-galactosidases giving galactose and glycerol.

Phospholipids are degraded by four specific types of phospholipases, each of which attacks the parent molecule in a distinct manner. For example, the C type phospholipases cleave the ester bond of carbon 3 of glycerol and phosphoric acid. In the case of phosphatidycholine this yields a 1,2-diacylglycerol and phosphorylcholine.

4.9. Postharvest Alterations

Postharvest changes in the lipid fraction of high moisture crops have not been thoroughly studied. In oleaginous plant products such as pecan kernels (approximately 74% lipid), changes in lipids are largely qualitative rather than quantitative. Approximately 98% of the lipid fraction is triacylglycerols of which 90% are unsaturated.[209,210] Oxidation represents the primary qualitative alteration in lipids during the storage and marketing period. Large quantitative and qualitative changes in oil seed crops do occur during germination when stored lipids are recycled, however, this is not considered a postharvest alteration.

In the fruit of avocado, the composition of the oil does not change during maturation and storage. While there is a large increase in fruit respiration during the ripening climacteric, mesocarp lipids do not appear to represent the source of carbon utilized.[17]

Substantial changes in the lipids of non-oleaginous tissues occur during senescence. Here there are significant alterations in both glycolipids and phospholipids. In cucumber cotyledons, a model system used for studying lipid changes during the onset and development of senescence, phosphatidyl choline, the major phospholipid present, begins to disappear once the cotyledons reach their maximum fresh weight. As senescence progresses, rapid desiccation of the tissue begins. By this time, 56% of the phosphatidyl choline has been broken down and phosphatidyl ethanolamine begins to be metabolized (Figure 4.21). The glycolipids begin to be lost concurrently with chlorophyll, approximately two weeks prior to initial weight loss of the tissue (Figure 4.22).[58] These changes in lipid composition may mediate alterations in the structure of the membranes resulting in abnormal permeability and decreased activity of membrane associated enzymes, thus accelerating senescence.

5. PLANT PIGMENTS

Our lives are surrounded and in many ways dominated by plant colors. These colors are due to the presence of pigments within the plant and their interaction with light striking them. Sunlight is composed of a number of different wavelengths, the composite of which is called the spectrum (Figure 4.23). Within the visible portion of the spectrum, specific ranges in wavelength display individual colors (e.g., red, blue, yellow). Some wavelengths (colors) are absorbed by the pigments in a plant while others are either reflected or transmitted through the tissue. What is seen as a specific plant color, such as the blue of grape-hyacinth flowers, is due to the absorption by pigments of all of the other wavelengths in the visible spectrum except the blue region that is reflected from the flower tissue.

In biological tissues there are two types of reflected light, surface (also called regular or spectacular) and body reflectance influence our visual image of the object. With surface reflectance, the light striking the product is reflected from the surface without penetrating the tissue. This represents only about 4% of the light striking most biological samples. Much more important is body reflectance where the light actually penetrates into the tissue, becomes diffused (spread out in all directions) upon interacting with internal surfaces and molecules and is eventually either absorbed or reaches the surface and escapes from the tissue. Part of

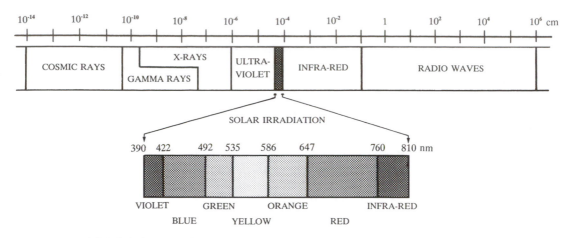

Figure 4.23. Visible light is a part of the spectrum of electromagnetic radiation from the sun. Radiation travels in waves; the length of each wave ranges from very short in gamma or cosmic rays (10^{-8} to 10^{-14}) to extremely long as in radio waves (1 to 10^6 cm). What is perceived as the color of a harvested product is due to the absorption of part of the visible spectrum by pigments in the surface cells and reflection (body reflectance) of the remainder, part of which is detected by the human eye.

this light is absorbed by the component pigments, while the remainder moves back out of the tissue becoming the color we perceive.

Plant pigments can be separated into four primary classes based on their chemistry, the chlorophylls, carotenoids, flavonoids and betalains, the latter being fairly limited in distribution (Table 4.13). There are, of course, additional pigments (e.g., quinones, phenalones, phyrones and others), however, these typically make only a minor contribution to the color of plants.

In nature, pigments serve a number of functions. The chlorophylls and carotenoids trap light energy in photosynthesis. The carotenoids are integral components of light harvesting complexes that increase the overall efficiency of photosynthesis[62,99] and also act as protecting agents for chlorophyll molecules to prevent photooxidation. Pigments are of paramount importance in many plant species for their role in facilitating pollination.[189] Flower colors are known to attract certain insects, birds and in some cases, bats, which pollinate the flower. In the wild, the color of some fruits and seeds also plays an important role in dispersal. Dispersal helps to minimize the potential for competition between the parent and its progeny, enhancing the potential for colonization of new areas.

For man, color, form and freedom from defects are three of the primary parameters used

Table 4.13. Plant Pigments.

Pigment	Color	Cellular Localization	Solubility
Chlorophyll	Blue-green, yellow-green	Chloroplasts	Insoluble in water, soluble in acetone, ether, alcohols
Carotenoids	Yellow, orange- red	Chloroplasts, chromoplasts	Insoluble in water, soluble in acetone, ether, alcohols
Flavonoids	Yellow, orange, red, blue	Vacuole	Water soluble
Betalains	Yellow, orange, red, violet	Vacuole, cytosol	Water soluble

to ascertain quality in postharvest plant products. Pigmentation provides us with quality information, such as the degree of ripeness of fruits (e.g., banana) or the mineral nutrition of ornamentals. It is especially important with ornamentals since our visual impressions are almost exclusively relied upon for quality judgment.

5.1. Classes of Pigment Compounds

5.1.1. Chlorophyll

The plant world is dominated by the color green due to the presence of the chlorophyll pigments that absorb red and blue wavelengths. The chlorophylls are the primary light-accepting pigments found in plants that carry out photosynthesis through the fixation of carbon dioxide and the release of oxygen. In nature, there are two predominate forms, chlorophyll a and chlorophyll b, differing only slightly in structure (Figure 4.24). Both are normally found concurrently within the same plant and usually at a 2.5–3.5:1 ratio between a and b. Two other chlorophylls, c and d, are found in a relatively limited number of species. For example, chlorophyll c is found in several marine plants.

Each form of chlorophyll is a magnesium containing porphyrin formed from 4 pyrrole rings (Figure 4.24). Attached to the propionyl group of a pyrrole ring of chlorophylls a, b and d is a 20 carbon phytol chain ($C_{20}H_{39}OH$). Chlorophylls a and b differ structurally only in the replacement of a methyl group on chlorophyll a with an aldehyde (-CHO). Unlike the flavonoid pigments, the chlorophylls are hydrophobic and hence, not soluble in water. The primary function of chlorophyll is to absorb light energy and convert it to chemical energy; a process that occurs in the chloroplasts.

5.1.2. Carotenoids

The carotenoids are a large group of pigments associated with chlorophyll in the chloroplasts and are also found in other chromoplasts. Their colors range from red, orange, and yellow to brown and are responsible for much of the autumn leaf pigmentation.

Chemically carotenoids are terpenoids comprised of eight isoprenoid units (Figure 4.25). Nearly all carotenoids are composed of 40 carbon atoms. They are divided into two subgroups, the carotenes and their oxygenated derivatives, the xanthophylls. Both are insoluble in water, although the xanthophylls tend to be less hydrophobic than the carotenes.

The nomenclature of the carotenoids is based on their 9 carbon end groups of which there are 7 primary types (Figure 4.25). These can be arranged in various combinations on the methylated straight chain portion of the molecule, for example α-carotene is β,ε-carotene while β-carotene is β,β-carotene. Lutein, a common xanthophyll, is structurally like α-carotene, differing only in the presence of a hydroxyl group on carbon 3 of both the β and ε end groups. Thus, a tremendous range in potential variation is possible and in some tissues, quite large assortments of specific compounds have been found. For example, in the juice of the 'Shamouti' oranges, 32 carotenoids have been identified (Table 4.14).

In photosynthetic tissue, carotenoids function both in the photosynthesis process *per se,* and as protectants, preventing the chlorophyll molecules from being oxidized (photooxidation) in the presence of light and oxygen. In flowers and fruits, carotenoids appear to act as attractants that aid in securing pollination or dispersal; however, in underground structures such as roots and tubers, their role is not understood.

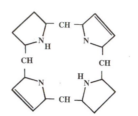

PORPHYRIN BASE STRUCTURE

PYRROLE RING

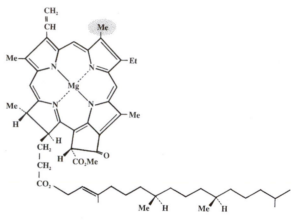

CHLOROPHYLL A

Figure 4.24. The chlorophylls are magnesium containing porphyrins derived from four pyrrole rings. The two prevalent chlorophylls, a and b, differ only in a single side group (shaded). Chlorophyll c, however, is structurally quite distinct from a, b, and d in that the phytol tail is not present.

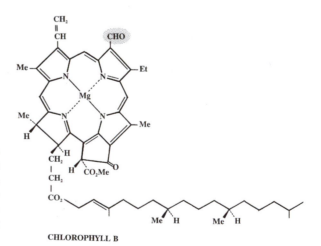

CHLOROPHYLL B

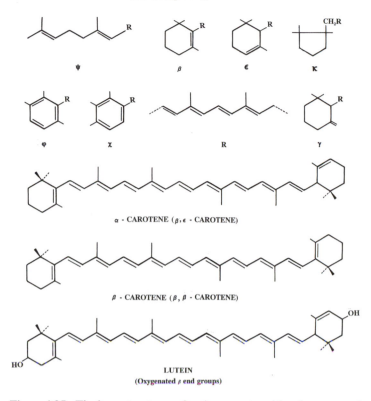

End Group Designation

ψ   β   ε   κ

φ   χ   R   γ

α - CAROTENE (β, ε - CAROTENE)

β - CAROTENE (β, β - CAROTENE)

LUTEIN
(Oxygenated β end groups)

Figure 4.25. The base structures of various carotenoid end groups and the representative structures of two carotenes (α-carotene and β-carotene). The figure shows the end group designation in naming. Xanthophylls, carotenes oxygenated on one or both end groups, are represented by the pigment lutein.

Table 4.14. Quantitative Composition of Carotenoids in the Juice of 'Shamouti' Oranges.*

Carotenoid	(% of total) Carotenoids	Carotenoid	(% of total) Carotenoids
Mutatoxanthin	15.13	α-Carotene	1.24
Cryptoxanthin	12.88	poli-cis-Cryptoxanthin	1.05
Trollixanthin	9.64	Carbonyl 422	0.78
Luteoxanthin-like	6.92	Cryptoflavin	0.70
Antheraxanthin	6.20	Auroxanthin	0.58
Phytoene	5.70	OH-α-Carotene	0.50
Lutein	5.20	Pigment 426	0.39
Isolutein	4.00	α-Carotene	0.36
Luteoxanthin	3.98	Citraurin	0.11
Neoxanthin	3.97	Cryptoxanthin diepoxide	0.10
Violaxanthin	3.00	Chrysanthemaxanthin	0.07
OH-Sintaxanthin	2.48	Sintaxanthin	0.07
Phytofluene	2.40	Rubixanthin	0.05
cis-Cryptoxanthin	2.00	β-Apo-10'-carotenal	0.03
Trollichrome	1.46	Mutatochrome	0.02
β-Carotene	1.40		

*Source: Gross et al.[76]

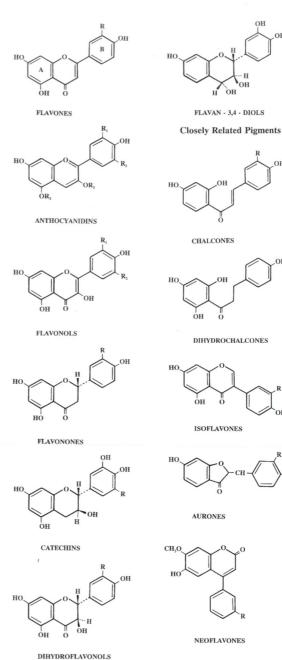

Figure 4.26. The basic structure of flavonoid pigments. Each consists of 2 benzene rings (A and B) joined by a 3 carbon link. The various classes of flavonoids differ only in the state of oxidation of the 3 carbon link. Within each class are a wide range of individual pigments varying in the number and position of groups (e.g., OH, CH_3, etc) attached to the 2 rings. Also illustrated are several closely related pigments (e.g., chalcones, dihydrochalcones, isoflavones, aurones and neoflavones).

5.1.3. *Flavonoids*

While green is the dominant color in plants, other colors have tremendous attraction both for man and other animals. Many of the intense colors of flowers, fruits, and some vegetables are the result of flavonoid pigments and closely related compounds. This large class of water-soluble compounds has a diverse range of colors. For example, there are yellows, reds, blues,

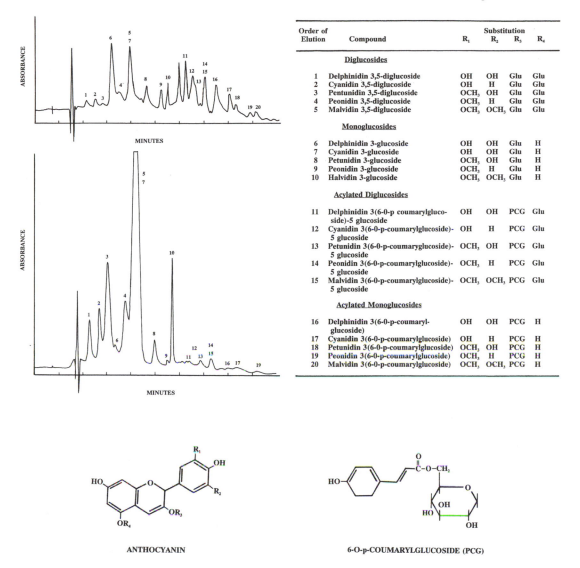

Order of Elution	Compound	Substitution			
		R₁	R₂	R₃	R₄
	Diglucosides				
1	Delphinidin 3,5-diglucoside	OH	OH	Glu	Glu
2	Cyanidin 3,5-diglucoside	OH	H	Glu	Glu
3	Pentunidin 3,5-diglucoside	OCH₃	OH	Glu	Glu
4	Peonidin 3,5-diglucoside	OCH₃	H	Glu	Glu
5	Malvidin 3,5-diglucoside	OCH₃	OCH₃	Glu	Glu
	Monoglucosides				
6	Delphinidin 3-glucoside	OH	OH	Glu	H
7	Cyanidin 3-glucoside	OH	OH	Glu	H
8	Petunidin 3-glucoside	OCH₃	OH	Glu	H
9	Peonidin 3-glucoside	OCH₃	H	Glu	H
10	Halvidin 3-glucoside	OCH₃	OCH₃	Glu	H
	Acylated Diglucosides				
11	Delphinidin 3(6-0-p coumarylglucoside)-5 glucoside	OH	OH	PCG	Glu
12	Cyanidin 3(6-0-p-coumarylglucoside)-5 glucoside	OH	H	PCG	Glu
13	Petunidin 3(6-0-p-coumaryglucoside)-5 glucoside	OCH₃	OH	PCG	Glu
14	Peonidin 3(6-0-p-coumarylglucoside)-5 glucoside	OCH₃	H	PCG	Glu
15	Malvidin 3(6-0-p-coumarylglucoside)-5 glucoside	OCH₃	OCH₃	PCG	Glu
	Acylated Monoglucosides				
16	Delphinidin 3(6-0-p-coumaryl-glucoside)	OH	OH	PCG	H
17	Cyanidin 3(6-0-p-coumarylglucoside)	OH	H	PCG	H
18	Petunidin 3(6-0-p-coumarylglucoside)	OCH₃	OH	PCG	H
19	Peonidin 3(6-0-p-coumarylglucoside)	OCH₃	H	PCG	H
20	Malvidin 3(6-0-p-coumarylglucoside)	OCH₃	OCH₃	PCG	H

ANTHOCYANIN

6-O-p-COUMARYLGLUCOSIDE (PCG)

Figure 4.27. A comparison of the anthocyanins of two grape cultivars (*top,* Concord and *bottom,* DeChaunac) separated by high preformance liquid chromatography. Substitutions in the basic structure of the anthocyanin molecule occur at any of the four positions, denoted as R₁–R₄. These substitutions are given for 20 of the anthocyanins found in the two grape cultivars. Part of the anthocyanins have an organic acid esterified to the hydroxyl group of their attached sugar (termed acylated). In this case, p-coumaric (PCG), a derivative of cinnamic acid, is present (*redrawn from Williams and Hrazdina*[247]).

and oranges. Numerous variations in color are derived both from structural differences between compounds and the relative concentration of specific pigments within the cells. The flavonoids are generally found in the vacuole.

The basic structure of flavonoid pigments is presented in Figure 4.26. It consists of 2 benzene rings (A and B) joined by a 3-carbon link which forms a γ-pyrone ring through oxygen. Various classes of flavonoids differ only in the state of oxidation of the 3-carbon link, while individual compounds within these classes differ mainly by the number and orientation of the hydroxy,

Betacyanins

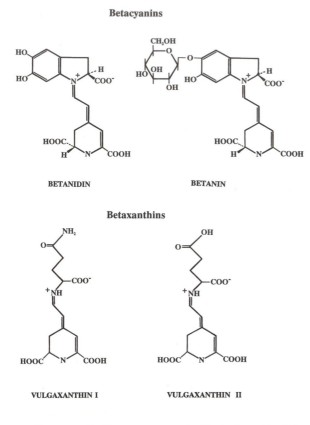

BETANIDIN   BETANIN

Betaxanthins

VULGAXANTHIN I   VULGAXANTHIN II

Figure 4.28. Examples of the two primary structural groups of betalains, the betacyanins and the betaxanthins. Betanidin, vulgaxanthin I, and vulgaxanthin II are found in the root of the common garden beet (*redrawn from Piattelli et al.*[183, 184]). Epimers of betacyanins are formed by alterations in the configuration, an example would be the epimer of betanin, isobetanin.

methoxy and other groups substituted on the 2 benzene rings. Individual classes of flavonoids include anthocyanidins, flavones, catechins, flavonols, flavanones, dihydroflavonols and the flavan-3,4-diols or proanthocyanidins (Figure 4.26). Most flavonoid pigments exist in live plant tissue as glycosides where one or more of their hydroxyl groups are joined to a sugar. In some anthocyanins, an organic acid may be esterified to one of the hydroxyls on the sugar, giving an acylated compound. This is the case with grapes where *p*-coumaric acid, a derivative of cinnamic acid, can be found attached to both the mono- and diglucoside anthocyanin pigments (Figure 4.27). The attached sugar confers higher solubility and stability (reduced reactivity).

Closely related to flavonoid compounds are the chalcones, dihydrochalcones, isoflavones, neoflavones and aurones (Figure 4.26). These do not have the 2-phenylchroman structure but are closely akin both chemically and in their biosynthesis.

5.1.4. Betalains

The betalains represent a fourth but substantially restricted group of plant pigments. They are found in the flower, fruits and in some cases in other plant parts, giving colors of yellow, orange, red and violet. Perhaps the best example is the red-violet pigment from the root of the beet, the first betalain isolated in crystalline form; hence the derivative name betalain.

As a group they are characterized by being water-soluble nitrogenous pigments that are found in the cytosol and in vacuoles. Chemically betalains are subdivided into two groups: the red-violet betacyanins illustrated by the structure of betanidin and betanin (Figure 4.28) and the yellow betaxanthins, characterized by vulgaranthin I and II. A number of the naturally occurring betaxanthins have the tail portion of the molecule either partially or totally closed into a ring structure as in dopaxanthin from the flowers of *Glottiphyllum longum* (Haw.) N.E. Br.[108]

While the precise function of the betalains is not known, it is possible that they may function like the anthocyanins in flowers and fruits enhancing insect or bird pollination and seed dispersal. No role has been presently ascribed for their presence in plant parts such as roots, leaves and stems though they may serve as antioxidants in the human diet.[116]

5.2. Pigment Biosynthesis and Degradation

5.2.1. Chlorophylls

Chlorophyll synthesis is modulated by a number of external influences, two of the most important being light and mineral nutrition. The initial pyrrole ring, porphobilinogen, is formed from two molecules of δ-amino levulinic acid derived from glycine and succinate (Figure 4.29). Four molecules of porphobilinogen are polymerized producing a ring structure, uroporphyrinogen, which has acetyl and propinoyl groups attached to each of the component pyrroles. After a series of decarboxylation reactions, protoporphyrin is formed and in subsequent steps, magnesium is inserted followed by the addition of the phytol tail. Chlorophyll a, which is blue-green in color, differs from chlorophyll b (yellow-green) only in the presence of a single methyl group instead of a formyl group.

Decomposition of chlorophyll may, in many cases, be quite rapid and dramatic in effect as in the autumn coloration of deciduous trees in the northern temperate zones or the ripening of bitter melon. In many tissues this loss of chlorophyll is part of a transition of the chloroplasts into chromoplasts containing yellow and red carotenoid pigments. The loss of chlorophyll can be mediated through several processes, such as the action of the enzyme chlorophyllase, enzymatic oxidation or photodegradation.[157] While the precise sequence of biochemical steps are not known, the initial reactions appear to be much like the reverse of the final steps in the synthesis pathway. Phytol may be removed to yield chlorophyllide or both magnesium and phytol to give pheophoride. In subsequent degradative steps, the low molecular weight products that are formed are colorless.

5.2.2. Carotenoids

Carotenoids, typically 40 carbon compounds, are built up from 5 carbon isoprene subunits, the most important of which is isopentenyl pyrophosphate. These initial subunits are formed in a series of steps from acetyl-CoA and acetoacetyl-CoA in the terpenoid pathway.[20,36,38] Isoprene subunits are sequentially added, build up to a 20 carbon intermediate, geranylgeranyl pyrophosphate, two of which condense to give phytoene with the typical carotenoid skeleton (Figure 4.30). Subsequent steps involve ring closure and, in the case of the xanthophylls, the addition of one or more oxygens.[199] Some carotenoids (e.g., carotenols) may be esterified to long chain fatty acids (e.g., oleate or palmitate) or other compounds. This occurs with leaf coloration in the fall[231] and in the peel of apple fruit during ripening. While carotenol esters were once thought to represent breakdown products formed during senescence (i.e., esterified with fatty acids formed during membrane degradation), current evidence points toward a controlled synthesis of the pigments.[128]

In xanthophyll synthesis, oxygen is added to the cyclic portions of the molecule, from carbon 1 to carbon 6 and often as hydroxyls. For example, the most common structural feature

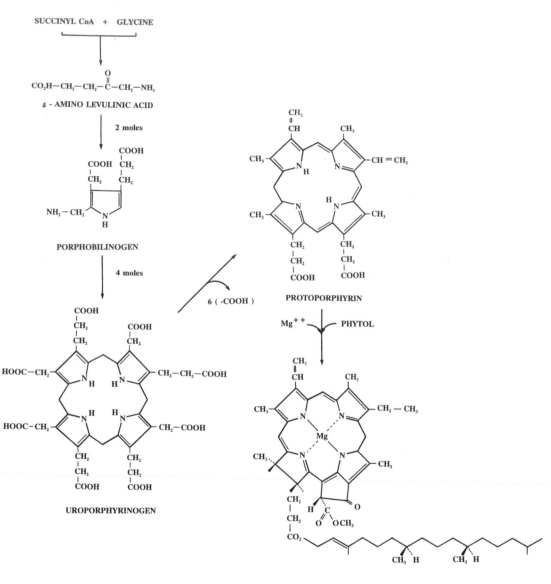

Figure 4.29. The general biosynthetic pathway for chlorophyll. Insertion of Mg^{++} and the addition of the phytol tail occur in the last series of steps.

of xanthophylls is the presence of a hydroxyl at the C-3 and C-3′ positions giving lutein (β,ε carotene-3,3′-diol) and zeaxanthin (β,β-carotene-3,3′-diol). Because of the large number of potential combinations for the placement of one or more oxygens, there are numerous xanthophylls. Of the over 300 naturally occurring carotenoids that have been identified, approximately 87% are xanthophylls.

The stability of carotenoids is highly variable. In some cases, such as in narcissus flowers, degradation occurs in only a few days,[19] while with dried corn kernels over 50% of the carotenoids remain after 3 years of storage.[191] A number of factors affect the rate of loss of carotenoids. These include the specific type of pigment, storage temperature, product moisture

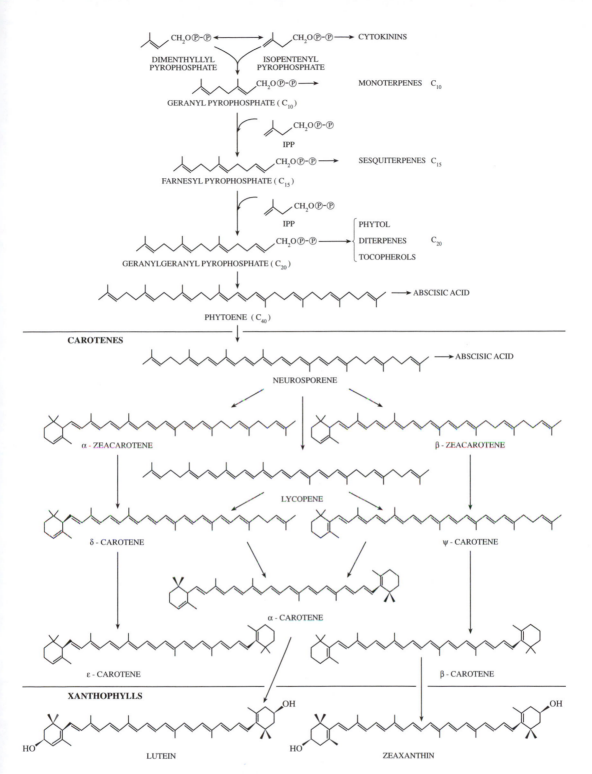

Figure 4.30. Generalized pathway for the biosynthesis of carotenes and their oxygenation to form xanthophylls.

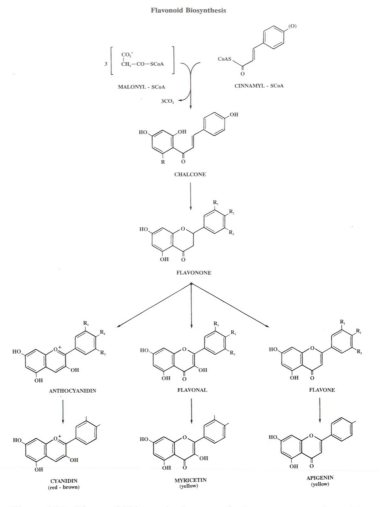

Figure 4.31. Flavonoid biosynthesis occurs in three stages starting with the combination of three malonyl-CoA molecules with one of cinnamyl-CoA. This is followed by ring closure forming a flavanone and conversion to the various classes of flavonoids. Final steps involve the formation of individual compounds with various additions to the two rings.

level, product type, prestorage treatments (e.g., drying of corn) and others. In edible products, the breakdown of β-carotene is of special concern due to its role as a precursor of vitamin A.

In the breakdown of carotenoids, the initial degradative steps involve oxygenation of the molecule. This differs from oxygenation reactions that occur largely on the ring structures during the formation of xanthophylls. Carotenoid degrading enzyme systems have been found in chloroplasts and mitochondria. The double bonds within the linear portion of the molecules are subject to attack by lipoxygenase[91] and activity is accelerated by the availability of oxygen, light and certain metals. A variety of shorter chain length terpenoids are formed, a number of which are volatile and in some cases, have distinct odors.[243]

As a group, xanthophylls are characteristically more stable than the carotenes. In leaf tissue in the autumn of the year, the xanthophylls are released into the cytoplasm upon the disruption of the chloroplasts. These molecules are subsequently esterified which appears to substantially enhance their stability, especially in contrast to the carotenes.

5.2.3. Flavonoids

Flavonoid biosynthesis begins with the formation of the basic $C_6C_3C_6$ skeleton through the combination of three malonyl-CoA molecules with one cinnamyl-CoA (Figure 4.31). This yields a chalcone and with ring closure, it is converted to a flavanone.[47] In the second series of steps, the flavanone may be converted to each of the different classes of flavonoids, e.g., flavone, flavonol, anthocyanidin. In the final stage these are converted to individual compounds, such as cyanidin, myricentin and apigenin.[61] The anthocyanins range from reds to blues in color. As the degree of methylation increases the individual compounds become increasingly red in color, while hydroxylation results in deeper blues. In addition, blue coloration can result from complexes formed by the chelation of Al^{+3} and Fe^{+3} to the hydroxyls of the A ring.

Of the flavonoids, the decomposition of anthocyanin has been studied in greatest detail.[47] The vulnerability of individual pigments to degradation tends to vary; substitutions at specific positions on the molecule can significantly affect its stability. For example, a hydroxyl at the 3' position enhances the pigments propensity for degradation.

Enzymes that have the potential to degrade anthocyanins have been isolated from a number of different tissues (e.g., flowers, fruits, and others). These tend to fall into two classes, glucosidases and polyphenol oxidases. Both have the ability to produce colorless products. Other possible mechanisms of pigment alteration and breakdown include pH alterations, which accompany ripening in some fruits, and attack of the charged portion of the molecule by naturally occurring nucleophiles (e.g., ascorbic acid).

5.2.4. Betalains

The betalains appear to be derived from 3,4-dihydroxyphenylalanine (L-DOPA). The betacyanins are formed from two of these molecules, one of which has an oxidative opening of the aromatic ring followed by subsequent closure to yield betalamic acid (Figure 4.32). The second forms cyclodopa and upon condensation with betalamic acid, yields betanidin, the base molecule for the formation of the various betacyanins. The betaxanthins are formed through the condensation of betalamic acid with an amino or imino group (other than cyclodopa).

Research on the breakdown of the betalains has centered to a large extent on color alterations that occur in beetroot, although these pigments may also be found in members of the Aizoaceae, Amaranthaceae, Basellaceae and Cactaceae families, which are of postharvest interest. This research was, in part, prompted by the elimination of the use of red dye number 2 in foods. Betalains were briefly considered as possible replacements. Nearly all of the studies to date have been done on either extracted pigments or processed tissue and, as a consequence, the extent to which these reactions can be implicated in color changes in stored fresh beetroot is not clear.

Beetroot tissue exposed to low pH (3.5–5.5) retains its color relatively well while at higher pH (7.5–8.5) discoloration occurs.[80] Action by β-glucosidase results in the removal of the sugar side group converting betanin and isobetanin to their aglucones, betanidin and isobetanidin. In addition, exposure to air and/or light results in the degradation of betalains, often causing a browning discoloration.

5.3. Postharvest Alteration in Pigmentation

During both the preharvest and postharvest period many products undergo significant changes in their pigment composition (Figures 4.33 and 4.34). These changes include both the

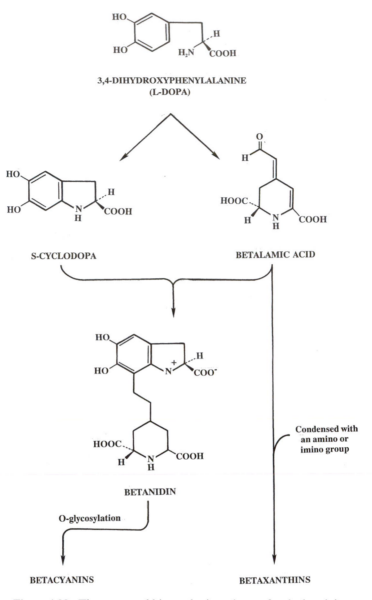

Figure 4.32. The proposed biosynthetic pathways for the betalains.

degradation of existing pigments and the synthesis of new pigments; in many cases, both processes may occur concurrently. Pigmentation changes are of paramount importance in many products in that they are used as a primary criterion for assessing quality.

The degradation of pigments can be subdivided into two general classes, pigments losses that are beneficial to quality and those that are detrimental. Many beneficial losses center around the degradation of chlorophyll with the concurrent synthesis of other pigments or the unmasking of pre-existing pigments already present with the tissue. Examples of this would be the degreening of oranges during which time carotenoids are being synthesized, and the loss of chlorophyll in banana, allowing the expression of the pigments that are already present. Detrimental losses of pigments after harvest can be seen in the color fading of flowers and in chlorophyll losses in broccoli florets or leaf crops. As with the degradation of pigments, the

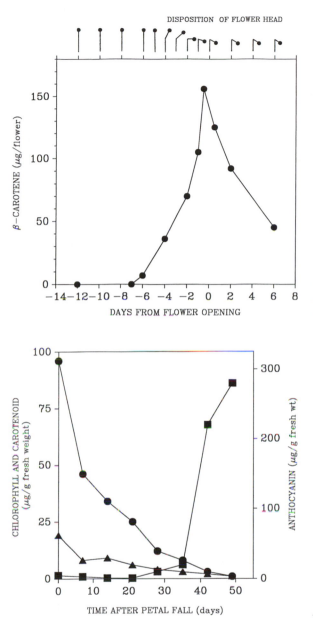

DISPOSITION OF FLOWER HEAD

Figure 4.33. Time course of the synthesis and degradation of β-carotene in the corona of *Narcissus majalis* Curtis flowers. Day 0 represents anthesis (*redrawn from Booth*[19]).

Figure 4.34. Changes in the pigment concentration in developing strawberry fruits (● chlorophyll; ▲ carotenoid; ■ anthocyanin (*redrawn from Woodward*[249]).

synthesis of pigments after harvest can be either beneficial or undesirable. The development of red coloration in the fruit of the tomato after harvest is highly desirable while the formation of chlorophyll in harvested potatoes or the synthesis of carotenoids in the bitter melon is undesirable.

Many postharvest factors affect the degree of change in pigmentation after harvest, the most important of which are light and temperature. Light is essential for the synthesis of chlorophyll and its presence delays the loss of these pigments in detached leaves. It also appears to be important in stimulating the synthesis of anthocyanins and lycopene in some products, however, not β-carotene in tomato fruits. Changes in the pigmentation of many tissues are temperature dependent. The effect of temperature, however, varies with the specific pigment, the tissue of interest and whether synthetic or degradative processes are operative. For example, in pink grapefruit high temperatures (30–35°C) favor the accumulation of lycopene but not carotene; at low temperatures (5 to 15°C) the opposite is true.[165] However, in the tomato temperatures above 30°C suppress the biosynthesis of lycopene, but not β-carotene.[232]

Several plant growth regulators have also been shown to have a significant effect on the pigmentation in some harvested products. The use of ethylene for the degreening of citrus and banana fruits and stimulating the synthesis of carotenoids (tomato) has become a major commercial postharvest tool in many areas of the world. Ethylene is also known to stimulate the formation of anthocyanins in the grape when applied prior to harvest.[82] In floral crops, ethylene emanating from the harvested product or adjacent products dramatically accelerate pigment degradation, color fading and senescence in many flowers (e.g., *Vanda* orchids). To prevent pigment degradation, chemicals that either inhibit the biosynthesis of ethylene (e.g., amino oxyacetic acid, aminoethoxyvinylglycine) or impede its activity (silver ions, l-methyl-cyclopropene) can be used.

Other growth regulators such as the cytokinins have pronounced effect on the retention of chlorophyll.[120] The potential use of synthetic cytokinins to retard chlorophyll loss has, as a consequence, been tested in several green vegetables (e.g., broccoli, Brussels sprouts, celery, endive and leaf lettuce). In broccoli stored at 13°C, the storage life can be doubled with a single application of zeatin and dihydrozeatin.[68] While many cases the results are positive, other postharvest techniques for color retention are commercially more acceptable.

6. VOLATILE COMPOUNDS

6.1. Classes of Volatile Compounds

Plants give off a wide range of volatile compounds, some of which are extremely important components of quality. Humans exhibit distinct patterns in the foods that they select to consume, and flavor is known to be a primary criterion in this selection. Most flavors are comprised of a combination of both taste and odor. Since taste is generally thought to be limited to four basic sensations—sweet, sour, salty and bitter—volatiles often play a very significant role in flavor. The four basic tastes can be contrasted with the potential perception of up to 10,000 distinct odors by the human nose olfactory epithelium. This difference in potential perception is also seen in the number of taste receptors that are in the thousands, while odor receptors are thought to be in the millions. Hence, aroma compounds are not only an extremely important part of flavor, they provide an almost unlimited potential for flavor diversity.

The importance of volatile compounds is not just limited to food products. The aroma of many ornamental and floral species complements our visual perception of these and often is an important part of the plant's aesthetic quality. The contribution of plant volatiles may impart either positive quality characteristics as in the fragrance of gardenia flowers or in the case of the carrion flowers a distinctly undesirable sensation, at least from the standpoint of humans. A primary requirement of plant volatiles is that they are present in a gaseous or vapor state as the molecules must be able to reach the olfactory epithelium in the roof of the nasal passages. In addition, some degree of water solubility is essential. Non-ionizable volatile compounds are typically only perceived by the olfactory system, while ionizable volatiles can contribute to the perception of both taste and aroma.

The volatiles emanating from living plant material after harvest represent a diverse array of chemical compounds. These include esters, lactones, alcohols, acids, aldehydes, ketones, acetals, hydrocarbons and some phenols, ethers and heterocyclic oxygen compounds. Over 300 volatile compounds have been identified from pear[193] and a comparable number from apple (Table 4.15). Superimposed on these raw product volatiles is a tremendous range to thermally generated volatile compounds formed during cooking.[119] The majority of edible plant material is consumed in a cooked form; for example, 370 of 390 commercially cultivated vegetables are routinely to intermittently cooked.[118] Exposure of foods to high temperatures can result in the synthesis of furans, furanones, pyranones, pyrroles, pyridines, pyrimidines, pyrazines, thiophenes, dihydrothiophenes, oxazoles, oxazolines, thiazoles, thiazolines, aldehydes, ketones, cyclopentanones, non-cyclic sulfur compounds and others that give rise to a tremendous range in flavors. Within each class of compounds is often a myriad of possible odors. For example, pyrazines range in aroma from roasted nut (2-methylpyrazine) to caramel-like (2,3-dimethyl-5-isopentylpyrazine) to bell pepper (2-propyl-3-methoxypyrazine). The specific volatile compounds formed vary with the chemistry of the plant material, cooking temperature, the length of cooking, and other factors.

The volatile substances generally are present in very small amounts, often only a fraction of a part per million. Of the large number of volatiles given off by a plant or plant organ, typically only a very small number of compounds impart the characteristic aroma. Of the approximately 330 volatile compounds from apple, only approximately a dozen are critical components of the characteristic apple aroma. Although present in very small amounts, ethyl-2-methylbutyrate is responsible for some of the characteristic apple aroma.

These critical volatiles are called character impact compounds (CIC) and plant products can be divided into the following four general groups based in part on the presence or absence of a character impact compound.[175]

   a. Those whose aroma is composed primarily of one character impact compound.
   b. Those whose aroma is due to a mixture of a small number compounds, of which one may be a character impact compound.
   c. Those whose aroma is due to a large number of compounds, none of which are character impact compounds, and with careful combination of these components the odor can be reproduced.
   d. Those whose aroma is made up of a complex mixture of compounds that cannot be reproduced.

Examples of character impact compounds are 2-methyl-3-ethyl-pyrazine in raw white potato tubers and ethylvinylketone in soybeans. Examples of food whose aroma is made up of a small number of compounds are boiled white potatoes (2-methyl-3-ethyl pyrazine (CIC) and methional), apple (ethyl-2-methylbutyrate (CIC), butyl acetate, 2-methylbutyl acetate, hexyl acetate, hexanal and trans-2-hexenal) and boiled cabbage (dimethyl disulfide (CIC) and 2-propenylisothiocyanate). Important aroma volatiles for a cross-section of fruits and vegetables are given in Table 4.16.

6.2. Synthesis and Degradation

Due to the extremely wide range in types of chemical compounds that are important in the aroma of harvested products,[198] their biosynthesis will not be discussed in detail. There are, however, several generalizations that can be made regarding biosynthesis.

Volatile compounds that are important in the aroma of postharvest products are formed by one of three general means in intact tissue. The first group is formed naturally by enzymes found within intact tissues. These include nearly all of the odors from fresh fruits, vegetables, and flowers. While biosynthesis of many of these compounds has not been stud-

Table 4.15. Volatile Compounds from Apple Fruit.*

Alcohols
  Methanol
  Ethanol
  Isopropanol
  Butanol
  2-Butanol
  Isobutanol
  Pentanol
  2-Methyl butanol
  2-Methyl-2-butanol
  3-Methyl butanol
  2-Pentanol
  3-Pentanol
  4-Pentanol
  2-Methyl-2-pentanol
  3-Methyl pentanol
  Hexanol
  cis-2-Hexenol
  cis-3-Hexenol
  trans-2-Hexenol
  trans-3-Hexenol
  1-Hexen 3-ol
  5 Hexenol
  Heptanol
  2-Heptanol
  Octanol
  2-Octanol
  3-Octanol
  an Ethylhexanol
  Nonanol
  2-Nonanol
  6-Methyl-5-heptenol
  Decanol
  3-Octenol
  Benzyl alcohol
  2-Phenethanol
  Terpinen-4-ol
  α-Terpineol
  Isobareol
  Citronellol
  Geraniol
Esters
(Formates)
  Methyl formate
  Ethyl formate
  Propyl formate
  Butyl formate
  2- or 3-Methyl butyl formate
  Pentyl formate
  i-Pentyl formate
  Hexyl formate

(Acetates)
  Methyl acetate
  Ethyl acetate
  Propyl acetate
  Butyl acetate
  Isobutyl acetate
  t-Butyl acetate
  Pentyl acetate
  2-Methyl butyl acetate
  3-Methyl butyl acetate
  Hexyl acetate
  Heptyl acetate
  Octyl acetate
  Benzyl acetate
  cis-3-Hexenyl acetate
  trans-2-Hexenyl acetate
  2-Phenyl ethyl acetate
  Nonyl acetate
  Decyl acetate
(Propionates)
  Methyl n-propionate
  Ethyl propionate
  Ethyl 2-methyl propionate
  Propyl propionate
  Butyl propionate
  Isobutyl propionate
  2- and/or 3-Methyl butyl
    propionate
  Hexyl propionate
(Butyrates)
  Methyl butyrate
  Ethyl butyrate
  Propyl butyrate
  Isopropyl butyrate
  Butyl butyrate
  Isobutyl butyrate
  Pentyl butyrate
  Isopentyl butyrate
  Hexyl butyrate
  Cinnamyl butyrate
  Ethyl crotonate
(Isobutyrates)
  Methyl isobutyrate
  Ethyl isobutyrate
  Butyl isobutyrate
  Isobutyl isobutyrate
  Pentyl isobutyrate
  Hexyl isobutyrate

(2 Methyl butyrates)
  Methyl-2-methyl butyrate
  Ethyl-2-methyl butyrate
  Propyl-2-methyl butyrate
  Butyl-2-methyl butyrate
  Isobutyl-2-methyl butyrate
  Pentyl-2-methyl butyrate
  Hexyl-2-methyl butyrate
(Pentanoates)
  Methyl pentanoate
  Ethyl pentanoate
  Propyl pentanoate
  Butyl pentanoate
  Amyl pentanoate
  Isoamyl pentanoate
(Isopentanoates)
  Methyl isopentanoate
  Ethyl isopentanoate
  Isopentyl isopentanoate
(Hexanoates)
  Methyl hexanoate
  Ethyl hexanoate
  Propyl hexanoate
  Butyl hexanoate
  Isobutyl hexanoate
  Pentyl hexanoate
  2- and/or 3-Methyl butyl
    hexanoate
(Hexenoates)
  Butyl trans-2 hexenoate
(Heptanoates)
  Ethyl heptanoate
  Propyl heptanoate
  Butyl heptanoate
(Octanoates)
  Ethyl octanoate
  Propyl octanoate
  Butyl octanoate
  Isobutyl octanoate
  Pentyl octanoate
  Isopentyl octanoate
  Hexyl octanoate
(Nonanoates)
  Ethyl nonanoate
(Decanoates)
  Ethyl decanoate
  Butyl decanoate
  Isobutyl decanoate
  Pentyl decanoate
  Isopentyl decanoate
  Hexyl decanoate

Table 4.15. (*continued*)

(Dodecanoates)
Ethyl dodecanoate
Butyl dodecanoate
Hexyl dodecanoate

(Other)
Diethyl succinate
Ethyl-2-phenylacetate
Dimethylphthalate
Diethylphthalate
Dipropylphthalate

Aldehydes
Formaldehyde
Acetaldehyde
Propanal
2 Propenal
2-Oxopropanal
Butanal
Isobutanal
2-Methyl butanal
trans-2-Butenal
Pentanal
Isopentanal
Hexanal
trans-2-Hexenal
cis-3-Hexenal
trans-3-Hexenal
Heptanal
trans-2-Heptenal
Octanal
Nonanal
Decanal
Undecanal
Dodecanal
Benzaldehyde
Phenyacetaldehyde

Ketones
2 Propanone
2 Butanone
3 Hydroxybutan-2-one
2,3-Butanedione
2-Pentanone
3-Pentanone
4-Methylpentan-2-one
2-Hexanone
2-Heptanone
3-Heptanone
4-Heptanone
2-Octanone
7-Methyloctan-4-one
Acetophenone

Ethers
Diethyl ether
Methyl propyl ether
Dibutyl ether
2- and/or 3-Methyl butyl ether
Dihexyl ether
Methylphenyl ether
4 Methoxyallybenzene
cis-Linalool oxide
trans-Linalool oxide

Acids
Formic
Acetic
Propanoic
Butanoic
Isobutanoic
2-Methyl butanoic
3-Methyl butanoic
Pentanoic
Pentenoic
4-Methyl pentanoic
Hexanoic
trans-2-Hexenoic
Heptanoic
cis-3-Heptenoic
Octanoic
cis-3-Octenoic
Nonanoic
cis-3-Nonenoic
Decanoic
Decenoic
Undecanoic
Undecenoic
Dodecanoic
Dodecenoic
Tridecanoic
Tridecenoic
Tetradecanoic
Tetradecenoic
Pentadecanoic
Pentadecenoic
Hexadecanoic
Hexadecenoic
Heptadecanoic
Heptadecenoic
Octadecanoic
9-Octadecenoic
9,12-Octadecadienoic
9,12,15-Octadecatrienoic

Nonadecanoic
Nonadecenoic
Eicosanoic
Benzoic

Bases
Ethylamine
Butylamine
Isoamylamine
Hexylamine

Acetals
Diethyoxymethane
Dibutoxymethane
Dihexoxymethane
1-Ethoxy-1-propoxyethane
1-Butoxy-1-ethoxyethane
1-Ethoxy-1-hexoxyethane
1-Ethoxy-1-octoxyethane
1,1-Diethoxyethane
1,1-Dibutoxyethane
1-Butoxy-1-2-methyl butoxy ethane
1-Butoxy-1-hexoxyethane
1,1-Di-2-methyl butoxy ethane
1,2-Methyl butoxy-1-hexoxy ethane
1,1-Di-hexoxyethane
1,1-Diethoxypropane
1,1-Dipthoxypentane
4-Methoxyally benzene
Furan
Furfural
5-Hydroxymethylfurfural
2,4,5-Trimethyl-1,3-dioxolane

Hydrocarbons
Ethane
Ethylene
α-Farnesene
β-Farnesene
Benzene
Ethyl benzene
1-Methylnaphthalene
2-Methylnaphthalene
Damascenone
α-Pinene

Source: Dimick and Hoskin[46]

Table 4.16. Volatile Compounds Responsible for the Characteristic Aroma of Selected Fruits, Vegetables and Nuts.*

Crop	Compounds
Vegetables	
Asparagus	Methyl-1,2-dithiolane-4-carboxylate; 1,2-diothiolane- 4-carboxylic acid
Beans	Oct-1-ene-3-ol, hex-*cis*-3-enol
Beet	2-*sec*-Butyl-3-methoxypyrazine, geosmin
Bell pepper	2-(2-methylpropyl)-3-methoxypyrazine, (*E,Z*)-2,6- nonadienal, (*E,E*)-2,4-decadienal
Brussel sprouts	2-Propenyl isothiocyanate, dimethyl sulfide, dimethyl disulfide, dimethyl trisulfide
Cabbage	2-Phenylethyl, 3-(methylthio) propyl, 4-(methylthio)-butyl isothiocyanates
Carrot	2-*sec*-Butyl-3-methoxy-pyrazine, sabinene, terpinolene, myrcene, octanal, 2-decenal
Celery	3-Butyl-phtalide, sedanolide (3-butyl-3*a*,4,5,6-tetra- hydrophthalide, β-selinene
Cucumber	(*E,Z*)-2,6-Nonadienal
Endive	Hexanal, (*Z*)-3-hexenol, (*Z*)-3-hexenal, β-ionone
Garden cress	4-Pentenyl isothiocyanate
Garlic	(*E*)- and (*Z*)-4,5,9-trithiadodeca-1,6,1-triene-9-oxide
Leek	(*E*)- and (*Z*)-3-Hexenol, dipropyl, trisulfide, propanethiol, methyl propyl disulfide, 2-propenyl propyl disulfide, 1-propenyl propyl disulfide, methyl propyl trisulfide, dipropyl trisulfide
Lettuce	2-Isopropyl-, 2-*sec*-butyl-, 2-(2-methyl-propyl)-3- methoxypyrazine
Mushrooms	1-Octen-3-one, 1-octen-3-ol, lenthionine
Onions	Thiopropanal S-oxide, 3,4-dimethyl-2, 5-dioxo-dihydro-thiophene, propyl methane-thiosulfonate
Parsley	β-Phellandrene, terpinolene, 1-methyl-4-isopropenyl benzene, *p*-mentha-1,3,8-triene, apiole
Parsnip	Myristicin
Peas	2-Alkenals, 2,4-alkadienals, 2,6-nonadienal, 3,5-octadien-2-ones, 2-alkyl-3-methoxypyrazines, hexanol
Radish	5-(Methylthio) pentyl isothiocyanate
Shallot	Methyl propyl trisulfide, dimethyl trisulfide, dipropyl trisulfide, (*E*)- and (*Z*)-1-propenyl propyl disulfide
Soybean	1-Penten-3-one, (*Z*)-3-hexenol, 2-pentylfuran, 2-pentylfuran, ethyl vinyl ketone
Sweetpotato	Maltol, phenylacetaldehyde, methyl geranate, 2-acetyl furan, 2-pentyl furan, 2-acetyl pyrrole, geraniol, β-ionone
Tomato	3-Methylbutanal, β-ionone, 1-penten-3-one, hexanal, (*Z*)-3-hexenol, 2- and 3-methylbutanol, 2- (2-methylpropyl) thiazole, eugenol, 6-methyl-5-hepten- 2-one, dimethyl trisulfide
Truffle	Dimethyl sulfide, methylenebis (methyl sulfide)
Watercress	2-Phenylethyl isothiocyanate
Fruits	
Apple	Ethyl-2-methylbutyrate, hexenal, butanol, (*E*)- 2-hexenol, hexyl acetate, (*E*)-2-hexenyl acetate
Apricot	Linalool, isobutyric acid, alkanolides
Banana	3-Methylbutyl acetate, butanoate, 3-methylbutanoate, 2-methylpropanol, 3-methylbutanol
Beli	3-Methylbutyl acetate, 3-methyl-2-buten-1-ol, α-phellandrene
Black currant	4-Methoxy-2-methyl-2-mercaptobutane, linalool, α-terpineol, 1-terpinen-4-ol, citronellol, *p*-cymene-8-ol, 2,3-butanedione
Blackberry	3, 4-Dimethoxyallybenzene, 2-heptanol, *p*-cymen-8-ol, (3,4,5-trimethoxyallybenzene), eugenol, isoeugenol, 4-hexanolide, 4-decanolide
Blueberry	Hydroxycitronellol, farnesyl acetate, farnesol, myristicine, linalool
Cashew apple	Hexanal, car-3-ene, limonene, (*E*)-2-hexenal, benzaldehyde

Table 4.16. (*continued*)

Crop	Compounds
Cherimoya	1-Butanol, 3-methyl-1-butanol, 1-hexanol, linalool, hexanoic acid, octanoic acid
Cherry	Benzaldehyde, (*E,Z*)-2, 6-nonadienal, linalool, hexanal, (*E*)-2-hexenol, benzaldehyde, phenylacetaldehyde, linalool, (*E,Z*)-2, 6-nonadienal, eugenol
Durian	Propanethiol, ethyl-2-methylbutanoate
Grape	Linalool, geraniol, nerol, linalool oxides
Grapefruit	Nootkatone, 1-*p*-menthene-8-thiol, limonene, acetaldehyde, decanal, ethyl acetate, methyl butanoate, ethyl butanoate
Guava	Myrcene, β-caryophyllene, α-humulene, α-selinene, α-copaene, benzaldehyde, 2-methylpropyl acetate, hexyl acetate, ethyl decanoate
Kiwi Fruit	Ethyl butanoate, (*E*)-2-hexenal
Lemon	Neral, geranial, geraniol, geranyl acetate, neryl acetate, bergamotene, caryophyllene, β-pinene, γ-terpinene, α-bisabolol
Litchi	2-phenylethanol, esters of cyclohexyl, hexyl, benzyl, citronellyl, and neryl alcohol, limonene, nonanal, decanal, citronellol, geraniol
Loquat	Phenylethanol, phenylacetaldehyde, hexenols, methyl cinnamate, β-ionone
Mango	β-carophyllene, limonene, myrcene, α-terpinolene, β-selinene
Mangosteen	Hexyl acetate, (*Z*)-3-hexenyl acetate, α-copaene
Melon	β-Ionone, benzaldehyde, (*Z*)-6-nonenal, (*Z,Z*)-3, 6- nonadienol
Orange	*d*-Limonene, ethyl butanoate, ethyl 2-methylbutanoate, ethyl propionate, methyl butanoate
Passion Fruit	2-Heptyl, 2-nonyl, (*Z*)-3-hexenyl esters, (*Z*)-3-octenyl acetate, geranyl, citronellyl esters, 3-methylthiohexanol, 2-methyl-4-propyl-1, 3-oxathianes
Papaya	Linalool, benzyl isothiocyanate
Peach	4-Decanolide, 3-methyl-butyl acetate, carvomenthenal, α-terpineol, linalool
Pear	Hexyl acetate, methyl and ethyl decadienoates, ethyl (*E*)- 2-octenoate, (*Z*)-4-decenoate, butyl acetate, ethyl butanoate
Pineapple	2-Methylbutanoates, hexanoates, methyl and ethyl 3-methylthiopropanoate
Plum	Benzaldehyde, linalool, ethyl nonanoate, methyl cinnamate, 4-decanolide
Pummelo	*d*-Limonene, myrcene, linalool, citronellal
Raspberry	1-(4-Hydroxyphenyl)-3-butanone, α- and β-ionones, linalool, gerniol
Rose-apple	Linalool, phenyl-1-propanol
Sapodilla fruit	Methyl benzoate, methyl salicylate, ethyl benzoate, phenylpropanone
Strawberry	Ethyl hexanoate, ethyl butanoate, (*E*)-2-hexenal, 2,5-dimethyl-4-methoxy-3(2*H*)-furanone, linalool
Nuts	
Brazil nut	Hexanal, methylbenzylfuran
Coconut	δ-Octalactone, δ-decalactone
Peanut	Methanol, acetaldehyde, ethanol, pentane, 2-propanol, propanal, acetone
Walnut	Hexanal, pentanal, 2, 3-pentanedione, and 2-methyl- 2-pentenal

* *Source:* Askar et al.;[5] Baldry et al.;[11] Bellina-Agostinone et al.;[12] Berger;[15] Berger et al.;[13] Berger et al.;[14] Boelens et al.;[18] Buttery et al.;[23,24,25,26] Carson;[31] Chitwood et al.;[32] Chung et al.;[33] Cronin;[37] Dürr and Röthlin;[48] Engel and Tressl;[51,52] Etievant et al.;[54] Flath and Forrey;[59] Flora and Wiley;[60] Fröhlich and Schreier;[67] Georgilopoulos and Gallois;[72] Guichard and Souty;[77] Haro and Faas;[85] Hayase et al.;[87] Hirvi et al.;[92] Hirvi;[93] Hirvi and Honkanen;[94,95] Horvat and Senter;[100] Idstein et al.;[104,106,107,109] Idstein and Schreier;[105] Jennings and Tressl;[112] Jennings;[113] Johnson and Vora;[114] Johnston et al.;[115] Kasting et al.;[117] Kemp et al.;[121,122] Kerslake et al.;[125] Kjaer et al.;[126] Kolor;[133] Koyasako and Bernhard;[134] Lee et al.;[135, 136] MacLeod and Islam;[139,140] MacLeod and Pikk;[141] MacLeod and Pieris;[142,143,147,149] MacLeod and de Troconis;[144, 145, 146,148] MacLeod et al.;[150] Marriott;[154] Mazza;[163] Morales and Duque;[170] Moshonas and Shaw;[171,172] Murray et al.;[173] Nyssen and Maarse;[176] Pyysalo;[190] Russell et al.;[197] Sawamura and Kuriyama;[202] Schreier et al.;[204] Schreier;[205] Schreyen et al.;[206,207] Shaw;[215] Shaw et al.;[213] Shaw and Wilson;[214] Spencer et al.;[219] Swords et al.;[223] Takeoka et al.;[226,227] Talou et al.;[228] Tressl and Drawert;[233] Tressl et al.;[234] Wallbank and Wheatley;[239] Wang and Kays;[241] Whitfield and Last;[245] Whitaker;[244] Winterhalter;[248] Wu et al.;[250] Yabumoto and Jennings;[252] Young and Paterson.[254]

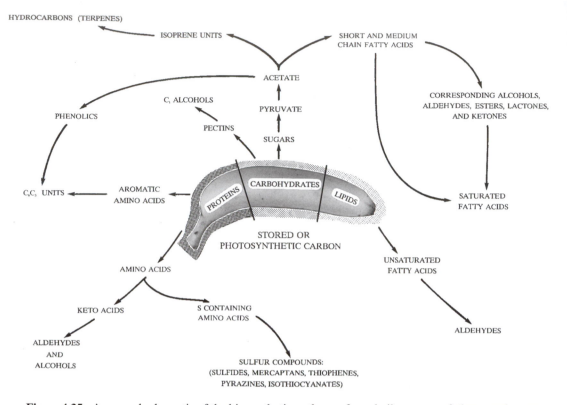

Figure 4.35. A general schematic of the biosynthetic pathways for volatile aromas of plant products.

ied in detail, three major pathways are known to be important (Figure 4.35). These are the isoprenoid pathway, the shikimic acid pathway and β-oxidation. The isoprenoid pathway contributes many of the terpenes (e.g., limonene in lemon aroma) that are found as multiples of 5 carbon isoprene subunits.[212] Over 200 monoterpenes (2 isoprene units, C_{10}) and 1,000 sesquiterpenes (5 isoprene units, C_{25}) have been identified.[229] The shikimic pathway provides benzyl alcohol, benzylaldehyde and many of the volatile phenolic compounds.[90] Beta-oxidation represents an important pathway for the production of volatiles throughout the oxidation of fatty acids. In Table 4.17, a list of potential oxidation products for three fatty acids, oleic, linoleic and linolenic, is given. Esters, an important class of fruit aromas, are formed by the enzyme acyl alcohol transferase that condenses acyl CoA with primary alcohols.[200]

A second group of volatiles are produced enzymatically after damage to the tissue. Examples would be part of the aroma of cucumbers (*cis*-3-nonenal and hexanal) formed during disruption of the intact cells[70] or the formation of methyl and propyl disulfides in onions. Cellular disruption allows enzymes and substrates, previously sequestered separately within the cells, to interact. The production of the aroma of onion is perhaps the most thoroughly studied example of this type of volatile flavor production.[192] Here the amino acids s-methyl-L-cysteine sulfoxide and s-*n*-propyl-L-cysteine sulfoxide are enzymatically degraded forming the characteristic onion volatiles. Other amino acids may also represent precursors for volatiles and as mentioned in the volatile production of undisturbed tissue, β-oxidation of fatty acids is also important in disrupted cells.

The third general means of flavor synthesis is through direct chemical reaction. This normally occurs with heating during processing or cooking. Since this alters the live product to a processed state, these volatiles are of less of a direct interest to postharvest biologists. They can

Table 4.17. Oxidation Products of Three Unsaturated Fatty Acids.*

Oleic	Propanal	Linoleic	2-Decenal
	Pentanal		Non-2,4-dienal
	Hexanal		Dec-2,4-dienal
	Heptanal		Undec-2,4-dienal
	Nonanal		Oct-1-en-3-ol
	2-Octenal		2-Heptenal
	2-Nonenal	Linolenic	Acetaldehyde
	2-Decenal		Propanal
Linoleic	Acetaldehyde		Butanal
	Propanal		2-Butenal
	Pentenal		2-Pentenal
	Hexanal		2-Hexenal
	2-Propenal		2-Heptenal
	2-Pentenal		2-Nonenal
	2-Hexenal		Hex-1,6-dienal
	2-Heptenal		Hept-2,4-dienal
	2-Octanal		Non-2,4-dienal
	2-Nonenal		Methyl ethyl ketone

*Source: Data from Hoffman.[98]

be important, however, when a particular postharvest handling practice alters the eventual flavor of a processed product.

In contrast to the biosynthesis of volatile compounds by plants, there has been much less interest in their degradation, due largely to the fate of the molecules once formed. Being volatile, most of these compounds simply dissipate into the atmosphere, eventually being degraded by biological, chemical or photochemical reactions.

6.3. Postharvest Alterations

The volatiles produced by harvested products can be altered by a wide range of preharvest and postharvest factors.[7] These include cultivar, maturity, season, production practices (e.g., nutrition), handling, storage, artificial ripening and eventual method of preparation.[95] Due to the importance of volatiles in the flavor quality of food crops and aesthetic appeal of many ornamentals, care must be taken during the postharvest period to minimize undesirable changes.

Early harvest is known to have detrimental effect on the synthesis of the volatile constituents of many fruits.[96] In the tomato, the production of volatiles increases with the development of the fruit and early harvest (breaker stage) with forced ripening (22–20°C) does not yield the same volatile profile as vine-ripened fruits (Figure 4.36). The concentrations of nonanal, decanal, dodecanal, neral, benzylaldehyde, citronellyl propionate, citronellyl butyrate, geranyl acetate and geranyl butyrate are higher in field ripened than artificially ripened fruits. As tomatoes develop from a ripe to an overripe state, the concentration of 2,3-butanedione, isopentyl butyrate, citronellyl butyrate 2,3-butanedione and geranyl butyrate increase while in general the concentration of alcohols, aldehydes, acetates and propionates tend to decrease. In non-climacteric fruits (e.g., oranges) that do not ripen normally if picked at a pre-ripe stage of development, the undesirable effect of early harvest on the flavor volatiles is even more pronounced.

Storage conditions and duration may also have a significant effect on the synthesis of volatiles after removal from storage.[218] Apple volatiles can be significantly influenced by genotype, cultural practices, ripening and storage atmospheres.[56] For example, apples stored under

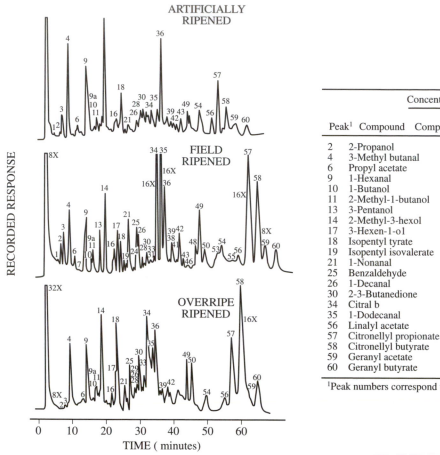

Peak[1]	Compound	Concentration (ppm in the fruit) Compound	Field Ripened	Overripe Red
2	2-Propanol	0.45	0.93	0.24
4	3-Methyl butanal	7.16	2.22	2.38
6	Propyl acetate	1.57	0.58	0.79
9	1-Hexanal	5.93	2.51	2.68
10	1-Butanol	1.12	0.35	0.76
11	2-Methyl-1-butanol	1.57	0.67	0.91
13	3-Pentanol	1.57	1.17	0.61
14	2-Methyl-3-hexol	7.94	2.51	2.68
17	3-Hexen-1-o1	0.78	1.28	0.79
18	Isopentyl tyrate	2.80	0.82	3.17
19	Isopentyl isovalerate	0.11	0.23	0.18
21	1-Nonanal	0.45	1.46	0.92
25	Benzaldehyde	0.22	1.17	1.40
26	1-Decanal	0.67	1.05	0.67
30	2-3-Butanedione	1.68	0.46	2.56
34	Citral b	1.56	5.95	3.66
35	1-Dodecanal	2.24	7.71	2.68
56	Linalyl acetate	2.80	2.28	1.22
57	Citronellyl propionate	5.82	17.87	7.14
58	Citronellyl butyrate	3.80	8.76	19.03
59	Geranyl acetate	2.78	2.92	2.38
60	Geranyl butyrate	1.68	2.80	4.02

[1]Peak numbers correspond to chromatogram

Figure 4.36. Effect of stage of maturity and artificial ripening on tomato fruit volatiles (*redrawn from Shah et al.*[211]).

controlled atmosphere conditions (2% oxygen, 3.5°C) have abnormal production of the critical flavor esters, butyl acetate and hexyl acetate upon removal from storage. Similar findings have been reported for hypobaric storage of fruits and the exposure of some fruits to chilling injury. Ozone (strawberry),[182] 1-methylcyclopropene (banana),[74] curing (sweetpotato)[240] and other treatments are known to alter the profile of volatiles after harvest.

Presently there is a very limited amount of information on the relationship between postharvest conditions and the beneficial or detrimental changes in the aroma of plant products. Much of the research to date has focused on determining the volatiles present and identifying specific character impact compounds. The relationship between aroma and flavor with regard to consumer preference is complex;[40,160] however, as more is learned about this area of postharvest biology, it will be possible to better control or perhaps even improve the aroma of many products.

7. PHENOLICS

Plant phenolics encompass a wide range of substances that have an aromatic ring and at least one hydroxyl group (Table 4.18). Included are derivatives of these aromatic hydroxyl compounds due to substitutions, for example, the presence of O-methylation instead of hydroxyls

Table 4.18. The Major Classes of Phenolics in Plants.*

Number of Carbon Atoms	Basic Skeleton	Class	Examples
6	C_6	Simple phenols Benzoquinones	2,6-Dimethoxybenzoquinone Catechol, hydroquinone,
7	C_6-C_1	Phenolic acids	p-Hydroxybenzoic, salicylic
8	C_6-C_2	Acetophenones Phenylacetic acids	3-Acetyl-6-methoxybenzaldehyde p-Hydroxyphenylacetic
9	C_6-C_3	Hydroxycinnamic acids Phenylpropenes Coumarins Isocoumarins Chromones	Caffeic, ferulic Myristicin, eugenol Umbelliferone, aesculetin Bergenin Eugenin
10	C_6-C_4	Naphthoquinones	Juglone, plumbagin
13	C_6-C_1-C_6	Xanthones	Mangiferin
14	C_6-C_2-C_6	Stilbenes Anthraquinones	Lunularic acid Emodin
15	C_6-C_3-C_6	Flavonoids Isoflavonoids	Quercetin, cyanidin Genistein
18	$(C_6$-$C_3)_2$	Lignans Neolignans	Pinoresinol Eusiderin
30	$(C_6$-C_3-$C_6)_2$	Biflavonoids	Amentoflavone
n	$(C_6$-$C_3)_n$ $(C_6)_n$ $(C_6$-C_3-$C_6)_n$	Lignins Catechol melanins Flavolans (Condensed tannins)	

Source: After Harborne.[84]

on methyleugenol. In this group, common phenolics are the flavonoids, lignin, the hormone abscisic acid, the amino acids tyrosine and dihydroxyphenylalanine, coenzyme Q and numerous end products of metabolism. Phenolics represent one of the most abundant groups of compounds found in nature and are of particular interest in postharvest physiology because of their role in color and flavor. The concentration of phenolics varies widely in postharvest products. For example, in ripe fruits it ranges from very slight to up to 8.5% (persimmon) of the dry weight (Table 4.19).

The general biological role of some phenolics in plants is readily apparent (e.g., pigments, abscisic acid, lignin, coenzyme Q), while others are involved in host plant defense, feeding deterrents, wood and bark characteristics, flower and fruit color and taste and aroma. However, for the majority of plant phenolics, their biological role has not been ascertained.

Plant phenolics are generally reactive acidic substances that rapidly form hydrogen bonds with other molecules. Often they will interact with the peptide bonds of proteins and when the protein is an enzyme this generally results in inactivation, a problem commonly encountered in the study of plant enzymes. The protein binding capacity of persimmon phenolics is so great that it is used to remove the protein from Japanese rice wine (sake), clarifying it. As a group, the phenols are susceptible to oxidation by the phenolases that convert monophenols to diphenols and subsequently to quinones. In addition, some phenols are capable of chelating metals.

Phenolic compounds rarely occur in a free state within the cell; rather they are commonly conjugated with other molecules. Many exist as glycosides linked to monosaccharides or disaccharides. This situation is especially true of the flavonoids, which are normally glycosylated. In addition, phenols may be conjugated to a number of other types of compounds. For example, hydroxycinnamic acid may be found esterified to organic acids, amino groups, lipids,

Table 4.19. Total Phenolic Content in Ripe Fruits*

Fruit	Phenolic Content (g · kg⁻¹ fresh weight)
Apple	
Various cultivars	1–10
'Cox's Orange Pippin'	0–55
'Baldwin'	2.5
Cider apple 'Launette'	11
Cider apple 'Waldhofler'	4.5
Banana	5.3
Date	5.0
Cherry	
'Montmorency'	5.0
Grape	
'Riesling', cluster	9.5
'Tokay', cluster	4.8
'Muscat', skin	3.5
'Muscat', pulp	1.0
'Muscat', seed	4.5
Passion fruit	0.014
Peach	
Mixed cultivars	0.3–1.4
'Elberta', flesh	0.7–1.8
'Elberta', skin	2.4
Pear	
'Muscachet'	4.0
Persimmon	85
Plum	
'Victoria', flesh	21
'Victoria', skin	57

*Source: van Buren[235]

terpenoids, phenolics and other groups, in addition to sugars. Within the cell, this serves to render mono- and diphenols less phytotoxic than when in the free state.

Phenolics are commonly divided into three classes based on the number of phenol rings present. The simplest class includes the monocyclic phenols composed of a single phenolic ring. Common examples found in plants are phenol, catechol, hydroquinone and *p*-hydroxycinnamic acid. Dicyclic phenols such as the flavonoids have two phenol rings while the remainder tend to be lumped into the polycyclic or polyphenol class. The structures of several common phenols of each class are illustrated in Figure 4.37. These general classes can be further divided into subclasses based upon the number of carbon atoms and the pattern of the basic carbon skeleton of the molecule (Table 4.18).

7.1. Biosynthesis of Phenols

Nearly all of the phenols are formed, initially from phosphoenolpyruvate and erythrose 4-phosphate through shikimate in the shikimic acid pathway (Figure 4.38). The aromatic amino acid phenylalanine is a central intermediate that is deaminated and hydroxylated in the para position on the phenol ring yielding *p*-hydroxycinnamic acid. As mentioned earlier in the section on pigments, malonate is essential in formation of the flavonoids. Three molecules of mal-

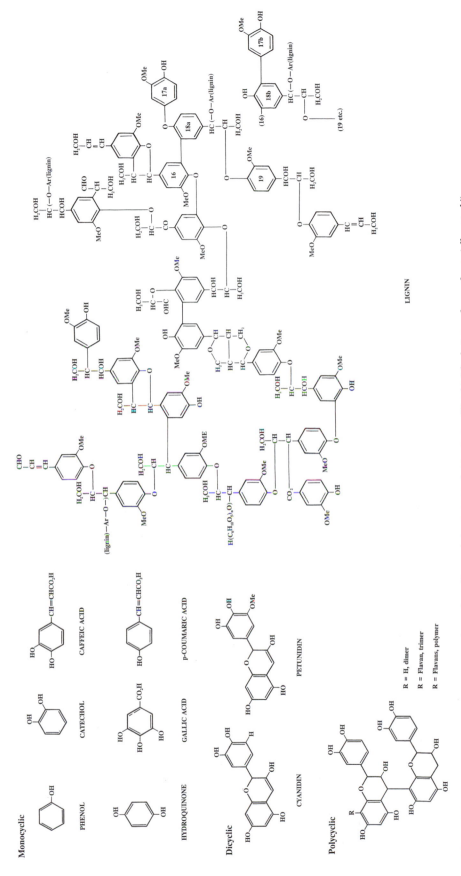

Figure 4.37. Structure of several common plant phenolics and the proposed structure of lignin (*the latter redrawn from Adler et al.*[1]).

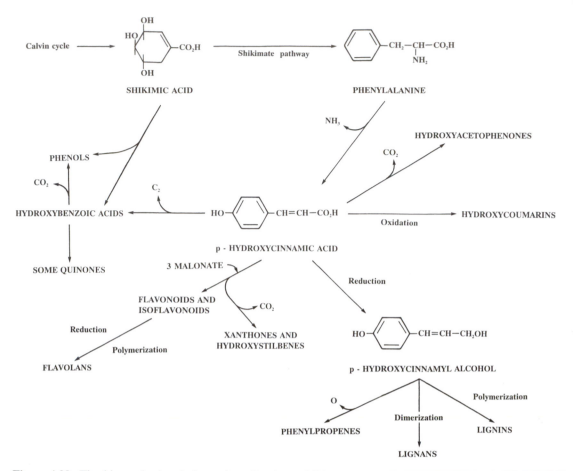

Figure 4.38. The biosynthesis of plant phenolics from shikimate and phenylalanine (*redrawn from Harborne*[84]).

onate in the form of malonyl-CoA combine with cinnamic acid (cinnamyl-CoA) to form a chalcone, which with ring closure gives the base structure for the flavonoids.

A number of physiological factors affect phenolic synthesis. These include light, temperature and cellular carbohydrate, mineral and water status. The response varies with the specific phenol and species in question. For example, at low temperature the synthesis of anthocyanin is favored while the synthesis of proanthocyanidin is repressed.

7.2. Decomposition of Phenols

For years many of the phenols were erroneously considered end products of metabolism, due in part to our inability to ascribe specific roles for these compounds within the cell. Some are now known to function as feeding deterrents to animals and insects and in disease resistance. Isotope labeling studies have shown that phenols do not simply accumulate unchanged in the plant, but in fact are in a constant state of flux, undergoing turnover *via* synthesis and degradation. The rate of turnover varies widely, from hours in some floral pigments to, in some cases, weeks. The actual sequence of catabolism varies with the diverse number of types of phenolic compounds. One common feature, however, is eventual ring cleavage and opening.

Higher plants have been shown to possess active aromatic ring-cleaving enzymes. With the simpler phenolics, β-oxidation appears to play a major role.

During the postharvest period of some plant products, changes in specific phenols are extremely important. This often entails the polymerization of the molecule into large, more complex substances.

7.3. Postharvest Alterations

Plant phenolics are of particular importance during the postharvest period due to their role in both flavor and color. While the majority of phenolics at the concentrations in which they are found in food products have no significant taste, several do substantially alter the flavor of certain products. The phenolic acids, when present in a sufficiently high concentration are distinctly sour and some of the flavonoids in citrus (e.g., naringin) are bitter. In addition, phenolics in the immature fruits of several species are highly astringent. For example, immature banana fruit contain 0.6% water soluble tannins (phenolics) on a fresh weight basis, the fruit of Chinese quince—0.4%, carob beans—1.7%, and persimmon fruit—2.0%.[159] These compounds are also found in other plant parts such as leaves, bark and seed coats of some species. In the persimmon, the water soluble tannins are composed of cathechin, catechin-3-gallate, gallocatechin, gallocatechin-3-gallate and an unknown terminal residue.[158]

During persimmon maturation and fruit ripening, the level of astringency and the concentration of water-soluble tannins diminish, eventually giving fruit of excellent flavor. This decrease in astringency is due to polymerization of the existing tannins, forming larger, water-insoluble molecules no longer capable of reacting with the taste receptors in the mouth. It has been known since the early part of this century that the astringency removal process could be substantially shortened by exposing the fruit to a high carbon dioxide atmosphere.[75] The rate of the induction response is temperature dependent; at 40°C the exposure may be as short as 6 hours. This initial induction process is followed by a period of astringency removal that occurs during a three-day air storage period (30°C). It is thought that the anaerobic conditions result in a small but significant build-up in acetaldehyde. Aldehydes react readily with phenols resulting in cross-linking between neighboring molecules, hence producing the insoluble non-astringent form. This same basic aldehyde-phenol reaction was used in the production of Bakelite plastic and when discovered in the early nineteen hundreds ushered in the plastics revolution.

Another important postharvest phenolic response is the discoloration (browning) of many products upon injury to the tissue. Injury may occur during harvest, handling or storage, resulting in the breakdown of organizational resistance between substrates and enzymes within the cell. Common examples would be bruising of fruits or broken-end discoloration of snap beans. When browning occurs, constituent phenols are oxidized to produce a quinone or quinone-like compound that polymerizes, forming brown pigments. These unsaturated brown polymers are sometimes referred to as melanins or melaninodins. Among the compounds believed to be important substrates are chlorogenic acid, neochlorogenic acid, catechol, tyrosine, caffeic acid, phenylalanine, protocatechin and dopamine.

Research on phenolic discoloration has focused on the cellular constituents responsible for browning and the handling and storage techniques that can be used to prevent browning. The success of attempts to correlate the concentration of total and specific phenols or the level of activity of the phenolase enzymes to tissue discoloration varies widely, depending on the tissue in question and in some cases, even with the cultivar studied. Several postharvest techniques, however, have been successful for inhibiting the browning of some products. For example, during the harvest of snap beans it is not uncommon for the ends of many of the bean pods to be broken. These brown fairly readily and upon processing yield a distinctly substandard product. The reactions involved in discoloration can be inhibited by exposing the beans to 7500 to 10,000 $\mu L \cdot L^{-1}$ sulfur dioxide for 30 seconds.[89] The use of controlled atmosphere storage and antioxidants has been successful with other products.

8. VITAMINS

Vitamins represent a group of organic compounds required in the human diet in relatively small amounts for normal metabolism and growth. Plant products provide a major source for many of the vitamins required by man. Exceptions to this would be vitamin B_{12} that appears to be synthesized only by microorganisms and vitamin D, obtained mainly from the exposure of our skin to ultraviolet light. In plants, many of the vitamins perform the same biochemical function as they do in animal cells. As a consequence, most have a vital role in plant metabolism, in addition to being a source of vitamins for animals.

Although vitamins are required in only small amounts in the diet, deficiencies have been a serious problem throughout the history of man. Even today deficiencies of certain vitamins result in major nutritional diseases in many areas of the world. Insufficient vitamin A in the diet is perhaps the most common, especially in parts of Asia. Prolonged deficiency, particularly in young children, results in blindness.

Typically, vitamins are separated into two classes based on their solubility: the water-soluble vitamins (thiamine, riboflavin, nicotinic acid, pantothenic acid, pyridoxine, biotin, folic acid and ascorbic acid) and those that are lipid-soluble (vitamins A, E and K). Normally the lipid-soluble vitamins are stored in the body in moderate amounts; as a consequence, a consistent daily intake is not essential. The water-soluble vitamins, however, tend not to be stored and a fairly constant day-to-day supply is required.

Vitamins function in a catalytic capacity as coenzymes—organic compounds that participate in the function of an enzyme. The role of the water-soluble vitamins (principally coenzymes) is for the most part fairly well understood, while the precise metabolic function of the lipid-soluble vitamins is not yet clear.

8.1. Water-Soluble Vitamins

8.1.1. Thiamine

Structurally thiamine or Vitamin B_1 is a substituted pyrimidine joined to a substituted thiazole group through a methylene bridge (Figure 4.39). It is water-soluble and is relatively stable at pH's below 7.0. The vitamin is found widely in both plants and animals, in several forms. In plant tissues it is found most abundantly as free thiamine; however, it may also be found as mono-, di- and triphosphoric esters and as mono- and disulfides in various biological materials. Thiamine functions in plants as the coenzyme thiamine pyrophosphate, which plays a major role in the glycolytic pathway (the decarboxylation of pyruvic acid), the tricarboxylic acid cycle (the decarboxylation of α-ketoglutaric acid) and in the pentose phosphate pathway (as a carboxyl group transferring enzyme).

In contrast to some vitamins such as ascorbic acid, the concentration of thiamine in various plants and plant parts is fairly uniform, generally varying only within a 20 to 40-fold range. Typically, dried beans and peas contain approximately 700 to 600 $\mu g \cdot 100\ g^{-1}$; nuts 500 to 600 μg; whole grain cereals, 300 to 400 μg; and fruits and vegetables 20 to 90 μg. In addition to variation among species, the absolute concentration of thiamine in edible plant products varies somewhat with cultivar and growing conditions. After harvest, the vitamin is relatively stable

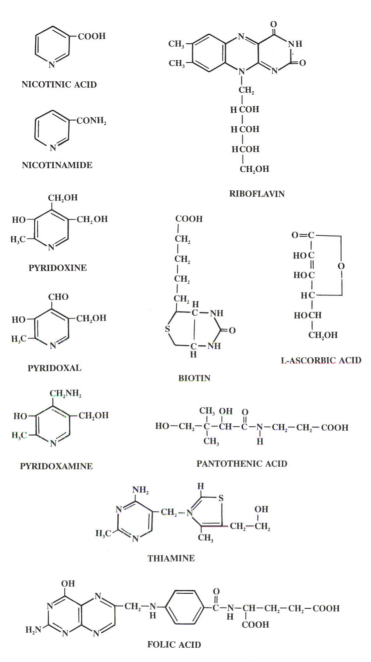

Figure 4.39. Structures of water soluble vitamins.

during storage. Primary losses incurred are most commonly the result of cooking, due largely to the molecules high water solubility.

8.1.2. *Riboflavin*

Riboflavin is a water-soluble derivative of D-ribose which contains an isoalloxazine ring [6,7-dimethyl-9-(D-1′-ribityl) isoalloxazine] (Figure 4.39). It is also know as vitamin B$_2$ and lactoflavin. Riboflavin or derivatives of riboflavin are found in all plants and many microorganisms; however, it is not synthesized by higher animals. Within plants it is found combined with other groups, largely as either flavin mononucleotide (FMN) or flavin adenine dinucleotide (FAD). This is also the case in most parts of the body of animals although in the retina of the eye it is found in its free form. Flavin nucleotides act as prothestic groups in oxidation-reduction enzymes. The isoalloxazine ring portion of the flavin nucleotide undergoes a reversible reduction to yield the reduced nucleotides $FMNH_2$ and $FADH_2$. Enzymes containing flavin nucleotides are essential for the oxidation of pyruvate and fatty acids and function in the electron transport system. Typically, the oxidized form of the molecule is colored yellow, red or green, while the reduced form is colorless.

Leafy vegetables represent a relatively good source of riboflavin, although vegetables that are high in riboflavin (e.g., pimento pepper, 0.46 mg $\cdot$ 100 g^{-1} fresh wt., mushrooms, 0.30 mg; lotus root, 0.22 mg; salsify, 0.22 mg) are consumed in only small amounts in most diets.

8.1.3. Niacin

Niacin or nicotinic acid is found widely in both plants and animals, either as the acid or amide (Figure 4.39). The name nicotinic acid comes from the molecules role as a component of the toxic alkaloid, nicotine from tobacco. Nicotinic acid is synthesized by animals, if their diet contains sufficient protein which is high in the amino acid tryptophan, the primary precursor of the vitamin. The coenzymes contain nicotinamide as an essential component. These are nicotinamide adenine dinucleotide (NAD) and nicotinamide adenine dinucleotide phosphate (NADP), known also as pyridine coenzymes. Both function as coenzymes in a large number of oxidation-reduction reactions catalyzed by what are known as pyridine-linked dehydrogenases. Most of the dehydrogenase enzymes are specific for either NAD or NADP, although several can utilize either form. In general, these reactions are reversible and are extremely important in many pathways within the cell.

8.1.4. Pyridoxine

Pyridoxine or vitamin B_6 is found in three forms, pyridoxine, pyridoxal and pyridoxamine (Figure 4.39)—pyridoxine typically being converted to the latter two forms, which are more efficacious. The active coenzyme forms of the vitamin are the phosphate derivatives: pyridoxal phosphate and pyridoxamine phosphate. Pyridoxine coenzymes function in a wide range of important reactions in amino acid metabolism such as transamination, decarboxylation and racemization reactions. Pyridoxal phosphate is also involved in the biosynthesis of ethylene, acting at the point of conversion of S-adenosyl methionine (SAM) to 1-aminocyclopropane-1-carboxylic acid (ACC). Ethylene synthesis is blocked at this step by 2-amino-4-aminoethoxy-*trans*-3-butenoic acid (AVG), a potent inhibitor of pyrixodal phosphate mediated enzyme reactions.

The three forms of the vitamin are found widely distributed in both the plant and animal kingdom; the predominant form, however, varies between sources. In vegetables, pyridoxial is the predominant form. Cereals (0.2 to 0.4 mg $\cdot$ 100 g^{-1}) and vegetables (e.g., Brussels sprouts, 0.28 mg $\cdot$ 100 g^{-1}; cauliflower, 0.2 mg; lima beans, 0.17 mg; spinach, 0.22 mg) represent good sources of the vitamin while many of the fruits are quite low (e.g., apple, 0.045 mg $\cdot$ 100 g^{-1}; orange, 0.05 mg). Glycosyl forms are also found.[224]

8.1.5. Pantothenic Acid

Pantothenic acid, formed from pantoic acid and the amino acid β-alanine, is found in limited quantities in most fruits and vegetables (Figure 4.39). The active form of the vitamin, coenzyme A (CoA), is synthesized from pantothenic acid in a series of steps. Coenzyme A functions as a carrier of acyl groups in enzymatic reactions during the synthesis and oxidation of fatty acids, pyruvate oxidation and a number of other acetylation reactions within the cell.

Deficiencies of the vitamin in animals are rare; a limited amount of storage of the molecule does occur in the heart, liver and kidneys. In diets with sufficient animal protein, most of the pantothenic acid is derived from this source. Dried peas and peanuts are considered good sources of the vitamin; walnuts, broccoli, peas, spinach, and rice, intermediate sources (0.5 to 2.0 mg · 100 g^{-1}); and onions, cabbage, lettuce, white potatoes, sweetpotatoes and most fruits, poor sources (0.1 to 0.5 mg · 100 g^{-1}).

8.1.6. Biotin

The vitamin biotin consists of fused imidazole and thiophene rings with an aliphatic side chain (Figure 4.39). Its structure, established in 1942, suggested the possible role of pimelic acid as the natural precursor of the molecule, which was later proven to be correct. Biotin is found widely in nature, usually in combined forms bound covalently to a protein through a peptide bond. When bound to a specific enzyme, it functions in carboxylation reactions. Here it acts as an intermediate in the transfer of a carboxyl group from either a donor molecule or carbon dioxide to an acceptor molecule. Examples of enzymes in which biotin acts as a carboxyl carrier are propionyl-CoA carboxylase and acetyl-CoA carboxylase.

Biotin is found widely distributed in foods and is also synthesized by bacteria in the intestine. As a consequence, deficiencies are extremely rare. When present they are normally associated with high intake of avidin, a protein found in raw egg whites that binds to the vitamin making it unavailable. Legumes, especially soybeans, represent an excellent plant source of biotin (61 µg · 100 g^{-1} edible product). In addition, nuts such as peanuts (34 µg), pecans (27 µg) and walnuts (37 µg) and a number of vegetables are good sources (e.g., southern peas, 21 µg; cauliflower, 17 µg; mushrooms, 16 µg), while most fruits and processed grains are consistently low in biotin.

8.1.7. Folic Acid

Folic acid is found widely distributed in plants, its name being derived from the Latin word *folium* for "leaf" from which it was first isolated. Structurally the molecule is composed of three basic subunits: 1) a substituted pteridine, 2) *p*-aminobenzoic acid, and 3) glutamic acid (Figure 4.39). The active coenzyme form of the vitamin is tetrahydrofolic acid, formed in a two-step reduction of the molecule. It functions as a carrier of one-carbon units (e.g., hydroxymethyl-CH$_2$OH, methyl-CH$_2$ and formyl-CHO groups) when these groups are transferred from one molecule to another. These reactions are critical steps in the synthesis of purines, pyrimidines and amino acids.

Folic acid is found widely in the plant kingdom and is also synthesized by microorganisms including intestinal bacteria. It is needed by humans and other animals in very small amounts (e.g., 0.4 mg · day^{-1} for humans) but is rapidly excreted from the body. Asparagus, spinach and dried beans are excellent sources of the vitamin; corn, snap beans, kale and many nuts, moderate sources (30 to 90 µg · 100 g^{-1}); and cabbage, carrots, rice, cucumbers, white potatoes, sweetpotatoes and most fruits, poor sources (0–30 µg · 100 g^{-1}).

8.1.8. Ascorbic Acid

Ascorbic acid is structurally one of the least complex vitamins found in plants.[217] It is a lactone of a sugar acid (Figure 4.39), synthesized in plants from glucose or another simple carbohydrate. It was first isolated in crystalline form in 1923. In spite of years of research, the precise physiological function of ascorbic acid in plant and animal cells remains unclear.[217] It is known to act as a cofactor in the hydroxylation of proline to hydroxyproline, however, other reducing agents can replace it.

Ascorbic acid is required in the diet of man and only a small number of other vertebrates and is supplied primarily by fruits and vegetables, although a small amount is found in animal products such as milk, liver and kidneys. In comparison with the other water-soluble vitamins in plants, ascorbic acid is found in relatively high concentrations. Guava (300 mg $\cdot 100$ g^{-1} fresh weight), black currants (210 mg), sweet peppers (125 mg) and several greens (kale, collards, turnips — 120 mg) are excellent sources. The West Indian cherry (acerola) contains approximately 1300 mg $\cdot$ 100 g^{-1} fresh weight.[179] Staples such as rice, wheat, corn and many of the starchy tubers tend to be extremely low. Fruits have a distinct advantage in the diet in that they are often served raw. During cooking a significant portion of the ascorbic acid of many vegetables is lost. This is due primarily to leaching of the water-soluble vitamin out of the tissue and to oxidation of the molecule. Losses from leaching tend to be greater in leafy vegetables due to the surface area in contrast to bulkier products.

The concentration of ascorbic acid often varies with location within a specific plant part and between different parts on the same plant. For example, in many fruits, the concentration in the skin is higher than in the pulp. The concentration of ascorbic acid declines fairly rapidly in many of the more perishable fruit and vegetables after harvest. Losses are greater with increasing storage temperature and duration.

8.2. Lipid-Soluble Vitamins

8.2.1. Vitamin A

Vitamin A or retinol is an isoprenoid compound with a 6 carbon cyclic ring and an 11 carbon side chain (Figure 4.40). It is formed in the intestinal mucosa by cleavage of carotene. Of the numerous naturally occurring carotenoids, only 10 have the potential to be converted into vitamin A and of these, β-, α- and γ-carotene are the most important. The presence of a β end group is essential for the formation of the molecule. Beta-carotene with two β end groups has twice the potential vitamin A as α-carotene, which is composed of a β and ε end group. Cleavage appears to be due to the presence of the enzyme β-carotene-15,15′-dioxygenase that oxidizes the central double bond; however, it is possible that other conversion mechanisms may occur.

Vitamin A is extremely important in human nutrition in that its synthesis is dependent upon carotene ultimately from plant sources. In contrast to ascorbic acid, only a small amount of vitamin A is needed in the diet. This ranges from 0.4 to 1.2 mg $\cdot$ day^{-1} depending on age and sex. Although it appears to be required in all of the tissues of the body, its general function in metabolism is not known, aside from its role in eyesight. A deficiency in young children results in permanent blindness, a common problem in many tropical areas of the world.

Since vitamin A per se is not present in plants, its potential concentration is measured in "international units" (IU), based on the concentration of α- and β-carotene in the tissue. One international unit of vitamin A is equal to 0.6 μg of β-carotene or 1.2 μg of α-carotene. Leafy vegetables average approximately 5000 IU $\cdot$ 100 g^{-1} fresh weight; and fruits 100–500 IU, although the mango (3000 IU) and papaya (2500 IU) are distinctly higher; while staple crops such as rice, peanuts and cassava have virtually none. An exception to this would be orange fleshed sweetpotatoes in which some of the high carotene cultivars contain up to 14,000 IU of vitamin A $\cdot$ 100 g^{-1} fresh weight.

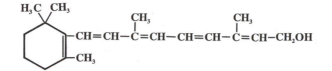

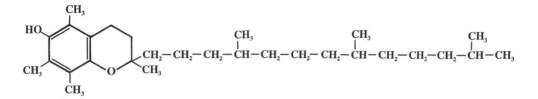

VITAMIN A

VITAMIN E

VITAMIN K

Figure 4.40. Structures of lipid soluble vitamins.

The concentration of carotene is known to vary widely among species and cultivars and less so due to production environmental conditions and cultural practices, although temperature and light intensity are known to have significant effects.[57, 212]

8.2.2. *Vitamin E*

Vitamin E or α-tocopherol is a molecule composed of a chromanol ring and a side chain formed from a phytol residue (Figure 4.40). In addition to α-tocopherol, β-, γ- and δ-tocopherol are also found in photosynthetic plants, although α-tocopherol is the most active form as a vitamin.

The biological role of α-tocopherol in animals, as in plants, is unclear. Vitamin E deficiency results in a number of symptoms in test animals, one of which is infant mortality, hence the derivation of the name tocopherol from the Greek word *tokos* meaning "childbirth." Tocopherols are known to have antioxidant activity, which prevents the autoxidation of unsaturated lipids. As a consequence, one function may be the protection of membrane lipids.

In plants, α-tocopherol is found associated with the chloroplast membrane and is thought to also be present in mitochondria. It also appears to be located in various plastids. Plant oils are an excellent source of tocopherols. Significant concentrations are found in wheat germ, corn and pecan oils. Pecans contain up to 600µg of tocopherol $\cdot$ g^{-1} of oil approximately 6 weeks prior to maturity. This amount declines to 100 to 200 µg $\cdot$ g^{-1} by maturation. In contrast to many other nuts (e.g., filberts, walnuts, Brazil nuts, almonds, chestnuts and peanuts), the γ-tocopherol isomer is found almost exclusively (>95%) in lieu of the α-tocopherol isomer.

The role of tocopherols in plants is thought to be related to their antioxidant properties.[39] This is supported by the correlation between the concentration of tocopherols in pecans, which tend to deteriorate in storage due to the oxidation of their component lipids, and the length of time they can be successfully stored. In pecan oils with a constant linolate concentration, keeping time increased in a linear fashion up to 800 µg of tocopherol $\cdot$ g^{-1} of oil.[196] The germination of wheat seeds is also correlated with tocopherol content.[45] Tocopherols are also thought to function as a structural component of chloroplast membranes and may in some way function in the initiation of flowering of certain species.

8.2.3. *Vitamin K*

Vitamin K or phylloquinone is a lipid soluble quinone, which is in many ways structurally very similar to α-tocopherol. Both are cyclic compounds with a phytol residue side chain composed of isoprene units (Figure 4.40). Two forms of the vitamin, K_1 and K_2, are known. A deficiency of vitamin K limits proper blood clotting in animals through the repressed formation of fibrin, the fibrous protein portion of blood clots. This in turn results in a tendency to hemorrhage. Aside from this specific function, its widespread occurrence in plants and microorganisms suggests a more general, but presently undefined biological role. As a quinone, it could possibly function as an electron carrier.

In plants, phylloquinone is present in most photosynthetic cells, hence leafy green tissues represent an excellent dietary source of the vitamin. Plant parts that normally do not contain chlorophyll have little vitamin K. Likewise, mineral deficiencies that repress chlorophyll synthesis (e.g., Fe) also appear to decrease the concentration of vitamin K.

9. PHYTOHORMONES

Five major groups of naturally occurring compounds, the phytohormones, are currently known to exist each of which exhibits strong plant growth regulating properties.[124] Included are ethylene, auxin, gibberellins, cytokinins and abscisic acid (Figure 4.41); each is structurally distinct and active in very low concentrations within plants.[41] Several new compounds have been identified (salicylic acid, jasmonate, and polyamines) which are involved in the plant's response to environmental stress and disease and insect resistance. In that their importance during the postharvest period is only beginning to be deciphered, the following critique will focus on the five "classical" plant hormones.

While each of the phytohormones has been implicated in a relatively diverse array of physiological roles in plants and detached plant parts, the precise mechanism in which they function is not yet completely understood.[42] During the postharvest period, ethylene is of major importance in that it is closely associated with the regulation of senescence in some products and the ripening of many fruits.[4] This section focuses primarily on the synthesis and deactivation of these molecules and changes in their concentration occurring during the postharvest period.

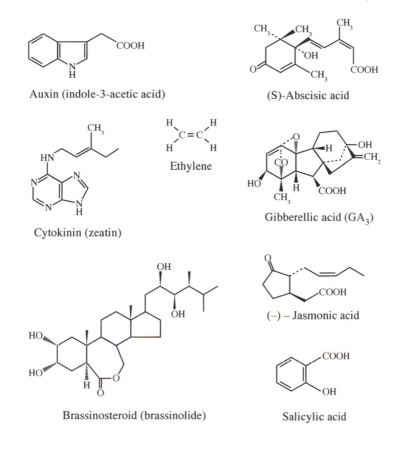

Figure 4.41. Structures of the phytohormones auxin, abscisic acid, ethylene, cytokinin, gibberellic acid, jasmonic acid, salicylic acid, brassinosteroid, and polyamines, represented by spermidine.

9.1. Classes, Synthesis and Degradation of Phytohormones

9.1.1. *Ethylene*

Ethylene, being a gaseous hydrocarbon, is unlike the other naturally occurring plant hormones. Although ethylene was known to elicit such responses as gravitropism and abscission early in the last century, it was not until the 1960's that it began to be accepted as a plant hormone.[4]

The effect of ethylene on plants and plant parts is known to vary widely. It has been implicated in ripening, abscission, senescence, dormancy, flowering and other responses. Ethylene appears to be produced by essentially all parts of higher plants, the rate of which varies with specific organ and tissue and their stage of growth and development. Rates of synthesis range from very low (0.04–0.05 µL · kg^{-1} · hr^{-1}) as in blueberries to extremely high (3400 µL · kg^{-1} · hr^{-1}) in fading blossoms of *Vanda* orchids. Alterations in the rate of synthesis of ethylene

ETHYLENE SYNTHESIS PATHWAY

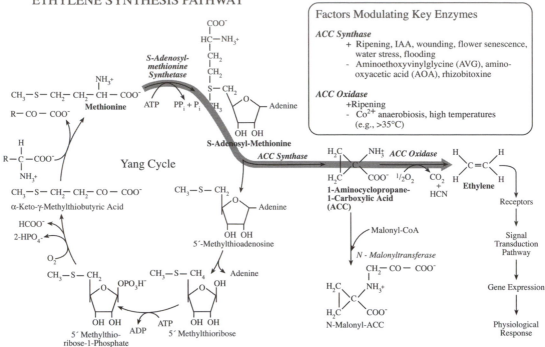

Figure 4.42. The ethylene synthesis pathway and Yang cycle. The initial precursor is the amino acid methionine and key regulatory enzymes in the pathway are ACC synthase and ACC oxidase which are positively and negatively modulated by a number of factors (see box). Hydrogen cyanide, formed in the last step of ethylene synthesis, is detoxified by cysteine. The ACC pool can be diminished *via* conversion to *N*-malonyl ACC. After synthesis ethylene may bind to receptors and through a series of steps instigate a physiological response or simply diffuse out of the tissue.

have been found, in some cases, to be closely correlated with the development of certain physiological responses in plants and plant organs, for example the ripening of climacteric fruits and the senescence of flowers.

Ethylene is synthesized from the sulfur-containing amino acid methionine which is first converted to s-adenosyl methionine (SAM) and then to the 4 carbon compound, 1-amino-cyclopropane-1-carboxylic acid (ACC) by the enzyme ACC synthase (Figure 4.42). During conversion to ACC, the sulfur-containing portion of the molecule, 5-methylthioadenosine, is cycled back to methionine *via* the formation of ribose and condensation with homoserine.[123,253] The final step in the synthesis pathway is the conversion of ACC to ethylene by ACC oxidase (ACO) which requires oxygen (previously called the "ethylene forming enzyme"). Stress (water, mechanical and others) is known to stimulate ethylene synthesis and, under some conditions, markedly so.

While ethylene appears to be synthesized in all cells, ACO is possibly associated with the tonoplast, vacuoles isolated from protoplasts were able to convert ACC to ethylene.[79, 166] Likewise, protoplasts that had their vacuoles removed (evacuolated) lose their capacity to produce ethylene from ACC; when the vacuoles were allowed to reform, synthesis is reinstated.[53]

Several potent inhibitors of ethylene synthesis have been found (rhizobitoxine and AVG, Figure 4.42) and were integral components in elucidating the pathway. Lieberman first showed that fungal metabolites from *Rhizobium japonicum, Streptomyces* sp. and *Pseudomonas aerug-*

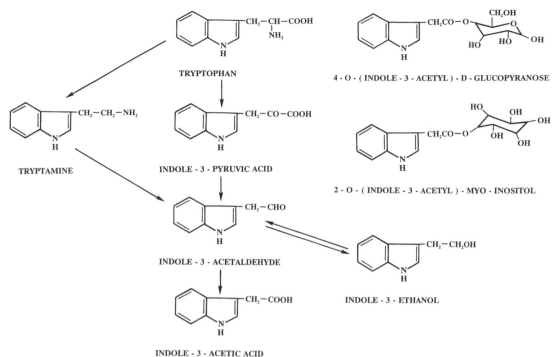

Figure 4.43. Biosynthesis of indoleacetic acid from tryptophan *via* both the indole-3-pyruvic acid and tryptamine pathways.

inosa inhibit the conversion of SAM to ACC. These unfortunately also inhibit other pyridoxal phosphate requiring enzymes in plants and animals and as a consequence are of little commercial value for postharvest products that are to be consumed.[4] Silver and 1-methylcyclopropene (1-MCP) are effective inhibitors of ethylene action *via* their interference with the binding site for ethylene. 1-Methylcyclopropene provides a more transitory (i.e., 5 to 7 days) inhibition than silver ions which exhibit a longer-term effect.

Since ethylene is continuously being produced by plant cells, some mechanism is essential to prevent the build-up of the hormone within the tissue. Unlike other hormones, gaseous ethylene diffuses readily out of the plant. This passive emanation of ethylene from the plant appears to be the primary means of eliminating the hormone. During the postharvest period, techniques such as ventilation and hypobaric conditions help to facilitate this phenomenon by maintaining a high diffusion gradient between the interior of the product and the surrounding environment. A passive elimination system of this nature would imply that the internal concentration of ethylene is controlled largely by the rate of synthesis rather than the rate of removal of the hormone.

Ethylene may also be metabolized within the cell, decreasing the internal concentration. Products such as ethylene oxide and ethylene glycol have been found, however, their importance in regulating the internal concentration of ethylene in most species appears to be very minimal.

9.1.2. Auxin

The name auxin, from the Greek word "*auxin*" meaning to increase, is given to a group of compounds that stimulate elongation. Indoleacetic acid (IAA) (Figure 4.43) is the prevalent

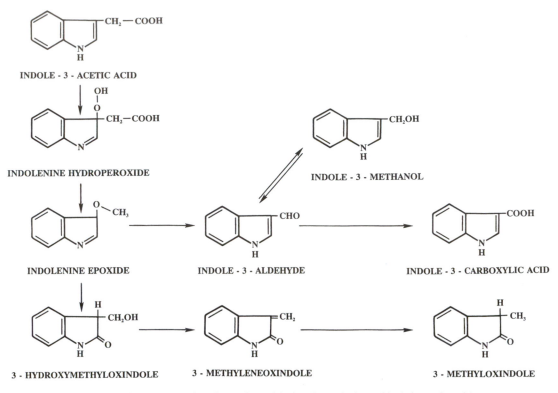

Figure 4.44. Proposed pathway for oxidative degradation of indoleacetic acid.

form, however, evidence suggests that there are other indolic auxins naturally occurring in plants.[9] Although auxin (indoleacetic acid) is found throughout the plant, the highest concentrations are localized in actively growing meristematic regions. It is found as both the free molecule and as inactive conjugated forms. When conjugated, auxin is metabolically bound to other low molecular weight compounds. This process appears to be reversible. The concentration of free auxin in plants ranges from 1 to 100 mg · kg^{-1} fresh weight. In contrast, the concentration of conjugated auxin can be substantially higher.

One striking characteristic of auxin is the strong polarity exhibited in its transport throughout the plant.[138] Auxin is transported *via* an energy dependent mechanism, basipetally-away from the apical tip of the plant toward the base. This flow of auxin represses the development of axillary lateral buds along the stem thus maintaining apical dominance. Movement of auxin out of the leaf blade toward the base of the petiole also appears to prevent leaf abscission.

Auxin has been implicated in the regulation of a number of physiological processes. For example, evidence for its role in cell growth and differentiation, fruit ripening, flowering, senescence, gravitropism, abscission, apical dominance and other responses has been given. The actual binding of the molecule, the signaling sequence and the means by which it instigates this diverse array of physiological events has not been fully elucidated. During auxin-induced cell elongation, it is thought to act both through a rapid direct effect on an ATPase proton pump mechanism in the plasma membrane and a secondary effect mediated through enzyme synthesis.

The obvious similarity between the amino acid tryptophan and indoleacetic acid (Figure 4.43) lead to the initial proposal that tryptophan represented the precursor of the hormone.

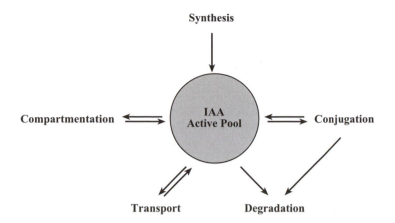

Figure 4.45. The concentration of IAA in the active pool within the cell is controlled by the rates of synthesis, conjugation, degradation, transport, and compartmentation.

Subsequent tests with labeled tryptophan substantiated its role as precursor and helped to elucidate the specific steps involved in the degradation of the side chain of the amino acid. This involves deamination, decarboxylation and two oxidation steps, by at least two general pathways, one through indole-3-pyruvic acid and a second by way of tryptamine (Figure 4.43). In addition to these two primary means of auxin synthesis, the identification of several chlorinated auxins, suggested the potential for alternate pathways.[9] Here the chlorine atom is found on the benzene ring, and appears to be added prior to alteration of the tryptophan side chain.

Auxin is active at very low levels in plant cells and, as a consequence, precise control over the internal concentration of the molecule is essential. As environmental conditions to which the plant is exposed change, relatively rapid and significant alterations in auxin concentration may be necessary. The concentration of auxin within a group of cells can be altered by: a) the rate of synthesis of the molecule; b) its rate of transport into or out of those cells; c) the rate of breakdown of the molecule; and d) the formation of conjugates or conversely auxin liberation from existing conjugates. Interconversion and catabolism of indoleacetic acid can be mediated by both enzymatic ("IAA oxidase," most probably a peroxidase) and non-enzymatic (e.g., H_2O_2 direct oxidation, light, UV radiation, and others) means. The proposed pathway for indoleacetic acid oxidation is presented in Figure 4.44. Peroxidases are found throughout the plant kingdom and some, in addition to exhibiting peroxidase activity, also appear to have the ability to oxidize auxin. Thus the endogenous concentration of auxin can be decreased by the action of these enzymes. Enzymatic control of the concentration of auxin, therefore, could represent a method of regulating certain physiological processes in which auxin is involved. In fact, the level of activity of auxin degrading enzymes has been correlated with the development of specific responses (e.g., fruit ripening). Another way of regulating response is to vary the tissues' sensitivity to auxin.

Conjugation of auxin to other low molecular weight compounds also represents a means of modulating the concentration of the hormones within a cell (Figure 4.45). This process does not exclude the potential for reversibility; thus the reaction can be reversed to yield the free active forms. At present, there are three major groups to which auxin has been found to be bound, in each case through the carboxyl group of the hormone. These include peptidyl IAA conjugated where auxin is linked to an amino acid through a peptide bond, glycosyl IAA conjugates where auxin is linked to a sugar through a glycosidic or an ester bond, and a myo-inositol conjugate where auxin is linked to myo-inositol through an ester bond.

9.1.3. Gibberellins

The gibberellins represent a group of acidic diterpenoids found in angiosperms, gymnosperms, ferns, algae and fungi; they do not however, appear to be present in bacteria. Over 70 different free and 16 conjugated gibberellins have been isolated, many of which represent intermediates in the synthesis pathway and lack hormonal activity. Typically the different gibberellins are designated with a number (e.g., GA_3, GA_4, GA_5, ...) based on their chronological order of isolation and identification. While gibberellins have been shown to induce stem elongation and other responses [e.g., increase radial diameter in stems (conifers), induce flowering, etc.] their precise role in plants remains unknown. Often several gibberellins are found in the same plant.

The base molecule for the various forms is gibberellin, a 20 carbon diterpeniod (Figure 4.46).[225] Some, however, are minus the carbon 20 methyl group and therefore only have 19 carbons (referred to as the C_{19}-GAs). Individual gibberellins differ from each other in the oxidation state of the ring structure and the carbon and hydroxyl groups present. In plants and plant

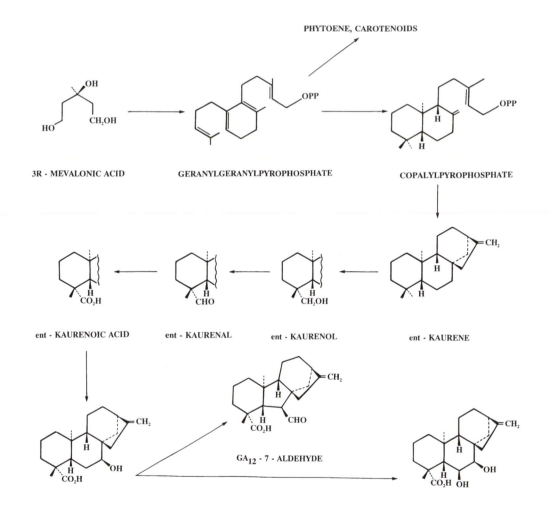

Figure 4.46. Pathway for gibberellic acid synthesis.

parts, gibberellins are also found as glycosides (typically of glucose) and other bound inactive forms. Several of these can be readily converted back to the free molecule upon hydrolysis.

Much of the data on the synthesis of gibberellins has come from studies on the fungus *Gibberella fujikuroi.* It appears that the same general pathway, at least until GA_{12}, is operative in higher plants. Since gibberellins are diterpenoids, labeled precursors of the terpenoid biosynthesis pathway (e.g., acetyl CoA and mevalonate) are monitored to establish the proposed pathway (Figure 4.46). The first complete cyclic compound in the pathway is kaurene. Kaurene is then, through a series of not clearly understood reactions, converted to various gibberellins.[86,88] Some interconversion between specific gibberellins occurs within plants. For example GA_1 can be converted to GA_3 or GA_5 and subsequently to GA_8.

Conjugates may represent an important means of modulating the internal concentrations of gibberellins within the plant. When gibberellins are applied to plant tissue there is often a rapid conversion to the inactive glucoside form. In addition, although more stable than auxins, gibberellins can be degraded to inactive compounds and some evidence for compartmentalization has been found.

Gibberellins are synthesized in the apical leaf primordia, root tips and in developing seeds. The hormone does not exhibit the same strongly polar transport as seen with auxin, although in some species there is basipetal movement in the stem. In addition to being found in the phloem, gibberellins have also been isolated from xylem exudate that suggests a more general, bi-directional movement of the molecule in the plant.

Several synthetic compounds have been shown to inhibit elongation of some plant species suggesting anti-gibberellin activity, a few of which are widely used in ornamental horticulture for producing more compact plants. They inhibit the normal synthesis of gibberellin within the plants, many of which through the inhibition of the enzyme kaurene synthase that catalyzes the synthesis step between geranylgeranyl pyrophosphate and copalyl pyrophosphate. Their use is of interest here in that inhibitors of GA synthesis appear to also significantly modify the postharvest response of many plants.

9.1.4. *Cytokinins*

Cytokinins are naturally occurring plant hormones that stimulate cell division. Initially they were called kinin, however, due to prior use of the name for a group of compounds in animal physiology, cytokinin (cyto kinesis or cell division) was adapted.

All of the naturally occurring cytokinins contain a N^6-substituted adenine moiety (Figure 4.47). Zeatin was the first naturally occurring cytokinin isolated and identified, however, since that time a number of others have been identified. They are found as the base molecule, a riboside (presence of ribosyl group at the R_3 position) or a ribotide.

The highest concentrations of cytokinins are found in embryos and young developing fruits, both of which are undergoing rapid cell division. The presence of high levels of cytokinins may facilitate their ability to act as a strong sink for nutrients. Cytokinins are also formed in the roots and are translocated *via* the xylem to the shoot. When in the leaf, however, the compounds are relatively immobile.

The precise mode of action of cytokinins is not known. While they do stimulate cell division, exogenous application is also known to cause several significant responses. When applied to detached leaves, cytokinins delay senescence, thus the rate at which degradative processes occur significantly decreases.[164] This is due in part to a facilitated movement of amino acids and other nutrients into the treated area. The site of response is localized to where the hormone is placed on the leaf, indicating little movement of cytokinin in the leaf. Considerable interest has been shown in this anti-senescence ability of cytokinins. Synthetic cytokinins

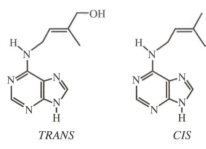

**BASE STRUCTURE
OF CYTOKININ**

STERIO ISOMERS OF ZEATIN

TRANS   *CIS*

SUBSTITUENTS			TRIVIAL NAME	SYSTEMATIC NAME	ABBREVIATIO
R_1	R_2	R_3*			
H	H		N^6-(Δ^2-ISOPENTENYL) ADENOSINE	6-(3-METHYLBUT-2-ENYLAMINO)-PURINE	i^6Ade
H	RIBOSYL		N^6-(Δ^2-ISOPENTENYL) ADENOSINE	6-(3-METHYLBUT-2-ENYLAMINO)-9-β-p-RIBOFURANOSYLPURINE	i^6A
H	H		*CIS*-ZEATIN	6-(4-HYDROXY-3-METHYL-*CIS*-BUT-2-ENYLAMINO)-PURINE	c-io^{6}Ade
H	RIBOSYL		*CIS*-ZEATIN RIBOSIDE	6-(4-HYDROXY-3-METHYL-*CIS*-BUT-2-ENYLAMINO)-9-β-p-RIBOFURANOSYLPURINE	c-io^{6}A
H	H		*TRANS*-ZEATIN	6-(4-HYDROXY-3-METHYL-*TRANS*-BUT-2-ENYLAMINO)-PURINE	t-io^{6}A
H	RIBOSYL		*TRANS*-ZEATIN RIBOSIDE	6-(4-HYDROXY-3-METHYL-*TRANS*-BUT-2-ENYLAMINO)-9-β-RIBOFURANOSYLPURINE	t-io^{6}A
H	H		DIHYDROZEATIN	6-(4-HYDROXY-3-METHYLBUTYLAMINO) PURINE	H$_2$-io^{6}Ade
H	RIBOSYL		DIHYDROZEATIN RIBOSIDE	6-(4-HYDROXY-3-METHYLBUTYLAMINO)-9-β-p-RIBOFURANOSYLPURINE	H$_2$-io^{6}A
CH_3-S	H			2-METHYLTHIO-6-(3-METHYLBUT-2-ENYLAMINO)-PURINE	ms^2-i^{6}Ade
CH_3-S	RIBOSYL			2-METHYLTHIO-6-(3-METHYLBUT-2-ENYLAMINO)-9-β-p-RIBOFURANOSYLPURINE	ms^2-i^{6}A
CH_3-S	H			2-METHYLTHIO-6-(4-HYDROXY-3-METHYL-*CIS*-BUT-2-ENYLAMINO)-PURINE	ms^2-c-io^{6}Ade
CH_3-S	RIBOSYL			2-METHYLTHIO-6-(4-HYDROXY-3-METHYL-*CIS*-BUT-2-ENYLAMINO)-9-β-p-RIBOFURANOSYLPURINE	ms^2-c-io^{6}A
CH_3-S	H			2-METHYLTHIO-6-(4-HYDROXY-3-METHYL-*TRANS*-BUT-2-ENYLAMINO)-PURINE	ms^2-t-io^{6}Ade
CH_3-S	RIBOSYL			2-METHYLTHIO-6-(4-HYDROXY-3-METHYL-*TRANS*-BUT-2-ENYLAMINO)-9-β-p-RIBOFURANOSYLPURINE	ms^2-t-io^{6}A

* In all cases, N^6 is linked to the C - 1 of the isoprenoid side chain

Figure 4.47. Naturally occurring cytokinins in plants (*redrawn from Sembdner et al.*[208]).

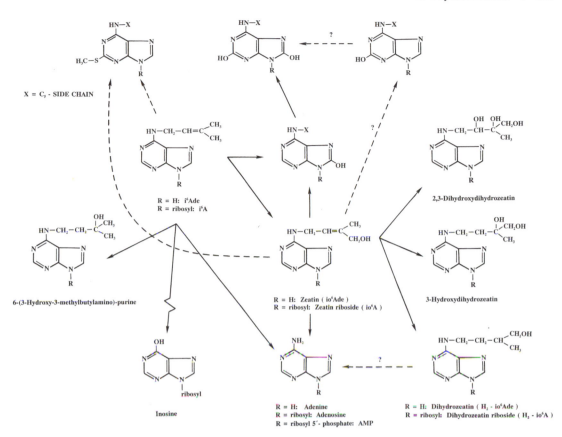

Figure 4.48. Proposed scheme for cytokinin interconversion (*redrawn from Sembdner et al.*[208]).

such as N⁶-benzyladenine have been applied to a number of postharvest products with varying degrees of success.

Other general effects of cytokinins on plants have been reported. These include: a) stimulation of seed germination; b) stimulation of the formation of seedless fruit; c) breaking seed dormancy; d) inducing bud formation; e) enhancing flowering; f) altering fruit growth; and g) breaking apical dominance. These responses tend to only be found in certain species and in some cases, cultivars, and are not widespread. It would appear, therefore, that they represent pharmacological responses, rather than precise physiological roles of the molecule in the plant. While these types of responses may not greatly expand our understanding of how cytokinins function in plants, some are of considerable interest in the commercial production and handling of agricultural plant products.

The biosynthesis of cytokinins is closely related to the metabolism of the purine adenine.[164] The isopentenyl side chain is added by the enzyme isopentenyl transferase. A general scheme for the interconversion of cytokinins, forming the various derivatives has been proposed (Figure 4.48). The level of activity is affected by the structure of these derivatives. The length of the side chain, degree of side chain unsaturation, and the stereochemistry of the double bond are important. Dihydroxyzeatin, without a double bond in the side chain, is only one-tenth as active as zeatin. Both the *cis* and *trans* stero-isomers of the side chain's double bond are found (Figure 4.47); however, the *trans* forms appear to be much more active.

Deactivation of cytokinins can occur through the conjugation of the molecule with a glycoside giving an inactive molecule. Cytokinins may also be degraded by the action of cytokinin

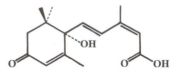

(+) - ABSCISIC ACID

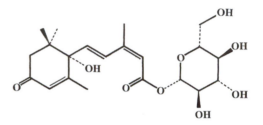

Figure 4.49. Structures of abscisic acid and its inactive glucose conjugate.

(+) - ABSCISYL - *β* - D - GLUCOPYRANOSIDE

oxidase that cleaves the side chain. The adenine portion of the molecule is then metabolized as a substrate or oxidized.

9.1.5. Abscisic Acid

Abscisic acid, previously known as dormin and abscisin, is a naturally occurring growth inhibitor in plants. Chemically, it is a terpenoid that is structurally very similar to the terminal portion of many carotenoids (Figure 4.49). Both *cis* and *trans* isomers are possible, however, only the *cis* form, designated (+)-ABA is active and is found almost exclusively in plants.

Abscisic acid is a potent growth inhibitor that has been proposed to play a regulatory role in such diverse physiological responses as dormancy, leaf and fruit abscission and water stress. Typically the concentration within plants is between 0.01 and 1 ppm, however, in wilted plants the concentration may increase as much as 40×. Abscisic acid is found in all parts of the plant; however, the highest concentrations appear to be localized in seeds and young fruits.

Two schemes for the synthesis of abscisic acid have been proposed.[137] The first or direct method involves the formation of the C_{15} carbon skeleton of abscisic acid from 3 isoprene units derived from mevalonic acid (Figure 4.50). The precise series of steps have yet to be fully elucidated. A second or indirect method was initially suggested based on the close similarity between the terminal ends of certain carotenoids, for example violaxanthin and abscisic acid. A lipoxygenase was subsequently isolated which would cleave carotenoids giving a range of compounds that were structurally similar to abscisic acid (e.g., xanthoxin). Exogenously applied xanthoxin was then converted to abscisic acid. In some cases this indirect scheme of synthesis may occur, however, its importance appears to be minimal. The major pathway for abscisic acid synthesis is the direct scheme from mevalonic acid.

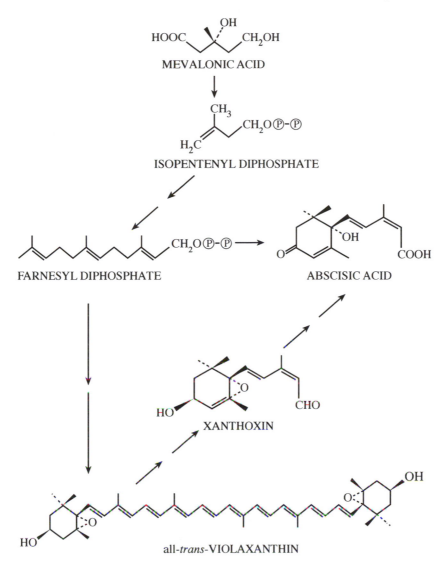

Figure 4.50. General pathway for the synthesis of abscisic acid from mevalonic acid (top) and a possible indirect scheme for the synthesis of abscisic acid from carotenoids (bottom). Violaxanthin is attacked by lipoxygenase yielding xanthoxin and other products. Xanthoxin is then converted to abscisic acid. This does not appear to represent a significant pathway *in vitro*.

Degradation of abscisic acid or loss of activity occurs through two primary mechanisms,[39] conjugation and metabolism (Figure 4.51). Abscisic acid rapidly forms an inactive conjugated with glucose. Glucosyl abscisate has been identified in a number of plants. Abscisic acid presumably may also form conjugates with other carbohydrates and types of compounds (i.e., proteins or lipids). It has been proposed that conjugation represents a means of interconversion between active and inactive forms of the molecule, thus controlling the internal concentration within the cell. Abscisic acid is rapidly metabolized in the plant. This results in a much less active derivative, (e.g., phaseic acid) or inactive compounds.

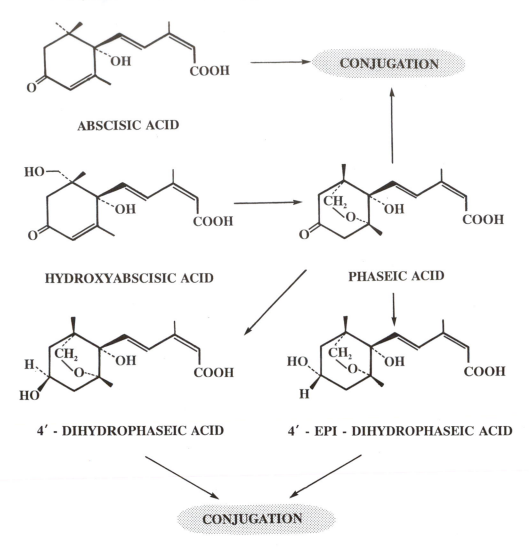

Figure 4.51. Conversion of abscisic acid to metabolites of low hormonal activity or devoid of activity.

9.2. Postharvest Alteration in the Concentration of Phytohormones

Due to the apparent effect of phytohormones on an array of physiological responses within the plant, the fate of these compounds after harvest is of considerable interest. Unfortunately, techniques for the measurement of phytohormones, especially auxin, gibberellins, cytokinins and abscisic acid are extremely complicated. In addition, few studies have monitored concurrently all five phytohormones and, as a consequence, a clear picture of gross postharvest alterations is not yet available. Typically, only one or occasionally two phytohormones are assessed in a single study. More importantly, the interpretation of the results of most studies is complicated by isolation and/or quantification procedures used. Often internal standards are not utilized during isolation and bioassays are relied upon to measure the relative activity of the impure isolates. Even when essential isolation and quantification prerequisites are met, one must assume that the hormone is not sequestered in a concentrated pool(s) somewhere within each cell (hence disruption during isolation would dilute the concentration) and that all the cell types making up an organ such as a fruit, have equal amounts of the hormone. Both are rather dubious assumptions. Thus, even when a close correlation is found between the gross concentration and a specific physiological event, the precise meaning remains in question. Interpretation of the response is further complicated by changes in tissue sensitivity during development.

The phytohormone ethylene, being a gas which diffuses readily out of a tissue, has the distinct advantage that tissue disruption is not required for isolation, and it can be analytically quantified relatively easily using gas chromatography at concentrations as low as several nL · L^{-1} of air (parts per billion). Because of this and the apparent relative importance of this phytohormone in senescence and fruit ripening, two major postharvest phenomena of both physiological and commercial interest, the following discussion will focus to a large extent on ethylene.

9.2.1. Flower Senescence

Distinctive alterations in the concentration of phytohormones are found during the senescence of flowers whether attached or detached from the parent plant. The synthesis of ethylene rises sharply as the result of pollination and senescence in the carnation and other flowers. Exogenous application of ethylene accelerates senescence while factors that inhibit its synthesis or action delay senescence. The rise in ethylene production occurs first in the carnation stigma followed by the ovary, receptacle and, lastly, the petal tissue. The chronological timing of the ethylene increase by the stigma would suggest that it is mediated by pollen tube penetration of the stigma rather than fertilization of the ovules. Thus, the initial response may be due partly to wounding.

With the increase in ethylene production upon pollination, there is a stimulation of ovary development with movement of carbohydrates out of the petals into the gynoecium. This transport effect can be stimulated without the occurrence of pollination by exposure of the flower to an appropriate concentration of ethylene. The close correlation between ethylene synthesis, ovary development and perianth senescence provides a strong argument for a natural role for ethylene in the flowering response.

Abscisic acid concentration also increases during carnation flower and rose petal senescence; its exogenous application can accelerate the process. The relationship between increases in abscisic acid concentration and senescence are not clear. In rose petals, the increase in abscisic acid occurs several days after an increase in ethylene synthesis, while the opposite appears to be the case with the carnation flower. Thus the nature of the interaction between these phytohormones is obscure.

Exogenous application of a synthetic cytokinin will extend the longevity of cut carnations[50] and anthuriums.[181] In addition, the endogenous concentration of cytokinin increases as rose flowers begin to open but declines substantially thereafter.[161] Based on this, it has been suggested that cytokinins are important during the opening and maturation response of the flower.

9.2.2. Fruit Ripening

Climacteric fruits exhibit an exponential increase in the rate of ethylene emanation near the onset of ripening.[2] During ripening of these fruits there is a large increase in the rate of respiration and carbon dioxide production. Since exogenous application of ethylene to preclimacteric fruits will induce ripening, the natural role of ethylene in inducing the ripening response has been a subject of considerable interest.

Many fruits exhibit a respiratory climacteric where the increase in ethylene synthesis precedes the increase in respiration; however, this sequence of events is not universal. In some species (e.g., mango and apple) the respiratory climacteric appears more or less simultaneously with, but not preceded by an increase in ethylene synthesis. In feijoa, cherimoya and avocado fruits, the respiratory climacteric significantly precedes the increase in ethylene synthesis. In addition, the fruit's sensitivity to the internal concentration of ethylene may change as ripening approaches. Thus, while ethylene appears to be present in all ripening fruits, its concentration does not necessarily need to rise in order to initiate ripening. In these cases, tissue sensitivity to ethylene could be changing.

Other phytohormones also appear to be important in the ripening response. Many avocado cultivars do not ripen while attached to the tree, suggesting that some inhibitor produced by the parent plant is operative. Likewise, when apple fruit are detached from the tree, the rate of ethylene synthesis increases.

Changes in auxin concentration have been proposed as part of a multi-hormonal scheme controlling ripening. This is supported by the retardation of ripening with the application of auxin and acceleration with an anti-auxin compound.[64,65] A decline in auxin concentration within some fruits as they approach the onset of ripening appears to be due to the action of "IAA oxidase."[63] However, a decline in IAA does not appear to be a universal phenomenon (e.g., apple). Thus there is insufficient evidence at this time to support IAA having a direct role in controlling the onset of ripening.

ADDITIONAL READING

Abel, S., and A. Theologis. 1996. Early genes and auxin action. *Plant Physiol.* 111:9–17.

Bandurski, R.S., J.D. Cohen, J. Slovin and D.M. Reinecke. 1995. Auxin biosynthesis and metabolism. Pp. 39–65. In: *Plant Hormones and Their Role in Plant Growth Development.* P.J. Davies (ed.). Kluwer, Dordrecht, Netherlands.

Binns, A. 1994. Cytokinin accumulation and action: Biochemical, genetic, and molecular approaches. *Annu. Rev. Plant Physiol. Plant Mol. Biol.* 45:173–196.

Birch, G.G., and M.G. Lindley (eds.). 1987. *Developments in Food Flavors.* Elsevier, New York.

Browse, J., and C. Somerville. 1991. Glycerolipid metabolism: Biochemistry and regulation. *Annu. Rev. Plant Physiol. Plant Mol. Biol.* 42:467–506.

Buchanan, B.B., W. Gruissem and R.L. Jones. 2000. *Biochemistry and Molecular Biology of Plants.* Amer. Society of Plant Physiologist, Rockville, MD.

Chappell, J. 1995. Biochemistry and molecular biology of the isoprenoid biosynthetic pathway in plants. *Annu. Rev. Plant Physiol. Plant Mol. Biol.* 46:521–547.

Chichester, C.O. (ed.). 1972. *The Chemistry of Plant Pigments.* Academic Press, New York.

Chory, J., and J. Li. 1997. Gibberellins, brassinosteroids and light-regulated development. *Plant Cell Environ.* 20:801–806.

Cordell, G. (ed.). 1997. *The Alkaloids.* Academic Press, San Diego, CA.

Cosgrove, D.J. 1997. Assembly and enlargement of the primary cell wall in plants. *Annu. Rev. Cell Dev. Biol.* 13:171–201.

Ellis, B.E., G.W. Kuroki and H.A. Stafford (eds.). 1994. *Genetic Engineering of Plant Secondary Metabolism.* Vol. 28 of *Recent Advances in Phytochemistry.* Plenum, New York.

Fischer, R.L., and A.B. Bennett. 1991. Role of cell wall hydrolases in fruit ripening. *Annu. Rev. Plant Physiol. Plant Mol. Biol.* 42:675–703.

Gershenzon, J., and Croteau, R. 1991. Terpenoids. Pp.165–219. In: *Herbivores: Their Interactions with Sec-*

ondary Plant Metabolites. Vol. 1 of *The Chemical Participants.* G.A. Rosenthal and M.R. Berenbaum (eds.). Academic Press, San Diego, CA.

Gianfagna, T.J. 1995. Natural and synthetic growth regulators and their use in horticultural and agronomic crops. Pp. 751–773. In: *Plant Hormones: Physiology, Biochemistry and Molecular Biology.* P.J. Davies (ed.). Kluwer, Boston.

Goodwin, T.W. (ed.). 1988. *Plant Pigments.* Academic Press, New York.

Graebe, J.E. 1987. Gibberellin biosynthesis and control. *Ann. Rev. Plant Physiol.* 38:419–465.

Harborne, J.B. (ed.). 1994. *The Flavonoids: Advances in Research since 1986.* Chapman and Hall, London.

Harborne, J.B., and F.A. Tomas-Barberan (eds.). 1991. *Ecological Chemistry and Biochemistry of Plant Terpenoids.* Clarendon Press, Oxford, England.

Hooykaas, P.J.J., M.A. Hall and K.R. Libbenga (eds.). 1999. *Biochemistry and Molecular Biology of Plant Hormones.* Elsevier, Amsterdam.

Horgan, R. 1995. Instrumental methods of plant hormone analysis. Pp.140–157. In: *Plant Hormones: Physiology, Biochemistry and Molecular Biology.* P.J. Davies (ed.). Kluwer, Dordrecht, The Netherlands.

Kieber, J.J. 1997. The ethylene response pathway in *Arabidopsis. Annu. Rev. Plant Physiol. Plant Mol. Biol.* 48:277–296.

Kolattukudy, P.E. 1996. Biosynthetic pathways of cutin and waxes, and their sensitivity to environmental stresses. Pp. 83–108. In: *Plant Cuticle.* G. Kerstiens (ed.). Biosis Scientific, Oxford, England.

Lange, T. 1998. The molecular biology of gibberellin synthesis. *Planta* 204:409–419.

Mayfield, S.P., C.B.Yohn, A. Cohen and A. Danon. 1995. Regulation of chloroplast gene expression. *Annu. Rev. Plant Physiol. Plant Mol. Biol.* 46:147–166.

Nishitani, K. 1997. The role of endoxyloglucan transferase in the organization of plant cell walls. *Int. Rev. Cytol.* 173:157–206.

Normanly, J. 1997. Auxin metabolism. *Physiol. Plant.* 100:431–442.

Post-Beittenmiller, D. 1996. Biochemistry and molecular biology of wax production in plants. *Annu. Rev. Plant Biochem. Mol. Biol.* 47:405–430.

Rose, J.K.C., H.H. Lee and A.B. Bennett. 1997. Expression of a divergent expansin gene is fruit-specific and ripening-regulated. *Proc. Natl. Acad. Sci.* USA 94:5955–5960.

Sakurai, A., T. Yokota and S.D. Clouse (eds.). 1999. *Brassinosteroids. Steroidal Plant Hormones.* Springer-Verlag, Tokyo.

Showalter, A.M. 1993. Structure and function of plant cell wall proteins. *Plant Cell* 5:9–23.

Teisseire, P.J. 1994. *Chemistry of Fragrant Substances.* VCH Pub., New York.

Trewavas, A.J., and R. Malho. 1997. Signal perception and transduction: the origin of the phenotype. *Plant Cell* 9:1181–1195.

Walden, R., and H. Lubenow. 1996. Genetic dissection of auxin action: More questions than answers? *Trends Plant Sci.* 1:335–339.

Zeevaart, J.A.D., and R.A. Creelman. 1988. Metabolism and physiology of abscisic acid. *Ann. Rev. Plant Physiol.* 39:439–473.

REFERENCES

1. Adler, E., S. Larsson, K. Lundquist and G.E. Miksche. 1969. Acidolytic, alkaline, and oxidative degradation of lignin. *Int. Wood. Chem. Symp.,* 260p.
2. Abeles, F.B., P.W. Morgan and M.E. Saltveit, Jr. 1992. *Ethylene in Plant Biology.* Academic Press, San Diego, CA.
3. Alscher, R.G., and J.L. Hess. 1993. *Antioxidants in Higher Plants.* CRC Press, Boca Raton, FL.
4. Argos, P., K. Pedersen, M.D. Marks and B.A. Larkins. 1982. A structural model for maize zein proteins. *J. Biol. Chem.* 257:9984–9990.
5. Askar, A., S.E. El-Nemr and S.S. Bassiouny. 1986. Aroma constituents in white and pink guava fruits. *Alimenta* 25:162–167.
6. Badings, H.T. 1960. Principles of autoxidation processes in lipids with special regard to the development of off-flavors. *Neth. Milk Dairy J.* 14:215–242.

7. Baldwin, E.A. 2002. Fruit flavor, volatile metabolism and consumer perception. Pp. 89–106. In: *Fruit Quality and its Biological Basis.* M. Knee (ed.). CRC Press, Boca Raton, FL.

8. Barrett, A., and D. Northcote. 1965. Apple fruit pectic substances. *Biochem. J.* 94:617–627.

9. Bartel, B. 1997. Auxin biosynthesis. *Annu. Rev. Plant Physiol. Plant Mol. Biol.* 48:51–66.

10. Bartley, I.M. 1976. Changes in the glucose of ripening apples. *Phytochemistry* 15:625–626.

11. Baldry, J., J. Dougan and G.E. Howard. 1972. Volatile flavouring constituents of durian. *Phytochemistry* 11:2081–2084.

12. Bellina-Agostinone, C., M. D'Antonio and G. Pacioni. 1987. Odour composition of the summer truffle, *Tuber aestivum. Trans. Br. Mycol. Soc.* 88:568–569.

13. Berger, R.G., F. Drawert, H. Kollmannsberger, S. Nitz and B. Schraufstetter. 1985. Novel volatiles in pineapple fruit and their sensory properties. *J. Agric. Food Chem.* 33:232–235.

14. Berger, R.G., F. Drawert and H. Kollmannsberger. 1989. The flavour of Cape gooseberry (*Physalis peruviana,* L.). *Z. Lebensm. Unters. Forsch.* 188:122–126.

15. Berger, R. G. 1991. Fruits I. Pp. 283–304. In: *Volatile Compounds in Foods and Beverages.* H. Maarse (ed.). Marcel Dekker, New York.

16. Biale, J.B. 1960. The postharvest biochemistry of tropical and subtropical fruits. *Adv. Food. Res.* 10:293–354.

17. Biale, J.B., and R.E. Young. 1971. The avocado pear. Pp.1–63. In: *The Biochemistry of Fruits and Their Products.* Vol. 2, A.C. Hulme (ed.). Academic Press, New York..

18. Boelens, M., P.J. De Valois, H.J. Wobben and A. van der Gen. 1971. Volatile flavor compounds from onion. *J. Agric. Food Chem.* 19:984–991.

19. Booth, V.H. 1963. Rapidity of carotene biosynthesis in *Narcissus. Biochem. J.* 87:238–239.

20. Britton, G. 1976. Biosynthesis of carotenoids. Pp. 262–327. In: *Chemistry and Biochemistry of Plant Pigments.* Vol. 1, T.W. Goodwin (ed.). Academic Press, London.

21. Buchanan, B.B., W. Gruissem and R.L. Jones. 2000. *Biochemistry and Molecular Biology of Plants.* Amer. Society of Plant Physiologist, Rockville, MD.

22. Buchanan-Wollaston, V. 1997. The molecular biology of leaf senescence. *J. Exp. Bot.* 48:181–189.

23. Buttery, R.G., R.M. Seifert, D.G. Guadagni, D.R. Black and L.C. Ling. 1968. Characterization of some volatile constituents of carrots. *J. Agric. Food Chem.* 16:1009–1015.

24. Buttery, R.G., R.M. Seifert and L.C. Ling. 1975. Characterization of some volatile constituents of dry red beans. *J. Agric. Food Chem.* 23:516–519.

25. Buttery, R.G., D.G. Guadagni, L.C. Ling, R.M. Seifert and W. Lipton. 1976. Additional volatile components of cabbage, broccoli, and cauliflower. *J. Agric. Food Chem.* 24:829–832.

26. Buttery, R.G., L.C. Ling and B.G. Chan. 1978. Volatiles of corn kernels and husks: possible corn ear worm attractants. *J. Agric. Food Chem.* 26:866–869.

27. Carpita, N.C. 1996. Structure and biosynthesis of plant cell walls. Pp.124–147. In: *Plant Metabolism.* D.T. Dennis, D.H. Turpin, D.D. Lefebvre and D.B. Layzell (eds.). Longman Press, Harlow, UK.

28. Carpita, N.C. 1997. Structure and biogenesis of the cell walls of grasses. *Annu. Rev. Plant Physiol. Plant Mol. Biol.* 47:445–476.

29. Carpita, N.C., and D.M. Gibeaut. 1993. Structural models of primary cell walls in flowering plants: Consistency of molecular structure with the physical properties of the walls during growth. *Plant J.* 3:1–30.

30. Carroll, D.E., M.W. Hoover and W.B. Nesbitt. 1971. Sugar and organic acid concentrations in cultivars of Muscadine grapes. *J. Am. Soc. Hort. Sci.* 96:737–740.

31. Carson, J.F. 1987. Chemistry and biological properties of onions and garlic. *Food Rev. Int.* 3:71–103.

32. Chitwood, R.L., R.M. Pangborn and W. Jennings. 1983. GC/MS and sensory analysis of volatiles from three cultivars of capsicum. *Food Chem.* 11:201–216.

33. Chung, T.Y., F. Hayase and H. Kato. 1983. Volatile components of ripe tomatoes and their juices, purées and pastes. *Agric. Biol. Chem.* 47:343–351.

34. Cockshull, K.E., and A.P. Hughes. 1967. Distribution of dry matter to flowers in *Chrysanthemum morifolium. Nature* (Lond.) 215:780–781.

35. Coggins, C.W., Jr., J.C.F. Knapp and A.L. Richer. 1968. Post-harvest softening studies of Deglet Noord dates: physical, chemical, and histological changes. *Date Growers Rept.* 45:3–6.

36. Colvin. J.R. 1980. The biosynthesis of cellulose. Pp. 544–570. In: *Carbohydrates: Structure and Function.* Vol. 3 of *The Biochemistry of Plants.* J. Preiss (ed.). Academic Press, New York.

37. Cronin, D.A. 1974. The major aroma constituents of the parsnip (*Pastinaca sativa* L.). *Proc. IV Intern. Congr. Food Sci. Technol.* 1:105–112.
38. Cunningham, F.X., and E. Goontt. 1998. Genes and enzymes of carotenoid biosynthesis in plants. *Annu. Rev. Plant Physiol. Plant Mol. Biol.* 49:557–583.
39. Cutler, A.J., and J.E. Krochko. 1999. Formation and breakdown of ABA. *Trends Plant Sci.* 12:472–478.
40. Daillant-Spinnler, B., H.J.H. MacFie, B.K. Beyts and D. Hedderley. 1996. Relationship between perceived sensory properties and major preference directions of 12 varieties of apples from the southern hemisphere. *Food Quality Pref.* 7:113–126.
41. Davies, P.J. (ed.). 1995. *Plant Hormones: Physiology, Biochemistry and Molecular Biology.* Kluwer, Dordrecht, The Netherlands.
42. Davies, P.J. 1995. The plant hormone concept: Concentration, sensitivity and transport. Pp. 13–38. In: *Plant Hormones: Physiology, Biochemistry and Molecular Biology.* P.J. Davies (ed.), Kluwer, Boston.
43. Delmer, D.P. 1999. Cellulose biosynthesis: exciting times for a difficult field of study. *Annu. Rev. Plant Physiol. Plant Mol. Biol.* 50:245–276.
44. Delmer, D.P., and Y. Amor. 1995. Cellulose biosynthesis. *Plant Cell* 7:987–1000.
45. Dicks, M. 1965. Vitamin E content of foods and feeds for human consumption. *Wyo. Agric. Exp. Sta. Bull.* 435.
46. Dimick, P.S., and J.C. Hoskin. 1983. Review of apple flavor—State of the art. *Crit. Rev. Food Sci. Nutr.* 18:387–409.
47. Dixon, R.A. 1999. Isoflavonoids: biochemistry, molecular biology and biological functions. Pp.773–824. In: *Polyketides and Other Secondary Metabolites Including Fatty Acids and Their Derivatives.* U. Sankawa (ed.). Vol. 1 of *Comprehensive Natural Products Chemistry.* D.H.R. Barton, K. Nakanishi and O. Meth-Cohn (eds.). Elsevier, Amsterdam.
48. Dürr, P., and M. Röthlin. 1981. Development of a synthetic apple juice odour. *Lebensm. Wiss. Technol.* 14:313–314.
49. Edelman, J., and T.G. Jefford. 1968. The mechanism of fructosan metabolism in higher plants as exemplified in *Helianthus tuberosus.* *New Phytol.* 67:517–531.
50. Eisinger, W. 1977. Role of cytokinins in carnation flower senescence. *Plant Physiol.* 59:707–709.
51. Engel, K.H., and R. Tressl. 1983a. Studies on the volatile components of two mango varieties. *J. Agric. Food Chem.* 31:796–801.
52. Engel, K.H., and R. Tressl. 1983b. Formation of aroma components from non-volatile precursors in passion fruit. *J. Agric. Food Chem.* 31:998–1002.
53. Erdmann, H., R.J. Griesbach, R.H. Lawson and A.K. Mattoo.1989. 1-Aminocyclopropane-1-carboxylic- acid-dependent ethylene production during re-formation of vacuoles in evacuolated protoplasts of *Petunia hybrida.* *Planta* 179:196–202.
54. Etievant, P.X., E.A. Guichard and S.N. Issanchou. 1986. The flavour components of Mirabelle plums. *Sci. Aliments* 6:417–432.
55. FAO. 1970. Amino-acid content of foods and biological data on proteins. *Nutritional Studies* No. 24.
56. Fellman, J.K., T.W. Miller, D.S. Mattinson and J.P. Mattheis. 2000. Factors that influence biosynthesis of volatile flavor compounds in apple fruits. *HortScience* 35:1026–1033.
57. Fernandez, M.C.C. 1954. The effect of environment on the carotene content of plants. Pp. 24–41. In: *Influence of Environment on the Chemical Composition of Plants.* Southern Coop. Ser. Bull. 36.
58. Ferguson, C.H.R., and E.W. Simon. 1973. Membrane lipids in senescing green tissues. *J. Exp. Bot.* 24:307–316.
59. Flath, R.A., and R.R. Forrey. 1977. Volatile components of papaya (*Carica papaya* L., Solo variety). *J. Agric. Food Chem.* 25:103–109.
60. Flora, L.F., and R.C. Wiley. 1974. Sweet corn aroma, chemical components and relative importance in the overall flavor response. *J. Food Sci.* 39:770–773.
61. Forkmann, G., and W. Heller. 1999. Biosynthesis of flavonoids. Pp.713–748. In: *Polyketides and Other Secondary Metabolites Including Fatty Acids and Their Derivatives.* U. Sankawa (ed.). Vol. 1 of *Comprehensive Natural Products Chemistry.* D.H.R. Barton, K. Nakanishi, and O. Meth-Cohn (eds.). Elsevier, Amsterdam.
62. Franks, H.A., A.J. Young, G. Britton and R.J. Cogdell (eds.). 1999. *The Photochemistry of Carotenoids.* Kluwer Academic, London.

63. Frenkel, C. 1972. Involvement of peroxidase and indole-3-acetic acid oxidase isozymes from pear, tomato, and blueberry fruit in ripening. *Plant Physiol.* 49:757–763.

64. Frenkel, C., and R. Dyck. 1973. Auxin inhibition of ripening in Bartlett pears. *Plant Physiol.* 51:6–9.

65. Frenkel, C., and N.F. Haard. 1973. Initiation of ripening in Bartlett pear with an antiauxin alpha (p-chlorophenoxy) isobutyric acid. *Plant Physiol.* 52:380–384.

66. Frenkel, C., I. Klein and D.R. Dilley. 1968. Protein synthesis in relation to ripening of pome fruits. *Plant Physiol.* 43:1146–1153.

67. Fröhlich, O., and P. Schreier. 1986. Additional neutral volatiles from litchi (*Litchi chinensis* Sonn.) fruit. *Flavour Fragr. J.* 1:149–153.

68. Fuller, G., J.A. Kuhnle, J.W. Corse and B.E. Mackey. 1977. Use of natural cytokinins to extend the storage life of broccoli (*Brassica oleracea,* Italica group). *J. Amer. Soc. Hort. Sci.* 102:480–484.

69. Galliard, T. 1975. Degradation of plant lipids by hydrolytic and oxidative enzymes. Pp. 319–357. In: *Recent Advances in the Chemistry and Biochemistry of Plant Lipids.* T. Galliard and E.I. Mercer (eds.). Academic Press, London.

70. Galliard, T., D.R. Phillips and J. Reynolds. 1976. The formation of *cis*-3-nonenal, *trans*-2-nonenal, and hexanal from linoleic acid hydroperoxide enzyme system in cucumber fruit. *Biochem. Biophys. Acta* 441:181–192.

71. Garwood, D.L., F.J. McArdle, S.F. Vanderslice and J.C. Shannon. 1976. Postharvest carbohydrate transformations and processed quality of high sugar maize genotypes. *J. Amer. Soc. Hort. Sci.* 101:400–404.

72. Georgilopoulos, D.N., and A.N. Gallois. 1987. Aroma compounds of fresh blackberries (*Rubus laciniata* L.). *Z. Lebensm. Unters. Forsch.* 184:1–7.

73. Gibeaut, D.M., and N.C. Carpita. 1994. The biosynthesis of plant cell wall polysaccharides. *FASEB J.* 8:904–915.

74. Golding, J.B., D. Shearer, W.B. McGlasson and S.G. Wyllie. 1999. Relationships between respiration, ethylene, and aroma production in ripening banana. *J. Agric. Food Chem.* 47:1646–1651.

75. Gore, H.C., and D. Fairchild. 1911. Experiments on processing of persimmons to render them nonastringent. *U.S. Dept. Agri. Bur. Chem., Bull.* 141, 31p.

76. Gross, J., M. Gabai and A. Lifshitz. 1971. Carotenoids in juice of Shamouti orange. *J. Food Sci.* 36:466–473.

77. Guichard, E., and M. Souty. 1988. Comparison of the relative quantities of aroma compounds in fresh apricot (*Prunus armeniaca*) from six different varieties. *Z. Lebensm. Unters. Forsch.* 186:301–307.

78. Guiraud, J.P., J. Daurelles and P. Galzy. 1981. Alcohol production from Jerusalem artichokes using yeasts with inulinase activity. *Biotechnol. Bioeng.* 23:1461–1466.

79. Guy, M., and H. Kende. 1984. Conversion of 1-aminocyclopropane-1-carboxylic acid to ethylene by isolated vacuoles of *Pisum sativum* L. *Planta* 160:281–287.

80. Habib, A.T., and H.D. Brown. 1956. The effect of oxygen and hydrogen ion concentration on color changes in processed beets, strawberries and raspberries. *Proc. Amer. Soc. Hort. Sci.* 68:482–490.

81. Hadfield, K.A., and A.B. Bennett. 1998. Polygalacturonases: many genes in search of a function. *Plant Physiol.* 117:337–343.

82. Hale, C.R., B.G. Coombe and J.S. Hawker. 1970. Effect of ethylene and 2-chloroethylphosphonic acid on the ripening of grapes. *Plant Physiol.* 45:620–623

83. Hane, M. 1962. Studies on the acid metabolism of strawberry fruits. *Gartenbauwissenschaft* 27:453–482.

84. Harborne, J.B. 1980. Plant phenolics. Pp. 329–402. In: *Secondary Plant Products.* E.A. Bell and B.V. Charlwood (eds.). Springer-Verlag, Berlin.

85. Haro, L., and W.E. Faas. 1985. Comparative study of the essential oils of Key and Persian limes. *Perfum. Flavor.* 10:67–72.

86. Harwood, J.L. 1996. Recent advances in the biosynthesis of plant fatty acids. *Biochim. Biophys. Acta* 1301:7–56.

87. Hayase, F., T.Y. Chung and H. Kato. 1984. Changes of volatile components of tomato fruits during ripening. *Food Chem.* 14:113–124.

88. Hedden, P., and Y. Kamiya. 1997. Gibberellin biosynthesis: enzymes, genes and their regulation. *Annu. Rev. Plant Physiol. Plant Mol. Biol.* 48:431–460.

89. Henderson, J.R., and B.W. Buescher. 1977. Effects of SO_2 and controlled atmospheres on broken-end discoloration and processed quality attributes in snap beans. *J. Amer. Soc. Hort. Sci.* 102:768–770.

90. Hermann, K.M. 1995. The shikimate pathway: early steps in the biosynthesis of aromatic compounds. *Plant Cell* 7:907–919.

91. Hildebrand, D.F., and T. Hymowitz. 1982. Carotene and chlorophyll bleaching by soybeans with and without seed lipoxygenase-1. *J. Agri. Food Chem.* 30:705–708.

92. Hirvi, T., E. Honkanen and T. Pyysalo. 1981. The aroma of cranberries. *Z. Lebensm. Unters. Forsch.* 172:365–367.

93. Hirvi, T. 1983. Mass fragmentographic and sensory analyses in the evaluation of the aroma of some strawberry varieties. *Lebensm. Wiss. Technol.* 16:157–161.

94. Hirvi, T., and E. Honkanen. 1983. The aroma of blueberries. *J. Sci. Food Agric.* 34:992–998.

95. Hirvi, T., and E. Honkanen. 1985. Analysis of the volatile constituents of black chokeberry (*Aronia melanocarpa* Ell.). *J. Sci. Food Agric.* 36:808–810.

96. Hobson, G.E. 1968. Cellulase activity during the maturation and ripening of tomato fruit. *J. Food Sci.* 33:588–592.

97. Hobson, G.E. 1975. Protein redistribution and tomato fruit ripening. *Coll. Intern. C.N.R.S.* 238:265–269.

98. Hoffman, G. 1962.1-Octen-3-ol and its relation to other oxidative cleavage products from esters of linoleic acid. *J. Amer. Oil Chemists* 39:439–444.

99. Hofman, E., P.M. Wrench, F.P. Sharples, R.G. Hiller, W. Welte and K. Diederichs. 2002. Structural basis of light harvesting by carotenoids: peridinin-chlorophyll-protein from *Amphidinium carterae*. *Science* 272:1788–1791.

100. Horvat, R.J., and S.D. Senter. 1987. Identification of additional volatile compounds from cantaloupe. *J. Food Sci.* 52:1097–1098.

101. Huber, S.C., and J.L. Huber. 1997. Role and regulation of sucrose-phosphate synthetase in higher plants. *Annu. Rev. Plant Physiol. Plant Mol. Biol.* 47:431–444.

102. Hulme, A.C. 1972. The proteins of fruits: Their involvement as enzyme in ripening. A review. *J. Food Tech.* 7:343–371.

103. Hulme, A.C., and L.S.C. Wooltorton. 1958. The acid content of cherries and strawberries. *Chem. Ind., Lond.* 22:659.

104. Idstein, H., W. Herres and P. Schreier. 1984. High-resolution gas chromatography—mass spectrometry and Fourier transform infrared analysis of cherimoya (*Annona cherimolia*, Mill.) volatiles. *J. Agric. Food Chem.* 32:383–389.

105. Idstein, H., and P. Schreier. 1985. Volatile constituents of Alphonso mango (*Mangifera indica*). *Phytochemistry* 24:2313–2316.

106. Idstein, H., C. Bauer and P. Schreier. 1985a. Volatile acids from tropical fruits: cherimoya (*Annona chermolia*, Mill.), guava (*Psidium guajava*, L.), mango (*Mangifera indica*, L., var. Alphonso), papaya (*Carica papaya*, L.). *Z. Lebensm. Unters. Forsch.* 180:394–397.

107. Idstein, H., T. Keller and P. Schreier. 1985b. Volatile constituents of mountain papaya (*Carica candamarcensis*, syn. *C. pubescens* Lenne et Koch) fruit. *J. Agric. Food Chem.* 33:663–666.

108. Impellizzeri , G., M. Piattelli and S. Sciuto. 1973. Acylated beta cyanins from *Drosanthemum foribundum*. *Phytochemistry* 12:2295–2296.

109. Ismail, H.M., A.A. Williams and O.G. Tucknott. 1981. The flavour of plums: an examination of the aroma components present in the headspace above four intact cultivars of intact plums, Marjorie's Seedling, Merton Gem, NA 10 and Victoria. *J. Sci. Food Agric.* 32:498–502.

110. Jacobi, G., B. Klemme and C. Postius. 1975. Dark starvation and metabolism. IV. The alteration of enzyme activities. *Biochem. Physiol. Pflanz.*168:247–256.

111. Jeffree, C.E. 1996. Structure and ontogeny of plant cuticles. Pp.33–85. In: *Plant Cuticles: An Integrated Functional Approach.* G. Kerstiens (ed.). BIOS Scientific, Oxford, England.

112. Jennings, W.G., and R. Tressl. 1974. Production of volatile compounds in the ripening Bartlett pear. *Chem. Microbiol. Technol. Lebensm.* 3:52–55.

113. Jennings, W.G. 1977. Volatile components of figs. *Food Chem.* 2:185–191.

114. Johnson, J.D., and J.D. Vora. 1983. Natural citrus essences. *Food Tech.* 37:92–93, 97.

115. Johnston, J.C., R.C. Welch and G.L.K. Hunter. 1980. Volatile constituents of litchi (*Litchi chinensis* Sonn.). *J. Agric. Food Chem.* 28:859–861.

116. Kanner, J., S. Harel and R. Granit. 2001. Betalains—A new class of dietary cationized antioxidants. *J. Agri. Food Chem.* 49:5178–5185.

117. Kasting, R., J. Andersson and E. Von Sydow. 1972. Volatile constituents in leaves of parsley. *Phytochemistry* 11:2277–2282.

118. Kays, S.J., and J.C. Silva Dias. 1996. *Cultivated Vegetables of the World: Latin Binomial, Common Name in 15 Languages, Edible Part, and Method of Preparation.* Exon Press, Athens, GA.

119. Kays, S.J., and Y. Wang. 2000. Thermally induced flavor compounds. *HortScience* 35:1002–1012.

120. Kefford, N.P., M.I. Burce and J.A Zwar. 1973. Retardation of leaf senescence by urea cytokinins in *Raphanus sativus. Phytochemistry* 12:995–1003.

121. Kemp, T.R., L.P. Stoltz and D.E. Knavel. 1972. Volatile components of muskmelon fruit. *J. Agric. Food Chem.* 20:196–198.

122. Kemp, T.R., D.E. Knavel and L.P. Stoltz. 1974. Identification of some volatile compounds from cucumber. *J. Agric. Food Chem.* 22:717–718.

123. Kende, H. 1993. Ethylene biosynthesis. *Annu. Rev. Plant Physiol. Plant Mol. Biol.* 44:283–307.

124. Kende, H., and J.A.D. Zeevart. 1997. The five "classical" plant hormones. *Plant Cell* 9:1197–1210.

125. Kerslake, M.F., A. Latrasse and J.L. Le Quéré. 1989. Hydrocarbon chemotypes of some varieties of blackcurrant (*Ribes* sp.). *J. Sci. Agric.* 47:43–51.

126. Kjaer, A., J. Ogaard Madsen, Y. Maeda, Y. Ozawa and Y. Uda. 1978. Volatiles in distillates of fresh radish of Japanese and Kenyan origin. *Agric. Biol. Chem.* 42:1715–1721.

127. Kliewer, W.M. 1966. Sugars and organic acids of *Vitis vinifera. Plant Physiol.,* Lancaster 41:923–931.

128. Knee, M, 1988. Carotenol esters in developing apple fruit. *Phytochemistry* 27:1005–1009.

129. Knee, M, and I.M. Bartley. 1981. Composition and metabolism of cell wall polysaccharides in ripening fruits. Pp. 133–148. In: *Recent Advances in the Biochemistry of Fruits and Vegetables.* J. Friend and M.J.C. Rhodes (eds.). Academic Press, New York.

130. Kolattukudy, P.E. 1980. Cutin, suberin and waxes. Pp. 571–645. In: *The Biochemistry of Plants.* Vol. 4, P.K. Stumpf and E.E. Conn (eds.). Academic Press, New York.

131. Kolattukudy, P.E., and T.J. Walton.1973. The biochemistry of plant cuticular lipids. *Prog. Chem. Fats Other Lipids.* 13:119–175.

132. Kollas, D.A. 1964. Preliminary investigation of the influence of controlled atmosphere storage on the organic acids of apples. *Nature* 204:758–759.

133. Kolor, M.G. 1983. Identification of an important new flavor compound in Concord grape. *J. Agric. Food Chem.* 31:1125–1127.

134. Koyasako, A., and R.A. Bernhard. 1983. Volatile constituents of the essential oil of kumquat. *J. Food Sci.* 48:1807–1812.

135. Lee, P.L., G. Swords and G.L.K. Hunter. 1975a. Volatile components of *Eugenia jambos* L., rose-apple. *J. Food Sci.* 40:421–422.

136. Lee, P.L., G. Swords and G.L.K. Hunter. 1975b. Volatile constituents of tamarind (*Tamarindus indica* L.). *J. Agric. Food Chem.* 23:1195–1199.

137. Leung, J. 1998. Abscisic acid signal transduction. *Annu. Rev. Plant Physiol. Plant Mol. Biol.* 49:199–222.

138. Lomax, T.L., G.K. Muday and P.H. Rubery. 1995. Auxin transport. Pp.509–530. In: *Plant Hormones and Their Role in Plant Growth Development.* P.J. Davies (ed.), Kluwer, Dordrecht, The Netherlands.

139. MacLeod, A.J., and R. Islam. 1975. Volatile flavour components of watercress. *J. Sci. Food Agric.* 26:1545–1550.

140. MacLeod, A.J., and R. Islam. 1976. Volatile flavour components of garden cress. *J. Sci. Food Agric.* 27:909–912.

141. MacLeod, A.J., and H.E. Pikk. 1979. Volatile flavor components of fresh and preserved Brussels sprouts grown at different crop spacings. *J. Food Sci.* 44:1183–1185, 1190.

142. MacLeod, A.J., and N.M. Pieris. 1981a. Volatile flavor components of beli fruit (*Aegle marmelos*) and a processed product. *J. Agric. Food Chem.* 29:1262–1264.

143. MacLeod, A.J., and N.M. Pieris. 1981b. Volatile flavor components of soursop (*Annona muricata*). *J. Agric. Food Chem.* 29:488–490.

144. MacLeod, A.J., and N.G. de Troconis. 1982a. Volatile flavour components of cashew "apple" (*Anacardium occidentale*). *Phytochemistry* 21:2527–2530.

145. MacLeod, A.J., and N.G. de Troconis. 1982b. Volatile flavour components of guava. *Phytochemistry* 21:1339–1342.

146. MacLeod, A.J., and N.G. de Troconis. 1982c. Volatile flavor components of sapodilla fruit (*Achras sapota* L.). *J. Agric. Food Chem.* 30:515–517.

147. MacLeod, A.J., and N.M. Pieris. 1982. Volatile flavour components of mangosteen, *Garcinia mangostana. Phytochemisry* 21:117–119.

148. MacLeod, A.J., and N.G. de Troconis. 1983. Aroma volatiles of aubergine (*Solanum melongena*). *Phytochemistry* 22:2077–2079.

149. MacLeod, A.J., and N.M. Pieris. 1984. Comparison of some mango cultivars. *Phytochemistry* 23:361–366.

150. MacLeod, A.J., C.H. Snyder and G. Subramanian. 1985. Volatile aroma constituents of parsley leaves. *Phytochemistry* 24:2623–2627.

151. Manners, D.J., and J.J. Marshall. 1971. Studies on carbohydrate-metabolizing enzymes. Part XXIV. The action of malted-rye alpha-amylase on amylopectin. *CarbohyDr. Res.* 18:203–209.

152. Marcus, A. (ed.). 1981. *Proteins and Nucleic Acids.* Vol. 6. of *The Biochemistry of Plants.* J. Preiss (ed.). Academic Press, New York.

153. Marks, M.D., and B.A. Larkins. 1982. Analysis of sequence microheterogeneity among zein messenger RNAs. *J. Biol. Chem.* 257:9976–9983.

154. Marriott, R.J. 1986. Biogenesis of blackcurrant (*Ribes nigrum* L.) aroma. *J. Amer. Chem. Soc.* 15:185–191.

155. Martin, C., and K.V. Thimann. 1972. The role of protein synthesis in the senescence of leaves. I. The formation of protease. *Plant Physiol.* 49:64–71.

156. Martin, C., and K.V. Thimann. 1973. The role of protein synthesis in the senescence of leaves. II. The influence of amino acids on senescence. *Plant Physiol.* 50:432–437.

157. Matile, P., S. Hortensteiner, H. Thomas and B. Krautler. 1996. Chlorophyll breakdown in senescent leaves. *Plant Physiol.* 112:1403–1409.

158. Matsuo, T., and S. Itoo. 1978. The chemical structure of Kaki-tannin from immature fruit of the persimmon (*Diosysyros Kaki* L.). *Agric. Biol. Chem.* 42:1637–1643.

159. Matsuo, T., and S. Itoo. 1981. Comparative studies of condensed tannins from severed young fruits. *J. Jpn. Soc. Hort. Sci.* 50:262–269.

160. Mattheis, J.P., and J.K. Fellman. 1999. Preharvest factors influencing flavor of fresh fruit and vegetables. *Postharv. Biol. Technol.* 15:227–232.

161. Mayak, S., A.H. Halevy and M. Katz. 1972. Correlative changes in phytohormones in relation to senescence processes in rose petals. *Physiol Plant.* 27:1–4.

162. Mazliak, P. 1968. Chemistry of plant cuticles. Pp. 49–111. In: *Progress in Phytochemistry.* Vol. 1, L. Reinhold and V. Liwschitz (eds.). J. Wiley and Sons, London.

163. Mazza, G. 1983. Gas chromatographic-mass spectrometric investigation of the volatile components of myrtle berries (*Myrtus communis* L.). *J. Chromatogr.* 264:304–311.

164. McGaw, B.A., and L.R. Burch.1995. Cytokinin biosynthesis and metabolism. Pp.98–117. In: *Plant Hormones: Physiology, Biochemistry and Molecular Biology.* P.J. Davies (ed.). Kluwer Academic Publishers, Dordrecht, The Netherlands.

165. Meredith, F.I., and R.H. Young. 1969. Effect of temperature on pigment development in Red Blush grapefruit and Ruby Blood oranges. Pp. 271–276. In: *Proceedings 1st International Citrus Symposium.* H.D. Chapman (ed.). Univ. of Calif., Riverside, CA.

166. Mitchell, T., A.J.R. Porter and P. John. 1988. Authentic activity of the ethylene-forming enzyme observed in membranes obtained from kiwkfruit (*Actinidia deliciosa*). *New Phytol.* 109:313–319.

167. Money, R.W. 1958. Analytical data on some common fruits. *J. Sci. Food Agric.* 9:19–20.

168. Money, R.W., and W.A. Christian. 1950. Analytical data on some common fruits. *J. Sci. Food Agr.* 1:8–12.

169. Moore, T.S., Jr. (ed). 1993. *Lipid Metabolism in Plants.* CRC Press, Boca Raton, FL.

170. Morales, A.L., and C. Duque. 1987. Aroma constituents of the fruit of the mountain papaya (*Carica pubescens*) from Colombia. *J. Agric. Food Chem.* 35:538–540.

171. Moshonas, M.G., and P.E. Shaw. 1986. Quantities of volatile flavor components in aqueous orange essence and in fresh orange juice. *Food Technol.* 40(11):100–103.

172. Moshonas, M.G., and P.E. Shaw. 1987. Quantitative analysis of orange juice flavor volatiles by direct-injection gas chromatography. *J. Agric. Food Chem.* 35:161–165.

173. Murray, K.E., P.A. Bannister and R.G. Buttery. 1975. Geosmin: an important volatile constituent of beetroot (*Beta vulgaris*). *Chem. Ind.* (London) 1975:973–974.
174. Nelson, O., and D. Pan. 1995. Starch synthesis in maize endosperms. *Annu. Rev. Plant Physiol. Plant Mol. Biol.* 46:475–496.
175. Nursten, H.E. 1970. Volatile compounds: The aroma of fruits. Pp. 239–268. In: *The Biochemistry of Fruits and Their Products*. Vol. 1, A.C. Hulme (ed.). Academic Press, New York.
176. Nyssen, L.M., and H. Maarse. 1986. Volatile compounds in blackcurrant products. *Flavour Fragr. J.* 1:143–148.
177. Ohlrogge, J.B., and J.A. Browse. 1995. Lipid biosynthesis. *Plant Cell* 7:957–970.
178. Ohlrogge, J.B., and J.G. Jaworski. 1997. Regulation of fatty acid synthesis. *Annu. Rev. Plant Physiol. Plant Mol. Biol.* 48:109–136.
179. Olliver, M. 1967. Ascorbic acid: Occurrence in foods. Pp. 359–367. In: *The Vitamins*. Vol. 1, W.H. Sebrell, Jr. and R.S. Harris (eds.). Academic Press, New York.
180. Paillard, N. 1968. Analyse de l'arome de pommes de la variete 'Calville blanc' par chromatographie sur colonne capillaire. *Fruits d'Outre Mer.* 23:393–38.
181. Paull, R.E., and T. Chantrachit. 2001. Benzyladenine and the vase life of tropical ornamentals. *Postharv. Biol. Tech.* 21:303–310.
182. Pérez, A.G., C. Sanz, J.J. Rios, R. Olías and J.M. Olías. 1999. Effects of ozone treatment on postharvest strawberry quality. *J. Agric. Food Chem.* 47:1652–1656.
183. Piattelli, M., and L. Minale. 1964. Pigments of Centrospermae -II. Distribution of betacyanins. *Phytochemistry* 3:547–557.
184. Piattelli, M., L. Minale and G. Prota. 1965. Pigments of Centrospermae—III. Betaxanthins from *Beta vulgaris* L. *Phytochemistry* 4:121–125.
185. Pinsky, A., S. Grossman and M. Trop.1971. Lipoxygenase content and antioxidant activity of some fruits and vegetables. *J. Food Sci.* 36:571–572.
186. Pollock, C.J., A.J. Cairns, H. Bohlmann and K. Apel. 1991. Fructan metabolism in grasses and cereals. *Annu. Rev. Plant Physiol. Plant Mol. Biol.* 42:77–107.
187. Preiss, J., and C. Levi. 1980. Starch biosynthesis and degradation. Pp. 371–423. In: *Carbohydrates: Structure and Function*. P.K. Stumpf and E.E. Conn (eds.). Vol. 3 of *The Biochemistry of Plants*. J. Preiss (ed.). Academic Press, New York.
188. Pressey, R. 1977. Enzymes involved in fruit softening. Pp. 172–191. In: *Enzymes in Food and Beverage Processing*. R.L. Ory and A.J. St. Angelo (eds.). ACS Symposium Series 47, Washington, DC.
189. Proctor, M., and P. Yeo. 1973. *The Pollination of Flowers*. Collins, Glasgow, Scotland.
190. Pyysalo, H. 1976. Identification of volatile compounds in seven edible fresh mushrooms. *Acta Chem. Scand.* 30:235–244.
191. Quackenbush, F.W. 1963. Corn carotenoids: Effects of temperature and moisture on losses during storage. *Cereal Chem.* 40:266–269.
192. Randle, W. 1997. Onion flavor chemistry and factors influencing flavor intensity. Pp 41–52. In: *Spices: Flavor Chemistry and Antioxidant Properties*. S.J. Risch, and C.T. Ho (eds.). ACS Symposium Series 660, Washington DC
193. Rapparini, R., and S. Predieri. 2002. Pear fruit volatiles. Hort. Rev. 28:237–324.
194. Richmond, A., and J.B. Biale. 1966. Protein synthesis in avocado fruit tissue. *Archs. Biochem. Biophys.* 115:211–214.
195. Richmond, A., and J.B. Biale. 1967. Protein and nucleic acid metabolism in fruits. II. RNA synthesis during the respiratory rise of the avocado. *Biochem. Biophys. Acta* 138:625–627.
196. Rudolph, C.J., G.V. Odell and H.A. Hinrichs. 1971. Chemical studies on pecan composition and factors responsible for the typical flavor of pecans. *Proc. 41st Okla. Pecan Grs. Assoc.,* pp. 63–70.
197. Russell, L.F., H.A. Quamme and J.I. Gray. 1981. Qualitative aspects of pear flavor. *J. Food Sci.* 46:1152–1158.
198. Salanke, D.K., and J.Y. Do. 1976. Biogenesis of aroma constituents of fruits and vegetables. *CRC Critical Rev. Food Tech.* 8:161–190.
199. Sandman, G. 2001. Genetic manipulation of carotenoid biosynthesis: strategies, problems and achievements. *Trends Plant Sci.* 6:14–17.
200. Sanz, C., Olias, J.M., Perez and A.G. 1997. Aroma biochemistry of fruits and vegetables. Pp. 125–156.

In: *Phytochemistry of Fruits and Vegetables.* F.A. Tomas-barberan and R.J. Robins, (eds.). Clarendon Press, Oxford, England.

201. Sasson, A., and S.P. Monselise. 1977. Organic acid composition of 'Shamouti' oranges at harvest and during prolonged postharvest storage. *J. Amer. Soc. Hort. Sci.* 102:331–336.

202. Sawamura, M., and T. Kuriyama. 1988. Quantitative determination of volatile constituents in the pummelo (*Citrus grandis* Osbeck forma Tosa-buntan). *J. Agric. Food Chem.* 36:567–569.

203. Schmid, K., and J.B. Ohlrogge. 1996. Lipid metabolism in plants. Pp. 363–390. In: *Biochemistry of Lipids, Lipoproteins and Membranes.* D.E. Vance and J. Vance (eds.). Elsevier Press, Amsterdam.

204. Schreier, P., F. Drawert and A. Junker. 1976. Identification of volatile constituents from grapes. *J. Agric. Food Chem.* 24:331–336.

205. Schreier, P. 1980. Quantitative composition of volatile constituents in cultivated strawberries *Fragaria ananassa* cv. Senga Sengana, Senga Litessa and Senga Gourmella. *J. Sci. Food Agric.* 31:487–494.

206. Schreyen, L., P. Dirinck, F. Van Wassenhove and N. Schamp. 1976. Analysis of leek volatiles by headspace condensation. *J. Agric. Food Chem.* 24:1147–1152.

207. Schreyen, L., P. Dirinch, P. Sandra and N. Schamp. 1979. Flavor analysis of quince. *J. Agric. Food Chem.* 27:872–876.

208. Sembdner, G., D. Gross, H.-W. Liehusch and G. Schneider. 1980. Biosynthesis and metabolism of plant hormones. Pp. 281–444. In: *Hormonal Regulation of Development. I. Molecular Aspects of Plant Hormones.* J. MacMillan (ed.). Springer-Verlag, Berlin.

209. Senter, S.D., and R.J. Horvat. 1976. Lipids of pecan nutmeats. *J. Food Sci.* 41:1201–1203.

210. Senter, S.D., and R.J. Horvat. 1978. Minor fatty acids from pecan kernel lipids. *J. Food Sci.* 43:1614–1615.

211. Shah, B.M., D.K. Salunkhe and L.E. Olsen. 1969. Effects of ripening processes on chemistry of tomato volatiles. *J. Amer. Soc. Hort. Sci.* 94:171–176.

212. Sharkey, T.D., and S. Yeh. 2001. Isoprene emission from plants. *Annu. Rev. Plant Physiol. Plant Mol. Biol.* 52:407–436.

213. Shaw, G.J., J.M. Allen and F.R. Visser. 1985. Volatile flavor components of babaco fruit (*Carica pentagona,* Heilborn). *J. Agric. Food Chem.* 33:795–797.

214. Shaw, P.E., and C.W. Wilson III. 1982. Volatile constituents of loquat (*Eriobotrya japonica* L.) fruit. *J. Food Sci.* 47:1743–1744.

215. Shaw, P.E. 1991. Fruits II. Pp. 305–327. In: *Volatile Compounds in Foods and Beverages.* H. Maarse (ed.). Marcel Dekker, New York.

216. Shine, W.E., and P.K. Stumpf. 1974. Fat metabolism in higher plants: Recent studies on plant oxidation systems. *Arch. Biochem. Biophys.* 162:147–157.

217. Smirnoff, N., P.C. Courklin and F.A. Loewus. 2001. Biosynthesis of ascorbic acid in plants: A renaissance. *Annu. Rev. Plant Physiol. Plant Mol. Biol.* 52:437–467.

218. Song, J., W. Dong, L. Fan, J. Verchoor and R. Beaudry. 1998. Aroma volatiles and quality changes in modified atmosphere packaging. Pp. 89–95. In: *Proc. Seventh International Controlled Atmosphere Research Confer. Fresh Fruit and Vegetable and MAP.* Gorny, J. (ed.). Univ. California, Davis, CA.

219. Spencer, M.D., R.M. Pangborn and W.G. Jennings. 1987. Gas chromatographic and sensory analysis of volatiles from Cling peaches. *J. Agric. Food. Chem.* 26:725–732.

220. Sterling, C. 1961. Physical state of cellulose during ripening of peach. *J. Food Sci.* 26:95–98.

221. Steward, F.C., A.C. Hulme, S.R. Freiberg, M.P. Hegarty, J.K. Pollard, R. Rabson and R.A. Barr. 1960. Physiological investigations on the banana plant. I. Biochemical constituents detected in the banana plant. *Ann. Bot.* 24:83–116.

222. Swisher, H.E., and L.H. Swisher.1980. Lemon and lime juices. Pp. 144–179. In: *Fruit and Vegetable Juice Processing Technology.* D.K. Tressler and M.A. Joslyn (eds.). AVI, Westport, CT.

223. Swords, G., P.A. Bobbio and G.L.K. Hunter. 1978. Volatile constituents of jack fruit (*Arthocarpus heterophyllus*). *J. Food Sci.* 43:639–640.

224. Tadera, K., T. Kaneko and F. Yagi. 1986. Evidence for the occurrence and distribution of a new type of vitamin B6 conjugate in plant foods. *Agri. Biol. Chem.* 50:2933–2934

225. Takahashi, N., I. Yamaguchi and H. Yamane.1986. Gibberellins. Pp. 57–151. In: *Chemistry of Plant Hormones.* N. Takahashi (ed.). CRC Press, Boca Raton, FL.

226. Takeoka, G.R., M. Güntert, R.A. Flath, R.E. Wurz and W. Jennings. 1986. Volatile constituents of kiwi fruit (*Actinidia chinensis,* Planch.). *J. Agric. Food Chem.* 34:576–578.

227. Takeoka, G.R., R.A. Flath, M. Güntert and W.G. Jennings. 1988. Nectarine volatiles: vacuum steam distillation versus headspace sampling. *J. Agric. Food Chem.* 36:553–560.

228. Talou, T., M. Delmas and A. Gaest. 1987. Principal constituents of black truffle (*Tuber melanosporum*) aroma. *J. Agric. Food Chem.* 35:774–777.

229. Teisseire, P.J. 1994. *Chemistry of Fragrant Substances.* VCH Pub., New York.

230. Tevini, M. 1977. Light, function, and lipids during plastid development. Pp. 121–145. In: *Lipids and Lipid Polymers in Higher Plants.* M. Tevini and H.K. Lichtenlhaler (eds.). Springer-Verlag, New York.

231. Tevini, M., and D. Steinmuller. 1985. Composition and function of plastogbuli. II. Lipid composition of leaves and plastoglobuli during beech leaf senescence. *Planta* 163:91–96.

232. Thomas, R.L., and J.J. Jen. 1975. Phytochrome-mediated carotenoids biosynthesis in ripening tomatoes. *Plant Physiol.* 56:452–453.

233. Tressl, R., and F. Drawert. 1973. Biogenesis of banana volatiles. *J. Agric. Food Chem.* 21:560–565.

234. Tressl, R., M. Holzer and M. Apetz. 1977. Formation of flavor components in asparagus. 1. Biosynthesis of sulfur-containing acids in asparagus. *J. Agric. Food Chem.* 25:455–459.

235. Ulrich, R. 1970. Organic acids. Pp. 89–118. In: *The Biochemistry of Fruits and Their Products.* Vol. 1. A.C. Hulme (ed). Academic Press, New York.

236. van Buren, J. 1970. Fruit phenolics. Pp. 269–304. In: *The Biochemistry of Fruit and Their Products.* Vol. 1. A.C. Hulme (ed). Academic Press, New York.

237. Vierstra, R. 1996. Proteolysis in plants. *Plant Mol. Biol.* 32:275–302.

238. von Wettstein-Knowles, P.M. 1993. Waxes, cutin and suberin. Pp.127–166. In: *Lipid Metabolism in Plants.* T.S. Moore, Jr. (ed.). CRC Press, Boca Raton, FL

239. Wallbank, B.E., and G.A. Wheatley. 1976. Volatile constituents from cauliflower and other crucifers. *Phytochemistry* 15:763–766.

240. Wang, Y., R.J. Horvat, R.A. White and S.J. Kays. 1998. Influence of curing treatment on the synthesis of the volatile flavor components of sweetpotato. *Acta Hort.* 464:207–212.

241. Wang, Y., and S.J. Kays. 2000. Contributions of volatile compounds to the characteristic aroma of baked 'Jewel' sweetpotatoes [*Ipomoea batatas* (L.) Lam.]. J. Amer. Soc. Hort. Sci. 125:638–643.

242. Watt, B.K., and A.L. Merrill. 1963. Composition of foods—raw, processed, prepared. *USDA Agri. Hbk.* 8.

243. Weeks, W.W. 1986. Carotenoids. A source of flavor and aroma. Pp. 156–166. In: *Biogeneration of Aromas.* T.H. Parliment and R. Croteau (eds.). Amer. Chem. Soc. Symp. Ser. 317, Washington, DC.

244. Whitaker, J.R. 1976. Development of flavor, odor, and pungency in onion and garlic. *Adv. Food Res.* 22:73–133.

245. Whitfield, F.B., and J.H. Last. 1991. Vegetables. Pp. 203–281. In: *Volatile Compounds in Foods and Beverages.* H. Maarse (ed.). Marcel Dekker, New York.

246. Widdowson, E.M., and R.A. McCance. 1935. The available carbohydrate of fruits. Determination of glucose, fructose, sucrose and starch. *Biochem. J.* 29:151–156.

247. Williams, M., G. Hrazdina, M.M. Wilkerson, J.G. Sweeney and G.A. Iacobucci. 1978. High-pressure liquid chromatographic separation of 3-glucosides, 3,5-diglucosides, 3-(6-O-p-coumaryl)glucosides, and 3-(6-O-p-coumaryl glucoside)-5-glucosides of anthocyanides. *J. Chromato.* 155:389–398.

248. Winterhalter, P. 1991. Fruits IV. Pp. 389–409. In: *Volatile Compounds in Foods and Beverages.* H. Maarse (ed.). Marcel Dekker, New York.

249. Woodward, J.R. 1972. Physical and chemical changes in developing strawberry fruits. *J. Sci. Food Agric.* 23:465–473.

250. Wu, J.L., C.C. Chou, M.H. Chen and C.M. Wu. 1982. Volatile flavor compounds from shallots. *J. Food Sci.* 47:606–608.

251. Wyman, H., and J.K. Palmer. 1964. Organic acids in the ripening banana fruit. *Plant Physiol.,* Lancaster. 39:630–633.

252. Yabumoto, K., and W.G. Jennings. 1977. Volatile constituents of cantaloupe, *Cucumis melo,* and their biogenesis. *J. Food. Sci.* 42:32–37.

253. Yang, S.F., and N.E. Hoffman. 1984. Ethylene biosynthesis and its regulation in higher plants. *Annu. Rev. Plant Physiol.* 35:155–189.

254. Young, H., and V.J. Paterson. 1985. The effect of harvest maturity, ripeness and storage on kiwi fruit aroma. *J. Sci. Food Agric.* 36:352–358.

5

MOLECULAR GENETICS, SIGNAL TRANSDUCTION AND RECOMBINANT DNA

Friar Gregor Mendel showed in the 1850's that there were heritable factors controlling plant traits such as flower color, stem length, pod shape, and seed color and shape. The expression of these factors can be modified by a wide range of environmental parameters (e.g., light quantity and quality, temperature, water and nutrient supply, pathogen and insect pressure) causing two genetically identical plants to develop into distinctly different phenotypes. In the early 1900's, it was postulated that these factors were located on the filamentous chromosomes in the nucleus. The factors were subsequently called "**genes**" which were thought to be consistently transmitted from one generation to the next. Over the next fifty years, using cytogenetic and crossing techniques, researchers confirmed that genes did occur on chromosomes and that some genes were linked, i.e., inherited together in a linkage group. Occasionally, the linkage was broken during division due to crossing-over, a physical exchange of genetic material that leads to a unique recombination.

Cytogeneticists were eventually able to ascribe certain genes to specific chromosomes. This was made possible due to variation in chromosome size and surface morphology. Maize, for instance, has 10 chromosomes varying in length from the longest (chromosome #1) to the shortest (chromosome #10). Differences in morphology led to mapping of a gene's relative position on a chromosome. Genes are said to be linked whenever 50% or more of the **gametes*** are the parental combination. When only 1% recombination occurs, the linkage distance is one map unit, whereas with 60% recombination, the linkage distance is 60 map units. A map unit is a relative value, in that factors other than physical chromosomal distance determine linkage, and interpretation problems arise when multiple crossovers occur. Multiple crossovers generally occur as the distance between two genes increases. *Gene-linkage mapping*† allowed the order of many genes to be determined on the chromosomes of certain widely studied species (e.g., corn, tomato).

Mapping of gene location requires a gene polymorphism (i.e., differences in physical structure) to be present. Polymorphisms occur when multiple versions (alleles) of a gene are present that vary in DNA sequence but produce a protein with the same function. For mapping, these alleles must produce a detectable phenotype or variation in the parent. An allele's

---

*A gamete is a cell or nucleus that may participate in sexual fusion to form a zygote.
†A linkage map is a diagrammatic representation of the order of genes on a chromosome.

phenotype can range from different grain color to variation in DNA polymorphism. Polymorphisms are now widely identified using molecular probes and are often in regions of the DNA that do not affect gene expression.

1. CHROMOSOMES AND GENES

There are relatively few chromosomes in the cells of most species. *Arabidopsis* has 5 chromosomes and corn 10, which contain approximately 20,000 and 30,000 genes, respectively. Therefore, many genes are found on each chromosome. The organization of information in the genes is a critical question being addressed by molecular geneticists. From the 1930s to 1960s, it was known that chromosomes were composed of proteins and deoxyribonucleic acids (DNA). Since proteins are large complex molecules composed of 20 essential amino acids and display a high level of specificity in activity, they were initially thought to contain sufficient information to function as genes. In contrast, DNA is a large molecule but composed of only four nucleotides, so it seemed unlikely that DNA could account for all of the genetic information required. However, mutation studies using bacteria and viruses eventually showed that genes consisted of DNA. Therefore, DNA must in some way contain the information needed for the thousands of genes present. The total of the genetic information contained in the DNA is called a species' genome.

1.1. DNA → RNA → Proteins

The nucleic acids that make up DNA and ribonucleic acid (RNA) polymers, store and transmit information. Information encoded in the DNA (genome) provides the blueprint for proteins required for the growth and development of all living organisms. Two major steps are involved in transferring the information contained in the DNA to proteins that carry out and control cellular functions. The nucleic acid sequence in DNA is first transcribed in the synthesis of an RNA molecule, referred to as a messenger RNA (mRNA). Messenger RNAs are translated into the amino acid sequence of a protein. The two steps are called **transcription** and **translation.** During cell division, DNA is replicated, and a copy of the entire genome is present in each new cell. Over the past decade, substantial progress has been made in identifying the genes present in plants, as evidenced by sequencing of the *Arabidopsis* genome. The function of many gene products, however, especially those involved in secondary metabolism, is unknown.

1.2. Nucleotides

Structurally, DNA is an unbranched double-stranded molecule, oriented in a double helical shape (Figure 5.1). Each rung in the double helix is made up of purine and pyrimidine nucleotides, consisting of a base (purine or pyrimidine), a pentose sugar (2-deoxyribose in DNA, ribose in RNA) and a phosphate. The ribose sugar in RNA is less stable than the 2-deoxyribose in DNA. The purine bases, adenine (A) and guanine (G), are present in both DNA and RNA. The pyrimidines, cytosine (C) and thymine (T), occur in DNA; however, in RNA thymine is replaced by uracil (U). Pyrimidine biosynthesis occurs principally in the plastids, while purine synthesis takes place in the cytoplasm. Each adjacent nucleic acid in DNA and RNA chains is linked by a phosphodiester bond through the 5′ and 3′ carbons of the pentose sugar.

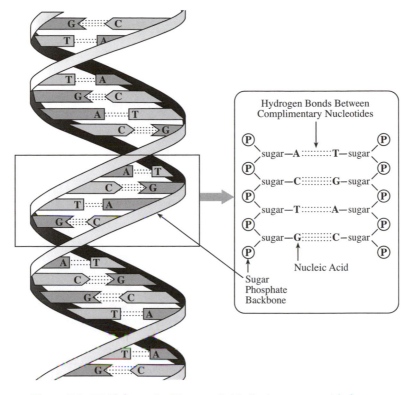

Figure 5.1. DNA has a double stranded helical structure with the two individual molecules oriented in opposite directions (i.e., 5′ to 3′ ends). Complementary nucleic acid base pairs are cross-linked by hydrogen bonds (three between cytosine and guanine and two between thymine and adenine). The sugar phosphate portion of each nucleotide forms the structural backbone of DNA *via* bonds through the phosphate of adjacent sugars (illustrated to the right). During RNA synthesis, the DNA strands separate, allowing one to be copied. As the RNA molecule is synthesized, the complementary base pair for each nucleic acid in the DNA molecule is added with the exception of thymine, which is replaced by uracil.

1.3. Nucleotide Sequence—The New Language

Rungs of the double helix are formed by pairing between complementary nucleotide bases. Hydrogen bonds are formed between a larger purine base (A or G) on one chain and a pyrimidine base (T or C) on the other chain. The **complementary pairing** of A-T involves two hydrogen bonds, while G-C pairing involves three hydrogen bonds (Figure 5.1). The two chains in the double helix are antiparallel with the sugar-phosphate backbones in opposite directions. RNA molecules are usually single stranded, but often fold upon themselves *via* hydrogen bonding of their bases.

The genetic information present in the nucleotide sequence of DNA is transcribed to a complementary mRNA. In turn, the mRNA encodes the sequence of amino acid in the protein (Figure 5.2). The translation to amino acids uses a three-nucleotide code. Each tri-nucleotide sequence (**codon**) encodes one amino acid (Table 5.1).

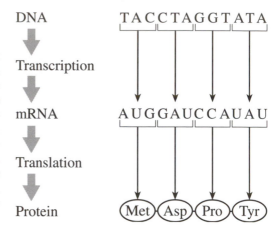

Nucleotide Sequence

Figure 5.2. The DNA nucleic acid sequence is first transcribed in the synthesis of an RNA copy, called transcription. Certain RNAs, called messenger RNAs (mRNA), are then decoded and translated into the amino acid sequence of a protein. The translation to amino acids utilizes a three nucleotides code. Each trinucleotide sequence (codon) encodes one amino acid and the codons do not overlap. If, during protein synthesis (translation), initiation starts at the U in the RNA sequence (AUGG.......) and not A, the codon UGG is read, which codes for the amino acid tryptophan instead of methionine (AUG).

Table 5.1. Individual Amino Acids Are Abbreviated Using Either a Three or Single Letter Code and Each is Specifically Coded for Within mRNA Using a Three Nucleotide Sequence Called a Codon.

Amino Acid	3-Letter Code	1-Letter Code	Codons
Alanine	Ala	A	GCC, GCU, GCG, GCA
Arginine	Arg	R	CGC, CGG, CGU, CGA, AGA, AGG
Asparagine	Asn	N	AAU, AAC
Aspartic acid	Asp	D	GAU, GAC
Cysteine	Cys	C	UGU, UGC
Glutamic acid	Glu	E	GAA, GAG
Glutamine	Gln	Q	CAA, CAG
Glycine	Gly	G	GGU, GGC, GGA, GGG
Histidine	His	H	CAU, CAC
Isoleucine	Ile	I	AUU, AUC, AUA
Leucine	Leu	L	UUA, UUG, CUA, CUG, CUU, CUC
Lysine	Lys	K	AAA, AAG
Methionine	Met	M	AUG
Phenylalanine	Phe	F	UUC, UUU
Proline	Pro	P	CCU, CCC, CCA, CCG
Serine	Ser	S	UCU, UCC, UCA, UCG, AGU, AGC
Threonine	Thr	T	ACU, ACC, ACA, ACG
Tyrosine	Tyr	Y	UAU, UAC
Tryptophan	Trp	W	UGG
Valine	Val	V	GUU, GUC, GUA, GUG
"Stop"	—	—	UAA, UAG, UGA

Codons do not overlap and are not separated by divisions. Therefore, if during protein synthesis (translation), initiation starts at the U in the RNA sequence (AUGG.......) and not A, the codon UGG is read which codes for the amino acid tryptophan instead of methionine (AUG). Incorrect initiation of protein synthesis can also lead to a nonsense codon, one for which there is no amino acid. Nonsense codons are possible, since there are four bases (A, U, C, G) and three nucleotides per codon; thus the number of possible codons is $4^3 = 64$. Since

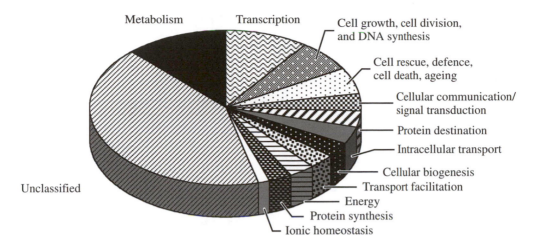

Figure 5.3. Functional analysis of *Arabidopsis* genes.[65]

there are 20 common amino acids and 64 possible codons, more than one codon can encode for each amino acid (Table 5.1). An exception is the codon AUG, the only sequence encoding methionine. AUG also signals the initiation of protein synthesis. Another single codon is UGG, which encodes tryptophan. Three codons (UAA, UAG, UGA) do not encode amino acids but signify the termination of protein synthesis or the end of the translated mRNA.

1.4. Chromosomes and Genes

The nucleus of a plant cell contains the largest amount of DNA and number of genes; therefore, it has the greatest complexity. By way of comparison, the bacterium *Escherichia coli* (*E. coli*) has 4.7×10^6 base pairs (bp), fruit flies 2×10^8 bp per haploid cell, humans 3×10^9 bp and plants from 1.5×10^8 bp in *Arabidopsis* to 1.6×10^{10} bp in wheat to the largest known genome in a member of the lily family, 1×10^{11} bp. The base pair measure for eukaryotes (fruit flies, humans, plants) is not a good indicator of the complexity of the organism, since not all the DNA encodes proteins. In prokaryotes (bacteria), nearly all the DNA encodes proteins or RNA. However, in eukaryotes non-coding DNA consists of multicopy sequences called repetitive DNA or single copy sequences called spacer DNA. In *Arabidopsis* only 10% of its nuclear genome is repetitive; therefore 90% of its genome encodes proteins or RNA. The function of only some of the genes is known or projected (Figure 5.3). Rice, with about five times the genome size, has a similar number of unique sequences as *Arabidopsis,* but it represents only about 20% of the total genome (versus 90% in *Arabidopsis*); the remainder is repetitive and spacer DNA. The *Arabidopsis* genome has been fully sequenced and is used as a model for studying gene expression and function in plants.[18,34,39,41,65,66]

Most plants, despite how much DNA they have or their genome size, are thought to possess approximately the same number of genes (20,000 to 30,000). Each gene comprises about 1,300 bases (1.3 kb); however, the bases need not be contiguous, since most eukaryotic genes contain sequences at either end and within the gene that do not contribute to the gene product. The internal non-coding sequences, called **introns** (Figure 5.4), are removed after transcription, as the mRNA is processed in preparation for translation.[7] The actual gene coding sequences subdivided by introns are called **exons.** The non-coding DNA sequences at the 5′

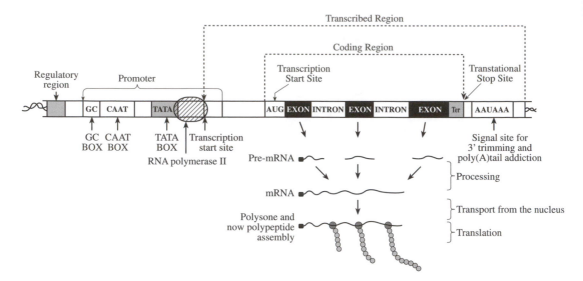

Figure 5.4. Structure and function of a plant gene that serves as a template for RNA synthesis. The gene is separated into a regulatory promoter region and a transcription region that codes for the protein. The transcribed region contains protein coding regions and regulatory elements (CAAT box, TATA box, GC box, promoter). In the protein coding region, there are sequences (introns) that are transcribed into RNA then removed prior to the messenger RNA (mRNA) being translated into a protein. Sequences that code for proteins are called exons.

end of the gene (flanking) are involved in gene regulation. These **promoter** regions control **transcription** of the gene. The flanking regions on the downstream end (3′) of the gene contain sequences that modify mRNA, a transcription stop site, and may contain regulatory elements.

1.5. Nuclear, Plastid and Mitochondria Genomes

Control within a plant cell is complicated by the presence of three different genomes found in the nucleus, mitochondria and plastids such as the chloroplast[23] (Figure 5.5). Mitochondria are thought to be the remains of a proteobacterium that became an endosymbiont in an anaerobic eukaryotic cell. Later, a cyanobacterium took up residence in an ancestral plant cell, evolving into the plastids.[27] In developing these endosymbiotic relationships, the proteobacteria and cyanobacteria gave up much of their original control over metabolism, division, development and gene expression, and became integrated with and coordinated by the nuclear genome. Information, therefore, is relayed among these intracellular compartments (organelles) and from environmental receptors.[23,33] Most protein and RNA synthesis needed for these organelles was transferred to the nuclear genome by deletions and DNA translocations from the organelle genome. Proteins originally encoded in the mitochondria and plastid genomes are now encoded by nuclear genes, assembled by ribosomes in the cytosol and directed to the appropriate organelle by **transit peptides**[7] attached to the protein. The genes that continue to be expressed in the mitochondria and plastids are thought to remain either because they are regulated by multi-component systems or the encoded protein must be synthesized in the organelle to function. Mitochondria and plastid DNA is inherited through the maternal (♀) contribution to the seed; consequently, it remains essentially unchanged from generation to generation. In contrast, inheritance within the nuclear genome is based on the contribution of both the male (pollen) and female (ovule) parent.

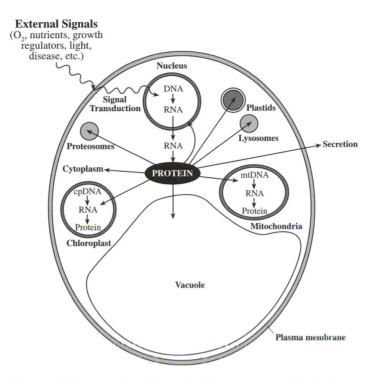

Figure 5.5. Genome interactions: The three genomes within plant cells interact to coordinate and integrate cellular activities during development and in response to environmental and cellular cues. In the maize (*Zea mays*) genome, there are 2.0×10^9 bp and 30,000 genes in the nucleus; 1.4×10^5 bp and 125 genes in the chloroplast and 5.7×10^5 bp and 40 genes in the mitochondria.

2. GENE EXPRESSION

The first step in eukaryotic gene expression is **transcription,** where mRNA is synthesized using the DNA base sequence for the gene (encoding the protein) as the template. The mRNA is capped at the 5′ end with 7-methyl-guanylate to prevent degradation by RNAses (enzymes that degrade RNA). The mRNA molecule is then cleaved at a specific site, and a poly-A tail, typically consisting of 100 to 200 adenylic acid subunits, is added. The poly-A tail protects the 3′ end from RNAses and is required for transit through the nuclear membrane. The second step involves processing the mRNA to remove the introns and splicing together the exons (coding regions). Splicing may also involve an exon being removed to produce a different polypeptide chain, thus allowing multiple proteins to be formed from a single gene.[7] The mRNA is then transported across the nuclear membrane into the cytosol. Protein synthesis occurs in the cytosol with multiple ribosomes attaching to the mRNA template, each translating the linear sequence of codons to the appropriate amino acid sequence and forming a series of identical proteins.

2.1. Prokaryote versus Eukaryote

Gene expression is more complex in eukaryotic than in prokaryotes (bacteria). In both, expression involves the DNA template being transcribed into mRNA. While prokaryotic transcription has been used as a model for the study of eukaryotic gene expression, the latter is very different and much more complex. In prokaryotes, translation is coupled to transcription. As the mRNA is synthesized on the DNA template, a ribosome binds to it and begins protein synthesis. In addition, the nuclear membrane or envelope is absent from prokaryotes, but in eukaryotes it has a substantial regulatory role. Prokaryote gene structure also differs from that of eukaryotes. Prokaryote genes are arranged in "operons," sets of linearly arranged genes that include both regulatory and structural gene sequences.

2.2. Eukaryote Gene Expression and RNA Processing

The relative simplicity of prokaryotes in their morphology and cellular structure contrasts sharply with the complexity of multicellular eukaryotes. Here regulation occurs at different levels, adding multiple layers of control over gene expression. The steps involved from transcription to translation illustrate the different levels of potential control. Control of gene expression can occur at the genome level by way of DNA methylation leading to gene silencing; chromatin condensation, making the gene inaccessible to RNA polymerase II; and in rare instances, gene amplification and rearrangement. RNA polymerase II is a multicomponent protein complex that synthesizes RNA using sections of the DNA as the template. If the gene's DNA is available for expression,[51] then the availability of necessary transcription factors (proteins) and RNA polymerase II determines expression and pre-mRNA synthesis. The pre-mRNA is then processed further or degraded. Processing involves the 5' end being capped with 7-methylguanylate, a poly-A tail added to the 3' end, splicing together of the exons and possible additional splicing.[5] Control over the mRNAs available for protein synthesis can also occur during splicing and *via* mRNA degradation. There are also 5' and 3' untranslated regions between the cap and the poly-A tail, the function of which is not yet clear. They may interact with proteins and be involved in the translation process. Each step in this stage occurs in the nucleus. Attachment of the 5' cap and poly-A tail, along with proper splicing, is completed before the mRNA exits the nucleus. Incomplete RNA transcripts are readily degraded in the nucleus.

The nucleus is bound by a double membrane, between which is the lumen or perinuclear space. The membrane acts as a boundary between the material (chromosomes, genes, proteins) in the nucleus and the cytoplasm, where protein synthesis occurs. RNA produced in the nucleus must pass through pores in the nuclear membrane, and proteins synthesized in the cytoplasm may be imported through similar pores. The pores make up 8 to 20% of the nuclear membrane surface and function as molecular sieves, controlling the outward and inward flow of molecules. Transport specificity is modulated by up to 120 different proteins (nucleoporins) arranged in pore complexes within the membrane. Proteins to be transported into the nucleus must have a specific sequence of amino acids on their tail to be recognized by the pore complex. Energy, as ATP, is required for protein importation. Some imported proteins include transcription factors and histones.

After exiting the nucleus, the mRNA associates with ribosomes, and protein synthesis occurs.[8] The life expectancy of a mRNA molecule is short; it varies from tissue to tissue.[20] The high turnover rate allows the cell to control how many copies of a particular protein are made, which can range from fewer than ten to thousands per cell. The mRNA may also vary in translatability due to folding, association with proteins and changes that influence accessibility of the start site for translation (AUG). In addition, whether the mRNA and ribosome complex is free or attached to the endoplasmic reticulum can affect the rate of protein synthesis.

Translation requires a number of small RNAs, 70 to 90 residues in length, called transfer or t-RNA, which bind the appropriate amino acid and transfer it to the mRNA. The number of different t-RNA's (31) is intermediate between the 64 codons and the 20 amino acids. This is a two-step process whereby the amino acid is first covalently bound to the 3' end of a specific t-RNA, then the complementary anticodon of the t-RNA binds to the matching exposed codon of the mRNA. The new amino acid is then bonded to the growing polypeptide chain. The t-RNA is released, and it can, following activation, transport another amino acid. Once the protein is synthesized (Figure 5.4), it undergoes folding, assembly and targeting to the correct site in the cell. The protein may also undergo post-translation modifications and cleavage. Thus, control over both the rate of synthesis and degradation (turnover) provides precision in the amount of a particular protein within a cell.[70]

Proteins are synthesized with a terminal amino acid sequence (signal peptide) that targets it to a specific location within the cell (chloroplast, mitochondria, proteosome, plastids, lysosome, vacuole, membrane system or the nucleus) or for secretion from the cell. Targeting is crucial for integrating cellular activities.

2.3. Eukaryotic Gene Regulation

Like prokaryotes, eukaryotic gene expression is regulated at the transcriptional level. Post-transcriptional control occurs; however, transcriptional regulation may be more important. Plant growth and cellular differentiation are due to regulated gene expression. An entire plant can be regenerated from a single root cell, demonstrating that all genes are retained and their expression depends upon the cell's environment. Genes can be separated into two general classes, **regulated** and **constitutive.** The latter are continually expressed and are needed for basic cellular metabolism. The regulated genes may be strictly controlled by a cell's developmental program and become active or quiescent with changing needs and developmental stages (vegetative to reproductive; juvenile to mature; growth to senescent). Besides developmental cues, many genes are environmentally regulated (e.g., light, temperature, water, pathogens, wounding, herbivore). The external environment is sensed by molecules on the cell surface (plasma membrane) and in the cytoplasm that signal the induction or termination of transcription of specific genes.

On either side of the genes are DNA regulatory sequences that interact with proteins (transcription factors) synthesized or activated during development or in response to stimuli. These regulatory sequences specify how much mRNA is produced and when the gene is turned "on" and "off." Therefore, DNA regulatory sequences are crucial for coordinated development.

2.3.1. Gene Promoters

As described previously, promoter sequences regulate gene expression and are found upstream (5') from the transcription start site (Figure 5.6). A nucleotide sequence called a TATA box, found in most eukaryotic genes about 30 nucleotides upstream (-30) from the transcription start site, is responsible for positioning RNA polymerase II. Certain constitutive genes, however, may not contain a recognizable TATA box. Two other proximal elements are the CAAT box (-80) and the GC box (-100), which are binding sites for transcription factors (i.e., proteins that influence the rate of transcription).* These two proximal elements, along with

---

*Cis-acting elements are sequences on the same DNA strand as the protein-coding region that regulate the amount and pattern of gene transcription. The most common cis-elements are the TATA and CAAT boxes; additional are found further away from the transcription start site. Trans-acting transcription factors are proteins that bind to the cis-acting elements and modulate the activity of common transcription factors that assemble with RNA polymerase II at the transcription start site.

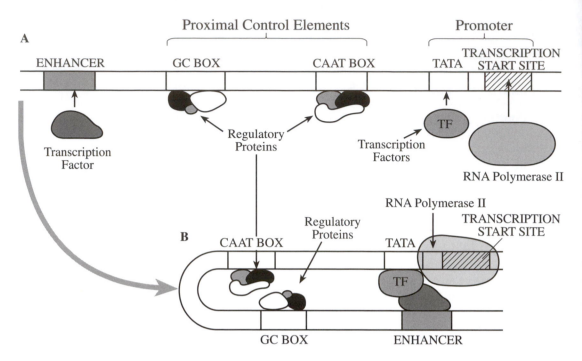

Figure 5.6. In eukaryotic gene expression, different levels of control occur, depending upon the gene being translated. (A) The minimum level of regulation is obtained with the TATA box, which binds to the RNA polymerase II. The transcription by RNA polymerase II can be enhanced by the CAAT box and the GC box, which are called proximal control elements, located 80 and 100 bp upstream of the start site. The binding of proteins to these sites enhances polymerase activity. (B) Further upstream from the proximal control elements are additional regulatory sequences to which transacting factors (proteins) bind and activate transcription by interacting with the transcription complex. Enhancer and prompter interaction can occur through the formation of a DNA loop, placing one or more enhancer binding regulatory proteins (transcription factors) in contact with RNA polymerase II.

the TATA box, comprise the minimum number of promoters for plant cells. The transcription factors that bind to these cis-acting sequences are also called trans-acting factors. Transcription factors are proteins that bind in a specific manner to a gene's promoter. The genes that encode these transcription factors are found elsewhere in the genome, i.e., at a different position on the same chromosome or on a different chromosome. The transcription factor genes are expressed in response to developmental and environmental signals.

2.3.2. Regulatory Sequences

Further upstream from the CAAT and GC boxes are distal regulatory sequences found about 1000 bp from the transcription start site (Figure 5.6). These sequences can function in either a positive (activator) or negative (repressor) manner to regulate gene expression once they have been triggered by transcription factors. Enhancer sequences, found some distance up- or down-stream from the promoter, may also be present.[62]

2.3.3. Trans-acting Transcription Factors

Transcription factors are proteins that generally have two structural features, a DNA binding domain and a domain that organizes the proteins involved in activating transcription by RNA polymerase II (Figure 5.6). The DNA binding domain is part of the transcription factor that interacts with a specific DNA sequence. Transcription factor proteins may interact with DNA as a single protein, a homodimer* or a heterodimer. They are often encoded by a family of genes that are responsible for a population of related proteins. Diversity of transcription factor proteins confers specificity to DNA binding and gene expression and allows coordinated expression of genes when appropriate.

2.3.4. RNA Polymerase II

Some transcription factors do not interact directly with a cis-acting element, but are required for the formation of a transcription complex with RNA polymerase II. RNA polymerase II cannot bind to the DNA template alone but requires the presence of several transcription factors (TF). The transcription factors include several TATA binding proteins, RNA polymerase II, TF protein H (TFII-H) that causes the enzyme to be phosphorylated and other TF proteins (TFII-A, B, D, E, F). The TF proteins form a multiprotein complex that interacts with other proteins that have bound to specific DNA template sequences, leading to mRNA synthesis from the transcription start site (Figure 5.6).

The binding of transcription factors and the RNA polymerase II complex to the DNA template does not occur unless the gene template is accessible.[17] Histone proteins regulate the degree of DNA condensation and, therefore, the access of transcription factors to gene promoter sequences. Regulation of condensation (accessibility) is dependent upon the degree of acetylation of the histone protein "tails."[61] Acetylated histone DNA-associated sequences are less compact. Two enzymes, acetyl transferase (adds acetyl groups) and deacetylase (removes acetyl groups), renders a gene accessible or less accessible to RNA polymerase. Other structural features of DNA also play a role in regulating gene expression. Eukaryote genomes are thought to be organized in loops, 5 to 200 kbp in size, which are anchored to attachment regions.[30] The attachment regions are AT rich and 200 to 1000 bp long. The topology of the loop and the matrix attachment region appear to influence gene regulation. DNA methylation can also influence gene expression.

3. INTEGRATION OF EXTRACELLULAR AND ENDOGENOUS SIGNALS WITH GENE EXPRESSION

As previously described, transcription is controlled at the molecular level by an intricate complex of molecules and nucleotide sequences (e.g., promoters, RNA polymerase II, regulatory sequences, transcription factors) that turn on or off the expression of specific genes. During cellular, tissue and plant development, gene expression is spatially and temporally coordinated in a very precise manner.[24] How then are developmental and environmental changes (e.g., photoperiod, oxygen concentration) recognized, assessed and integrated—and the appropriate genes from the 25,000 to 30,000 present, turned on or off?[63,67] Genes involved in "housekeeping" or maintenance roles (constitutive genes) are continually expressed with the number of

---

*A dimer is a condensation product consisting of two molecules, in this case proteins. A homodimer is comprised of two identical proteins, while a heterodimer is comprised of two different proteins.

copies of each protein being regulated by both its rate of synthesis and degradation (turnover). For many genes, however, the timing, location and level of expression are critical. This section examines how environmental information is passed through a signaling system that leads to expression of the appropriate gene(s).

In animals, signaling is accomplished *via* the nervous and endocrine systems. Plants, lacking a nervous system, have developed a chemical signaling system that allows precise assessment of environmental factors such as the quantity and quality of light.[64]

During a plant's life cycle, it responds to internal and external stimuli. These provide the plant with information about light, mineral nutrients, organic molecules, water availability, soil pH, gravity, temperature, wind, growth regulators, cellular pH, atmosphere gas composition, bacterial, fungal and virus pathogens, nematodes, herbivores and other factors.[16,36] The information, which varies in both quantity and quality, may be received *via* plasmodesmata between adjacent cells[29] and over longer distances through the xylem and phloem. The final response will often vary due to the plant's ability to integrate experience, developmental stage, internal clocks and other factors, the net effect of which is to optimize growth and development.

For many years, ideas on intercellular signaling were confined to the five plant hormones—auxin, cytokinins, gibberellins, abscisic acid and ethylene.[32,67] This small number of growth regulators were shown to interact in signal transduction (i.e., auxin and ethylene, cytokinin and abscisic acid),[44,48,53] and respective combinations could explain a very small number of coordinated developmental responses. More recently other chemicals have been shown to have significant signaling properties.[13,40,44] These have parallels to the chemicals found in animal systems, such as steroids, sterols and lipid derivatives, nitric oxide, jasmonic acid, salicylic acid, proteins, peptides and possibly small RNA sequences.[42,45,52,54,71,72] Besides movement within the plant *via* the phloem and xylem, cell-to-cell movement can occur in a regulated way through plasmodesmata. Plasmodesmata provides symplastic (cytoplasmic) continuity between cells. These intercellular pores are plasma membrane lined with a core of endoplasmic reticulum in the center. The pores regulate the passage of proteins and ribonucleoproteins between cells that signal cell development in adjoining and distant cells. Cell development in root and shoot meristems are coordinated *via* such signals. Other signals that can be transferred long distances through the phloem sometimes play a role in plant defense.

3.1. Receptors

Considerable progress has been made on animal signal transduction and its role in gene expression. While animal systems are thought to provide some parallels with plants, they differ in significant ways. In plants, as in animals, common elements exist in the signal transduction pathways. The first step is **signal recognition** that may occur extracellularly, at the plasma membrane or within the cell. Receptor proteins are involved in recognition that may involve catalytic activity. The second step is **signal transduction.** Signal transduction can directly or indirectly affect cellular protein activities and membrane permeability, alter the cell's metabolism and be amplified and transferred to the nucleus to alter gene expression.

Most receptors are found on the plasma membrane, though a few are in the cytoplasm and other cellular compartments. Other receptors, such as red and far red light and plant growth regulators, modify membrane potential.[53,64] The selective binding of the signal (ligand) to the receptor is reversible, and signal concentration can exceed receptor concentration and lead to the saturation of a response. Receptors have different ligand binding strengths (e.g., an ethylene molecule binds for only around ten minutes before dissociation from the site). Receptors are also subject to turnover. In certain species, the ethylene response can be recovered even after treatment with a potent inhibitor (1-methylcyclopropene) that binds very tightly to the receptor, blocking ethylene responses.

The gene and amino acid sequences are known for several plant receptors. Though little sequence conservation exists, all receptors have certain similarities in structure. The two structural elements are seven hydrophobic domains that apparently span the membrane and a ligand binding site (receptor site) present in one of the extracellular hydrophobic domains. Other receptors, such as the c-terminal receptor, are phosphorylated by protein kinases on a serine/threonine or tyrosine residue. Another class of receptors has a single membrane spanning domain with a large cytoplasmic domain containing a protein kinase activity. This receptor-like protein kinase responds by dimerization after the ligand binds to the receptor. Other receptors located intracellularly act as Ca^{2+} channels and involve inositol 1,4,5-triphosphate (IP_3) and cyclic AMP. Both signaling molecules are synthesized by enzymes on the plasma membrane and translocated to the endoplasmic reticulum and vacuole where their receptors are located. Binding to the receptor leads to Ca^{2+} channels opening and an influx of Ca^{2+} into the cytoplasm. The receptors have four membrane spanning domains.

Our understanding of the receptors for the phytohormones is rapidly expanding. Receptors have been identified for auxin, phytochrome, cytokinin, abscisic acid and ethylene. For example *Arabidopsis* has 5 ethylene receptors (ETR1, ETR2, ERS1, ERS2 and EIN4).[31] While the possibility of multiple binding sites had been proposed for some time,[25] not until recently has our technical ability been such that the genes for the sites could be identified. The ethylene binding receptor (Figure 5.7) has three hydrophobic stretches that span the membrane[11] and a copper atom [(Cu(I)] at the binding site.[55] Questions about differences in sensitivity among sites, their turnover rate and location within the plant are now starting to be addressed.

Ethylene receptors are negative regulators in that ethylene responses are repressed in the absence of the molecule. Binding of ethylene removes the block to downstream signal transduction. Transgenic *etr1-1* tomato plants are insensitive to ethylene and give phenotypes with fruit that fail to ripen and roots that have difficulty penetrating the soil. 1-Methylcyclopropene competitively binds to the receptor site with a high affinity, blocking the plant's response to ethylene. Transgenic plants with disrupted ethylene binding and/or signal transduction and the use of 1-MCP are viable commercial options when an ethylene response is unwanted (e.g., flower senescence); however, when the objective is to delay rather than prevent ripening, transgenic repression of the genes for ethylene synthesis (ACC synthase and ACC oxidase) is the appropriate approach in that the ethylene response (e.g., ripening) can be instigated with exogenous ethylene treatment. Equally important, constitutive repression of ethylene binding and/or signal transduction generally results in phenotypes with little or no commercial utility due to the inhibition of other critical ethylene mediated systems in the plant. Thus transgenic alterations must be tissue specific and not extend across the entire plant.

3.2. Signal Transduction

The principal elements in plant cell signal transduction are the secondary messengers, intracellular (Ca^{2+}) and protein kinases. These messengers can lead to the amplification of a signal that causes changes in ion flux, regulation of metabolic pathways and altered gene expression (Figure 5.8).

A critical question is how does a transitory increase in calcium provide the needed level of response specificity? It is known that the increase in calcium varies, which could influence the location or the distance into the cytoplasm it moves. The calcium signal mechanism is only found in eukaryotes and may have developed initially as a detoxification mechanism to remove calcium.

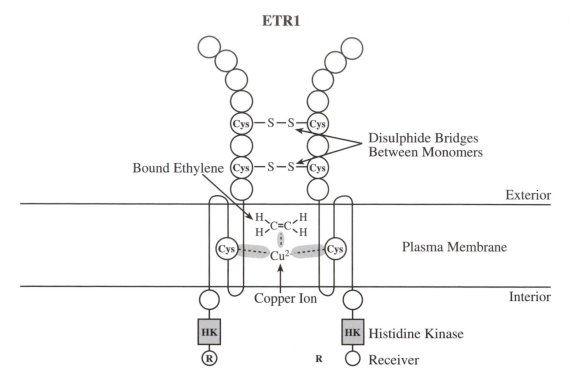

Figure 5.7. The ethylene receptor, as illustrated by ETR1, is comprised of two identical proteins (monomers) that span the plasma membrane and are bound together *via* disulfide bridges between cysteine residues near the N terminal of each molecule (exterior to the membrane; internal amino acids are not displayed, thus the illustration is not drawn to scale). Ethylene binds to a copper cofactor that is thought to be positioned between a cysteine (or histidine) residue in each monomer within the membrane. On the interior of the membrane are a histidine kinase domain (HK) and a receiver domain (R) which are thought to comprise the transmitter portion of the receptor. Binding of ethylene inactivates the transmitter, blocking subsequent steps in the signaling pathway (a form of negative regulation). Five different ethylene binding receptors have been identified in *Arabidopsis (after Rodríguez et al.)*.[55]

3.3. Secondary Messenger—Calcium and Calmodulin

The short-term influx of calcium is a major cell signal transduction mechanism. The signals sensed are thought to include wind, temperature stress, red and blue light, wounding, anaerobiosis, abscisic acid, osmotic stress and mineral nutrition. The signal depends on the Ca^{2+} electrochemical gradient across the plasma membrane.[56] Cells normally maintain a very low Ca^{2+} concentration in the cytoplasm (100 to 200 nM), while in the vacuole, endoplasmic reticulum lumen and external to the plasma membrane, the concentration may be 10,000 fold greater (0.5 to 1 mM). Other potential calcium stores can be in the chloroplast, mitochondria and nucleus. High concentrations of Ca^{2+} in the cytoplasm are toxic, interfering with cellular metabolism. When the concentration is high, calcium "pumps" are activated in the tonoplast and mitochondria membranes and Ca^{2+} is pumped out of the cytoplasm to reestablish the appropriate concentration.

The primary sensor (receptor) of calcium is calmodulin, a small protein (15 to 17 kDa)[75] with four Ca^{2+} binding sites. Calmodulin is found in both the cytoplasm and nucleus and attached to the plasma membrane. The concentration of calmodulin within a cell varies widely,

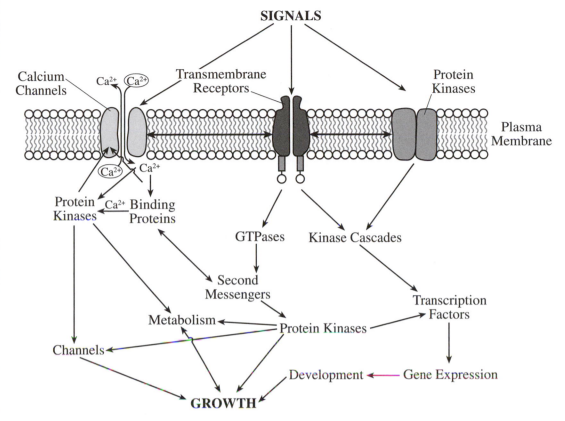

Figure 5.8. Plants respond to a multiplicity of internal and external signals. These responses are necessary for coordinated growth, development and survival. The signals are received by all cells in the plant; however, they may respond in different ways. Following amplification of the signal, changes may occur in ion fluxes, metabolic pathways, gene expression and the cytoskeleton.

depending on stage of development and tissue type. Upon binding calcium, the molecule undergoes a structural change, exposing a hydrophobic region of methionine, leucine and phenylalanine. The hydrophobic region is recognized by target proteins, which combine with calmodulin, resulting in activation. The calcium signal transduction pathway involves hundreds of different proteins, while there are at least 1000 protein kinases that phosphorylate proteins. The signals can interact cooperatively and synergistically with each other, giving a wide range of possible final responses.

3.4. Phosphorylation — Kinases

Phosphorylation, the addition of a phosphate, is one of the most common covalent modifications of plant proteins.[60] Phosphorylation can alter a protein's biological activity, subcellular localization, half-life and protein-protein interactions.[3] Phosphorylation by kinases and dephosphorylation by phosphatases is seen as a crucial factor in the integration of signaling within plant cells, influencing both the extent and duration of a response (Figure 5.9). *Arabidopsis* has about 1000 protein kinases and 200 phosphatases. Important groups are the mitogen-activated protein kinases (MAPK—23 genes), MAPK kinases (MAPKK—9 genes),

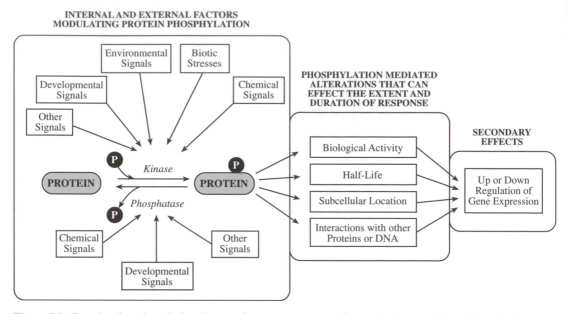

Figure 5.9. Protein phosphorylation is central to many processes in regulating protein activity, development and responses to environmental and biotic stress. Phosphorylation alters a protein's activity, cellular localization, half-life and interaction with other proteins or DNA. Kinases that phosphorylate proteins play a crucial role, integrating various signals and coordinating the cell's response *via* the MAPK pathway.

MAKK kinases (MAKKK—>25 genes) and calcium-dependent protein kinases (CDPK—29 genes). These kinases are involved in modulating and controlling cell division, cell metabolism (including RNA polymerase activity), transport and hormone responses, and the response to environmental and pathogen stresses in plants. Many receptors have kinase activity that starts the signal amplification process leading to gene expression.

Protein kinases transfer phosphate from ATP to the amino acid serine, threonine or tyrosine on a targeted protein. Conversely, specific phosphatases remove the phosphate. A single protein kinase can phosphorylate hundreds of individual proteins, thereby amplifying a signal. Some of these kinases are located in the plasma membrane; they are automatically phosphorylated upon ligand binding *via* a homologous dimerization. Other protein kinases are Ca^{2+} dependent with a calmodulin-like domain. Protein kinase cascades are thought to be involved in the ethylene, gibberellin, auxin and abscisic acid responses.

3.5. Ethylene Signal Transduction Pathway

The phytohormone ethylene instigates its action within the plant by first attaching to an ethylene-specific receptor site. Alterations in the configuration of the binding protein trigger a signal transduction pathway where the message "ethylene is bound to the receptor" is propagated through a series of molecules, eventually triggering the expression of the genes that lead to the physiological response (e.g., flower senescence). While the overall sequence is not yet understood, ethylene response mutants and transgenic clones are starting to provide an insight into the ethylene signal transduction pathway. The first step occurs at the ethylene receptor on the plasma membrane, a membrane-spanning histidine kinase with a copper cofac-

ETHYLENE SIGNAL TRANSDUCTION PATHWAY

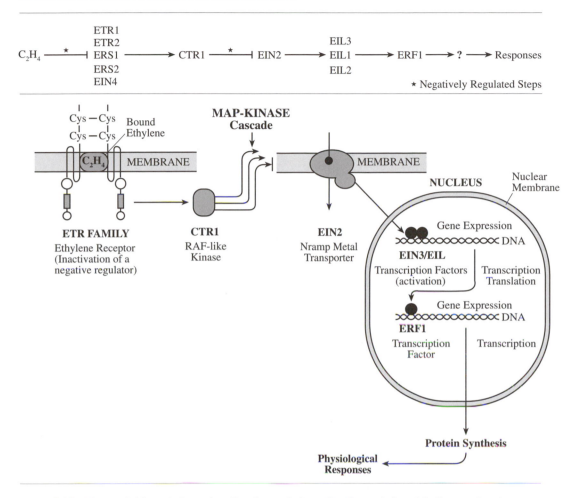

Figure 5.10. The model for ethylene signaling in *Arabidopsis* involves ethylene binding to the plasma membrane embedded receptor such as ETR1. This binding deactivates the receptor molecule, and the CTR1 protein becomes inactive. EIN2 is an integral membrane protein that occurs before the EIN3 transcription factors.[5,6,55]

tor. Upon binding, the molecule associates as a homodimer *via* a disulfite linkage (Figure 5.10). The receptor is a two-component signaling system[10] where an amino-terminal input domain perceives ethylene and a carboxyl terminal histidine kinase transmits the signal. Upon ethylene binding, the receptor molecule is deactivated through the phosphorylation of the histidine. In bacterial two-component systems, the next step is the transfer of the phosphoryl group to the asparagine residue in the receiver domain at the carboxyl end thereby activating the output domain. In *Arabidopsis,* there are five ethylene receptors (ETR1, ERS1, ETR2, EIN4, ERS2) having similar ethylene domains; however, two vary in their receiver domains. Without ethylene, the receptors are predicted to be functionally active histidine kinases.[59]

Following the receptor, the next known component in the pathway is a protein kinase, CTR1 (Figure 5.10). CRT1 has a sequence similar to an MAPKK kinase that may act in a protein phosphorylation cascade and is thought to be a negative regulator of ethylene responses. Downstream to CTR1 is the integral membrane EIN2, a positive regulator, thought to be central to ethylene responses[9] in that loss-of-function mutants are highly ethylene insensitive. The mutants also do not response to certain other plant growth regulators. EIN2 is followed by ethylene-responsive transcription proteins that bind to specific gene promoters in the nucleus to instigate the expression of the genes required for the physiological response.

4. RECOMBINANT DNA

Approximately 10,000 years ago, man began to cultivate several wild annual grain species, an endeavor that required saving seed from one growing season to the next for planting. At some point, a few early farmers began to save seed from superior plants and in so doing, began the genetic modification of plants by humans. Plant selection eventually led to plant breeding, a relatively recent event in the history of agriculture. With breeding, desirable parent plants are intentionally crossed, increasing the rate and effectiveness of genetically modifying crops. During the past 50 years, exceptional quantitative and qualitative advances in crop plants have been made with the yield of most major food crops doubling and sometimes, tripling. Conventional plant breeding, however, is limited to a relatively narrow genetic base—only plants of the same species (or in a few cases, very closely related species) can be successfully crossed. Likewise, significant genetic advances often require years to accomplish. With conventional plant breeding, the entire genome of both parent plants is combined, creating a myriad of possible combinations that must be pragmatically examined to find the best and then often backcrossed to eliminate undesirable genes.

The most recent advance in crop improvement is recombinant DNA technology, which allows the introduction of one or a few genes into an existing superior clone. In addition, recombinant DNA opens the door to a vast array of beneficial genes that are not restricted to a given species but can be obtained from any organism or artificially created.

Plants created using recombinant DNA are called genetically engineered, transgenic or genetically modified. The latter term, genetically modified (GM), is the least accurate in that classical plant breeding also genetically modifies plants. The term GM, however, is widely used in the popular literature on the subject to denote organisms developed with recombinant DNA. The spread of genetically modified crops has been remarkable since the first commercial plantings in 1996. It is estimated that in 2000, more than 44 million hectares of genetically modified crops were grown worldwide, 30 million hectares in the United States alone[47] (Table 5.2). Many foods we consume are either genetically modified or contain ingredients derived from gene modification technology.[68]

The objective of genetic engineering is to insert new genes into the nuclear genomes of targeted plants and have it expressed in the desired manner. The insertion must be genetically stable, so that it can be transferred to subsequent generations *via* normal sexual reproduction. Decisions that must be made include determining: 1) what is to be inserted; 2) how it is to be inserted; 3) how transformed cells are to be separated from non-transformed cells; and 4) how the transformed cells are to be converted back into plants?

4.1. Transferred DNA

Besides the DNA encoding the gene of interest, a promoter region and a reporter gene must be attached to the genetic material to be transferred to the nuclear DNA of the plant cell. Collectively the inserted material is termed a **gene construct.** The promoter region controls the ex-

Table 5.2. Transformed plants released in the United States and Western Europe.*

African violet	Eggplant	pine
aspen, quaking	*Eucalyptus grandis*	pineapple
alfalfa	European aspen	plum
American chestnut	European aspen (*alba* ×	*Poa pratensis* × *Poa arachnifera*
Anthurium andraeanum	*tremula*)	poplar
apple	European plum	*Populus deltoides*
Arabidopsis thaliana	*Festuca arundinacea*	potato
avocado	fodder beet	rapeseed
barley	gladiolus	raspberry
barrelclover	grape	rhododendron
beet, leaf, sugar and garden	grape (*berlandieri* × *riparia*)	rice
Begonia semperflorens	grape (*berlandieri* × *rupestris*)	rose gum
belladonna	grapefruit	*Rosa hybrida*
bermudagrass	Kentucky bluegrass	*Rubus idaeus*
broccoli	kiwi	Russian wild rye
Brassica oleracea	lettuce	rye grass
Brassica rapa	lily	safflower
cabbage	limonium	sand grape
cantaloupe	marigold	sorghum
carrot	melon	soybean
cassava	mustard	spruce
cauliflower	*Metaseiulus occidentalis*	squash
cherry	*Nicotiana attenuata*	St. Augustine grass
chicory	oat	strawberry
chrysanthemum	onion	sugarcane
Cichorium intybus	orange	sunflower
Citrus sinensis × *Poncirus*	papaya	sweetpotato
trifoliatus	*Paspalum notatum*	sweetgum
clary	pea	Tasmanian blue gum
coffee	peanut	tobacco
corn	pear	tomato
cotton	*Pelargonium*	turnip
cranberry	pepper	velvet bentgrass
creeping bentgrass	peppermint	walnut
cress, thale	perennial ryegrass	watermelon
cucumber	persimmon	wheat
Dendrobium	petunia	

*As of August, 2003.

pression of the gene within the plant; without expression, the gene would be of no value. A promoter from the cauliflower mosaic virus (CaMV35S) is widely used for plant genetic engineering; it is positioned up-stream from the transcription start site for the inserted gene. Once in the plant, DNA-binding proteins (transcription factors) interact with the promoter, triggering the expression of the genes fused to it. DNA-binding proteins allow expression of genes in specific cell types or at a precise time. Also attached to the promoter and gene of interest is a reporter gene. Since T-DNA is inserted at random within the plant genome, some cells are not transformed, while in others expression may be very low. Consequently, a method must be employed to separate the successfully transformed cells from the remainder. A reporter or marker gene is added to the T-DNA segment that provides a means of identifying the transformed cells. With some reporter genes, non-transformed cells can be efficiently screened and eliminated.

4.2. T-DNA Delivery

Transformation of a wide cross-section of plants has exploited a sophisticated technique that several members of the plant bacterial genus *Agrobacterium* evolved millions of years ago that allows them to force a diverse range of host plants to produce essential bacterial nutrients. This is accomplished through the organism's ability to insert several of its genes into the plant's nuclear genome. What is remarkable is that the gene structure of bacteria (prokaryotes) is distinctly different from that of plants (eukaryotes), so that its DNA would not be expressed even if successfully inserted in the plant's genome. *A. tumefaciens,* however, has evolved a segment of its DNA that is eukaryote-like in structure and thus able to be expressed by the plant.

The inserted genes contain regulatory sequences that encode proteins involved in the biosynthesis of plant growth factors and bacterial nutrients.[22] With *A. tumefaciens,* the plant cells over-produce auxin and cytokinin, which leads to the formation of an undifferentiated clump of cells (crown gall) on the surface of the plant that acts as an ecological niche for the organism. The organism, along with *A. rhizogenes,* infects a wide range of dicotyledonous plants.

The transferred DNA (T-DNA) is located on a large plasmid—pTi (tumor inducing plasmid) found within the bacteria (Figure 5.11). The T-DNA actually represents only about 10% of the plasmid's genes; there are number of other genes present, some of which (i.e., virulence genes *VirA, B, C, D, E, G, H*) are critical for the transfer of the T-DNA into the plant.

Transfer requires a complex series of closely choreographed steps (Figure 5.11). Initially a wound to the plant triggers the synthesis of response compounds that are detected by a virulence protein on the organism's surface, triggering plasmid gene expression. A single stranded copy of the T-DNA is made and transported through a membrane channel formed by proteins (*VirB* locus) between the bacterial and plant cell. Once in the cytosol, the T-DNA is coated with a protective protein (*VirE* locus), also transported into the cell, targeted to the nucleus by another protein (*VirD$_2$*) and then transported through a nuclear pore complex and integrated at random into a plant chromosome. Transcripts of the inserted T-DNA are subsequently expressed by the plant and their mRNAs exported to the cytosol, where they are translated and proteins are formed.

Molecular geneticists have found that by removing the portion of the T-DNA that produces the tumor symptoms and replacing it with a gene or genes of interest, they can utilize *Agrobacterium* to insert new genes into the plant's nuclear genome. Disarmed *Agrobacterium*-based transformation has been used to genetically alter a wide range of dicots and, more recently, monocots such as rice, maize and barley.

DNA can also be directly inserted into plant cells using microprojectiles, vortexing, or one of several other techniques that physically puncture the cellular membrane.[58] In this case, the same DNA construct is required (promoter, gene of interest and the reporter gene); however, they are not placed in a plasmid but are inserted "naked." Since *Agrobacterium* typically only infects dicots, direct injection allows modification of monocots and certain dicots that the organism does not normally infect. The most widely used of these direct injection techniques is microprojectile bombardment (i.e., gene gun or biolistic). The DNA to be transferred is coated on very small diameter (1–3 μm) gold or tungsten particles and literally shot, using a compressed gas or electrical discharge, into the cells. Once inside, the DNA separates from the particle and becomes incorporated into the plant's chromosomal DNA.

4.3. Identification of Transformed Cells

Generally only a very small percentage of the cells treated are successfully transformed (i.e., contain the DNA in a functional position). The problem becomes how to isolate the transformed cells from the thousands of untransformed cells. Identification is accomplished using

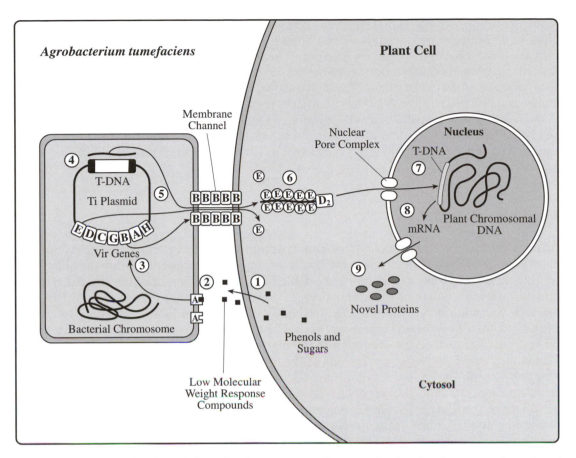

Figure 5.11. The transfer of DNA from *Agrobacterium tumefaciens* to the plant involves a complex series of closely choreographed steps. The plant responds to wounding by producing several low molecular weight response compounds (1) that diffuse out of the cell, activating receptors on the surface of *A. tumefaciens* (2) which, in turn, activate virulence genes (Vir) on the plasmid (3). Activation triggers the production of a copy of the T-DNA section (4), which is transported through a special membrane channel (5) of B proteins into the cytosol of the plant's cell, where it is coated with protective proteins [E] and has a directional protein [D2] attached to one end (6)—both of which are synthesized in the microorganism. The T-DNA then moves through a nuclear pore complex into the nucleus, where it is incorporated into the plant's chromosomal DNA (7). The inserted genes are subsequently transcribed producing mRNAs (8) that move through pores into the cytoplasm where translation takes place and the novel *Agrobacterium* derived proteins are formed.

a reporter gene that is also included with the transferred DNA. Expression of the reporter gene allows identification of transformed cells that are then separated and exposed to conditions conducive to the regeneration of the new, genetically modified plants. Initially antibiotic resistance genes were used as reporters since they allow both identification and isolation from untransformed cells. The cells are cultured on a media containing the antibiotic so that the cells expressing the antibiotic resistance (transformed) grow, while those without die. The surviving cells are then cultured under conditions that result in the regeneration of intact plants.

4.4. Regeneration

An individual plant cell, under the appropriate conditions, has the genetic potential to develop into an entire plant. The cells are grown in flasks containing nutrients, hormones and other requisites, using techniques that have been being developed over the past 40+ years. As the individual cells divide, they produce a mass of undifferentiated cells that can be used for transformation *via* a bacterial vector or gene gun. The transformed cells are then separated from the untransformed cells, and the former are placed in an appropriate medium for the regeneration of many small clonal plants. Through a sequence of steps, the individual plants are separated and allowed to grow until they eventually reach a size that will allow them to be transplanted into potting media. Since the exogenous DNA is inserted randomly and its location can be critical to expression, it is necessary to regenerate plants from many transformed cells and then screen the progeny to select clones that have the appropriate timing and degree of expression.

4.5. Gene Silencing

Altering the genetic make up of a plant can be accomplished by inserting new genes into the plant's nuclear DNA. An alternative method is turning off existing genes within the plant, preventing them from being expressed—**gene silencing.** Gene silencing causes decreased gene expression (down-regulation) to varying degrees, including complete suppression of the gene product. Silencing is accomplished by preventing mRNA from being formed or by inactivating it before it can begin protein synthesis. Expression is altered using either antisense or sense constructs of the gene inserted into the plant's DNA. An antisense gene has the complementary nucleic acid sequence to that of the targeted endogenous gene. Upon expression, an mRNA is produced that is complementary to the mRNA produced by the endogenous gene, causing it to hybridize, producing a nonfunctional double stranded molecule that can no longer be translated. The mode of action of a sense construct in turning a gene off is less well understood. Unlike antisense constructs, which simply represent the complimentary code, sense constructs are identical to the endogenous gene. These are similarly inserted into the plant's DNA; however, their effectiveness depends upon where in the genome they are inserted. Expression of the gene can range from a reduction to complete inactivated. The actual mechanism of this co-suppression is not yet known.

4.6. DNA and Protein Microarrays

Microarray technology has provided a very useful tool for studying which genes (DNA arrays) are expressed and what proteins (protein arrays) are produced. DNA microarrays use hundreds of different short single-stranded oligonucleotide probes or cDNA clones attached as microdots on a solid surface (chip). The technique works on the basis of the hybridization of the isolated mRNA with a complementary sequence attached to the chip. The population of mRNAs is first converted to cDNAs and then to cRNAs labeled with a fluorescent dye. These are then allowed to hybridize with their respective complementary oligonucleotide probes on the chip. By knowing the location of each oligonucleotide probe on the chip and scanning the fluorescence signal at each position, a snapshot of the total mRNA pool can be obtained.[37] Miniaturization has allowed monitoring thousands of genes simultaneously and comparing the relative levels of RNA transcripts formed, so that differences in transcripts (or proteins in protein microarrays) between two samples can be determined by quantifying the fluorescent signal intensity of each dot. Therefore, at the onset of ripening, genes up-regulated, down-regulated or not expressed can be monitored in a single DNA microarray. Protein arrays function similarly, showing the presence and relative level of gene products. While both types of microarrays are having a significant impact on our understanding of the coordination of changes occurring at the onset of shifts in metabolism (e.g., circadian clock, plant defense, ripening)[1] they do not identify the function of uncharacterized protein or genes.

4.7. Benefits and Controversy over the Use of Recombinant DNA

The ability to up- or down-regulate specific genes opens the door to a wide range of scientific possibilities. For example, recombinant DNA can be used to discover the biological role of individual genes within plants. From a commercial standpoint, it can be used to create new cultivars that have improved production, quality or other attributes. While an exceedingly diverse range of positive benefits could be derived from the use of recombinant DNA, transgenic organisms have not been uniformly accepted by consumers. Currently the lack of adequate scientific knowledge of the risks versus the benefits of introducing genetically modified organisms into the environment has resulted in a high level of polarization between pro- and anti-biotechnology groups. The following section briefly critiques a cross-section of the benefits and criticisms of introducing genetically modified organisms. The scientific information available, however, is rapidly changing, so a pragmatic effort should be made to seek out the most recent publications.

4.7.1. Scientific Benefits

Recombinant DNA represents a very effective tool for understanding the molecular control of plant development and senescence and allows posing questions about the functional control and organization of the genome that were either extremely difficult or impossible to answer previously. "Knock-outs," insertions that eliminate the expression of a gene or the synthesis of its product, offer an extremely powerful tool. Sometimes, the effect of a knock-out may be unanticipated, opening the door to new areas of inquiry. For instance, when a gene for a critical protein in the ethylene signaling pathway (essential for the action of ethylene) was blocked, it not only inhibited fruit ripening (anticipated) but created transgenic plants that were much more susceptible to microorganisms (unanticipated). This result showed a role for the phytohormone in the plant's response to pathogen invasions (e.g., four genes encode ethylene-response element binding factor expressed in response to invasion by bacterial blight/ bacterial leaf spot pathogen *Pseudomonas syringae*[49,50]). Knock-outs are particularly valuable for identifying and exploring genes that regulate the expression of other genes.

Recombinant DNA has been used to begin to explain several postharvest processes. For example, the first commercial agricultural use for recombinant DNA was the insertion of an antisense gene for polygalacturonase into tomato. Most tomatoes are harvested virtually green and then artificially ripened with ethylene. This gives a very firm fruit that can be shipped with little damage; however, the flavor of the fruit is decidedly inferior to that of fruit that ripens naturally. At the onset of the program, the generally held theory for the control of fruit softening during ripening was that there is an increase in the expression of a gene for the enzyme polygalacturonase. Polygalacturonase then mediates the degradation of pectin, the key cementing agent between neighboring cells, which gives the tissue much of its structural rigidity and was therefore thought to have a dominant role in the softening process. If the gene could be blocked while other ripening genes functioned normally (e.g., color and flavor development), the fruit could be left on the plant longer to ripen without it becoming too soft, increasing the overall quality of the product. However, while the transgenic fruit (antisense gene for polygalacturonase) softened a little more slowly, it did soften, undermining the accepted dogma of the time that polygalacturonase was the dominant enzyme in the control of softening. This led to a reexamination of the softening process and the production of plants with antisense genes for other possible softening associated enzymes (see chapter 6, section 1.3). It is now thought that softening is mediated by several enzymes acting synergistically. Blocking the synthesis of one enzyme may slow softening; however, the remaining enzymes can complete the task.

An antisense gene for aminocyclopropane-1-carboxylic acid (ACC) oxidase has also allowed the exploration of the role of ethylene in the overall fruit ripening process. When expressed in cantaloupe fruit,[4] it reduced ethylene synthesis and blocked ripening. Likewise, the gene for the ethylene receptor protein (*etr1-1*), isolated from *Arabidopsis* and inserted into petunia, made the flowers insensitive to ethylene, significantly extending their functional existence before senescence. Therefore, the role of ethylene in the overall flower senescence process can be studied.

Another approach is to randomly insert T-DNA; if inserted within an existing gene, it can result in the disruption of the gene. When the disrupted gene is of biological interest, it can be isolated using the T-DNA as a tag. The gene can then be identified and its role in the plant explored, contrasting the wild type with the transgenic clone.

4.7.2. Agricultural Benefits

At the onset of genetically modifying organisms, considerable publicity touted the potential impact of the technology for solving problems confronting humans. In crop plants, higher yields, expanding production to more marginal lands, removal of environmental contaminants, healthier and more nutritious foods,[21] reduced pesticide use, and myriad other proposed benefits elevated the expectations of the public. While these expectations remain valid, their rate of delivery has been far slower than originally anticipated. Evidently, the genetic control of plants and our ability to manipulate it is much more complex than initially thought. For example, transgenic alterations in the genome are often transitory. The transgene may be expressed for a generation or two and then for some yet-to-be-explained reason silenced, reiterating the fact that our existing understanding remains superficial. However, even with the current impediments to the routine genetic alteration of plants, some significant successes have been realized.

A diverse and rapidly expanding range of crops have been altered (Table 5.2) and new phenotypes derived *via* genetic engineering. The most widely altered traits are herbicide, insect and virus resistance and the control of ripening/senescence [collectively these four categories represent nearly 90% of the cultivars deregulated (or applied for deregulation) in the United States as of October, 2001].[28] Over 399 phenotypes have been field-tested in the United States as of 2002 (Appendix I).

Genes for resistance to several herbicides have been introduced into crop plants.[15] Using genes for 2,4-D degradation and resistance to glyphosate and glufsinate has allowed the engineering crops in which weeds can be readily controlled. As a herbicide, glyphosate has a short half-life once applied and a low toxicity to animals and it controls a diverse cross-section of weed species, underscoring possible ecological advantages of transgenic crops with tolerance to the herbicide.

Insects cause extensive quantitative and qualitative losses in crop plants each year requiring the application of astronomical amounts of pesticides (over 10 billion dollars are spent annually in the United States alone for insecticides). In developing countries, the application of insecticides is rarely economical and often represents a serious safety hazard to farm workers. Consequently, the genetic engineering of insect resistance has very significant advantages. The two most prevalent approaches used to date have been the incorporation of genes for proteinase inhibitors and the genes for the production of bacterial proteins from *Bacillus thuringiensis* (Bt) that are toxic to a cross-section of insects. Proteinase inhibitors are found in many plant species, including several crop plants (e.g., cowpea, sweetpotato); they appear to have evolved naturally as part of the plant's insect defense system. They work by inhibiting the degradation of proteins within the insect's digestive system, preventing the acquisition of essential amino acids and leading to larvae mortality. Proteinase inhibitors are readily degraded by cooking and in the uncooked state seldom represent a significant health factor in the diet of humans due to their low level of toxicity. The Bt toxin is a protein that when ingested by Lepidoptera and Coleoptera insects results in the inhibition of the activity of epithelial cells in the digestive tract. The toxin is not toxic to mammals and is rapidly degraded by them.

Viruses result in significant losses during production and diminish the quality of many crops. The insertion into plants of genes for viral coat proteins can block virus replication. The inhibition apparently occurs through the interference of the plant's expressed coat protein in the cytoplasm with the uncoating and replication of nucleic acids from the viral particles. When crops are susceptible to more than one virus, multiple coat protein genes are required (e.g., cantaloupe is susceptible to cucumber, zucchini yellow and watermelon mosaic viruses).[19]

To date, ripening and senescence related transgenic traits of commercial interest have largely focused on altering the synthesis or action of ethylene. Antisense genes for ACC oxidase and ACC synthase have been inserted in carnation, tomato and cantaloupe; ACC deaminase, which shunts ACC away from ACC oxidase and ethylene formation, has been up-regulated in tomato. Likewise, an antisense gene for ethylene binding (*etr1-1*) has been placed in petunia flowers, significantly slowing senescense.[73] Transgenic crops, with up or down regulation of various genes, give phenotypes with increased flower pigmentation and unique colors, impeded fruit softening and decreased bruising susceptibility. As additional public laboratories develop transgenic crops, it is anticipated that many unique and desirable phenotypes, though possibly less profitable than those developed by commercial interests, will become available.

4.7.3. Controversy Over Recombinant DNA

Acceptance of genetically modified crops by the public has varied from substantial resistance (Western Europe) to general acceptance (United States). Consumer understanding of the technology and their confidence in government regulatory systems have a distinct impact on the position adopted, which in turn often impedes focusing upon the critical questions: 1) food safety and public health; 2) environmental safety; and 3) impact upon agricultural practices and society.[38,47] Several specific concerns voiced by groups opposing the genetic engineering of plants follows.

Food safety and human health concerns. Allergic reactions represent a very serious, even potentially fatal, risk to individuals who are sensitive to minute quantities of an allergen.[43] The possibility of inadvertently transferring a potentially fatal allergen was illustrated by the incorporation of a storage protein from Brazil nut into soybean to enhance its protein's sulfur containing amino acid complement.[26] Individuals sensitive to the Brazil nut allergen would have been sensitive to the transgenic soybean.[46] Consequently, the soybean line was never released.

For similar reasons, Star Link, a transgenic corn cultivar containing a Bt gene (*Cry9C*) for insect resistance, was cleared for use as animal feed but not for human consumption. It contains the insecticidal Bt protein that shares some properties with a known allergen. When StarLink corn was inadvertently mixed with corn for human consumption, a number of products containing the contaminated material had to be recalled. It has been estimated that as much as 5% of the U.S. corn crop in 2000 may have been contaminated.[35]

Knowledge of the potential liability and health risks of allergens in transgenic crops has made the biotechnology industry and regulatory agencies very careful in screening new clones. It should be noted that the incorporation of toxins, antinutrients and allergens also occurs with conventional plant breeding.[69] For example, conventionally bred insect-resistant celery accumulated psoralens that caused severe skin burns to workers handling the cultivar.[2]

Ecological risks. Due to the inherent complexity of ecosystems, it is extremely difficult to predict the occurrence and extent of the long-term environmental effects of introducing novel organisms.[74] New crops, whether derived from traditional breeding programs or genetic engineering, may be altered in a manner that enhances their potential to become an invasive species. This is because the altered organism is novel to the existing network of ecological relationships. Thus a genetically modified crop might be invasive, becoming a weed in other crops or in uncultivated habitats. An initial assessment of transgenic oilseed rape, potato, corn, and sugar beet grown in 12 habitats more than 10 years, however, has not found genetically modified plants to be more invasive or persistent that their non-genetically modified counterparts.[14]

Genetic pollution. A potentially significant concern is that genes from genetically modified plants will escape into the ecosystem *via* interspecific hybridization with wild relatives, causing weedy hybrids that are much more difficult to control or have a negative impact on the ecosystem. For example, oil seed rape can hybridize with at least 14 different weed species.[57] An alternative means of contamination is through the pollination of non-transgenic cultivars with pollen from genetically modified plants. Growers of organic or non-transgenic crops near transgenic crops of the same species face inadvertent contamination of their material. The European Commission has set a threshold of 1% for accidental contamination of foods with genetically modified organisms, above which the product must be labeled as "containing genetically modified material" (labeling is not currently required in the United States).[26] The transfer of genes to bacteria, fungi or other microorganisms is also a concern.[47] The development of technology to prevent the transfer of genes to non-target organisms, therefore, would be advantageous.

Impact on agricultural practices and society. a) Decisions on what transgenic products to develop are generally based upon short-range economic considerations made by commercial companies. Frequently, the criticism is directly aimed at large multinational seed companies which are seen as controlling the fate of small growers, preventing growers from saving seed for next years crop. b) Certain transgenic crops will impede the future development of newer, safer pesticides (e.g., herbicide resistant cultivars where the transgenic crop is inseparably linked to the chemical, minimizing the incentive to develop a new herbicide). c) The long-term consequences of the introduction of new genes into the environment are not known. d) Widespread adoption and growth in monoculture of a small number of cultivars decreases crop diversity, increasing the chances for unprecedented losses.

Perhaps potentially the most serious problem associated with the genetic modification of organisms is that the technology is neither particularly difficult nor expensive and can be accomplished by one or a few individuals. To date, transgenic organisms have been designed to achieve some "beneficial effect" and have been assessed and approved by appropriate government monitoring agencies. It is possible, however, that harmful transgenic alterations could be intentionally developed and hidden from regulatory oversight, with catastrophic consequences. The frequency of introduction of new, destructive computer viruses suggests that individuals are willing to do so.

While there are two very divergent sides to the genetically modified organism question, transgenic crops are rapidly becoming a fixture in agriculture. It is essential that the development and introduction of transgenic plants be carefully scrutinized and controlled by appropriate government regulatory agencies. Recombinant DNA technology opens the door to understanding genome structure and control in plants and a wealth of solutions to agricultural problems.

ADDITIONAL READING

Bevan, M., K. Mayer, O. White, J.A. Eisen, D. Preuss, T. Bureau, S.L. Salzberg and H.W. Mewes. 2001. Sequence and analysis of *Arabidopsis* genome. *Curr. Opin. Plant Biol.* 4:105–110.

Browning, K. 1996. The plant translational apparatus. *Plant Mol. Biol.* 32:107–144.

Chopra, V.L., V.S. Malik and S.R. Bhat (eds.). 1999. *Applied Plant Biotechnology.* Science Pub., Enfield, NJ.

Gillham, N.W. 1994. *Organelle Genes and Genomes.* Oxford Univ. Press, Oxford, England.

Gilroy, S., and A.J. Trewavas. 2001. Signal processing and transduction in plant cells: the end of the beginning? *Nat. Rev. Mol. Cell Biol.* 2:307–314.

Johri, M.M., and D. Mitra. 2001. Action of plant hormones. *Cur. Sci.* 80:199–205.

Kanellis, A.K., C. Chang, H. Klee, A.B. Bleecker, J.C. Pech, and D. Grierson (eds.). 1999. *Biology and Biotechnology of the Plant Hormone Ethylene II.* Kluwer Acad. Pub., London.

Khachatourians, G.G., A. McHughen, R. Scorza, W.-K. Nip and Y.H. Hui (eds.). 2002. *Transgenic Plants and Crops.* M. Dekker, New York.

Lindsey, K. (ed.). 1998. *Transgenic Plant Research.* Harwood Acad. Pub., Amsterdam.

McCourt, P. 2001. Plant hormone signaling: getting the message out. *Mol. Cell* 8:1157–1158.

Meyer, P. (ed.). 1995. *Gene Silencing in Higher Plants and Related Phenomena in other Eukaryotes.* Springer-Verlag, New York.

Miglani, G.S. 1998. *Dictionary of Plant Genetics and Molecular Biology.* Food Prod. Press, New York.

Nottingham, S. 2002. *Genescapes. The Ecology of Genetic Engineering.* Zed Books, London.

Reymand, P., and E.E. Farmer. 1998. Jasmonate and salicylate as global signals for defense gene expression. *Curr. Opin. Plant Biol.* 1:404–411.

Scheel, D., and C. Wasternack (eds.). 2002. *Plant Signal Transduction.* Oxford Univ. Press, Oxford, England.

Sobral, B.W.S. (ed.). 1966. *The Impact of Plant Molecular Genetics.* Birkhauser, Cambridge, MA.

Stepanova A.N., and J.R. Ecker. 2000. Ethylene signaling: from mutants to molecules. *Cur. Opin. Plant Biol.* 3: 353–360.

Trewavas, A.J., and R. Malho. 1997. Signal perception and transduction: the origin of the phenotype. *Plant Cell* 9:1181–1195.

Watanabe, K.N., and E. Pehu. 1997. *Plant Biotechnology and Plant Genetic Resources for Sustainability and Productivity.* Academic Press, San Diego, CA.

Zielinski, R.E. 1998. Calmodulin and calmodulin-binding proteins in plants. *Annu. Rev. Plant Physiol. Plant Mol. Biol.* 49:697–727.

REFERENCES

1. Aharoni, A., and O. Vorst. 2002. DNA microarrays for functional plant genomics. *Plant Mol. Biol.* 48:99–118.

2. Ames, B.N., and L.S. Good. 1999. Pp.18–38. In: *Fearing Food: Risk, Health and Environment.* J. Morris and R. Bates (eds.). Butterworth Heinemann, Oxford, England.

3. Asai, T., G. Tena, J. Plotnikova, M.R. Willman, W.L. Chiu, L. Gomez-Gomez, T. Boller, F.M. Ausubel and J. Sheen. 2002. MAP kinase signaling cascade in *Arabidopsis* innate immunity. *Nature* 415:977–983.

4. Ayub, R., M. Guis, M. Benamor, L.Gillot, J.P. Roustan, A. Latche, M. Bouzayen and J.C. Pech. 1996. Expression of ACC oxidase antisense gene inhibits ripening of cantaloupe melon fruits. *Nature Biotech.* 14:862–866.

5. Bleecker, A.B., A.E. Hall, F.I. Rodriguez, J.J. Esch and B. Binder. 1999. The ethylene signal transduction pathway. Pp. 51–57. In: *Biology and Biotechnology of the Plant Hormone Ethylene II.* A.K. Kanellis, C. Chang, H. Klee, A.B. Bleecker, J.C. Pech and D. Grierson (eds.). Kluwer Acad. Pub., London.

6. Bleecker, A.B., and H. Kende. 2000. Ethylene: a gaseous signal molecule in plants. *Annu. Rev. Cell. Dev. Biol.* 16:1–18.

7. Brown, J.W.S., and C.G. Simpson. 1998. Splice site selection in plant pre-mRNA splicing. *Annu. Rev. Plant Physiol. Plant Mol. Biol.* 49:77–95.

8. Browning, K.S. 1996. The plant translational apparatus. *Plant Mol. Biol.* 32:107–144.

9. Chang, C., and R. Stradler. 2001. Ethylene hormones receptor action in *Arabidopsis. BioEssays* 23:619–627.

10. Chang, C., and R.C. Stewart.1998. The two-component system. Regulation of diverse signaling pathways in prokaryotes and eukaryotes. *Plant Physiol.* 117:723–731.

11. Chang, C., S.F. Kwok, A.B. Bleecker and E.M. Meyerowitz. 1993. *Arabidopsis* ethylene-response gene ETR1: similarity of product to two-component regulators. *Science* 262:539–545.

12. Chen, Y.F., M.D. Randlett, J.L. Findell and G.E. Schaller. 2002. Localization of the ethylene receptor ETRI to the endoplasmic reticulum of *Arabidopsis. J. Biol.Chem.* 277:19861–19866.

13. Clouse, S.D. 2001. Integration of light and brassinosteroids signals in etiolated seedling growth. *Trends Plant Sci.* 6:443–445.

14. Crawley, M.J., S.L. Brown, R.S. Hails, D.D. Kohn and M. Rees. 2001. Transgenic crops in natural habitats. *Nature* 409:682–683.

15. Dale, P.J. 1995. R and D regulation and field trialing of transgenic crops. *Trends Biotech.* 13:398–403.

16. Dangl, J.L., and J.D.G. Jones. 2001. Plant pathogens and integrated defense responses to infection. *Nature* 411:826–833.

17. Dean, C., and R. Schmidt. 1995. Plant genomes: A current description. *Annu. Rev. Plant Physiol. Plant Mol. Biol.* 46:395–418.

18. European Union Chromosome 3 Arabidopsis Sequencing Consortium, The Institute for Genome Research, and Kazuma DNA Research Institute. 2000. Sequence and analysis of chromosome 3 of the plant *Arabidopsis thaliana. Nature* 408:820–822.

19. Fuchs, M., J.R. McFerson, D.M. Tricoli, R. McMaster, R.Z. Deng, M.L. Boeshore, J.R. Reynolds, P.F. Russell, D.D. Quemada and D. Gonsalves. 1997. Cantaloupe line CZW-30 containing coat protein genes for cucumber mosaic virus, zucchini yellow mosaic virus and watermelon mosaic virus-2 is resistant to these viruses in the field. *Mol. Breed.* 3:279–290.

20. Futterer, J., and T. Hohn. 1996. Translation in plants-rules and exceptions. *Plant Mol. Biol.* 32:159–189.

21. Galili, G., S. Galili, E. Lewinsohn and Y. Tadmor. 2002. Genetic, molecular, and genomic approaches to improve the value of plant foods and feeds. *Crit. Rev. Plant Sci.* 21:167–204.

22. Gelvin, S.B. 2003. *Agrobacterium*-mediated plant transformation: the biology behind the "gene-jockeying" tool. *Microbiol. Mol. Biol. Rev.* 67:16–37.

23. Gillham, N.W. 1994. *Organelle Genes and Genomes.* Oxford Univ. Press, Oxford, England.

24. Gilroy, S., and A. Trewavas. 2001. Signal processing and transduction in plant cells: the end of the beginning? *Nature. Rev. Mol. Cell Biol.* 2:307–314.

25. Goeschl, J.D., and S.J. Kays. 1975. Concentration dependencies of some effects of ethylene on etiolated pea, peanut, bean and cotton seedlings. *Plant Physiol.* 55:670–677.

26. Goldman, K.A. 2000. Bioengineered food—safety and labeling. *Science* 290:457, 459.

27. Gray, M.W. 1999. Evolution of organelle genomes. *Cur. Opin. Genetic Develop.* 9:678–687.

28. Grumet, R. 2002. Plant biotechnology in the field—a snapshot with emphasis on horticultural crops. *HortScience* 37:435–436

29. Haywood, V., F. Kragler and J.J. Lucas. 2002. Plasmodesmata: pathways for protein and ribonucleoprotein signaling. *Plant Cell* 14:supplement S303–S325.

30. Holmes-Davis, R., and L. Comai. 1998. Nuclear matrix attachment regions and plant gene expression. *Trends Plant Sci.* 3:91–97.

31. Hua, J., and E.M. Meyerowitz. 1998. Ethylene responses are negatively regulated by a receptor gene family in *Arabidopsis thaliana. Cell* 94:261–271.

32. Johri, M.M., and D. Mitra. 2001. Acton of plant hormones. *Cur. Sci.* 80:199–205.

33. Jarvis, P. 2001. Intracellular signaling: the chloroplast talks! *Cur. Biol.* 11:R307–R310.

34. Kazusa DNA Research Institute, Cold Spring Harbor and Washington University in St. Louis Sequencing Consortium, and The European Union Arabidopsis Genome Sequencing Consortium. 2000. Sequence and analysis of chromosome 5 of the plant *Arabidopsis thaliana. Nature* 408:823–826.

35. Kleiner, K. 2000. Unfit for humans corn has been contaminated with a potentially harmful protein. *New Scientist,* Dec. 2, p. 11.

36. Knight, H., and M.R. Knight. 2001. Abiotic stress signaling pathways: specificity and cross-talk. *Trends Plant Sci.* 6:262–267.

37. Knudsen, S. 2002. *A Biologist's Guide to Analysis of DNA Microarray Data.* Wiley-Interscience, New York.

38. Lappé, M., and B. Bailey. 1999. *Against the Grain: The Genetic Transformation of Global Agriculture.* Earthscan, London.
39. Lin, X.Y., S. Kaul, S. Rounsley, T.P. Shea, M.I. Benito, C.D. Town, C.Y. Fujii, T. Mason, C.L. Bowman and M. Barnstead. 1999. Sequence and analysis of chromosome 2 of the plant *Arabidopsis thaliana. Nature* 402:761–768.
40. Lindsey, K., S. Cassan and P. Chilley. 2002. Peptides: new signaling molecules in plants. *Trends Plant Sci.* 7:78–83.
41. Mayer, K., C. Schuller, R. Wambutt, G. Murphy, G. Volckaert, T. Pohl, A. Dusterhoft, W. Stiekema, K.D. Entian, N. Terryn, B. Harris, W. Ansorge, P. Brandt, L. Grivell, M. Rieger, M. Weichselgartner, V. de Simone, B. Obermaier, R. Mache, M. Muller, M. Kreis, M. Delseny, P. Puigdomenech, M. Watson, W.R. McCombie, et al. 1999. Sequence and analysis of chromosome 4 of the plant *Arabidopsis thaliana. Nature* 402:769–777.
42. Memelink, J., R. Verpoorte and J.W. Kijne. 2001. Organization of jasmonate-responsive gene expression in alkaloid metabolism. *Trends Plant Sci.* 6:212–219.
43. Metcalfe, D.D., J.D. Astwood, R. Townsend, H.A. Sampson, S.L. Taylor and R.L. Fuchs. 1996. Assessment of the allergenic potential of foods from genetically engineered crop plants. *Crit. Rev. Food Sci. Nutr.* 36:S165–186.
44. McCourt, P. 2001. Plant hormone signaling: Getting the message out. *Mol. Cell* 8:1157–1158.
45. Munnik, T. 2001. Phosphatidic acid: an emerging plant lipid second messenger. *Trends Plant Sci.* 6:227–233.
46. Nordlee, J.A., S.L. Taylor, J.A. Townsend, L.A. Thomas and R.K. Bush. 1996. Identification of a Brazil-nut allergen in transgenic soybeans. *New Eng. J. Med.* 334(11):688–692.
47. Nottingham, S. 2002. *Genescapes. The Ecology of Genetic Engineering.* Zed Books, London.
48. O'Donnell, P.J., C. Calvert, R. Atzorn, C. Wasternack, H.M.O. Leyser and D.J. Bowles. 1996. Ethylene as a signal mediating the wound response in tomato plants. *Science* 274:1914–1917.
49. Ohme Takagi, M., K. Suzuki and H. Shinshi. 2000. Regulation of ethylene-induced transcription of defense genes. *Plant Cell Physiol.* 41:1187–1192.
50. Onate-Sanchez, L., and K.B. Singh. 2002. Identification of *Arabidopsis* ethylene-responsive element binding factor with distinct induction kinetics after pathogen infection. *Plant Physiol.* 128:1313–1322.
51. Pikaard, C.S. 1998. Chromosome topology-organizing genes by loops and bounds. *Plant Cell* 10:1229–1232.
52. Preston, C.A., G. Laue and I.T. Baldwin. 2001. Methyl jasmonate is blowing in the wind, but can it act as a plant-plant airborne signal? *Biochem. Syst. Ecol.* 29:1007–1023.
53. Reed, J.W. 2001. Roles and activities of Aux/IAH proteins in *Arabidopsis. Trends Plant Sci.* 6:420–425.
54. Reymand, P., and E.E. Farmer. 1998. Jasmonate and salicylate as global signals for defense gene expression. *Cur. Opin. Plant Biol.* 1:404–411.
55. Rodríguez, F.I., J.J. Esch, A.E. Hall, B.M. Binder, G.E. Schaller and A.B. Bleecker. 1999. A copper cofactor for the ETR1 receptor from *Arabidopsis. Science* 283:996–998.
56. Sanders, D., C. Brownlee and J.F. Harper. 1999. Communicating with calcium. *Plant Cell* 11:691–706.
57. Scheffler, J.A., and P.J. Dale. 1994. Opportunities for gene transfer from transgenic oilseed rape (*Brassica napus*) to related species. *Transgenic Res.* 3:263–278.
58. Songstadd, D.D., D.A. Somers and R.J. Griesbach. 1995. Advances in alternative DNA delivery techniques. *Plant Cell Tissue Organ Cult.* 40:1–15.
59. Stepanova, A.N., and J.R. Ecker. 2000. Ethylene signaling: from mutants to molecules. *Cur. Opin. Plant Biol.* 3:353–360.
60. Stone, J.M., and J.C. Walker. 1995. Plant protein kinase families and signal transduction. *Plant Physiol.* 108:451–457.
61. Struhl, K. 1998. Histone acetylation and transcriptional regulatory mechanisms. *Genes and Development* 12:599–606.
62. Sundaresan, V., P. Springer, T. Volpe, S. Howard, J.D.G. Jones, C. Dean, H. Ma and R. Martienssen. 1995. Patterns of gene action in plant development revealed by enhancer trap and gene trap transposal elements. *Genes Dev.* 9:1797–1810.
63. Swarup, J., G. Parry, N. Graham, T. Allen and M. Bennett. Auxin cross-talk: integration of signaling pathways to control plant development. *Plant Mol. Biol.* 49:411–426.

64. Tepperman, J.M., T. Zhu, H.-S. Chang, X. Wang and P.H. Quail. 2001. Multiple transcription-factor genes are early targets of phytochrome A signaling. *Proc. Nat. Acad. Sci.* USA 98:9437–9442.

65. The Arabidopsis Genome Initiative. 2000. Analysis of the genome sequence of the flowering plant *Arabidopsis thaliana. Nature* 408:796–813.

66. Theologis, A., J.R. Ecker, C.J. Palm, N.A. Federspiel, S. Kaul, O. White, J. Alonso, H. Altafi, R. Araujo and C.L. Bowman. 2000. Sequence and analysis of chromosome 1 of the plant *Arabidopsis thaliana. Nature* 408:816–820.

67. Trewavas, A.J., and R. Malho. 1997. Signal perception and transduction—the origin of the phenotype. *Plant Cell* 9:1181–1195.

68. Uzogana, S.G. 2000. The impact of genetic modification of human foods in the 21st century: a review. *Biotech. Adv.* 18:179–206.

69. Van Gelder, W.M.J., J.H. Vinke and J.J.C. Scheffer. 1988. Steroidal glycoalkaloids in tubers and leaves of solanum species used in potato breeding. *Euphytica* 38:147–158.

70. Vierstra, R. 1996. Proteolysis in plants. *Plant Mol. Biol.* 32:275–302.

71. Weber, H. 2002. Fatty acid-derived signals in plants. *Trends Plant Sci.* 7:217–224.

72. Wendehenne, D., A. Pugin, D.F. Klessig and J. Durner. 2001. Nitric oxide: comparative synthesis and signaling in animals and plant cells. *Trends Plant Sci.* 6:177–183.

73. Wilkinson, J.Q., M.B. Lanahan, D.G. Clark, A.B. Bleecker, C. Chang, E.M. Meyerowitz and H.J. Knee. 1997. A dominant mutant receptor from *Arabidopsis* confers ethylene insensitivity in heterologous plants. *Nature Biotech.* 15:444–447.

74. Wolfenbarger, L.L., and P.R. Phifer. 2000. The ecological risks and benefits of genetically engineered plants. *Science* 290:2088–2093.

75. Zielinski, R.E. 1998. Calmodulin and calmodulin-binding proteins in plants. *Annu. Rev. Plant Physiol. Plant Mol. Biol.* 49:697–725.

6

DEVELOPMENT OF PLANTS AND PLANT ORGANS

Plants and plant organs progress through a dynamic series of genetically controlled developmental processes terminating in their eventual senescence and death. Their development is the combination of both growth (an irreversible increase in size or volume which is accompanied by the biosynthesis of new protoplasmic constituents) and differentiation (qualitative changes in the cells) and can be viewed at either the whole plant or individual organ level.

Unlike animals, during development, plants display a remarkable degree of variability in form that is strongly influenced by the environment in which they are grown. An animal, for example a primate, will develop only four limbs regardless of variation in environment during the development period. Thus in animals, morphological development is very tightly controlled. This is in sharp contrast with the tremendous diversity found in the plant kingdom. Responding to light, temperature, soil nutrient's and other factors, two genetically identical plants may develop into structurally distinct mature plants. Environment, therefore, has a pronounced influence on the development of plants and plant parts, and this influence carries over into the postharvest period. Variations in product composition and structure can significantly alter the way a product responds after harvest and, as a consequence, how it must be handled. If we are to understand the physical and chemical changes occurring during the postharvest period, it is essential that we first understand how the postharvest period fits into the entire developmental cycle of the plant.

The developmental period encompasses the entire time frame from the first initiation of growth to the eventual death of the plant or plant organ. In the past, senescence was not generally accepted as part of development. We now know that many of the initial processes of senescence are precisely regulated. Thus the onset and initial developments in senescence do not represent a complete collapse of organization with ensuing disorder; rather this period represents a distinct portion of the overall developmental cycle of the plant or plant organ.

In many crop plants this natural developmental cycle is interrupted prior to its completion by harvest. For example, lettuce plants are harvested very early in their developmental cycle, bean sprouts almost at the beginning. Grains and dried pulse crops, on the other hand, reach the end of their developmental cycle before being harvested.

In its life cycle, a plant passes through a number of distinct developmental stages. These stages are closely synchronized with the development of the plant (internal control) and often with the environment in which the plant or plant organ is held (external control). Many species have built-in control mechanisms that restrict a particular developmental step (e.g., germination, flowering). When the requirements for the control mechanism are satisfactorily met by an internal or external signal, development then proceeds.

1. SPECIFIC DEVELOPMENTAL STAGES

While there are numerous stages and substages in the developmental cycle of a plant, in this section we focus on processes that can be and often are significantly modulated during the postharvest period. Developmental steps or stages such as dormancy, flowering, fruit ripening, abscission and senescence are of particular importance for many postharvest products.

1.1. Dormancy

Dormancy is a period of suspended growth common to many plants and plant parts (e.g., buds, tubers, seeds). In nature, dormancy serves to synchronize subsequent growth and development with desirable environmental conditions. Many plants or plant parts are dormant during extended cold or dry periods which are unfavorable for growth. In some seed-bearing species, the duration of dormancy within the seeds from a single parent plant may vary widely. This tends to enhance the long-term survival potential for the species. Other dormancy mechanisms appear to aid in the dispersal of the reproductive unit.

Dormancy can be separated into three general classes based upon the site or cause of the inhibition of growth.[250] **Ecodormancy** occurs when one or more factors in the basic growth environment are unsuitable for overall growth. These are external requisites such as moisture, light or temperature. Ecodormancy is more or less synonymous to the older dormancy term quiescence. **Endodormancy** occurs when an internal mechanism prevents growth even though the external conditions may be ideal. The initial reaction leading to growth control is perception of an environmental or endogenous signal by the affected structure. **Paradormancy** is similar to endodormancy; however, the signal originates in or is initially perceived by a structure other than the one in which the dormancy is manifested (e.g., perceived by the seed coat while manifested in the embryo).

An often confusing plethora of dormancy terms is found in the literature and within dormancy classes, a diverse array of subclasses has been proposed. For example, within endodormancy the number and make-up of subclasses varies depending upon the plant part (seed, tubers, buds), species and other factors. For seeds alone, seven classes of dormancy have been proposed, which, in turn, contain numerous subdivisions.[29] Seed dormancy can also be separated into primary and secondary (induced) dormancy, the former occurring during the natural developmental maturation of the seed, the latter after the seed physically separates from the parent plant.[180]

The internal mechanisms controlling dormancy are not fully understood.[287] In dormant seeds and buds, roles for the phytohormones abscisic acid and gibberellic acid have been proposed.[238] Abscisic acid is thought to impose dormancy while the seed is attached to the plant, preventing premature sprouting (**vivipary**), with the concentration declining in tandem with an increase in germination potential. Abscisic acid-deficient mutants of *Arabidopsis,* which are nondormant, appear to support this hypothesis, as do treatments that alter endogenous abscisic acid levels. Based upon the number of variations in dormancy across species, it is evident the control mechanism(s), when finally elucidated, will no doubt be much more complex than simply a balance between abscisic acid and gibberellic acid levels.

The postharvest period can dramatically affect dormancy and may take the form of prolonging, breaking or preventing the establishment of dormancy. For many species of domesticated plants or plant parts, the storage environment is used to satisfy specific dormancy requirements. With the yam[457] and the white potato,[77] the dormancy period determines the length of time the tubers can be successfully stored. Thus, care must be taken to prolong the dormant period. In contrast, storage temperatures can be used to break seed dormancy in a diverse cross-section of species and in potato (seed) tubers to be used for planting material.[196] For certain other species, it is essential that the plant or plant propagule not be stored under conditions that will induce dormancy. Spruce seeds that have been stored under conditions to alleviate dormancy undergo a reimposition of dormancy if the seeds are redried to 4–10% moisture.[206] Precise postharvest requirements, therefore, vary widely, depending upon species and, in some cases, even the cultivar in question.

The specific conditions for fulfillment of the requirements of the dormancy mechanism, thus allowing growth to proceed, vary widely. In the buds of many temperate perennials, sufficient exposure to cold temperatures above freezing and below a specific maximum is required.[98] Many seeds also respond to low temperatures. Holding seeds in cold, moist conditions (**stratification**) has been a widely used agriculture practice for centuries. Other species fulfill their dormancy requirements with environmental signals such as temperature fluctuations (*Lycopus europaeus* L.), light quality (certain lettuce cultivars), photoperiod (*Vertonica persica* Poir.) or moisture (many desert species). The dormancy mechanism may respond to the summation of an external signal (e.g., hours of chilling) or it may require several types of signals simultaneously (e.g., photoperiod and temperature) to insure against aberrant environmental conditions inadvertently triggering the continuation of growth.

1.2. Flowering

The ability to reproduce is a unifying and essential characteristic of all organisms. Although there are a number of reproductive strategies found in the plant kingdom, sexual reproduction by way of flower and seed production is one of the most common. Many flowering species have evolved beautiful and elaborate floral appendages that appear to facilitate successful reproduction. Because of their beauty, many of these, in turn, have been domesticated by man, adding immeasurably to the aesthetic quality of life.

Flowering represents a distinct stage in the overall developmental cycle of most plant species. Because of this, the factors controlling flowering, from initiation to anthesis and eventually senescence, are of considerable interest to plant scientists. For postharvest biologists, species in which the flowering process is in some way modulated during storage are of particular interest (e.g., cut flowers, flowering bulbs, biennial transplants). In this section, the relationship between flowering and storage is examined.

Cut flowers held in refrigerated storage are the products most commonly associated with the postharvest handling of floral crops. Considerable research has been devoted to expanding the longevity of cut flowers both during and after storage. While cut flowers represent a substantial part of the total volume of flowers and flowering products handled, the range of products is actually much broader. Many flowers are sold attached to the parent plant (e.g., potted chrysanthemums). The handling techniques used for intact plants and plant propagules often vary considerably from those used for detached flowers of the same species. In some cases, flowering may be induced during storage, with subsequent floral development occurring after storage (e.g., flowering bulbs such as tulips or Easter lilies). For some biennial species, it may be essential to store the plants or plant propagules under conditions that will not induce flowering.

The apical meristem of the shoot is the source of all of the aerial parts of the plant. As the apical cells divide, they displace previously formed cells to the periphery where they differentiate into leaf or flower primordia (Laufs et al., 1998).[255] The fate of these cells is controlled by a group of genes* in *Arabidopsis* that appear to be involved in the communication between

---

CLAVATA (*CLV1, 2,* and *3*).

neighboring cells.[141] When an apical meristem becomes a floral meristem, new genes are expressed that control this developmental process. What controls the decision to switch from a vegetative phase (formation of leaves) to a reproductive phase (formation of flowers)? Control of genes regulating the fate of the apical meristem is modulated by a series of developmental signals. Included are internal (endogenous) controls such as circadian rhythms that interface with external (environment-sensing) controls such as photoperiod and temperature. When the appropriate developmental signals are present, a floral stimulus is translocated from the leaves to the apical meristem of the shoot initiating flower development. Flowering, therefore, can be separated into two distinct physiological processes: induction and subsequent development. For some plants, flower induction can occur during the postharvest period and may be beneficial or detrimental, depending upon the species and its intended use.

1.2.1. Flower Induction

a. Internal control

One essential internal requirement for floral induction is that the shoot apical meristem must have made the transition from a juvenile stage to an adult vegetative phase. At this point, the plant has the potential to form reproductive structures if the appropriate environmental signals are present. The transition, or phase change, between the juvenile phase and adult vegetative phase tends to be gradual, varying in the length of time required to be attained and the extent of development of the plant. A few species (e.g., *Pharbitis nil* Choisy) can make the transition from juvenile to adult vegetative and then flower, under optimum inductive conditions, as soon as their cotyledons are above the ground. Other species require longer periods, extending in length from days to years [e.g., 25–30 years for English oak].[91] Several species of bamboo require 50–100 years before flowering and then do so only once. Genetic and chemical controls over phase changes are not well understood.

Circadian rhythms are a ubiquitous internal mechanism that is important not only within plants but animals and microorganisms as well. Circadian rhythms act as a biological clock, allowing the organism to determine the time of day (photoperiod, in contrast, makes it possible for the plant to determine the time of year). The clock is an endogenous, self-sustaining oscillator that maintains rhythms of about 24 hours and controls flowering in many species and a wide variety of other biological processes [e.g., leaf movements, cyclic flower opening and closing, emission of volatile floral aroma compounds].[174] Circadian clocks are synchronized (entrained) by environmental clues such as light and temperature cycles. Suspension of the rhythm, for example by continuous light or its alteration during the postharvest period, can have potentially significant consequences.

b. External control

Flower induction is known to be controlled or modulated by three primary external mechanisms: vernalization, photoperiodism, and thermoperiodism. Some species are influenced by only one of these mechanisms, whereas the flowering of others is controlled by more than one.

Vernalization is the promotion of flowering by low temperature preconditioning of plants.[351] The dominant gene(s) for flowering time in *Arabidopsis* normally confers late flowering, which is reversed by vernalization (3–8 weeks of cold). Early flowering ecotypes are due to 1 or 2 mutations that cause a loss of function.[201] Individual species may be receptive to vernalization at a very specific growth stage or throughout development. For example, seeds of many winter annuals can be induced to flower by increasing their moisture content to approximately 40% while maintaining them at a low temperature. The actual flowering process does not proceed, however, until germination and subsequent development of the plant. Most biennials that grow as a rosette the first year must reach a minimum size before becoming receptive to low temperatures. Intact plants such as *Matthiola* are induced to flower by a 2- to 3-week cold period. Bulbs of Dutch iris,[208] tulip,[353] hyacinths, muscari,[367] and Easter lilies also require a cold period for flower induction.[106]

Perception of the vernalization stimulus occurs within the meristematic zone of the shoot apex. Temperature requirements are generally from 1 to 7°C but range from just below freezing to approximately 10°C.[249] The minimum duration of cold exposure and the effective temperature varies widely among species. With many bulbs, part of the postharvest storage period may be used for flower induction. Proper timing of vernalization allows synchronizing flowering with specific sales periods. For many biennial species, it is essential to store the plants or plant propagules under conditions that will not result in flower induction. Exposure to low temperatures at the transplant stage can be sufficient to cause premature flowering after field planting and loss of a significant portion of the crop.[126,297]

Photoperiodism, the length of the daily light and dark periods, provides a means by which many species synchronize the timing of their reproductive phase during the growing season.[268] In addition to flowering, photoperiod also influences asexual reproduction, the formation of storage organs and the onset of dormancy in certain species. Though called *photo*period, the length of the dark period is actually more important than that of the light period (especially so in short-day plants).

Flowering plants can be divided into three general groups based on photoperiodic response: short-day (flowering is promoted when the day length is less than a certain critical value), long-day (flowering is promoted when the day length is longer than a critical value), and day-neutral (flowering is not regulated by photoperiod). Within both short- and long-day plants are **qualitative** (flowering occurs only under the appropriate photoperiod) and **quantitative** (flowering is accelerated by the appropriate photoperiod) photoperiodic responses. For a few very sensitive species, extremely short exposures to the proper day-night time sequence are sufficient to induce flowering. Storage conditions of some species, then, may be modified to enhance or, conversely, prevent flower induction depending on needs.

Perception of the photoperiod occurs within leaves that are generally located some distance from the shoot apex. In some instances, exposure of a single leaf, as in the short-day plant *Xanthium,* is sufficient to trigger flowering. Photoperiod assessment by the plant involves a circadian timekeeping mechanism that synchronizes the perception of the environmental clue. The photoperiodic response appears to be controlled through the function of the clock.[407] Both phytochrome and cryptochromes are known to function as photoreceptors of the light signal.[401] Once the photoperiodic requirement has been fulfilled, the signal that regulates the transition to flowering is translocated from the leaves *via* the phloem to the floral meristem in the shoot apex. To date, the chemical signal that induces floral meristem development has eluded identification though sought by a number of research groups since the discovery of photoperiodism in the 1920's.

The influence of daily temperature fluctuations on plant growth and development is known as thermoperiodism. The effects of thermoperiodism on flowering tend to be more quantitative than inductive. Optimum temperature fluctuations may result in more flowers per plant and other growth effects, rather than an all-or-nothing response such as found with photoperiodism. Thermoperiodism may be a diurnal or seasonal phenomenon. In the tulip, the optimum temperature varies during the growth of the plant. Therefore, thermoperiodism is much more important during periods of plant growth than during periods of suspended growth in storage.

Flower induction is thus modulated or controlled by several environmental parameters in addition to the stage of development of the plant. In some instances, the storage period for a crop can be utilized to induce flowering. This is the case with many bulbs. It may also be essential to store plants that have already been induced to flower under conditions where the

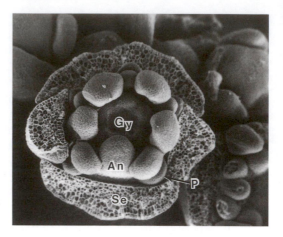

Figure 6.1. Most perfect flowers develop their parts in a distinct sequence, forming whorls from the outside to the inside, i.e. sepals (Se) → petals (P) → androecium (An) → gynoecium (Gy), as illustrated by this geranium flower bud. The androecium is composed of the anthers and filaments, while the gynoecium is made up of the stigma, style and ovary) (*photograph courtesy of H. Y. Wetzstein*).

flowering stimulus is not diminished or eliminated. When poststorage flowering is not desirable, storage conditions must be selected that will prevent induction.

1.2.2 Flower Development

After receiving the floral induction stimulus that results in the conversion of the vegetative meristem into a floral one, the meristem undergoes a series of developmental changes. These can be grouped into the following general stages: cell division, cell expansion and flower maturation, anthesis, and senescence. The initial stage involves extensive cell division with the sequential development of the individual floral organs. The floral organs are initiated by periclinal division of cells, usually found deep beneath the protoderm of the apical meristem. Generally they develop in a distinct sequence, arising from the outside inward, that is, sepals → petals → stamens → carpels (Figure 6.1). Development of the individual floral organs is regulated by homeotic genes which are expressed in the organ primordia of the developing flower.[454] Occasionally aberrant environmental conditions or pharmacological treatments can disrupt the gene expression sequence, resulting in the absence of a floral organ or its incorrect placement (e.g., petals at the tip of the stigma). The duration and timing of the cell division phase varies widely among individual floral organs. For example, the petals and stamen often cease cell division much earlier than the carpels, which may continue to divide in some species even after fruit set. The initial organogenesis of the floral organs has been widely studied by plant anatomists.[384] These studies tend to focus on the initial developmental stage of each floral organ through completion of cell division.

The cell division phase of flower development is followed by cell expansion. The timing and duration of the cell expansion phase varies widely among the individual floral organs, as well as with species and environmental conditions. Each flower part appears to go through the same general sequence of developmental events as the complete flower, that is, cell division, cell expansion, maturation and senescence. As with flower induction and cell division, the cell expansion phase is also strongly modulated by environmental factors. Unlike induction and division, however, cell expansion and subsequent flower opening stages have been little studied. These stages are of critical importance during postharvest handling and storage.

During the cell expansion phase, the individual floral parts develop, often approaching their final size. The rate at which development occurs varies widely with species. In many flowers, petal growth proceeds very rapidly just prior to opening (Figure 6.2), with a differential elongation rate between the interior and exterior portion of the petals mediating the actual opening response.[334] Osmotic changes within the cells provide the driving force for cell expan-

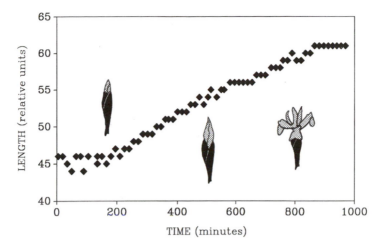

Figure 6.2. The rapid elongation of the petals is the driving force in flower opening in many species. This elongation phenomenon is illustrated above by the increase in bud length of iris flowers during anthesis (*redrawn from Reid and Evans*[358]).

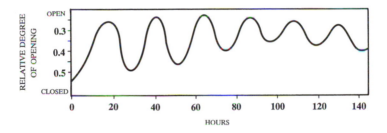

Figure 6.3. Although virtually all of the flower species that are used as cut flowers in the floral trade remain open after anthesis, a few exhibit distinct opening and closing cycles. This is illustrated by the cyclic movement of *Kalanchoë blossfeldiana* flowers (*redrawn from Engelmann et al.*[130]).

sion. The movement of sugars from the leaves and stem into the flower is a critical component in opening in that sugars provide energy[411] and alter the osmotic potential of the vacuole, causing water to move into the cells resulting in their expansion.[132] When the stem to which the flower is attached is severed from the parent plant, flower opening is generally impeded because the carbohydrate and water supply is restricted to that which can be recycled from the attached vegetative portion (longer rose stems represent more carbohydrate for recycling, resulting in larger flowers upon opening).[194] As a consequence, holding cut flowers in aqueous solutions containing sugars is highly beneficial.[61,430] Elongation-induced flower opening differs from petal movements that are turgor driven, which mediate cyclic flower opening-closing responses found in some species. The basic control mechanisms for cyclic opening-closing responses are not yet fully understood.

Anthesis, especially in species with large, brightly colored flowers, represents one of the most spectacular phenomena in the plant kingdom. The actual flower opening response, however, varies widely. Many flowers open and remain open until death, the duration dependent upon species and environmental conditions. Others open and close at certain times of the day (Figure 6.3), the number of cycles of which depending on the longevity of the flower.[130] The

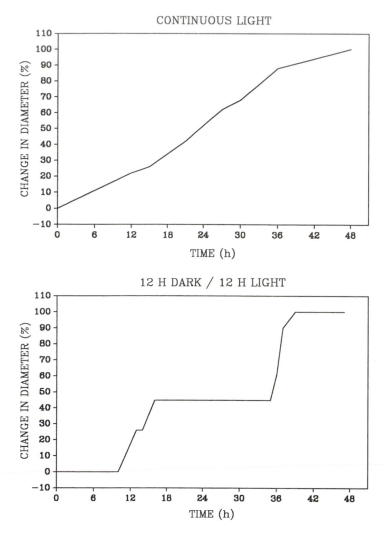

Figure 6.4. The environment used for holding flowers after harvest can have a pronounced effect on bud opening. This figure illustrates the effect of continuous light versus a 12-hour dark/light cycle on the opening of cut rose flowers (*redrawn from Evans and Reid*[131]).

former tend to be the most commonly used species in the floral trade. The actual opening response may be modulated by light conditions (Figure 6.4), temperature, moisture and carbohydrate supply. Likewise, whether the flower is attached to the parent plant can significantly modulate the handling practices needed to maximize flower opening and longevity.

Petal and, in some cases, sepal pigmentation are critical factors in flower quality and like other events in the development of flowers, pigmentation can be affected by environmental conditions. The timing of pigment synthesis relative to other developmental events in the flower varies with species. Most flowers develop their full complement of pigments prior to opening. In some cases, however, major changes in pigments occur after anthesis. For example, the flowers of *Victoria amazonica* Sowerby* are pure white the night of anthesis. Pig-

---

*Synonymous to *V. regia* Lindl.

Figure 6.5. Pollination, in this case by bees (right), can trigger senescence and abscission of some species of flowers (e.g., snapdragon).

ments are synthesized while the flowers are closed on the subsequent day, and upon reopening that night, the flowers are red.[143]

Floral scents, likewise, represent important quality attributes in many species. In nature floral scents may be important as both long- and short-distance attractants and nectar guides to a cross-section of animal pollinators,[112] as well as a key ingredient in their attractiveness to humans. Floral volatiles are often a complex mixture of relatively low molecular weight compounds,[115] only a portion of which are responsible for the characteristic scent. Quantitative and qualitative differences in emission of critical odorants can be due to circadian rhythms,[174] the length of time after anthesis, or other environmental factors. Severing the flower from the parent plant can also impact scent production. The biosynthesis and accumulation of volatile compounds are compartmentalized and restricted to specific tissues in the flower.[335]

In nature, pollination follows anthesis in the normal sequence of developmental events, setting off a dramatic series of chemical and morphological changes within the flower. In some species, pollination greatly accelerates the senescence of certain flower parts (e.g., calyx and corolla) and thus can be undesirable. For example, pollination results in rapid flower fading in orchids[20] and petunia[152] and abscission in the snapdragon, sweet pea and other flowers (Figure 6.5).

1.2.2 Perianth/Flower Senescence

While the term *flower senescence* is widely used, upon fertilization typically only certain parts of the flower, the perianth (calyx and corolla), actually senesce (a more detailed treatment of senescence is presented later in this chapter). The perianth of flowers used in floriculture is almost invariably enlarged and showy, to attract animal pollinators. Wind pollinated species, in contrast, tend to have much less conspicuous flowers. From an evolutionary standpoint, removal of these floral parts after pollination makes sense; with pollination their function has been fulfilled. Maintenance of the perianth after pollination, or the entire flower in the absence of pollination, has distinct evolutionary disadvantages [e.g., additional maintenance costs (carbon skeletons and energy) and the potential for excess pollination or herbivore damage to the flower]. Therefore the perianth has a genetically programmed functional existence, the duration of which varies widely. The range in longevity of the perianth is illustrated by two orchid species—from less than 60 minutes in *Desmotrichum appendiculatum* Blume,*[,244] to approximately 80 days in *Odontoglossum rossii* Lindl.[223] The perianth of most species has a relatively short functional existence. For example in one diverse cross-section of species (280), approximately two thirds had longevities of 4 days or less.[23,349] An ephemeral functional existence would be unacceptable in the cut flower trade; however, for intact ornamentals that produce a relatively large numbers of flowers over an extended period (e.g., evening-primrose or daylilies), an abbreviated longevity is acceptable. Interestingly, two of the species that have been used to develop floral senescence models are very short-lived (day lilies and morning glory), allowing scientists to do numerous experiments over a short time interval.

The presence of a genetically controlled program triggering senescence of the flower and/or perianth is readily evident; how it functions and what factors modulate it are just beginning to emerge. Fertilization represents a critical branch point in the fate of the flower. Without fertilization, genes are expressed that instigate the senescence of the entire flower. When fertilized, however, genes are expressed that instigate the senescence of just the perianth, leaving the ovary to develop and form seeds. In both instances, perianth senescence proceeds *via* two very general mechanisms: petal wilting and petal abscission.[†] Within both mechanisms, pollination can significantly impact the longevity of the perianth. In species with a very short-lived perianth, senescence does not appear to be accelerated by pollination. However, in species in which the flower persists for several days or longer, their longevity is often shortened appreciably by pollination or fertilization. In these species, a separate program appears to be instigated with pollination that overrides or accelerates the normal perianth senescence sequence.

a. Flowers displaying petal wilting

The petals of these species typically begin to curl inward with the onset of wilting, which is accompanied by a loss of cellular integrity. The phytohormone ethylene may (morning glory)[217] or may not (daylily)[256] be involved in the response. During the onset of petal senescence, inhibitors of protein synthesis prevent wilting, indicating that the synthesis of new proteins, presumably hydrolytic enzymes needed to dismantle the cells, is required.[267] In short-lived flowers such as morning glory and daylily, the onset of senescence actually begins before the flower is completely open. As the petals or tepals enlarge, cell size and intercellular space increases markedly and continues to do so through flower opening. With cell enlargement, the attachment between neighboring cells erodes[405] and by the day after anthesis, cellular integrity has been lost and the petals have collapsed. Based upon the role of abscisic acid in certain stress-induced osmotic responses, preliminary evidence suggests that phytochromes may also be involved in petal wilting.[322]

b. Flowers displaying petal abscission

In a number of species, the petals abscise from the flower (a more detailed treatment of abscission is presented later in the chapter). Abscission may occur with virtually no change in the

---

*Synonymous to *Dendrobium appendiculatum* Lindl.
[†]See Appendix II for senesence mechanism and ethylene sensitivity of a cross-section of flower species.[429,431,462]

petals *per se* (without loss of fresh or dry weight) or may occur after significant cellular alterations and wilting. In abscising species, a preformed abscission zone of cells is present at the base of the corolla, similar to that found in leaves. With the onset of the abscission process, the cells undergo pronounced biochemical and physical alterations,[133] leading to a decrease in cell-to-cell bonding strength, precipitating the eventual structural failure and shedding of the organ.

Ethylene has been implicated in the abscission process, though it may function in more of a coordination than inductive capacity. There is typically an increase in ethylene synthesis during or just prior to abscission. For example, in *Cyclamen* flowers the rate of ethylene synthesis increases 60-fold with pollination,[165] while in foxglove there is a 10-fold increase.[406]

c. Pollination signal transduction

Pollination in many species sets off a dramatic sequence of events leading to early petal senescence, which can be seen in both wilting and abscising species. The pollen grains germinate on the stigma and elongate down the style, organs that are some distance from the petals. Thus the flower's pollination status must be signaled by some means to the petals in order to instigate their senescence. The carnation, whose flowers exhibit a climacteric-like increase in ethylene synthesis with the onset of senescence,[310] has been used as a model system for studying floral interorgan communication. Existing evidence points to ethylene, triggered by pollination, diffusing through the style into the base of the petals as the signal, though other molecules have been proposed. An increase in ethylene in the petal activates autocatalytic ethylene synthesis (system II ethylene), which in turn triggers the expression of a cross-section of senescence-related genes. For example, the induction of mRNAs encoding for ACC synthase and ACC oxidase, key enzymes in the ethylene biosynthesis pathway, are expressed with concomitant increases in the activity of both enzymes.[323] There are multiple copies of the ACC synthase gene present in carnation that are expressed in a tissue specific manner (e.g., the *DCACS1* gene mRNAs are found primarily in the petals).[205] Ethylene synthesis and action are both required, in that the inhibition of synthesis *via* an antisense gene for ACC oxidase delays petal senescence,[296] and blocking ethylene binding with 2,5-norbornadiene prevents petal senescence.[450] The requirement for the ethylene synthesis pathway to be turned on for petal senescence to proceed has been further substantiated with the transformation of carnation with the *etr1-1* mutant allele from *Arabidopsis,* conferring ethylene insensitivity. The inability to perceive ethylene, which blocks autocatalytic induction of ethylene synthesis, increases the vase life of transgenic flowers 3-fold over non-transgenic controls.[65]

In keeping with the tremendous variation among flowers in the plant kingdom, other mechanisms controlling senescence are probable. As more is learned about the key genes controlling senescence, postharvest biologists will be able to manipulate the process more effectively.

1.3. Fruit Ripening

As fruits near the end of their growth phase, a series of qualitative and in some cases quantitative transformations occur with ripening. The fruit of agricultural products vary widely in both chemical composition and physical structure. However, these may be divided into two very general classes based on the structure of their fruit wall: parenchymatic or fleshy fruits (e.g., banana, citrus, apple, tomato); and sclerenchymatic or dry fruits (e.g., rice, wheat, peas, corn, nuts). With fleshy fruits, the fruit tissue rather than the actual reproductive organ, the embryo, represents the object of commercial importance.* In nature, the fruit tissue, in fact,

---

*There are a few exceptions to this (e.g., almond, coffee).

has only an indirect role in the perpetuation of the species. With most dry fruits, the individual seed (fertilized mature ovule), rather than the surrounding tissue, is the article of postharvest interest. Many one-seeded dry indehiscent fruits, however, are regarded as seeds in a functional sense, a unit of dissemination.

In fleshy fruits, ripening refers to the transition period between maturation of the fruit and senescence. During ripening, fleshy fruits undergo a series of distinct and, in some cases, very dramatic changes in their physical and chemical condition. With dried fruits, however, ripening is less well defined and may refer, depending upon species, to all the final stages of development and maturation. For example, the ripening period in wheat is divided into six stages: watery-ripe, milky-ripe, mealy-ripe, waxy-ripe, fully-ripe and dead-ripe.[333] During the water-ripe stage, cell division in both the endosperm and embryo is still occurring. Similar but not necessarily identical stages have been ascribed to many other dried fruits.

Ripening, therefore, describes distinctly different events in fleshy and dry fruits. Therefore, the concept of ripening and the changes occurring during this time period will be treated separately.

1.3.1. Fleshy Fruits

Botanically, a fruit is a mature ovary which contains one or more seeds and may include accessory floral parts. In fleshy fruits, the fruit wall, which may consist of the ovary wall or the ovary wall fused with noncarpellary tissue(s), is notably developed. Depending upon the type of fruit, the entire ovary wall or various parts differentiate into the fleshy parenchymatic tissue. The individual "fruit" of commerce may be formed from a single, enlarged ovary (simple fruit, e.g., avocado), a number of ovaries belonging to a single flower (aggregate fruit, e.g., strawberry) or enlarged ovaries of several flowers including accessory parts fused to form the fruit (multiple fruit, e.g., pineapple) (Table 6.1).

While immature, the parenchymatic tissues, which comprise the main body of the fruit, have a firm texture and the cells retain their protoplasts. At the transition phase following maturation, the fruit undergoes a series of generally irreversible qualitative changes that render it attractive for consumption. These diverse physical and chemical alterations, termed "ripening" in fleshy fruits, lead to the senescence phase of development, culminating in the ultimate death of the fruit.[45] Changes in texture, storage materials, pigments and flavor components may occur. Due to the great diversity in the origin and composition of the parenchymatic tis-

Table 6.1. Classification of Fruits.

Dry Fruits		Fleshy Fruits		
Indehiscent	*Dehiscent*	*Simple*	*Aggregate*	*Multiple*
grains	**legume**	**berry**	**berry**	**berry**
(wheat, rice)	(pea, bean)	(tomato)	(strawberry)	(mulberry)
nuts	**silique**	**drupe**	(raspberry)	(pineapple)
(chestnut, pecan)	(mustard)	(peach)		
caryopsis	**capsule**	**pome**		
(corn)	(snapdragon)	(apple)		
achene	**follicle**	**hesperidium**		
(sunflower)	(columbine)	(orange)		
samara				
(maple, elm)				
schizocarp				
(celery)				

sue and the number of ovaries that constitute the individual fruit, considerable diversity is found in the ripening behavior among species.

Fruits vary widely in the length of time required to ripen once the ripening process begins and their longevity after ripening before the tissue deteriorates to an unacceptable level. One major responsibility of postharvest biologists and technologists is to develop the information and methods needed to maximize the duration of the period between ripening and deterioration.

Fruits of different species vary in their ability to ripen once detached from the parent plant. Many fruits must be harvested only when fully ripe (e.g., grape, cherry, bramble fruits); they are not capable of ripening after detachment (non-climacteric fruits, Table 6.2). These fruits do undergo many physical and chemical changes after harvest however, these alterations are largely degradative and generally do not enhance the products' aesthetic qualities. Other fruits can be harvested unripe but if properly handled will undergo normal ripening even though detached from the plant (climacteric fruit): apples, bananas, tomatoes and many others. These fruits often undergo hydrolytic conversions in storage materials and synthesize the pigments and flavors associated with a ripe fruit. With the exception of certain cultivars of several species (e.g., avocado), detachment is not a prerequisite for ripening; each will ripen while attached to the parent plant.

This latter class of fruits, which can ripen normally after harvest, has been widely studied since many offer a much greater degree of storage and marketing flexibility. If harvested unripe and held under conditions that prevent ripening, the ripening process can then be induced, allowing synchronization of ripening and marketing. The quality attributes that make up the aesthetic appeal of the ripe fruit are not present during the storage period and, as a consequence, are not subject to loss. Thus, consistently high quality fruit can be marketed over extended periods. Many of the fruits that are available in retail stores during periods well beyond their normal field production time belong to this latter class.

A relatively large number of fruits that fall into both of the preceding classes are marketed and utilized in an unripe state. Ripening results in undesirable changes which, in fact, decrease the aesthetic quality of the fruit. Many of the fruits that are eaten as vegetables are examples (e.g., bitter melon, cucumber, squash, okra). Normally these fruits are harvested in an immature or mature unripe condition. Inhibiting the ripening process, therefore, may be essential during the postharvest period.

a. Changes that occur with ripening

During ripening, fleshy fruits undergo major changes in their chemical and physical state. These changes represent a wide spectrum of synthetic and degradative biochemical processes, many of which, although not all, occur concurrently or sequentially within the fruit. Table 6.3 lists a cross-section of these alterations. Those that represent changes in quality attributes of the fruit can be grouped into three general categories: 1) textural changes, 2) changes in pigmentation, and 3) changes in flavor.

The induction and development of ripening are tightly controlled steps in the overall developmental cycle of a fruit. During this period, specific enzymes are synthesized or activated triggering or accelerating specific metabolic events. For example in some fruit (like apples), protein synthesis has been shown to increase dramatically with the beginning of the ripening response.[192] These proteins are thought to be enzymes and other polypeptides required for ripening.

i. Softening

Softening is one of the most significant quality alterations consistently associated with the ripening of fleshy fruits. Alterations in texture affect both the edibility of the fruit and the length of time the fruit may be held. In many fruits that are consumed in an unripe state (e.g.,

Table 6.2. Classification of Fleshy Fruits According to Their Respiratory Pattern.

Common Name	Scientific Name	Reference
Climacteric Fruits		
Apple	*Malus sylvestris* Mill.	44
Apricot	*Prunus armeniaca* L.	44
Atemoya	*Annona squamosa* L.	70
Avocado	*Persea americana* Mill.	44
Banana	*Musa* spp.	44
Biriba	*Rollinia deliciosa* Safford	47
Bitter melon	*Momordica charantia* L.	213
Blueberry, highbush	*Vaccinium corymbosum* L.	195
Blueberry, lowbush	*Vaccinium angustifolium* Ait.	195
Blueberry, rabbiteye	*Vaccinium ashei* Reade	264
Breadfruit	*Artocarpus altilis* (Parkins.) Fosb.	47, 465
Cantaloupe	*Cucumis melo* L. Cantalupensis group	269
Cape Gooseberry	*Physalis peruviana* L.	426
Cherimoya	*Annona cherimola* Mill.	44
Chiku	*Manilkara zapota* (L.) von Royen	69, 253
Chinese gooseberry	(see Kiwifruit)	
Corossol sauvage	*Rollinia orthopetala* A. DC.	43
Durian	*Duiro zibelhinus* Murr.	59, 423
Feijoa	*Feijoa sellowiana* O. Berg.	44
Fig, common	*Ficus carica* L.	275
Goldenberry	(see Cape gooseberry)	
Guava, 'Purple Strawberry'	*Psidium littorale* var. *longipes* (O. Berg.) Fosb.	14
Guava, 'Strawberry'	*Psidium littorale* Raddi	14
Guava, 'Yellow Strawberry'	*Psidium littorale* var. *littorale* Fosb.	14
Guava	*Psidium guajava* L.	14
Honeydew melon	*Cucumis melo* L. Inodorus group	341
Japanese pear	*Pyrus serotina* Rehder	177
Jujube, Chinese	*Zizyphus jujuba* Mill.	1
Jujube	*Zizyphus mauritiana* Lamk.	1
Jujube	*Zizyphus spina-christi* (L.) Willd.	16
Kiwifruit	*Actinidia chinensis* Planch.	343
Mammee-apple	*Mammea americana* L.	13
Mango	*Mangifera indica* L.	44
Mangosteen	*Garcinia mangostana* L.	311
Papaw	*Asimina triloba* (L.) Dunal.	44
Papaya	*Carica papaya* L.	44
Passion fruit	*Passiflora edulis* Sims.	44
Peach	*Prunus persica* (L.) Batsch.	44
Pear	*Pyrus communis* L.	44
Persimmon	*Diospyros kaki* L.f.	356
Plum	*Prunus americana* Marsh	44
Sapodilla	(see Chiku)	
Sapote	*Casimiroa edulis* Llave.	44
Soursop	*Annona muricata* L.	47
Tomato	*Lycopersicum esculentum* Mill.	44
Sapota	*Mauilkara achras* (Mill.) Forsberg	352
Zapote	(see Chiku)	

Table 6.2. (*continued*)

Common Name	Scientific Name	Reference
Nonclimacteric Fruits		
Blackberry	*Rubus* L. subgenus *Rubus* Watson	264, 332
Cacao	*Theobroma cacao* L.	47
Carambola	*Averrhoa carambola* L.	248, 320
Cashew	*Anacardium occidentale* L.	47
Cherry, sour	*Prunus cerasus* L.	57
Cherry, sweet	*Prunus avium* L.	44
Coconut	*Cocos nicifera* L.	327
Cucumber	*Cucumis sativus* L.	44
Dragon fruit	*Hylocereus polyrhizus* (Weber) Britton & Rose	307
Dragon fruit	*Hylocereus undatus* (Weber) Britton & Rose	257, 307
Duku-Lanson	(see Langsat)	
Grape	*Vitis vinifera* L.	44
Grapefruit	*Citrus* × *paradisi* Macfady	44
Java plum	*Syzygium cumini* (L.) Skeels*	14
Lemon	*Citrus limon* (L.) Burm. f.	44
Langsat	*Lansium domesticum* Correa.	325
Lanzone	(see Langsat)	
Litchi	(see Lychee)	
Longkong	(see Langsat)	
Loquat	*Eriobotrya japonica* Lindl.	58, 476
Longan	*Euphoria longana* Lamk.	326, 422
Lychee	*Litchi chinensis* Sonn.	13, 85
Mountain apple	*Syzygium malaccense* (L.) Merrill & Perry[†]	14
Olive	*Olea europaea* L.	285
Orange	*Citrus sinensis* (L.) Osbeck.	44
Pepino	*Solanum muricatum* Ait.	12, 178
Pepper	*Capsicum annuum* L.	376
Pineapple	*Ananas comosus* (L.) Merrill	44
Pitaya, Yellow	*Selenicereus megalanthus* (Scum. ex Vaupel) Moran	308
Pitaya	(see Dragon fruit)	
Pomegranate	*Punica granatum* L.	129, 392
Rambutan	*Nephelium lappaceum* L.	290
Raspberry	*Rubus idaeus* L.	331
Rose apple	*Syzygium jambos* (L.) Alston[‡]	14
Satsuma mandarin	*Citrus reticulata* Blanco	356
Star apple	*Chrysophyllum cainito* L.	342
Strawberry	*Fragaria* × *Ananassa* Duchesne	44
Strawberry pear	(see Dragon fruit)	
Surinam cherry	*Eugenia uniflora* L.	14
Tree tomato	*Cyphomandra betacea* (Cav.) Sendtu.	344
Watermelon	*Citrullus lunatus* (Thunb.) Mansf.	128, 299
Wax apple	*Syzygium samarangense* (Blume) Merrill & L.M. Perry[§]	14

*Originally cited as *Eugenia cumini* (L.) Druce.
[†]Originally cited as *Eugenia malaccensis* L.
[‡]Originally cited as *Eugenia jambos* L.
[§]Fruit is thought to be of the indicated classification.

Table 6.3. Physical and Chemical Alterations That Occur During the Ripening of Fleshy Fruits.*

	Important Quality Attributes
1. Seed maturation	
2. Changes in pigmentation	
a. degradation of chlorophyll	
b. unmasking of existing pigments	Color
c. synthesis of carotenoids	
d. synthesis of anthocyanins	
3. Softening	
a. changes in pectin composition	
b. possible alterations in other cell wall components	Texture
c. hydrolysis of storage materials	
4. Changes in carbohydrate composition	
a. starch conversion to sugar	
b. sugar interconversions	Flavor
5. Production of aromatic volatiles	
6. Changes in organic acids	
7. Fruit abscission	
8. Changes in respiration rate	
9. Changes in the rate of ethylene synthesis	
10. Changes in tissue permeability	
11. Changes in proteins	
a. Quantative	
b. Qualitative	
1. enzyme synthesis	
12. Development of surface waxes	

*Note: The order does not represent the sequence of occurrences during ripening.

cucumbers, squash), softening may be detrimental. In others, it is an essential component in the development of optimum quality. Therefore, during the postharvest handling and storage period, when feasible, we try to maximize our control over textural changes, whether it is to prevent, synchronize or accelerate the process.

Once the softening process is initiated, the rate of textural change is a function of the type of fruit and the conditions under which the product is held. With the exception of some turgor-mediated textural alterations, softening in fruit is an irreversible process once initiated. This does not mean that all synthetic processes enhancing wall rigidity cease; rather the degradative reactions proceed at a more rapid rate.

The textural properties of fruit can have a pronounced influence on their acceptability to consumers, and often acceptable texture has a very narrow range which can be rapidly exceeded, diminishing the quality of the product. For example, in apples when the adjacent cell walls are tightly bonded together, the cells are crushed during the process of chewing, giving a crisp, juicy sensation. However, as the apple begins to soften, the walls between neighboring cells become less tightly bound and when pressure is applied the neighboring cells now simply slide past each other, resulting in an unacceptable mealy, non-juicy sensation. Understanding the biochemical and structural factors controlling texture is critical to maintaining the desired textural properties during the postharvest period.

Our understanding of the control of fruit softening has changed markedly over the past 10 years. For some time it was thought that softening was controlled by the status of the pectin molecules found in the middle lamella between neighboring cell walls. Pectins act as cementing agents, holding cells together and giving the tissue structural rigidity. This view was supported by increases in the enzymes that degrade pectin (polygalacturonase and pectin esterase) and increases in the solubility of the pectin component correlating with loss of firmness. Subsequently, however, transgenic tomatoes were developed in which polygalacturonase, a key enzyme responsible for degrading the pectins, was eliminated. The fruit softened just the same as normal fruit, indicating that softening in tomato is not controlled simply by alterations in the pectins but by a much more complex process.

The texture of fleshy fruits is affected by the composition of their cell walls and the degree of hydration. During ripening, enzymatically mediated degradative changes in the cell walls have been found in most fruits. The enzymes may be either synthesized, activated or a combination of both, at or near the onset of the ripening process. Changes in cellular osmotic properties are generally not associated with normal textural changes during ripening. In many fruits, however, desiccation-induced osmotic alterations represent a potentially critical path for undesirable textural changes, which may occur throughout the postharvest period.

During cell division and enlargement, the cell wall is assembled to precise specifications in a highly complex and coordinate manner. As the cell enlarges, its volume increases from 10- to 1000-fold, a seemingly improbable task when encased within a relatively rigid wall. Increases in volume, therefore, require some means of relaxing the wall to allow expansion, a process that occurs through alterations in the wall structure.[95] As the cell reaches its final size, subsequent alterations occur that increase the wall's final strength and rigidity. In contrast, softening during ripening can be viewed as a controlled disassembly of the wall and likewise entails a highly coordinated sequence of events. While our understanding of both processes has increased substantially, neither has been adequately explained.

The parenchymatic tissue of fleshy fruits is composed of polyhedral or elongated cells with primary cell walls and considerable space between neighboring cells. The wall is a heterogeneous polymeric structure[71] in which cellulose microfibrils are embedded in a complex polysaccharide matrix with smaller quantities of structural proteins, enzymes, phenolic and hydrophobic compounds, and inorganic molecules (e.g., ~30% cellulose, 30% hemicellulose, 35% pectin, and 1–5% structural proteins on a dry weight basis). In addition, the wall matrix is typically ~75% water, giving the composite the physical properties of a dense gel interlaced with a fibrous support structure.

Cellulose is present as long, unbranched β-1,4-linked glucan chains (i.e., 3000 to 5000 D-glucose subunits) that are arranged into tightly packed, parallel strands forming a microfibril (30–100 chains per microfibril) (Figure 2.23). Much of the microfibril is rigid, insoluble and crystalline in structure and is not subject to disassembly. Interspersed along the microfibril are thought to be areas that are more amorphous and less crystalline (paracrystalline), which may be more labile.

The hemicellulose component is comprised of β-1,4-linked glucosyl residues interspersed with mono-, di-, and trisaccharide side-chains (consisting of xylose, galactose, fucose or arabinose) which alter the properties of the polymer. Xyloglucans represent a major form of hemicellulose in dicots. They coat the surface of the microfibrils and cross-link neighboring microfibrils,[173] giving a three-dimensional structure. These cross-linking xyloglucans represent sites where a relatively small amount of enzymatic action could significantly alter the structural rigidity of the wall.

Pectins are comprised of D-galacturonic acid, L-rhamnose, L-arabinose, D-galactose, and to a lesser extent, D-xylose, 2-O-methyl-L-fucose, D-apiose, and D-galacturonic acid bonded to form polymers such as rhamnogalacturonans, arabinans, homogalacturonans, galactans, and arabinogalactans. Rhamnogalacturonan I, for example, consists of galacturonic acid, interspersed with rhamnose, reaching a total of approximately 2000 subunits. In contrast, rhamnogalacturonan II is substantially smaller (~25 to 50 subunits) and consists of galacturonic acid interspersed with monosaccharides such as 2-O-methyl-D-xylose and D-apiose esterified to the molecule. The contribution of pectin to the structural rigidity of the wall is enhanced through non-covalent bonds formed when Ca^{++} ions cross-link free carboxyl groups on adjacent pectin molecules.[251] The high affinity of pectin for water gives it a dense gel-like structure that embeds the cellulose-hemicellulose network, preventing its collapse.[199]

Phenolic components, primarily ferulic and *p*-coumaric acids are esterified to arabinose or galactose subunits in pectin molecules. These compounds may increase the rigidity of the wall by cross-linking neighboring pectin molecules.[445] In Chinese water chestnuts, diferuloyl bridges may be responsible for their thermal stability, allowing the corms to remain crisp during cooking.[442]

The wall also contains both structural and enzymatic proteins. Extensin, for example, is a structural protein that is thought to add mechanical strength and assist in the assembly of other wall components. As a wall protein, it is distinct due to its high content of the amino acid hydroxyproline and level of glycosylation (~50% carbohydrate, principally arabinofuranose). Whether or not alterations in extensin occur during ripening-mediated softening is not known.

Other proteins include a wide cross-section of enzymes found in the cell wall, a number of which catalyze structural and compositional changes occurring as the fruit softens during ripening.[67] Included are β-1,4-glucanases, xyloglucan endotransglycosylases, glycosidases, hydrolases, expansins and other enzymes. They may be present throughout fruit development or synthesized with the onset of ripening. It is thought that certain enzymes instigate critical alterations mediating or facilitate softening while others cause general compositional changes in the wall.

Several expansin genes are expressed during tomato fruit ripening,[369] apparently in response to ethylene. The enzyme is thought to contribute to the disassembly of the cellulose-xyloglucan structural network, disrupting the tethering between molecules and in so doing diminishing cell wall strength.[368] Expansins work in concert with other enzymes that attack wall components, leading to fruit softening.

In some fruit, softening is compressed into a relatively short time period (e.g., several days in banana); while in others it occurs more slowly (e.g., several weeks in carambola[87]). Differences in rate may be due to wall compositional and structural differences or variation in the abundance, type, isoform and timing of expression of specific enzymes. For example, the β-1,4-glucanases that are thought to degrade amorphous cellulose (the majority of the cellulose is in a crystalline microfibrillar form that is unaffected by the enzyme) are encoded by a multigene family allowing tissue specific expression of various isoforms. The enzyme undergoes substantial increases in fruit such as avocado, tomato, pepper, and strawberry during ripening, and isoforms *Cel1* and *Cel2* increase in tomato in response to ethylene. *Cel2* dismantles the cell walls in the abscission zone of the fruit, but neither *Cel1* nor *Cel2* attacks the cell walls of the fruit.[72]

The pectin fraction of the cell wall does undergo significant modification during ripening, changes which appear to parallel softening (Figure 6.6). Soluble pectins increase during this period while insoluble pectin declines. In unripe peaches, approximately 25% of the pectin fraction is water soluble; upon ripening, this increases to approximately 70% of the total.[340] Enzyme-mediated changes in the insoluble pectin fraction, resulting in solubilization, may occur through the cleavage of linkages between the pectin molecule and other cell wall components or through the direct hydrolysis of the pectin molecule. The former appears to occur in the apple, while hydrolysis is the predominate mechanism in a variety of other fruits.

The hydrolysis of pectin molecules involves the action of pectin esterases, polygalacturonases and a number of glycosidases.[397] Pectin esterases catalyze the removal of methyl esters, proceeding in a linear manner down the pectin molecule, leaving free carboxyl groups, an apparent prerequisite for further degradation of the molecule by polygalacturonase.

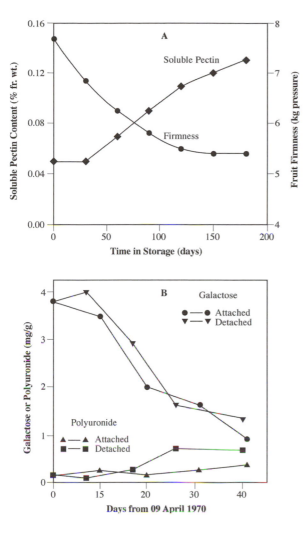

Figure 6.6. During fruit ripening there are pronounced changes in texture. In many fruits, softening involves degrading the large pectin molecules found in the middle lamella that bind together the walls of neighboring cells. (A) As the size of the pectin molecules decreases their water solubility increases as illustrated here (a) in Delicious apples (*redrawn from Gerhardt and Smith*[151]). (B) A decline in cell wall galactose content and an increase in polyuronides are also indicative of pectin hydrolysis (*redrawn from Knee*[231]).

Generally two forms of polygalacturonase are present, an endopolygalacturonase which attacks the pectin molecule at various sites within the chain and an exopolygalacturonase which sequentially removes galacturonic acid residues from the end of the molecule. Of the two, endopolygalacturonase is much more important in that it cleaves the pectin molecule toward the center, yielding two molecules approximately one-half the size of the original and greatly increasing their solubility. Removal of a terminal subunit by exopolygalacturonase, however, results in only a very slight alteration in the properties of the pectin molecule. Both mediate changes in solubility; however, the two enzymes differ greatly in the rate in which they may effect change. The levels of the two polygalacturonases vary among and within species. For example, apples appear to lack endopolygalacturonase while it is present in pears, peaches, strawberries, tomatoes and many other fruit.[345] Freestone peaches, which soften extensively during ripening, have high activities of both endo- and exopolygalacturonase. Clingstone cultivars, however, have low activities of endopolygalacturonase and display much less softening and pectin solubilization with ripening.[346]

Even with the very marked correlations that have been reported between polygalacturonase activity and fruit softening, the precise role of polygalacturonase in softening is far from clear. Analysis of softening in transgenic plants has renewed the debate on the relative importance of polygalacturonase. When transgenic tomato plants were produced in which the level of polygalacturonase synthesis in the fruit was virtually eliminated (~99%), fruit softening was not significantly different from normal non-transformed fruit.[398] Likewise, when polygalacturonase is over-expressed in fruit lacking polygalacturonase, softening was not significantly altered.[154] These results suggest that polygalacturonase, along with at least two previously mentioned β-1,4-glucanases, are not critical factors in fruit softening.

It is likely that no single enzyme is responsible for softening, but rather textural changes involve the concerted and synergistic action among a variety of enzymes.[368] Likewise, the mechanism of softening may vary among different types of fruit. It is evident that our understanding of the overall dismantling process is in its infancy. Using antisense suppression of various enzymes and isoforms up-regulated during ripening will make it possible to identify their respective roles and critical alterations essential for softening.

Alteration in cellular hydration is known to result in marked changes in the texture of plant products. Often the percent water loss needed before realizing undesirable textural changes is quite small. Although changes in the rate of transpiration have been found during the ripening of fruits such as the banana,[394] changes in osmotic conditions do not appear to be a significant part of the normal softening process during ripening. Only small changes in hydration normally occur during ripening when fruits are held under proper relative humidity conditions. With improper handling, however, water loss may yield significant losses in textural quality.

ii. Hydrolytic conversion of storage materials

Many fruits that have the potential to ripen after being detached from the plant undergo significant changes in their storage form of carbon during this process. Typically starch is the prevalent carbon storage compound, and it undergoes hydrolytic conversion during ripening, yielding free sugar. In other fruits, carbon may be stored as lipids or organic acids, which may or may not be altered during ripening.

The banana fruit represents an excellent example of hydrolytic alterations during ripening (Figure 6.7). The edible mesocarp tissue contains 20 to 30% starch when the fruit is at a mature green stage. During ripening, the starch concentration decreases to only 1 to 2% while the concentration of sugar increases from 1% to 14–15%.[472] Along with changes in the textural properties of the tissue, this conversion of starch to sugar enhances the palatability of the fruit.

The rate and extent of hydrolysis of carbon storage compounds during ripening varies widely among fruits of different species. In banana, there is a dramatic conversion of starch to sugar, and this generally occurs over a relatively short period (e.g., 8 to 10 days).[439] In avocado,

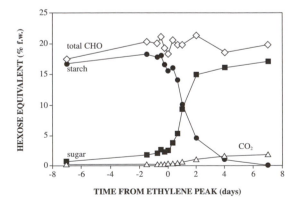

Figure 6.7. In banana fruit, there is a very rapid hydrolysis of starch and a corresponding increase in sugar concentration during ripening (*redrawn from Beaudry et al.*[33]). In other fruits (e.g., oranges), changes in composition may occur at a much slower rate.

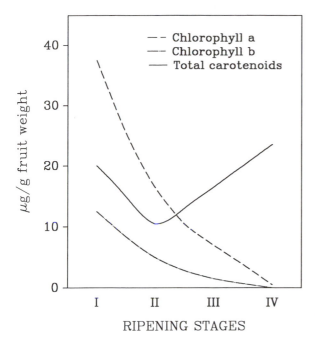

Figure 6.8. In some cases (e.g., banana), the degradation of chlorophyll unmasks largely pre-existing pigments, accounting for the rapid change in fruit color (*redrawn from Gross and Flugel*[163]).

lipids represent the predominate storage form of carbon; however, they do not appear to furnish a respiratory source of carbon during the ripening process.

In fruits that ripen normally only when attached to the parent plant, there is often also a significant increase in sugars during this period. In this case, the sugars are not derived *via* the hydrolysis of carbon compounds stored within the fruit. Rather, the sugars are composed of carbon transported from the parent plant into the fruit during the ripening process. As a consequence, the accumulation of sugar is dependent upon attachment; fruit removal prior to ripening prevents further accumulation. Although postharvest softening will generally occur in these fruits if prematurely harvested, they do not develop the full complement of other quality characteristics associated with a ripe fruit.

iii. Changes in pigmentation

Changes in the coloration of fruits during ripening are often spectacular. In nature, these changes facilitate the natural dispersal of the seeds of many species by attracting animals. With fruits consumed by man, color changes that occur during the ripening period are often used as an index to the degree of ripeness. Thus, the timing of harvest for some fruits (e.g., tomato) may be determined by fruit color. Color is also the primary criterion used by consumers in assessing ripeness of many fruits. So color changes during ripening and storage are of primary importance. It is necessary to be cognizant, therefore, of the conditions that mediate both desirable and undesirable changes in color after harvest.

Alteration in the coloration of fruits normally involves the loss of chlorophyll and the synthesis of other pigments such as carotenoids and anthocyanins (Figure 6.8) or the unmasking of these pigments which were formed earlier in the development of the fruit. Some fruits do, however, retain their green coloration throughout the ripening period (e.g., avocado, kiwifruit, honeydew melons, and some apple cultivars).

Changes in color are instigated by the expression, or increased expression, of genes encoding key enzymes in pigment synthesis or degradation pathways. Over the past few years, a better understanding has emerged as to the individual pigments making up the coloration of a broad cross-section of fruits, the synthesis pathway for these pigments, and factors affecting the timing and expression of critical genes controlling pigment synthesis.

The timing, rate and extent of change in fruit color vary widely among different species and cultivars of the same species. The timing of color change can be assessed relative to the time of harvest or to the actual ripening of the fruit. Many red apple cultivars develop much of their coloration during the final stages of development prior to ripening. Synthesis of pigments may continue, however, throughout the ripening period. A number of climacteric fruits have the potential to develop their normal coloration after removal from the parent plant (e.g., banana, tomato). Many fruits, however, only develop their normal coloration while attached to the parent plant. If picked prior to ripening, these fruits may lose much of their chlorophyll but will seldom develop more than a small portion of their normal complement of pigments.

Changes in fruit color may or may not coincide with the development of the other quality criteria associated with ripening. With apples, color development does not closely parallel the respiratory climacteric. Color, therefore, is not generally an acceptable means of assessing ripeness of this fruit. There is, however, a relatively close association between color changes and ripening in climacteric fruit such as the banana, tomato and bitter melon and non-climacteric fruits such as the cherry, blueberry and strawberry.

The rate of color change also varies widely. In many fruits, there is a relatively slow but steady synthesis of the pigments making up the final coloration of the product (e.g., grapes). In others, this change is compressed into a short time period, giving a spectacular effect. For example, the fruit of the bitter melon when exposed to ethylene can proceed from green to bright orange in a 24-hour period.[213] Likewise, the color change from green to yellow in the banana can be compressed into several days with appropriate treatment. Many of these very rapid changes in coloration are mediated to a large extent by the rapid degradation of chlorophyll and the exposure of existing pigments that were previously masked rather than *de novo* pigment synthesis alone (Figure 6.8).

Most color changes in fruit are associated with a decrease in the concentration of chlorophyll molecules in the chloroplasts. The chloroplasts are transformed into chromoplasts during ripening, through extensive changes in their internal membranes and the synthesis of carotenoids (yellow to red in color). The loss in chlorophyll is instigated by an increase in the activity of the enzyme chlorophyllase.[*,266] Concurrent synthesis of carotenoids does not occur in all fruits, however. Significant color changes may be mediated in some fruits through the degradation of chlorophyll and the exposure of preexisting carotenoids. For example, during the transformation of three cultivars of banana from green to yellow during ripening, the mean chlorophyll content of the peel decreased from 77 to 0 $\mu g \cdot g^{-1}$ while the carotenoid concentration increased from 10.5 to only 11.3 $\mu g \cdot g^{-1}$.[109]

Unlike chlorophyll and carotenoids, which are sequestered in chloroplasts or chromoplasts, anthocyanins accumulate in the vacuoles and give the pink, red, purple and blue colors of fruits. Cells with high concentrations of anthocyanins may be found distributed throughout the fleshy portion of the fruit (e.g., sweet cherry cultivars) or found in abundance only in the epidermal or subepidermal tissues of the fruit (e.g., cranberry,[381] apples, plums, pears[105]) (Figure 6.9). Additional details on the synthesis and degradation of pigments are covered in section 5 of chapter 4.

Our understanding of the mechanisms controlling pigment alterations at the genomic level has expanded substantially over the past several years. Dramatic shifts in fruit color involve the expression of genes encoding key enzymes in the synthesis and/or degradation path-

---

*Two other enzymes (chlorophyll oxidase and peroxidase) have been shown to degrade chlorophyll.[7,278]

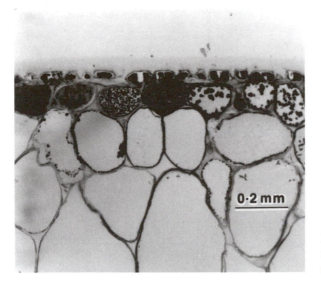

Figure 6.9. Anthocyanin pigments are found in the outer two layers of cells of cranberry fruit (cv. Pilgrim) (*from Sapers et al.*[381]).

ways for pigments. Genes encoding enzymes mediating alterations in pigment synthesis/ degradation are progressively being identified and the timing and control of their expression elucidated.

A central feature in the color alterations of many fruits during ripening is the degradation of chlorophyll, perhaps the single most abundant pigment in the biosphere.[280] Chlorophyllase catalyzes the removal of the phytol side-chain, the first step in chlorophyll degradation (see review by Matile et al.[279]). Exogenous ethylene has long been known to stimulate the degradation of chlorophyll, apparently through increasing the expression of the gene encoding chlorophyllase.[350] Hence, an *in vivo* role for ethylene in triggering the onset of degradation of chlorophyll during ripening has been postulated for both nonclimacteric and climacteric fruits. In the peel of citrus, a nonclimacteric fruit in which there is no notable increase in ethylene during ripening, the gene for chlorophyllase is expressed throughout development and does not increase with the onset of chlorophyll degradation during ripening.[198] The assumption, therefore, would be that ethylene is not involved; however, inhibitors of ethylene action (2,5-norbornadiene, silver thiosulphate, 1-methylcyclopropene) inhibit chlorophyll breakdown,[157,339] indicating that ethylene's role in nonclimacteric fruit may be more complex.

The development of antisense clones, in which ethylene synthesis has been blocked, has shown that the degradation of chlorophyll in climacteric fruit can be either ethylene-dependent or independent. In tomato, the absence of the normal climacteric increase in ethylene did not delay chlorophyll loss (i.e., ethylene-independent chlorophyll degradation),[319,227] while in melon the loss of rind chlorophyll is totally prevented (i.e., ethylene-dependent chlorophyll degradation).[164] The precise role of ethylene in chlorophyll catabolism during ripening in both climacteric and nonclimacteric fruits requires further elucidation.

Chlorophyll degradation does appear to be independent of changes in the synthesis of other pigments, even though both may occur concurrently. The regulatory separation of chlorophyll degradation is apparent in mutants in which chlorophyll is retained, while carotenoid synthesis develops normally during ripening (e.g., the pepper cultivar Negral).

Fruits such as tomato undergo a tremendous increase in carotenoids, in particular lycopene, which increases 500-fold during ripening.[146] Carotenoid synthesis occurs within the plastids,[100] and the pigments accumulate with the transition of chloroplasts into chromoplasts.[160] While the enzymes controlling carotenoid synthesis are found in the plastids, their encoding genes are nuclear,[226] with transcriptional regulation the predominant mechanism controlling carotenoid synthesis.

The transcription of phytoene synthase and phytoene desaturase, key enzymes in the synthesis pathway, is developmentally regulated,[146] with their expression considerably elevated at the onset of ripening. However, not all genes in the synthesis pathway are expressed simultaneously. In contrast to genes that encode enzymes of early steps in the pathway, the genes for lycopene β- and ε-cyclase are down-regulated at the breaker stage, forcing the pathway toward the accumulation of lycopene.[329,366]

Flavonoids, such as anthocyanins, accumulate during ripening of a number of fruits. A diverse cross-section of pigments is synthesized,[286] involving a number of structural and regulatory genes.[186] The expression of seven genes in grapes, encoding enzymes in the anthocyanin biosynthetic pathway, coincides with the onset of pigment synthesis, indicating a coordination probably involving regulatory genes. One of the structural genes monitored was expressed independently, however, apparently controlled by a separate regulatory mechanism.[63] In strawberries, expression of the gene for dihydroflavonol-4-reductase, which catalyzes the last common step in the flavonoid biosynthetic pathway, likewise is up-regulated at the onset of ripening.[303]

Color alterations within a fruit during ripening are affected by a number of factors. As mentioned previously, the time of harvest may significantly alter the development of the normal complement of pigments in many fruits. Light, temperature and oxygen concentration may also have a pronounced affect on color development. Light is not essential for the synthesis of carotenoids and in some cases (e.g., peach and apricot) color development has been shown to be greater in the absence of light. Although not an obligatory requirement for synthesis, light has also been shown to enhance the synthesis of carotenoids by as much as 2.3-fold in tomatoes harvested prior to ripening. Interestingly, light-induced lycopene accumulation in tomato pericarp occurs independently of ethylene biosynthesis.[15] Taken together with data that demonstrate ethylene does in fact regulate some components of tomato pigmentation, it now appears that color development in tomato consists of both light-dependent and light-independent components. Ethylene presumably regulates the light-independent component.

Anthocyanin synthesis is markedly stimulated by light in many fruits (e.g., apple). Red apple cultivars allowed to mature in total darkness remain green.[393] Light mediated synthesis appears to involve two photoreactions: a low-energy reversible phytochrome controlled reaction and a high-energy reaction. Thus, both light quantity and quality are important in anthocyanin accumulation in apples.[246] Postharvest irradiation can improve the color of the fruit of some species (e.g., tomato and apple, but not pear)[15,230,274] and may be a viable means of improving the fruit color of some species during the postharvest period.

Temperature can affect both the rate of synthesis of specific pigments and their final concentration within the fruit. The optimum and maximum temperature for synthesis of a specific pigment varies among species. For example, lycopene synthesis in the tomato is inhibited above 30°C,[158] while in the watermelon synthesis is not prevented until the fruit temperature rises above 37°C.[438]

Oxygen is essential for carotenoid synthesis,[438] and increasing the oxygen concentration enhances the synthesis of the pigments.[107]

iv. Changes in flavor

Flavor is generally considered to consists of two major attributes: taste and odor. Taste is perceived by taste buds in the mouth, while odor is detected by olfactory receptors in the nose. Fruit flavor is primarily a composite of sugars, acids and volatile compounds, and with ripening there are very dramatic changes in the flavor of most fruits.

Alterations in taste during ripening center predominately on changes in sugars and or-

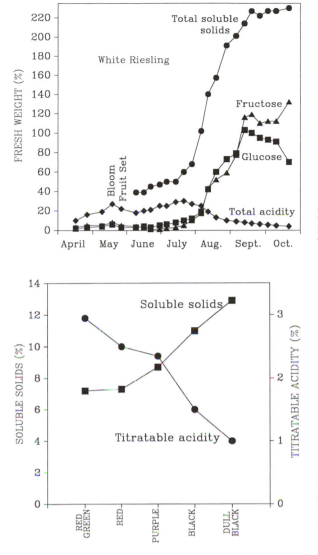

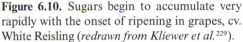

Figure 6.10. Sugars begin to accumulate very rapidly with the onset of ripening in grapes, cv. White Reisling (*redrawn from Kliewer et al.*[229]).

Figure 6.11. Sweetness (as indicated by soluble solids) and acidity are important components in the taste of many fruits. Both commonly change with the onset of ripening, illustrated by blackberry fruit (*redrawn from Walsh et al.*[444]).

ganic acids. With fruits that must ripen while attached to the parent plant, sugars increase *via* translocation of sucrose from the leaves. Upon arrival, sucrose in grapes is hydrolyzed by invertase, forming glucose and fructose. The total sugar concentration (as measured by soluble solids) in White Riesling grapes begins to accumulate rapidly with the onset of ripening (Figure 6.10). In some, although not all climacteric fruits, changes in internal sugars result from the hydrolysis of carbohydrate reserves within the tissue. For example, the sugar concentration in the banana increases 12 to 15 fold during ripening (Figure 6.7). Increases in free sugars are due to hydrolysis of starch reserves by α- and β-amylase and/or starch phosphorylase. The activity of these enzymes increases markedly during the ripening of many fruits. Lipids may also be hydrolyzed and converted to sugars.

Some fruits that do not accumulate significant reserve carbohydrates (e.g., honeydew melons) will also ripen after harvest. Their internal sugar concentration is dependent upon accumulation of sugars prior to harvest; thus early harvest can greatly compromise final quality.

Changes in acidity are also important in the development of the characteristic taste in many fruits (Figure 6.11). Although there are a number of organic acids found in plants, gen-

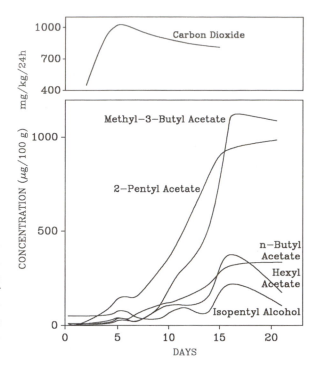

Figure 6.12. In addition to taste, perceived by the taste buds in the mouth, aromatic compounds are an extremely important component of the flavor of edible products. The synthesis of critical volatiles typically increases markedly with ripening, illustrated by changes in volatile compounds emanating from banana fruit (*redrawn from Drawert*[114]).

erally only one or two accumulate in the fruit of a species. For example, malic acid predominates in apples, bananas and cherries; citric acid in oranges, lemons and currents; citric and malic acids in tomatoes and gooseberries; and malic and tartaric acids in the grapes. During ripening, there is a decrease in organic acids in most fruits. This loss is due largely to the utilization of these compounds as respiratory substrates and as carbon skeletons for the synthesis of new compounds during ripening. The decrease in total acidity in the grape tends to coincide with the onset of ripening and the accumulation of sugars (Figure 6.10). The concentration of organic acids does not, however, decline in all fruits during ripening. In the banana, there is a significant increase in the concentration of malic acid and a decrease in pH (e.g., 5.4 to 4.5).[321]

The aroma of a fruit is an extremely important quality criterion, and as fruits ripen there is an increase in the rate of synthesis of volatile compounds (Figure 6.12). Over 400 different compounds have been identified emanating from tomato fruit,[78,79,80,81] and relatively large numbers of volatile compounds have been identified in various other fruits.[317] Typically only a small number of the total complement of volatile compounds tend to make-up the characteristic aroma we perceive for a specific fruit. In some instances, one (e.g., 3-methylbutyl acetate in banana)[38] or a small number of compounds have a dominating influence on the characteristic aroma of a fruit (character impact compounds). In other instances, it is more complex (e.g., approximately 16 compounds appear to be important in tomato flavor).[78]

Fruits vary considerably in the concentration and type of volatile compounds they produce. In the banana, odor concentrates between 65–338 ppm were collected, while the strawberry yielded only 5–10 ppm. In addition, our ability to perceive the odor of specific compounds varies widely. For example, vanillin can be sensed at a concentration of 0.000,001 mg · m^{-3} of air, while we can detect diethyl ether only when the concentration has reached 1 mg · m^{-3} or greater.

There are distinct quantitative and qualitative differences in the composition of the volatiles produced by individual fruits as they ripen. Over 85 compounds are known to emanate from ripe peaches, none of which exhibits a characteristic peach-like aroma by itself. Rather, the aroma of a ripe peach appears to be consist of γ- and δ-lactones, esters, aldehydes, benzyl alcohol and d-limonene.[111] The concentration of these compounds increases as the fruit ripens. If picked at a pre-ripe stage of development (hard-mature) and artificially ripened at room temperature, the production of these volatiles is markedly depressed. This appears to be due to the fact that some of the aromatic compounds are apparently synthesized in the leaves and translocated into the fruit.[187] Even with fruits (e.g., tomato) which ripen "normally" after detachment, the characteristic odor may be significantly diminished, although still within an acceptable range.

The complexity of volatile flavor chemistry stems from the fact that there is an exceedingly wide range of pathways,[371,380,468] precursors,[254,284,330,459] and eventual compounds involved. As a consequence, identifying the key genes involved has progressed slowly. Initial data on the effect of down-regulating genes involve in the general ripening process (e.g., polygalacturonase, 1-aminocyclopropane-1-carboxylic acid synthase, 1-aminocyclopropane-1-carboxylic acid oxidase) has indicated significant quantitative alterations in the volatiles formed.[25,26,30,319] Reports describing quantitative and qualitative alterations in precursors such as carotenoids and fatty acids have shown a significant impact on the fruits' volatile chemistry.[11,82,347,403,451] As a better understanding of the genetic control over the synthesis of volatile flavor compounds emerges, it will eventually increase our ability to positively impact flavor quality during the postharvest period.

In addition to preharvest factors,[74] stage of maturity at harvest, handling[134] and storage conditions[318] may also significantly alter the synthesis of volatiles. Fruit held in controlled atmosphere[282] and hypobaric storage are often incapable of synthesizing normal quantities of volatile flavor compounds upon removal. This does not appear to be due to an inhibition of the over-all ripening process, since with apples chlorophyll degradation and ethylene synthesis proceed normally.[172]

b. Changes in the respiration rate of certain fruits

The fruit of a number of species exhibits a spectacular increase in the rate of respiration, called the **respiratory climacteric,** during ripening. During the final stages of maturation, fruit respiration declines to a very low level (the preclimacteric minimum, see Figure 3.25 in chapter 3). This is followed by a tremendous increase in the rate of respiration (as much as 4 to 5 times) with the onset of ripening. The rate of respiration eventually peaks (climacteric peak) and then declines during the post-climacteric period. Climacteric fruits are also distinguished from fruit not exhibiting this burst in respiration (nonclimacteric) in that they increase their rate of synthesis of the phytohormone ethylene in response to exposure to low levels of the hormone (termed "autocatalytic", "autostimulatory" or "ethylene-dependent" production of ethylene).

The respiratory climacteric was first studied in detail by Kidd and West.[224] Since this early work, postharvest biologists have had a fascination with the respiratory climacteric which has translated into an impressive accumulation of research data.

The respiratory climacteric does not occur in all fruits, however. Rather, a relatively low, consistent rate of respiration is maintained in some fruits during ripening (Figure 3.25). While lacking the respiratory climacteric and ethylene-dependent synthesis of ethylene, nonclimacteric fruits obviously do ripen. While the fundamental differences in ripening between climacteric and nonclimacteric fruit are not fully understood, recent advances in molecular genetics have begun to clarify the control mechanisms in climacteric fruit.

The list of climacteric species that have been identified is longer than that of nonclimacteric (Table 6.2); however this is misleading, in that it is much easier to establish that the respiratory climacteric exists in a species than to prove its nonexistence. Thus, there are probably more nonclimacteric fruits than the existing ratio of species reported thus far.

Why has there been such an interest in the respiratory climacteric among postharvest biologists? From a scientific basis, the respiratory climacteric represents a convenient map or timetable on which other changes occurring during ripening can be placed and related to each other. Early work also indicated that environmental factors that inhibited the respiratory climacteric likewise inhibited ripening. Thus, fruit could be held in storage and induced to ripen as needed for individual studies. This allowed greatly extending the time period in which the ripening process could be studied. In addition, many economically important fruit crops (e.g., apples, pears, bananas, tomatoes) are also climacteric. Both the level of grower interest and support for research on these crops have enhanced their attractiveness as research projects.

Commercially, our ability to harvest many climacteric fruits when unripe, store them for extended periods under conditions that prevent ripening and subsequently ripen them, has greatly expanded the length of time these fruits are available fresh for consumption. Sales are no longer limited to just the harvest season and yearly consumption is similarly expanded.

The timing of the respiratory climacteric relative to optimum eating quality of the fruit varies with species. In pear fruits, the climacteric peak more or less coincides with this optimum. In the apple and banana, it occurs slightly before and in the tomato, well before optimum ripeness.

Individual species also vary in the speed in which the climacteric proceeds and the maximum rate of respiration at the climacteric peak (Figure 6.13). The climacteric occurs quite rapidly in the banana and breadfruit; at intermediate rates in apple and mango; slowly in the fig and not at all in nonclimacteric fruits, e.g., lemon, grape and strawberry. At the climacteric peak of the breadfruit and cherimoya,[47,242] respiration is proceeding at 170–180 mg CO_2 · kg^{-1}· hr^{-1}; in the mango[75] the rate is approximately 80 mg CO_2 · kg^{-1}· hr^{-1}; and the apple[400] and tomato[270] 10–20 mg CO_2 · kg^{-1}· hr^{-1}.

c. Requisites for ripening

Essential requisites for ripening include gene expression, synthesis of key enzymes, availability of respiratory substrate and energy. Ripening is a genetically programmed event involving the regulated expression of specific genes[160] that control a complex of synthetic and degradative changes (Table 6.3). Certain developmentally regulated genes are expressed at the onset of ripening, and these, in turn, lead to the expression of additional genes that regulate a host of genes essential for ripening. The majority of the gene products are enzymes that control the quantitative and qualitative alterations (texture, flavor, color) that occur during ripening. The need for *de novo* synthesis of enzymes during ripening was initially illustrated with the incorporation of radioactive amino acids into fruit proteins. Incorporation increased substantially from the pre-climacteric period to the early climacteric rise but dropped drastically at the climacteric peak.[359] More recently, antisense genes have allowed blocking the formation of specific gene products and thus assessing their role in ripening. Likewise, inhibitors of protein synthesis block softening and pigmentation changes without affecting the climacteric rise in respiration.[147] This suggests that the respiratory rise is not linked to the synthesis of new enzymes, while many of the ripening changes apparently are.

Other requisites for ripening include a source of respiratory substrate to provide the energy to drive the reactions occurring during ripening. The substrate(s) also provides carbon skeletons for many of the new compounds that are formed. We have seen that in many fruits the respiratory substrate comes from the parent plant as the fruit ripens while still attached. In others, it is recycled from stored carbon within the fruit itself.

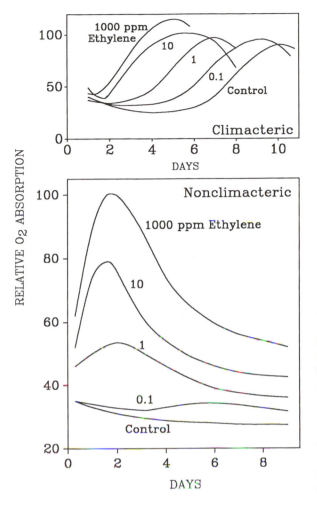

Figure 6.13. The respiratory rate of both climacteric (top) and nonclimacteric (bottom) fruit increases when exposed to an external source of ethylene; however, the response differs in several ways. For example, the increase in respiration (as indicated by oxygen absorption) of nonclimacteric fruit is concentration dependent, while increases in ethylene concentration only shift the time of the respiratory increase in climacteric fruits (*redrawn from Biale*[45]). See Table 6.4 for other differences in response of climacteric and nonclimacteric fruit to exogenous ethylene.

The importance of respiration to ripening was shown by using inhibitors of the metabolic production of energy (e.g., dinitrophenol, arsenite).[37,277] When energy production is inhibited, ripening is usually inhibited.

d. Initiation of ripening

The initiation of ripening in both climacteric and nonclimacteric fruit is developmentally controlled. Immature fruit will not ripen, and their propensity to ripen increases with advancing maturity. In climacteric fruit, ethylene is a central feature in the ripening process; however, the onset of ripening is not initiated by ethylene. Rather, specific changes at the genomic level trigger an increase in ethylene synthesis, which in turn instigates the expression of the broad cross-section of genes required for many of the changes in color, flavor, texture and other biochemical and physiological processes. A global regulator gene* that encodes ripening transcription factors is expressed well before an increase in ethylene synthesis or other ripening alterations.[440] The transcription factors, in turn, instigate the expression of ripening-specific genes. The transcrip-

---

*The *LeMSDS-RIN* gene is part of a MADS-box family of transcription factors.

tion factor gene is also expressed during the ripening of strawberry, a nonclimacteric fruit, indicating that the developmental control may be universal across all fruits, both fleshy and dried.

What then, is the role of ethylene if the events leading to the initiation of ripening are set into motion prior to an increase in ethylene synthesis? Since fruits comprise a diversity of cell types, and these cells are in varying stages of maturity, it seems unlikely that the decision to initiate ripening would be made in all of the cells simultaneously. Rather, the onset of ripening in climacteric fruit probably occurs in one area of the fruit and spreads, *via* the diffusion of ethylene through the intercellular air space to the other cells, triggering the ripening of the entire organ. Thus, while essential for ripening in climacteric fruit, the role of ethylene might be more accurately described as one of synchronizing ripening, triggering a complex and coordinated transformation.[153] A synchronizing role for ethylene is indicated by the fact that when preclimacteric bananas are wounded in one relatively small area of the fruit, increasing ethylene synthesis, ripening is often initiated.

When climacteric fruits are artificially exposed to ethylene at a sufficient concentration and length of time, the respiratory climacteric of the fruit is initiated and respiration increases, often dramatically. The fruit's sensitivity to ethylene is greatest just prior to the respiratory rise. After respiration has been stimulated, additional ethylene has no stimulatory effect. Above a threshold level of ethylene, the respiratory increase is independent of the concentration of ethylene to which the fruit is exposed. If the ambient ethylene is removed after the respiratory rise has begun, the response is irreversible and the respiratory rate continues to increase.

The initiation of ripening in nonclimacteric fruit is also developmentally regulated;[440] however, ethylene does not play a central regulatory role. Ethylene, exogenously applied, can stimulate the rate of respiration throughout the postharvest life of the tissue, but the response appears to be pharmacological. The magnitude of the response depends upon the concentration of ethylene to which the fruit is exposed (peak respiratory rates are generally proportional to the logarithm of the ethylene concentration applied). Continued high respiration is dependent upon the continuous exposure to ethylene. Removal of the exogenous ethylene results in a resumption of the respiratory rate found prior to exposure. Whether or not ethylene plays a role in the ripening of nonclimacteric fruit remains a topic of debate.[258]

Climacteric and non-climacteric fruits also differ in three additional ways (Table 6.4). Climacteric fruits exhibit ethylene-stimulated production of ethylene (ethylene-induced ethylene synthesis), while in nonclimacteric fruits it is absent. Furthermore, in nonclimacteric fruit the endogenous concentration of ethylene is generally quite low and remains so throughout the ripening period. In climacteric fruits, the internal concentration is highly variable, being quite low prior to ripening to very high during the respiratory climacteric. Finally, the onset of ripening in climacteric fruit results from the turning on of a gene for ethylene synthesis, while in nonclimacteric fruit, the onset may follow the turning off of the gene for auxin synthesis.

Table 6.4. Differences Between Climacteric and Nonclimacteric Fruits in the Synthesis of Ethylene and Their Response to Exogenous Ethylene.

	Climacteric	*Nonclimacteric*
Response to exogenous application of ethylene	Stimulates respiration only prior to respiratory rise	Stimulates respiration throughout postharvest postharvest life
Magnitude of respiration response	Independent of ethylene concentration	Dependent on ethylene concentration
Reversibility of ethylene mediated respiratory rise	Irreversible	Reversible, dependent on continued exposure
Autocatalytic production ethylene	Present	Absent
Endogenous concentration of ethylene	Highly variable, ranging from low to very high	Low

In addition to questions concerning the fundamental differences in control over the ripening process between climacteric and nonclimacteric fruit, a myriad of evolutionary questions arise. For example, which ripening method represents the more primitive form? What evolutionary advantage, if any, does one form have over the other? What selective pressures lead to the development of the more recent form? Both climacteric and nonclimacteric fruit are confronted with the same problem—how to synchronize the ripening of the diverse cross-section of cells within the fruit to yield a uniformly ripe product. Synchronized ripening maximizes the fruit's attractiveness to herbivores, facilitating dispersal of the seed away from the parent plant.

i. Climacteric fruit

It has been known for a number of years that exogenous application of ethylene to climacteric fruit would induce their ripening.[232] In fact, in the mid-1940's a book describing the use of ethylene to ripen tomatoes, a climacteric fruit, was published.[183] The discovery of ethylene as the causal agent led from the observation that fruits held in rooms warmed by oil heaters ripened much earlier than those not similarly exposed.* Sufficient ethylene was given-off in the exhaust fumes to initiate the ripening process. Later research established that ethylene also emanated from the fruits themselves; the rate of production being much greater in ripe than non-ripe fruits.[49,234] Hence, ethylene emanating from ripe fruits would initiate ripening of unripe fruits if they were stored together. Nonclimacteric fruits also produce ethylene, however, only a very low background level of ethylene and their response to the hormone differs from climacteric fruits (Figure 6.13). While exposure of the fruit to ethylene, does increase respiration,† it does not initiate the ripening process in nonclimacteric fruits. Other differences between the two fruit types center on their respiratory response to applied ethylene and their synthesis of the hormone (Table 6.4).

Ethylene is synthesized in and evolves from the cells of both climacteric and nonclimacteric fruit throughout their growth and development. During the ripening phase of climacteric fruit, ethylene appears to assume a much more dominant regulatory role. This ethylene associated with ripening appears to represent a separate system from the normal background levels of ethylene synthesized by the plant. The two sources of ethylene have been designated system 1 and system 2.[291] System 1 ethylene, found throughout development in both climacteric and nonclimacteric fruits, is responsible for the low basal and wound-induced ethylene production. System 2 ethylene is activated in climacteric fruit during ripening and is responsible for the upsurge in ethylene production. The synthesis pathway is identical between system 1 and system 2 ethylene; however, control over the expression of the key rate limiting enzyme in the pathway [1-aminocyclopropane-1-carboxylic acid synthase (ACC synthase)] differs. The gene family is composed of at least nine copies of genes for ACC synthase,[475] which differ in the location within the plant where they are expressed and in what controls their expression. System 1 ethylene is inhibited by ethylene (autoinhibition) which maintains the internal concentration at a low level, whereas system 2 ethylene is stimulated by ethylene (autostimulation). As the ripening of climacteric fruit switches on and system 2 ethylene synthesis is engaged, a gene‡ is activated that is responsible for the initiation and maintenance of ethylene-dependent synthesis.[27]

Thus ethylene sets in motion a diverse range of ripening-associated alterations that result in the quality attributes associated with a ripe fruit. The key role of ethylene in triggering the cascade

---

*For a review of the chronological sequence of ethylene use in agriculture, see Kays and Beaudry.[212]
†With the exception of cherry.[262]
‡(*LEACS2* in tomato).

of alterations in climacteric fruit has been demonstrated using antisense genes for critical steps in the ethylene synthesis pathway. In melon fruit, inhibition of ethylene synthesis completely blocks fruit softening;[164] however, not all changes in climacteric fruit appear to be controlled by ethylene. For example, certain pigments can be either ethylene-dependent or ethylene-independent. In tomato, chlorophyll degradation is not inhibited by reduced ethylene synthesis while lycopene synthesis is thoroughly retarded.[227,304,319] In melons, however, carotenoid accumulation in the flesh does not appear to be regulated by ethylene,[24,164] while chlorophyll degradation in the rind is dependant upon ethylene.[209] The relationship between ethylene and alterations in texture are less clear. Expression of two genes* coding for endo-β-1,4-glucanase is regulated by ethylene;[252] however, for endopolygalacturonase the importance of ethylene is less definitive.[415]

The application of ethylene to a number of climacteric fruits is known to initiate the early onset of ripening. The minimum concentration of ethylene and the length of exposure required to accelerate ripening varies with different fruits. For example, $10 \, \mu L \cdot L^{-1}$ of ethylene is required for the induction of ripening in the avocado while only $1 \, \mu L \cdot L^{-1}$ is sufficient for the banana.[45] Likewise, the internal concentration of ethylene at the onset of normal ripening is known to vary (e.g., $0.44 \, \mu L \cdot L^{-1}$ in persimmon,[356] 0.1 to $1.0 \, \mu L \cdot L^{-1}$ in banana,[44] and $3.0 \, \mu L \cdot L^{-1}$ in honeydew melons[341]).

In addition to variation between types of fruits for the minimum concentration of ethylene required to initiate ripening, fruits of the same cultivar change in their sensitivity to ethylene as they approach the time of natural ripening. In Anjou pears, the minimum concentration and time required to initiate ripening decreased from a 14-day exposure of $1 \, \mu L$ of ethylene $\cdot L^{-1}$ of air in fruit that are 57% mature, to $0.5 \, \mu L \cdot L^{-1}$ for 15 days when 71% mature, to $0.2 \, \mu L \cdot L^{-1}$ for 17 days when 86% mature.[449] The resistance of young fruit to ripening by exogenous ethylene has been documented in a number of other species.

If ethylene does, in fact, synchronize ripening in climacteric fruits, it was initially felt that the internal concentration of ethylene should increase prior to the first signs of ripening within the fruit (generally the onset of the respiratory climacteric). In some species this is the case. The synthesis of ethylene by banana (Figure 6.14),[75] cantaloupe,[269] avocado[75] and honeydew melon[156] precedes the onset of the respiratory climacteric. In the mango[75] and feijoa,[356] however, the increase in ethylene synthesis occurs after the initial onset of the respiratory climacteric. In these fruits the concentration of ethylene already present appears to be sufficient to stimulate ripening; therefore an increase in internal ethylene need not necessarily precede the climacteric. The timing of the onset of ripening in these fruits must then be controlled by factors other than an increase in ethylene synthesis. A change in the number of ethylene binding sites as the fruit approaches the time to ripen,[469] increasing the cells' sensitivity to existing ethylene or a decline in factors that impede the onset of ripening may be involved.

It is important to note that while ethylene triggers a cascade of gene expression, the products of which are responsible for controlling a cross-section of alterations in the fruit, ethylene does not control all of the changes associated with ripening. Certain ethylene-independent alterations appear to be controlled by other mechanisms.

ii. Nonclimacteric fruit

As with climacteric fruit, the ripening of nonclimacteric fruit is developmentally regulated,[440] the onset of which instigates specific changes in gene expression and cellular metabolism that lead to alterations in firmness, texture, coloration and flavor. In contrast to climacteric fruit where ethylene plays a central role in ripening, nonclimacteric fruit do not display an increase in respiration or ethylene synthesis during ripening, nor is autocatalytic ethylene synthesis

---

*(Cel1 and Cel2)

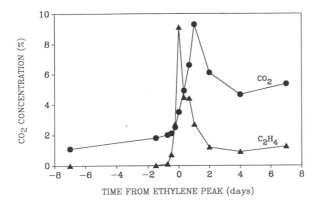

Figure 6.14. Changes in the rate of ethylene synthesis precede the onset of the respiratory climacteric in some fruits, as illustrated above in banana (*redrawn from Beaudry et al.*[33]). In other climacteric fruit (e.g., mango and feiyoa), however, the increase in ethylene synthesis may occur after the increase in respiration.

present. While far less in known about the control of ripening in nonclimacteric fruit, evidence to date indicates that other phytohormones, in particular auxin, may play a central role. Of the nonclimacteric fruits (Table 6.2), the ripening of strawberries and grapes has been the most widely studied, though collectively the research on these two crops represents only a small fraction of that published on apples. Both strawberries and grapes produce relatively low levels of ethylene during ripening[8,235] and exposure of strawberries to high levels of exogenous ethylene has no apparent effect on the ripening process.

The onset of ripening in the fruit of both species appears to be regulated by auxin, a decline in which instigates the process.[103,155,273] Auxin is synthesized in the achenes of the strawberry and their surgical removal results in a decline in auxin in the underlying fruit tissue while instigating anthocyanin accumulation and a decline in chlorophyll and firmness. A central role for auxin is also supported by the fact that the application of synthetic auxin inhibits ripening and the expression of genes associated with ripening.[171,273] Therefore, auxin acts as a negative regulator, and its decline triggers ripening and the derepression of a cross-section of ripening-related genes. However, not all ripening-associated genes are coordinated by auxin.[89] Thus other signals (e.g., abscisic acid, sugar status) may be operative in nonclimacteric fruits.[104,139]

In many nonclimacteric fruits, ripening occurs over a relatively long time interval. For example, in grapes alterations associated with ripening often occur over an extended period (e.g., 8 weeks)[139] with the initial inception of ripening occurring quite rapidly (1–2 days).[94] In contrast, the signal controlling the synchronization of ripening in strawberries appears to propagate from cell to cell and at a much slower rate than the diffusion of ethylene in climacteric fruit. This can be seen in the sequential development of pigment synthesis, which often begins in one location and propagates outward into unpigmented areas. In both climacteric and nonclimacteric fruit, the control over ripening is no doubt much more complex that simply altering the synthesis of the two respective phytohormones. In both instances alterations occur in the fruit that are independent of ethylene or auxin.

1.3.2. Dried Fruits*

Like their fleshy counterparts, dried fruits such as grains, legumes and nuts also have a segment of their development which is commonly, although not universally, referred to as the ripening phase. Dried fruits differ in that while the fruit's accessory tissues undergo changes in

---

*The term "dried fruit" is used here to denote grains, legumes and nuts rather than dried fleshy fruits such as raisins or prunes.

some ways similar to those in fleshy fruits during ripening (e.g., alterations in pigmentation), the focus in dried fruit is almost exclusively on changes that occur in the seed rather that the accessory tissues. Therefore, in dried fruit there are ripening changes that involve the entire fruit and, more importantly, ripening changes within the individual seeds.

While the fruit undergoes ripening changes that lead to the eventual senescence of the accessory tissues, changes in the seed do not lead to senescence; therefore, the biochemical and physiological alterations there are distinctly different from those in the accessory tissues. Ripening within the seed involves changes that occur during the latter part of development that enhance the seed's reproductive potential, which are the focus of the following section, along with the acquisition of storage compounds essential for a viable reproductive propagule.

The accessory tissue, such as the pericarp (e.g., pod wall of the soybean), has a cross-section of functions. In the case of legumes, the pericarp encloses and protects the developing seeds, photosynthetically fixes carbon that is utilized by the pod and developing seed,[99,225] accumulates nutrients which are remobilized during the seed filling stage,[404,418] and facilitaties seed dispersal.[419] So there is a close relationship between the seed and accessory tissue that involves a level of control essential for synchronizing the events during development and ripening. The cross-communication between the developing seeds and their surrounding tissues is just starting to be elucidated; however, the presence of an interaction has long been apparent. Pecan kernels, for instance, trigger the dehiscence and release of the nut from the shuck in which it is enclosed through the enhanced biosynthesis of the ethylene.[264] In canola, the silique wall may contribute 50–60% of the dry matter that is imported into the embryo during seed-filling.[260] Likewise, in field peas there is a coordinated remobilization of dry matter and nitrogen (~50%) during the developmental period in which a linear increase in dry seed weight occurs.[142]

The biochemical and physiological changes associated with ripening within seeds generally begin toward the end of the logarithmic period of dry weight accumulation. Unfortunately, in seed biology the terminology utilized for seed development and maturation is not universal. Whether or not the term "ripening" is utilized, and, if used, the time period assigned in the overall development of an individual seed, depends upon the species and the individual investigator. In some instance, ripening changes are simply considered as part of the overall maturation and desiccation stages. Even in instances when the term "ripening" is not used, the terminology is further complicated by the general use of the term "after-ripening" for the chemical and physical changes occurring in mature seeds of certain species that fulfill a dormancy requirement.[204]

a. Seed development

Seeds generally contain a fully formed embryo and extra-embryonic tissues that are desiccated and in a quiescent state at maturity. While it is advantageous to divide seed development prior to germination into distinct stages during which specific developmental events predominate, a cross-section of stages have been proposed without any one system being universally accepted (e.g., formative—storage—ripening;[436] embryo morphogenesis—seed maturation—desiccation;[169] cotyledon—embryo maturation—post-abscission—desiccation[191]). A number of the terms used for developmental stages are very general (e.g., maturation, embryogenesis, late-embryogenesis) and lack the specificity needed to ascertain when in the overall developmental process they apply. As a consequence, in the following discussion, seed development is separated into five general stages: (1) fertilization, (2) cell division and elongation, (3) seed-filling, (4) ripening and (5) germination; these are utilized because they temporally correlate relatively well with morphological, biochemical, physiological and gene expression patterns (Figure 6.15A).

Stages of Seed Development

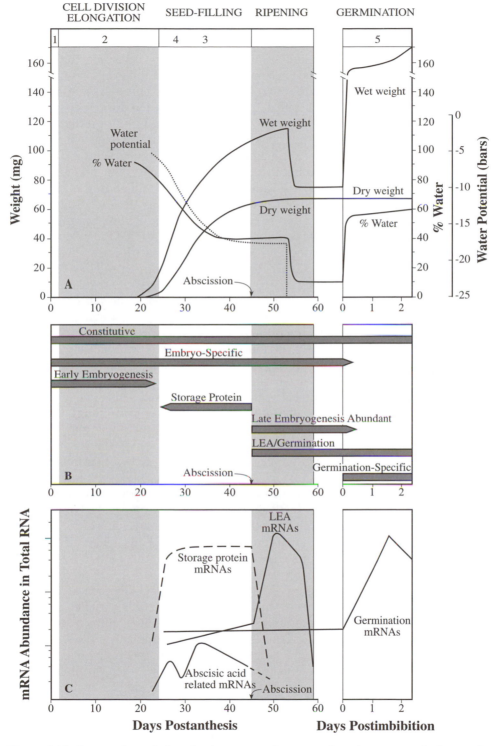

Figure 6.15. Stages of cotyledon development and expressed against temporal changes in (A) composition, (B) general classes of mRNA and (C) the relative abundance of selected mRNAs of *Gossypium hirsutum* L. (*after Dure,*[119] *Galau et al.,*[148] *and Hughes and Galau*[191]).

The initial stage (1) involves fertilization and the formation of a single-celled zygote. Once fertilization has taken place, there is a tremendous increase in synthetic activity as many of the structural components of the seed are formed (stage 2). This is a period of rapid cell division, cell elongation and organelle biogenesis. At this time an excised embryo, held under the appropriate conditions, will readily germinate; however, later in the developmental sequence it is inhibited. The acquisition of embryo dormancy is a means of inhibiting premature germination of the seed while it is attached to the parent plant (**vivipary**). Through the study of mutants, abscisic acid had been shown to be a critical factor in this acquired dormancy (see 'inhibition of germination' below). Stage 2 is followed by a period in which photosynthates are translocated into the young cells and reserve forms of carbon and nitrogen (e.g., starch, lipids, storage proteins) are synthesized. The seed-filling period (stage 3) is characterized by a rapid increase in dry weight and a gradual decline in seed moisture content (Figure 6.15A). During this time, the seed also begins to develop desiccation tolerance that will allow it to dry at maturity without dying. Seed-filling is followed by ripening (stage 4) where the seed progresses through a series of physical and chemical alterations essential for the production of a propagule capable of germination and growth. Ripening is characterized by changes that are independent of carbon loading into the seed and, under normal conditions, begins with ovule abscission, though the ripening process can be initiated earlier with certain treatments. Thus seed-filling and ripening can be uncoupled, indicating control by at least two separate regulatory systems. At the onset of ripening, the fresh weight of the seed is often at its peak and subsequently declines. With ripening, there is a pronounced shift in the pattern of protein synthesis. Storage protein synthesis has been terminated, however, proteins involved in ripening, conferring desiccation tolerance and those needed for eventual germination, increase. The imbibition of water (stage 5) triggers the onset of germination during which germination-specific proteins are synthesized, which instigate recycling of stored carbon and the development of new growth.

In some instances, the transition between stages may be well delineated (e.g., seed-filling → ripening) while in others (e.g., cell division and elongation → seed-filling), there is no precise point at which one stops and the next begins. In the latter, the stages may overlap as one stage declines in activity and the subsequent one proceeds. This is in part due to a gradual transition in developmental processes within the cells, but also to the fact that not all of the cells within a seed are identical; they may vary in type, age and degree of development. In addition, different aspects of development (e.g., dormancy, desiccation tolerance, storage protein synthesis) appear to be controlled by distinctly different regulatory programs. For example, morphogenesis is arrested shortly after the onset of the seed-filling stage; however, seed-filling can be initiated independently of the termination of cell division. Thus, while there is an overall coordination of the developmental processes occurring from fertilization to final maturity, no single master regulatory control appears to exist.

Variation in the synchronization of development can be readily seen at the intact fruit level in fruits that have multiple seeds in which the individual seeds vary in age.[125] Typically, seeds at the base of the ear or pod are the least mature,* and their ripening also lags behind.[22]

The focus in this section is on the final stages in the development of the seed, seed-filling and ripening. The length of these stages varies considerably among individual species and in some cases may differ significantly within a species. For example, when germination potential is used as a measure of the completion of the ripening process, seed of wheat cultivars 'Holdfast' and 'Atle' of the same age and weight differ considerably.[455] Seed of 'Holdfast' harvested 7 weeks after anthesis had 80–90% germination, while similar seed of 'Atle' had only 7–10% germination.

---

*The opposite is true in maize (corn).

While it has been common to delineate the end of the ripening period and hence matura-tion of the seed as occurring at harvest, this distinction in some instances is not warranted.* In certain species, biochemical and physical changes ("after-ripening") remain that are es-sential to produce a seed capable of germination. These alterations are not complete when the grower decides it is time for harvest and the seed is considered dormant.[123] After-ripening may represent the final stages of the ripening process that in these species has not been completed.

b. Regulation of seed development and ripening

Seed development, from fertilization through ripening, requires a system of coordination that interfaces environmental, parent plant, fruit and individual seed factors. Expression of the thousands of genes that encode proteins operative during seed development is subject to com-plex patterns. Regulatory control appears to involve a number of distinct programs in which temporal and spatial patterns in gene expression are activated. Genes are subject to both on/off and up/down regulation, and the predominate mechanism for controlling expression is at the transcription level. Transcription factors activate or repress transcription, often through binding to DNA at specific locations, with a single transcription factor or transcription fac-tor complex, in some cases, controlling the transcription of more than one gene. Transcription factors, in turn, are also subject to a variety of regulatory controls.[388] How the signaling sys-tem for developmental and other clues are structured and integrated to effect gene expression during seed development is just starting to be elucidated.

It is estimated that between 12,000 and 18,000 genes are expressed during seed develop-ment, with a majority of the mRNAs being formed over the entire developmental period (con-stitutive). Therefore, shifts in metabolism during development (e.g., the transition from seed-filling to ripening) are thought to be controlled by a relatively small number of genes.[156] Seven distinct patterns of mRNA accumulation during seed development have been identified, sev-eral of which correspond to specific categories of gene products (Figure 6.15B,C).[122,119] Type 1 represents constitutively produced mRNAs—those that are present throughout seed devel-opment and subsequent germination. Type 2 mRNAs are produced during the entire devel-opmental period of the seed but terminate prior to germination. Types 3–6 are expressed only during segments of the developmental period [e.g., 3—during the cell division and elongation phase (early embryogenesis); 4—mRNAs that predominate during the seed-filling phase, con-trolling the synthesis of storage proteins and proteins needed for the synthesis of storage car-bohydrates and lipids; 5—late embryogenesis abundant (LEA) mRNAs that are expressed during the ripening phase, some of the proteins of which are thought to be involved in confer-ring desiccation tolerance to the seed; and 6—LEA and mRNAs needed for the eventual ger-mination of the seed]. The final type (7) are germination-specific mRNAs that are not tran-scribed until the imbibition of water, triggering the onset of germination.

The simultaneous monitoring of approximately 3,500 genes during seed-filling using mi-croarray technology has allowed characterizing the relative abundance of mRNAs for specific proteins (Figure 6.16A), giving an indication of the degree of regulatory coordination among and within pathways.[374] Only about 35% of the *Arabidopsis* genes display significant changes (i.e., ≥ twofold) during seed-filling; the majority are expressed constitutively. Genes linked to the biosynthesis of storage lipids do not display closely coordinated regulation (a similar situ-ation is found for storage proteins). For example, certain lipid synthesis mRNAs increase early in the seed-filling stage and then decline markedly (e.g., oleate desaturase, Figure 6.16B); oth-

---

*A similar situation can be seen in fleshy fruits, especially so for climacteric species that almost exclusively in-stigate or continue, depending upon stage of maturity at harvest, their ripening process after harvest.

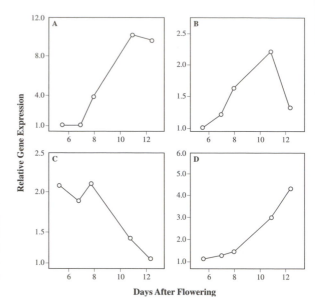

Figure 6.16. Temporal microarray mRNA expression profiles during maximum seed-filling (8–13 days after flowering) for (A) 12s storage protein, (B) oleate desaturate, (C) chloroplast glucose-6-phosphate transporter and (D) sucrose synthase in *Arabidopsis* cotyledons (*after Ruuska et al.*[374]).

ers reach a maximum toward the latter part of seed-filling; and the rest remain essentially unchanged, indicating that distinctly different mechanisms regulate their expression.

Likewise, during this time period, carbon metabolism genes display a cross-section of temporal responses in mRNA accumulation. Chloroplast glucose-6-phosphate transporter mRNA declines progressively over the seed-filling period (Figure 6.16C) while sucrose synthase, which cleaves sucrose upon its entry into the cell, providing carbon skeletons for the synthesis of storage compounds, increases markedly over the same interval (Figure 6.16D). Messenger RNAs associated with photosynthesis increases rapidly and then declines during the latter part of seed-filling. Many seeds contain chloroplasts, and photosynthesis therein may be an important source of the oxygen and energy required for seed development and the refixation of respiratory carbon.[362] Likewise, genes for an amino acid transporter are up-regulated prior to the appearance of storage protein transcripts.[298] Collectively, the data indicate that control over gene expression is much more complex than might initially be supposed.

It is important to note that gene expression can vary substantially over very short distances within the seed. For example, the epidermis of the embryo differentiates into a transfer cell layer containing sucrose and other transporters that mediate nutrient uptake during seed-filling.[60] In loss-of-function mutants for the sucrose transporter, embryo development is severely compromised. It is essential, therefore, to monitor gene expression within uniform regions of cells to have an accurate assessment of temporal changes occurring.

It is evident from the temporal expression of mRNAs that distinct sets of genes are expressed at different stages in the development of the seed. The factors that coordinate and control the expression of both regulatory and structural genes are just starting to be identified.[460] For example, abscisic acid is thought to play a significant role in certain aspects of seed development. Three genes identified thus far (*ABI3, FUS3* and *LEC1*) affect a wide spectrum of seed characteristics and appear to be of critical importance in regulation of development. They act synergistically, possibly as transcription factors, regulating developmental processes.[460] Their expression patterns broadly overlap during seed-filling and ripening.

Abscisic acid is thought to play a substantial role in seed development; however, the expression of seed protein genes can occur in the absence of abscisic acid, indicating that it is not the only factor involved in initiating and maintaining this phase of development. As the abscisic acid signal transduction pathway is further identified,[295] its interaction with other systems modulating gene expression will be ascertained more clearly.

c. Morphological, chemical and physiological alterations during seed filling and ripening

i. Inhibition of germination

The ability to germinate is an essential requisite for a viable seed, thus seeds possess a series of genes responsible for the onset and development of germination. As a dried, non-dormant seed imbibes water, the high moisture environment triggers the expression of genes essential for germination (Figure 6.15). However, early in the development of the seed, while attached to the parent plant, the embryo is also in a high moisture environment conducive to germination. At this time, immature seeds of a number of species can be surgically removed and, if placed under the appropriate environmental conditions, the embryo will germinate and develop into a normal seedling.[218] To prevent premature germination during development, plants have evolved dormancy mechanisms to inhibit precocious germination; these may also be operative in mature seeds.[40]

The central role abscisic acid plays in suppressing germination during seed development has been established using plants possessing gene mutations for abscisic acid synthesis[238,241,305,361] and perception[240,305] and temporal changes in the abundance of abscisic mRNAs (Figure 6.15C) in relation to dormancy. When either synthesis or perception of the phytohormone is inhibited, the seeds germinate within the pod or ear midway through their development. Dormancy is instigated by abscisic acid formed in the embryo itself rather than within maternal tissues since neither maternally derived nor applied abscisic acid is able to induce dormancy.[216,237] Gibberellic acid may also be involved, acting as a positive stimulus for germination which abscisic acid negates early in the seed's development.[239,456]

ii. Seed-filling

The primary forms of storage reserves in seeds are proteins, carbohydrates and lipids. The most abundant seed proteins are sequestered in storage protein vacuoles or protein bodies. Lipids, primarily in the form of triacylglycerols, are stored in oil bodies. A variety of storage carbohydrates, ranging from starch which accumulates in amyloplasts to the hemicellulose, mannans and xyloglucans which accumulate in the cell walls are also formed in the developing seed. These reserves are synthesized largely during the seed-filling period and accumulate primarily in either the endosperm or the embryonic cotyledons. While the seed of all species will contain carbohydrates, storage proteins and lipids, some species preferentially accumulate substantially more of one type of storage reserve. Stored reserves represent essential sources of carbon and nitrogen that is recycled and utilized during germination and early seedling development.

1. Proteins

Plants provide over 70% of the protein consumed by humans, the majority of which comes from the seed of cereals and legumes (i.e., 8–15% and up to 40% of the total dry weight, respectively). Seed proteins, which include storage, late embryogenesis abundant and other proteins, some with roles in plant defense such as protease inhibitors, accumulate to high levels in developing seeds. Typically the bulk of the protein component is comprised of a relatively small number of individual proteins, collectively referred to as storage proteins.[17,181,182] Storage proteins are "any protein accumulated in significant quantities in the developing seed which on germination is rapidly hydrolyzed to provide a source of reduced nitrogen for the early stages of seedling growth."[179]

Storage proteins are separated into four very general classes (albumins, globulins, prolamins and glutelins) based upon their solubility. In cereals, prolamins are the primary form of storage protein, and individual prolamins within various cereal species are given trivial names (e.g., hordein in barley, avenin in oats, zein in maize, gluten in wheat). Prolamins are the predominant protein form in cereals, as are globulins in legumes, though other forms may be present in each.* For example, globulins contribute about 10–20% of the total storage protein in maize.[446]

Seed proteins accumulate largely during the seed-filling period.[34] In soybeans, two storage proteins (conglycinin and glycinin) constitute approximately 70% of the seed protein at maturity and 30 to 40% of the total seed weight. There are temporal differences in the accumulation of the protein fractions during development.[102] Messenger RNAs for glycinin and conglycinin can be first detected 14–18 days after pollination.[293] However, during the final developmental period (ripening) these mRNAs are no longer detected, indicating a cessation of storage protein synthesis. A similar decline in storage protein synthesis is seen in castor bean[222] and most other seeds.

Storage proteins are synthesized on polysomes attached to the rough endoplasmic reticulum and subsequently stored in numerous small protein bodies or storage protein vacuoles. In dicots, protein bodies are formed directly from the endoplasmic reticulum, while sequestering in storage protein vacuoles involves the movement of the proteins from the endoplasmic reticulum to the Golgi and then to the vacuole.[88] After initial assembly, the proteins typically undergo a broad cross-section of structural modifications, which can occur in the endoplasmic reticulum, Golgi or storage protein vacuole. In grain of starch-storing monocots, however, storage proteins formed during the latter period of seed-filling (stage 3) are assembled in the endoplasmic reticulum, segments of which form protein bodies, bypassing the Golgi apparatus and vacuole.

Seed protein gene expression is regulated largely by transcriptionally controlled processes such as DNA binding proteins that may directly activate genes or function as transcription factors controlling the expression of a number of genes. A number of different DNA elements and their associated factors act in combination to control the transcriptional activity of a gene. Different seed protein genes appear to be regulated by distinct sets of factors with multiple factors being involved in controlling gene expression.

Protease inhibitors also accumulate during seed-filling, and in a number of species they may reach 5–10% of the total seed protein.[375] They are thought to have both regulatory (e.g., repressing protease activity in the storage organ) and protective roles (e.g., against herbivores). They may also be involved in protecting enzymes during dehydration of the seed.[247] Regulation of protease inhibitor expression is thought to be at the transcriptional level[135] or by a differentially regulated multigene family.

While storage protein synthesis terminates with ovule abscission and the onset of ripening, other proteins, presumably those required for ripening and subsequent germination, begin to be synthesized.[221] The decline in abundance of the 12s storage protein mRNA late in the seed-filling stage of *Arabidopsis* cotyledons is indicative of the approaching transition to ripening (Figure 6.16A).

2. Carbohydrates

Deposition of reserve carbohydrates, typically as starch, begins in earnest with the onset of the seed-filling period. Starch synthesis is a crucial process in many seeds in that carbohydrates

---

*The amino acid sequences for a number of globulin storage proteins have been delineated. Existing evidence indicates that all of the globulin storage proteins in flowering plants descended from two genes that existed at the beginning of angiosperm evolution.[62]

represent the primary source of energy and carbon skeletons utilized by the developing seedling upon germination. In starch storing seeds, this form of carbohydrate may make up as much as two-thirds of the final seed dry weight.

Starch is comprised of amylose and amylopectin molecules that are synthesized primarily from sucrose translocated into the developing embryo from adjacent photosynthesizing tissues and carbon recycled from temporary storage sites. The overall process is relatively complex, involving at least 21 enzymes[390] and is controlled by both structural and regulatory genes. Gene expression for starch synthesis and deposition varies during the development of the seed, indicating that not all genes are coordinately regulated. Likewise, there are significant changes in the activity of specific enzymes controlling carbohydrate interconversion during the latter stages of development. In soybean seed, for example, galactinol synthase activity increases, often peaking around the time the seed reaches its maximum dry weight[382] and appears to be important in the conversion of sucrose to raffinose and stachyose.

The level of sucrose translocated into the embryo is thought to be an important signal triggering the expression of certain genes required for carbohydrate synthesis and deposition.[271] This is supported by the fact that decreasing the sucrose levels in the seed *via* the enhanced expression of invertase (which cleaves sucrose into glucose and fructose) impairs carbohydrate storage.[309]

Starch synthesis occurs initially in proplastids which, with granule development, mature into amyloplasts. Biosynthesis involves several starch synthases and starch branching enzymes found within the plastid.[108] As the starch grains develop, they exhibit concentric rings that form from the interior outward. Amyloplast size varies; however, they are typically separated into two general size categories. In cereals, the smaller amyloplasts are numerically greater (~10×), but comprise only about 30% of the total starch. Variation in amyloplast size facilitates tighter packing within the starchy endosperm.

Other carbohydrates such as glactomannans, which are commonly accumulated by leguminous seeds, may be deposited in the cell wall region.[355] In the field pea, for example, hemicellulose in the cell wall can comprise up to 40% of the seeds final dry weight. The hemicelluloses are largely glactomannans which are synthesized from sucrose entering the seed. Deposition occurs first in the cell walls adjacent to the embryo and with time, progresses outward toward the periphery of the endosperm.[124]

3. Lipids

Oleogenic seeds contain carbon stored primarily as lipids such as triacylglycerols, which are synthesized during the seed-filling stage. Synthesis involves the production of the glycerol backbone, formation of the fatty acids and then their esterification to the glycerol molecule. Sucrose translocated into the seed serves as the carbon source, which is initially converted to acetyl-CoA. The fatty acid composition in oil-seed crops varies among species; thus there are distinct genetic controls over the relative amounts of the fatty acids formed and available for incorporation into triacylglycerols.

In crambe, there is little lipid synthesis during the early stages of development, however, with seed-filling (between 8th and 30th day after flowering), triacylglycerols begin to increase along with changes in their fatty acid composition.[19] Initially erucic acid is very low but increases markedly after day 15 and by ripening becomes the predominate fatty acid form. Lipid composition is strongly modulated by genotype and environmental conditions during formation (e.g., light, temperature).[425] The deposition of lipids in seeds typically follows a sigmoid pattern,[436] plateauing by ovule abscission and the onset of ripening.

Reserve triacylglycerols are sequestered in oil bodies that originate in the endoplasmic reticulum. The lipids initially begin to accumulate between the two layers of the membrane, and as the vesicle increases in size, it separates, either partially or completely. Formation between the two membrane layers of the endoplasmic reticulum causes the mature lipid body to be surrounded by a single (half-unit) membrane. The single membrane creates a potential problem in that it would allow other oil bodies to readily coalesce into a single large body. The presence of a class of low molecular weight proteins, oleosins, incorporated within the membrane appear to be important in stabilizing the structure. They give the membrane surface a negative charge so that adjacent oil bodies repel each other rather than coalesce.

iii. Synthesis of flavor precursors and compounds

Many staple seed crops have a relatively low level of flavor compounds that contribute to the taste and aroma of the product, a significant portion of which are not formed until cooking. Crops such as wheat, corn and rice have low flavor intensities, which is a distinct advantage in that they often are used as flavor carriers (i.e., staples to which flavors from other sources are added during preparation or before consumption). This is true for most grains, although in Asia, rice is commonly eaten without flavor augmentation. In a number of countries, however, the addition of flavors to rice is common.

In many instances, the characteristic aroma volatiles in seeds are not present in the raw product but are derived *via* thermal reactions that occur during cooking. A diverse range of volatile compounds are formed, a number of which may contribute to the aroma. In rice, for example, over 100 compounds contribute to the aroma, some of which have a positive (e.g., 2-acetyl-1-pyrroline) and others a negative (rancid odors) impact on consumer acceptance.[39,81] 2-Acetyl-1-pyrroline is also a major contributor to the characteristic baked crust aroma of wheat breads[162,386] that is formed through the reaction of the amino acid proline with 2-oxopropanal, a thermal degradation product of sugars. In rice, however, the chemical is an endogenous component, found in uncooked grains, leaves and other plant parts and is not formed during cooking.[470]

Depending upon the species, nut crops may be consumed either raw or cooked in some manner (e.g., roasted). The aroma of raw pecan kernels comes from a series of alcohols, aldehydes and one lactone;[373] however, upon roasting, pyrazines form the backbone of the characteristic aroma.[452]

A wide range of compounds can act as precursors for the flavors formed during thermal reactions.[215] In addition, age of the seed, postharvest conditions and the method of preparation can have a substantial impact on the flavor formed. Due to the large number of possible compounds and the complexity of flavor formation, the deposition of flavor compounds or their precursors during seed development has been little studied in seed crops consumed by humans. In rice, where there are aromatic and nonaromatic cultivars, the aromatic trait is controlled by either 1 or 2 recessive genes.[113] The focus, thus far, has been predominately to breed for qualitative or quantitative differences in specific compounds (e.g., certain unsaturated fatty acids). As a consequence, little is known about the genetic regulation of the synthesis of aroma compounds in rice and most cereals.

iv. Morphological alterations during seed ripening

As lima bean seeds approach the ripening stage, the cotyledons, consisting largely of parenchyma cells, are well vacuolated, containing chloroplasts with well developed grana, starch grains and polysomes associated with the endoplasmic reticulum (Figure 6.17A).* During ripening the cotyledons undergo significant changes in chloroplast structure, with the internal membranes being dismantled and the grana disappearing.[279] The loss of polysomes

---

*Klien and Pollock[228] term what is herein referred to as the seed-filling stage as "ripening" and ripening as the "maturation stage".

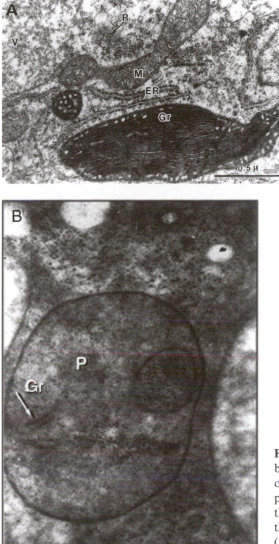

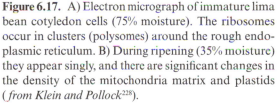

Figure 6.17. A) Electron micrograph of immature lima bean cotyledon cells (75% moisture). The ribosomes occur in clusters (polysomes) around the rough endoplasmic reticulum. B) During ripening (35% moisture) they appear singly, and there are significant changes in the density of the mitochondria matrix and plastids (*from Klein and Pollock*[228]).

(Figure 6.17B) during this period appears to coincide with a sharp decline in protein synthesis (largely storage proteins) with the onset of ripening. Similarly, there is a disappearance of Golgi bodies. For many seeds, chlorophyll degradation occurs during the ripening period. For example, the concentration of total chlorophyll declines rapidly as soybean seeds approach 60–70% moisture,[413] and this tends to coincide with a rapid decline in seed respiratory rate (Figure 6.18). In pecan kernels, there is a linear logarithmic decline in respiratory rate with seed moisture content (between 13 and 3%).[32] For many seeds there is a distinct transition to a low level of respiration around the time that seed moisture content is typically considered safe for harvest.

v. Acquisition of desiccation tolerance

Desiccation is one of the most visibly pronounced terminal events in the development of seeds, transforming the embryo into a state of metabolic quiescence. Seeds seldom are able to survive desiccation during the cell division, elongation and early seed-filling stages; however, tol-

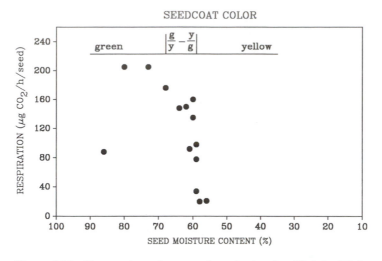

Figure 6.18. Changes in soybean seed respiration (cv. 'Fiskeky V') in relation to changes in moisture content and seed coat color (*redrawn from Tekrony et al.*[413]).

erance to desiccation is acquired as the seed matures.[41] Seeds are grouped into two very general classes, orthodox and recalcitrant, based upon their ability to withstand desiccation. The moisture content of **orthodox** seeds may decline to 5–10% (fwt) during ripening on the parent plant and may be further reduced to 1–5% without damage if adequate tolerance has been achieved. **Recalcitrant** species, however, are not capable of withstanding drying during development or after shedding. To remain viable they must maintain a high moisture content.

The acquisition of desiccation tolerance in orthodox seeds generally occurs well before the onset of drying during the ripening stage.[219,220] Immature seed can actually survive drying long before the normal decline in water content if they are desiccated slowly.[9,220] Slow drying allows the seed to instigate the necessary physical and chemical alterations required for tolerance. Mature seed that have acquired desiccation tolerance, however, are able to survive rapid drying. Thus, the degree of desiccation tolerance increases as the seed matures. How do orthodox seed survive extreme water loss and, upon imbibition, resume normal metabolic activity? Two changes in the seed's metabolism instigated early in the developmental sequence are thought to be critical in attaining desiccation tolerance: 1) the expression of late embryogenesis abundant (LEA) genes and the accumulation of LEA proteins;[121] and 2) accumulation of certain carbohydrates.

LEA proteins, made up of several protein families, represent about 30% of the nonstorage proteins in cotton seeds,[190] a subset of which are called **dehydrins.** Dehydrins exhibit common structural features and are thought to function in a protective role, preparing the embryo for desiccation, although the evidence at this time is largely circumstantial. The expression of dehydrin genes is triggered by both developmental and environmental clues. Expression can also be triggered prematurely in immature seeds with exogenous abscisic acid, indicating that the level of the hormone within the seed may be a factor controlling the onset of desiccation tolerance.[336] How intimately abscisic acid is connected with the expression of LEA/dehydrin genes is not yet known. It is apparent that other regulators of desiccation tolerance genes are operative.[68,92] Since a cross-section of genes are being expressed, their regulatory control is thought to be relatively complex.

Proteins formed during the ripening stage are largely found in the cytosol.[93] Both dehydrin and certain LEA proteins, being hydrophilic, are thought to protect cellular membranes and proteins from damage when water is lost from the cell.[120] As the free water within the cell evaporates, only a thin hydration monolayer remains around proteins and membrane surfaces. Loss of this monolayer layer with further desiccation can be lethal. These specialized proteins are thought to function by binding water within the cell or mimicking the properties of water within the cell[101] and, in so doing, confer resistance to the damage.

The role of dehydrin proteins in recalcitrant seeds is not yet clear; they are found in species of temperate origin but are absent in those of tropical origin.[136,140] It is possible that since desiccation tolerance is a quantitative trait,[437] the amount of LEA/dehydrin proteins or the timing and rate of accumulation may be critical features in the acquisition of desiccation tolerance.

The accumulation of certain sugars (sucrose, stachyose, raffinose) also appears to be a central feature in the acquisition of desiccation tolerance. Sugars are thought to stabilize cellular membranes through the interaction of their hydroxyl groups with the hydrophilic polar heads of membrane phospholipids[101] and may function similarly in stabilizing proteins, preventing irreversible denaturation. Stachyose accumulates in excised immature soybean seeds when subjected to slow drying but does not if water loss is prevented;[56] thus its synthesis can be instigated by water stress.

vi. Abscission

Abscission of the ovule at the onset of ripening sets in motion a diverse cross-section of alterations within the seed (a more detailed critique of abscission is covered later in this chapter). While it has been argued that desiccation triggers the onset of gene expression during this developmental stage, clearly (Figure 6.15A,C) increased expression begins well before (i.e., 8 to 10 days) marked changes in seed moisture content. The termination of the flow of carbon and other requisites into the seed marks a pronounced shift in the prevailing status. Coupled with the discovery of sucrose sensors and the timing of significant alterations in water status, it would appear that the latter is a secondary factor, at least with regard to the changes in gene expression at the onset of ripening. Changes in water status, however, may well instigate further alterations in gene expression later in the ripening process.

vii. Desiccation

During the seed-filling stage, the water content of the seed begins to decline. Initially the decrease is gradual, as the water within the seed is replaced by starch or other storage materials. This period of gradual decline is terminated during the ripening stage, but, interestingly, not immediately after ovule abscission (Figure 6.15A) as one might anticipate due to the absence of the ability to replenish losses. The loss of water from the seed occurs through evaporation, the rate of which is a function of the seed's existing water content, its physical and chemical make-up and the environmental conditions to which it is exposed. What mediates this delay in water loss in not clear; however, the delay is significant and the eventual onset and rate of water loss pronounced.

The essential role of desiccation in instigating gene expression during the latter part of ripening is supported by the fact that detached immature seeds maintained in a fully hydrated state do not become viable reproductive structures.[170] Only partial, rather than complete, drying is required to make this transition.[42] Since the seeds are detached form their source of carbon and other requisites, drying must activate additional genes that are essential for producing a viable reproductive propagule.[370]

The requirement for desiccation is not universal, since the seed of certain fleshy fruits (e.g., tomatoes) do not require drying for completion of the ripening process, and in recalcitrant seeds, such as mangrove, dehydration results in a rapid loss of viability.[408] In most seeds, however, drying appears to be required. Drying, likewise, confers a state of metabolic quiescence which greatly extends the seed's functional existence.

viii. Dormancy

Seeds that are dormant at maturity lack the ability to germinate even when exposed to the appropriate environmental conditions. Dormant seeds can be separated into two very general classes—primary and secondary. Seeds exhibiting **primary dormancy** are dormant when they emerge from the parent plant. In contrast, seeds that were not initially dormant can be induced to become dormant if conditions are not favorable for germination—termed **secondary dormancy.** Seeds exhibiting primary dormancy may have the condition imposed by the seed coat due to its impermeability to water, restricted gas exchange or other factors (**coat imposed dormancy**). Alternatively, they may exhibit a dormancy that is inherent to the embryo (**embryo dormancy**). The latter form can be released by a variety of external treatments (e.g., the slow drying of dormant seeds, chilling of imbibed seeds, and exposure to specific light treatments).[41]

Though the seeds of a number of crops are not dormant at maturity, most seeds require a desiccation period before they are capable of germination (e.g., in barley some cultivars are dormant while others are not).[363] The dormancy mechanism is instigated during the early stages of development (i.e., around the third week after anthesis in barley)[50] to prevent precocious germination while the seeds are still developing on the parent plant. Abscisic acid is thought to be a key factor in dormancy due to its role in inhibiting germination. The role of abscisic acid has been explored using mutants that exhibit repressed synthesis or sensitivity. For example, seeds of *Arabidopsis* and tomato mutants that are deficient in abscisic acid are not dormant,[161,238] and mutants with defects in abscisic acid perception have reduced dormancy.[240] It is highly probably that other genes are also be involved.

Is the dormancy operative at seed maturity the same as in immature seeds or are different dormancy mechanisms operative? Based upon the range in environmental clues that fulfill primary dormancy requirements (e.g., drying, cold-moist conditions, light) in various species, it is probable that there are additional mechanisms operative. This is supported by the genetic analysis of dormancy in mature barley seeds. Four regions of the genome on chromosomes 1, 4 and 7 are involved,[167,428] accounting for approximately 75% of the dormancy expressed. The barley dormancy loci appear to operate *via* the release of dormancy at the end of the developmental process during after-ripening.[363] Without release, dormancy is carried over into the normal germination period.

ix. After-ripening

Seeds of a number of species are dormant when shed from the parent plant. The dormancy appears to be a mechanism that prevents the seed from germinating at the end of the growing season, ensuring that germination does not occur until the next season. Little is known about this type of dormancy. Does it represent a negative control of germination or simply the lack of completion of the ripening process while attached to the parent? Our knowledge of this type of dormancy is largely restricted to the environmental conditions that facilitate its removal. Under appropriate conditions, the seeds undergo chemical and physical alterations, collectively termed after-ripening,[204] which alleviate the dormancy. The standard method to facilitate after-ripening and confer germination potential is to dry the seed, followed by dry storage, the time interval varying with species. The amount of time required to fulfill this type of dormancy is temperature-dependent; increasing temperature increases the rate at which after-ripening is completed.[245] The length of dry storage can range from two weeks (*Lepidium virginicum* L.) to several years (*Sporobolus cryptandrus* Gray)[28] depending upon species, temperature and other factors. Environmental conditions during seed development are known to influence the degree of dormancy; for example, short-day, cool temperatures and high moisture during the seed-filling phase of barley increase dormancy.[387]

1.4. Abscission

The shedding of plant parts is a natural process, affording plants a number of survival and thus evolutionary advantages. Shedding allows plants to remove injured, infected or senescent organs, recycle nutrients, adjust leaf/flower/fruit numbers when under stress conditions, disperse seeds, and other benefits. Most shedding responses are precipitated by physical and chemical changes occurring in a specialized abscission zone. Abscission *per se* represents the separation of cells, tissue or organs from the remainder of the plant at one of these zones. During the developmental phase, naturally occurring abscission responses are part of normal growth under non-stress conditions. After harvest, however, abscission responses generally are detrimental to product quality. Examples are the abscission of florets in cut and potted flowers,[83] the leaves of foliage plants[276] and food crops such as cabbage,[289] and celery. As a consequence, considerable research effort has been directed toward the inhibition of abscission.

Abscission zones are comprised of specialized cells, only a few cells wide, that transverse a major portion of the cross-sectional area of the organ (Figure 6.19). Prior to the onset of abscission, the cells in the abscission zone are relatively indistinguishable from neighboring cells but somewhat smaller. The presence and location of the abscission zone on a plant is genetically controlled.

A wide range of plant parts are abscised by various species. The most commonly shed parts are leaves, flowers, fruits and seeds; however, branches, bark, roots, spines and other parts may be shed, depending upon the species. Leaf abscission can involve individual leaves, individual leaflets on compound leaves and, in a limited number of species, entire leaf branchlets (e.g., *Larix*). Entire flowers abscise in response to stress conditions and individual floral parts (e.g., calyx, petals, stamen) after anthesis. In some species, petal abscission occurs very quickly after pollination (e.g., *Digitalis, Clarkia* and *Antirrhinum*). Others shed their petals at varying intervals after anthesis; for example, in some species of *Linum* and *Geranium,* petal abscission occurs during the afternoon of the day of anthesis; for *Gossypium hirsutum* L., the day after anthesis; and for *Eschscholzia californica* Cham., the fifth day after anthesis.[10]

Individual fruits may be shed due to the lack of pollination, to adjust the fruit load on the plant or due to physical injury to the organ or subtending branch. Seeds abscise from the placenta of most fruits. In dry dehiscent fruits, this allows the propagule to be dispersed from the parent plant *via* a diverse array of species-dependent mechanisms. The actual dehiscence process is a modification of abscission. Rather than detachment, dehiscence involves the opening of a structure, with the separation occurring along a defined zone of cells.[402]

The location of the abscission zone is adjacent to the site of detachment. For simple leaves, this is normally at the base of the petiole. Fruit abscission zones may lie between the ovary and receptacle, the ovary and pedicel, the nut and receptacle or other locations depending upon the species.

Abscission entails a highly coordinated series of alterations under the control of several developmental, environmental and hormonal factors. Three general processes are involved: 1) changes in the physical and chemical condition of the cells within the abscission zone that

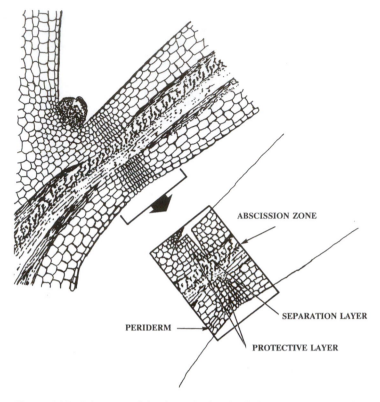

ABSCISSION ZONE

SEPARATION LAYER

PERIDERM

PROTECTIVE LAYER

Figure 6.19. Diagram of the tissue in the abscission zone transversing the base of a leaf petiole. Abscission zone cells are typically smaller and less differentiated than neighboring cells. As the chemical and physical processes leading to abscission develop (box), the juncture between neighboring cells within the zone weakens and subsequently separates. A layer of cells then seals the leaf scar, protecting the plant from pathogen invasion and water loss (*redrawn from Addicott*[10]).

result in a progressive weakening of the juncture; 2) synthesis of compounds in the abscission zone that function in protecting of the fracture surface against microbial attack; and 3) the mechanical separation of the plant part. Abscission is coordinated by the phytohormone ethylene which regulates the expression of a cross-section of genes encoding essential enzymes that mediate the physical and chemical alterations. Ethylene, however, does not function alone but appears to act in tandem with auxin. During normal growth, auxin synthesized in the organ diffuses outward through the zone, inhibiting abscission.[302] Once conditions (e.g., disease, stress, reaching the terminal stage of development) activate the abscission process, auxin synthesis decreases and ethylene synthesis in the abscission zone is stimulated. Reduced auxin in the abscission zone increases the sensitivity of the cells to ethylene. Thus, ethylene accelerates abscission, while auxin impedes it. A shift in the balance, whether natural or artificial in origin (e.g., exposure to exogenous ethylene), can instigate or impede the process.

After induction, there is a lag phase during which mRNA and protein synthesis occur,[6] and, if the synthesis is blocked, there is a marked inhibition of the weakening process. During this time there are no changes in the physical strength of the abscission zone. Following the lag phase, the abscission zone undergoes a progressive weakening (a decline in break-strength) due to the enzymatic hydrolysis of the primary cell wall and the middle lamella, reducing the strength of the molecules holding the cells together along the separation zone. Fracture typically follows a line through the middle lamella between neighboring cells. Physical forces are important in the eventual mechanical separation of the organ. In addition to gravity, wind, rain, ice accumulation and other factors often facilitate final separation.

Alterations in the abscission zone cell walls are mediated by two types of hydrolytic enzymes, cellulases (endo-β-1,4-glucanases) and polygalacturonases. The cellulases belong to a multigene family[412] whose expression is tissue-specific. Of the six cellulase genes in tomatoes, *Cel2* mRNA increases in the abscission zone.[72] In transgenic plants with antisense constructs that repress 80% of the *Cel2* cellulase, abscission is significantly impeded. Multiple genes and their differential expression indicate that shedding of various organs from the plant may be controlled by different isozymes of cellulase.

The pectin-degrading enzyme polygalacturonase also increases during abscission with the first signs of cell separation occurring in the pectin-rich middle lamella region of the walls.[474] Three different polygalacturonases are expressed in tomato leaf and fruit abscission zones,[207] which differ from those of the fruit,[412] indicating a high level of specificity and control over this developmental step.

Though organ abscission leaves a ruptured surface, it seldom results in microbial invasion. In addition to the hydrolytic enzymes, genes expressed during abscission result in the synthesis of a cross-section of pathogen-related proteins that appear to facilitate the plants defense against invasion. For example, there is an increase in peroxidase activity in the proximal cells of the abscission zone,[176] in addition to several pathogen-related proteins, chitinases, a proteinase inhibitor, a metallothionein-like protein and polyphenyl oxidase.[96] The accumulation of polyphenyl oxidase, which can oxidize chlorogenic acid and related compounds to fungistatic molecules, and chitinase is ethylene-dependent in *Sambucus nigra* L.[97]

Environmental and other factors that accelerate abscission (e.g., mineral deficiency, drought, low light, pollination, the lack of pollination, pollution) have been well documented. However, how these signals are translated into the induction of the abscission response is not well understood. During the postharvest period, ethylene, low light intensity and water stress are the most common causes of abscission in ornamental plants. Physiologically active levels of ethylene may be produced by the plants themselves[214] or may come from an external source (e.g., growth regulators, internal combustion engines, improperly adjusted gas heaters). Preventing losses caused by ethylene mediated abscission after harvest has been approached three ways: 1) avoiding exposure to exogenous sources of ethylene; 2) treatments to inhibit the synthesis of ethylene by the plant, and 3) treatments to reduce the sensitivity of the plant to ethylene. Proper ventilation and other management practices (e.g., sanitation, proper product selection for mixed storage, use of ethylene scrubbers) should be routinely used to prevent the concentration of ethylene from reaching a detrimental level. While the inhibition of ethylene synthesis with various compounds (e.g., aminoethoxyvinylglycine) may be beneficial, it does not protect the plant material from other sources of ethylene. The use of silver ions, a potent inhibitor of ethylene binding,[43] has proven to be an extremely effective way of preventing abscission in many ornamentals*,[357] When formulated as silver thiosulphate, it is relatively non-phytotoxic and stable while being readily translocated within the plant. An earlier technique, the application of synthetic auxins, has been found to be effective for the inhibition of leaf abscission of harvested English holly[360] and several other ornamental foliages. For food crops, high levels of carbon dioxide, a putative inhibitor of ethylene binding,[76] can reduce ethylene mediated abscission in some crops.

---

*Silver is a toxic ion that should not be used on crops to be consumed.

1.5. Aging and Senescence

Living organisms share several universal traits, one of which is that their existence is finite. All organisms eventually die. The question thus becomes not will they die, but when will they die and how will they die? The life expectancy of plants and plant parts varies widely, ranging from only minutes to hundreds of years. When moving live harvested plant material from the producer to the eventual consumer, we are necessarily concerned not just with the impending death of the product but often even more so with the general degradative changes in the product that lead up to its eventual death. Our interest in these changes arises from the fact that many of these alterations diminish product quality. Our objective during the postharvest period is not to prevent the eventual death of the product or, for that matter, extend the storage life to its theoretical maximum duration. Economic constraints make the latter an unrealistic commercial option. Thus, if an adequate supply of lettuce is being harvested for the market every week of the year, there is little incentive to maintain lettuce in storage for 6 months, even though it may be theoretically possible.

Many of the changes that occur after harvest, especially those in highly perishable products, are part of the process of senescence. **Senescence** can be defined as a series of endogenously controlled deteriorative changes which result in the natural death of cells, tissues, organs or organisms. This differs from **aging,** which entails changes that accumulate over time without reference to death as an eventual consequence.[292] Non-living objects, as well as living, therefore can age. In living organisms, aging is not a direct cause of death, but it may increase the probability of death by decreasing the product's resistance to stress.

Both aging and senescence are of interest from a postharvest standpoint, though aging has generally been less studied. With highly perishable products, the life expectancy is so brief that aging is generally of only minor consequence relative to the onset of senescence. In more stable products (e.g., dried corn kernels), however, the physical and chemical changes associated with aging are of considerable interest. Aging alterations often impact quality, and when the product is a reproductive propagule, these alterations can translate into losses in productivity in the subsequent crop. In humans, the disease progeria results in tremendously accelerated aging, where very young children (≥ 10 yr) have aged to a point normally reached in 70 to 80 years. Clues from gene expression patterns have pointed toward genetic instability and disturbances in gene function as being central components of the disease. Analysis of accelerated aging in both plants and animals may possibly provide clues as to the causes of normal aging.

1.5.1. Aging

Aging is a passive, non-programmed process; the factors precipitating it are largely external. Loss of seed viability with time is often a function of aging. With aging, viability declines and progressively more seedlings have developmental and pigmentation abnormalities.[348] Thus, age-related lesions accumulate with time, the rate of occurrence strongly modulated by temperature, moisture content and oxygen concentration. Higher levels of each accelerate the rate.

Cells are dynamic homeostatic systems* that have the ability to undergo repair. Aging lesions accumulate when the cells are inable to maintain homeostasis.[364] In many situations repair reactions occur, ranging from DNA synthesis to the replacement of defective cells. For example, harvested fruit sensitive to chilling injury can largely counteract low temperature injury during storage if exposed to periods of intermittent warming.[36] Likewise, DNA repair is

---

*Homeostasis—the tendency to maintain internal stability *via* a coordinated response to the disruption of normal condition or function.

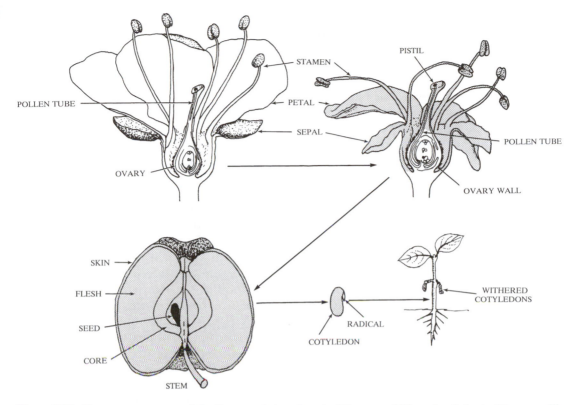

Figure 6.20. Organ senescence and death are an integral part of the normal life cycle of plants. Flower pollination, for example, sets off a dramatic series of developmental events within the ovary. Coupled to this is the onset of senescence in the floral parts that are no longer essential (e.g., petals, stamen, stigma and style). Carbon and nitrogen compounds may be recycled out of these organs into the developing ovaries. The fruit tissue subsequently decomposes, leaving the seeds, which in turn undergo senescence and death of a major portion of their cells.

a routine phenomenon.[443] In low moisture reproductive propagules, such as seeds, the concurrent low level of metabolic activity exacerbates the accumulation of lesions through the cells' inability to instigate necessary repairs.

1.5.2. Senescence

Senescence, an integral part of the normal developmental cycle of plants, can be viewed on an organism, organ, tissue or cellular level.[259,316] Monocarpic annuals such as wheat complete an entire cycle from germination to death within a growing season. The sight of miles upon miles of wheat in the Great Plains of North America simultaneously dying in the early summer dramatically illustrates the presence of an internal program controlling plant death responses. A variation of the death syndrome can likewise be seen in herbaceous perennials, where the above-ground portion of the plant dies while the below-ground system remains alive. Organ and cellular death are essential components of the normal developmental process in plants. Organs such as fruit, leaves and flower parts eventually undergo senescence and death. For example, at anthesis a simple fruit is comprised of the immature ovary and the various flower parts. Within a week or so after flower opening and pollination, most of the floral parts (petals, pistil, stigma, style), having fulfilled their biological function, senesce and drop to ground (Figure 6.20). With the exception of the seeds, the tissues making up the fruit in turn senesce. Thus, of the thousands of cells that make up the fruit, nearly all have by now died. Those remaining are largely concentrated in the cotyledons which, with germination, will also senesce. Thus organ senescence and death are integral parts of the normal cycle from seed to seed.

Organ senescence is what we most often encounter during the postharvest period. Many of our handling and storage techniques have been designed to retard the rate of development of senescence in organs such as fruits, flowers, leaves, roots, tubers, etc. In nature, organ senescence allows the removal of plant parts that have already fulfilled their biological function or are no longer functioning properly. This eliminates the maintenance costs (energy) to the parent plant for that organ, and may allow recycling nutrients back into the plant, increasing the plant's overall efficiency.

Groups of cells may also senesce while the cells around them remain alive, for example the thin layer of cells making up abscission and dehiscence zones. These cells undergo a form of programmed death. Although the cells are not dead at the time of separation, many of the physical and chemical changes are similar to those occurring in cells during senescence.[312]

Death also occurs at the cellular level, with certain cells dying while surrounding cells remain alive. An example of cellular death can be seen in xylem cells, which are not alive at maturity and in single-celled root hairs, formed just behind the growing root tip, which have a relatively short life expectancy.[300] The death of these cells represents part of the normal development of the plant. Our understanding the control of death in plants is at this time very limited, since the genes expressed during senescence are just starting to be identified. In that death at the cellular level represents the key to understanding the death of tissues, organs and organisms, the following paragraphs focus on programmed cellular death.

In a typical day, 60 to 70 billion cells perish in our bodies; much of this turnover involves **apoptosis,** one form of programmed cellular death, characterized by a unique set of symptoms. After years of relative neglect, cellular death has become an *avant garde* topic in animal physiology with a plethora of research papers (e.g., between 1995 and 1999, over 20,000 papers were published on cellular death). Do programmed cell death mechanisms differ in plants and animals, or is there homology in the basic programs? In animal cells, death typically involves the removal of the corpse; however, in plant cells, with their very stable walls, removal is a much more difficult task. There is, however, some indication that certain types of plant-cell death possess the fundamental elements of apoptosis in animal cells.[451]

Cellular death in plants can be mediated *via* external or internal means. For example, severe trauma causing a very rapid, uncontrolled cellular death occurs frequently during the postharvest handling of some products. Death is caused by an external force, such as mechanical wounding, where the cells are crushed or very rapidly killed in some other manner. Severe trauma, however, does not represent a form of senescence, since the cells are not an active participant in the instigation and development of their own death, nor is there a controlled disassembly of cellular constituents.

Is there a genetically controlled cell death program? The presence of genetic control over senescence has been established through four lines of evidence: a) the removal of the nucleus (i.e., the primary source of genetic information) inhibits senescence;[471] b) inhibitors of RNA synthesis inhibit senescence, indicating that *de novo* gene expression is required;[313] c) senescence retarding mutations have been identified in a wide cross-section of crops,[315] thus defective genes can impede senescence; and d) there are distinct alterations in gene expression during senescence, when some genes are turned on and others are turned off, the former probably being more important in the control process. Thus, cellular death is controlled at the genomic level and is tightly programmed.

Programmed cellular death can be classified a number of ways, e.g., based upon the presence or absence of a controlled disassembly and recycling of cellular constituents, the speed of death, or the type of program instigating death. Are there different death programs or sub-

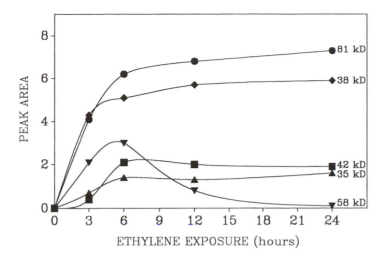

Figure 6.21. Changes in the relative abundance of selected mRNAs in carnation (*Dianthus caryophyllus,* L.) petals with time, after the onset of senescence mediated *via* exposure to ethylene (7.5 μL · L⁻¹)(*redrawn from Woodson and Lawton*[464]). Thus genes for specific proteins appear to be turned on during the early stages of senescence, and their duration of transcription varies.

programs operative, and, if so, how should they be categorized? Typically the categorization of death has been based either on morphological alterations taking place within the cell as senescence develops or on the general mechanism controlling the onset of death. Presently, three general classes of cellular death mechanism in plants have been documented: 1) developmental control—such as that instigated in flower petal cells after pollination; 2) perturbed development—when cellular death is caused by genetic lesions; and 3) environmental stress— where stress is monitored *via* some internal mechanism, and once it reaches a predetermined level, death is instigated.

The presence or absence of a controlled disassembly of the cell is likewise an important distinction. In many instances, the cell is progressively dismantled of its cellular components, and, as might be expected, there is a fairly distinct priority in cellular components for dismantling. The mitochondria, needed for continued energy production, the nucleus and the plasma membrane are maintained until the cell nears a final organizational collapse and death. In instances of very rapid death (e.g., hypersensitive responses with the invasion of certain pathogens), disassembly may be largely nonexistent, with alterations limited to the very basic essential reactions involved in the death response. Controlled disassembly, in contrast, allows carbohydrates, proteins and nutrients such as nitrogen, phosphorous and metals present in the organ to be reallocated to actively growing tissues or stored for subsequent use.

a. Changes in Gene Expression with Senescence

In contrast to the once-held view that senescence is simply an organizational collapse of the cell, we now know that senescence is initially a tightly controlled developmental event which normally occurs in a highly ordered sequence of steps. The onset of programmed cell death is controlled by a series of external and internal signals that direct the cell on how to instigate its death and how the cellular constituents will be processed. Thus, specific genes controlling the death mechanism(s) are up-regulated (Figure 6.21), while others are down-regulated.[281,328,464] Genes whose transcripts are up-regulated with the onset of senescence are termed **senescence-**

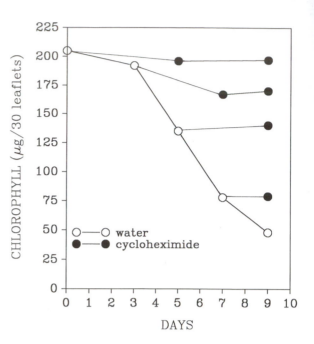

Figure 6.22. Inhibition of chlorophyll degradation in *Anacharis* leaflets using cycloheximide, an inhibitor of protein synthesis (*redrawn from Makovetski and Goldschmidt*[272]).

associated genes. The mRNAs from over 30 such genes have been reported to be up-regulated during senescence.[73,202] For example, genes encoding certain hydrolases appear to be up-regulated in all programmed cell-death mechanisms. However, other genes that are expressed during senescence are unique to the cell-death program activated. Thus, pathogen-mediated hypersensitive responses will instigate genes that are unique to that means of death. With environmental stress-induced death, different genes are activated. Interestingly, there is some overlap where certain defense-related genes, though not pathogen-related, are expressed in both instances. Defense-related gene products may be present to combat opportunistic infections that commonly accompany senescence.

The requirement for senescence-associated gene products for the instigation of senescence is underscored by the inhibition of senescence by cycloheximide, an inhibitor of protein synthesis (Figure 6.22).[236] Cycloheximide inhibits protein synthesis on 80s ribosomes found largely in the cytoplasm. In contrast, the inhibition of protein synthesis with chloramphenicol, which acts on 70s ribosomes found in the plastids and mitochondria, does not inhibit senescence.[461] Thus the control of senescence clearly resides in the nuclear genome rather than in DNA found in individual organelles.

The overall process of cellular senescence can be separated into three very general stages: 1) an initiation phase—where internal or external factors are sensed in some manner and, upon reaching an appropriate level, trigger a signal transduction mechanism that instigates senescence; 2) a coordinated degradation phase—where genes are activated or inactivated that control cellular disassembly; and 3) a terminal phase—where changes are instigated that precipitate the death of the cell.[314] Since a diverse range of environmental and internal factors can initiate senescence, a range of signal transduction mechanisms may be operative, but whether these lead to a single, universal death program is not known. The mitochondria, due to its crucial role in cellular energy metabolism, has been proposed as a possible stress sensor and subsequent dispatcher of programmed cell death.[203]

Total protein content, while characteristically declining in senescing leaves, does not necessarily decline in all senescing organs. During fruit ripening total protein concentration has been shown to remain essentially unchanged (e.g., avocado,[48] banana,[441] citrus[261]) or in some cases increase (e.g., apple,[192] pear,[168] cantaloupe[372]). In each case, the concentration of certain enzymes increases. These typically are degradative in nature, involved in dismantling large molecules. Genes encoding degrading enzymes such as proteases and nucleases and those involved in lipid, carbohydrate and nitrogen metabolism are activated. For example in fruit, increases in ribonuclease, invertase, acid phosphatase, α-1,3-glucanase, α-1,4-glucanase, cellulase and polygalacturonase have been reported, while in leaves protease, ribonuclease, invertase, acid phosphatase and α-1,3-glucanase increase.[66]

The idea of a controlled series of events is further underscored by the fact that senescence is an active process, requiring energy. The disruption of energy availability, such as through depressed respiration *via* low oxygen,[463] retards the rate of development of senescence.

b. Chemical and Ultrastructural Changes During Senescence

The cellular membrane system (plasma membrane, endoplasmic reticulum, vacuolar membrane, the highly specialized thylakoid membranes of the chloroplasts, etc.) represent selective barriers to the movement of compounds within and between cells. Control *via* access is one important means of regulation for a number of biochemical and physiological events within the cell. During senescence, however, there is a progressive loss of membrane integrity and, consequently, regulatory control.[417] As one might anticipate, the chemistry and structure of the various membranes within the cell differ, and those, in turn, affect what senescence processes operate and their rate.

A relatively consistent event in the senescence of chloroplast-containing plant organs is the degradation of chlorophyll and the corresponding transition in color of the product.* In leaves, the chloroplasts contain most of the protein in the cell (and thus nitrogen), so they are one of the first organelles targeted for recycling. Changes in chlorophyll are dramatically illustrated with leaf senescence in the autumn when an estimated 300 million tons of chlorophyll is degraded.[175] While chlorophyll degradation is in most products closely tied with senescence, its dismantling does not appear to be an essential requisite for senescence to proceed, as in a mutant genotype of fescue, in which death occurs without chlorophyll degradation.[416]

Chlorophyll loss occurs in tandem with significant changes in the chloroplast *per se,* which undergo a sequential series of alterations in their ultrastructure.[150] Ultrastructural changes center on the thylakoids, the internal membrane system within the chloroplast and the site of the light reactions in photosynthesis (photosystem I and II). Initially the stroma thylakoids lose their integrity, followed by swelling and disintegration of the grana thylakoids.[189] Interestingly, the double membrane envelope that encloses the chloroplast does not begin to lose its integrity until very late in the dismantling of the chloroplast. This allows the process to be reversed until quite late in the degradative sequence. Regreening is a common occurrence in leaves and some fruits exposed to inductive conditions. Likewise, retention of the outer membrane system would account for the ability of the chloroplasts in some products (e.g., tomato fruits) to be transformed into other types of plastids.

In keeping with the idea of an orderly dismantling of the cell, one would anticipate that cellular structures and organelles that are absolutely essential would be retained until the cell enters the very final stages of senescence. The mitochondrial,[127,365] plasmalemma, nuclear and vacuolar membranes persist to the very late stages.

---

*The loss of chlorophyll can, depending upon the product, unmask a range of pigments and, therefore, colors.

c. Endogenous Regulators of Senescence

Ethylene is known to stimulate senescence or senescence-associated processes in a variety of organs. For example, the exogenous application of ethylene stimulates the ripening of climacteric fruit, senescence of flowers and leaves and other senescence processes. In climacteric fruit, treatments that inhibit the synthesis, action or internal concentration of ethylene tend to delay though not prevent ripening.[110,449] Likewise, the leaves of antisense tomato clones with reduced ethylene synthesis and ethylene insensitive *Arabidopsis* plants both display a delayed onset of foliar senescence. Is ethylene an integral controlling signal of both physiological processes? Closely tied to ethylene-stimulated fruit ripening and leaf senescence are age-dependent factors. In both instances, as the respective organ matures, its susceptibility to ethylene increases. The role of ethylene and age-dependent factors appears, however, to differ between fruit ripening and leaf senescence. Both are essential for ripening; however, with leaf senescence, ethylene is neither necessary, nor acts as a senescence-promoting signal. Ethylene function in leaf senescence appears to result from its effect on age-related factors, with its role limited to modulating the timing of senescence.[159]

In contrast to ethylene, cytokinin and calcium ions have been shown to delay senescence. In general, exogenous application of cytokinin or synthetic forms of cytokinin at the appropriate concentration and timing have been shown to delay senescence in many tissues.[433] In addition, a decline in internal concentration of cytokinins coincides with the onset and development of senescence in some tissues (e.g., petals[434] and leaves[432]). In fruits, the transition of chloroplasts to non-chlorophyll-containing chromoplasts is accompanied by a decrease in endogenous cytokinin concentration.[288] The biological role of cytokinin has been further underscored by the development of a transgenic tobacco line in which the gene that catalyzes the rate-limiting step in cytokinin biosynthesis is over-expressed,[149] increasing the internal concentration of the hormone. Leaf senescence was significantly retarded suggesting that cytokinin level within the leaf is an integral factor in the induction or progression of senescence.

Exogenous Ca^{2+} application delays senescence in a number of organs (e.g., leaves and fruits).[138,263,338] This effect is thought to be related to the extracellular role of Ca^{2+} in stabilizing the cell wall and external surface of the plasmalemma.[138] Within the cell, the concentration of Ca^{2+} is substantially lower than that found in the extracellular region, and its role appears to be much more complex. Within the cell, Ca^{2+} is thought to act as part of an information transduction system where extracellular signals are translated into changes in metabolism.[337]

The precise role of ethylene, cytokinin and Ca^{2+} in the promotion or inhibition of senescence and their interaction among themselves and with other endogenous growth regulators is not presently known. As our understanding of this complex relationship improves, so should our ability to manipulate these factors to retard senescence.

d. Environmental Factors Modulating the Rate of Senescence

Stress can significantly modulate the rate of senescence of harvested plants and plant organs. In some cases, stress can induce the onset of senescence. Environmental stresses such as temperature (high and low), composition of the gas atmosphere surrounding the product, water deficit or excess, pathogens, herbivores, irradiation, mechanical damage, mineral imbalances, salinity, and air pollutants have been shown to accelerate senescence in various organs. Some environmental stresses (e.g., light depravation) can trigger the onset of senescence.[55] During the postharvest period, the regulation of temperature and gas concentration (refrigerated and controlled atmosphere storage) is routinely used to decrease the rate of development of senescence.

2. MATURATION

During development, plants pass through a series of distinct but often overlapping stages. With monocarpic plants, development begins with germination, passing through the juvenile stage and progressing to maturity and finally senescence. Maturity, as viewed from the natural reproductive biology of the plant, is generally considered to be the stage of development where the plant is capable of shifting from vegetative to reproductive growth. In agriculture, however, maturity is much more arbitrary. Generally, maturity is seen as a stage of development we superimpose upon the plant or plant part relative to our needs. The object in question is considered mature when it meets our requirements for harvest (i.e., **harvestable maturity**). This does not necessarily imply that the product meets our maturity requirements for immediate utilization. Many products are sufficiently mature for harvest but not for utilization. For example, apples, bananas and persimmons that are to be held in storage for considerable time are harvested prior to having developed sufficiently for immediate consumption. With proper handling, they will continue to develop after harvest, reaching an acceptable level of culinary quality.

Harvest maturity (Figure 6.23) varies widely with the plant product involved. Plants that are sold or consumed at the seedling stage reach a harvestable maturity very early in the natural development cycle of the plant. Plant parts such as inflorescences (artichoke, cauliflower) or partially developed fruits (cucumber, bitter melon, sweet corn) progress through a significant portion of their developmental cycle. Fruits such as apples, bananas and citrus are nearly fully developed, while most nut and seed crops are fully developed at harvest. Harvestable maturity, therefore, can occur throughout the developmental cycle, with the precise time varying with the product in question.

Harvestable maturity for a number of crops occurs over a relatively wide time frame. For example, with cassava or taro this time period may be several months. With the giant taro, the period of harvestable maturity extends for several years without an appreciable loss of quality. Many crops, however, have relatively short, precise time periods for harvest, and exceeding this period results in the impairment of quality or product loss. The length of this period ranges from weeks (e.g., oranges) to days (e.g., apples) to hours within a single day (e.g., pollen, gherkins).

The timing and duration of this period of harvestable maturity can be modulated by a number of factors. For example, cultivar can have a significant influence. Environmental conditions during development can also have a pronounced effect on the timing and length of the period of harvestable maturity. Substantial losses are encountered each year due to environmentally induced alterations in the maturation time period.

Harvesting at the proper stage of maturity is essential for optimum quality and often for the maintenance of this quality after harvest. When is the optimum point for harvest in the period of harvestable maturity? Within a given crop, optimum maturity is a highly subjective determination. One critical variable is who in the production-harvest-storage-marketing-utilization chain determines the criteria to be used. In that the needs vary at each stage, their criteria for optimum maturity may also vary. For example, optimum harvest maturity for the grower is a function of both product and marketing conditions. When can the crop be harvested to maximize profits? When supply is low and price high, lettuce is often harvested very early in the normal harvestable maturity time period. Thus for an individual grower, optimum maturity may vary with each successive crop. Growers must also consider how the crop is to be harvested. Optimum harvest maturity for a crop of tomatoes that is to be harvested using a destructive once-over mechanical harvest is going to be substantially later than when multiple hand harvests are used.

The method in which the product is to be handled after harvest may also influence what

Developmental Stages

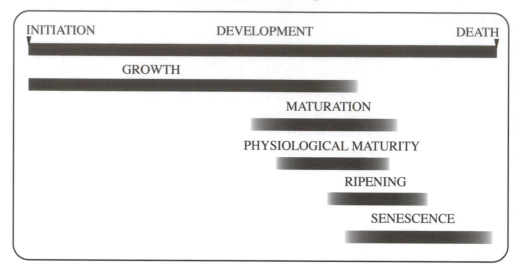

Harvestable Maturity

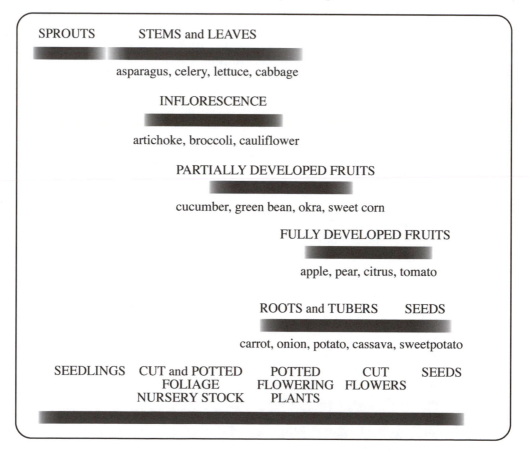

Figure 6.23. The timing of harvestable maturity within the normal developmental cycle of a plant varies widely among individual products. Some (e.g., bean sprouts) have just begun development when they reach a harvestable maturity, while others (e.g., grains, seeds) complete the cycle from seed to seed (*redrawn from Watada et al.*[453]).

is considered the optimum degree of maturity for harvest. Apples that are to be held in cold storage for extended periods are generally harvested at a less mature stage than those destined for the immediate fresh market. The optimum maturity for spinach that is to be hand-harvested for fresh market often does not coincide with that of spinach to be processed. Likewise, spinach that is to be processed and packed as a chopped product can be harvested later than that which is packed with the leaves intact.

Optimum harvest maturity is not a fixed point in the developmental cycle of a plant or plant part; but varies depending upon the criteria utilized to determine it. These criteria may vary substantially among the various individuals in the production-harvest-storage-marketing-utilization chain and the constraints under which they are operating at that particular time.

How then do we determine when optimum harvestable maturity has been reached? With most crops, optimum maturity is determined by specific physical and/or chemical characteristics of the plants or plant organ to be harvested. Implicit in the use of single or multiple physical or chemical characteristics to determine optimum maturity is the principle that changes in the selected parameter correlate with the attainment of the general composite of quality characteristic of the product.

2.1. Physical Measurements of Maturity

Physical attributes such as size, color and texture are commonly used to determine crop maturity. With individual crops, these physical characteristics may be very specific. For example, muskmelon maturity is determined by the ease with which the stem separates from the fruit. Although very closely related, honeydew melons have totally different criteria for indentifying maturity. Spinach for processing is harvested shortly after a hollow area develops within the interior of the base of the plant where the stem and tap root meet. With cut roses, optimum maturity for harvest varies with individual cultivar; yellow cultivars generally can be cut at a slightly tighter bud stage than pink or red cultivars. The latter are allowed to develop until one or two of the outer petals begin to unfurl and the sepals are either at a right angle to the stem axis or pointing downward.

Maturity criteria based on physical characteristics can be made either objectively or subjectively. Experienced growers commonly determine harvest date based on a subjective assessment of the crop's physical characteristics. Although lacking precise analytical measurements, many growers have developed the ability to determine with a high degree of precision the optimum maturity for a specific crop. Not all growers, however, are equally proficient at determining maturity. Futhermore, with some crops it is not possible to accurately determine optimum maturity with a subjective evaluation. Therefore, objective or analytical measurements of maturity have become more widely used, and, if properly carried out, they tend to be highly consistent. This procedure requires measurement, using a standard instrument, of some physical characteristic that is known to change in relation to the maturation of the product in question. For example, the maturation of peas is determined by the force required to shear the seed. Several instruments have been developed for this test (e.g., Tenderometer, Maturometer, and others). Flesh firmness is commonly used to determine the maturity of apples. As the fruit matures and ripens, dissolution of the middle lamella of the cell wall results in softening of the edible accessory tissue. Firmness is measured using the resistance to penetration of the tissue by a plunger of a standard shape and size. Several relatively inexpensive pressure testers are commercially available (e.g., Magness-Taylor, Effegi pressure testers).[64] Recent work has been directed toward developing non-destructive techniques for measuring firmness.[86,294]

Optical measurements may also be used to assess the maturity of many crops. These may require removing a representative sample of fruit and analyzing it under laboratory conditions (e.g., starch content in apple assessed by iodine color reaction). Color changes for many fruits are associated with ripening and may be used to determine the optimum maturity for harvest. The simplest types of equipment measure the surface color of the sample,[145] and a number of on-line commercial grading systems are available.

2.2. Chemical Measurements of Maturity

During maturation, many plant products undergo distinct chemical alterations which are correlated with maturity. Consequently, it is possible to use either a subjective or objective measure of these changes to determine maturity. The most common chemical parameters measured are water, soluble solids, specific sugars, starch and acidity. Numerous other compounds or groups of compounds, however, can be used for specific crops.

Subjective evaluations of the basic chemistry of the product are made by feel, taste and aroma. The concentration of water, the predominant chemical in many fleshy plant products, is often judged subjectively by the feel of the product. The moisture content of many dried seeds and nuts is commonly estimated in this manner and often with reasonable accuracy.

Taste and odor are also utilized as subjective measures of chemical composition. Humans can perceive four basic taste sensations (sweet, sour, salty and bitter) and up to an estimated 10,000 distinct odors. Thus, sweetness (sugars), sourness (acidity) and aroma are sometimes used as criteria. Numerous factors can, however, significantly alter our estimation of maturity using these parameters, so they are less commonly used than subjective physical measurements.

Objective measurements of maturity are widely used for many crops. Water content can be quickly measured by weighing the sample before and after drying. Instruments such as the Ohaus Moisture Determination Balance simplify this process; however, only one sample can be dried at a time. Water content is extremely important in the postharvest handling and storage of seed and nut crops that need to be relatively dry at harvest. Soluble solids or specific sugars are also tested extensively (e.g., processing grapes). Soluble solids can be determined on samples of juice squeezed from the tissue by use of a refractometer or hydrometer. Acidity may also be readily measured by titration. In some crops (e.g., citrus), the ratio of soluble solids to titratable acidity accurately reflects the proper maturity for harvest.

The conversion of starch to sugar is also used as an index of maturity in some products. The reaction of starch with iodine produces a bluish-to-purple color, and the pattern of staining and intensity can be used as a measure of the amount of starch remaining in the tissue.

3. QUALITY

In harvested plant products, quality is the composite of those characteristics that differentiate individual units of the product and have significance in determining the unit's degree of acceptability to the user.[21,243] These characteristics or components of quality differ among types of products, according to where within the production-storage-marketing-utilization chain they are assessed, and with the expertise or purpose of individuals assessing quality.[391,233]

3.1. Components of Quality

The ultimate consumer attempts to assess the general quality of a product at the time of purchase. Generally their assessment is a composite of several characteristics of the product that they have found or believe to be good indices of overall quality. These components of quality can be separated into two major classifications, nutritional and sensory. Nutritional components are primarily applicable to plant products that are to be consumed; however, nutrition may also strongly modulate the quality of other products, for example those used as reproductive propagules (seeds, small plants) or for their ornamental value (foliage plants). Sensory criteria, including appearance, texture, aroma and taste, may be modulated by both preharvest[211] and postharvest[31,324,378] factors. Not all of these criteria will necessarily be applicable to all products. Taste is not a quality consideration for ornamental flowers, while odor may or may not be, depending upon the species and cultivar.

Quality, therefore, is assessed from the relative values of several characteristics considered together (seldom is only a single parameter used). While it is possible to develop a precise list of what factors make-up the quality of a particular product, ascertaining their relative importance is difficult, since not all characteristics contribute equally to the over-all quality of the product. Likewise, the value of one parameter may change relative to other parameters. For some products, numerical scores have been assigned for individual characteristics, where each represents part of a total numeric value. Schemes have been developed for which the composite score of all components represents the grade of the product. Unfortunately, even when the maximum possible value for each individual component of quality is weighed relative to its estimated contribution, the scores are rarely additive; the composite decreases the amount of information available. For example, a very low score for one quality component (e.g., number of flowers on a potted chrysanthemum plant) may render the product unacceptable to the consumer while lowering the overall grade only slightly. Likewise, one quality characteristic may or may not be closely tied to another. Apple taste and aroma are important quality components, but they only significantly improve the product's overall quality when the fruit's textural properties are at or near optimum.

3.1.1. Size

The size of individual units of a product can significantly affect consumer appeal, handling practices, storage potential, market selection and final use. With many agricultural products, consumers discriminate on the basis of size. Often exceedingly small and large individual units are considered undesirable. Small water chestnuts are less desirable due to the greater losses incurred during peeling and the additional labor required. In contrast, large peas are generally considered lower in quality than small or *petit pois*. Seldom are large apples selected for long-term storage, because they are much more susceptible to postharvest physiological disorders such as internal breakdown. Size may also determine what markets are available to the producer for the sale of the product. Early season peaches grown in Georgia must be at least 4.1 cm in diameter to be shipped out of the state. Likewise, large sweetpotatoes are used for processing or animal feed rather than fresh market sales.

Size may be determined by one of three general means: 1) dimension (length, width, diameter or circumference); 2) weight; or 3) volume. In some instances, multiple measurements of size for a single product are utilized. For example, the Economic Commission for Europe uses both stem diameter and length for grading asparagus. Standards for trees and shrubs in the United States (American Association of Nurserymen) include such measurements as overall height, trunk diameter, height of branching, number of branches and either diameter and depth of the ball or size of the container.

3.1.2. Shape and Form

Shape (the general outline of the product) and form (the arrangement of individual parts) are also important components of the overall quality of many postharvest products. Shape can be determined precisely by specific measurements or their mathematical relationships; however, more often than not, it is ascertained subjectively. Shape is an important factor in distinguishing between individual cultivars (e.g., among carrots, pears, apples, pecans). Likewise, unacceptable products are often eliminated based on their lack of conformity to a predetermined shape. Thus, like size, shape and form can alter acceptability, potential markets and final use. With ornamental products, both form and shape are important quality criteria. The number, placement and orientation of individual branches/flowers/leaves may be important in ascertaining quality. For example, the British Standard Institution recognizes 11 different forms of fruit trees.

Both shape and form can be altered by a number of factors, of which species, cultivar and production conditions are particularly important. Improper postharvest handling and storage practices may also significantly affect shape and form. Incorrect sleeving of poinsettias and leaf shedding (e.g., in *Ficus benjamina* L.) due to low light are examples of postharvest situations where shape and form can be negatively altered.

3.1.3. Color

The color of an agricultural product probably contributes more to our assessment of quality than any other single factor. Consumers have developed distinct correlations between color and the overall quality of specific products. Tomatoes should be red, bananas yellow. Reverse this (i.e. yellow tomatoes and red bananas) and it would be difficult to give the products away, even if their flavor is superior. Hence, on our first visual assessment of the quality of a product, color is critical.

Color is a function of the light striking the product, the differential reflection of certain wavelengths and our visual perception of those wavelengths. The color we perceive is based on the absorption of some wavelengths and the reflection of others in the visible portion of the electromagnetic spectrum (400 to 770 nm). Reflected light in the 422 to 492 nm range is blue (Figure 4.23), the 535 to 586 nm range yellow and the 647 to 760 nm range red (the division of wave lengths into subsections is rather arbitrary). If there is equal reflection of all wavelengths, the sensation is white, while complete absorption of all wavelengths gives black. What wavelengths are absorbed is determined primarily by the pigmentation of the product (see chapter 4, section 5).

Color can be described by three basic properties: 1) **hue,** the actual color which is a function of the dominant wavelength reflected; 2) **lightness,** the amount of light reflected (this depends not only upon the product but also the intensity of light from the source); and 3) **saturation,** the portion of the total light having a given wavelength. Our perception of a product's color can be altered by changing any of these three properties.

Products with smooth polished surfaces tend to be shiny while those with irregular surfaces are flat in color. Flat coloration comes from the irregular surface reflecting the light at different angles. A bright luster is added to some products (e.g., apples, oranges) by waxing and polishing the surface. Our perception of color can also be altered by the quality of the light striking the product. Artificial lights that do not display a spectrum that closely coincides with that of sunlight distort our perception of the product's color. While color is used as a primary criterion to assess the general quality of many products, quality and color do not necessarily correlate closely with each other. In some cases, the association between what is perceived as optimum color and optimum quality is not at all valid. For example, a number of orange cultivars have fruit that are quite green when at their peak of quality. Since most consumers believe that oranges should be orange, the marketability of green fruit is much diminished. As a consequence, when destined for the fresh market, the fruit are often either dyed or gassed with ethylene to remove the chlorophyll pigments imparting the green color. In pecans, dark brown kernel coloration is considered undesirable. This preference is based upon a gradual increase in darkness with age and rancidity. However, some cultivars of excellent quality are quite dark at optimum harvest. In both cases, color does not accurately reflect the true quality of the product.

A number of factors can affect the color of harvested products. For example, cultivar may have a tremendous effect on color and color stability after harvest. Apple cultivars have been selected whose fruit are primarily red, yellow or green. Likewise, a wide range of carnation cultivars have been selected because of their flower color. This is often the case for flowering ornamentals, where flower color is an extremely important component of over-all quality.

The color of many products changes during development. Banana fruits make the transition from green to yellow upon ripening, indicating an increase in acceptability for consumption. Conversely, the outer leaves of Brussels sprouts change from green to yellow as they begin to deteriorate. Maturity, therefore, may have a significant effect upon product color. Additional preharvest and postharvest factors that can alter the color of agricultural products include nutrition, light, moisture content, season, weather, improper handling, chilling injury and physical damage.

3.1.4. Condition and Absence of Defects

Condition is a somewhat nebulous quality consideration which appears to encompass a wide range of the properties of the product in question. Our assessment of condition may include general visible quality parameters such as color, shape and the freedom from defects. It may also include less readily definable parameters such as freshness, cleanliness and maturity. Freshness often includes the general physical condition of the product, for example the absence of wilting of lettuce or shriveling of fruits. Other characteristics of freshness may be more elusive. For example, aroma may be lost or an undesirable aroma gained as the product's freshness deteriorates. Oil seed crops and nuts develop a rancid odor with the loss of freshness. Freshness may also carry a time association, i.e. recently harvested. Considerable emphasis is now placed on freshness in the purchase of flowers, fruits and vegetables.

Variation is an inherent factor in biological systems and thus in the production of agricultural products. Because of variation, some portion of the total of each harvested commodity will deviate from what is considered optimum for one or more quality components (shape, size, color, etc.). These products display quality defects, the presence of a fault that is undesirable and prevents them from being optimal in quality. Defects can be classified under the following general headings: 1) Biological factors—pathological, entomological and animal; 2) Physiological factors—physiological disorders, nutritional imbalance and maturity; 3) Environmental factors—climate and weather, soils and water supply; 4) Mechanical damage; 5) Extraneous matter—growing medium, vegetable matter and chemical residues; 6) Genetic aberrations.

Defects have both production[137] and post-production causes. Of primary interest from the perspective of postharvest biology are post-production causes of quality loss. The most important of these, from the standpoint of losses incurred, are harvest and postharvest mechanical damage and the occurrence of insect and pathogen problems. In addition, we can exert a much greater control over some causes of defects than others. For example, the weather during production can be little influenced. After harvest, however, it is possible to precisely control the temperature and humidity to which many products are exposed. Understanding the causes for defects, especially those we can influence, often allows the prevention of their occurrence.

3.1.5. Odor

Collectively, a keen sense of taste and smell are extremely important in the survival of animals, in that these senses aid in the identification of desirable foods but, more importantly, avoiding unwholesome or toxic ones. The relative importance of taste and smell can be seen in the fact that of the estimated 40,000 to 75,000 genes in the human genome,[184,466] approximately 1,000 encode chemoreceptors. Odor is perceived through the chemical stimulation of chemoreceptor sites on the olfactory epithelium. In contrast to the four primary taste sensations, a trained person can distinguish over 10,000 distinct odors, some in very minute quantities (e.g., 10^{-9} mg $\cdot$ m^{-3}). Unlike taste, odor may be a primary quality criterion assessed in selecting a product. The aroma of gardenia flowers or other aromatic ornamentals often contributes significantly to the decision to purchase the product.

The chemical compounds that make-up the aromatic properties of plant products must be volatile at the temperature at which the product is utilized, and exhibit at least some degree of water solubility. In recent years, considerable research has been directed toward isolating and identifying the volatile compounds produced by plants (for additional details see chapter 4, section 6). Most volatile compounds do not contribute to the characteristic aroma of a product because their odor cannot be sensed or it occurs in too minute a quantity to be significant. Generally only a small number of volatile compounds are important in the aroma. Volatile compounds may also make-up distinctly undesirable odors. These are especially evident when some products are improperly handled after harvest (e.g., the exposure of Brussels sprouts to anaerobic conditions) or held for too long. Other factors that can influence both desirable and undesirable odors include species, cultivar, maturity and ripening.

Although it is possible to precisely measure the concentration of many of the desirable and undesirable compounds that impart the characteristic odor of a product, this is in fact rarely done in assessing quality. Generally, odor is determined subjectively by simply smelling the product.

3.1.6. Taste

The taste of an edible product is perceived by specialized taste buds on the tongue. Although there are a great many tastes, most appear to involve combinations of four dominant chemical sensations—sweet, sour, bitter and salty. Of these, sweet and sour predominate, with bitterness being important in some products. Saltiness, on the other hand, is seldom a factor in fresh products. Taste represents one of the quality attributes we as consumers try to correlate with visual parameters of the product (e.g., maturity, color, cultivar). This is because taste is seldom a direct quality consideration in deciding whether to purchase a product. Rather, the assessment of taste is made after purchase.

Sweetness due to sugars and sourness from organic acids are dominant components in the taste of many fruits. In some cases, wholesale purchases are made based upon sugar concentration (as indicated by soluble solids) or sugar/acid ratio. Likewise, bitterness may be an important component of the taste of some products (e.g., grapefruit, bitter melon). Many phenolic compounds, in varying levels of polymerization, are major contributors to bitterness. In some products, bitterness is considered a desirable attribute while in others it is undesirable, avoided with cultivar selection (e.g., in cucumbers) and maturity. Of the many preharvest[283] and postharvest factors that can affect the taste quality of products, ripeness, maturity, cultivar, irrigation and fertilization are particularly important.

3.1.7. Texture

Agricultural plant products display a seemingly unlimited array of textural characteristics. These may be external properties, e.g., the surface geometry of an Ice-Plant leaf, or those internal properties that are of critical importance in edible products. Texture comprises those properties of a product that can be appraised visually or by touch. With edible products, texture may also be assessed by skin and muscle senses in the mouth. Textural properties can be divided into three major classifications, with subclasses within each.[409] 1) Mechanical characteristics: hardness—the force necessary to attain a given deformation; cohesiveness—the strength of internal bonds of the product; viscosity—the rate of flow per unit force; elasticity—after deformation, the speed at which the material returns to its original shape; adhesiveness—the work required to overcome the attraction between food and mouth; brittleness—the force required to fracture a product; chewiness—the energy required to masticate a solid food until it can be swallowed; gumminess—the energy required to masticate a semi-solid food until it can be swallowed. 2) Geometrical characteristics: particle size and shape; particle shape and orientation. 3) Other characteristics: moisture content; lipid content.

The internal textural properties of plant products result from the composition and the structure of the cells and their supporting tissues. These structural properties may involve the cell walls *per se* or non-cell wall materials such as storage carbohydrates. Turgor pressure is an extremely important parameter affecting texture in many fleshy products. Physical and chemical differences in structure result in the many dissimilarities found among species and products and within individual products.

Texture may be measured both subjectively and objectively. We subjectively scrutinize the texture of an apple when we bite into it. The textural properties of apples represent one of the single most important components of over-all quality (e.g., crisp to mealy). Texture can also be determined using a more objective means. In recent years, a number of instruments have been developed to objectively measure discrete differences in textural properties among and within individual products. These include shear and compression instruments and instruments that assess texture indirectly by determining the moisture, fiber or lipid content of the product.

As with other quality attributes, a number of preharvest[379] and postharvest factors can affect the texture of plant products. Among these, maturity is one of the most important for a significant cross-section of products. For example, asparagus, beans and peas become fibrous or harden with advancing maturity, while many of the fleshy fruits (e.g., peaches, apples, pears) soften. The handling and storage conditions to which products are exposed after harvest may also significantly alter their textural properties. The loss of water due to improper control of the relative humidity in storage can result in serious textural quality losses. These losses may be reversible, as in leaf lettuce, or largely irreversible (e.g., apples). Exposure of some products to chilling temperatures, even for short periods, may result in textural alterations. For example, holding sweetpotatoes at temperatures below 10°C can result in a condition called "hardcore," where the center of the root becomes woody and inedible. Several other factors that can significantly affect the textural properties of harvested products are cultivar and production practices such as nutrition and irrigation.

3.1.8. Other Components of Quality

Not all of the critical components of quality for all products are described by the characteristics just discussed. Some plant products, due to their end use or unique nature, require specialized quality criteria. For example, with products which are to be used as propagules (e.g., seeds, seedlings, transplants), viability, vigor and purity are of critical importance. With seeds, the freedom from seed-born diseases is also a consideration.

Nutritional quality may be important for some food crops. For example, losses in vitamins during handling and storage are known to occur and may be significant. Although there is a growing interest in nutrition, it is at present rarely utilized by consumers as a quality criterion in discriminating among possible choices within a given commodity.

3.2. Measurement of Quality

Most commercially grown crops are graded for quality at one or more points between harvest and final retail sales. For many, precisely defined quality standards (grades) have been established and are enforced by various government agencies or grower organizations. Assessment of quality may be subjective or objective, depending upon the technique utilized and may be made on each product unit or on only a small sample (i.e., sampling). Sampling, however, cannot be used to segregate undesirable from superior product units; rather it can only give an estimate of the overall quality of the entire lot. Likewise, assessment can be either nondestructive or destructive, the latter being done on only a very small sample. Deficiencies in subjective methods and in sub-sample analysis have made the automation of objective, nondestructive methods a major preoccupation of agricultural engineers, since it ideally allows the accurate assessment of every product unit. Objective techniques also tend to increase the consistency and uniformity of quality evaluation in comparison with subjective techniques.

For commercial crops, generally more than one criterion is utilized in assessing quality. For example, a subjective visual evaluation may be made by trained personnel for surface characteristics such as shape, disease, insect damage, discoloration and wilting. This type of evaluation is limited to characteristics that can be determined visually. Subsequently, the product may be screened using one of many possible automated objective measurements.[2] These include characteristics such as size, weight, specific gravity, mechanical properties, electromagnetic properties, electrochemical properties, and chemistry. Since the methods for subjective quality evaluation are very diverse, the following critique focuses on objective measures of quality.

3.2.1. Size

Size separation of individual product units may be accomplished using a number of techniques depending upon the product in question. Screens of perforated plates or wire cloth with various hole sizes and shapes are used for the separation of small products such as seeds. Size separation may also be made using belts with holes, cups, rollers or diverging rollers. For example, Chinese water chestnuts are graded for size using two rollers placed in a V configuration. The corms are fed onto the narrow end of the rollers, as they proceed toward the base, the distance between the rollers becomes greater. Eventually the chestnut drops onto one of several conveyers beneath, each collecting corms of a certain size.

3.2.2. Weight

As with size, a variety of techniques have been developed for the separation of products based upon their weight. Often weight is a more precise measure for separation than size (diameter) in that weight generally varies with the cube of the diameter of regularly-shaped products.

Typically a cup or pocket holding a single unit moves along a path where the weight required to drop the product into its appropriate class changes with distance. The actual weight required to activate the tripping device is generally adjustable so that the machine may be used for more than one type of product.

3.2.3. Shape

Instrumentation for shape separation has been developed for some plant products. Mechanical means of this type of quality assessment are especially prevalent in the grading of seeds. Disc, cylinder, incline belt and spiral separators may be used to select for or against length or roundness. Separators may also be used for cleaning seed lots. More recently, the widespread availability of computers and the development of computer imaging technology has stimulated the development of automated shape analysis.[385] It is currently possible to compute the area, perimeter, maximum length, width, centroid, horizontal and vertical fret, curvature, circularity and form factor of harvested products.[188] Techniques for sorting products such as tomatoes,[383] apples,[354] sweetpotatoes[466] and other agricultural products for shape have been developed. Interfacing computer vision with robotics will have a pronounced effect on the grading of harvested products in the future.

3.2.4. Density/Specific Gravity

The density or specific gravity of individual product units can be an important quality criterion for some crops.[473] Three general techniques utilized for the measurement of density are flotation, fluidized-beds and machine vision + weighing. Flotation separation uses a liquid medium in which the specific gravity may be controlled (e.g., salt solutions). A specific gravity is selected that allows the heavier products to sink or the lighter products to float to the surface. Specific gravity, measured in this manner, has been shown to correlate with the quality of processed tomatoes.[458]

When it is not desirable to wet the product, fluidized-bed technology may be applicable. A continuous stream of air is forced through a bed of granular particles to produce enough force to counterbalance the particle weight. The particles, in this case the product, perform like a liquid, such that very light and very dense product units can be separated. In addition to density, however, the product's aerodynamic characteristics can significantly influence the separation causing irregularly shaped product units to perform aberrantly. Fluidized-bed separation is especially useful for seeds of species in which density is correlated with vigor.[399] Air is also extensively used for the removal of foreign material from the product.

The combination of machine vision and automated weighing can be used to determine the density of products such as fruits for which flotation and fluidized-bed techniques are not applicable. Some commercial citrus packinghouses utilize this technique, which can grade 5–6 fruit $\cdot$ sec^{-1} $\cdot$ lane^{-1}.[427]

3.2.5. Mechanical Properties

Mechanical properties such as firmness represent critical quality attributes in many products. Consumers commonly test fruit firmness subjectively by squeezing the product and assessing the level of resistance. Many of the objective methods suitable for packinghouse operations involve applying a sufficient load (force) to the fruit or vegetable for the tissue to begin to de-

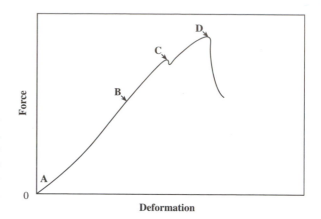

Figure 6.24. Textural properties can be ascertained using a force-deformation curve of a cylindrical sample of the product. In this example of apple tissue under compression, individual points on the curve are: A. onset of deformation; B. point of inflection; C. yield point; and D. rupture point (*after Abbott et al.*[2]).

form and monitoring the resulting force-deformation curve (Figure 6.24). In the first region of the curve (A-B), the force and deformation increase more or less linearly. When the force is released, due to the tissue's elasticity, it reverts back to its original shape with no damage. In region B-C, the deformation increases until reaching the yield point (C), where the deformation increases without a corresponding increase in force. Here the tissue begins to rupture, and the yield point is indicative of mechanical failure of the microstructure. Rupture occurs at D when the macrostructure fails. Each region on the curve can yield information about the mechanical properties of the tissue; however, only the first region (A-B) is nondestructive. Past this point, irreversible damage to the tissue occurs. Thus, it is possible to measure the mechanical properties either nondestructively or destructively, the latter being far more accurate at this time; therefore, it has become the standard in commercial settings. Destructive measurements, in contrast, can be used only on a very small sample of the product; hence it is not possible to measure the texture of all product units and grade them into firmness categories. Several types of measurement techniques have been tested (i.e., quasi-static or dynamic loading, impact, low-frequency vibrations, sonic vibrations and ultrasonics).[2]

Numerous destructive mechanical instruments have been developed for agricultural products that measure force, deformation and dynamic force-deformation. Loading types include puncture, shear, compression, crushing, bending, extrusion, twisting and tension, with puncture[5] and shear[18,22] being the most common. Force-type testers, principally used for firm fruits (e.g., apples, pears), are less accurate for fruits having softer tissues (e.g., tomatoes, cherries). Deformation-type instruments monitor the deformation of the tissue under a constant load for a specific time interval. Variations have been developed for nondestructive measurements.[410] Dynamic force-deformation instruments have been tested on apples, kiwifruit[3] and other fruit. One nondestructive mechanical technique utilizes air pressure as the mechanical force,[193] causing a slight deformation in the tissue that is correlated with the product's firmness.

Impact techniques analyze the change in energy that occurs when two objects collide.[301] Impact response curves, obtained by the use of drop, falling mass or other techniques, are related to firmness. Impact techniques that are nondestructive, however, remain largely experimental.

Low frequency vibration techniques were initially developed for the separation of seeds. They are based upon the principle that the bounce of individual product units, caused by an imposed vibrational energy, is dependent upon the vibrational frequency. The product's response is related to its mechanical properties. Prototypes have been tested on several high moisture crops (e.g., muscadine grapes, tomatoes)[166,185] with varying success.

Sonic techniques utilize sound waves within the audible range to apply a force to the product, which, in turn, vibrates at the same frequency as the applied force. The resonance is related to the mechanical properties of the product. Ultrasonic sensing, in contrast, utilizes sound waves above the audible frequency to assess the effect of the product on attenuating the sound. While sound, vibrational, impact and certain force-deformation techniques are non-destructive, none are sufficiently accurate at this time for measuring the textural properties of high moisture products.

3.2.6. *Electromagnetic properties*

The electromagnetic spectrum exhibits an exceptionally wide range in wavelengths, from radio waves to gamma rays.* Certain segments of the spectrum can be used to measure both external and internal quality attributes of plant products.

a. *Optical*

The use of light in quality evaluation has increased substantially in recent years due to the amount of information that can be derived, the speed of which it is accomplished and its non-destructive nature. Among optical techniques are devices that use visible, near infrared or ultraviolet regions of the spectrum.

Light striking a plant can undergo several possible fates (Figure 6.25). It may be reflected from the surface of the product without entering the surface cells, such as light striking a polished metal surface. However, generally only a very small amount of light is reflected. Most of the light enters the product and goes through a series of interactions which randomly include refraction, reflection, absorption and transmission.† Some of the light will exit the sample at points from 0 to 360 degrees from the point of illumination. The character of the light exiting the sample will have been changed by the absorption process, and these changes can be directly related to the chemical make up of the absorbing substance. Therefore, light exiting the product can be used to measure internal characteristics, such as the color, the level of specific chemical components, the presence of certain internal disorders, and other quality attributes.

When the light is composed of a cross-section of wavelengths in the visible spectrum, pigments found below the surface absorb specific wavelengths. For example, chlorophyll absorbs red light and blue light but very little in the green region of the spectrum. The green color we perceive for chlorophyll is from this non-absorbed light moving back out of the tissue. Wavelengths that are not absorbed can be readily measured,[117] the basis for electronic color sorters. Absorbed wavelengths can be determined by the difference between the spectral composition of the light entering versus that leaving the sample.

In some instances, the amount of light absorbed at specific wavelengths correlates with the quantity of an absorbing component. Analysis of the light exiting the product (not absorbed) can be used to measure internal characteristics. For example, the presence of hollow heart in potato tubers can be non-destructively detected with light.[51] Since the disorder is found in the interior of the tuber, while the exterior is devoid of any physical manifestations,

---

*Wavelength range for various regions of the electromagnetic spectrum (radiowave 3mm—30,000m; microwave 3mm–300mm; infrared 0.75μm–1000μm; visible 400nm–770nm; ultraviolet 4nm–400nm; X-ray 0.002nm–100nm; gamma ray 0.00005nm–0.002nm). Some regions overlap or form subregions of others.
†Other terms used for the various fates of light are: spectacular = surface reflectance; reflectance = body reflectance; body transmittance = transmitted.

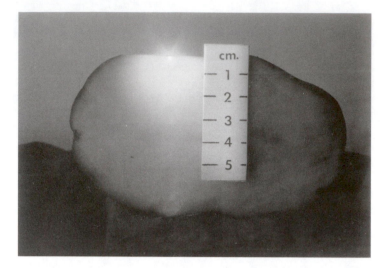

Figure 6.25. As illustrated with this longitudinal section of potato tuber, much of the light striking a plant product moves into the tissue, with only a very small amount being reflected from the surface. Within the product, the light undergoes a series of random interactions (refraction, reflection, absorption and transmission) with some of the light exiting the sample 0 to 360 degrees from the point of illumination. The character of the light exiting has been changed by the absorption process and these changes can be directly related to the chemical makeup of the absorbing substance (*Photograph courtesy of G.G. Dull*).

dissection was the only means of detection prior to optical analysis. Likewise, light in the near infrared portion of the spectrum can be used to non-destructively determine the soluble solids,[116] acidity, dry matter, and sugar content in a wide variety of fruits and vegetables.[210] The use of light, therefore, is a very powerful, nondestructive tool for determining not only the stage of maturity but the quality of harvested products.

Instruments have now been developed that identify potato tubers with hollow heart[51] and screen onion bulbs[53] and potato tubers[116] for percent dry matter, papayas for internal color,[52] honeydew and cantaloupes for soluble solids[118] and apples for water-core[54] using near infrared spectroscopy. It may eventually be possible to nondestructively measure the internal concentration of specific sugars within harvested products. Commercial on-line instruments with high through-put rates are now available for certain applications. The technology allows grading each individual product unit nondestructively into quality categories.

Machine vision represents a variation of optical techniques that allows performing visual-based inspection of the product where both the spatial content and the intensity of the color can be monitored objectively. With the use of cameras, the output at one or multiple wavelengths is digitized for analysis. The type of light used to illuminate the sample depends upon the quality attributes of interest. Products can be graded on-line on the basis of size, shape, maturity and surfaced defects.

b. Fluorescence and delayed light emission

Both fluorescence and delayed light emission have been studied as a means of assessing chlorophyll content, which in some products is related to maturity or certain quality attributes.

Fluorescence light is emitted by the chlorophyll molecule following excitation by high-energy light. Upon relaxation, there is an instantaneous emission of lower-energy light from the molecule. In delayed light emission, some of the energy initially trapped by the chlorophyll molecule is transferred back through reverse reactions to reexcite the molecule, resulting in the emission of fluorescence light. Excitation and emission spectra between fluorescence and delayed light emission are similar; however, the timing between excitation and emission differs. Fluorescence has been used to monitor ripening and chilling injury,[396] and delayed light emission for maturity[144,197] and mechanical injury.[4] Both techniques remain experimental at this time.

c. *X-ray and gamma ray*

The high energy of gamma and X-rays, much of which is capable of passing completely through a product, has been utilized to assess conditions within the interior. The product is exposed to gamma or X-rays (typically the latter), and the pattern of the energy exiting the product is indicative of certain internal quality attributes (e.g., cellular breakdown, bruising, softening, hydration, dehydration, internal decay, insect presence or injury, internal structure). Therefore, internal disorders such as cork spot, bitter pit, watercore[421] and brown core can be detected. X-ray instruments are currently used commercially for hollow-heart detection in potato and freeze damage in citrus. The primary deficits preventing wider use of the technology are equipment and operating costs and safety considerations.

d. *Magnetic resonance and magnetic resonance imaging*

In a magnetic field, certain nuclei (e.g., ^1H, ^{13}C, ^{31}P, ^{23}Na) interact with radiation in the radio frequency range. The nuclei orient to the magnetic field, and upon exposure to radio frequency radiation, they become excited to a higher energy level, reorienting the nuclei. With the removal of the radio frequency excitation, the nuclei return to their original energy level, realigning (relaxation) through the release of energy, which can be detected. Magnetic resonance images can be constructed to visualize the interior of products.[90] The technique has been used to determine maturity[377] and detect watercore,[420] core breakdown, chilling injury, insect damage,[414] bruises, dry regions and other internal defects. At present, the acceptance of magnetic resonance for quality assessment of agricultural products has been hindered by its high cost and the relatively slow through-put rate of product.

e. *Dielectric and electrical properties*

The resistance or conductance of a wide cross-section of fruits and vegetables has been studied.[306] In general, electrical properties are not sufficiently sensitive for quality evaluation.

3.2.7. *Electrochemical Properties*

Ripening and maturity have been assessed with the use of electronic sensors that respond to volatiles emitted from fruits. As the fruit ripens or matures, quantitative and qualitative changes in the volatile profile occur. Certain volatiles bind to different receptors, altering their electrical potential, which is monitored. Electrochemical techniques have been used to monitor ripening and maturity of several fruits.[35,395] The primary deficits at this time are precision and the length of time required between individual measurements.

3.2.8. Surface Characteristics

Air trapped on the surface of the product alters the rate at which the product settles in a solution. Chemicals are added to the solution to enhance air trapping. Generally, this technique is used for separating damaged fruits, which tend to have more air adhering to their surface, causing them to float. Machine vision, likewise, can be used to detect surface characteristics, especially deficits due to abrasions and cuts.

3.2.9. Chemical Analyses

Individual products can be measured for various chemical characteristics (e.g., moisture, sugar, soluble solids, acidity, pH, impurities, rancidity). With the exception of spectrophotometric measurements, chemical analyses tend to be destructive. Therefore, normally only a small representative sample is measured. This information is then used to infer the condition of the larger quantity from which the sample was drawn. Chemical analyses are not used to determine the quality of each individual unit of the product.

3.3. Quality Standards

While most individuals have a subjective set of quality standards for the products with which they are familiar, these tend to be highly arbitrary. What is acceptable quality to one consumer may be totally unacceptable to another. Likewise, what is considered to be acceptable quality to the seller may not coincide with the opinion of the buyer. Our individual quality criteria are also subject to variation due to changes in supply and demand. When a particular product is scarce, we may accept a level of quality that would be unacceptable when the supply is abundant. The need for a well-defined set of standards to characterize quality was evident even in the Roman era, when the Emperor Diocletian decreed standards for many food products.[424]

The advantages of standards are twofold. First, standards protect the purchaser by providing a uniform quality product. This is especially important when one purchases products without having seen them; the buyers need to know what they will receive. Secondly, standards typically bring about a general upgrading of the overall quality of a commodity. Producers, aware of the economic advantages of providing a top quality product, have the incentive to alter their production practices to achieve as high of standard as possible.

Standards are established by individual countries or groups of countries (e.g., the European Economic Community), by provinces, states, counties or municipal authorities within individual countries, by grower organizations and by buyers. There is an increasing impetus for worldwide standards for many agricultural products. These standards are being established by the Codex Alimentarius Commission.

Compliance with standards may be voluntary or mandatory, depending upon the organization establishing them and the authorization under which it operates. Standards established or operated by government organizations are generally mandatory.

An important requisite for quality standards is a clear, concise and accurate description of each criterion used to determine quality and the acceptable range for each criterion within a specific grade. For example, the upper and lower diameter limits for each grade of lemon and orange fruits is defined, as well as the ranges for other quality components. Rigorous characterization of grades leads to uniformity in grading between individuals, locations and time during the season.

ADDITIONAL READING

Abeles, F.B., P.W. Morgan and M.E. Saltveit, Jr.1992. *Ethylene in Plant Biology.* Academic Press, New York.

Bartz, J.A., and J.K. Brecht. 2003. *Postharvest Physiology and Pathology of Vegetables.* Dekker, New York.

Coen, E.S. 1992. The role of homeotic genes in flower development and evolution. *Annu. Rev. Plant Physiol. Plant Mol. Biol.* 42:241–279.

Florkowski, W.J., S.E. Prussia and R.L. Shewfelt (eds.). 2000. *Integrated View of Fruit and Vegetable Quality.* Technomic Pub., Lancaster, PA.

Fry, S.C. 1995. Polysaccharide-modifying enzymes in the plant cell wall. *Annu. Rev. Plant Physiol. Plant Mol. Biol.* 46:497–520.

Galili, G., and J. Kigel (eds.). 1995. *Seed Development and Germination.* Dekker, New York.

Kende, H. 1993. Ethylene biosynthesis. *Annu. Rev. Plant Physiol. Plant Mol. Biol.* 44:283–307.

Kieber, J.J. 1997. The ethylene response pathway in *Arabidopsis. Annu. Rev. Plant Physiol. Plant Mol. Biol.* 48:277–296.

Knee, M. (ed.). 2002. *Fruit Quality and Its Biological Basis.* CRC Press, Boca Raton, FL.

Koornneef, M., C. Alonso-Blanco, A.J.M. Peeters and W. Soppe. 1998. Genetic control of flowering time in *Arabidopsis. Annu. Rev. Plant Physiol. Plant Mol. Biol.* 49:345–370.

Larkins, B.A., and I.K. Vasil (eds.). 1997. *Cellular and Molecular Biology of Plant Seed Development.* Kluwer Acad., Boston, MA.

Lumsden, P.J. 1992. Circadian rhythms and phytochrome. *Annu. Rev. Plant Physiol. Plant Mol. Biol.* 42:351–371.

Matile, P., S. Hörtensteiner and H. Thomas. 1999. Chlorophyll degradation. *Annu. Rev. Plant Physiol. Plant Mol. Biol.* 50:67–95.

McCarty, D.R. 1995. Genetic control and integration of maturation and germination pathways in seed development. *Annu. Rev. Plant Physiol. Plant Mol. Biol.* 46:71–93.

Nelson, O., and D. Pan. 1995. Starch synthesis in maize endosperms. *Annu. Rev. Plant Physiol. Plant Mol. Biol.* 46:475–496.

Noodén, L.D., and A.C. Leopold (eds.). 1988. *Senescence and Aging in Plants.* Academic Press, New York.

Shewfelt, R.L., and S.E. Prussia (eds.). 1993. *Postharvest Handling: A Systems Approach.* Academic Press, New York.

REFERENCES

1. Abbas, M.F., and R.A.M. Saggar. 1989. Respiration rate, ethylene production, and certain chemical changes during the ripening of jujube fruit. *J. Hort. Sci.* 64:223–225.

2. Abbott, J.A., R. Lu, B.L. Upchurch and R.L. Stroshine. 1997. Technologies for nondestructive quality evaluation of fruits and vegetables. *Hort. Rev.* 20:1–120.

3. Abbott, J.A., and D.R. Massie. 1993. Nondestructive firmness measurement of apples. *Amer. Soc. Agr. Eng. Paper* 93–6025.

4. Abbott, J.A., A.R. Miller and T.A. Campbell. 1991. Detection of mechanical injury and physiological breakdown of cucumbers using delayed light emission. *J. Amer. Soc. Hort. Sci.* 116:52–57.

5. Abbott, J.A., A.E. Watada and D.R. Massie. 1976. Effe-gi, Magness-Taylor, and Instron fruit pressure testing devices for apples, peaches, and nectarines. *J. Amer. Soc. Hort. Sci.* 101:698–700.

6. Abeles, F.B. 1968. Role of RNA and protein synthesis in abscission. *Plant Physiol.* 43:1577–1586.

7. Abeles, F.B., L.J. Dunn, P. Morgans, A. Callahan, R.E. Dinterman and J. Schmidt. 1988. Induction of 33-kD and 60-kD peroxidases during ethylene induced senescence of cucumber cotyledens. *Plant Physiol.* 87:609–615.

8. Abeles, F.B., and F. Takeda. 1990. Cellulase activity and ethylene in ripening strawberry and apple fruits. *Scient. Hort.* 42:269–275.

9. Adams, C.A., and R.W. Rinne. 1981. Seed maturation in soybeans (*Glycine max,* L. Merr.) is independent of seed mass and of the parent plant, yet is necessary for production of viable seeds. *J. Exp. Bot.* 32:615–620.

10. Addicott, F.T. 1982. *Abscission.* University Calif. Press, Berkeley, CA.

11. Aharoni, A., L.C. P. Keizer, H.J. Bouwmeester, Z. Sun, M. Alvarez-Huerta, H.A. Verhoeven, J. Blaas, A.M.M.L. von Houwelingen, R.C.H. De Vos, H. van der Voet, R.C. Jansen, M. Guis, J. Mol, R.W. Davis, M. Schena, A.J. van Turen and A.P. O'Connell. 2000. Identification of the *SAAT* gene involved in strawberry flavor biogenesis by use of DNA microarrays. *Plant Cell* 12:647–661.

12. Ahumada, M., and M. Cantwell. 1996. Postharvest studies on pepino dulce (*Solanum muricatum* Ait.)— maturity at harvest and storage behavior. *Postharv. Biol. Tech.* 7:129–136.

13. Akamine, E.T., and T. Goo. 1978. Respiration and ethylene production in mammee apple (*Mammea americana* L.). *J. Amer. Soc. Hort. Sci.* 103:308–310.

14. Akamine, E.K., and T. Goo. 1979. Respiration and ethylene production in fruits of species and cultivars of *Psidium* and species of *Eugenia. J. Amer. Soc. Hort. Sci.* 104:632–635.

15. Alba, R., M.M. Cordonnier-Pratt and L.H. Pratt. 2002. Fruit-localized phytochromes regulate lycopene accumulation independent of ethylene production in tomato. *Plant Physiol.* 123:363–370.

16. Al-Niami, J.H., R.A.M. Saggar and M.F. Abbas. 1992. The effect of temperature on some chemical constituents and storage behavior of jujube fruit cv. Zaytoni. *Basrah J. Agric. Sci.* 2:31–36.

17. Altschul, A.M., L.Y. Yatsu, R.L. Ory and E.M. Engleman. 1966. Seed proteins. *Annu. Rev. Plant Physiol.* 17:113–136.

18. Anthistel, M.J. 1961. The composition of peas in relation to texture. *Fruit Veg. Cann. Quick Freeze Res. Assoc., Sci. Bull.,* No. 4.

19. Appleby, R.S., M.I. Gurr and B.W. Nichols. 1974. Studies on seed-oil triglycerides. Factors controlling the biosynthesis of fatty acids and acyl lipids in subcellular organelles of maturing *Crambe abyssinica* seeds. *Eur. J. Biochem.* 48:209–216.

20. Arditti, J., and H. Flick. 1976. Post-pollination phenomena in orchid flowers. VI. Excised floral segments of Cymbidium. *Amer. J. Bot.* 63:201–211.

21. Arthey, V.D. 1975. *Quality of Horticultural Products.* Butterworth, London.

22. Arthey, V.D., and C. Webb. 1969. The relationship between maturity and quality of canned broad beans (*Vicia faba* L.). *J. Food Technol.* 4:61–74.

23. Ashman, T.-L., and D.J. Schoen. 1996. Floral longevity: Fitness consequences and resource costs. Pp. 112–139. In: *Floral Biology.* D.G. Lloyd and S.C.H. Barrett (eds.). Chapman and Hall, New York.

24. Ayub, R., M. Guis, M. Ben Amor, L. Gillot, J.P. Roustan, A. Latché, M. Bouzayen and J.C. Pech. 1996. Expression of ACC oxidase antisense gene inhibits ripening of cantaloupe melon fruits. *Nature Biotechnol.* 14:862–866.

25. Baldwin, E.A., M.O Nisperos-Carriedo and J.W. Scott. 1992. Levels of flavor volatiles in a normal cultivar, ripening inhibitor and their hybrid. *Proc. Fla. State Hort. Soc.* 104:86–89.

26. Baldwin, E.A., J.W. Scott, C.K. Shewmaker and W. Schuch. 2000. Flavor trivia and tomato aroma: Biochemistry and possible mechanisms for control of important aroma components. *HortScience* 35:1013–1022.

27. Barry, C.S., M. Immaculada Llop-Tous and D. Grierson. 2000. The regulation of 1-aminocyclopropane-1-carboxylic acid synthase gene expression during the transition from system-1 to system-2 ethylene synthesis in tomato. *Plant Physiol.* 123:979–986.

28. Barton, L.V. 1965. Seed dormancy: General survey of dormancy types in seeds. Pp. 699–720. In: *Handbook of Plant Physiology.* Vol. 15a(2). W. Rukland (ed.). Springer-Verlag, Berlin.

29. Baskin, C.C., and J.M. Baskin, 1998. *Seeds. Ecology, Biogeography and Evolution of Dormancy and Germination.* Academic Press, San Diego, CA.

30. Bauchot, A.D., D.S. Mottram, A.T. Dodson and P. John. 1998. Effect of aminocyclopropane-1-carboxylic acid oxidase antisense gene on the formation of volatile esters in cantaloupe Charentais melon (cv. Védrandais). *J. Agri. Food Chem.* 46:4787–4792.

31. Beaudry, R.M. 1999. Effect of O_2 and CO_2 partial pressure on selected phenomena affecting fruit and vegetable quality. *Postharv. Biol. Tech.* 15:293–304.

32. Beaudry, R.M., J.A. Payne and S.J. Kays. 1985. Variation in the respiration of harvested pecans due to genotype and kernel moisture level. *HortScience* 20:752–754.

33. Beaudry, R.M., R.F. Severson, C.C. Black and S.J. Kays. 1989. Banana ripening: Implications of changes in glycolytic intermediate concentrations, glycolytic and gluconeogenic carbon flux, and fructose 2,6-bisphosphate concentration. *Plant Physiol.* 91:1436–1444.

34. Beevers, L., and R. Poulson. 1972. Protein synthesis in cotyledons of *Pisum sativum* L. I. Changes in cell-free amino acid incorporation capacity during seed development and maturation. *Plant Physiol.* 49:476–481.

35. Benady, M., J.E. Simon, D.J. Charles and G.E. Miles. 1995. Fruit ripeness determination by electronic sensing of aromatic volatiles. *Trans. Amer. Soc. Agr. Eng.* 38:251–257.

36. Ben-Aire, R., S. Lavee and S. Guelfat-Reich. 1970. Control of woolly breakdown of 'Elberta' peaches in cold storage by intermittent exposure to room temperature. *Proc. Amer. Soc. Hort. Sci.* 95:801–802.

37. Ben-Yehoshua, S. 1964. Respiration and ripening of discs of the avocado fruit. *Plant Physiol.* 17:71–80.

38. Berger, R.G. 1991. Fruits I. Pp. 283–304. In: *Volatile Compounds in Foods and Beverages.* H. Maarse (ed.). Marcel Decker, New York.

39. Bergman, C.J., J.T. Delgado, R. Bryant, C.Grimm, K.R. Cadwallader and B.D. Webb. 2000. Rapid gas chromatographic technique for quantifying 1-acetyl-1-pyrroline and hexanal in rice (*Oryza sativa,* L.). *Cereal Chem.* 77:454–458.

40. Bewley, J.D. 1997. Seed germination and dormancy. *Plant Cell* 9:1055–1066.

41. Bewley, J.D., and M. Black. 1995. *Seeds. Physiology of Development and Germination.* Plenum, New York.

42. Bewley, J.D., A.R. Kermode and S. Misra. 1989. Desiccation and minimal drying ("undrying") treatments of seeds of caster bean and *Phaseolus vulgaris* which terminate development and promote germination cause changes in protein and messenger RNA synthesis. *Ann. Bot.* 63:3–17.

43. Beyer, E.M. 1976. A potent inhibitor of ethylene action in plants. *Plant Physiol.* 58:268–271.

44. Biale, J.B. 1960. Respiration of fruits. Pp. 536–592. In: *Encyclopedia of Plant Physiology.* Vol. 12(2). W. Ruhland (ed.), Springer, Berlin.

45. Biale, J.B. 1964. Growth, maturation, and senescence in fruits. *Science* 146:880–888.

46. Biale, J.B. 1976. Recent advances in postharvest physiology of tropical and subtropical fruits. *Acta Hort.* 57:179–187.

47. Biale, J.B., and D.E. Barcus. 1970. Respiratory patterns in tropical fruits of the Amazon Basin. *Trop. Sci.* 12:93–104.

48. Biale, J.B., and R.E. Young. 1971. The avocado pear. Pp. 65–105. In: *The Biochemistry of Fruits and Their Products.* Vol 1. A.C. Hulme (ed). Academic Press, London.

49. Biale, J.B., and R.E. Young. 1981. Respiration and ripening in fruits _ retrospect and prospect. Pp. 1–39. In: *Recent Advances in the Biochemistry of Fruits and Vegetables.* J. Friend and M.J.C. Rhodes (eds.), Academic Press, New York.

50. Bianco, J., G. Garello and M.T. Le Page-Degivry. 1994. Release of dormancy in sunflower embryos by dry storage: involvement of gibberellins and abscisic acid. *Seed Sci. Res.* 4:57–62.

51. Birth, G.S. 1960. A nondestructive technique for detecting internal discoloration in potatoes. *Amer. Potato J.* 37:53–60.

52. Birth, G.S., G.G. Dull, J.B. Magee, H.T. Chan and C.G. Cavaletto. 1984. An optical method for estimating papaya maturity. *J. Amer. Soc. Hort. Sci.* 109:62–66.

53. Birth, G.S., G.G. Dull, W.T. Renfroe and S.J. Kays. 1985. Nondestructive spectrophotometric determination of dry matter in onion. *J. Amer. Soc. Hort. Sci.* 110:297–303.

54. Birth, G.S., and K.L. Olsen. 1964. Nondestructive detection of watercore in Delicious apples. *Proc. Amer. Soc. Hort. Sci.* 85:74–84.

55. Biswal, U.C., and B. Biswal. 1984. Photocontrol of leaf senescence. *Photochem. Photobiol.* 39:875–879.

56. Blackman, S.A., R.L. Obendorf and A.C. Leopold. 1992. Maturation proteins and sugars in desiccation tolerance of developing soybean seeds. *Plant Physiol.* 100:225–230.

57. Blanpied, G.D. 1972. A study of ethylene in apple, red raspberry and cherry. *Plant Physiol.* 48:627–630.

58. Blumenfeld, A. 1980. Fruit growth of loquat. *J. Amer. Soc. Hort. Sci.* 105:747–750.

59. Booncherm, P. 1990. Physiological changes in durians (*Durio zibethinus* Murr.) cv. Chanee after harvest. MS. Thesis, Kasetsart Univ., Bangkok, 57p.

60. Borisjuk, L., S. Walenta, H. Rolletschek, W. Mueller-Klieser, U. Wobus and H. Weber. 2002. Spatial analysis of plant metabolism: sucrose imaging with *Vicia faba* cotyledons reveals specific developmental patterns. *Plant J.* 29:521–530.

61. Borochov, A., and W.R. Woodson. 1989. Physiology and biochemistry of flower petal senescence. *Hort. Rev.* 11:15–43.

62. Borrolo, K., and L. Dure, III. 1987. The globulin seed storage proteins of flowering plants are derived from two ancestral genes. *Plant Mol. Biol.* 8:113–131.

63. Boss, P.K., C. Davies and S.P. Robinson. 1996. Analysis of the expression of anthocyanin pathway genes in developing *Vitis vinifera* L. cv. Shiraz grape berries and the implications for pathway regulation. *Plant Physiol.* 111:1059–1066.

64. Bourne, M.C. 1980. Texture evaluation of horticultural crops. *HortScience* 15:51–57.

65. Bovy, A.G., G.C. Angenent, H.J.M. Dons and A.C. von Altvorst. 1999. Heterologous expression of the *Arabidopsis* etr1-1 allele inhibits the senescence of carnation flowers. *Mol. Breed.* 5:301–308.

66. Brady, C. 1988. Nucleic acid and protein synthesis. Pp. 147–179. In: *Senescence and Aging in Plants.* L.D. Noodén and A.C. Leopold (eds.). Academic Press, New York.

67. Brady, C.J. 1987. Fruit ripening. *Annu. Rev. Plant Physiol.* 38:155–178.

68. Bray, E.A. 1991. Regulation of gene expression by endogenous ABA during drought stress. Pp. 81–98. In: *Abscisic Acid Physiology and Biochemistry.* W.J. Davies and H.G. Jones (eds.). Bios, Oxford, England.

69. Broughton, W.J., and H.C. Wong. 1979. Storage conditions and ripening of chiku fruits, *Achrssapota* L. *Scientia Hort.* 10:377–385.

70. Brown, B.I., L.S. Wong, A.P. George and R.J. Nissen. 1988. Comparative studies on the postharvest physiology of fruit from different species of Annona (custard apple). *J. Hort. Sci.* 63:521–528.

71. Brownleader, M.D., P. Jackson, A. Mobasheri, A.T. Pantelides, S. Sumar, M. Trevan and P.M. Dey. 1999. Molecular aspects of cell wall modifications during fruit ripening. *Crit. Rev. Food Sci. Nutr.* 39:149–164.

72. Brummell, D.A., B.D. Hall and A.B. Bennett. 1999. Antisense suppression of tomato endo-1,4-β-glucanase Cel2 mRNA accumulation increases the force required to break fruit abscission zones but does not affect fruit softening. *Plant Mol. Biol.* 40:615–622.

73. Buchanan-Wollaston, V. 1997. The molecular biology of leaf senescence. *J. Exp. Bot.* 48:181–199.

74. Bureau, S.M., R.L. Baumes and A.J. Razungles. 2000. Effects of vine or bunch shading on the glycosylated flavor precursors in grapes of *Vitis vinifera* L. cv. Syrah. *J. Agri. Food Chem.* 48:1290–1297.

75. Burg, S.P., and E.A. Burg. 1962. Role of ethylene in fruit ripening. *Plant Physiol.* 37:179–189.

76. Burg, S.P., and E.A. Burg. 1967. Molecular requirements for the biological activity of ethylene. *Plant Physiol.* 42:144–152.

77. Burton, W.G. 1957. The dormancy and sprouting of potatoes. *Food Sci. Abstr.* 29:1–12.

78. Buttery, R.G. 1993. Quantitative and sensory aspects of flavor of tomato and other vegetables and fruits. Pp. 259–286. In: *Flavor Science: Sensible Principles and Techniques.* T.E. Acree and R. Teranishi (eds.). Amer. Chem. Soc., Washington, DC.

79. Buttery, R.G., and L.C. Ling. 1993a. Enzymatic production of volatiles in tomatoes. Pp.137–146. In: *Flavor Precursors.* P. Schreier and P. Winterhalter (eds.). Allured Publ., Wheaton, IL.

80. Buttery, R.G., and L.C. Ling. 1993b. Volatiles of tomato fruit and plant parts: Relationship and biogenesis. Pp. 23–34. In: *Bioactive Volatile Compounds from Plants.* R. Teranishi, R. Buttery and H Sugisawa (eds.). Amer. Chem. Soc., Washington, DC.

81. Buttery, R.G., L.C. Ling, B.O. Juliano and J.G. Turnbaugh. 1983. Cooked rice aroma and 2-acetyl-1-pyrroline. *J. Agri. Food Chem.* 31:823–826.

82. Buttery, R.G., R. Teranishi, L.C. Ling, R.A. Flath and D.J. Stern. 1988. Quantitative studies on origins of fresh tomato aroma volatiles. *J. Agri. Food Chem.* 36:1247–1250.

83. Cameron, A.C., and M.S. Reid. 1983. Use of silver thiosulfate to prevent flower abscission from potted plants. *Sci. Hort.* 19:373–378.

84. Carpita, N.C., and D.M. Gibeaut. 1993. Structural models of primary cell walls in flowering plants: consistency of molecular structure with the physical properties of the walls during growth. *Plant J.* 3:1–30.

85. Chan, Y.K., Y.H. Yan and N. Li. 1998. Low-temperature storage elicits ethylene production in nonclimacteric lychee (*Litchi chinensis* Sonn.) fruit. *J. Amer. Soc. Hort. Sci.* 33:1228–1230.

86. Chen, P., and Z. Sun. 1989. Nondestructive methods for quality evaluation and sorting of agricultural products. *Proc. 4th Intern. Conf. on Properties of Agric. Materials,* Rostock, FDR.

87. Chin, L.-H., Z.M. Ali and H. Lazan. 1999. Cell wall modifications, degrading enzymes and softening of carambola fruit during ripening. *J. Exp. Bot.* 50:767–775.

88. Chrispeels, M.J. 1991. Sorting of proteins in the secretory system. *Annu. Rev. Plant Physiol. Plant Mol. Biol.* 42:21–53.

89. Civello, P.M., A.L.T. Powell, A. Sabehat and A.B. Bennett. 1999. An expansin gene expressed in ripening strawberry fruit. *Plant Physiol.* 121:1273–1279.

90. Clark, C.J., P.D. Hockings, D.C. Joyce and R.A. Mazucco. 1997. Application of magnetic resonance imaging to pre- and post-harvest studies of fruits and vegetables. *Postharv. Biol. Tech.* 11:1–21.

91. Clark, J.R. 1983. Age-related changes in trees. *J. Arboriculture* 9:201–205.

92. Close, T.J., and P.M. Chandler. 1990. Cereal dehydrins: serology, gene mapping and potential functional roles. *Aust. J. Plant Physiol.* 17:333–344.

93. Close, T.J., R.D. Fenton, A.Yang, R. Asghar, D.A. DeMason, D.E. Crone, N.C. Meyer and F. Moonan. 1993. Dehydrin: the protein. Pp. 104–118. In: *Responses of Plants to Cellular Dehydration during Environmental Stress.* T.J. Close and E.A. Bray (eds.). Amer. Soc. Plant Physiol., Rockville, MD.

94. Coombe, B.G. 1992. Research on development and ripening of the grape berry. *Amer. J. Enol. Vitic.* 43:101–110.

95. Cosgrove, D.J. 1997. Relaxation in a high-stress environment: The molecular bases of extensible cell walls and cell enlargement. *Plant Cell* 9:1031–1041.

96. Coupe, S.A., J.E. Taylor and J.A. Roberts. 1995. Characterization of an mRNA encoding a metallothionein-like protein that accumulates during ethylene-promoted abscission of *Sambucus nigra* L. leaflets. *Planta* 197:442–447.

97. Coupe, S.A., J.E. Taylor and J.A. Roberts. 1997. Temporal and spatial expression of mRNAs encoding pathogenesis-related proteins during ethylene-promoted leaflet abscission in *Sambucus nigra. Plant, Cell Environ.* 20:1517–1524.

98. Couvillon, G.A. 1995. Temperature and stress effects on rest in fruit trees: A review. *Acta Hort.* 395:11–19.

99. Crookston, R.K., J. O'Toole and J.L. Ozbun. 1974. Characterization of the bean pod as a photosynthetic organ. *Crop Sci.* 14:708–712.

100. Cunningham, F.X., Jr., and E. Gantt. 1998. Genes and enzymes of carotenoid biosynthesis in plants. *Annu. Rev. Plant Physiol. Plant Mol. Biol.* 49:557–583.

101. Crowe, J.H., F.A. Hoekstra and L.M. Crowe. 1992. Anhydrobiosis: The water replacement hypothesis. Pp. 440–455. In: *The Properties of Water in Foods.* D.S. Reid (ed.). Blackie, London.

102. Danielson, C.E. 1952. A contribution to the study of the synthesis of reserve proteins of ripening pea seeds. *Acta Chem. Scand.* 6:149–159.

103. Davies, C., P.K. Boss and S.P. Robinson. 1997. Treatment of grape berries, a nonclimacteric fruit with a synthetic auxin, retards ripening and alters the expression of developmentally regulated genes. *Plant Physiol.* 115:1155–1161.

104. Davies, C., and S.P. Robinson. 2000. Differential screening indicates a dramatic change in mRNA profiles during grape berry ripening. Cloning and characterization of cDNAs encoding putative cell wall and stress response proteins. *Plant Physiol.* 122:803–812.

105. Dayton, D.F. 1966. The pattern and inheritance of anthocyanin distribution in red pears. *Proc. Amer. Soc. Hort. Sci.* 89:110–116.

106. DeHertogh, A.A. 1977. *Holland Bulb Forcer's Guide.* Netherlands Flower-Bulb Institute, New York.

107. Denisen, E.L. 1951. Carotenoid content of tomato fruits. I. Effect of temperature and light. II. Effects of nutrients, storage and variety. *Iowa State Coll. J. Sci.* 25:549–574.

108. Denyer, K., C.M. Hylton, C.F. Jenner and A.M. Smith. 1995. Identification of multiple isoforms of soluble and granule-bound starch synthase in developing wheat endosperm. *Planta* 196:256–265.

109. Desai, B.B., and P.B. Deslipande. 1975. Chemical transformations in three cultivars of banana (*Musa paradisica* Linn.) fruit stored at 20°C. *Mysore J. Agri. Sci.* 9:634–643.

110. Dilley, D.R. 1977. Hypobaric storage of perishable commodities—fruits, vegetables, flowers, and seedlings. *Acta Hort.* 62:61–70.

111. Do, J.Y., D.K. Salunkhe and L.E. Olson. 1979. Isolation, identification and comparison of the volatiles of peach fruit as related to harvest maturity and artificial ripening. *J. Food Sci.* 34:618–621.

112. Dobson, H.E.M. 1993. Floral volatiles in insect biology. Pp. 47–81. In: *Insect-Plant Interactions.* Vol. V., E. Bernays (ed.). CRC Press, Boca Raton, FL.

113. Dong, Y.J., E. Tsuzuki and H. Terao. 2000. Inheritance of aroma in four rice cultivars (*Oryza sativa* L.). *Intern. Rice Res. Notes* 25.2:15.

114. Drawert, F. 1975. Formation des aromes a différents stades de l'évolution du fruit; enzyme intervenant

dans cette formation. Pp. 309–319. In: *Facteurs et Régulation de la Maturation des Fruits.* No. 238, CNRS, Paris.

115. Dudareva, N., B. Piechulla and E. Pichersky. 2000. Biogenesis of floral scents. *Hort. Rev.* 24:31–54.

116. Dull, G.G., G.S. Birth and R.G. Leffler. 1989. Use of near infrared analysis for the nondestructive measurement of dry matter in potatoes. *Amer. Pot. J.* 66:215–225.

117. Dull, G.G., G.S. Birth and J.B. Magee. 1980. Nondestructive evaluation of internal quality. *HortScience* 15:60–63.

118. Dull, G.G., G.S. Birth, D.A. Smittle and R.G. Leffler. 1989. Near infrared analysis of soluble solids in intact cantaloupe. *J. Food Sci.* 54:393–395.

119. Dure, L., III. 1985. Embryogenesis and gene expression during seed formation. *Oxford Surveys Plant Mol. Cell Biol.* 2:179–197.

120. Dure, L., III. 1997. Lea proteins and the desiccation tolerance in seeds. Pp. 525–543. In: *Cellular and Molecular Biology of Plant Seed Development.* B.A. Larkins and I.K. Vasil (eds.). Kluwer Acad., Boston, MA.

121. Dure, L., III. 1993. The Lea proteins of higher plants. Pp. 325–335. In: *Control of Plant Gene Expression.* D.P.S. Verma (ed.). CRC Press, Boca Raton, FL.

122. Dure, L., III. and C.A. Chlan. 1985. Cottonseed storage proteins: Products of three gene families. Pp. 179–197. In: *Molecular Form and Function of the Plant Genome.* L. Van Vloten-Doting, G.S.P. Groot, and T.C. Hall (eds.). Plenum Press, New York.

123. Eckerson, S. 1913. A physiological and chemical study of after-ripening. *Bot. Gaz.* 55:286–299.

124. Edwards, M., C. Scott, M.J. Gidley and J.S.G. Reid. 1992. Control of mannose/galactose ratio during glactomannan formation in developing legume seeds. *Planta* 187:67–74

125. Egli, D.B., J.E. Leggett and J.M. Wood. 1978. Influences of soybean seed size and position on the rate and duration of filling. *Agron. J.* 70:127–130.

126. Eguchi, T., T. Matsumura and T. Koyama. 1963. The effect of low temperatures on flower and seed formation in Japanese radish and Chinese cabbage. *Proc. Amer. Soc. Hort. Sci.* 82:322–331.

127. Eisenburg, B.A., and G.L. Staby. 1985. Mitochondrial changes in harvested carnation flowers (*Dianthus caryophyllus* L.) during senescence. *Plant Cell Physiol.* 26:829–837.

128. Elkashif, M.E., D.J. Huber and J.K. Brecht. 1989. Respiration and ethylene production in harvested watermelon fruit: evidence for nonclimacteric respiratory behavior. *J. Amer. Soc. Hort. Sci.* 114:81–85.

129. Elyatem, S.M., and A.A. Kader. 1984. Post harvest physiology and storage behavior of pomegranate fruits. *Scient. Hort.* 24:287–298.

130. Engelmann, W., I. Eger, A. Johnsson and H.G. Karlsson. 1974. Effect of temperature pulses on the petal rhythm of *Kalanchoe:* an experimental and theoretical study. *Int. J. Chronobiol.* 2:347–358.

131. Evans, R.Y., and M.S. Reid. 1985. Control of petal expansion during diurnal opening of roses. *Acta Hort.* 181:55–63.

132. Evans, R.Y., and M.S. Reid. 1988. Changes in carbohydrates and osmotic potential during rhythmic expansion of rose petals. *J. Amer. Soc. Hort. Sci.* 113:884–888.

133. Evensen, K.B., A.M. Page and A.D. Stead. 1993. Anatomy of ethylene-induced petal abscission in *Pelargonium* × *hortorum. Ann. Bot.* 71:559–566.

134. Fallik, E., D.D. Archbold, T.R. Hamilton-Kemp, J.H. Loughrin and R.W. Collins.1997. Heat treatment temporarily inhibits aroma volatile compound emission from Golden Delicious apples. *J. Agri. Food Chem.* 45:4038–4041.

135. Farmer, E.E., R.R. Johnson and C.A. Ryan. 1992. Regulation of expression of proteinase inhibitor genes by methyl jasmonate and jasmonic acid. *Plant Physiol.* 98:995–1002.

136. Farrant, J.M., N.W. Pammenter, P. Berjak, E.J. Farnsworth and C.W. Vertucci. 1996. Presence of dehydrin-like proteins and levels of abscisic acid in recalcitrant (desiccation sensitive) seeds may be related to habitat. *Seed Sci. Res.* 6:175–182.

137. Ferguson, I., R. Volz and A. Wolf. 1999. Preharvest factors affecting physiological disorders of fruit. *Postharv. Biol. Tech.* 15:255–262.

138. Ferguson, I.B. 1984. Calcium in plant senescence and fruit ripening. *Plant Cell Environ.* 7:477–489.

139. Fillion, L., A. Ageorges, S. Picaud, P. Coutos-Thévenot, R. Lemoine, C. Romieu and S. Delrot. 1999. Cloning and expression of a hexose transporter gene expressed during the ripening of grape berry. *Plant Physiol.* 120:1083–1093.

140. Firich-Savage, W.E., S.K. Pramanik and J.D. Bewley.1994. The expression of dehydrin proteins in desiccation-sensitive (recalcitrant) seeds of temperate trees. *Planta* 193:478–485.

141. Fletcher, J.C., U. Brand, M.P. Running, R. Simon and E.M. Meyerowitz. 1999. Signaling of cell fate decisions by *CLAVATA3* in *Arabidopsis* shoot meristems. *Science* 283:1911–1914.

142. Flinn, A.M., and J.S. Pate. 1968. Biochemical and physiological changes during maturation of fruit of the field pea (*Pisum arvense* L.). *Ann. Bot.* 32:479–495.

143. Floren, G. 1941. Untersuchungen über Blütenfärbmuster und Blütenfärbunger. *Flora* 135:65–100.

144. Forbus, W.R., Jr. and G.G. Dull. 1990. Delayed light emission as an indicator of peach maturity. *J. Food Sci.* 55:1581–1584.

145. Francis, F.J. 1980. Color quality evaluation of horticultural crops. *HortScience* 15:58–59.

146. Fraser, P.D., M.R. Truesdale, C.R. Bird, W. Schuch and P.M. Bramley. 1994. Carotenoid biosynthesis during tomato fruit development. Evidence for tissue-specific gene expression. *Plant Physiol.* 105:405–413.

147. Frenkel, C., I. Klein and D.R. Dilley. 1968. Protein synthesis in relation to ripening of pome fruits. *Plant Physiol.* 43:1146–1153.

148. Galau, G.A., N. Bijaisoradat and D.W. Hughes. 1987. Accumulation kinetics of cotton late embryogenesis-abundant mRNAs and storage protein mRNAs: Coordinate regulation during embryogenesis and the role of abscisic acid. *Dev. Biol.* 123:198–212.

149. Gan, S., and R.M. Amasino. 1995. Inhibition of leaf senescence by autoregulated production of cytokinin. *Science* 270:1966–1967.

150. Gepstein, S. 1988. Photosynthesis. Pp. 85–109. In: *Senescence and Aging in Plants.* L.D. Noodén and A.C. Leopold (eds.). Academic Press, New York.

151. Gerhardt, F., and E. Smith. 1946. Physiology and dessert quality of Delicious apples as influenced by handling, storage and simulated marketing practice. *Proc. Wash. State Hort. Assoc.* 1945:151–172.

152. Gilissen, L.J.W. 1977. Style controlled wilting of the flower. *Planta* 133:275–280.

153. Giovannoni, J. 2001. Molecular biology of fruit maturation and ripening. *Annu. Rev. Plant Physiol. Plant Mol. Biol.* 52:725–749.

154. Giovannoni, J.J., D. Della Penna, A.B. Bennett and R.L. Fischer. 1989. Expression of a chimeric polygalacturonase gene in transgenic *rin* (ripening inhibitor) tomato fruit results in polyuronide degradation but not fruit softening. *Plant Cell* 1:53–63.

155. Given, N.K., M.A. Venis and D. Grierson. 1988. Hormonal regulation of ripening in the strawberry, a non-climacteric fruit. *Planta* 174:402–406.

156. Goldberg, R.B., G. Hoschek, S.H. Tam, G.S. Ditta and R.W. Breidenbach. 1981. Abundance, diversity, and regulation of mRNA sequence sets in soybean embryogenesis. *Dev. Biol.* 83:201–217.

157. Goldschmidt, E.E., M. Huberman and R. Goren. 1993. Probing the role of endogenous ethylene in the degreening of citrus fruit with ethylene antagonists. *Plant Growth Regul.* 12:325–329.

158. Goodwin, T.W., and M. Jamikorn. 1952. Biosynthesis of carotenes in ripening tomatoes. *Nature* 170:104–105.

159. Grbić, V., and A.B. Bleecker. 1995. Ethylene regulates the timing of leaf senescence in Arabidopsis. *Plant J.* 8:595–602.

160. Grierson, D., and W. Schuch. 1993. Control of ripening. *Phil. Trans. Roy. Soc.,* London (Biol.) 342:241–250.

161. Groot, S.P.C., and C.M. Karssen. 1992. Dormancy and germination of abscisic acid deficient tomato seeds: Studies with the *sitiens* mutant. *Plant Physiol.* 99:952–958.

162. Grosch, W., and P. Schieberle. 1997. Flavor of cereal products—A review. *Cereal Chem.* 74:91–97.

163. Gross, J., and M. Flugel. 1982. Pigment changes in peel of the ripening banana *Musa cavendish. Gartenbauwis* 44:134–135.

164. Guis, M., R. Botondi, M. Ben-Amor, R. Ayub, M. Bouzayen, J.C. Pech and A. Latché. 1997. Ripening-associated biochemical traits of cantaloupe Charentais melons expressing an antisense ACC oxidase transgene. *J. Amer. Soc. Hort. Sci.* 122:748–751

165. Halevy, A.H., C.S. Whitehead and A.M. Kofranek. 1984. Does pollination induce corolla abscission of cyclamen flowers by promoting ethylene production? *Plant Physiol.* 75:1090–1093.

166. Hamann, D.D., and D.E. Carroll. 1971. Ripeness sorting of muscadine grapes by use of low-frequency vibrational energy. *J. Food Sci.* 36:1049–1051.

167. Han, F., S.E. Ullrich, J.A. Clancy, V. Jitkov, A. Kilian and I. Romagosa. 1996. Verification of barley seed dormancy loci via linked molecular markers. *Theor. Appl. Genet.* 92:87–91.

168. Hansen, E. 1967. Ethylene-stimulated metabolism of immature "Bartlett" pears. *Proc. Amer. Soc. Hort. Sci.* 91:863–867.

169. Harada, J. 1997. Seed maturation and control of germination. Pp. 545–592. In: *Cellular and Molecular Biology of Plant Seed Development.* B.A. Larkins and I.K. Vasil (eds.). Kluwer, Boston, MA.

170. Harlan, H.V., and M.N. Pope. 1922. The germination of barley seeds harvested at different stages of growth. *J. Heredity* 13:72–75.

171. Harpster, M.H., D.A. Brummell and P. Dunsmuir. 1998. Expression analysis of a ripening-specific, auxin-repressed endo-1,4-β-glucanase gene in strawberry. *Plant Physiol.* 118:1307–1316.

172. Hatfield, S.G.S., and B.D. Patterson. 1975. Abnormal volatile production by apples during ripening after controlled atmosphere storage. Pp. 57–62. In: *Facteurs et Régulation de la Maturation des Fruits,* No. 238, CNRS, Paris.

173. Hayashi, T. 1989. Xyloglucans in the primary cell wall. *Annu. Rev. Plant Physiol. Plant Mol. Biol.* 40:139–168.

174. Helsper, J.P.F.G., J.A. Davies, H.J. Bouwmeester, A.F. Krol and M.H. van Kampen. 1998. Circadian rhythimicity in emission of volatile compounds by flowers of *Rosa hybrida* L. cv. Honesty. *Planta* 207: 88–95.

175. Hendry, G. 1988.Where does all the green go? *New Scientist* 120:38–42.

176. Henry, E.W., J.G. Valdovinos and T.E. Jensen. 1974. Peroxidases in tobacco abscission zone tissue. II. Time-course of peroxidase activity during ethylene-induced abscission. *Plant Physiol.* 54:192–196.

177. Hewett, E.W., and R.E. Lill. 1992. Hosui fruit (*Pyrus serotina*) grown in New Zealand is nonclimacteric. *New Zealand J. Crop Hort. Sci.* 20:371–375.

178. Heyes, J.A., F.H. Blaikie, C.G. Downs and D.F. Sealey. 1994. Textural and physiological changes during pepino (*Solanum muricatum* Ait.) ripening. *Scient. Hort.* 58:1–15.

179. Higgins, T.J.V. 1984. Synthesis and regulation of major proteins in seeds. *Annu. Rev. Plant Physiol.* 35:191–221

180. Hilhorst, H.W.M., and P.E. Toorop. 1997. Review on dormancy, germinability, and germination in crop and weed seeds. *Adv. Agron.* 61:111–165.

181. Hill, J.E., and R.W. Breidenbach. 1974a. Proteins of soybean seeds. I. Isolation and characterization of the major components. *Plant Physiol.* 53:742–746.

182. Hill, J.E., and R.W. Breidenbach. 1974b. Proteins of soybean seeds. II. Accumulation of the major components during seed development and maturation. *Plant Physiol.* 53:747–751.

183. Hills, L.D., and E.H. Haywood. 1946. *Rapid Tomato Ripening.* Faber and Faber, London.

184. Hogenesch, J.B., K.A. Ching, S. Batalov, S.A. Kay, P.G. Schultz and M.P. Cooke. 2001. A comparison of the Celera and ensemble predicted gene sets reveals little overlap in novel genes. *Cell* 106:413–415.

185. Holmes, R.G. 1979. Vibratory sorting of process tomatoes. *Amer. Soc. Agr. Eng. Paper* 79–6543.

186. Holton, T.A., and E.C. Cornish. 1995. Genetics and biochemistry of anthocyanin biosynthesis. *Plant Cell* 7:1071–1083.

187. Horvat, R.J., and G.W. Chapman. 1990. Comparison of volatile compounds from peach fruit and leaves (cv. Monroe) during maturation. *J. Agri. Food Chem.* 38:1442–1444.

188. Howarth, M.S., and W.F. McClure. 1987. Agricultural product analysis by computer vision. *Amer. Soc. Agri. Eng. Paper* No. 87–3043.

189. Huber, D.J., and D.W. Newman. 1976. Relation between lipid changes and plastid ultra-structural changes in senescing and regreening soybean cotyledons. *J. Exp. Bot.* 27:490–511.

190. Hughes, D.W., and G.A. Galau.1987. Translation efficiency of *Lea* mRNAs in cotton embryos: minor changes during embryogenesis and germination. *Plant Mol. Biol.* 9:301–313.

191. Hughes, D.W., and G.A. Galau. 1989. Temporally modular gene expression during cotyledon development. *Genes Dev.* 3:358–369.

192. Hulme, A.C. 1954. Studies in the nitrogen metabolism of apple fruit. *J. Exp. Bot.* 5:159–172.

193. Hung, Y.C., and S.E. Prussia. 1995. Firmness measurement using a nondestructive laser-puff detector. *Proc. Food Processing Automation IV Conf.,* Pp. 145–154.

194. Ichimura, K., Y. Mukasa, T. Fujiwara, K. Kohata, R. Goto and K. Suto. 1999. Possible roles of methyl glucoside and *myo*-inositol in the opening of cut rose flowers. *Ann. Bot.* 83:551–557.

195. Ismael, A.A., and W.T. Kender. 1969. Evidence of a respiratory climacteric in highbush and lowbush blueberry fruit. *HortScience* 4:342–344.

196. Ittersum, M.K. van. 1992. *Dormancy and Growth Vigour of Seed Potatoes.* Landbouwuniversiteitte Wageningen, Wageningen, The Netherlands.

197. Jacob, R.C., R.J. Romani and C.M. Sprock. 1965. Fruit sorting by delayed light emission. *Trans. Amer. Soc. Agr. Eng.* 8:18–19, 24.

198. Jacob-Wilk, D., D. Holland, E.E. Goldschmidt, J. Riov and Y. Eyal. 1999. Chlorophyll breakdown by chlorophyllase: isolation and functional expression of the *Chlase1* gene from ethylene-treated fruit and its regulation during development. *Plant J.* 20:653–661.

199. Jarvis, M.C. 1992. Control of thickness of collenchyma cell walls by pectin. *Planta* 187:218–220.

200. Jen, J.J. 1974. Influence of spectral quality of light on pigment systems of ripening tomatoes. *J. Food Sci.* 39:907–910.

201. Johanson, U., J. West, C. Lister, S. Michaels, R. Amasino and C. Dean. 2000. Molecular analysis of FRIGIDA, a major determinant of natural variation in *Arabidopsis* flowering time. *Science* 290:344–347.

202. John, I., R. Hackett, W. Cooper, R. Drake, A. Farrell and D. Grierson. 1997. Cloning and characterization of tomato leaf senescence-related cDNAs. *Plant Mol. Biol.* 33:641–651.

203. Jones, A. 2000. Does the plant mitochondria integrate cellular stress and regulate programmed cell death? *Trends Plant Sci.* 5:225–230.

204. Jones, J.A. 1920. Physiological study of maple seeds. *Bot. Gaz.* 69:127–152.

205. Jones, M.L., and W.R. Woodson. 1999. Differential expression of three members of the 1-aminocyclo-propane-1- carboxylate synthase gene family in carnation. *Plant Physiol.* 119:755–764.

206. Jones, S.K., P.G. Gosling and R.H. Ellis. 1998. Reimposition of conditional dormancy during air-dry storage of prechilled Sitka spruce seeds. *Seed Sci. Res.* 8:113–122.

207. Kalaitzis, P., T. Solomos and M.L. Tucker. 1997. Three different polygalacturonases are expressed in tomato leaf and flower abscission, each with a different temporal expression pattern. *Plant Physiol.* 113:1303–1308.

208. Kamerbeek, G.A., and J.J. Beijer. 1964. Vroege bloei van Iris 'Wedgwood'. *Meded. Dir. Turinb.* 27:598–604.

209. Karvouni, Z., I. John, J.E. Taylor, C.F. Watson, A.J. Turner and D. Grierson.1995. Isolation and characterization of a melon cDNA clone encoding phytoene synthase. *Plant Mol. Biol.* 27:1153–1162.

210. Kays, S.J. 1999. Non-destructive quality evaluation of intact, high-moisture products. *Near Infrared News* 10(3):12–15.

211. Kays, S.J. 1999. Preharvest factors affecting appearance. *Postharv. Biol. Tech.* 15:233–248.

212. Kays, S.J., and R.M. Beaudry. 1987. Techniques for inducing ethylene effects. *Acta Hort.* 210:77–116.

213. Kays, S.J., and M.J. Hayes. 1978. Induction of ripening in the fruits of *Momordica charactia,* L. by ethylene. *Trop. Agric.* 55:167–172.

214. Kays, S.J., C.A. Jaworski and H.C. Price. 1976. Defoliation of pepper transplants in transit by endogenously evolved ethylene. *J. Amer. Soc. Hort. Sci.* 101:449–451.

215. Kays, S.J., and Y. Wang. 2000. Thermally induced flavor compounds. *HortScience* 35:1002–1012.

216. Karssen, C.M., D.L.C. Brinkhorst-van der Swan, A.E. Breekland and M. Koornneef. 1983. Induction of dormancy during seed development by endogenous abscisic acid: studies on abscisic acid deficient genotypes of *Arabidopsis thaliana* (L.) Heynh. *Planta* 157:158–165.

217. Kende, H., and A.D. Hanson. 1976. Relationship between ethylene evolution and senescence in Morning Glory flower tissue. *Plant Physiol.* 57:523–527.

218. Kermode, A.R. 1990. Regulatory mechanisms involved in the transition from seed development to germination. *Crit. Rev. Plant Sci.* 9:155–195.

219. Kermode, A.R.1997. Approaches to elucidate the basis of desiccation-tolerance in seeds. *Seed Sci. Res.* 7:75–95.

220. Kermode, A.R., and J.D. Bewley. 1985. The role of maturation drying in the transition from seed development to germination. I. Acquisition of desiccation-tolerance and germinability during development of *Ricinus communis* L. seeds. *J. Exp. Bot.* 36:1906–1915.

221. Kermode, A.R., and J.D. Bewley. 1985. The role of maturation drying in the transition from seed development to germination. II. Post-germinative enzyme production and soluble protein synthetic pattern changes within the endosperm of *Ricinus communis* L. seeds. *J. Exp. Bot.* 36:1916–1927.

222. Kermode, A.R., D.J. Gifford and J.D. Bewley. 1985. The role of maturation drying in the transition from seed development to germination. III. Insoluble protein synthetic pattern changes within the endosperm of *Ricinus commurus* L. seeds. *J. Exp. Bot.* 36:1928–1936.

223. Kerner von Marilaun, A. 1891. *Pflanzenleben.* Band 2. Verlag des Biblographisches Instituts, Leipzig, Germany.

224. Kidd, F., and C. West. 1930. Physiology of fruit. I. Changes in the respiratory activity of apples during their senescence at different temperatures. *Proc. Roy. Soc.* (London) B, 106:93–109.

225. King, S.P., J.E. Lunn and R.T. Furbank. 1997. Carbohydrate content and enzyme metabolism in developing canola siliques. *Plant Physiol.* 114:153–160.

226. Kirk, J.T.O., and R. Tilney-Bassett. 1978. *The Plastids.* Freeman, San Francisco, CA.

227. Klee, H.J. 1993. Ripening physiology of fruit from transgenic tomato (*Lycopersicon esculentum*) plants with reduced ethylene synthesis. *Plant Physiol.* 102:911–916.

228. Klein, S., and B.M. Pollock. 1968. Cell fine structure of developing lima bean seeds related to seed desiccation. *Amer. J. Bot.* 55:658–672.

229. Kliewer, W.M. 1967. The glucose-fructose ratio of *Vitis vinifera* grapes. *Amer. J. Enol. Vitic.* 18:33–41.

230. Knee, M. 1972. Anthocyanin, carotenoid, and chlorophyll changes in the peel of 'Cox's Orange Pippin' apples during ripening on and off the tree. *J. Exp. Bot.* 23:184–196.

231. Knee, M. 1973. Polysaccharide changes in cell walls of ripening apples. *Phytochemistry* 12:1543–1549.

232. Knee, M. 1985. Evaluating the practical significance of ethylene in fruit storage. Pp. 297–315. In: *Ethylene and Plant Development.* J.A Roberts and G.A. Tucker (eds.). Butterworths, London.

233. Knee, M. (ed.). 2002. *Fruit Quality and Its Biological Basis.* CRC Press, Boca Raton, FL.

234. Knee, M., F.J. Proctor and C.J. Dover. 1985. The technology of ethylene control: use and removal in post-harvest handling of horticultural commodities. *Ann. Appl. Biol.* 107:581–595.

235. Knee, M., J.A. Sargent and D.J. Osborne. 1977. Cell wall metabolism in developing strawberry fruits. *J. Exp. Bot.* 28:337–396.

236. Knypl, J.S., and W. Mazurczyk. 1971. Arrest of chlorophyll and protein breakdown in senescing leaf discs of kale by cycloheximide and vanillin. *Curr. Sci.* 40:294–295.

237. Koornneef, M., C.J. Hanhart, H.W.M. Hilhorst and C.M. Karssen. 1989. In vivo inhibition of seed development and reserve protein accumulation in recombinants of abscisic acid biosynthesis and responsiveness mutants of *Arabidopsis thaliana. Plant Physiol.* 90:463–469.

238. Koornneef, M., M.L. Jorna, D.L.C. Brinkhorst-van der Swan and C.M. Karssen. 1982. The isolation of abscisic acid (ABA) deficient mutants by selection of induced revertants in non-germinating gibberellin sensitive lines of *Arabidopsis thaliana* (L.) Heynh. *Theor. Appl. Genet.* 61:385–393.

239. Koornneef, M., K.M. Léon-Kloosterziel, S.H. Schwartz and J.A.D. Zeevaart. 1998. The genetic and molecular dissection of abscisic acid biosynthesis and signal transduction in *Arabidopsis. Plant Physiol. Biochem.* 36:83–89.

240. Koornneef, M., G. Reuling and C.M. Karssen. 1984. The isolation and characterization of abscisic acid-insensitive mutants of *Arabidopsis thaliana. Physiol. Plant.* 61:377–383.

241. Koornneff, M., and J.H. Ven der Veen. 1980. Induction and analysis of gibberellin sensitive mutants in *Arabidopsis thaliana* (L.) Heynh. *Theor. Appl. Genet.* 58:257–263.

242. Kosiyachinda, S., and R.E. Young. 1975. Ethylene production in relation to the initiation of respiratory climacteric in fruit. *Plant Cell Physiol.* 16:595–602.

243. Kramer, A., and B.A.Twigg. 1970. *Quality Control for the Food Industry.* Vol. 1, AVI, Westport, CT.

244. Kränzlin, F. 1910. Orchidaceae, Monandrae, Dendrobiinae. Pars 1. In: *Das Pflanzenreich.* A. Engler (ed.). Wilhelm Engelmann Verlag, Leipzig, Germany.

245. Kust, C.A. 1963. Dormancy and viability of witchweed seeds as affected by temperature and relative humidity during storage. *Weeds* 11:247–250.

246. Ku, P.K., and A.L. Mancinelli. 1972. Photocontrol of anthocyanin synthesis. *Plant Physiol.* 49:212–217.

247. Lam, J.-M., K.-H. Pwee, W.Q. Sun, Y.-L. Chua and X.-J. Wang. 1999. Enzyme-stabilizing activity of seed trypsin inhibitors during desiccation. *Plant Sci.* 142:209–218.

248. Lam, P.F., and C.K. Wan. 1983. Climacteric nature of the carambola (*Averrhoa carambola* L.) fruit. *Pertanika* 6:44–47.

249. Lang, A. 1965. Physiology of flower induction. Pp. 1380–1535, Vol. 15. In: *Encyclopedia of Plant Physiology.* W. Ruhland (ed.). Springer, Berlin.

250. Lang, G.A. 1987. Dormancy: A new universal terminology. *HortScience* 22:817–820.

251. Lapasin, R., and S. Pricl. 1995. *The Rheology of Industrial Polysaccharides: Theory and Applications.* Blackie Academic, London.

252. Lashbrook, C., C. Gonzalez-Bosch and A.B. Bennett. 1994. Two divergent endo-β-1,4-gluconase genes exhibit overlapping expression in ripening fruit and abscising flowers. *Plant Cell* 6:1485–1493.

253. Latifah, M.N. 1996. Effect of exogenous ethylene in the ripening of chiku (*Achras sapota* L.). Pp. 367–376. In: *Proc. Intern. Conf. Trop. Fruits.* Kuala Lumpur, Malaysia.

254. Latza, S., D. Gansser and R.G. Berger. 1996. Identification and accumulation of 1-*O-trans*-cinnamoyl-β-D-glucopyranose in developing strawberry fruit (*Fragaria ananassa* Duch. cv. Kent). *J. Agri. Food Chem.* 44:1367–1370.

255. Laufs, P., O. Grandjean, C. Jonak, K. Kieu and J. Traas. 1998. Cellular parameters of the shoot apical meristem in *Arabidopsis. Plant Cell* 10:1375–1389.

256. Lay-Yee, M., A.D. Stead and M.S. Reid. 1992. Flower senescence in daylily (*Hemerocallis*). *Physiol. Plant.* 86:308–314.

257. Le, V.T., N. Nguyen, D.D. Nguyen and T.T.H. Ha. 2002. Dragon fruit quality and storage life: effect of harvest time, use of plant growth regulators and modified atmosphere packaging. *Acta Hort.* 575 (Vol. 2):611–621.

258. Leliévre, J.-M., A. Latché, B. Jones, M. Bouzayen and J.-C. Pech. 1997. Ethylene and fruit ripening. *Physiol. Plant.* 101:727–739.

259. Leopold, A.C. 1961. Senescence in plant development. *Science* 134:1727–1732.

260. Lewis, G.J., and N. Thurling. 1994. Growth, development, and yield of three oilseed *Brassica* species in a water-limited environment. *Aust. J. Exp. Agric.* 34:93–103.

261. Lewis, L.N., C.W. Coggins, Jr., C.K. Labanauskas and W.M. Dugger, Jr. 1967. Biochemical changes associated with natural and gibberellin A$_3$ delayed senescence in the navel orange rind. *Plant Cell Physiol.* 8:151–160.

262. Li, S., P.K. Andrews and M.E. Patterson. 1994. Effects of ethephon on the respiration and ethylene evolution of sweet cherry (*Prunus avium* L.) fruit at different development stages. *Postharv. Biol. Tech.* 4:235–243.

263. Lieberman, M., and S.Y. Wang. 1982. Influence of calcium and magnesium on ethylene production by apple tissue slices. *Plant Physiol.* 69:1150–1155.

264. Lipe, J.A. 1978. Ethylene in fruits of blackberry and rabbiteye blueberry. *J. Amer. Soc. Hort. Sci.* 103:76–77.

265. Lipe, J.A., and P.W. Morgan. 1970. Ethylene: Involvement in shuck dehiscence in pecan fruits (*Carya illinoensis* [Wang.] K. Koch.). *HortScience* 5:266–267.

266. Looney, N.E., and M.E. Patterson. 1967. Chlorophyllase activity in apples and bananas during the climacteric phase. *Nature* 214:1245–1246.

267. Lukaszewski, T.A., and M.S. Reid. 1989. Bulb-type flower senescence. *Acta Hort.* 261:59–62.

268. Lumsden, P., and A.J. Millar (eds.). 1998. *Biological Rhythms and Photoperception in Plants.* Bios, Oxford, England.

269. Lyons, J.M., W.B. McGlasson and H.K. Pratt. 1962. Ethylene production, respiration and internal gas concentrations in cantaloupe fruits at various stages of maturity. *Plant Physiol.* 37:31–36.

270. Lyons, J.M., and H.K. Pratt. 1964. Effect of stage of maturity and ethylene treatments on respiration and ripening of tomato fruits. *Proc. Amer. Soc. Hort. Sci.* 84:491–500.

271. Maas, C., S. Schaal and W. Werr. 1990. A feedback control element near the transcription start site of the maize *Shunken* gene determines promoter activity. *Eur. Mol. Biol. Org. J.* 9:3447–3452.

272. Makovetski, S., and E.E. Goldschmidt. 1976. A requirement for cytoplasmic protein synthesis during chloroplast senescence in the aquatic plant *Anacharis canadensis. Plant Cell Physiol.* 17:859–862.

273. Manning, K. 1998. Isolation of a set of ripening-related genes from strawberry: their identification and possible relationship to fruit quality traits. *Planta* 205:622–631.

274. Marais, E., G. Jacobs and D.M. Holcroft. 2001. Postharvest irradiation enhances anthocyanin synthesis in apples but not in pears. *HortScience* 36:738–740.

275. Marei, N., and J.C. Crane. 1971. Growth and respiratory response of fig (*Ficus carica* L. cv. Mission) fruits to ethylene. *Plant Physiol.* 48:249–254.

276. Marousky, F.J., and B.K. Harbough. 1979. Interactions of ethylene, temperature, light and carbon diox-

ide on leaf and stipule abscission and chlorosis in *Philodendron scandens* spp. *oxycardium. J. Amer. Soc. Hort. Sci.* 104:876–880.

277. Marks, J.P., R. Bernlohr and J.P. Varner. 1957. Esterification of phosphate in ripening fruit. *Plant Physiol.* 32:259–262.

278. Martinoia, E., M.J. Dalling and P. Matile. 1982. Catabolism of chlorophyll: demonstration of chloroplast-localized peroxidative and oxidative activities. *Z. Pflanzenyphysiol.* 107:269–279.

279. Matile, P., S. Hortensteiner and H. Thomas. 1999. Chlorophyll degradation. *Annu. Rev. Plant Physiol. Plant Mol. Biol.* 50:67–95.

280. Matile, P., S. Hortensteiner, H. Thomas and B. Krautler. 1996. Chlorophyll breakdown in senescent leaves. *Plant Physiol.* 112:1403–1409.

281. Matile, P., and F. Winkenbach. 1971. Function of lysosomes and lysosomal enzymes in senescing corolla of the morning glory (*Ipomoea purpurea*). *J. Exp. Bot.* 122:759–771.

282. Mattheis, J.P., D.A. Buchanan and J.K. Fellman. 1997. Volatile constituents of bing cherry fruit following controlled atmosphere storage. *J. Agri. Food Chem.* 45:212–216.

283. Mattheis, J.P., and J.K. Fellman. 1999. Preharvest factors influencing flavor of fresh fruit and vegetables. *Postharv. Biol. Tech.* 15:227–232.

284. Mayorga, H., H. Knapp, P. Winterhalter and C. Duque. 2001. Glycosidically bound flavor compounds of Cape gooseberry (*Physalis peruviana* L.). *J. Agri. Food Chem.* 49:1904–1908.

285. Maxie, E.C., P.B. Catlin and H.T. Hartman. 1960. Respiration and ripening of olive fruits. *Proc. Amer. Soc. Hort. Sci.* 75:275–291.

286. Mazza, G., and E. Miniati. 1993. *Anthocyanins in Fruits, Vegetables, and Grains.* CRC Press, Boca Raton, FL.

287. McCarty, D.R. 1995. Genetic control and integration of maturation and germination pathways in seed development. *Annu. Rev. Plant Physiol. Plant Mol. Biol.* 46:71–93.

288. McGlasson, W.B., N.L. Wade and I. Adato. 1978. Phytohormones and fruit ripening. Pp. 447–493. In: *Phytohormones and Related Compounds: A Comprehensive Treatise.* Vol. 2. D.S. Letham, P.B. Goodwin, and T.J.V. Higgens (eds.). Elsevier/North-Holland, Amsterdam.

289. McKeown, A.W., E.C. Lougheed and D.P. Murr. 1978. Compatibility of cabbage, carrots, and apples in low pressure storage. *J. Amer. Soc. Hort. Sci.* 103:749–752.

290. McLauchlan, R.L., L.R. Barker and A. Prasad. 1994. Temperature effects on respiration of rambutan and carambola. Pp. 38. In: *Hort. Postharvest Group Biennial Rev. 1992–4.* McLauchlan, R., G. Meiburg and J. Bagshaw (eds.). Queensland Dept. Primary Ind., Australia.

291. McMurchie, E.J., W.B. McGlasson and I.L. Eaks. 1972. Treatment of fruit with propylene gives information about the biogenesis of ethylene. *Nature* 237:235–236.

292. Medawar, P.B. 1957. *The Uniqueness of the Individual.* Methuen, London.

293. Meinke, D.W., J. Chen and R.W. Beachy. 1981. Expression of storage-protein genes during soybean seed development. *Planta* 153:130–139.

294. Merideth, F.I., R.G. Leffler and C.E. Lyons. 1988. A firmness detector for peaches using impact forces analysis. *Amer. Soc. Agric. Eng.* paper No. 88–6570.

295. Merlot, S., and G. Giraudat. 1997. Genetic analysis of abscisic acid signal transduction. *Plant Physiol.* 114:751–757.

296. Michael, M.Z., K.W. Savin, S.C. Baudinette, M.W. Graham, S.F. Chandler, C.-Y. Lu, C. Caesar, I. Gautrais, R. Young, G.D. Nugent, D.R. Stevenson, E.L.-J. O'Conner, C.S. Cobbett and E.C. Cornish. 1993. Cloning of ethylene biosynthetic genes involved in petal senescence of carnation and petunia, and their antisense expression in transgenic plants. Pp. 298–303. In: *Cellular and Molecular Aspects of the Plant Hormone Ethylene.* J.C. Pech, A. Latché and C. Balagué, (eds.). Kluwer Acad., Dordrecht, The Netherlands.

297. Miller, J.C. 1929. A study of some factors affecting seed-stalk development in cabbage. *Cornell Univ. Agr. Expt. Sta. Bull.* 488.

298. Miranda, M., L. Borisjuk, A. Tewes, D. Dietrich, D. Rentsch, H. Weber and U. Wobus. 2003. Peptide and amino acid transporters are differentially regulated during seed development and germination in faba bean. *Plant Physiol.* 132:1950–1960.

299. Mizano, S., and H.K. Pratt. 1973. Relations of respiration and ethylene production to maturity in the watermelon. *J. Amer. Soc. Hort. Sci.* 98:614–617.

300. Molisch, H. 1938. *The Longevity of Plants.* H. Fullington, transl., Science Press, Lancaster, PA.

301. Molsenin, N.N. 1986. *Physical Properties of Plant and Animal Materials.* Gordon Breach Sci., New York.

302. Morgan, P.W. 1984. Is ethylene the natural regulator of abscission? Pp. 231–240. In: *Ethylene: Biochemical, Physiological and Applied Aspects.* Y. Fuchs and E. Chalutz (eds.). Martinus Hijhoff, The Hague, The Netherlands.

303. Moyano, E., I. Portero-Robles, N. Medina-Escobar, V. Valpuesta, J. Muñoz-Blanco and J.L. Caballero. 1998. A fruit-specific putative dihydroflavonol 4-reductase gene is differentially expressed in strawberry during the ripening process. *Plant Physiol.* 117:711–716.

304. Murray, A.J., G.E. Hobson, W. Schuch and C. Bird. 1993. Reduced ethylene synthesis in EFE antisense tomatoes has differential effects on fruit ripening processes. *Postharv. Biol. Tech.* 2:301–303.

305. Neill, S.J., R. Horgan and A.D. Perry. 1986. The carotenoid and abscisic acid content of viviparous kernels and seedlings of *Zea mays* L. *Planta* 169:87–96.

306. Nelson, S.O. 1973. Electrical properties of agricultural products—a critical review. *Trans. Amer. Soc. Agr. Eng.* 16:384–400.

307. Nerd, A., F. Gutman and Y. Mizrahi. 1999. Ripening and postharvest behavior of fruits of two *Hylocereus* species (Cactaceae). *Postharv. Biol. Tech.* 15:99–105.

308. Nerd, A., and Y. Mizrahi. 1998. Fruit development and ripening in yellow pitaya. *J. Amer. Soc. Hort. Sci.* 123:560–562.

309. Neubohn, B., S. Gubatz, U. Wobus and H. Weber. 2000. Sugar levels altered by ectopic expression of a yeast-derived invertase affect cellular differentiation of developing cotyledons of *Vicia narbonensis* L. *Planta* 211:325–334.

310. Nichols, R. 1977. Sites of ethylene production in the pollinated and unpollinated senescing carnation (*Dianthus caroyophyllus*) inflorescence. *Planta* 135:155–159.

311. Noichinda, S. 1992. Effect of modified atmosphere conditions on quality and storage life of mangosteen (*Garcinia mangostana* L.) fruit. M.S. Thesis, Kasetsat Univ., Bangkok.

312. Noodén, L.D. 1988. The phenomena of senescence and aging. Pp. 1–50. In: *Senescence and Aging in Plants.* L.D. Noodén and A.C. Leopold (eds.). Academic Press, New York.

313. Noodén, L.D. 1988. Postlude and prospects. Pp. 499–517. In: *Senescence and Aging in Plants.* L.D. Noodén and A.C. Leopold (eds.). Academic Press, New York.

314. Noodén, L.D., J.J. Guiamét and I. John. 1997. Senescence mechanisms. *Physiol. Plant.* 101:746–753.

315. Noodén, L.D., and J.J. Guiamét. 1996. Genetic control of senescence and aging in plants. Pp. 94–118. In: *Handbook of the Biology of Aging.* T.E. Johnson, N.J. Holbrook and J.H. Morrison (eds.), Academic Press, San Diego, CA.

316. Noodén, L.D., and J.W. Thompson. 1985. Aging and senescence in plants. Pp. 105–127. In: *Handbook of the Biology of Aging.* C.E. Finch and E.L. Schneider (eds.). Van Nostrand-Reinhold, New York.

317. Nursten, H.E. 1970. Volatile compounds: The aroma of fruits. Pp. 239–268. In: *The Biochemistry of Fruits and Their Products.* Vol. 1. A.C. Hulme (ed.). Academic Press, New York.

318. Obenland, D.M., M.L. Arpaia, R.K. Austin and B.E. MacKey.1999. High-temperature forced-air treatment alters the quantity of flavor-related, volatile constituents present in Navel and Valencia oranges. *J. Agri. Food Chem.* 47:5184–5188.

319. Oeller, P.W., L. Min-Wong, L.P. Taylor, D.A. Pike and A. Theologis. 1991. Reversible inhibition of tomato fruit senescence by antisense RNA. *Science* 254:437–439.

320. Oslund, C.R., and T.L. Davenport. 1983. Ethylene and carbon dioxide in ripening fruit of *Averrhoa carambola. HortScience* 18:229–230.

321. Palmer, J.K. 1971. The banana. Pp. 65–105. In: *The Biochemistry of Fruits and Their Products.* Vol. 2. A.C. Hulme (ed), Academic Press, London.

322. Panavas, T., E.L. Walker and B. Rubinstein. 1998. Possible involvement of abscisic acid in senescence of daylily petals. *J. Exp. Bot.* 49:1987–1997.

323. Park, K.Y., A. Drory and W.R. Woodson. 1992. Molecular cloning of a 1-aminocyclopropane-1- carboxylate synthase from senescing carnation flower petals. *Plant Mol. Biol.* 18:377–386.

324. Paull, R.E. 1999. Effect of temperature and relative humidity on fresh commodity quality. *Postharv. Biol. Tech.* 15:263–278.

325. Paull, R.E. 2001. Langsat-Longkong-Duku-Lanson. *Postharvest Quality Maintenance Guidelines,* Univ. Hawaii, Honolulu, HI.

326. Paull, R.E., and N.J. Chen. 1987. Changes in longan and rambutan during postharvest storage. *HortScience* 22:1303–1304.

327. Paull, R.E., and S. Ketsa. 2001. Coconut. *Postharvest Quality Maintenance Guidelines.* Univ. Hawaii, Honolulu, HI.

328. Pech, J.C., and R.J. Romani. 1979. Senescence in pear *Pyrus communis* fruit cells cultured in a continuously renewed auxin-deprived medium. *Plant Physiol.* 63:814–817.

329. Pecker, I., R. Gabbay, F.X. Cunningham, Jr. and J. Hirschberg. 1996. Cloning and characterization of the cDNA for lycopene β-cyclase from tomato reveals decrease in its expression during fruit ripening. *Plant Mol. Biol.* 30:807–819.

330. Pérez, A.G., A. Cert, J.J. Ríos and J.M. Olías. 1997. Free and glycosidically bound volatile compounds from two banana cultivars: Valery and Pequeña Enana. *J. Agric. Food Chem.* 45:4393–4397.

331. Perkins-Veazie, P., and G. Nonnecke. 1992. Physiological changes during ripening of raspberry fruit. *J. Amer. Soc. Hort. Sci.* 27:331–333.

332. Perkins-Veazie, P., J.R. Clark, D.J. Huber and E.A. Baldwin. 2000. Ripening physiology in 'Navaho' thornless blackberries: color, respiration, ethylene production, softening, and compositional changes. *J. Amer. Soc. Hort. Sci.* 125:357–363.

333. Peterson, R.F. 1965. *Wheat. Botany, Cultivation and Utilization.* Leonard Hill, London.

334. Pfeffer, W. 1873. *Physiologische Untersuchungen.* W. Engelmann, Leipzig, Germany.

335. Pichersky, E., R.A. Lewinsohn and E. Croteau. 1994. Floral scent production in *Clarkia* (Onagraceae). I. Localization and developmental modulation of monoterpene emission and linalool synthase activity. *Plant Physiol.* 106:1533–1540.

336. Pla, M., J. Gómez, A. Goday and M. Pagés. 1991. Regulation of the abscisic acid-responsive gene *rab* 28 in maize *viviparous* mutants. *Mol. Gen. Genet.* 230:394–400.

337. Poovaiah, B.W. 1988. Calcium and senescence. Pp. 369–389. In: *Senescence and Aging in Plants.* L.D. Noodén and A.C. Leopold (eds.). Academic Press, New York.

338. Poovaiah, B.W., and A.C. Leopold. 1973. Deferral of leaf senescence with calcium. *Plant Physiol.* 52:236–239.

339. Porat, R., B. Weiss, L. Cohen, A. Daus, R. Goren and S. Droby. 1999. Effects of ethylene and 1-methyl-cyclopropene on the post-harvest qualities of 'Shamouti' oranges. *Postharv. Biol. Tech.* 15:155–163.

340. Postlmayer, H.L., B.S. Luk and S.J. Leonard. 1956. Characterization of pectin changes in freestone and clingstone peaches during ripening and processing. *Food Technol.* 10:618–625.

341. Pratt, H.K., and J.D. Goeschl. 1968. The role of ethylene in fruit ripening. Pp. 1295–1302. In: *Biochemistry and Physiology of Plant Growth Substances.* F. Wightman and G. Setterfield (eds.). Runge, Ottawa, Canada.

342. Pratt, H.K., and D.B. Mendoza. 1980. Fruit development and ripening of the star apple (*Chrysophyllum cainito* L.). *HortScience* 15:721–722.

343. Pratt, H.K., and M.S. Reid. 1974. Chinese gooseberry: seasonal patterns in fruit growth and maturation, ripening, respiration and the role of ethylene. *J. Sci. Food. Agri.* 25:747–753.

344. Pratt, H.K., and M.S. Reid. 1976. The tamarillo: fruit growth and maturation, ripening, respiration, and the role of ethylene. *J. Sci. Food. Agri.* 27:399–404.

345. Pressey, R. 1977. Enzymes involved in fruit softening. *Amer. Soc. Chem. Symp.* Ser. 47:172–191.

346. Pressey, R., and J.K. Avants. 1978. Difference in polygalacturonase composition of clingstone and freestone peaches. *J. Food Sci.* 43:1415–1423.

347. Prestage, S., R.S.T. Linforth, A.J. Taylor, E. Lee, J. Speirs and W. Schuch. 1999. Volatile production in tomato fruit with modified alcohol dehydrogenase activity. *J. Sci. Food Agri.* 79:131–136.

348. Priestly, D.A. 1986. *Seed Aging.* Comstock Publ. Assoc., Ithaca, NY.

349. Primack, R.B. 1985. Longevity of individual flowers. *Annu. Rev. Ecol. Systematics* 16:15–37.

350. Purvis, A.C., and C.R. Barmore. 1981. Involvement of ethylene in chlorophyll degradation in peel of citrus fruit. *Plant Physiol.* 68:854–856.

351. Purvis, O.N. 1961. The physiological analysis of vernalization. Pp. 76–122. In: *Encyclopedia of Plant Physiology.* Vol. XVI. W. Rukland (ed.). Springer-Verlag, Berlin.

352. Rao, D.V.R., and B.S. Chundawat. 1988. Ripening changes in sapota cv. Kalipatti at ambient temperature. *Indian J. Plant Physiol.* 31:205–208.

353. Rees, A.R. 1977. The cold requirements of tulip cultivars. *Scientia Hort.* 7:383–389.

354. Rehkugler, G.E., and J.A. Throop. 1986. Apple sorting with machine vision. *Trans. Amer. Soc. Agri. Eng.* 29:1388–1397.

355. Reid, J.S.G. 1985. Cell wall storage carbohydrates in seeds. Biochemistry of seed "gums" and "hemicelluloses". *Adv. Bot. Res.* 11:125–155.

356. Reid, M.S. 1975. The role of ethylene in the ripening of some unusual fruits. Pp. 177–182. In: *Facteurs et Régulation de la Maturation des Fruits,* No. 238, CNRS, Paris.

357. Reid, M.S. 1985. Ethylene and abscission. *HortScience* 20:45–50.

358. Reid, M.S., and R.Y. Evans. 1985. Control of cut flower opening. *Acta Hort.* 181:45–54.

359. Richmond, A., and J.B. Biale. 1966. Protein and nucleic acid metabolism in fruits: I. Studies of amino acid incorporation during the climacteric rise in respiration of the avocado. *Plant Physiol.* 41:1247–1253.

360. Roberts, A.N., and R.L. Ticknor. 1970. Commercial production of English holly in the Pacific Northwest. *Amer. Hort.* 49:301–314.

361. Robertson, D.S. 1955. The genetics of vivipary in maize. *Genetics* 40:745–760.

362. Rolletschek, H., H. Weber and L. Borisjuk. 2003. Energy status and its control on embryogenesis of legumes. Embryo photosynthesis contributes to oxygen supply and is coupled to biosynthetic fluxes. *Plant Physiol.* 132:1196–1206.

363. Romagosa, I., F. Han, J.A. Clancy and S.E. Ullrich. 1999. Individual locus effects on dormancy during seed development and after-ripening in barley. *Crop Sci.* 39:74–79.

364. Romani, R.J. 1987. Senescence and homeostasis in postharvest research. *HortScience* 22:865–868.

365. Romani, R. 1978. Long term maintenance of mitochondria function *in vitro* and the course of cyanide-insensitive respiration. Pp. 3–10. In: *Plant Mitochondria.* G. Ducet and C. Lance (eds.). Elsevier/North Holland, Amsterdam.

366. Ronen, G., M. Cohen, D. Zamir and J. Hirschberg. 1999. Regulation of carotenoid biosynthesis during tomato fruit development: expression of the gene for lycopene epsilon-cyclase is down-regulated during ripening and is elevated in the mutant *Delta. Plant J.* 17:341–351.

367. Roozen, F.M. (ed.). 1980. *Forcing Flowerbulbs.* International Flower-Bulb Centre, Hillegom, The Netherlands.

368. Rose, J.K.C., and A.B. Bennett. 1999. Cooperative disassembly of the cellulose-xyloglucan network of plant cell walls: parallels between cell expansion and fruit ripening. *Trends Plant Sci.* 4:176–183.

369. Rose, J.K.C., H.H. Lee and A.B. Bennett. 1997. Expression of a divergent expansin gene is fruit-specific and ripening-regulated. *Proc. Natl. Acad. Sci.* USA 94:5955–5960.

370. Rosenburg, L.A., and R.W. Rinne. 1986. Water loss as a prerequisite for seedling growth in soybean seeds (*Glycine max* L. Merr.). *J. Exp. Bot.* 37:1663–1674.

371. Rowan, D.D., H.P. Lane, J.M. Allen, S. Fielder and M.B. Hunt. 1996. Biosynthesis of 2-methylbutyl, 2-methyl-2-butenyl, and 2-methylbutanoate esters in Red Delicious and Granny Smith apples using deuterium-labeled substrates. *J. Agri. Food Chem.* 44:3276–3285.

372. Rowan, K.S., W.B. McGlasson and H.K. Pratt. 1969. Changes in adenosine pyrophosphates in cantaloupe fruit ripening normally and after treatment with ethylene. *J Exp. Bot.* 20:145–155.

373. Rudolph, C.J., Jr. 1971. Factors responsible for flavor and off-flavor development in pecans. Ph.D. Dissertation, Okla. State Univ., Stillwater, OK.

374. Ruuska, S.A., T. Girke, C. Benning and J.B. Ohlrogge. 2002. Contrapuntal networks of gene expression during Arabidopsis seed filling. *Plant Cell* 14:1191–1206.

375. Ryan, C.A. 1981. Proteinase inhibitors. Pp. 351–370. In: *The Biochemistry of Plants.* Vol 6. A. Marcus (ed.). Academic Press, New York.

376. Saltveit, M.E. 1977. Carbon dioxide, ethylene, and color development in ripening mature green bell peppers. *J. Amer. Soc. Hort. Sci.* 102:523–525.

377. Saltveit, M.E. 1991. Determining tomato maturity with nondestructive in vivo nuclear magnetic resonance imaging. *Postharv. Biol. Tech.* 1:153–159.

378. Saltveit, M.E. 1999. Effect of ethylene on quality of fresh fruits and vegetables. *Postharv. Biol. Tech.* 15:279–292.

379. Sams, C.E. 1999. Preharvest factors affecting postharvest textures. *Postharv. Biol. Tech.* 15:249–254.

380. Sanz, C., J.M. Olias and A.G. Perez. 1997. Aroma biochemistry of fruits and vegetables. Pp. 125–156. In: *Phytochemistry of Fruits and Vegetables.* F.A. Tomas-Barberan and R.J. Robins (eds.). Clarendon Press, Oxford, England.

381. Sapers, G.M., S.B. Jones and G.T. Maher. 1983. Factors affecting the recovery of juice and anthocyanin from cranberries. *J. Amer. Soc. Hort. Sci.* 108:246–249.

382. Saraviz, D.M., D.M. Pharr and T.E. Carter, Jr. 1987. Galactinol synthesis activity and soluble sugars in developing seeds of four soybean genotypes. *Plant Physiol.* 83:185–189.

383. Sarkar, N., and R.R. Wolfe. 1985. Future extraction techniques for sorting tomatoes by computer vision. *Trans. Amer. Soc. Agri. Eng.* 28:970–974,979.

384. Sattler, R. 1973. *Organogenesis of Flowers; A Photographic Text-atlas.* University of Toronto Press, Toronto, Canada.

385. Schatzki, T.F., A. Grossman and R. Young. 1983. Recognition of agricultural objects by shape. *IEEE Trans. on Pattern Anal. Artif. Intell.* 5:645–653.

386. Schieberle, P., and W. Grosch. 1985. Identification of the volatile flavour compounds of wheat bread crust. Comparison with rye bread crust. *Z. Lebensm. Unters. Forsch.* 180:474–478.

387. Schurink, R.C., J.M.M. van Beckum and F. Heidekamp. 1992. Modulation of grain dormancy in barley by variation of plant growth conditions. *Hereditas* 117:137–143.

388. Schwechheimer, C., and M. Bevan. 1998. The regulation of transcription factor activity in plants. *Trends Plant Sci.* 3:378–383.

389. Sexton, R., and H.W. Woolhouse. 1984. Senescence and abscission. Pp. 469–497. In: *Advanced Plant Physiology.* M.B. Wilkins (ed.). Pitman, London.

390. Shannon, J.C., and D.L. Garwood. 1984. Genetics and physiology of starch development. Pp. 25–86. In: *Starch: Chemistry and Technology.* R.L. Whistler, J.N. Bemiller and E.F. Paschall (eds.). Academic Press, New York.

391. Shewfelt, R.L. 1999. What is quality? *Postharv. Biol. Tech.* 15:197–200.

392. Shulman, Y., L. Fainberstein and S. Levee. 1984. Pomegranate fruit development and maturation. *J. Hort. Sci.* 59:265–274.

393. Siegelman, H.W. 1964. Pp. 437–456. In: *Biochemistry of Phenolic Compounds.* J.B. Harborne (ed.). Academic Press, New York.

394. Simmonds, N.W. 1966. Fruit biochemistry. Pp. 223–225. In: *Bananas.* Longman, London.

395. Simon, J.E., A. Hetzroni, B. Bordelon, G.E. Miles and D.J. Charles. 1996. Electronic sensing of aromatic volatiles for quality sorting of blueberry fruit. *J. Food Sci.* 61:967–969.

396. Smillie, R.M., S.E. Hetherington, R. Nott, G.R. Chaplin and N.L. Wade. 1987. Applications of chlorophyll fluorescence to the postharvest physiology and storage of mango and banana fruit and the chilling tolerance of mango cultivars. *ASEAN Food J.* 3:55–59.

397. Smith, D.L., and K.C. Gross. 2000. A family of at least seven beta-galactosidase genes is expressed during tomato fruit development. *Plant Physiol.* 123:1172–1183.

398. Smith, C.J.S., C.F. Watson, P. Morris, R.C. Bird, G.B. Seymour, J.E. Grey, C. Arnold, G.A. Tucker, W. Schuch, S. Hardings and D. Grierson. 1990. Inheritance and effect of ripening of antisense polygalacturonase genes in transgenic tomatoes. *Plant Mol. Biol.* 12:369–379.

399. Smittle, D.A., R.E. Williamson and J.R. Stansell. 1976. Response of snap beans to seed separation by aerodynamic properties. *HortScience* 11:469–471.

400. Smock, R.M. 1972. Influence of detachment from the tree on the respiration of apples. *J. Amer. Soc. Hort. Sci.* 97:509–511.

401. Somers, D.E., P.F. Devlin and S.A. Kay. 1998. Phytochromes and cryptochromes in the entrainment of the Arabidopsis circadian clock. *Science* 282:1488–1490.

402. Sparks, D., and L.E. Yates. 1995. Anatomy of shuck abscission in 'Desirable' pecan. *J. Amer. Soc. Hort. Sci.* 120:790–797.

403. Speirs, J., E. Lee, K. Holt, K. Yong-Duk, N.S. Scott, B. Loveys and W. Schuch.1998. Genetic manipulation of alcohol dehydrogenase levels in ripening tomato fruit affects the balance of some flavor aldehydes and alcohols. *Plant Physiol.* 117:1047–1958.

404. Staswick, P.E. 1989. Preferential loss of an abundant storage protein from soybean pods during seed development. *Plant Physiol.* 90:1252–1255.

405. Stead, A.D., and W.G. Van Doorn. 1994. Strategies of flower senescence—a review. Pp. 215–237. In: *Molecular and Cellular Aspects of Plant Reproduction.* R.J. Scott and A.D. Stead (eds.). Soc. Exp. Biol. Seminar Ser. 55, Cambridge University Press, Cambridge.

406. Stead, A.D. and K.G. Moore. 1983. Studies on flower longevity in *Digitalis:* II. The role of ethylene in corolla abscission. *Planta* 157:15–21.

407. Strayer, C., T. Oyama, T.F. Schultz, R. Raman, D.E. Somers, P. Más, S. Panda, J.A. Kreps and S.A. Kay. 2000. Cloning of the *Arabidopsis* clock gene TOC1, an autoregulatory response regulator homolog. *Science* 289:768–771.

408. Sussex, I.M. 1975. Growth and metabolism of the embryo and attached seedlings of the viviparous mangrove, *Rhizophora mangle. Amer. J. Bot.* 62:948–953.

409. Szczesnick, A.S. 1963. Classification of external characteristics. *J. Food Sci.* 28:385–389.

410. Takao, H., and S. Ohmori. 1994. Development of device for nondestructive evaluation of fruit firmness. *Jpn. Agri. Res. Quart.* 28:36–43.

411. Tanner, W., and T. Caspari. 1996. Membrane transport carriers. *Annu. Rev. Plant Physiol. Plant Mol. Biol.* 47:595–626.

412. Taylor, J.E., G.A. Tucker, Y. Lasslett, C.J.S. Smith, C.M. Arnold, C.F. Watson, W. Achuch, D. Grierson and J.A. Roberts. 1990. Polygalacturonase expression during leaf abscission of normal and transgenic tomato plants. *Planta* 183:133–138.

413. Tekrony, D.M., D.B. Egli, J. Balles, T. Pfeiffer and R. J. Fellows. 1979. Physiological maturity in soybean. *Agron. J.* 71:771–775.

414. Thai, C.N., E.W. Tollner, K. Morita and S.J. Kays. 1997. X-ray characterization of sweetpotato weevil larvae development and subsequent damage in infested roots. Pp. 361–368. In: *Sensors for Nondestructive Testing—Measuring the Quality of Fresh Fruits and Vegetables.* NE Reg. Agr. Eng. Serv., Ithaca, NY.

415. Theologis, A., P.W. Oeller, L.M. Wong, W.H. Rottmann and D. Gantz. 1993. Use of a tomato mutant constructed with reverse genetics to study ripening, a complex developmental process. *Dev. Genet.* 14:282–295.

416. Thomas, H., and J.L. Stoddart. 1975. Separation of chlorophyll degradation from other senescence processes in leaves of a mutant genotype of meadow fescue (*Festuca pratensis*). *Plant Physiol.* 56:438–441.

417. Thompson, J.E. 1988. The molecular basis for membrane deterioration during senescence. Pp. 51–83. In: *Senescence and Aging in Plants.* L.D. Noodén and A.C. Leopold (eds.). Academic Press, New York.

418. Throne, J.H. 1979. Assimilate redistribution from soybean pod wall during seed development. *Agron. J.* 71:812–816.

419. Tiwari, S., and V.S. Bhatia. 1995. Characterization of pod anatomy associated with resistance to pod-shattering in soybeans. *Ann. Bot.* 72:483–485.

420. Tollner, E.W. 1990. Magnetic resonance for non-contact measurement of weight and pit presence in tart cherries. Pp. 184–193. *Proc. Food Automation Conf. Amer. Soc. Agr. Eng.,* St. Joseph, MI.

421. Tollner, E.W., Y.C. Hung, B.L. Upchurch and S.E. Prussia. 1992. Relating X-ray absorption to density and water content in apples. *Trans. Amer. Soc. Agr. Eng.* 35:1921–1928.

422. Tongdee, S.C. 1977. Study on the characteristics of longan during storage. *Kasikorn* 50(2):95–97.

423. Tongdee, S.C., A. Chayasombat and S. Neamprem. 1988. Respiration, ethylene production and changes in the internal atmospheres of durian (*Durio zibethinus* Murray). Pp. 22–30. In: *Proc. Seminar on Durian.* Thailand Inst. Scientific & Tech. Res., Bangkok.

424. Townshend, F. 1967. Food standards. Pp. 185–365, Vol. 1. In: *Quality Control in the Food Industry.* S.M. Herschdoerfer (ed.). Academic Press, London.

425. Trénolières, A., J.P. Dubacq and D Drapier. 1982. Unsaturated fatty acids in maturing seeds of sunflower and rape: regulation by temperature and light intensity. *Phytochemistry* 21:41–45.

426. Trinchero, G.D., G.O. Sozzi, A.M. Cerri, R. Vilella and A.A. Fraschina. 1999. Ripening-related changes in ethylene production, respiration rate and cell-wall enzyme activity in goldenberry (*Physalis peruviana* L.), a solanaceous species. *Postharv. Biol. Tech.* 16:139–145.

427. Upchurch, B.L., H.A. Affeldt, D.J. Aneshansley, G.S. Birth, R.P. Cavalieri, P. Chen, W.M. Miller, Y. Sarig, Z. Schmilovitch, J.A. Throop and B.W. Tollner. 1994. Detection of internal disorders. Pp. 80–85. In: *Nondestructive Technologies for Quality Evaluation of Fruits and Vegetables.* G. Brown and Y. Sarig (eds.). Amer. Soc. Agri. Eng., St. Joseph, MI.

428. Ullrich, S.E., P.M. Hayes, W.E. Dyer, T.K. Blake and J.A. Clancy. 1993. Quantitative trait locus analysis of seed dormancy in 'Steptoe' barley. Pp. 136–145. In: *Pre-harvest Sprouting in Cereals.* N. Noda and D.J. Mares (eds.). Amer. Assoc. Cereal Chem., St. Paul, MN.

429. Van Doorn, W.G. 2001. Categories of petal senescence and abscission: A re-evaluation. *Ann. Bot.* 87:447–456.

430. Van Doorn, W.G., G. Groenewegen, P.A. Van de Pol and C.E.M. Berkholst. 1991. Effects of carbohydrate and water status on flower opening of cut Madelon roses. *Postharv. Biol. Tech.* 1:47–57.

431. Van Doorn, W.G., and A.D. Stead. 1997. Abscission of flowers and floral parts. *J. Exp. Bot.* 48:821–837.

432. Van Staden, J. 1976. Season changes in the cytokinin context of *Ginkgo biloba* leaves. *Plant Physiol.* 38:1–5.

433. Van Staden, J., E.L. Cook and L.D. Noodén. 1988. Cytokinins and senescence. Pp. 281–328. In: *Senescence and Aging in Plants.* L.D. Noodén and A.C. Leopold (eds.). Academic Press, New York.

434. Van Staden, J., and G.G. Dimalla. 1980. The effect of silver thiosulphate preservative on the physiology of cut carnations. II. Influence on endogenous cytokinin. *Z. Pflangenphysiol.* 99:19–26.

435. Vendrell, M. 1969. Reversion of senescence: Effects of 2,4-dichloro-phenoxyacetic acid and indoleacetic acid on respiration, ethylene production, and ripening of banana fruit slices. *Aust. J. Biol. Sci.* 22:601–610.

436. Vereshchagin, A.G. 1992. Kinetics of content of oil and nonlipid materials in ripening seeds. *Soviet. Plant Physiol.* 39:242–250.

437. Vertucci, C.W., and J.M. Farrant. 1995. Acquisition and loss of desiccation tolerance. pp. 701–746. In: *Seed Development and Germination.* G. Galili and J. Kigel (eds.). M. Dekker, New York.

438. Vogele, A.C. 1937. Effect of environmental factors upon the color of the tomato and the watermelon. *Plant Physiol.* 12:929–955.

439. von Loeseche, H.W. 1950. *Bananas.* Interscience, New York.

440. Vrebalov, J., D. Ruezinsky, V. Padmanabhan, R. White, D. Medrano, R. Drake, W. Schuch and J. Giovannoni. 2002. A MADS-box gene necessary for fruit ripening at the tomato *ripening-inhibitor* (*rin*) locus. *Science* 296:343–346.

441. Wade, N.L., P.B.H. O'Connell and C.J. Brady. 1972. Content of RNA and protein of the ripening of banana. *Phytochemistry* 11:975–979.

442. Waldron, K.W., A.C. Smith, A. Ng and M.L. Parker. 1997. Approaches to understanding and controlling cell separation in relation to fruit and vegetable texture. *Trend Food Sci. Technol.* 8:213–221.

443. Walker, G.C. 1985. Inducible DNA repair systems. *Annu. Rev. Biochem.* 54:425–475.

444. Walsh, C.S., J. Popenoe and T. Solomos. 1983. Thornless blackberry is a climacteric fruit. *HortScience* 18:482–483.

445. Wallace, G., and S.C. Fry. 1994. Phenolic components of the plant cell wall. *Intern. Rev. Cytol.* 151:229–268.

446. Wallace, N.H., and A.L. Kriz. 1991. Nucleotide sequence of a cDNA clone corresponding to the maize globulin-1 gene. *Plant Physiol.* 95:973–975.

447. Wang, C., C.-K. Chin, C.-T. Ho, C.-F. Hwang, J.J. Polashock and C.E. Martin.1996. Changes of fatty acids and fatty acid-derived flavor compounds by expressing the yeast △-9 desaturase gene in tomato. *J. Agri. Food Chem.* 44:3399–3402.

448. Wang, C.Y., and W.M. Mellenthin. 1977. Effect of aminothoxy analog of 2-rhizobitoxine on ripening of pears. *Plant Physiol.* 59:546–549.

449. Wang, C.Y., W.M. Mellenthin and E. Hausen. 1972. Maturation of 'Anjou' pears in relation to chemical composition and reaction to ethylene. *J. Amer. Soc. Hort. Sci.* 97:9–12.

450. Wang, H., and W.R. Woodson. 1989. Reversible inhibition of ethylene action and interruption of petal senescence in carnation flowers by norbornadiene. *Plant Physiol.* 89:434–438.

451. Wang, H., J. Li, R.M. Bostock and D.G. Gilchrist. 1996. Apoptosis: a functional paradigm for programmed plant cell death induced by a host-selective phytotoxin and invoked during development. *Plant Cell* 8:375–391.

452. Wang, P.-S., and G.V. Odell. 1972. Characterization of some volatile constituents of roasted pecans. *J. Agri. Food Chem.* 20:206–210.

453. Watada, A.E., R.C. Herner, A.A. Kader, R.J. Romani and G.L. Staby. 1984. Terminology for the description of developmental stages of horticultural crops. *HortScience* 19:20–21.

454. Weigel, D., and E.M. Meyerowitz. 1993. Activation of floral homeotic genes in *Arabidopsis. Science* 261:1723–1726.

455. Wellington, P.S. 1956. Studies on the germination of cereals. I. The germination of wheat grains in the ear during development, ripening, and after-ripening. *Ann. Bot.* 20:105–120.

456. White, C.N., W.M. Proebsting, P. Hedden and C.J. Rivin. 2000. Gibberellin and seed development in maize. I. Evidence that gibberellin/abscisic acid balance governs germination versus maturation pathways. *Plant Physiol.* 122:1081–1088.

457. Wickham, L.D., H.C. Passam and L.A. Wilson. 1984. Dormancy responses to postharvest application of growth regulators in *Dioscorea* species. 2. Dormancy responses in ware tubers of *D. alata* and *D. esculeuta. J. Agric. Sci.* (Camb.) 102:433–436.

458. Williams, J.W., and W.A. Sistrunk. 1979. Effects of cultivar, irrigation, ethephon, and harvest date on the yield and quality of processing tomatoes. *J. Amer. Soc. Hort. Sci.* 104:435–439.

459. Williams, P.J. 1993. Hydrolytic flavor release in fruit and wines through hydrolysis of nonvolatile precursors. Pp. 287–308. In: *Flavor Science, Sensible Principles and Techniques.* T.E. Acree and R. Teranishi (eds.). Amer. Chem. Soc., Washington, DC.

460. Wobus, U., and H. Webber. 1999. Seed maturation: genetic programmes and control signals. *Curr. Opin. Plant Biol.* 2:33–38.

461. Wollgrehn, R., and B. Parthier. 1964. Der Einfluss des Kinetins auf den NS-und Protein-Stoffwechsel in abgeschnittenen mit Hemmstoffen behandelten Tabakblättern. *Phytochemistry* 3:241–248.

462. Woltering, E.J., and W.G. van Doorn. 1988. Role of ethylene in senescence of petals—morphological and taxonomic relationships. *J. Exp. Bot.* 39:1605–1616.

463. Wood, J.G., and D.H. Cruickshank. 1944. The metabolism of starving leaves. 5. Changes in amounts of some amino acids during starvation of grass leaves and their bearing on the nature of the relationship between proteins and amino acids. *Aust. J. Exp. Biol. Med. Sci.* 22:111–123.

464. Woodson, W.R., and K.A. Lawton. 1988. Ethylene-induced gene expression in carnation petals. Relationship to autocatalytic ethylene production and senescence. *Plant Physiol.* 87:498–503.

465. Worrell, D.B., and C.M. Sean Carrington. 1997. Breadfruit. Pp. 347–363. In: *Postharvest Physiology and Storage of Tropical and Subtropical Fruit.* S.K. Mitra (ed.). CAB Intern., Wallingford, England.

466. Wright, M.E., and J.H. Tappen. 1985. The size and shape of typical sweet potatoes. *Amer. Soc. Agric. Eng.* Paper No. 85–6017, pp. 207–225.

467. Wright, F.A., W.J. Lemon, W.D. Zhao, R. Sears, D. Zhuo, J.-P. Wang, H.-Y. Yang, T. Baer, D. Stredney, J. Spitzner, A. Stutz, R. Krahe and B. Yuan. 2001. A draft annotation and overview of the human genome. *Genome Biol.* 2:1–18.

468. Wyllie, S.G., and J.K. Fellman. 2000. Formation of volatile branched chain esters in bananas (*Musa sapientum* L.). *J. Agri. Food Chem.* 48:34–93–3496.

469. Yen, H.C., S. Lee, S.D. Tanksley, M.B. Lanahan, H.J. Klee and J.J. Giovannoni. 1995. The tomato *never-ripe* locus regulates ethylene-inducible gene expression and is linked to a homolog of the Arabidopsis *ETR1* gene. *Plant Physiol.* 107:1343–1353.

470. Yoshihashi, T. 2002. Quantitative analysis of 2-acetyl-1-pyrroline of an aromatic rice by stable isotope dilution method and model studies on its formation during cooking. *J. Food Sci.* 67:619–622.

471. Yoshida, Y. 1961. Nuclear control of chloroplast activity in *Elodea* leaf cells. *Protoplasma* 54:476–492.

472. Young, R.E., S. Salminen and P. Sornsrivichai. 1975. Enzyme regulation associated with ripening of banana fruit. *Facteurs et Régulations de la Maturation des Fruits* No. 238:271–280.

473. Zaltzman, A., B.P. Verma and Z. Schmilovitch. 1987. Potential of quality sorting of fruits and vegetables using fluidized bed medium. *Trans. Amer. Soc. Agr. Eng.* 30:823–831.

474. Zanchin, A., C. Macato, L. Trainotti, G. Casadoro and N. Rascio. 1995. Characterization of abscission zones in the flower and fruits of peach [*Prunus persica* (L.) Batsch]. *New Phytol.* 129:345–354.

475. Zarembinski, T.I., and A. Theologis. 1994. Ethylene biosynthesis and action: A case of conservation. *Plant Mol. Biol.* 26:1579–1597.

476. Zheng, Y., Y. Xi and T. Ying. 1993. Studies on postharvest respiration and ethylene production in loquat fruits. *Acta Hort. Sinica* 20:111–115.

7

STRESS IN HARVESTED PRODUCTS

1. NATURE OF STRESS IN RELATION TO HARVESTED PRODUCTS

Plants have evolved over many millennia, adapting to a wide range of environmental conditions. Relative to the tremendous environmental diversity encountered over the earth's surface, the range of conditions in which plants can survive is remarkably wide. The precise conditions and potential duration of exposure that a plant can withstand before the onset of serious injury and death, however, vary substantially among species. For example, most species of tropical origin (e.g., citrus, bananas) sustain significant damage from low, nonfreezing temperatures (i.e., >10°C), considered mild for temperate species. Likewise, different parts of the same plant may display a significant disparity in the precise conditions under which injury occurs.

Within this broad range of conditions, there is a relatively narrow span under which the plant or plant part functions at its optimum. When the plant is exposed to conditions outside the boundaries of this narrow optimum span, it is subject to stress. Stress is any environmental factor that can induce potentially injurious alterations in a living system. A stress, therefore, is an external factor (or succession of factors) of such magnitude that it tends to interrupt, restrict or accelerate normal metabolic processes and does so in an adverse or negative manner. The extent of injury sustained is determined by the severity of the stress, the length of time the plant is exposed and the plant's general constitutive resistance to the stress. There are circumstances where stress is beneficial; for example, many tropical plants require a period of water deficit before flowering is induced.

Most recommended postharvest environmental conditions represent a stress. While the storage of an apple fruit at 0°C represents a significant stress, it also represents the optimum storage temperature for preservation of the fruit. Implicit in the definition of stress is the question of whether stress is regarded as relative to the plant *per se* or relative to the eventual use for the plant or plant part. From the postharvest biologist's position, stress is an external factor that, if the plant or plant part is exposed to it for sufficient duration or magnitude, will result in undesirable quality changes. Therefore, while recommended apple storage conditions represent a stress to the apple fruit, to the postharvest biologist these conditions are the optimum for the maintenance of product quality.

While attached to the parent plant, an individual plant part is continually replenished with energy, nutrients, hormones, water and other requisites. This availability of essential substances enhances the plant part's ability to withstand environmental stresses and recover from the changes they cause. Plant products that are severed from the parent plant at harvest, however, are deprived of this replenishing process. This loss can significantly alter the commodity's ability to withstand certain stresses and recover from damage.

Commodities are subjected to a significant stress when harvested, but are also subjected to a progressive series of stresses during the postharvest period. As a consequence, the postharvest period requires stress management. In this context, stress is defined relative to the end use of the product.

The importance of stress management during the postharvest period necessitates an understanding of how plants and plant parts respond to stress. The plant's stress response depends upon the plant part affected, its stage of development and the magnitude and duration of a stress. Plant tissue can generally withstand short durations of stress, depending upon the magnitude; however, after the stress exceeds a critical time × magnitude threshold, injury occurs. For example, papaya fruit can be held for 10 days at 5°C before symptoms of chilling injury occur, but injury requires 21 days at 10°C. Bananas, in contrast, are damaged after 2 days at 5°C or minutes at freezing temperatures. Before reaching the threshold for irreversible damage, the changes induced by the stress in the physiology and metabolism of the tissue can be "repaired" upon return to non-stress conditions.

The importance of stress management during the postharvest period necessitates understanding how plants or plant parts respond to stress. Stress results in a physical or chemical change in the cells, of which there are two types.[153]

The alteration may be reversible upon removal of the stress or irreversible, producing a permanent injury to the tissue. Physical manifestations of stress-induced alterations may be seen, for example, as changes in the shape of a cell or the cessation of protoplasmic streaming. A shift in metabolism would be an example of a chemical change precipitated by a stress.

Stress can produce an injurious effect on a tissue in one of three principal ways. When a single stress is imposed, a direct, indirect or secondary stress injury may occur. A **direct stress injury** generally results in a relatively rapid manifestation of the symptoms (e.g., freeze damage). An **indirect stress injury** is due to a normally reversible alteration that, if imposed for a sufficient duration, becomes irreversible (e.g., compression of a fruit for an extended period that results in an irreversible shape change). A **secondary stress injury** is due to the induction of a secondary stress, which in turn causes the injury. High-temperature-induced desiccation is an example of a secondary stress injury (i.e., high temperature increases evaporation, which results in a water stress from which the injury is derived).

Plants have an inherent adaptive potential that they use to resist stress. This resistance to stress can occur *via* two possible means. The plant can avoid the stress by preventing or decreasing the penetration of the stress into its tissues (stress avoidance). Avoidance results in an increase in the amount of stress required to produce a given injury. The plant may also respond by decreasing or eliminating the alterations caused by the stress (stress tolerance).

2. TYPES OF STRESS

During the postharvest period, potential stresses include temperature, water, gas composition, radiation, chemical, mechanical, gravitational, herbivory and pathological. These are normally grouped into abiotic (e.g., temperature, water) and biotic (e.g., insects, pathogens, herbivores) stresses. An understanding of the effect of these stresses on the plant and its quality, and the response of the plant to stress, is needed to effectively design postharvest practices that will minimize their detrimental impact.

2.1. Temperature Stress

2.1.1. High Temperature Stress

Temperature represents perhaps the most critical single factor in the maintenance of quality in harvested plant products. The conversion of water from a liquid to a gas (evaporation) and its loss from the plant provides the primary method for maintaining the plant's temperature within a biological safe range. Due to evaporation (transpiration), sweetpotato leaves growing under conditions where the soil surface temperature can frequently exceed 50°C during the day were able to maintain their maximum leaf temperature under 33°C.[150] Control of temperature by transpiration, however, necessitates a continual replenishment of the water lost. When plant parts are harvested or prevented from replacing this water, they lose the potential to prevent a buildup in temperature. As a consequence, the air temperature, thermal energy from the sun striking the product and metabolic heat produced by the product often result in a progressive increase in product temperature. If the temperature is not reduced, significant quality losses occur. Consequently, one of the primary concerns after harvest is the removal of field and metabolic heat. Postharvest biologists and agricultural engineers have determined the effect of prolonged exposure to high temperature on the loss of product quality, the most desirable method for heat removal and the optimum temperature for subsequent storage for a diverse group of species.

Tolerance to high temperatures varies with species, cultivar, stage of development, condition and plant part in question. Some thermophilic bacteria can grow and develop at temperatures up to the boiling point of water.[24] In contrast, apple and tomato fruits have temperature maximums of 49–52°C and 45°C, respectively.[113] Generally tissues that are in a state of rest are more stable than those that are actively growing. Low moisture content increases the thermotolerance of some commodities (Figure 7.1), for example, dry wheat grains (-9% mois-

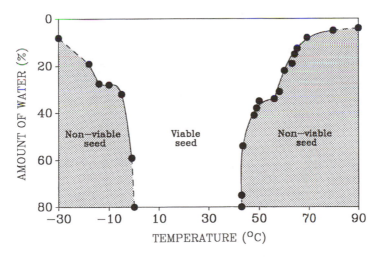

Figure 7.1. The relationship between corn (grain) moisture content and death caused by temperature extremes. High moisture increases the susceptibility of grain to both low and high temperature stress (*redrawn from Robbins and Petsch;*[220] *low temperature data from Kisselbach and Ratcliff*[142]).

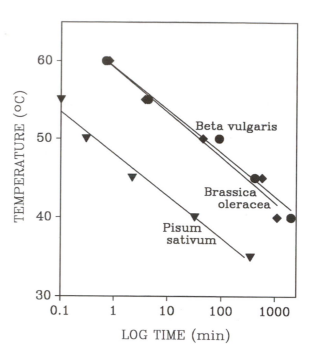

Figure 7.2. The relationship between temperature and the log of the duration of exposure required to cause death (*data from Collander*[49]).

ture) can withstand temperatures of approximately 90°C,[96] while moist grains are injured at much lower temperatures.

The level of injury sustained by the plant or plant part due to high temperature stress is dependent on temperature and exposure duration (Figure 7.2). Generally, plant products withstand periods of high temperature for a relatively short duration better than they withstand exposure to somewhat lower temperatures for longer periods. The temperature at which a product suffers severe injury and possibly death varies inversely with exposure duration (duration being an exponential function) and the tissue's physiological state.[200,203]

The actual injury incurred due to acute high temperature stress occurs at several levels. For example, primary direct injury such as the loss of membrane integrity, cellular leakage, and the disruption of protein and nucleic acid synthesis can occur.[165,203] The integrity of membrane structure and function is essential for normal cellular function;[19] thus its disruption can mediate widespread secondary responses. The effects of chronic exposures to moderate heat stress, however, are not as well understood. Injury may be due to changes in the rate of synthesis or breakdown of enzymes or metabolites, metabolic imbalances or other processes.

Indirect effects of high temperature stress can be seen in the inhibition of pigment synthesis and the formation of surface burns (e.g., sun scald of grape, papaya, pineapple, cherry and tomato fruits) and lesions. In tomato fruit, high temperatures can significantly alter the synthesis of pigments during ripening. At 32 to 38°C, only yellow pigmentation develops while the synthesis of lycopene, the principal red pigment, is inhibited.[274] The softening of tomato and papaya fruit during ripening is also inhibited by high temperatures,[101,203] possibly due to decreased activity of pectolytic and other wall degrading enzymes.[195] Bananas stored at 40°C fail to ripen[294] while at 35°C the production of volatiles is greatly reduced.[295] Exposure to 40 to 45°C for 24 to 48 hours, however, can extend postharvest life and reduce subsequent chilling injury.[165,203] Likewise, very short exposures (e.g., 30 to 90 sec) to similar temperatures can increase the chilling tolerance of seedlings (see below) and delay browning of fresh-cut lettuce and celery.

High temperatures can greatly increase the rate of transpiration. This is due to both a direct effect on the free energy of water and an increase in the vapor pressure gradient between the product and its surrounding environment. When the leaf temperature is 5°C above the atmospheric temperature, the vapor pressure gradient is equivalent to a 30% reduction in the surrounding relative humidity.[56] Thus, when *Ficus* plants were held at 39°C for 4 to 12 days, leaf desiccation occurs, greatly decreasing the quality of the plants.[50]

Plants respond to heat stress through heat avoidance and/or changes in heat tolerance. Avoidance can occur through a decrease in respiratory rate, the synthesis of compounds that increase the reflectance of radiant energy and an increase in transpiration. The plant may effect changes in heat tolerance through changes in its protein complement, chemical composition or moisture content. For example, many species are known to respond to heat stress by synthesizing a new set of proteins (called **heat shock proteins**) while at the same time repressing the synthesis of other proteins.[165] Heat shock proteins, formed during exposure to a 39–41°C treatment, provide thermal protection to a subsequent thermal stress treatment (45–48°C) which would otherwise be lethal.[136,137] There is a rapid synthesis of mRNA's for heat shock proteins, which can be detected 3 to 10 minutes after exposure to the high temperature.[5] The ability of a heat stress treatment to suppress the synthesis of a stress-induced protein may also be important in alleviating the deleterious responses to subsequent stress.[162,227]

Single-celled nonphotosynthesizing *Tetrahymena pyriformis* can alter the fatty acid composition of the lipid portion of their membranes in response to high or low temperature stress,[87] allowing the organism to adapt rapidly to changes in its environment. Since the fluidity of the lipid component of the membrane increases with increasing temperature, this can significantly alter membrane function and integrity.[166] Specific enzymes, synthesized in response to thermal stress, modify certain fatty acids that are incorporated into the lipid portion of the membrane. By decreasing the unsaturated to saturated fatty acid ratio, they decrease the fluidity of the membrane at high temperatures.[60,192] This heat stress survival mechanism is found widely in the plant kingdom and may be important in the thermotolerance of harvested plant products.[165]

2.1.2. *Low Temperature Stress–Chilling Injury*

Most species that evolved in the tropics and subtropics and a few of temperate origin are injured by temperatures slightly above freezing, in the 0 to 10°C range,[199] a phenomenon especially important in the postharvest handling and storage of tropical plant products (e.g., bananas, cucumber, pineapple, papayas, tomato). Exposure of these commodities to low temperatures (<10°C) to extend their storage life must be done with great care to avoid injury. Chilling temperatures may also be encountered in the field or during handling, transit, wholesale distribution, retail sales or in the home.

The degree of chilling injury incurred by a plant or plant part depends upon the temperature, the exposure duration and the species sensitivity to chilling temperatures (Figure 7.3). The lower the exposure temperature below a threshold chilling temperature, the greater the severity of the eventual injury.[271] The rate of development of injury symptoms in storage is also generally decreased with temperature; however, upon removal to non-chilling conditions the full manifestation of the injury incurred becomes apparent (Figure 7.4). Likewise, the longer the product is exposed to temperatures below the threshold for injury, the greater the injury. Only a few hours of exposure of banana fruit to chilling temperatures is sufficient to cause injury, while weeks or months may elapse without harm to some cultivars of apple and grapefruit.

The sensitivity of a plant or plant part to chilling stress varies with species, cultivar, plant part and its morphological and physiological condition at time of exposure.[80,241] Likewise, cel-

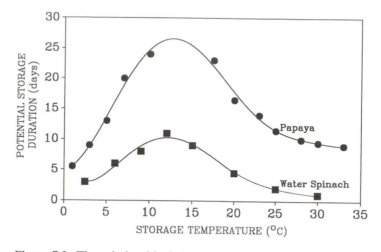

Figure 7.3. The relationship between storage temperature and the maximum potential storage duration for papaya fruit and water spinach (*redrawn from Paull*[199]).

Figure 7.4. Effect of duration of exposure to chilling temperatures on the respiratory rate of okra fruit (*redrawn from Ilker*[120]). The longer the exposure to chilling (●), the greater the respiration rate of the product once transferred to room temperature (■). Elevated respiration is indicative of damage to the tissue.

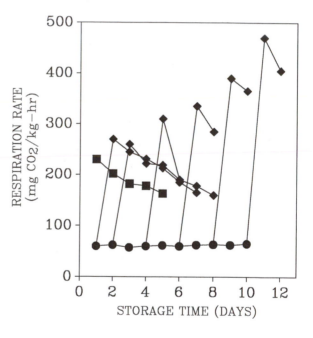

lular organelles within a product may also vary in susceptibility.[43] Some of the more common chill sensitive postharvest products are avocado, beans, banana, cassava, various species of citrus, corn, cucumber, eggplant, okra, mango, melon, papaya, pepper, pineapple, some cultivars of potato, pumpkin, sweetpotato, yam and numerous ornamentals. The fruits of some temperate species, but not the plant, show chilling injury. These include nectarines, peach and plum. In temperate species, the condition is often called low temperature injury.

The critical threshold temperature below which chilling stress occurs varies widely among species. For example, banana fruit is injured by temperatures in the 12–13°C range; mango 10–13°C; lime, muskmelon and pineapple 7–10°C; cucumber, eggplant and papaya 7°C; and orange 5°C. Fruit ripening of some species can significantly alter chilling sensitivity.

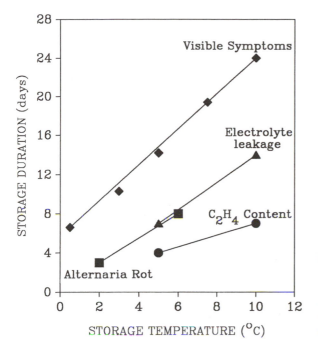

Figure 7.5. The relationship between storage duration and temperature on incipient chilling injury symptoms in papaya fruit (*redrawn from Paull*[199]). Symptoms include electrolyte leakage, fruit cavity ethylene accumulation, *Alternaria* rot and visible signs of injury.

Avocado, papaya and tomato fruit are very sensitive at the preclimacteric stage, increase to a maximum that coincides with the climacteric peak, and then decline rapidly as the fruit ripens further.[147] Susceptibility to chilling injury can also be modulated by field production conditions[83] such as mineral nutrition.

Direct stress injury, the primary response to a chilling temperature, is generally considered to be physical in nature. Chilling temperatures are thought to result in changes in the physical properties of the cellular membranes' lipids and proteins that result in a series of possible indirect injuries or dysfunctions.[120,256] A transition can occur in the fluidity of the membrane lipids that coincides with the threshold chilling temperature in some chill sensitive species.[215] Since membranes vary substantially over their surface, injury symptoms are not uniform. For example, fluidity changes are thought to occur more in certain areas within the membrane (microdomains) that are more susceptable.[6,207] The area affected may be very limited, since membrane permeability does not immediately increase upon exposure of sensitive tissue to chilling temperatures.[229] Alternatively, low temperatures may have a direct effect on specific critical proteins or protein synthesis and degradation rates leading to metabolic disruption upon return to nonchilling temperatures.

After sufficient exposure of sensitive species to chilling, changes in the membranes result in a number of possible secondary effects, e.g., loss of membrane integrity, leakage of solutes (Figure 7.5), loss of compartmentation and changes in enzyme activity (Figure 7.6).[280] These secondary effects lead to the eventual manifestation of chilling injury symptoms. Specific physical and chemical symptoms vary widely among chilling-sensitive species. Injury may take the form of surface lesions (e.g., cucumber fruit[72]) (Figure 7.7), inhibition of ripening (e.g., tomatoes[224]), discoloration (e.g., banana fruit[125]) (Figure 7.7), low temperature breakdown,[289] inhibited growth and increased susceptibility to decay.

Dysfunctions resulting from primary molecular changes induced by chilling temperatures can be repaired or even reversed if the tissue is returned to non-chilling conditions before permanent injury occurs. By using temperature cycling (also known as intermittent warning), 'Bramley's Seedling' apples can be stored at 0°C, using 5 day exposures to 15°C at intervals

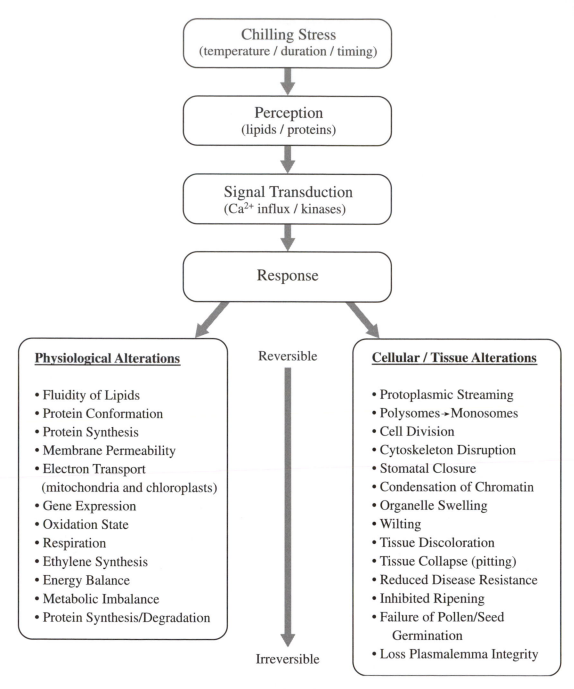

Figure 7.6. Scheme for chilling.

during the storage period, without the development of the characteristic low temperature breakdown at 0°C.[247] Similar improvements with return to non-chilling temperatures have been shown for citrus, cucumbers, peppers and several other crops. The potential for reversal varies with species, condition of the product, length of time under chilling conditions and the exposure temperature.

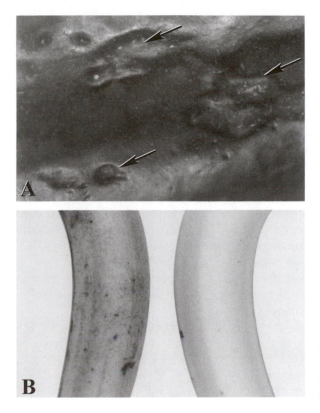

Figure 7.7. Chilling injury is manifested in an array of physical and chemical symptoms. For example in cucumber (A) injury is seen as localized surface pitting, while bananas (B) display a generalized surface discoloration

In addition to direct and indirect stresses, chilling temperatures may also lead to secondary stress injury that result in quality loss and death. When the root systems of sensitive plants (e.g., tobacco, corn, sugarcane) are exposed to temperatures just above freezing, the plants subsequently wilt.[188] If exposed for sufficient time, desiccation induced death occurs, apparently due to the inability of the plant to absorb water *via* a failure of the root's cellular membranes to transfer water to the vascular system.[133] Water loss and wilting may also occur with exposure of the aerial portion of the plant to chilling temperatures. In some species, the stomata are unable to close and the plant's ability to transport water is impaired.

As with high temperature stress, *Tetrahymena pyriformis* displays a remarkable adaptive mechanism, responding to low temperatures *via* rapid alterations in membrane physical and chemical characteristics.[63] Acyl groups within the membrane lipids undergo selective replacement that increases the level of unsaturation with a commensurate increase in fluidity. Chill sensitive products may lack this adaptive mechanism, or the rate of membrane alteration maybe too slow to circumvent injury. Alternatively, injury may be through a non-lipid mechanism (e.g., specific proteins). Regardless of the mechanism resulting in injury, chilling sensitive tissues display undesirable alterations in quality when exposed to a low temperature stress. Aside from avoiding chilling temperatures, few techniques have been successful for circumventing chilling injury. Perhaps the most successful method has been intermittent warming of peaches and nectarines, especially when coupled with controlled atmosphere storage.[4] There is some indication that the development of low temperature injury in apples can be reduced using lower relative humidities in storage.[238] Likewise, postharvest dips of apples in $CaCl_2$ solutions have been shown to reduce low temperature breakdown of 'Jonathan' apples.[239]

2.1.3. Low Temperature Stress—Freezing Injury

Postharvest plant products vary in the temperature at which they freeze and the damage incurred due to freezing.[180] For most fleshy flowers, fruits and vegetables, freezing occurs one to several degrees below the freezing point of water (Table 7.1 and Appendix V and VI); the roots of woody ornamentals are killed by significantly lower temperatures (Appendix IV). The depression in freezing point is due to the presence of solutes within the aqueous medium of the cell. Freezing usually has disastrous consequences upon quality. Generally, even brief exposure of fleshy products to freezing temperatures renders them useless, especially for fresh market sales. Many non-fleshy products (e.g., seeds and nuts) and some acclimatized ornamentals can withstand temperatures well below freezing with little or no damage. For example, wheat seeds (10.6% moisture[163]) and *Ranunculus* tubers (9% moisture[15]) can withstand temperatures of -196°C with no loss in survival. The same tissues similarly exposed, when the moisture content is elevated (25% and 30–50%, respectively), are killed. The ability of a tissue to withstand freezing temperatures is strongly dependent upon its moisture level. Tissues that have very low moisture content have the majority of their water present in a "bound" state. Bound water is not converted to ice crystals even at very low temperatures. The penetration of cellular membranes by growing ice crystals is thought to be the primary cause of damage along with cellular desiccation mediated by ice crystal formation.

Some cells may freeze without loss of viability if the freezing process occurs under the appropriate conditions.[167] Ice crystals may be formed extracelluarly or intracellularly depending upon the rate of cooling. Normally ice forms first in the extracellular space outside of the cells due to a lower solute concentration. When ice crystals begin to form in the extracellular medium, solutes within the freezing water are excluded, increasing the chemical potential of the unfrozen solution. This results in an osmotic gradient between the intracellular and extracellular water that leads to the movement of intracellular water outwards, through the plasma membrane. Freezing of extracellular water, therefore, results in an increase in solute

Table 7.1. Freezing Points of Selected Postharvest Products.* For a more detailed list and the killing point of young and mature roots of selected woody ornamentals see Appendixes IV–VI.

Commodity	Highest Freezing Point (°C)	Commodity	Highest Freezing Point (°C)
Fruit		*Nuts*	
Apple	−1.1	Chestnuts, European	−1.7
Avocado	−0.3	Coconut	−0.8
Banana	−0.6	Pecan	−6.7
Cherimoya	−2.2	Walnut, Persian	−5.5
Date	−15.7	*Cut Flowers*	
Mango	−0.9	Acacia	−3.6
Peach	−0.9	Carnation	−0.7
Vegetables		Chrysanthemum	−0.8
Bean, green, or snap	−0.7	Daisy, Shasta	−1.1
Carrot, mature (topped)	−1.4	Narcissus (daffodils)	−0.1
Lettuce	−0.2	Poinsettia	−1.1
Potato, late-crop	−0.6	Snapdragon	−0.9
Rutabaga	−1.1	*Florist greens*	
Squash, summer	−0.5	Asparagus (*Plumosus*)	−3.3
Water chestnuts	−2.8		

*Source: Havis, 1976;[105] Lutz and Hardenburg, 1968;[167] Studer et al., 1978;[263] Whiteman, 1957;[285] Wright, 1942.[292]

Figure 7.8. Freeze damage in the woody ornamental *Magnolia grandiflora* L. In the intact plant (A), injury (right) is seen as discoloration and altered leaf and shoot orientation after thawing. Leaves (B) display darkening (right) and in many species, a water soaked appearance.

concentration of both the extracellular and intracellular solutions and, in turn, further decreasing the freezing point of the remaining water in a liquid state. Water movement continues until the chemical potential of the unfrozen extracellular water is in equilibrium with that within the cell.

When the rate of cooling is relatively slow, the efflux of water through the plasma membrane is rapid enough that excessive supercooling of the intracellular medium does not occur and intracellular ice crystals do not form. If, however, cooling is quite rapid, excessive supercooling of the intracellular solution occurs and ice crystals form within the cell. This type of ice formation, paradoxically, causes foliage sun-scalding of evergreen ornamentals during the winter (Figure 7.8).[284] The plants are not impaired by the normal slow-freezing process during the winter; for example some species may withstand temperatures of -87°C. However, when leaves are in very bright, direct sunlight, the leaf tissue may thaw even though the surrounding air temperature is below freezing. Intermittent blockage of the sunlight by clouds can result in very rapid cooling (8–10°C · min^{-1}) and refreezing of the leaf, causing intercellular ice formation. Thus, the same plant that can withstand very low winter temperature extremes may be killed by temperatures of only a few degrees below freezing.

After thawing, freeze injured tissue exhibits a flaccid, water soaked appearance (Figure 7.9) with plasma membrane damage a central feature. The rapidity of the response indicates that injury is not due to a metabolic dysfunction such as occurs with chill injury. Injury can also occur when the cooling rate is relatively slow due to concentrating or dehydrating of the intercellular solution.[174,175] Manifestation of freezing injury may result from: a) mechanical penetration of plasma membranes by growing ice crystals; b) expansion-induced lysis or split-

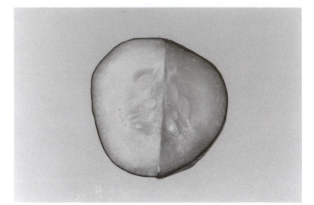

Figure 7.9. Freeze damage in cucumber fruit (right half). Upon thawing, frozen tissue of succulent species typically becomes translucent, flaccid and darker in color.

ting of the plasma membrane during warming; c) loss of osmotic responsiveness during cooling; d) altered osmotic behavior during warming; and e) intracellular ice formation.[256]

Many plants, and in some cases plant parts, may be acclimatized under appropriate conditions, greatly decreasing their susceptibility to freeze injury. Based on the central role of the plasma membrane in damage symptomology, it is inferred that acclimatization may involve the limitation of ice crystal growth or cellular alterations in the membrane that enhance its ability to withstand freezing temperatures. This ability may be due to changes that lead to numerous ice nucleation sites in the extracellular space, altered permeability of the membrane to water, increased membrane stability or the ability of the cellular metabolic system to withstand desiccation during freezing and recovery during thawing.

2.2. Water Stress

Water stress is a universal problem for agricultural plants. Even under optimum soil moisture conditions, plants may undergo a daily cycle of mild but significant water stress. Water stress begins when the tissue's moisture content deviates from the optimum. When cells lose water, their turgor pressure drops, which has a pronounced effect on growth; the expansion of new cells is largely driven by turgor pressure. Cell division is also disrupted by water stress, and it is division that determines the potential for growth and hence final size. The economic consequences of water stress on growth (yield) each year are astronomical.

The potential for injury due to water stress increases sharply after harvest. When plants are actively growing, they are able to maintain a dynamic equilibrium between water uptake by the roots and loss due to transpiration. When individual parts are severed from the parent plant at harvest, their ability to replace water lost through transpiration is eliminated making them much more susceptible to water stress. In the case of root crops (e.g., carrot, parsnip), a portion of the root system is the plant part harvested, however, it is removed from its source of moisture. The situation is different for decapitated plant parts in which transpiration losses can be replaced without the presence of the root system. Harvested fleshy products, however, are generally dependent upon their existing internal moisture supply. Thus, the conservation of internal water is of primary concern after harvest.

Container grown ornamentals represent a somewhat similar situation. The container's growing medium contains moisture that represents the plants limited water resources. The potential contribution of container water during the postharvest period is a function of the medium's moisture content at harvest and the container's volume. Containerized ornamentals

Table 7.2. Generalized sensitivity to water stress of plant processes or parameters (after Hsiao[112]).

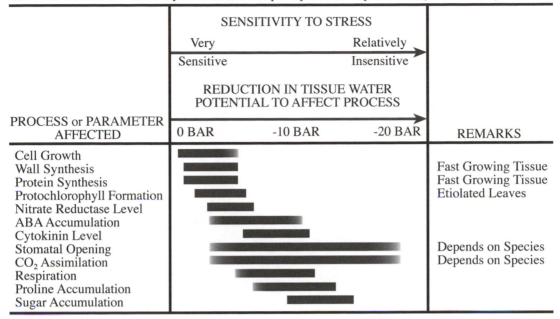

PROCESS or PARAMETER AFFECTED	SENSITIVITY TO STRESS	REMARKS
Cell Growth		
Wall Synthesis		Fast Growing Tissue
Protein Synthesis		Fast Growing Tissue
Protochlorophyll Formation		Etiolated Leaves
Nitrate Reductase Level		
ABA Accumulation		
Cytokinin Level		
Stomatal Opening		Depends on Species
CO$_2$ Assimilation		Depends on Species
Respiration		
Proline Accumulation		
Sugar Accumulation		

have the distinct advantage over many harvested products in that moisture losses can be readily replenished if applied in time.

Water stress can be described using cell water potential or water content relative to the full turgor level as indexes (for a more detailed discussion of cell water relations, see chapter 8). Hsiao[112] has separated water stress into three somewhat arbitrary classes based on cell water potential. Mild stress occurs when the cell water potential is lowered by only a few bars. This represents a small loss of turgor pressure. Moderate stress occurs between a few bars and -12 to -15 bars and mediates a sufficient decline in turgor pressure to result in the wilting of leaves of most species. Severe stress is found below -15 bars when the cells are subjected to serious dehydration and mechanical stress.

Water stress leads to dehydration of the cells that in turn mediates a number of indirect responses (e.g., metabolic alterations, changes in enzyme activation, altered ion flux). Using leaves as a model, the effect of water stress on the functioning of the tissue is illustrated in Table 7.2. Cell growth appears to be the plant process most sensitive to water stress, with a turgor decline of less than one bar being sufficient to decrease the rate of cell elongation. A decline of this magnitude does not, however, stop growth. In fact, some detached products (e.g., cabbage heads) continue to grow in storage, producing stems of considerable length. In this case, there is not a net increase in weight but in fact a decline, with requisite water and nutrients being recycled from existing sources within the head.

Closely following growth in the degree of sensitivity to water deficit stress are cell wall and protein synthesis, chlorophyll formation and nitrate reductase levels, followed by changes in the concentration of abscisic acid and cytokinin, stomatal closure and a sequential drop in the rate of photosynthesis. Changes in each of these parameters occur at relatively mild levels of stress (i.e., <-6 bar). At higher levels of stress (cell water potentials of -10 to -20 bars), changes in respiration and proline and sugar accumulation occur.

Harvested plant products vary widely in their ability to withstand water deficiency. Primary factors include species, surface to volume ratio, surface characteristics and stage of development. There are three general categories of seeds based on their desiccation tolerance

Figure 7.10. Water deficit stress as displayed in: a pepper fruit (A, right); leaf lettuce (B); tomato transplants (C); a Chinese waterchestnut corm (D, right); and (E). sweet corn (bottom). A shriveled or wilted appearance is common in most succulent products undergoing a water deficit stress. In sweet corn (E, bottom), dehydration is first seen as separation between individual kernels (1) and indentation on the top of the individual kernels (2).

and the time they remain viable in storage. a) Seed, spores, and other reproductive bodies of many species are desiccation tolerant (i.e., orthodox) and can be air dried without loss of viability. For example, Kentucky blue grass seed is capable of withstanding a moisture content of approximately 0.1%,[104] while other seeds die at higher moisture levels. Most annual and biennial crops produce desiccant-tolerant seeds that can easily be stored for many years in a dry, cool storage environment. These crops include grains, legumes, vegetables, floral crops and temperate fruit trees. b) By contrast, desiccant-intolerant (i.e., recalcitrant) seeds are difficult to store and usually remain viable for only a few weeks or months. Desiccant-intolerant seeds do not enter dormancy after maturing and tend to have relatively high rates of respiration and physiological activity that lead to rapid deterioration. Desiccant-intolerant seeds must be planted while still fresh or stored moist at low temperatures. Even if properly stored, they can be kept only for short periods of time before they succumb to infection or exhaust their energy reserves. Desiccant-intolerant species include tropical perennials (e.g., avocado, rambutan, litchi) and some temperate deciduous trees (e.g., chestnut, buckeye, maple, oak). c) Between these two extremes are the intermediate seeds that can be stored for a few years if maintained

Table 7.3. Maximum Acceptable Loss of Water as a Percent of Original Fresh Weight for Selected Harvested Products Before the Commodity is Considered Unmarketable.

Commodity	Maximum Acceptable Loss (%)*	Commodity	Maximum Acceptable Loss (%)*
Apple	5	Onion, 'Bedfordshire Champion'	10
Asparagus	8	Orange	5
Bean, broad	6	Parsnip, 'Hollow Crown'	7
Bean, runner	5	Potato	
Beetroot		Maincrop	7
Storing	7	New	7
Bunching with leaves	5	Peas in pod	
Blackberries, 'Bedford Giant'	6	Early	5
Brussels sprouts	8	Maincrop	5
Cabbage, 'Primo'	7	Pepper, green	7
Cabbage, 'January King'	7	Raspberries, 'Malling Jewel'	6
Cabbage, 'Decema'	10	Rhubarb, forced	5
Carrot		Spinach, 'Prickly True'	3
Storing	8	Sprouting broccoli	4
Bunching with leaves	4	Strawberries, 'Cambridge Favourite'	6
Cauliflower, 'April Glory'	7	Sweet corn	7
Celery, white	10	Tomato, 'Eurocross BB'	7
Cucumber, 'Femdam'	7	Turnip, bunching with leaves	5
Leeks, 'Musselburgh'	7	Watercress	7
Lettuce			
'Unrivalled'	5		
'Kordaat'	3		
'Kloek'	3		

Source: Kaufman et al., 1956;[132] Pieniazeh, 1942;[205] Robinson et al., 1975.[221]
*Percent of original fresh weight

under proper conditions of temperature and humidity. They include tropical and subtropical perennials (e.g., coffee, citrus, macadamia, papaya) and some tree nuts (e.g., hazelnut, hickory, pecan, walnut). Soybean seed can be readily air dried when mature with no loss in viability; however, when harvested at less than the mature stage, rapid drying is lethal. If, however, the seeds are held under the appropriate temperature and relative humidity conditions for a sufficient period and then allowed to slowly decrease in moisture content, viability is maintained. Apparently the seeds are capable of undergoing the essential physical and chemical changes needed to maintain viability even though prematurely severed from an external source of moisture and nutrients.

In contrast to mature reproductive structures, most fruits, vegetables and flowers rapidly decline in quality with only minor moisture losses (Figure 7.10). Generally the loss of only 5–10% moisture renders a wide range of products unmarketable (Table 7.3).[132,205,221] There is generally a loss of crispness and turgidity and, for green vegetables, coloration. The rate of water loss also varies widely among different products. Organs such as potato tubers and onion bulbs, capable of long-term storage, lose water less readily than pepper and avocado fruits and much less readily than immature okra and leaf lettuce. Leafy vegetables with large surface-to-volume ratios are particularly vulnerable to the rapid loss of water.

The loss of moisture results in a reduction in the fresh weight of the harvested product, which when sold on a weight basis translates into a loss in monetary return. In addition to losses in turgidity and firmness, there are other symptoms, depending upon the product in question. Significant quality diminutions occur due to undesirable changes caused by water loss, e.g., discoloration of flowers, leafy vegetables and some fruits (e.g., oranges), reduced production of flavor and aroma compounds, a decline in nutritional value due to metabolism of vitamins, an increase in susceptibility to chilling injury, increased incidence of pathogen invasion and an accelerated rate of senescence.

Plants and plant parts vary widely in their ability to withstand water stress. Many intact plants exposed to periods of low water availability become more resistant to subsequent stress, a process called **acclimatization** or drought hardening. Some species have also evolved adaptive mechanisms such as increased cuticle formation, shedding of leaves, rapid stomatal closure, and alterations in shape and size to combat drought situations.

Excessive moisture can also result in water stress conditions in plants. In intact plants, the excess moisture replaces the gas phase of the soil or root media with water, leading to several possible secondary stresses. Excess water may cause leaching of nutrients from the plant or more importantly, impose a gas stress upon the plant (see section 3 of this chapter for additional details). The latter can be due to insufficient oxygen, excessive carbon dioxide or excessive production or accumulation of the hormone ethylene. These changes may result in increased invasions by pathogens. One common flooding symptom in many plants is wilting of the above-ground portion of the plant due to impaired water uptake by the root system. Under waterlogged conditions the lack of oxygen also promotes the accumulation of 1-aminocyclo-propane-1-carboxylic acid (ACC) that is translocated to the aerial portion. When the ACC arrives in cells with adequate oxygen, it is converted to ethylene, which can cause epinastic bending of the leaves (e.g., tomato) and other undesirable alterations.

Seeds and other postharvest plant parts may also sustain stress due to excessive moisture. The moisture optima for seed storage are much lower than that for the normal germination, growth, development and reproduction cycle. As a consequence, low moisture is a desirable condition for long-term quality maintenance in most seeds. There are of course notable exceptions, e.g., the seed of *Acer saccharinum* L. and many tropical tree species will not withstand even air drying. In most seeds, however, increases in moisture above the optimum for storage results in undesirable changes in quality which, from a postharvest standpoint, represents a stress (Figure 7.11).

Different seeds have different optimum storage moisture levels, and the maximum acceptable moisture content varies with storage temperature. Higher moisture content seeds must be stored at lower temperatures to prevent deterioration. Generally the moisture content that leads to a significant increase in respiration coincides with the level at which spoilage begins to occur in storage. The most common cause for increases in seed moisture after drying is storage at an excessively high relative humidity. Seeds absorb or lose water from the atmosphere until they reach equilibrium. Under these conditions, the moisture content of the grain increases until equilibrium is reestablished between the seed and its storage environment.

The composition of the seed also influences the optimum storage relative humidity. For example, at a relative humidity of between 20 and 70%, the seed-moisture content of a carbohydrate containing wheat seeds is around 30% more than that of oil-containing soybean seeds at the same relative humidity. Therefore the relative humidity must be adjusted to give the same moisture level in seeds with differing compositions. Seed moisture level also has a pronounced effect upon the activity of storage pathogens. For example, fungal attack of corn stored at 18% moisture was significantly greater after 7.5 months storage than that stored at 14 to 15.5% moisture.[44]

Water-excess stress is also a common occurrence in other postharvest products when moisture is allowed to remain or collect on the surface of the product. Orchid petals become

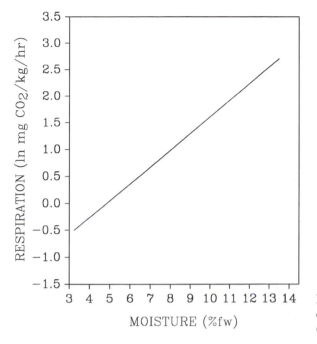

Figure 7.11. The relationship between water content in pecan seed and respiratory activity (*redrawn from Beaudry, et al.*[14]).

translucent if free water remains in contact with them for a few hours. The formation of surface water occurs when the temperature of the product is below the dew point of the water vapor in the surrounding air. Condensation forms on the surface of the product, which impairs gas diffusion and greatly enhances the potential for pathogen invasion. In some products, the damage incurred is relatively minor, while in others (e.g., Chinese chestnuts) it is devastating. Even in Roman times great care was taken to prevent this occurrence.

2.3. Gas Composition

Postharvest stress due to the storage gas atmosphere can have a pronounced effect upon quality changes. A gas stress that slows down the normal maturation and senescence phases of the life cycle of a plant or plant part can be a positive attribute; it is a basic component in the handling and storage of certain crops. Of primary concern during storage is the concentration of oxygen, carbon dioxide and ethylene within the tissue. Postharvest biologists have found that appropriate alterations in the storage gas composition decrease the rate of metabolism without causing permanent injury. An appropriately altered storage gas atmosphere can greatly extend the storage life and maintain quality in some products, and in some instances repress or kill storage insects. However, exceedingly low oxygen, high carbon dioxide or high ethylene concentrations may have a disastrous impact on quality.

2.3.1. Oxygen Stress

Lowering the availability of molecular oxygen causes a decline in a cell's general metabolism. Large changes in metabolic rates, however, do not take place prior to reaching relatively low oxygen concentrations (i.e., <5%). The critical factor is the actual concentration of oxygen within the cells which is controlled by the product's resistance to oxygen diffusion, its rate of oxygen utilization and the difference in oxygen partial pressure between the product exterior and interior. Due to differences in the rate of use and diffusion, there is considerable variation in the optimum external oxygen concentration for various products. Large dense products (e.g., sweetpotato) generally have a higher external oxygen requirement than smaller or less dense products. Likewise, some products must be stored at higher temperatures than others to prevent chilling injury, which increases their oxygen utilization rate. Examples of recommended oxygen concentrations for several postharvest products under optimum storage conditions are: 'Bramley's Seedling' apples, 11–13% O_2 (3–4°C, 8–10% CO_2); 'Golden Delicious' apples, 3% O_2 (2.5°C, 5% CO_2); 'William's Bon Chrétien' pear, 2% O_2 (0°C, 2% CO_2), blackcurrant 5–6% O_2 (2–4°C, 40–50% CO_2); and red cabbage, 3% O_2 (0°C, 3% CO_2).[81,259]

Low oxygen stress can be separated into three general classes: a) Severe stress occurs when anaerobic conditions are reached within the product. This situation, if it continues for a sufficient period, generally results in a rapid and irreversible decline in product quality (Figure 7.12A). b) Moderate stress occurs at oxygen concentrations above that required for anaerobiosis; however, it can impair the quality of the product. Particularly notable are undesirable changes in flavor and aroma. c) Mild stress does not result in injury and can, in some products, greatly increase longevity and maintain quality (Figure 7.12B).

Not all quality components are similarly affected. The optimum oxygen concentration is therefore above that which results in a decline in any critical quality component. It should be noted that optimum oxygen conditions are not necessarily economically practical. Although small improvements in longevity can be achieved for a number of products with mild oxygen stress, the additional expense often does not warrant its use.

Under very low internal oxygen conditions, the terminal electron acceptor in the electron transport system, cytochrome oxidase, ceases to function. This enzyme has a high affinity for oxygen, so inhibition does not occur until the internal oxygen concentration is less than 0.2%, though the external oxygen concentration may be 2 to 3%. When cytochrome oxidase inhibition does occur, $NADH_2$ can no longer be oxidized to NAD and cycled back to the tricarboxylic acid cycle. As a consequence, the tricarboxylic acid cycle is inhibited. Anaerobic conditions in most products, therefore, result in an increase in glycolysis and a low energy yield (i.e., 2 ATPs per molecule of glucose versus 32 under aerobic conditions). Since the cell still requires energy when anaerobic and the efficiency of ATP production declines, respiration must be significantly accelerated to meet even minimal cellular requirements. The glycolytic pathway is not blocked by the absence of oxygen. Its rate is actually accelerated to meet the cell's energy needs. This leads to a build-up of acetaldehyde, ethanol and lactic acid, which can be toxic to the cell by disrupting cellular organization. For example ethanol, the least toxic, may act upon cellular membranes,[143] leading to death. Acetaldehyde is the most toxic of the three. Induction of alcohol dehydrogenase, which catalyzes the conversion of acetaldehyde to ethanol, is often the difference between survival and death in anaerobic tissue in that ethanol and to a lesser extent lactic acid are better tolerated by the cell.

Susceptibility to detrimental low oxygen stress is related to the nature of the product, e.g., its size, stage of development, anatomy, morphology and general condition. Considerable variation in the susceptibility to low oxygen stress can be seen among cultivars. For example, the injury sustained by the storage roots of sweetpotato cultivars exposed to flood induced low oxygen conditions varies widely but does not correlate closely with the absolute alcohol concentration attained nor the activity of alcohol dehydrogenase, the enzyme controlling its synthesis.[42] Other factors, such as differences in ability to metabolize accumulated ethanol upon return to aerobic conditions may be important in determining the extent of injury incurred.

Low oxygen conditions after harvest often develop when air exchange is inadequate. This

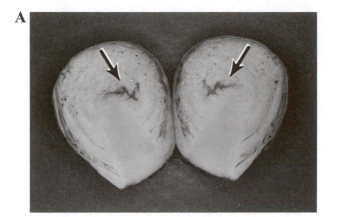

A

B   **% O$_2$**

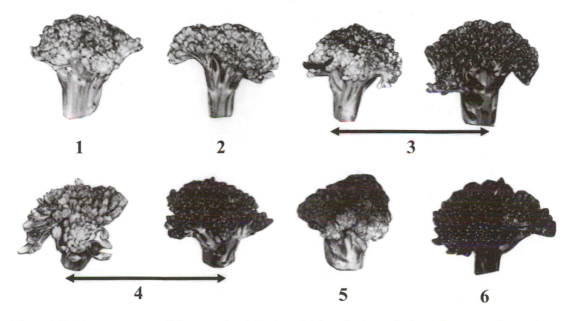

Figure 7.12. Low oxygen conditions can be either beneficial or detrimental, depending upon the product, oxygen concentration, temperature and other factors. The detrimental effect of too little oxygen on stored Brussels sprouts is seen as internal discoloration (A). Both a positive and negative effect of low oxygen on stored broccoli florets is displayed in (B). Inadequate oxygen results in loss of chlorophyll and subsequent lightening of the florets, while low but adequate oxygen minimizes chlorophyll degradation (*photographs courtesy of W. Lipton*).

may be due to excessively tight packaging (especially when product temperature is high), harvesting and bulk storage of grains when their moisture level has not been sufficiently reduced, over-watering of containerized plants, prolonged submergence in or coverage with water, as well as improper controlled gas atmosphere storage conditions. Low oxygen stress may also occur prior to harvest, with the majority of the injury symptoms developing during the postharvest period. Sweetpotato storage roots exposed to flooding prior to harvest are predisposed to spoilage in storage.

The extent of the development of low oxygen injury is highly dependent upon the duration of the exposure to anaerobic conditions. Generally, the longer the exposure, the greater the accumulation of stress metabolites (acetaldehyde, ethanol, lactic acid) and subsequent injury. However, since the reactions for the conversion of pyruvate to ethanol are reversible upon return to aerobic conditions, most plants, if not extensively injured, are capable of recycling the metabolites formed, thus preventing further injury.[58]

Plant products exposed to anaerobic conditions sustain significant losses in quality and if exposure is sufficiently long, death will occur. Injury symptomology and subsequent quality losses vary widely among products. Anaerobic conditions can also result in the formation of undesirable flavors (in particular odor), altered texture, discoloration, increased incidence of physiological disorders and changes in composition.[82,141,157,158,249]

Mild oxygen stress is utilized to decrease the metabolic rate of a number of harvested plant products, decreasing the rate at which quality is lost and increasing the products functional life expectancy. Unlike severe and moderate low oxygen stresses, the effects of mild stress can be beneficial. Examples of several beneficial effects of mild oxygen stress after harvest are decreases in the rate of softening, compositional changes and pigmentation loss. Low oxygen storage conditions are often commercially used to retard the onset of climacteric fruit ripening.[140]

Harvested plant products can also be injured by excess oxygen in the postharvest environment. Several studies have monitored the effects of extremely high oxygen concentrations (i.e., up to 100%); however, the absence of their occurrence outside of the laboratory makes the results academic.[11,138] High oxygen stress may possibly occur under ambient storage conditions for some products that are produced in low oxygen environments, e.g., aquatic crops such as lotus root or Chinese water chestnut. This area of research has been little explored.

2.3.2. Carbon Dioxide Stress

A central component in extended quality maintenance by use of elevated carbon dioxide is a decrease in product respiration. Carbon dioxide, however, may also impose a significant stress on harvested products. The response produced maybe beneficial or detrimental to product quality depending upon the nature of the product in question, the concentration of carbon dioxide within the tissue, the duration of exposure, the internal oxygen concentration and other factors. Increased levels of carbon dioxide significantly decrease the respiratory rate of a number of products. The molecule's primary action appears to be on the rate of a reversible reaction within the tricarboxylic acid cycle catalyzed by succinate dehydrogenase. As the rate of the enzyme is inhibited, the concentration of succinate increases, often to toxic levels.[115,145] Other enzymes in the cycle may also be affected, but to a lesser extent.[189] Carbon dioxide inhibition may also lead to the accumulation of toxic quantities of acetaldehyde and ethanol,[243] a disorder referred to as 'zymasis' in early research on carbon dioxide toxicity. At concentrations above 20% carbon dioxide, zymasis leads to acetaldehyde poisoning due to the inhibition of alcohol dehydrogenase.[266]

Plant products vary widely in their susceptibility to carbon dioxide injury. Variation can be found between individual cultivars,[114,139] with maturity of the fruit[54,61] and with location within the product. For example, the non-chlorophyll containing rib portion of lettuce leaves ('Crisp Head') is much more susceptible to carbon dioxide injury than sections of the chlorophyll containing leaf tissue.[156]

Injury from excessively high carbon dioxide is seen in many products as an increased incidence of various internal and external physiological disorders (Figure 7.13). These include: brown heart[139,249] and core flush of apples;[152] internal browning of cabbage;[121] low temperature

1%O₂   **2%O₂**   **21%O₂**

0.% CO₂

10% CO₂

Figure 7.13. As with oxygen, modified storage carbon dioxide gas environments can be either beneficial or detrimental. A positive effect of high carbon dioxide on a stored product is illustrated by the improved quality of broccoli florets stored in 10% CO_2 (*photographs courtesy of W. Lipton*).

breakdown of apples;[80] brown stain of lettuce;[258] discoloration of mushrooms;[248] calyx discoloration and internal browning of bell peppers;[190] and surface blemishes on tomatoes and pitting of broad bean pods.[268] Quality losses may also be due to accelerated softening;[157] the formation of off-flavors (e.g., strawberries,[53] broccoli,[157,246] kiwi fruit[103] and spinach[177]); inhibited or uneven ripening of tomatoes after removal from storage;[190] decreased resistance to pathogens;[259] and inhibited wound healing in potatoes.[30]

Low levels of supplemental carbon dioxide can result in a number of positive benefits during storage of some crops, and as a consequence, carbon dioxide is an important component of the controlled atmosphere storage of these species (Figure 7.13). In some cases, the elevated carbon dioxide treatment may represent only a short-term exposure. In most instances, however, the product is maintained at the elevated carbon dioxide level during storage. This appears to be due primarily to a progressive decline in succinate dehydrogenase activity with increasing carbon dioxide concentration.[85] Inhibited respiratory rates have been demonstrated in a wide cross-section of harvested products (e.g., spinach, avocado, lettuce, cherries, apples and numerous others). Other positive benefits include: a decreased rate of softening of kiwi and apple fruit;[103,152] inhibition of chlorophyll synthesis in potato tubers held in the light[10,212] and degradation in chlorophyll-containing products such as green beans and broccoli;[231] stimulated loss of astringency in persimmons;[71] decreased ethylene synthesis;[179,269] also decreased ethylene action and inhibition of some storage pathogens.[26,27,47,48,73] The differential in carbon dioxide concentration between where maximum storage benefits are obtained and the onset of detrimental injuries is often quite narrow. Consequently, the level of carbon dioxide in the storage atmosphere must be very closely monitored and controlled.

2.3.3. *Ethylene Stress*

Ethylene within the storage environment, whether produced by the stored product, microorganisms or other sources, is known to cause a significant stress to many harvested products.[1]

The hormone affects the rate of metabolism of many succulent plant products and is generally active at very low concentrations. While the primary mode of action of the molecule is not known, ethylene has been shown to increase the rate of respiration, alter the activity of a number of enzymes, increase membrane permeability, alter cellular compartmentalization, and alter auxin transport and metabolism.[210] Cellular changes induced by ethylene cause an acceleration of senescence and the deterioration that accompanies it. Ethylene within the storage environment results in a significant decline in the level of a number of quality attributes and is responsible for the induction or aggravation of several physiological disorders.[127]

Primary losses in quality in response to ethylene are due to the induction of abscission (e.g., leaves, flower, leaf, fruit) in an extremely wide cross-section of harvested products (Figure 7.14A). Abscission occurs not only in intact products such as ornamental species[169] and vegetable transplants,[134] but also in decapitated products such as cabbage and cauliflower (see review by Reid[216]). There are also changes in flavor, color and texture. The exposure of carrots to ethylene increases the concentration of phenolics[232] and isocoumarin,[41] which is very bitter in taste. Likewise, undesirable flavors develop in cabbage[267] and sweetpotatoes[28] as well as general changes in sugars and acids in climacteric fruits upon exposure to ethylene. In green tissues, ethylene accelerates the degradation of chlorophyll, resulting in undesirable yellowing. This has been shown in cucumber fruit,[208] cabbage,[107] Brussels sprouts, broccoli, cauliflower[267] and acorn squash,[196] as well as in the leaves of many ornamentals. Ethylene also accelerates undesirable changes in coloration of many flowers. When carotenoid pigments are prevalent, degradation of the chromoplasts occurs at a much more rapid rate. In flowers where anthocyanins predominate, changes in vacuole pH hastened by ethylene can result in pronounced alterations in color.[257] Exposure of some products to ethylene stress results in significant changes in texture. Watermelons display a decrease in firmness[219] as do sweetpotato roots after cooking.[28] In some products there is a marked acceleration of lignin synthesis which increases fiber formation and toughness. This acceleration has been shown in asparagus spears[97] and rutabaga roots.[217]

Ethylene in the transit or storage environment has also been shown to induce a number of physiological disorders which impair quality in harvested products. These include flower petal fading, wilting and closure,[100] leaf epinasty[131,169] (Figure 7.14B), russet spotting of lettuce (Figure 7.14C)[118] and gummosis, necrosis and flower-bud blasting in some flowering bulbs.[130] In addition, many products exposed to ethylene stress display an increased susceptibility to storage pathogens.[73]

Not all of the postharvest effects of ethylene stress are undesirable. Ethylene is used widely to initiate the ripening of several climacteric fruits held in storage (e.g., banana and tomato). This synchronizes ripening of all of the fruit and increases the uniformity of ripening among and within individual fruit. Ethylene may also be used to induce the selective abscission of leaves, flowers and fruits of certain ornamentals. In the citrus industry, ethylene is utilized to enhance the degradation of fruit chlorophyll (degreening) of some cultivars of lemons and oranges prior to sale (Figure 7.14D).[281]

2.4. Radiation Stress

2.4.1. Visible Light

Light stress after harvest significantly alters the quality and marketability of plants and plant parts. Stress can be due to either a deficiency or an excess of light depending upon the product and the exposure conditions. The injury sustained from excessive light may be due to the development of a secondary stress (e.g., heat, water deficit) or a direct effect on a light-sensitive

Figure 7.14. Ethylene mediated responses in harvested products. (A) A beneficial effect of ethylene after harvest is illustrated by degreening of citrus; (B) Russet spot in head lettuce is a distinctly undesirable effect of ethylene; (C) Untreated transplants of tomato and pepper and (D) the same plants after exposure to ethylene during simulated transit. Leaves of the tomato exhibit epinasty (downward curvature of the petioles) while pepper leaves abscise at the base of the petioles.

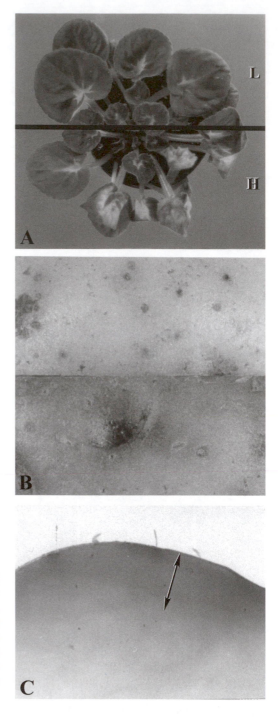

Figure 7.15. The detrimental effect of light stress during the postharvest period is demonstrated by: (A). Sunscald in African violet during retail sales (L—low light, H—high light); (B). Greening of potato tubers during marketing caused by light induced chlorophyll synthesis (bottom half); (C). Unlike the pigmentation found is some products that is limited to the surface layer or two of cells, chlorophyll synthesis in the potato can occur some distance into the tuber.

system within the tissue. Light entering the product increases the object's energy level resulting in a buildup of heat that, if not adequately dissipated, can cause a thermal stress. Heat stress, in turn, may result in an increase in water loss from the tissue. In fruits, foliage and young bark of some species, water loss develops into the classical symptoms of sunscald. Plants produced under low light conditions, such as shade tolerant species, are particularly susceptible to excess light during the postharvest period. Injury is unfortunately a common occurrence for ornamental plants, such as African violet (Figure 7.15A) and gloxinia, during retail sales.

In chlorophyll containing tissues, the symptoms of excess light include a decline in chlorophyll concentration within the exposed leaves and an increase in a yellow or whitish coloration. Normally leaf chlorophylls are adequately protected from photooxidation by carotenoids and other pigments. However, when the light intensity exceeds the capacity of the protective molecules, photooxidation and subsequent breakdown of chlorophyll begins. The degree of injury due to light stress depends upon species, cultivar, conditions of growth, and the light intensity and duration to which the product is exposed. The level of injury can range from slight discoloration to plant death.

Another form of light stress occurs in products that are grown in the absence of light (e.g., roots and tubers). Even very low light intensities after harvest (e.g., 25 lux)[7] can induce the formation of chlorophyll (Figure 7.15B,C). In potato tubers, chlorophyll synthesis is usually accompanied by the accumulation of toxic glycoalkaloids.[122] Root crops may also undergo light stress prior to harvest when the upper portion of the tap root, protruding from the soil, is exposed to light (e.g., green shoulder in carrots). Bamboo shoots are bitter if exposed to light and must be harvested before they break through the soil. Susceptibility to excess light stress varies widely with species, cultivar, light intensity, duration of exposure and other factors.

Light deficiency stress occurs when plants are held under conditions where the daily light exposure results in undesirable changes in quality. This is particularly evident when illumination falls below the light compensation point of the plant. Indirect injury occurs due to the plant's inability to synthesize sufficient carbohydrates for essential respiratory and maintenance reactions. Low light is also known to decrease the activity of certain enzymes[161] and mediate significant alterations in chloroplast structure.[193] Therefore, during the postharvest period both light intensity and duration may be critical. Short exposures to light well above the light compensation point may be insufficient to meet the respiratory requirements of the plant over the remainder of the day. Many plants harvested intact, however, are held for prolonged periods in environments devoid of light.

Transferring ornamental plants, such as *Ficus benjamina* L., produced under conditions of full sunlight, into a low light postharvest environment generally results in excessive leaf abscission (Figure 7.16). If the plants are first acclimatized to low light conditions, however, leaf shedding is minimized.[51] The acclimatization process results in significant changes in anatomy, morphology[76] and composition[182] of the leaves. Acclimatization or hardening of plants has also been shown to result in a lower respiratory rate[176] and light compensation point.[123] As a consequence, these plants tend to survive better in low-light interior environments than plants that have not been acclimatized.[58]

2.4.2. Ultraviolet Radiation

Interest in ultraviolet radiation as a source of stress during plant production has increased significantly since the first report of a decline in the earth's protective ozone layer. Ultraviolet radiation can produce significant injury and death to plants if the exposure is sufficiently high.[68,164] Much of the ultraviolet radiation from the sun is filtered out by ozone and oxygen molecules in the upper atmosphere before reaching the earth's surface. Plants have also evolved organic compounds (e.g., flavonoids, anthocyanins) that absorb ultraviolet radiation and act as protectants. At present, there is little evidence that would suggest that ultraviolet radiation presents a significant stress (deficit or excess) during production or postharvest phases of agriculture.

Figure 7.16. Low light stress during the postharvest period is illustrated by leaf shedding in a Ficus plant during simulated transportation to market. Symptoms include extensive leaf abscission and may include stem die-back. Along with light intensity and the duration of light depravation, preharvest production conditions are known to affect the degree of leaf shedding.

2.4.3. *Ionizing Radiation*

Ionizing radiation represents a form of energy that, when absorbed by an organism, ionizes or electronically excites the molecules with which it comes in contact. Generally this results in chemical changes and damage to the tissue if the radiation is of sufficient magnitude (Figure 7.17). The primary types of ionizing radiation are alpha, gamma and x-rays. These, however, are found only at very low levels in nature. Thus, exposure of plant products to high levels of ionizing radiation represents an artificial situation imposed upon certain harvested products to control microorganisms or insects, or to prevent sprouting (e.g., potato and onion).[60,253]

When used to control insects or pathogens, the radiation causes damage to both the cells of the product and of the target organism. Ionizing radiation has been tested on a wide range

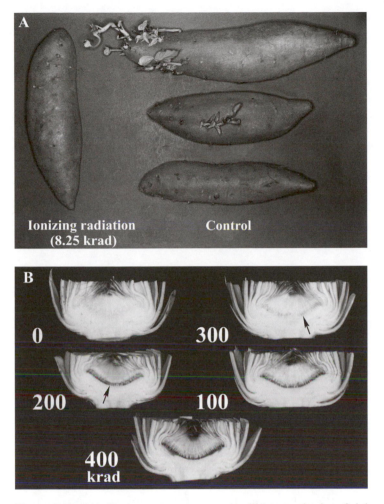

Figure 7.17. (A) Sweetpotato storage roots illustrate the beneficial effect of the use of ionizing radiation after harvest to control sprouting. The vertical root (left), exposed to 82.5 Grays, displays a pronounced inhibition of sprouting in contrast to control roots. When used for insect and microorganism control, success is dependent upon causing more damage to the target organism than to the host plant. (B) The lower photograph illustrates the undesirable effect of radiation on a susceptible artichoke florets (*photographs courtesy of W. Lipton*).

of postharvest products (e.g., strawberry,[172] mango,[62] papaya,[67] potato,[29] several grains) but has not proven commercially viable. Minimum exposure needed for most products to obtain a positive benefit ranges from 1,000 to 10,000 Grays.[25] This, however, is generally greater than the maximum dose the plant tissue will tolerate before injury occurs.[173] At present, irradiation of harvested plant products is used on a commercial scale in Japan to prevent potato sprouting (i.e., low doses, 100 to 200 Grays), Hawaii for insect disinfestation (minimum 250 Grays) and the continental United States for treatment of meat products. The objective in disinfestation is not to kill the insect but to sterilize the eggs and larvae, as a lethal dose would cause significant product quality loss.[201] Dry herbs and spices, however, can be irradiated (see Maxie and Kader[171] and Romani[223] for reviews).

The dose requirements depend upon the sensitivity of the target organism and the ability of the host to withstand undesirable changes. The susceptibility of both the target organism and host vary widely. Since the host product is damaged during exposure, products such as seeds or other propagules that are intended for future growth are not viable candidates for irradiation treatment.

Ionizing radiation stress occurs at any level of exposure; however, in postharvest products our concern is primarily with levels of radiation that produce significant changes in product quality. Injury may be direct, altering a critical type of molecule, for example, a component of the cell membrane whose alteration leads to a loss of semipermeability.[282] Symptoms of this type of injury occur very rapidly. Injury may also be indirect, such as through the formation of free radicals which cause metabolic disruptions. Sub-lethal, but excessive, doses of radiation to the host product have been shown to cause changes in texture, undesirable changes in flavor, the formation of lesions and discoloration of the tissue. Optimum exposures are those which are high enough to facilitate the desirable effects without causing significant changes in product quality or potential longevity.

2.5. Chemical Stress

Plants and plant parts are subjected to a wide range of chemical stresses that may impair postharvest quality. These include salts, ions, air pollutants and agricultural chemicals.

2.5.1. Salt Stress

Excess salt represents a potentially significant stress during the postharvest life of products such as ornamental plants. High levels of salt can cause a direct toxic effect upon the cells of the plant. More often, however, excess salts result in secondary stresses (e.g., osmotic and/or nutritional),[18] that lead to decreased product quality[196] (Figure 7.18). The addition of salts to the aqueous environment of the plant decreases the osmotic potential of the water. In addition, high salts in the growing media can suppress the uptake of other nutrients, e.g., sodium has a competitive effect on potassium uptake.[250]

Secondary osmotic stress represents potentially the most critical concern in the handling, storage and retail sales of ornamentals. During production, the plants are irrigated on a relatively precise and frequent schedule. However, during the postharvest period the responsibility for this task changes hands, generally to personnel who are less familiar with the requirements of the plants. As conditions change from those of the production zone, water use patterns also change, as does the precision of replenishing water lost due to evapotranspiration. High salts in the potting media greatly increase the osmotic stress as the limited water resources of the container are depleted. This decreases the rate of growth, photosynthesis and other metabolic events within the plant. Symptoms of salt stress that ensue range from wilting, leaf burn, nutrient deficiency and leaf drop to death of the plant.

2.5.2. Ion Stress

Unlike salts, ions do not result in a significant decrease in water potential. The strain produced by ion stress is due to an alteration in the normal ionic balance within the cell. Spencer[255] divides ions into three groups based upon their potential toxicity: nontoxic; toxic at intermediate concentrations; and toxic at low concentrations. Of the rather large number of ions that

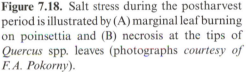

Figure 7.18. Salt stress during the postharvest period is illustrated by (A) marginal leaf burning on poinsettia and (B) necrosis at the tips of *Quercus* spp. leaves (photographs *courtesy of F.A. Pokorny*).

can cause injury, the most critical that are commonly encountered are Ag, Cd, Co, Hg, Mn, Ni, and Zn. These may be introduced as contaminants in the root media of intact container-ized plants, from the water used, or in postharvest treatments.

Ion stress during plant production may result in no visible symptoms. During this period the plants are grown under near optimum conditions. However, upon movement to postharvest conditions, other stresses are imposed upon the plant (e.g., osmotic) that may increase the plant's susceptibility and subsequent damage from ion stress.

2.5.3. Air Pollutant Stress

Air pollutants can cause extensive damage to plants during the production phase and may also be important during the postharvest period. The most common air pollutants encountered are sulfur dioxide, ozone and ammonia (Figure 7.19).[168,191]

Sulfur dioxide injury varies with species, gas concentration, length of exposure and the conditions of the plant and its environment at the time of exposure.[283] The molecule may be either absorbed on the surfaces of the product or absorbed through the stomates.[74] Upon entry into the plant, sulfur dioxide dissolves to sulfurous acid, which forms sulfite salts. When the concentration is sufficiently low, the plant can oxidize these sulfites to sulfates, a normal constituent of plants. In fact, some species of plants assimilate sulfur dioxide with carbon dioxide[77] to be used in various sulfur-containing metabolic products (e.g., the amino acids methionine and cysteine). It is the sulfites not converted to sulfates that appear to cause injury to the tissue.

A relatively wide range of agricultural crops is sensitive to sulfur dioxide injury. Typical symptoms are a marginal and interveinal chlorosis of leaves[111] (Figure 7.19B). The leaves of

Figure 7.19. In addition to ethylene, a number of gases can cause quality loss during the postharvest period. (A) The bottom three kernels illustrate the effect of ammonia vapor on pecans. (B) Sulfur dioxide under appropriate conditions can cause extensive damage to bedding plants (e.g. petunia) during marketing (*photograph from Howe and Woltz*[111]).

some sensitive species may develop a water-soaked appearance due to the effect of sulfur dioxide on membrane permeability. While crop response to sulfur dioxide has been studied extensively in the production phase, little information is available on the effects of postharvest exposure. Sulfur dioxide is used widely for postharvest disease control in grapes and for litchi and longan skin color retention.

Ozone, produced through the photochemical action of sunlight on nitrogen oxides and reactive hydrocarbons from fuel combustion emissions, is considered one of the most serious of the gaseous pollutants. Susceptibility varies widely with species, cultivar and plant condition.[33] Newly expanded leaves appear to be the most sensitive, and rapid luxuriant growth intensifies sensitivity. Susceptible species develop interveinal chlorosis, blisters and necrotic areas on the leaves. Blistering, apparently due to swelling of guard and epidermal cells, is followed by tissue collapse and dehydration.[283] Several chemicals, daminozide[32] and ethylene diurea,[33] have been shown to act as anti-air pollution compounds, decreasing the effect of ozone on certain species. As with sulfur dioxide, the postharvest effects of these chemicals have been little studied.

Ammonia toxicity to plants has been known for many years. Exposure to the gas during postharvest handling and storage has come primarily from ammonia refrigerant leaks. The injury response is a very rapid darkening of the tissue. In pecans, exposure is most dramatic, shifting the surface color from a light brown to black in a few minutes (Figure 7.19A).[135] Ammonia appears to mediate a change in the oxidation state of iron atoms in the surface testa of the kernels. Strong reducing agents can significantly, although not totally, reverse this color change;[277] however, due to toxicity or flavor alterations, these chemicals do not represent a commercially viable method for circumventing injury. Fortunately, the incidence of postharvest exposure to ammonia has declined substantially with the increased utilization of newer refrigerants.

Other gaseous pollutants, e.g. hydrogen fluoride,[95] are also a problem in the production phase of agriculture. Their importance in the postharvest biology of agricultural crops, however, has yet to be established.

2.5.4. Agricultural Chemicals

The number of chemicals used in agriculture has increased dramatically in the past 40 years. Each can have a pronounced effect on yield or product quality attributes. Although most are used during the production phase (herbicides, insecticides, fungicides, growth regulators), some have been found to exhibit distinct advantages when used after harvest. A unifying characteristic of all of these compounds is some form of direct or indirect biological activity. A chemical is selected for a particular type of activity that is advantageous (e.g., the control of one type of insect or group of insects). Its actual biological activity, however, often extends outside of its target role. Plant growth regulators, for example, have been selected for certain benefits they impart to the plant; however, they may also mediate a number of secondary responses some of which occur after harvest.[9]

The effects of biologically active chemicals can be separated into two categories, the beneficial effects as intended and unintentional effects. These unintentional effects may be positive or negative to the plant or plant part in question. Biologically active chemicals, therefore, may result in a stress to the plant (altering normal metabolism); they may also prevent or retard the effect of other stresses.

Inadvertent secondary effects of agricultural chemicals on plant processes have been known for many years. Apple trees treated with lime-sulfur fungicidal sprays undergo a marked reduction (i.e., 50%) in the rate of photosynthesis after application.[106,109] If applied in hot weather, they can cause severe phytotoxicity and death of leaves and scald on fruit. Other indirect effects include changes in respiratory rate,[233] fruit shape and color, and the occurrence of physiological disorders.

The time of application of chemicals may subsequently alter the postharvest biology of a product. These changes range from very early in the development of the product, to just prior to storage. For example, sprays applied to induce parthenocarpy often result in faster ripening fruit, higher respiratory rates, lower acidity, firmness and vitamin C and an increased susceptibility to postharvest physiological disorders many months later.[91,148,185] Chemicals used to thin fruits early in the season can result in increased fruit color, soluble solids and titratable acidity.[153] Likewise, postharvest but prestorage fumigation with ethylene dibromide decreases color development of the outer pericarp of tomato fruit upon eventual ripening,[218] increases the respiratory rate of some deciduous fruits[45] and alters the normal ripening pattern of banana.[78]

Daminozide, one of the most widely studied chemicals applied preharvest (generally to prevent fruit drop), results in both positive and detrimental indirect effects during storage. In some fruits (especially apples) it significantly retards ripening, while in stone fruits ripening is accelerated.[79,159,240] Other effects include changes in texture, color and the susceptibility to certain storage disorders. Questions about the safety of daminozide residues to consumers have led to its removal from use on food crops.

The extent of injury caused by these compounds ranges widely, varying with the chemical in question, its concentration, the plant species, timing of application and other factors. Generally, the undesirable effects do not result in death of the product or render it totally unacceptable for sale. If this were the case, the chemical's positive benefits would be sufficiently

Table 7.4. Postharvest Losses of Selected Fruits and Vegetables Sampled at the Retail and Consumer Level in the New York Marketing Area.

| Commodity | | Cause of Loss % | | |
		Mechanical Injury	Parasitic Disease	Nonparasitic Disorder
Apple	'Red Delicious'*	1.8	0.5	1.3
Cucumber		1.2	3.3	3.4
Grape	'Emperor'*	4.2	0.4	0.9
	'Thompson'*	8.3	0.6	1.6
Lettuce	Iceberg	5.8	2.7	3.2
Orange	Navel	0.8	3.1	0.3
	Valencia	0.2	2.6	0.4
Peach		6.4	6.2	—
Pear	'Bartlett'	2.1	3.1	0.7
	'Bosc'	4.1	3.8	2.2
	'd'Anjou'	1.6	1.7	0.8
Pepper	Bell	2.2	4.0	4.4
Potato	'Katahdin'†	2.5	1.4	1.0
	'White Rose'†	1.5	2.4	0.4
Strawberry		7.7	15.2	—
Sweetpotato		1.7	9.2	4.2

Source: Ceponis and Butterfield[34,35,36,37,38,39]

*Retail losses only

†Wholesale and consumer losses only.

overshadowed by the detrimental effects to preclude its use. More often, undesirable effects are seen as some degree of quality loss, e.g., increased susceptibility to a physiological disorder[20] or more rapid softening.[13]

The cause of the chemical stress from biologically active compounds is not well understood. For many compounds, the response is not seen until quite some time after exposure and therefore does not appear to be a direct effect. The response may affect cell division, assimilate partitioning or occur more directly, for example, a metabolic imbalance leading to the formation of lesions.

2.6. Mechanical Stress

During growth, plants are subjected to mechanical stresses from a variety of sources. Wind, rain, hail, herbivores and soil compaction are but a few of the many sources of mechanical stress by which the growth and development of plants or plant parts may be altered. In addition to these forms of mechanical stress, the effect of mechanical perturbation of plants is also important and has been known for many years. Charles Darwin, in a series of experiments first published in 1881,[57] showed that very gentle tactile stimulation of pea roots caused a change in both the rate and direction of growth. Subsequently the effects of various forms of pre- and postharvest mechanical stress have been studied by a number of scientists. For example, some fruits[245,273] and nuts[254] split while on the tree when subjected to certain conditions.

Mechanical stress is a dominant factor throughout the postharvest handling, transport and storage phases of agriculture. During this period, mechanical stresses initiate a wide range of physical injuries to harvested products which decrease their value, increase their susceptibility to disease and water loss, and often significantly shorten their longevity. Losses due to postharvest mechanical stress can be considerable. For example, retail and consumer losses

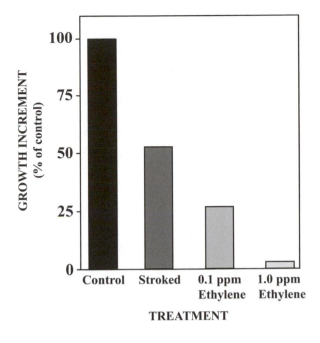

Figure 7.20. The effect of mechanical perturbment on the vertical growth of lily plants *(after Hiraki and Ota[108])*. The leaf blades were lightly stroked five times a day during the course of the experiment (22 days) with a dusting brush. Similar plants were exposed to low levels of ethylene which also significantly inhibited stem elongation.

alone due to mechanical stress for a cross-section of fruits and vegetables sampled in the New York marketing area ranged from 0.2 to more than 8% (Table 7.4).[31,39] These figures do not include losses that occur during and after harvest and prior to packaging.

Mechanical stresses can be separated into two general classes based upon the injury produced. The first, mechanical perturbment, originates from a wide range of sources and is characterized by the absence of a direct physical wounding of the tissue. Mechanical perturbment does, however, result in a decrease in the rate of growth and alters the developmental pattern of many species. The second type of mechanical stress results in a direct physical wounding to the tissue, e.g., cuts, impact and abrasion.

Generally, some degree of mechanical perturbment accompanies all wounding responses. In addition, the same source of the mechanical stress can result in disturbance at one level and wounding at another and, in some cases, a force causes perturbment that may facilitate mechanical wounding, e.g., wind-blown sand.[211] Mechanical stress may also arise as a secondary stress induced by a primary stress such as freezing.

2.6.1. Mechanical Perturbment

Mechanical perturbment of plants whether from stroking,[108] bending,[222] flexing, vibrating, shaking, rubbing[230] or other actions, exerts a pronounced influence on growth and development (Figure 7.20). Subgroups of mechanical perturbment have even been identified based upon the type of stress [e.g., sesimomorphogenesis (shaking), vibromorphogenesis (vibrational)]. Some plants are extremely sensitive to mechanical perturbment with very light touch (e.g., with a camel hair brush) being sufficient to instigate the response. Mechanical perturbment or stimulation has evolved in a few species as part of the function of certain organs (e.g., tendrils) or the trapping movements of certain insectivorous plants (e.g., *Dionaea muscipula Ellis.*).[244]

Mechanical perturbment does not result in physical injury to the tissue. Rather, it instigates significant changes in the development pattern of the plant or plant part. In addition to an inhibition of elongation and an increase in stem radial diameter, mechanical perturbment has been shown to decrease the number of nodes and leaves, increase greening of new leaves, increase lateral branch development,[184] inhibit flowering (*Mimosia pudica* L.)[117] and induce epinasty.[184,225] The mechanism by which mechanical perturbment results in these effects is not yet clear. Reception of the stimulus is quite rapid, while recovery is gradual, often requiring several days before normal growth resumes. Perturbment is known to increase the level of ethylene synthesis, which may be the cause for growth inhibition.[114,222] There are also, however, significant changes in the synthesis of gibberellic acid.[264] Both appear to represent sequential responses occurring well after the initial sensory reception and action. It is thought that the early stages of the response to perturbment involve a rapid change in membrane permeability and bioelectrical potential.

An example of a postharvest mechanical perturbment that alters the value of a crop is the effect of sleeving on potted poinsettia plants. To prevent leaf and bract breakage during handling and transport, open-ended plastic sleeves are placed around each individual plant. This, however, often slightly bends the petioles of the leaves and results in a mechanical stress that is sufficient to increase the level of ethylene synthesis by the plants, which in turn causes leaf and bract epinasty.[209] Foliar sprays of an inhibitor of ethylene synthesis or action, 24 hours prior to sleeving, has been shown to significantly reduce the development of epinasty after the sleeves are removed.[230]

2.6.2. *Physical Wounding*

Mechanical stresses that cause physical injury (mechanical failure) represent one of the most serious sources of quality loss during the postharvest period. Wounding causes increases in respiration and ethylene production and provides entry sites for decay organisms. Although there are also many preharvest sources of mechanical injury to plant tissue (e.g., hail, growth cracks, insect feeding), of particular interest are those that occur from harvest until final utilization. Injury may occur at any point. For example, skin punctures of apple fruits increased progressively as the product moved from the orchard (26%), to the packinghouse (30%), the retail store (36%) and the display bin (50%).[40] A major postharvest source of wounds are injuries imposed during the preparation of fresh-cut fruits and vegetables. The responses to these injuries are similar to those occurring naturally with physical wounds; however, the conditions during imposition and subsequent tissue response can be effectively controlled to minimize deleterious changes. In some products (e.g., grapes, lettuce, apples, and 'Katahden' potatoes) mechanical injury has been shown to be the primary cause of postharvest losses.[31,32] The actual loss may be: a) physical (loss of part of the product); b) physiological (loss in weight from increased water and respiratory losses); c) pathological, due to facilitated entry of microorganisms; or d) qualitative.

Three of the most important types of mechanical stress are friction (abrasion), impact and compression. **Friction**, which results from movement of the product against an adjacent object, can result in surface abrasion. This type of damage occurs on grading belts and during transit where irregularities in the roadbed or wheels of the transport vehicle result in vibrations that are transmitted to the product. Friction induced injury in fleshy products (e.g., pears)[251,252] is generally limited to the epidermis and a few underlying layers of cells. The surface tissue darkens quickly due to the enzymatic oxidation of the constituents of the injured cells (Figure 7.21A). This also provides sites for entry of fungi and other microorganisms. In non-fleshy products such as seeds and nuts,[69] friction results in an abrasion of the surface. Generally this does not result in sufficient damage to decrease product quality.

Impact stress occurs when the product is dropped from a sufficient distance to cause in-

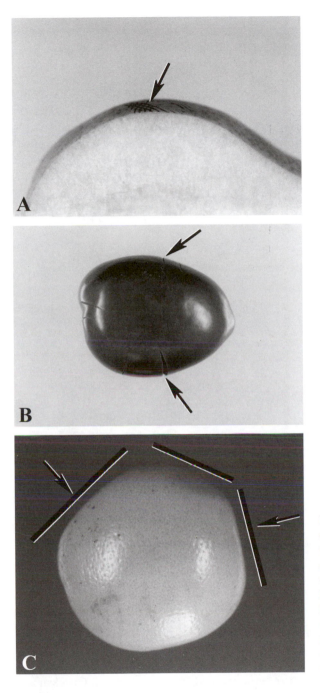

Figure 7.21. The effect of different types of mechanical stress on harvested products. (A) Friction damage to the surface of a pear fruit. In contrast to bruising, the injury incurred with surface abrasions due to friction is normally limited to the surface cells. (B) Impact stress is illustrated by surface splitting of a tomato fruit. (C) Compression stress commonly occurs when products are stacked excessively high. The flat deformed sides of grapefruit are typical of this type of mechanical injury.

jury. In very soft fruits, this may be only a few inches, with the injury generally seen as bruising (Figure 7.21B). The volume injured is usually proportional to the energy of the impact.[226] With bruising, the injury is restricted to the interior flesh and may or may not be seen on the surface (Figure 7.22). In some products the bruise may be initially seen as a water-soaked area after peeling. With time, exposure of the internal contents of the damaged cells to air in the intercellular spaces results in the typical browning symptoms. The damaged tissue may eventually become desiccated.[170]

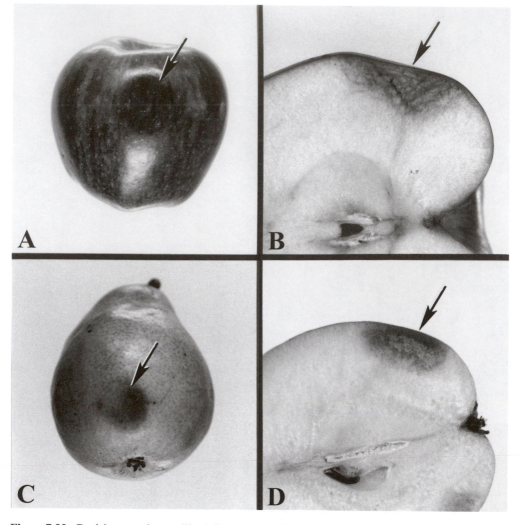

Figure 7.22. Bruising may be readily visible at the surface as in light colored pear fruit (C), or virtually invisible as in intensely pigmented 'Red Delicious' apple fruit (A). In both cases, the damage to cells below the surface is extensive (B and D).

Impact often causes tissue failure where the surface splits (Figure 7.21B) separating partially or entirely into two pieces (Figure 7.23A) or the tissue undergoes slip failure (Figure 7.23B), where the cells rupture or separate along defined surfaces within the tissue. With slip failure, the tissue along either side of the fracture remains relatively undamaged. Cracking or splitting is common in potato tubers, watermelons, cabbages, tomatoes and other large fleshy fruit and vegetable products.[186,187]

Compression injuries are frequently incurred during handling, transport and storage. Stacking the product too deep is the most common cause of compression injury (Figure 7.21C). The injury may take the form of splitting as in tomatoes where the fruit diameter is expanded to a point of failure, internal shear in potatoes[242] or permanent deformation as in citrus. In one study,[99] 33 to 60% of the grapefruit shipped to Japan arrived seriously deformed. The most extensive injury was found in fruit located at the base of the load.

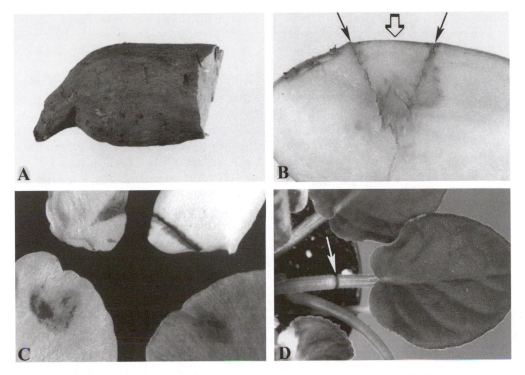

Figure 7.23. Tissue failure occurs *via* four mechanisms: cleavage (A), illustrated by a sweetpotato root; slip (B), where the impact force causes the neighboring cells of the potato tuber to break apart at a 45° angle; bruising (C), illustrated by petal bruising (clockwise from top left: *Impatiens walleriana* Hook. f.; *Gardenia zasminoceides* Ellis; *Mandelvillia laxa* Woodson; *Hedychium coronarium* Hoenig.); and buckling (D), illustrated by the mechanical failure of an African violet petiole.

Edible plant tissues have three primary components that determine their mechanical behavior.[206] These mechanical properties determine the ability of the product tissue to absorb and dissipate mechanical energy. They are the strength of the parenchyma cell walls, the degree of intercellular bonding between neighboring cells and the amount extracellular space between the cells. Turgor pressure and cell wall strength control the mechanical properties of parenchyma cells. Turgor pressure within the cell is modulated by the bidirectional movement of water through the plasma membrane. Water movement allows the extracellular and internal water potentials to equilibrate. Restraining the outward stretch of the protoplasm is the cell wall, structurally the most rigid single component of the cell. In parenchyma cells, there is only a thin wall; no thickening of the cell wall has occurred as in thickened and lignified secondary cell walls of sclerenchyma cells in more rigid woody products. Stiffness of the primary cell wall is imparted by the cellulose microfibrils; however, since the wall is quite thin, it will bend fairly easily, and it is readily deformed.

Analysis of the sequence of events occurring when a load is applied to a cell illustrates how mechanical stress damages plant tissue. Initially the cell begins to change in shape, decreasing in diameter in the direction of the force. Since the contents of the cells do not compress or compress very little, the change in shape alters the surface-to-volume ratio of the cell and increases its turgor pressure. To equilibrate the internal and external water potentials, water must now move out of the cell to compensate for the increased turgor pressure. If the magnitude of the stress is sufficiently low, not applied rapidly and of a short duration, the deformation will be largely elastic, with the cell recovering most of its original shape. Longer periods of compression result in a greater net efflux of water and less recovery. If the magnitude to the stress is great enough or rapidly applied as in impact, the strength of the wall is exceeded and the cell ruptures.

Intracellular bonding adds another dimension to the mechanical properties of tissue. Neighboring cells are bonded together by a pectin layer called the middle lamella. Physically the middle lamella is plastic in nature with the pectin bonds allowing the cells to change position slowly during compression. If the shear stress in the middle lamella exceeds its strength, then the cells will separate, without necessarily rupturing. When debonding occurs, the separation plane usually proceeds at a 45° angle to the direction of the force applied.

Finally, the extracellular space is also an important component of the tissue's rheological and strength properties. Part of the area between neighboring cells is not middle lamella but air and extracellular fluid. When the extracellular space is sufficiently large, as in peach fruits, it provides room for the cells to reorient when compressed; thus, the tissue volume can change significantly. In very dense products (i.e., potato tubers or sweetpotato storage roots), the extracellular space is low and there is little compression of the tissue.

When a plant tissue is subjected to sufficient friction, impact or compression stress, it undergoes mechanical failure. Tissue failure may occur by one of four possible modes: cleavage, slip, bruising and buckling (Figure 7.23).[110] **Cleavage** occurs when failure is along the line of maximum shear stress, leaving two essentially intact pieces (Figure 7.23A).[65] When an impact force results in failure along planes at approximately a 45° angle to the direction of loading, the two sections of tissue slip or slide relative to each other (Figure 7.23B). **Slip** failure occurs in apples,[181] potatoes[64] and a number of other products. The most common failure mode in harvested products is bruising. **Bruising** occurs when the cells rupture, allowing the interaction of previously sequestered components (e.g., enzymes, substrates) and exposing their internal contents to the air (Figure 7.23C). Both enzymatic and non-enzymatic reactions occur that cause the characteristic dark discoloration of the bruise. **Buckling** failure is illustrated by petiole breakage in African violet (Figure 7.23D). Stems and petioles of intact plants often sustain this type of failure during transit and marketing.

Plants and their component parts vary widely in their susceptibility to mechanical damage. Several important factors affect the degree of injury sustained from a given stress, e.g., cultivar, degree of cell hydration, stage of maturity, product weight and size, skin characteristics, and environmental conditions (e.g., temperature). Differences in the susceptibility to mechanical damage may be substantial even among individual cultivars. For example in a damage survey conducted by the Potato Marketing Board in Great Britain, total flesh damage between potato cultivars ranged from 13.3% in the cultivar 'Golden Wonder' to 36.7% in 'Kerrs Pink'.[5] Potato firmness markedly influences the degree of damage within a cultivar. When the tubers are very turgid, the incidence of splitting and cracking increases during harvesting and handling. Conversely, when flaccid they are more susceptible to black spot, a disorder caused by mechanically induced cell rupture and subsequent discoloration due to the oxidation of the amino acid tyrosine. In dried seeds such as beans,[12] cracking decreases with increasing seed moisture content. Pecans and several other nuts are either soaked in water or steam conditioned prior to shelling to minimize kernel breakage.

The stage of maturity is likewise an important parameter in a number of products. Many fruits undergo distinct textural changes as they ripen, with softening of the flesh greatly increasing their potential for mechanical damage. Because of this, many fruit that are to be harvested for commercial sales are gathered before they reach complete ripeness. Peaches harvested at the green ripe stage, underwent significantly less bruising than those of more advanced maturity.[116] Papayas are susceptible to friction injury when less than half ripe and to impact injury as ripening increases.[214] Susceptibility to mechanical damage increases with increasing size and weight of the product.

Skin strength and other mechanical properties are important factors governing differences in damage among types of products and cultivars. For example, tomato skin strength is a critical factor in the resistance of the fruit to cracking.[275] Both the shape of the surface cells and the deposition of cutin appear to affect skin strength.[102] With potato tubers, both periderm thickness and maturity modulate the degree of abrasion damage sustained during harvesting and handling.[287]

Environmental conditions such as temperature, relative humidity and oxygen concentration can also significantly alter the degree of damage sustained from a given level of mechanical stress. This may be due to a direct effect on the mechanical properties of the tissue or the development of injury symptoms after wounding. Potato tubers are more prone to bruising when cold than when warm.[197] Damage can be reduced, therefore, by starting harvest later in the day if night temperatures have been cool and by warming tubers removed from storage before handling. The volume of bruises produced increases with increasing temperature at impact and holding temperature as the bruise develops.[226]

The wounding of plant tissue starts a dramatic series of changes that culminate in either the rapid deterioration and death of the product or healing of the damaged surfaces. The sequence of events that occur during wound healing depends upon the nature of the damaged tissue. Healing may involve complete replacement of the damaged tissue (e.g., as in apical meristems), repair through the induction of a secondary meristem (e.g., wound periderm formation in tubers) or changes in the physical and chemical composition of the tissue near the damaged cells.[21,22,129,155] Younger, more meristematic tissues tend to heal wounds more successfully than older tissues.[89] Apple fruit attached to the tree can heal mechanically induced wounds completely until mid-July; thereafter wound-healing ability declines rapidly.[245] Stored apples, in contrast, display virtually no wound-healing ability.

The chronological sequence of anatomical changes occurring in response to wounding includes: desiccation of several of the outer layers of parenchyma cells adjacent to the wound; suberization[146] and in some cases lignification[279] of the cells below the desiccated cells; and stimulation of cell division in the cells below this to form a cambium (phellogen) and with continued cell division, the formation of cork (phellem). In some tissue, there may be no phellogen formation, just suberization of the cells adjacent to the wound. In response to mechanical wounding caused by leaf miners, species with thin leaves tended to form callus while species with fleshy or evergreen leaves form periderm.[291]

Changes in anatomy are mediated through a series of pronounced biochemical alterations in the cells adjacent to the wound. There is an increase in respiration, synthesis of nucleic acids, RNA, protein and ethylene, transformation of carbohydrates, and activation of certain enzymes and repression of others.[46,119,128,129,178] In tomato fruit, the activity of hydroxycinnamate-CoA ligase and O-methyltransferase increases sharply, an increase that appears to be involved in the formation of monomeric units from which, through the activity of peroxidases, lignins are formed.[84] Phytohormones, especially auxin, cytokinin and ethylene, are thought to play a role in the wound response.[204]

In addition to age, species and even cultivar,[88] a number of other factors can influence the response of plants to wounding. Variation due to temperature, relative humidity, light and oxygen concentration strongly modulates both qualitative and quantitative responses.[155,228] High humidity is conductive to wound healing in many tissues (e.g., 85 to 95% RH is recommended for sweetpotatoes); however, excess moisture can impede the process. Likewise, moderate to warm temperatures generally facilitate wound healing. In Kalanchoe, an inverse linear relationship exists between temperature (21–36°C) and wound healing.[154]

2.7. Gravitational Stress

Plants and many of their organs exhibit a precise gravitationally controlled orientation. Subterranean organs such as primary roots display a positive gravitropism;* they grow in the direction of gravity. Stems and flower stalks, however, are negatively gravitropic and develop at an 180° orientation to the pull of gravity. Secondary roots may grow at intermediate angles, e.g., 45° (termed plagiotropism), while many petioles, stems, rhizomes, roots and stolons grow more or less horizontally.

When a gravitationally sensitive plant or plant part is changed from its normal orientation, it responds by orienting its new growth in the direction required to reestablish its original plane. Therefore, snapdragon racemes (Figure 7.24A) or asparagus spears laid on their sides after harvest (90° to the orientation of gravity) begin to grow in a vertical direction. Potted narcissus plants also undergo gravitropic curvature when their orientation to the gravitational field is altered (Figure 7.24B). Unlike the case with snapdragons, their curvature is a turgor-driven response that is normally reversible upon returning the plant to the proper orientation. This change in orientation results in stress which can be seen as a small decline in the overall growth rate.[202] Much more important, from a postharvest standpoint, is the effect of the plant's new orientation upon product quality. Bent stems of gladiolus and asparagus spears are considered inferior in quality, so care must be taken to keep gravitropically sensitive products upright during transit and storage.

The product's response to gravity can be separated into three distinct components: perception of the gravitational stimulus; transduction the stimulus to the zone of response; and the growth response.[276,288] Perception is believed to occur in specialized cells called **statocytes,** found in the apical tip of roots, shoots and flower stalks. These cells contain gravitropically sensitive masses called **statoliths,** which move in response to a change in orientation to the bottom of the cell. Starch-containing amyloplasts are the only subcellular particles of sufficient mass and density to sediment in the cytoplasm at a rapid enough rate to account for graviperception, and indeed their movement has been correlated with the plant's perception of an alteration in its gravitational orientation.

The response zone in higher plants is often several cells removed from where gravitational perception takes place. The tips of coleoptiles detect gravity, while the cells below respond in a differential growth pattern that results in upward growth. If the tip is removed, the shoot loses its graviperception, even though it continues to elongate. For many years the signal was thought to be transmitted through a differential movement of auxin (Cholodny-Went hypothesis). Auxin accumulated on the lower side of the organ, promoting the growth of those cells in stems while inhibiting them in roots. An alternative explanation is that auxin does not elicit the response, but a growth inhibitor accumulates on the basal side, at least in the roots.[66,126]

The response of the stem and root involves a differential in growth over the cross-section of the organ. Curvature is caused by a growth inhibition; for roots the inhibition is only slight on the upper side but pronounced on the lower and continues until the organ reestablishes its original gravitational orientation.

The occurrence of gravitropic-mediated quality losses is relatively infrequent. While many containerized ornamentals will display curvature if left in a non-vertical position for sufficient time, their response rate is generally slow enough to preclude significant damage during normal handling, transport and storage periods. Several crops [e.g., seedling cress, asparagus spears, potted flowering bulbs, and red ginger, gladiolus and snapdragon inflores-

---

*Previously termed "geotropism."

Figure 7.24. The effect of gravitational stress on harvested products: (A) a cut snapdragon raceme; and (B) a potted *Narcissus* plant. The examples illustrate two different types of gravitropic response. The former (A) represents an irreversible curvature due to growth, while the latter (B) is a turgor driven response that will normally reorient in a vertical position once the pot is placed upright.

cences (Figure 6-24B)] do, however, display a high level of sensitivity and relatively short response time. Curvature due to improper stacking of boxes during transit or storage can result in significant losses in product quality.

2.8. Herbivore Stress

Feeding by an extremely diverse array of herbivores results in extensive damage to many plants and, in some cases, even death. These range from invertebrate (e.g., protozoa to mites and insects) to vertebrate herbivores (e.g., fish to mammals).[55] In nature, essentially all anatomical

Table 7.5. Plant Tissues and the Herbivores That Feed on Them.*

Tissue	Mode of Feeding	Examples of Feeders
Leaves	Clipping	Ungulates, slugs, sawflies, butterflies, etc.
	Skeletonizing	Beetles, sawflies, capsid bugs
	Holing	Moths, weevils, pigeons, slugs, etc.
	Rolling	Microlepidoptera, aphids
	Spinning	Lepidoptera, sawflies
	Mining	Microlepidoptera, Diptera
	Rasping	Slugs, snails
	Sucking	Aphids, psyllids, hoppers, whitefly, mites, etc.
Buds	Removal	Finches, browsing ungulates
	Boring	Hymenoptera, Lepidoptera, Diptera
	Deforming	Aphids, moths
Herbaceous stems	Removal	Ungulates, sawflies, etc.
	Boring	Weevils, flies, moths
	Sucking	Aphids, scales, cochineals, bugs
Bark	Tunneling	Beetles, wasps
	Stripping	Squirrels, deer, goats, voles
	Sucking	Scales, bark lice
Wood	Felling	Beavers, large ungulates
	Tunneling	Beetles, wasps
	Chewing	Termites
Flowers	Nectar drinking	Bats, hummingbirds, butterflies, etc.
	Pollen eating	Bees, butterflies, mice
	Receptacle eating	Diptera, Microlepidoptera, thrips
	Spinning	Microlepidoptera
Fruits	Beneficial	Monkey, thrushes, ungulates, elephants
	Destructive	Wasps, moths, rodents, finches, pigeons
Seeds	Predation	Deer, squirrels, mice, finches, pigeons
	Boring	Weevils, moths, bruchids
	Sucking	Lygaeid bugs
Sap	Phloem	Aphids, whitefly, hoppers
	Xylem	Spittlebugs, cicadas
	Cell contents	Bugs, hoppers, mites, tardigrades, etc.
Roots	Clipping	Beetles, flies, rodents, ungulates, etc.
	Tunneling	Nematodes, flies
	Sucking	Aphids, cicadas, nematodes, etc.
Galls	Leaves	Hymenoptera, Diptera, aphids, mites
	Fruit	Hymenoptera
	Stems	Hymenoptera, Diptera
	Roots	Aphids, weevils, Hymenoptera

*After Crawley.[55]

parts of plants are subject to consumption by herbivores (Table 7.5). Stems, leaves, roots, buds, flowers, seeds and other plant parts are consumed or damaged by the various modes of feeding. Significant postharvest losses are incurred, especially due to the action of the insect, mite, bat, and rodent groups. In Southeast Asia, even monkeys and elephants are a problem. Obtaining an accurate estimate of worldwide postharvest losses caused by herbivores is difficult, however, losses are known to be extensive. For example, maize storage losses in Honduras were estimated at 50%.[270] Often the value of the losses sustained *via* storage herbivores is several orders of magnitude greater than the cost of appropriate control measures.

Feeding by herbivores results in a distinct stress on both pre- and postharvest plant products.[198] Due to the diversity of herbivores consuming plant parts, several potential stresses may be operative. These include mechanical stress, the introduction of biologically toxic or active chemicals (chemical stress) and the development of secondary stresses (e.g., pathogen,[31] water stress).

The action of herbivores can be separated into three general classes, based upon their effects on the plant or plant part: those where the action of the herbivore causes a) only a small effect on the general growth of the plant or plant part, b) significant metabolic changes in the plant part attacked, and c) both metabolic and developmental changes. These classes are not mutually exclusive; e.g., even minor damage results in some localized metabolic changes. The mechanical removal of leaf tissue by phytophagous insects results in short-term effects on growth roughly proportional to the volume of the tissue removed.[149] Leaf area loss decreases the potential photosynthetic area of the plant, hence reducing the total carbon fixed. If the insect population is small relative to the size of the plant, dry matter losses may represent only a very small portion of the total carbon turnover by the plant. However, when the insect population is large, losses may be extensive and significantly decrease subsequent growth.[272] During storage, the loss of foliage decreases the quality of the product, and with intact plants or plant propagules, it can significantly reduce subsequent growth.

The action of some herbivores initiates significant metabolic changes within the affected tissue. Aphid feeding has been shown to increase the respiratory rate of leaf tissue, while photosynthetic activity declines.[144] Leaf miners (e.g., *Stigmella argentipedella* A.) in birch leaves increase the cytokinin and protein concentrations in the tissue surrounding their tunnels (up to $20\times$ the cytokinin concentration of normal leaves). Because of this, the affected tissue remains greener much longer in the fall.[75] When larvae of the sweetpotato weevil (*Cylas* spp.), tunnel through storage roots (pre- and postharvest), they release a terpene-inducing factor which triggers the synthesis of the toxic furanoterpenoid ipomoeamerone by the roots. Although toxic to animals,[23] it does not appear to affect the insect. The precise biological function of the chemical is not known. The terpene inducing factor consists of a glycoprotein, a protein and a heat-stable low molecular weight compound.[234,235]

Other herbivores are known to cause both metabolic and developmental changes in the plant part attacked. Perhaps one of the clearest examples is the effect of gall-forming insects. Various gall wasps deposit their eggs in buds, stems, leaves or roots of the host species. The resulting larva alters the metabolism and development of the tissue, producing a malformed growth. These changes appear to be due to chemicals released by the larvae. In the Oriental chestnut gall wasp (*Dryocosmus kuriphilus* Yasumatsu), the larvae instigate a preferential movement of carbohydrates and other requisites to the infested bud *in lieu* of the normal buds on the stem. The resulting distorted growth prevents normal leaf and stem formation, decreasing the photosynthetic area of the plant. When infestations are sufficiently high, tree death often occurs within several seasons.

Losses arising from the action of herbivores on postharvest products can be viewed at several levels. Feeding decreases the net volume of the stored product. For example, under appropriate storage conditions weevils (e.g., *Oryzaephilus surinamenis* L.) can cause tremendous losses. Herbivore activity also has a major effect on product quality. The external or internal appearance of the product may be impaired by tissue damage (e.g., tunneling insects in fleshy fruits and vegetables) (Figure 7.25). In addition to the physical damage to the product, the presence of excreta, dead insects and webbing decreases quality and product cleanliness. Quality may be impaired by the mere presence of the insect prior to actual damage to the product. The action of herbivores may also decrease the nutrient value of the product. The production of toxic furanoterpenoids by sweetpotato storage roots in response to the activity of sweetpotato weevil larvae minimizes its desirability as food. Nutrient losses in plant propagules, likewise, can reduce vigor and subsequent growth.

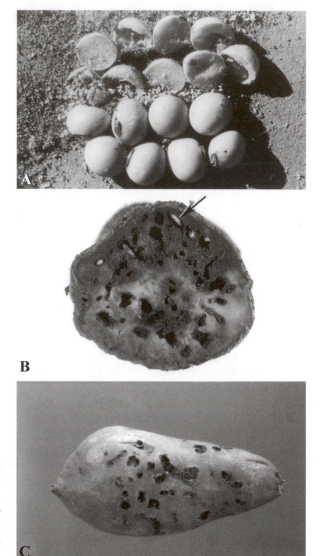

Figure 7.25. Herbivory stress on harvested products: (A) Cigarette beetle (*Lasioderma serricorne*) damage to stored soybean; (B) Sweetpotato weevil (*Cylas formicarius elegantulus* Summers) damage to a stored sweetpotato root; and (C) Preharvest damage to sweetpotato caused by wireworms (*Conoderus falli* Lane).

2.9. Pathological Stress

Harvested plants and plant parts undergo substantial losses due to the stress imposed by invading microorganisms (Figure 7.26). A wide range of such organisms (fungi, bacteria, viruses, mycoplasms and nematodes) affect plant products and cause disease. Disease is the result of a successful infection. Some infections (termed **latent, incipient** or **quiescent**) occur prior to harvest with the disease not developing until the postharvest period, for example, anthracnose diseases of tropical fruit caused by *Colletotrichum gloeosporioides*.[212] Not all fungi and bacteria found on fresh produce cause disease. Many actually inhibit the growth and development of pathogenic fungi and bacteria. In some instances, these organisms can be used as biocontrol agents, though they generally are less effective than chemical controls.[290]

Development of a pathogen within the tissue causes marked metabolic changes in the host plant. In contrast with what little is known about induced metabolic changes in plants

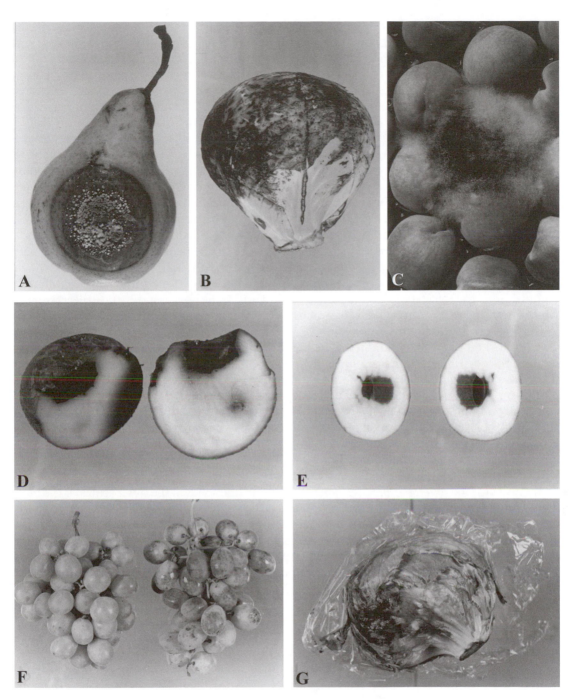

Figure 7.26. Illustrated are a cross-section of postharvest diseases: (A) Blue mold rot on pear—*Penicillium expansium,* Link; (B) Black rot on Brussels sprouts—*Xanthomonas campestris* (Pam.) Dows.; (C) Rhizopus rot on peach—caused by both *Rhizopus oryzae* Went & Prinsen Geerligs and *R. stolonifer* (Ehreub. ex Fr.) Lind.; (D) Gangrene on potato—*Phoma exigua* Desm. var. *foveata,* (Foister) Boerema; (E) Black heart of potato—a physiological disorder thought to be caused by low oxygen; (F) Grey mold rot on grapes—*Botryotinia fucheliana* (de Bary) Whetzel; and (G) Bacterial soft rot on head lettuce—caused by both *Erwinia carotovora* (Jones) Holland and *Pseudomonas marginalis* (Brown) Stevens (*photographs courtesy of A. Snowdon*[250]).

due to herbivory,[128] the effect of pathogens has received considerable attention. This is due in part to the low mobility of plant pathogens after invasion relative to that of most herbivores. As a plant is altered by the action of a herbivore, it becomes a less desirable food source; most herbivores can simply move to adjacent plants or a different part of the same plant. This is rarely the case with pathogens, and as a consequence, there has evolved a very close interrelationship between the host and the disease organisms. Plants have evolved a number of defenses, both preexisting and induced, against the interaction with disease organisms.[90]

Infection by microorganisms is a complex process.[122] The interaction between pathogen and host can be separated into two general processes, the infection sequence of the organism and the various types and degrees of stress imposed upon the host by the pathogen's colonization. While the former is of critical importance in the overall pathogenicity of the organism, it is the stresses imposed upon the host during colonization that result in product losses. Some pathogens directly penetrate the plant's surface with mechanical pressure or enzymes, while others enter through natural openings (e.g., stomata, lenticels) or mechanical wounds (e.g., cuts). The epidermis represents both a physical and chemical barrier to invasion. Soft rots of stored fruits caused by *Sclerotinia fructigena* secrete polygalacturonase enzymes, which degrade the pectic substances of the cell wall. After infection, various pathogens use different colonization strategies; some kill the plant cells and utilize the host as a substrate (**necrotrophs**) while others feed on the living cells (**biotrophs**). The infection and colonization phases are followed by reproduction of the pathogen.

Plants and plant parts exhibit a wide range of structural and biochemical barriers to infection. Successful infection is an infrequent event and depends upon a failure or breach in these barriers. The barriers may be preexisting or formed in response to invasion by the pathogen. Preexisting structural defenses include substances such as waxes, cuticle and other compounds found in or on the epidermal cell walls. In response to infection, the tissue may also form cork layers, deposit gums or gels, form abscission zones or limit the infection through the formation of necrotic areas. Likewise, plants may possess a number of preexisting biochemical inhibitors to infection. For example, the level of phenolics has been correlated with resistance of certain plants to invasion and colonization by some pathogens.[194] Onion cultivars resistant to *Colletotrichum circinans* accumulate phenolics in the outer scales of the bulb which inhibit germination and penetration of the fungi.[278] Sweetpotato external tissue (periderm and outer cortex) contain antifungal compounds that resist soft rot caused by *Rhizopus stolonifer,* while internal tissue accumulate antifungal compounds when cut.[260] Other examples include an antifungal diene in avocado epicarp, oil glands in citrus and phenolics in mango. Plants may also respond to infection by accumulating antimicrobial stress metabolites called **phytoalexins**.[262] These compounds are usually of a low molecular weight and are general, rather less specific, in their activity. Other induced biochemical defenses include mechanisms for the detoxification of pathogen toxins, altered respiration and metabolic pathways, and the induction of hypersensitive reactions.[4,16,98]

Stress to plant cells caused by the infection process activates a cascade of defense responses[122] which may overlap with those to other stresses (e.g., wounding, temperature, drought).[297] The first step in the plant's response is activation of one or more receptors that recognize the stimuli and transmits the information to the interior (Figure 7.27) instigating defense responses.[92,240] The plant must be able to discriminate between self-generated signals and those instigated by pathogens (elicitors) and abiotic stresses. Likewise, signals can be specific to a pathogen or general in nature. Non-specific signals include cell wall breakdown fragments and non-plant derived proteins.[122] The plant responds through the synthesis of antipathogen compounds, increased cell wall cross-linking and lignification, development of systemic resistance and the death of the cells around the infection site.[92] Localized cell death (**hypersensitive response**) is a genetically programmed response in which the cells adjacent to the site of in-

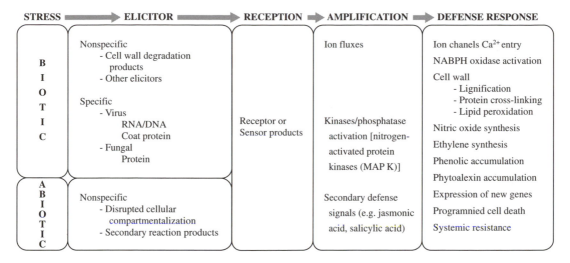

Figure 7.27. Biotic (pathogens: bacteria, fungi, virus, nematode) and abiotic (wounding, salt and temperature stress, etc.) stress stimuli and plant defense responses. The general process is stimulus → reception and recognition → amplification → defense reaction.

vasion are sacrificed to block the development of the pathogen into other parts of the plant.[122] The object of these defense responses is to limit the pathogen's infection and colonization.

Pathogens have, in turn, developed mechanisms to avoid these plant defense processes; for example, biotrophs that are dependent upon living plant cells for survival have evolved mechanisms to avoid triggering the programmed cell death response. The interaction between pathogens and their host is currently the subject of considerable research interest, and similarities have been found with certain animal defense systems.[17] The involvement of protein kinases/phosphatases in the amplification of the response is one area of active interest.[297] Kinase cascades are activated by phosphorylation with signal specificity conferred by multiple kinases.[286]

Plant pathogens are also known to alter the metabolism of nitrogenous compounds, phenols and hormones of the infected plant.[3,59,86,94] A number of the effects on cellular metabolism can be modified by postharvest treatments (e.g., heat treatments);[237] some, however, represent defense mechanisms of the plant. Studies of the nature, metabolism and mode of action of these defense mechanisms is currently a rapidly expanding area of research.[7,122]

When plants are infected by pathogens, their rate of respiration generally increases shortly after invasion and increases further as the invading organism multiplies.[183] Elevated respiration is a very general response that occurs with a wide range of microorganisms in both susceptible and resistant cultivars.[122] Susceptible and resistant cultivars differ, however, in that respiratory increases with the entry of the pathogen into a resistant line are generally more rapid and pronounced than in comparable susceptible lines.[3,122,151] It is thought that much of the increase in respiration is by an increased energy demand for the synthesis of compounds needed to combat the infection.

Some plant pathogens reduce the rate of photosynthesis in intact plants through a) the destruction of leaf tissue, b) the induction of processes leading to chlorophyll degradation, or c) alteration of the actual rate of the photosynthetic process.[160,293] Chlorosis may develop through an effect on chlorophyll synthesis, chlorophyll degradation or degeneration of the chloroplasts *per se*. Most virus infections that reduce photosynthesis appear to alter chlorophyll synthesis. The importance of postharvest pathogens that reduce photosynthesis is normally minimal since photosynthesis is generally either much reduced or completely inhibited due to the low level or absence of light. Infection prior to harvest may, however, increases the level of leaf shedding of infected leaves during storage.

The action of some plant pathogens results in blockage of the plant's translocation system. Blockage may involve inhibition of the acropetal movement of water and inorganic nutrients in the xylem or interference in the translocation of organic compounds basipetally in the phloem.[70] Xylem vessels may be blocked by bacterial bodies or water transport impaired by compounds produced by the microorganism (e.g., slime molds such as *Pseudomonas solanacearum*).[117] Stem blockages are often cited as the cause for decreased vase life of cut flowers. Vase water typically contains a broad cross-section of microorganisms; in carnations ('Improved White Sim'), 3 of 25 microorganisms present reduced the vase life of the flowers.[296] Several of these microorganisms, when transferred to the vase solution of other flower species, also reduced vase life.

One important aspect of the postharvest stress mediated by plant pathogens is that maturation and harvest often result in an increase in susceptibility of the tissue.[212] This increase in susceptibility is illustrated by the resistance of immature fruits to colonization after initial infection by quiescent pathogens.[212] Although present, the organism is not able to proceed with further development. As the fruit matures, inhibition of development appears to be lost and colonization proceeds. Several possible reasons have been shown for this increased resistance in immature tissue: a) young tissues may contain toxic compounds that are inactivated as the tissue ages and/or ripens; b) as most products mature, their composition changes which may now meet the energy and nutritional requirements of the pathogen; c) young tissues may exhibit a greater propensity for the production of phytoalexins than mature tissues; and d) the enzyme potential of the invading organism may not be sufficient for colonization of immature tissues.[212,265] Therefore, as the length of time a product is in storage increases, its ability to resist pathogen invasion decreases. This decrease in resistance during storage is a common feature of chilling sensitive commodities held just below or at the chilling injury threshold. Similarly, carrot roots inoculated with *Botrytis cinerea* at varying intervals after harvest displayed an increase in pathogen development as the interval between harvest and inoculation increased.[93] These changes in resistance appear to be largely associated with changes in the effectiveness of the plant's defense mechanisms.

ADDITIONAL READING

Abeles, F.B., P.W. Morgan and M.E. Saltveit. 1992. *Ethylene in Plant Biology.* Academic Press, New York.

Agrawal, S.B., and M. Agrawal. 2000. *Environmental Pollution and Plant Response.* Lewis Pub., Boca Raton, FL.

Barkai-Golan, R. 2001. *Postharvest Diseases of Fruits and Vegetables.* Elsevier, Amsterdam.

Bartz, J.A., and J.K. Brecht. 2003. *Postharvest Physiology and Pathology of Vegetables.* M. Dekker, New York.

Basra, A.S., and R.K. Basra. 1997. *Mechanisms of Environmental Stress Resistance in Plants.* Harwood Intern. Pub., Amsterdam.

Bell, J.N.B., and M. Treshow. 2002. *Air Pollution and Plant Life.* Wiley, New York.

DeVries, J.W., M.W. Trucksess and L.S. Jackson (eds.). 2002. *Mycotoxins and Food Safety.* Kluwer Academic, New York.

Grillo, S., and A. Leone. 1996. *Physical Stresses in Plants: Genes and Their Products for Tolerance.* Springer, Berlin.

Inzé, D., and M. Van Montagu. 2002. *Oxidative Stress in Plants.* Taylor & Francis, London.

Karban, R., and I.T. Baldwin. 1997. *Induced Responses to Herbivory.* University Chicago Press, Chicago, IL.

Koval, T.M. 1997. *Stress-Inducible Processes in Higher Eukaryotic Cells.* Plenum Press, New York.

Larcher, W. 2003. *Physiological Plant Ecology: Ecophysiology and Stress Physiology of Functional Groups.* Springer-Verlag, Berlin.

Lerner, H.R. (ed.). 1999. *Plant Responses to Environmental Stresses.* M. Dekker, New York.

Li, P.H., and E.T. Palva (eds.). 2002. *Plant Cold Hardiness: Gene Regulation and Genetic Engineering.* Kluwer Academic, New York.

Lumsden, P.J. 1997. *Plants and UV-B: Responses to Environmental Change.* Cambridge University Press, Cambridge.

Luo, Y., and H.A. Mooney. 1999. *Carbon Dioxide and Environmental Stress.* Academic Press, San Diego, CA.

Mattoo, A.K., and J.C. Stuttle. 1991. *The Plant Hormone Ethylene.* CRC Press, Boca Raton, FL.

Muhammad, A., and W.T. Frankenberger. 2002. *Ethylene: Agricultural Sources and Applications.* Kluwer Academic, New York.

Nilsen, E.T., D.M. Orcutt and M.G. Hale. 1996–2000. *The Physiology of Plant Under Stress.* 2 vols. Wiley, New York.

Omasa, K.. H. Saji, S. Youssefian, and N.Kondo (eds.). *Air Pollution and Plant Biotechnology: Prospects for Phytomonitoring and Phytoremediation.* Springer, New York.

Percy, K.E., J.N. Cape, R. Jagels and C.J. Simpson (eds.). 1994. *Air Pollutants and the Leaf Cuticle.* Springer-Verlag, Berlin.

Pessarakli, M. 1999. *Handbook of Plant and Crop Stress.* M. Dekker, New York.

Polak, M. 2003. *Developmental Instability.* Oxford University Press, Oxford, England.

Rosen, E., R. Chen and P.H. Masson. 1999. Root gravitropism: a complex response to a simple stimulus? Trends *Plant Sci.* 4:407–412.

Scandalios, J.G. 1997. *Oxidative Stress and the Molecular Biology of Antioxidant Defenses.* Cold Spring Harbor Laboratory Press, Plainview, NY.

Shinozaki, K., and K. Yamaguchi-Shinozaki. 1999. *Molecular Responses to Cold, Drought, Heat, and Salt Stress in Higher Plants.* R.G. Landes Co., Austin, TX.

Sinha, K.K., and D. Bhatnagar. 1998. *Mycotoxins in Agriculture and Food Safety.* M. Dekker, New York.

Tasaka, M., T. Kato and H. Fukaki. 1999. The endodermis and shoot gravitropism. *Trends Plant Sci.* 4:104–107.

Yunus, M., N. Singh and L.J. De Kok. 2000. *Environmental Stress: Induction, Mitigation, and Exo-conservation.* Kluwer Academic, Boston, MA.

REFERENCES

1. Abeles, F.B., P.W. Morgan and M.E. Saltveit. 1992. *Ethylene in Plant Biology.* Academic Press, New York.

2. Agrawal, A.A. 2000. Overcompensation of plants in response to herbivory and the by-product benefits of mutualism. *Trends Plant Sci.* 5:309–313.

3. Akai, S., and S. Ouchi (eds.). 1971. *Morphological and Biochemical Events in Plant-Parasite Interaction.* Phytopath. Soc. Japan, Tokyo.

4. Allen, P.J. 1954. Physiological aspects of fungus diseases of plants. *Annu. Rev. Plant Physiol.* 5:225–248.

5. Anderson, R.E., and R.W. Penney. 1975. Intermittent warming of peaches and nectarines stored in controlled atmosphere or air. *J. Amer. Soc. Hort. Sci.* 100:151–153.

6. Anon. 1974. National damage survey, 1973. *U.K. Potato Marketing Board,* London.

7. Baerug, R. 1962. Influence of different rates and intensities of light on solanine content and cooking quality of potato tubers. *Eur. Potato J.* 5:242–251.

8. Bailey, J.A., and B.J. Deverall. 1983. *The Dynamics of Host Defense.* Academic Press, New York.

9. Bangerth, F. 1983. Hormonal and chemical preharvest treatments which influence postharvest quality, maturity and storability of fruit. Pp. 331–354. In: *Post-Harvest Physiology and Crop Preservation.* M. Lieberman (ed.). Plenum Press, New York.

10. Banks, N.H. 1985. Coating and modified atmosphere effects on potato tuber greening. *J. Agric. Sci.* 105:59–62.

11. Barker, J., and L.W. Mapson. 1955. Studies in the respiratory and carbohydrate metabolism of plant tissues. VII. Experimental studies with potato tubers of an inhibition of the respiration and of a 'block' in the tricarboxylic acid cycle induced by 'oxygen poisoning'. *Proc. Roy. Soc. B.* 143:523–549.

12. Barriga, C. 1961. The effects of mechanical abuse of Navy bean seed at various moisture levels. *Agron. J.* 53:250–251.

13. Baumgardner, R.A., G.E. Stembridge, L.O. Van Blaricom and C.E. Gambrell, Jr. 1971. Effects of succinic acid-2,2-dimethylhydrazide on the color, firmness, and uniformity of processing peaches. *J. Amer. Soc. Hort. Sci.* 97:485–488.

14. Beaudry, R.M., J.A. Payne and S.J. Kays. 1985. Variation in the respiration of harvested pecans due to genotype and kernel moisture. *HortScience* 20:752–754,

15. Becquerel, P. 1932. L'anhydrobiose des tubercules des Renoncules dans l'azote liquid. *C.R. Acad. Sci. Paris.* 194:1974–1976.

16. Bell, A.A. 1981. Biochemical mechanisms of disease resistance. *Annu. Rev. Plant Physiol.* 32:21–81.

17. Bergey, D.R., G.A. Howe, and C.A. Ryan. 1996. Polypeptide signaling for plant defensive genes exhibits analogies to defense signaling in animals. *Proc. Nat. Acad. Sci.* (USA) 3:12053–12058.

18. Bernstein, L. 1964. Effects of salinity on mineral composition and growth of plants. *Plant Anal. Fert. Probl. Colloq.* 4:25–45.

19. Bjorkman, O., M. Badger and P. Armond. 1980. Adaptation of plants to water and high temperature stress. Pp. 233–249. In: *Response and Adaptation of Photosynthesis to High Temperature.* N.C. Turner and P.J. Kramer (eds.). Wiley-Interscience, New York.

20. Blanpied, G.D. 1978. The soft McIntosh problem. *Proc. New York State Hort. Soc.* 123:122–124.

21. Bloch, R. 1941. Wound healing in higher plants. *Bot. Rev.* 7:110–146.

22. Bloch, R. 1952. Wound healing in higher plants. II. *Bot. Rev.* 18:655–679.

23. Boyd, M.R., L.T. Burka, T.M. Harris and B.J. Wilson. 1973. Lung toxic furanoterpenoids produced by sweet potatoes. *Biochem. Biophys. Acta* 337:184–195.

24. Brock, T.D., and G.K. Darland. 1970. Limits of microbial existence: Temperature and pH. *Science* 169:1316–1318.

25. Brodrick, H.T., and R. Thord-Gray. 1982. Irradiation in perspective—the significance for the mango industry. *Res. Rpt. S. African Mango Grs. Assoc.* 2:23–26.

26. Brooks, C., E.V. Miller, C.O. Bratley, J.S. Cooley, P.V. Mook and H.B. Johnson. 1937. Effect of solid and gaseous carbon dioxide upon transit diseases of certain fruits and vegetables. *USDA Tech. Bull.* 318.

27. Brown, W. 1922. On the germination and growth of fungi at various temperatures and in various concentrations of oxygen and of carbon dioxide. *Ann. Bot.* 36:257–283.

28. Buescher, R.W., W.A. Sistrunk and P.L. Brady. 1975. Effects of ethylene on metabolic and quality attributes in sweet potato roots. *J. Food Sci.* 40:1018–1020.

29. Burton, W.G. 1978. Post-harvest behavior and storage of potatoes. *Appl. Biol.* 3:86–228.

30. Butchbaker, A.F., D.C. Nelson and R. Shaw. 1967. Controlled atmosphere storage of potatoes. *Trans. Amer. Soc. Agr. Eng.* 10:534–538.

31. Carter, W. 1973. *Insects in Relations to Plant Disease.* Wiley, New York.

32. Cathey, H.M., J. Halperin and A.A. Piringer. 1965. Relations of N-dimethylamino- succinamic acid to photoperiod, kind of supplemental light, and night temperature, and its effects on the growth and flowering of garden annuals. *Hort. Res.* 5:1–12.

33. Cathey, H.M., and H.E. Heggestad. 1982. Ozone and sulfur dioxide sensitivity of petunia: modification by ethylene diurea. *J. Amer. Soc. Hort. Sci.* 107:1028–1035.

34. Ceponis, M.J., and J.E. Butterfield. 1973. The nature and extent of retail and consumer losses in apples, oranges, lettuce, peaches, strawberries, and potatoes marketed in greater New York. *USDA Marketing Res. Rept.* 996.

35. Ceponis, M.J., and J.E. Butterfield. 1974. Causes of cullage of Florida bell peppers in New York wholesale and retail markets. *Plant Dis. Rept.* 58:367–369.

36. Ceponis, M.J., and J.E. Butterfield. 1974. Market losses in Florida cucumbers and bell peppers in metropolitan New York. *Plant Dis. Rept.* 58:558–560.

37. Ceponis, M.J., and J.E. Butterfield. 1974. Retail and consumer losses in sweet potatoes in metropolitan New York. *HortScience* 9:393–394.

38. Ceponis, M.J., and J.E. Butterfield. 1974. Retail and consumer losses of western pears in metropolitan New York. *HortScience* 9:447–448.

39. Ceponis, M.J., and J.E. Butterfield. 1974. Retail losses in California grapes marketed in metropolitan New York. *USDA, ARS-NE-53.*

40. Ceponis, M.J., J. Kaufman and S.M. Ringel. 1962. Quality of prepackaged apples in New York City retail stores. *USDA, AMS-461.*

41. Chalutz, E., J.E. DeVay and E.C. Maxie. 1969. Ethylene-induced isocoumarin formation in carrot root tissue. *Plant Physiol.* 44:235–241.

42. Chang, L.A., L.K. Hammett and D.M. Pharr. 1982. Ethanol, alcohol dehydrogenase, and pyruvate decarboxylase in storage roots of four sweet potato cultivars during simulated flood-damage and storage. *J. Amer. Soc. Hort. Sci.* 107:674–677.

43. Chang, T.-S., and R.L. Shewfelt. 1988. Effect of chilling exposure of tomatoes during subsequent ripening. *J. Food Sci.* 53:1160–1162.

44. Christensen, C.M. (ed.). 1982. *Storage of Cereal Grains and Their Products.* Amer. Assoc. Cereal Chemists, St. Paul, MN.

45. Claypool, L.L., and H.M. Vines. 1956. Commodity tolerance studies of deciduous fruits to moist heat and fumigants. *Hilgardia* 24:297–355.

46. Click, R.E., and D.P. Hackett. 1963. The role of protein and nucleic acid synthesis in the development of respiration in potato tuber slices. *Proc. Nat. Acad. Sci.* (USA) 50:243–250.

47. Cochrane, V.W. 1958. *Physiology of Fungi.* Wiley, New York.

48. Cochrane, J.C., V.W. Cochrane, F.G. Simon and J. Spaeth. 1963. Spore germination and carbon metabolism in *Fusarum solani.* I. Requirements for spore germination. *Phytopathology* 53:1155–1106.

49. Collander, R.1924. Beobachtungen über die quantitativen Beziehungen zwischen Tötungsgeschwindigkeit und Temperatur beim Wärmetod pflanzlicher Zellen. *Commentat. Biol. Soc. Sci. Fenn.* 11–12.

50. Collins, P.C., and T.M. Blessington. 1983. Postharvest effects of shipping temperatures and subsequent interior keeping quality of *Ficus benjamina. HortScience* 18:757–758.

51. Conover, C.A., and R.T. Poole. 1975. Acclimatization of tropical trees for interior use. *HortScience* 10:600–681.

52. Conover, C.A., and R.T. Poole. 1977. Effects of cultural practices on acclimatization of *Ficus benjamina* L. *J. Amer. Soc. Hort. Sci.* 102:529–531.

53. Couey, H.M., and J.M. Wells. 1970. Low oxygen and high carbon dioxide atmospheres to control postharvest decay of strawberries. *Phytopathology* 60:47–49.

54. Crane, W.M., and D. Martin. 1938. Apple investigation in Tasmania: Miscellaneous notes. *J. Council Sci. Ind. Res. Austr.* 11:47–60.

55. Crawley, M.J. 1983. *Herbivory the Dynamics of Animal-Plant Interactions.* Blackwell Scientific, Oxford, England.

56. Curtis, O.F. 1936. Comparative effects of altering leaf temperatures and air humidities on vapor pressure gradients. *Plant Physiol.* 11:595–603.

57. Darwin, C. 1881. *The Power of Movement in Plants.* John Murry, London.

58. Davies, D.D., S. Grego and P. Kenworthy. 1974. The control of the production of lactate and ethanol by higher plants. *Planta* 118:297–310.

59. Dennis, C. (ed.). 1983. *Post-harvest Pathology of Fruits and Vegetables.* Academic Press, New York.

60. Dennison, R.A., and E.M. Ahmed. 1975. Irradiation treatment of fruits and vegetables. Pp. 118–129. In: *Symposium: Postharvest Biology and Handling of Fruits and Vegetables.* N.F. Haard and D.K. Salunkhe (eds.). AVI, Westport, CT.

61. Dewey, D.H. 1962. Factors affecting the quality of Jonathan apples stored in controlled atmospheres. *Proc. 16th Intern. Hort. Congr.* 1:278.

62. Dharkar, S.D., K.A. Savagaon, A.N. Spirangarajan and A. Sreenivasan. 1966. Irradiation of mangoes. I. Radiation-induced delay in ripening of Alphonso mangoes. *J. Food. Sci.* 31:863–869.

63. Dickens, B.F., and G.A. Thompson, Jr. 1982. Phospholipid molecular species alterations in microsomal membranes as an initial key step during cellular acclimation to low temperature. *Biochemistry* 21:3604–3611.

64. Diehl, K.C., and D.D. Hamann. 1980. Relationships between sensory profile parameters and fundamental mechanical parameters for potatoes, melons and apples. *J. Texture Studies* 10:401–420.

65. Diehl, K.C., D.D. Hamann and J.K. Whitfield. 1980. Structural failure in selected raw fruit and vegetables. *J. Texture Studies* 10:371–400.

66. Digby, J., and R.D. Firn. 1976. A critical assessment of the Cholodny-Went theory of shoot gravitropism. *Curr. Adv. Plant Sci.* 8:953–960.

67. Dollar, A.M., M. Hanaoka, G.A. McClish and J.H. Moy. 1970. Semi-commercial scale studies on irradiated papaya. *Rpt. U. S. Atomic Energy Comm.* 1970, Contract No. AT-(26-1)-374.

68. Dubrov, A.P. 1968. *The Genetic and Physiological Effect of the Action of Ultraviolet Radiation on Higher Plants.* Nauka, Moscow.

69. Dull, G.G., and S.J. Kays. 1988. Quality and mechanical stability of pecan kernels with different packaging protocols. *J. Food Sci.* 53:565–567.

70. Durbin, R.D. 1967. Obligate parasites: Effect on the movement of solutes and water. Pp. 80–99. In: *The Dynamic Role of Molecular Constituents in Plant-Parasite Interaction.* C. J. Mirocha and I. Uritani (eds.). Bruce, St. Paul, MN.

71. Eaks, I.L. 1967. Ripening and astringency removal in persimmon fruits. *Proc. Amer. Soc. Hort. Sci.* 91:868–875.

72. Eaks, I.L., and L.L. Morris. 1962. Deterioration of cucumbers at chilling and non-chilling temperatures. *Proc. Amer. Soc. Hort. Sci.* 69:388–399.

73. El-Goorani, M.A., and N.F. Sommer. 1981. Effects of modified atmospheres on postharvest pathogens of fruits and vegetables. *Hort. Rev.* 3:412–461.

74. Elkiey, T., D.P. Ormrod and B. Marie. 1982. Foliar sorption of sulfur dioxide, nitrogen dioxide, and ozone by ornamental woody plants. *HortScience* 17:358–360.

75. Engelbrecht, L., U. Arban and W. Heese. 1969. Leaf-miner caterpillars and cytokinins in the 'green islands' of autumn leaves. *Nature* 223:319–321.

76. Fails, B.S., A.J. Lewis and J.A. Barden. 1982. Anatomy and morphology of sun- and shade-grown *Ficus benjamina* L. *J. Amer. Soc. Hort. Sci.*104:410–413.

77. Faller, N. 1976. Simultaneous assimilation of sulfur and carbon dioxides by some plants. *Acta. Bot. Croat.* 35:87–95.

78. Faroogi, W.A., and E.G. Hall. 1972. Effects of ethylene dibromide on the respiration and ripening of bananas (*Musa cavendishii* L.). *Nucleus* 9:22–28.

79. Faust, M. 1973. Effect of growth regulators on firmness and red color of fruit. *Acta Hort.* 34:407–420.

80. Fidler, J.C. 1968. Low temperature injury to fruits and vegetables. *Recent Adv. Food Sci.* 4:271–283.

81. Fidler J.C., and G. Mann. 1972. *Refrigerated Storage of Apples and Pears—A Practical Guide.* Commonwealth Agric. Bur., Slough, England.

82. Fidler, J.C., and C.J. North. 1971. The effect of periods of anaerobiosis on the storage of apples. *J. Hort. Sci.* 46:213–221.

83. Fidler, J.C., B.G. Wilkinson, K.L. Edney and R.O. Sharples. 1973. *The Biology of Apple and Pear Storage.* Commonwealth Agric. Bur., Slough, England.

84. Fleuriet, A., and J.J. Macheix. 1984. Orientation nouvelle du métabolisme des acides hydroxycinnamiques dans les fruits de tomates blessés (*Lycopersicon esculentum*). *Physiol. Plant* 61:64–68.

85. Frenkel, C., and M.E. Patterson. 1973. Effect of carbon dioxide on activity of succinic dehydrogenase in 'Bartlett' pears during cold storage. *HortScience* 8:395–396.

86. Friend, J., and D.R. Threlfall (eds.). 1976. *Biochemical Aspects of Plant-Parasite Relationships.* Academic Press, New York.

87. Fukushima, H., C.E. Martin, H. Iida, Y. Kitajima, G.A. Thompson, Jr. and Y. Nozawa. 1976. Changes in membrane lipid composition during temperature adaptation by a thermotolerant strain of *Tetrahymena pyriformis*. *Biochem. Biophys. Acta* 431:165–179.

88. Gallaghen, P.W., and T.D. Sydnor. 1983. Variation in wound response among cultivars of red maple. *J. Amer. Soc. Hort. Sci.* 108:744–746.

89. Garms, H. 1933. Untersuchungen über Wundhieling an Früchten. *Beih. Bot. Centralbl. I.* 51:437–516.

90. Genoud, T., and J.P. Metraux. 1999. Crosstalk in plant cell signaling: structure and function of a genetic network. *Trends Plant Sci.* 4:503–507.

91. Gil, G.F., W.H. Griggs and G.C. Martin. 1972. Gibberellin-induced parthenocarpy in 'Winter Nelis' pear. *HortScience* 7:559–561.

92. Glazebrook, J. 1999. Genes controlling expression of defense responses in *Arabidopsis. Cur. Opin. Plant Biol.* 2:280–286.

93. Goodliffe, J.P., and J.B. Heale. 1977. Factors affecting the resistance of stored carrot roots to *Botrytis cinerea. Ann. Appl. Biol.* 85:163.

94. Goodman, R.N., Z. Kiraly and M. Zaithin. 1967. *The Biochemistry and Physiology of Infectious Plant Disease.* Van Nostrand, New York.

95. Granett, A.L. 1982. Pictorial keys to evaluate foliar injury caused by hydrogen fluoride. *HortScience* 17:587–588.
96. Groves, J.F. 1917. Temperature and life duration of seeds. *Bot Gaz.* 63:169–189.
97. Haard, N.F., S.C. Sharma, R. Wolfe and C. Frenkel. 1974. Ethylene induced isoperoxidase changes during fiber formation in postharvest asparagus. *J. Food Sci.* 39:452–456.
98. Hartleb, H., R. Heitefuss and H.H. Hoppe. 1997. *Resistance of Crop Plants Against Fungi.* Gustav Fischer, Jena, Germany.
99. Hale, P.W., and J.J. Smoot. 1973. Exporting Florida grapefruit to Japan. An evaluation of new shipping cartons and decay control treatments. *Citrus & Veg. Mag.* 37(3):20–23, 45.
100. Halevy, A.H., and S. Mayak. 1981. Senescence and postharvest physiology of cut flowers—part 2. *Hort. Rev.* 3:59–143.
101. Hall, C.B. 1964. The effects of short periods of high temperature on the ripening of detached tomato fruits. *Proc. Amer. Soc. Hort. Sci.* 84:501–506.
102. Hankinson, B., and V.N.M. Rao. 1979. Histological and physical behavior of tomato skins susceptible to cracking. *J. Amer. Soc. Hort. Sci.* 104:577–581.
103. Harman, J.E., and B. McDonald. 1983. Controlled atmosphere storage of kiwi fruit: Effects on storage life and fruit quality. *Acta Hort.* 138:195–201.
104. Harrington, G.T., and W. Crocker. 1918. Resistance of seeds to desiccation. *J. Agr. Res.* 14:525–532.
105. Havis, J.R. 1976. Root hardiness of woody ornamentals. *HortScience* 11:385–386.
106. Heiniche, A.J. 1937. How lime/sulphur spray effects the photosynthesis of an entire ten year old apple tree. *Proc. Amer. Soc. Hort. Sci.* 35:256–259.
107. Hicks, J.R., and P.M. Ludford. 1981. Effects of low ethylene levels on storage of cabbage. *Acta Hort.* 116:65–73.
108. Hiraki, Y., and Y. Ota. 1975. The relationships between growth inhibition and ethylene production by mechanical stimulation in *Lillium longiflorum. Plant Cell Physiol.* 16:185–189.
109. Hoffman, M.B. 1933. The effect of certain spray materials on the carbon dioxide assimilation by McIntosh apple leaves. *Proc. Amer. Soc. Hort. Sci.* 29:389–398.
110. Holt, J.E., and D. Schoorl. 1982. Mechanics of failure in fruits and vegetables. *J. Texture Studies* 13:83–97.
111. Howe, T.K., and S.S. Woltz. 1981. Symptomology and relative susceptibility of various ornamental plants to acute airborne sulfur dioxide exposure. *Proc. Fl. St. Hort. Soc.* 94:121–123.
112. Hsiao, T.C. 1973. Plant responses to water stress. *Annu. Rev. Plant Physiol.* 24:519–570.
113. Huber, H. 1935. Der Wärmehaushalt der Pflanzen. *Naturwiss. Landwirtsch.* 17:148–152.
114. Huelin, F.E., and G.B. Tindale. 1942. Investigations on the gas storage of Victorian pears. *J. Agr. Vict.* 40:594–606.
115. Hulme, A.C. 1956. Carbon dioxide injury and the presence of succinic acid in apples. *Nature* 178:218–219.
116. Hung, Y.-C., and S.E. Prussia.1989. Effect of maturity and storage on the bruise susceptibility of peaches (cv. Red Globe). *Trans. Amer. Soc. Agri. Eng.* 32:1377–1382.
117. Husain, A., and A. Kelman. 1958. Relation of slime production to mechanism of wilting and pathogenicity of *Pseudomonas solanacearum. Phytopathology* 48:155–165.
118. Hyodo, H., H. Kuroda and S.F. Yang. 1978. Induction of phenylalanine ammonia-lyase and increase in phenolics in lettuce leaves in relation to the development of russet spotting caused by ethylene. *Plant Physiol.* 62:31–35.
119. Hyodo, H., K. Tanaka and K. Watanake. 1983. Wound-induced ethylene production and 1-aminocyclopropane-1-carboxylic acid synthase in mesocarp tissue of winter squash fruit. *Plant Cell Physiol.* 24:963–969.
120. Ilker, Y. 1976. Physiological manifestations of chilling injury and its alleviation in okra plants (*Abelmoschus esculentus* (L.) Moench). Ph.D. Thesis, Univ. Calif., Davis, CA.
121. Isenberg, F.M., and R.M. Sayles. 1969. Modified atmosphere storage of Danish cabbage. *J. Amer. Soc. Hort. Sci.* 94:447–449.
122. Jackson, A.O., and C.B. Taylor. 1996. Plants-microbe interactions. *Plant Cell* 8:1651–191
123. Jadhov, S.J., and D.K. Salunkhe. 1975. Formation and control of chlorophyll and glycoalkaloids in tubers of *Solanum tuberosum* L. and evaluation of glycoalkaloid toxicity. *Adv. Food. Res.* 21:307–354.
124. Johnson, C.R., J.K. Krantz, J.N. Joiner and C.A. Conover. 1979. Light compensation point and leaf dis-

tribution of *Ficus benjamina* as affected by light intensity and nitrogen-potassium nutrition. *J. Amer. Soc. Hort. Sci.* 104:335–338.

125. Jones, R.L., H.T. Freebairn and J.F. McConnell. 1978. The prevention of chilling injury, weight loss reduction, and ripening retardation in banana. *J. Amer. Soc. Hort. Sci.* 103:129–221.

126. Juniper, B.E. 1976. Geotropism. *Annu. Rev. Plant Physiol.* 27:385–406.

127. Kader, A.A. 1985. Ethylene-induced senescence and physiological disorders in harvested horticultural crops. *HortScience* 20:54–57.

128. Kahl, G. 1974. Metabolism in plant storage tissue slices. *Bot. Rev.* 40:263–314.

129. Kahl, G. (ed.). 1978. *Biochemistry of Wounded Plant Tissues.* Walter de Gruyter, New York.

130. Kamerbeek, G.A., and W.J. DeMunk. 1976. A review of ethylene effects in bulbous plants. *Scientia Hort.* 4:101–115.

131. Kang, B.G. 1979. Epinasty. Pp. 657–667. In: *Physiology of Movements.* W. Haupt and M.E. Feinleik (eds.). vol. 7, *Encylo. Plant Physiol.* Springer-Verlag, Berlin.

132. Kaufman, J., R.E. Hardenburg and J.M. Lutz. 1956. Weight losses and decay of Florida and California oranges in mesh and perforated polyethylene consumer bags. *Proc. Amer. Soc. Hort. Sci.* 67:244–250.

133. Kaufmann, M.R. 1975. Leaf water stress in Engelmann spruce: Influence of the root and shoot environments. *Plant Physiol.* 56:841–844.

134. Kays, S.J., C.A. Jaworski and H.C. Price. 1976. Defoliation of pepper transplants in transit by endogenously evolved ethylene. *J. Amer. Soc. Hort. Sci.* 101:449–451.

135. Kays. S.J., and D.M. Wilson. 1977. Alteration of pecan kernel color. *J. Food Sci.* 42:982–988.

136. Key, J.L., C.Y. Lin, E. Ceglarz and F. Schöffl. 1982. The heat shock response in soybean seedlings. In: *Structure and Function of Plant Genomes.* L. Dure, (ed.). NATO Advanced Studies Series, Plenum Press, New York.

137. Key, J.L., C.Y. Lin and Y.M. Chen. 1981. Heat shock proteins of higher plants. *Proc. Nat. Acad. Sci.* (USA) 78:3526–3530.

138. Kidd, F. 1919. Laboratory experiments on the sprouting of potatoes in various gas mixtures (nitrogen, oxygen and carbon dioxide). *New Phytol.* 18:248–252.

139. Kidd, F., and C. West. 1923. Brown heart a functional disease of apples and pears. *Gt. Brit. Dept. Sci. Ind. Res. Food Invest. Board Special Rpt.* Number 12.

140. Kidd, F., and C. West. 1927. A relation between the concentration of oxygen and carbon dioxide in the atmosphere, rate of respiration and length of storage life in apples. *Gt. Brit. Dept. Sci. Ind. Res. Rept. Food Invest. Board,* 1925, 1926, pp. 41–42.

141. Kidd, F., and C. West. 1939. The gas-storage of Cox's Orange Pippin apples on a commercial scale. *Rept. Food Invest. Board,* 1938. pp. 153–156.

142. Kiesselbach, T.A., and J.A. Ratcliff. 1920. Freeze injury of seed corn. *Univ. Neb. Agric. Exp. Sta. Res. Bull.* 16.

143. Kiyosawa, K. 1975. Studies on the effects of alcohols on membrane water permeability of *Nitella. Protoplasma* 86:243–252.

144. Kloft, W., and P. Ehrhardt. 1959. Untersuchungen uber Saugtatigkeit und Schadwirkung der Sitkafichtenlaus *Liosomaphis abietina* (Walk.) (*Neomyzaphis abietina* Walk.). *Phytopath. Z.* 35:401–410.

145. Knee, M. 1973. Effects of controlled atmosphere storage on respiratory metabolism of apple fruit tissue. *J. Sci. Food Agric.* 24:1289–1298.

146. Kolattukudy, P.E., and V.P. Agrawal. 1974. Structure and composition of aliphatic components of potato tuber skin (suberin). *Lipids* 9:682–691.

147. Kosiyachinada, S., and R.E. Young. 1976. Chilling sensitivity of avocado fruit at different stages of respiratory climacteric. *J. Amer. Soc. Hort. Sci.* 101:665–667.

148. Kotob, M.A., and W.W. Schwake. 1975. Respiration rate and acidity in parthenocarpic and seeded Conference pears. *J. Hort. Sci.* 50:435–445.

149. Kulman, H.M. 1971. Effects of insect defoliation on growth and mortality of trees. *Annu. Rev. Ento.* 16:289–324.

150. Kuraiski, S., and N. Nito. 1980. The maximum leaf surface temperatures of higher plants observed in the Inland Sea area. *Bot. Mag.* 93:209–228.

151. Lamb, C., and R.A. Dixon. 1997. Oxidative burst in plant disease resistance. *Annu. Rev. Plant Physiol. Plant Mol. Biol.* 48:251–275.

152. Lau, O.L. 1983. Effects of storage procedures and low oxygen and carbon dioxide atmospheres on storage quality of 'Spartan' apples. *J. Amer. Soc. Hort. Sci.* 108:953–957.

153. Link, H. 1967. Der Einfuss der Ausdünnung auf Fruchtqualität und Erntemenge bei der Apfelsorte 'Golden Delicious'. *Gartenbauwiss* 32:423–444.

154. Lipetz, J. 1966. Crown gall tumorigenesis II. Relations between wound healing and the tumorgenic response. *Cancer Res.* 26:1597–1605.

155. Lipetz, J. 1970. Wound-healing in higher plants. *Intern Rev. Cytol.* 27:1–28.

156. Lipton, W.J. 1977. Toward an explanation of disorders of vegetables induced by high CO_2 or low O_2? *Proc. 2nd Nat. CA Res. Conf.,* Mich. St. Univ. Hort. Rept. 28:137–141.

157. Lipton, W.J., and C.M. Harris. 1974. Controlled atmosphere effects on the market quality of stored broccoli (*Brassica oleracea* L., Italica group). *J. Amer. Soc. Hort. Sci.* 99:200–205.

158. Little, C.R., J.D. Faragher and H.J. Taylor. 1982. Effects of initial oxygen stress treatments in low oxygen modified atmosphere storage of 'Granny Smith' apples. *J. Amer. Soc. Hort. Sci.* 107:320–323.

159. Liu, F.W. 1979. Interaction of daminozide, harvesting date, and ethylene in CA storage on 'McIntosh' apple quality. *J. Amer. Soc. Hort. Sci.* 104:599–601.

160. Livne, A. 1964. Photosynthesis in healthy and rust-affected plants. *Plant Physiol.* 39:614–621.

161. Lloyd, E.J. 1976. The influence of shading on enzyme activity in seedling leaves of barley. *Z. Pflanzenphysiol.* 78:1–12.

162. Loaiza-Velarde, J.G., and M.E. Saltveit. 2001. Heat shocks applied either before or after wounding reduce browning of lettuce leaf tissue. *J. Amer. Soc. Hort. Sci.* 126:227–234.

163. Lockett, M.C., and B.J. Luyet.1951. Survival of frozen seed of various water contents. *Biodynamica* 7(134):67–76.

164. Lockhart, J.A., and U. Brodführer-Franzgrote. 1961. The effects of ultraviolet radiation on plants. *Encyclo. Plant Physiol.* 16:532–554. Springer-Verlag, Berlin.

165. Lurie, S. 1998. Postharvest heat treatments. *Postharv. Biol. Tech.* 14:257–269.

166. Lurie, S., S. Othman and A. Borochov. 1995. Effect of heat treatment on plasma membrane of apple fruit. *Postharv. Biol. Tech.* 5:29–38.

167. Lutz, J.M., and R.E. Hardenburg. 1968. The commercial storage of fruits, vegetables, and florist and nursery stocks. *U.S. Dept. Agri., Agr. Hbk.* 66.

168. Mansfield, T.A. (ed.). 1976. *Effects of Air Pollutants on Plants.* Soc. Exp. Biol. Sem. Ser., vol. 1. Cambridge Univ. Press, Cambridge.

169. Marousky, F.J., and B.K. Harbaugh. 1982. Responses of certain flowering and foliage plants to exogenous ethylene. *Proc. Fla. St. Hort. Soc.* 95:159–162.

170. Mattus, G.E., L.E. Scott and L.L. Claypool. 1959. Brown spot of Bartlett pears. *Calif. Agric.* 13(7):8, 13.

171. Maxie, E.C., and A. Abel-Kader. 1966. Food irradiation—physiology of fruits as related to feasibility of the technology. *Adv. Food Res.* 15:105–145.

172. Maxie, E.C., and N.F. Sommer. 1971. Radiation technology in conjunction with postharvest procedures as a means of extending the shelf-life of fruits and vegetables. *Rpt. U. S. At. Energy Comm.* 1971, Contract No. AT-(11-1)-34.

173. Maxie, E.C., N.F. Sommer and F.G. Mitchell. 1971. Infeasibility of irradiating fresh fruits and vegetables. *HortScience* 6:202–204.

174. Mazur, P. 1969. Freezing injury in plants. *Annu. Rev. Plant Physiol.* 20:419–448.

175. Mazur, P. 1970. Cryobiology: The freezing of biological systems. *Science* 168:939–949.

176. McCree, K.J., and J.H. Troughton. 1966. Prediction of growth rate at different light levels from measured photosynthesis and respiration rates. *Plant Physiol.* 41:559–566.

177. McGill, J.N., A.I. Nelson and M.P. Steinberg. 1966. Effects of modified storage atmosphere on ascorbic acid and other quality characteristics of spinach. *J. Food Sci.* 31:510–517.

178. McGlasson, W.B., and H.K. Pratt. 1964. Effects on wounding on respiration and ethylene production by cantaloupe fruit tissue. *Plant Physiol.* 39:128–132.

179. Meigh, D.F. 1960. Use of gas chromatography in measuring the ethylene production of stored apples. *J. Sci. Food Agr.* 11:381–385.

180. Meryman, H.T. 1966. *Cryobiology.* Academic Press, New York.

181. Miles, J.A., and G.E. Rehkugler. 1973. A failure criterion for apple flesh. *Trans. Amer. Soc. Agr. Eng.* 16:1148–1153.

182. Milks, R.R., J.N. Joiner, L.A. Garard, C.A. Conover and B. Tjia. 1979. Influence of acclimatization on carbohydrate production and translocation of *Ficus benjamina* L. *J. Amer. Hort. Sci.* 104:410–413.

183. Millerd, A., and K.J. Scott. 1962. Respiration of the diseased plant. *Annu. Rev. Plant Physiol.* 13:550–574.

184. Mitchell, C.A., C.J. Severson, J.A. Wott and P.A. Hammer. 1975. Seismorphogenetic regulation of plant growth. *J. Amer. Soc. Hort. Sci.* 100:161–165.

185. Modlibowska, J. 1966. Inducing precocious cropping on young Dr. Jules Guyot pear trees with gibberellic acid. *J. Hort. Sci.* 41:137–144.

186. Mohsenin, N.N. 1970. *Physical Properties of Plant and Animal Materials.* 2 vols. Gordon and Breach, New York.

187. Mohsenin, N.N. 1977. Characterization and failure in solid food with particular reference to fruits and vegetables. *J. Texture Studies* 8:169–193.

188. Molisch, H. 1896. Das Erfrieren von Pflanzen bei Temperaturen über dem Eispuckt. *Sitzber. Akad. Wiss. Wien. Math. Naturwis. Kl.* 105:1–14.

189. Monning, A. 1983. Studies on the reaction of Krebs cycle enzymes from apple tissue (cv. Cox Orange) to increased levels of CO_2. *Acta Hort.* 138:113–119.

190. Morris, L.L., and A.A. Kader. 1977. Physiological disorders of certain vegetables in relation to modified atmosphere. *Proc. 2nd Nat. CA Res. Conf.,* Mich. St. Univ. Hort. Rept. 28:266–267.

191. Mudd, J.B., and T.T. Kozlowski (eds.). 1975. *Responses of Plants to Air Pollution.* Academic Press, New York.

192. Murakami, Y., M. Tsuyama, Y. Kobayashi, H. Kodama, and K. Iba. 2000. Trienoic fatty acids and plant tolerance of high temperatures. *Science* 287:476–479.

193. Nagarojuah, S. 1976. The effects of increased illumination and shading on the low-light-induced decline of photosynthesis in cotton leaves. *Physiol. Plant.* 36:338–342.

194. Nicholson, R.L., and R. Hammerschmidt. 1992. Phenolic compounds and their role in disease resistance. *Annu. Rev. Phytopath.* 30:369–389.

195. Ogura, N., N. Nakayawa and H.E. Takehana. 1975. Studies on the storage temperature of tomato fruit. I. Effect of high temperature-short term storage of mature green tomato fruits on changes of their chemical constituents after ripening at room temperature. *J. Agric. Soc. Japan* 49:189–196.

196. Olorunda, A.O., and N.E. Looney. 1977. Response of squash to ethylene and chilling injury. *Ann. Appl. Biol.* 87:465–469.

197. Ophuis, B.G., J.C. Hesen and E. Kroesbergen. 1958. The influence of temperature during handling on the occurrence of blue discolorations inside potato tubers. *European Pot. J.* 1:48–65.

198. Osborne, D.J. 1973. Mutual regulation of growth and development in plants and insects. Pp. 33–42. In: *Insect/Plant Relationships.* H.F. van Emden, (ed.). Blackwell Scientific Pub., London.

199. Paull, R.E. 1990. Chilling injury of crops of tropical and subtropical origin. Pp. 17–36. In: *Chilling Injury of Horticultural Crops.* C.-Y. Wang (ed.). CRC Press, Boca Raton, FL.

200. Paull, R.E. 1995. Preharvest factors and the heat sensitivity of field grown ripening papaya fruit. *Postharv. Biol. Tech.* 6:167–175.

201. Paull, R.E., and J. Armstrong. 1994. Insect pests and fresh horticultural products: Introduction treatments and responses. Pp. 1–36. In: *Insect Pests and Fresh Horticultural Products: Treatments and Responses.* R.E. Paull and J.W. Armstrong (eds.). CABI, Wallingford, England.

202. Paull, R.E., and N.J. Chen. 1999. Heat treatment prevents postharvest geotropic curvature of asparagus (*Asparagus officinalis* L.). *Postharv. Biol. Tech.* 16:37–41.

203. Paull, R.E., and N.J. Chen. 2000. Heat treatments and fruit ripening. *Postharv. Biol. Tech.* 21:21–37.

204. Pena-Cortes, H., and L. Willmitzer. 1995. The role of hormones in gene activation in response to wounding. Pp. 395–414. In: *Plant Hormones.* P.J. Davies (ed.). Kluwer Academic Press, Dordrecht, The Netherlands.

205. Pieniazeh, S.A. 1942. External factors affecting water loss from apples in cold storage. *Refrig. Eng.* 44:171–173.

206. Pitt, R.E. 1982. Models for the rheology and statistical strength of uniformity stressed vegetative tissue. *Trans. Amer. Soc. Agri. Eng.* 25:1776–1784.

207. Platt-Aloia, K.A., and W.W. Thomson.1987. Freeze fracture evidence for lateral phase separations in the plasmalemma of chilling-injured avocado fruit. *Protoplasma* 136:71–80.

208. Poenicke, E.F., S.J. Kays, D.A. Smittle and R.E. Williamson. 1977. Ethylene in relation to postharvest quality deterioration in processing cucumbers. *J. Amer. Soc. Hort. Sci.* 102:303–306.

209. Poljakoff-Mayber, A., and J. Gale (eds.). 1975. *Plants in Saline Environments.* Ecological Studies, Vol. 15, Springer-Verlag, New York.

210. Pratt, H.K., and J.D. Goeschl. 1969. Physiological roles of ethylene in plants. *Annu. Rev. Plant Physiol.* 20:541–584.

211. Precheur, R., J.K. Greig and D.V. Armbrust. 1978. The effects of wind and wind-plus-sand on tomato plants. *J. Amer. Soc. Hort. Sci.* 103:351–355.

212. Proapst, P.A., and F.R. Forsyth. 1973. The role of internally produced carbon dioxide in the prevention of greening in potato tubers. *Acta Hort.* 38:277–290.

213. Prusky, D. 1996. Pathogen quiescence in postharvest diseases. *Annu. Rev. Phytopathology* 34:413–434.

214. Quintana, M.U., and R.E. Paull. 1993. Mechanical injury during postharvest handling of Solo papaya fruit. *J. Amer. Soc. Hort. Sci.* 118:618–622.

215. Raison, J.K., J.M. Lyons, R.J. Mahlhoin and A.D. Keith. 1971. Temperature-induced phase changes in mitochondrial membranes detected by spin labeling. *J. Biol. Chem.* 246:4036–4040.

216. Reid, M.S. 1985. Ethylene and abscission. *HortScience* 20:45–50.

217. Rhodes, M.J.C., and L.S.C. Wooltorton. 1973. Changes in phenolic acid and lignin biosynthesis in response to treatment of root tissue of the Swedish turnip (*Brassica napo-brassica*) with ethylene. *Qual. Plant.* 23:145–155.

218. Rigney, C.J., D. Graham and T.H. Lee. 1978. Changes in tomato fruit ripening caused by ethylene dibromide fumigation. *J. Amer. Soc. Hort. Sci.* 103:402–423.

219. Risse, L.A., and T.T. Hatton. 1982. Sensitivity of watermelons to ethylene during storage. *HortScience* 17:946–948.

220. Robbins, J.W., and K.F. Petsch. 1932. Moisture content and high temperature in relation to the germination of corn and wheat grains. *Bot. Gaz.* 93:85–92.

221. Robinson, J.E., K.M. Browne and W.G. Burton. 1975. Storage characteristics of some vegetables and soft fruits. *Ann. Appl. Biol.* 81:399–408.

222. Robitaille, H.A., and A.C. Leopold. 1974. Ethylene and regulation of apple stem growth under stress. *Physiol. Plant.* 32:301–304.

223. Romani, R.J. 1966. Radiobiological parameters in the irradiation of fruits and vegetables. *Adv. Food Res.* 15:57–103.

224. Rosa, J.R. 1926. Ripening and storage of tomatoes. *Proc. Amer. Soc. Hort. Sci.* 23:233–242.

225. Sacalis, J.N. 1978. Ethylene evolution by petioles of sleeved poinsettia plants. *HortScience* 13:594–596.

226. Saltveit, M.E. 1984. Effect of temperature on firmness and bruising of 'Starkcrimson Delicious' and 'Golden Delicious' apples. *HortScience* 19:550–551.

227. Saltveit, M.E. 1996. Physical and physiological changes in minimally processed fruits and vegetables. Pp. 205–220. In: *Phytochemistry of Fruit and Vegetables.* F.A. Tomás-Barberán (ed.). Oxford University Press, Oxford, England.

228. Saltveit, M.E. 2000. Wound induced changes in phenolic metabolism and tissue browning are altered by heat shock. *Postharv. Biol. Tech.* 21:61–69.

229. Saltveit, M.E. 2002. The rate of ion leakage from chilling-sensitive tissue does not immediately increase upon exposure to chilling temperatures. *Postharv. Biol. Tech.* 26:295–304.

230. Saltveit, M.E., and R.A. Larson. 1981. Reduced leaf epinasty in mechanically stressed poinsettia (*Euphorbia pulcherrima* cultivar Annette Hegg Diva) plants. *J. Amer. Soc. Hort. Sci.* 106:156–159.

231. Salunkhe, D.K. (ed.). 1974. *Storage, Processing and Nutritional Quality of Fruits and Vegetables.* CRC Press, Cleveland, OH.

232. Sarkar, S.K., and C.T. Phan. 1979. Naturally-occurring and ethylene-induced compounds in the carrot root. *J. Food Prod.* 42:526–534.

233. Sasaki, S., and T.T. Kozlowski. 1968. Effects of herbicides on respiration of red pine (seedlings) II. Monuron, Diuron, DCPA, Dalapon, CDEC, CDAA, EPTC, and NPA. *Bot. Gaz.* 129:268–293.

234. Sato, K., I. Uritani and T. Saito. 1981. Characterization of the terpene-inducing factor isolated from larvae of the sweet potato weevil, *Cylas formicarius* FABRICUS (Coleoptera:Brenthidae). *Appl. Ento. Zoology* 16:103–112.

235. Sato, K., I. Uritani and T. Saito. 1982. Properties of terpene-inducing factor extracted from adults of the

sweet potato weevil, *Cylas formicarius* FABRICIUS (Coleoptera: Brenthidae). *Appl. Ento. Zoology* 17:368–374.

236. Scheel, D. 1998. Resistance response physiology and signal transduction. *Cur. Opinions Plant Biol.* 1:305–310.

237. Schirra, M., G. D'hallewin, S. Ben-Yehoshua and E. Fallik. 2000. Host-pathogen interactions modulated by heat treatments. *Postharv. Biol. Tech.* 21:71–85.

238. Scott, K.J., and E.A. Roberts. 1968. The importance of weight loss in reducing breakdown of 'Jonathan' apples. *Aust. J. Expt. Agr. Animal Husb.* 8:377–380.

239. Scott, K.J., and R.B.H. Wills. 1975. Postharvest application of calcium as control for storage breakdown of apples. *HortScience* 10:75–76.

240. Sharples, R.O. 1973. Orchard sprays. Pp.194–203. In: *The Biology of Apple and Pear Storage.* J.C. Fidler, B.G. Wilkinson, K.L. Edney and R.O. Sharples (eds.). Commonwealth Agr. Bur., London.

241. Sharples, R.O. 1980. The influence of orchard nutrition on the storage quality of apples and pears grown in the United Kingdom. Pp.17–28. In: *Mineral Nutrition of Fruit Trees.* D. Atkinson, J.E. Jackson, R.O. Sharples and W.M. Waller (eds.). Butterworths, London.

242. Sherif, S.M. 1976. The quasi-static contact problem for nearly incompressible agricultural products. Ph.D. Thesis, Mich. State University, East Lansing, MI.

243. Shipway, M.R., and W.J. Bramlage. 1973. Effects of carbon dioxide on activity of apple mitochondria. *Plant Physiol.* 51:1095–1098.

244. Silsaoka, T. 1969. Physiology of rapid movements in higher plants. *Annu. Rev. Plant Physiol.* 20:165–184.

245. Skene, D.S. 1980. Growth stresses during fruit development in Cox's Orange Pippin apples. *J. Hort Sci.* 55:27–32.

246. Smith, W.H. 1938. The storage of broccoli. *Rept. Fd. Invest. Bd.,* 1937, pp.185–187.

247. Smith, W.H. 1958. Reduction of low-temperature injury to stored apples by modulation of environmental conditions. *Nature* 191:275–276.

248. Smith, W.H. 1965. Storage of mushrooms. *Rpt. Ditton Lab.* 1964–1965, p. 25.

249. Smock, R.M. 1977. Nomenclature for internal storage disorders of apples. *HortScience* 12:306–308.

250. Solov'ev, V.A. 1969. Plant growth and water and mineral nutrient supply under NaCl salinization conditions. *Fiziol. Rast.* 16:870–876.

251. Somner, N.F. 1957. Pear transit simulated in test. *Calif. Agric.* 11(9):3–5,16.

252. Somner, N.F. 1957. Surface discoloration of pears. *Calif. Agric.* 11(1):3–4.

253. Somner, N.F., and R.J. Fortlage. 1966. Ionizing radiation for control of postharvest diseases of fruits and vegetables. *Adv. Food Res.* 15:147–193.

254. Sparks, D. 1986. Pecan. Pp.323–339. In: *Handbook of Fruit Set and Development.* S.P. Monselise (ed.). CRC, Boca Raton, FL.

255. Spencer, E.L. 1937. Frenching of tobacco and thallium toxicity. *Amer. J. Bot.* 24:16–24.

256. Steponkus, P.L. 1984. Role of the plasma membrane in freezing injury and cold acclimation. *Annu. Rev. Plant Physiol.* 35:543–484.

257. Steward, R.N., K.H. Norris and S. Asen. 1975. Microspectrophotometric measurement of pH and pH effect on color of petal epidermal cells. *Phytochemistry* 14:937–942.

258. Stewart, J.K., and M. Uota. 1971. Carbon dioxide injury and market quality of lettuce held in controlled atmospheres. *J. Amer. Soc. Hort. Sci.* 96:27–30.

259. Stoll, K. 1972. Lagerung von Früchten und Gemusen in kontrollierter Atmosphere. *Mitt. Eidg. Forsch. Anst. Obst. Wein. Gartenbau, Wädenswil, Flugschrift Schweiz Obst. Weinbau* 107:614–623, 648–652, 711–714, 741–745.

260. Stange, R.R., S.L. Midland, G.J. Holmes, J.E. Sims and R.T. Mayer. 2001. Constituents from the periderm and outer cortex of *Ipomoea batatas* with anti-fungal activity against *Rhizopus stolonifer. Postharv. Biol. Tech.* 23:85–92.

261. Strobel, G.A. 1974. Phytotoxins produced by plant parasites. *Annu. Rev. Plant Physiol.* 25:541–566.

262. Strauss, S.Y., and A.A. Agrawal. 1999. The ecology and evolution of plant tolerance to herbivory. *Trends Ecol. Evol.* 14:179–185.

263. Studer, E.J., P.L. Stepoukus, G.L. Good and S.C. Wiest. 1978. Root hardiness of container-grown ornamentals. *HortScience* 13:172–174.

264. Suge, H. 1978. Growth and gibberellin production in *Phaseolus vulgaris* as affected by mechanical stress. *Plant and Cell Physiol.* 19:1557–1560.

265. Swinburne, T.R. 1983. Quiescent infections in post-harvest diseases. Pp.1–21. In: *Post-harvest Pathology of Fruits and Vegetables.* C. Dennis (ed.). Academic Press, New York.

266. Thomas, M. 1925. The controlling influence of carbon dioxide. V. A quantitative study of the production of ethyl alcohol and acetaldehyde by cells of the higher plants in relation to the concentration of oxygen and carbon dioxide. *Biochem. J.* 19:927–947.

267. Toivonen, P., J. Walsh, E.C. Lougheed and D.P. Murr. 1982. Ethylene relationships in storage of some vegetables. Pp. 299–307. In: *Controlled Atmospheres for Storage and Transport of Perishable Agricultural Commodities.* D.G. Richardson and M. Meheriuk (eds.). Timber Press, Beaverton, OR.

268. Tomkins, R.G. 1965. The storage of broad beams. *Annu. Rpt. Ditton and Covent Garden Labs,* 1964–1965, p. 24.

269. Tomkins, R.G., and D.F. Meigh. 1968. The concentration of ethylene found in controlled atmosphere stores. *Ditton Lab. Annu. Rpt.* 1967–68. pp.33–36.

270. United Nations. 1950. *Dept. Econ. Affairs,* II. G.I., 65–66. Washington, D.C.

271. Van der Plank, J.E., and R. Davies. 1938. Temperature-cold injury curves of fruit. *J. Pomol Hort. Soc.* 15:226–247.

272. Varley, G.C., and G.R. Gradwell. 1962. The effect of partial defoliation by caterpillars on the timber production of oak trees in England. *Proc. 11th Intern. Cong. Ent.,* Vienna 1960, 2:211–214.

273. Verner, L., and E.C. Blodgett. 1931. Physiological studies of the cracking of sweet cherries. *Idaho Agr. Exp. Sta. Bull.* 184.

274. Vogele, A.C. 1937. Effect of environmental factors upon the color of the tomato and the watermelon. *Plant Physiol.* 12:929–955.

275. Voisey, P.W., L.H. Lyall and M. Klock. 1970. Tomato skin strength—its measurement and relation to cracking. *J. Amer. Soc. Hort. Sci.* 95:485–488.

276. Volkmann, D., and A. Sievers. 1979. Graviperception in multicellular organs. Pp.353–367. In: *Physiology of Movements.* W. Haupt and M.E. Feinleik (eds.). Vol. 7. *Encyclopedia of Plant Physiology.* Springer-Verlag, Berlin.

277. Von Wandruszka, R.M.A., C.A. Smith and S.J. Kays. 1980. The role of iron in pecan kernel color. *Lebensm. Wiss. U. Technol.* 13:38–39.

278. Walker, J.C., and K.P. Link. 1935. Toxicity of phenolic compounds to certain onion bulb parasites. *Bot. Gaz.* 96:468–484.

279. Walter, W.M., Jr., and W.E. Schadel. 1983. Structure and composition of normal skin (periderm) and wound tissue from cured sweet potatoes. *J. Amer. Soc. Hort. Sci.* 108:909–914.

280. Wang, C.Y. 1982. Physiological and biochemical responses of plants to chilling stress. *HortScience* 17:173–186.

281. Wardowski, W.F., and A.A. McCormack. 1973. Degreening Florida citrus fruits. *Fla. Agri. Exp. Sta. Ext. Ser. Cir.* 389.

282. Wattendorf, J. 1970. Permeabilität für Wasser, Malonsäurediamid und Harnstoff nach α-Bestrahlung von Convallaria-Zellen. *Ber. Deut. Bot. Ges.* 83:3–17.

283. Webster, C.C. 1967. The effects of air pollution on plant and soil. *Agri. Res. Counc.,* London.

284. Weiser, C.J. 1970. Cold resistance and injury in woody plants. *Science* 169:1269–1278.

285. Whiteman, T.M. 1957. Freezing points of fruits, vegetables and florist stocks. *U.S. Dept. Agr., Market Res. Rpt.* 196.

286. Widmann, C., S. Gibson, M.B. Jarpe and G.L. Johnson. 1999. Mitogen-activated protein kinases: conservation of a three-kinase module from yeast to human. *Physiological Rev.* 70:143–180.

287. Wilcockson, S.J., R.L. Griffith and E.J. Allen. 1980. Effects of maturity on susceptibility to damage. *Ann. Appl. Biol.* 96:349–353.

288. Wilkin, M.B. 1979. Growth-control mechanisms in gravitropism. Pp.601–621, In: *Physiology of Movements.* W. Haupt and M.E. Feinleib (eds.). Vol. 7. *Encyclopedia of Plant Physiology.* Springer-Verlag, Berlin.

289. Wills, R.B.H., and K.J. Scott. 1971. Chemical induction of low temperature breakdown in apples. *Phytochemistry* 10:1783–1785.

290. Wilson, C.L., and M.E. Wisniewski. 1989. Biological control of postharvest diseases of fruits and vegetables: An emerging technology. *Annu. Rev. Phytopathology* 27:425–441.

291. Woit, M. 1925. Über Wundreaktionen an Blättern und den Anatomischen ban der Klattminen. *Mitt. Dent. DenDr. Ges.* 35:163–187.

292. Wright, R.C. 1942. The freezing temperatures of some fruits, vegetables and florists' stocks. *U.S. Dept. Agri. Circ.* 447.

293. Wynn, W.K., Jr. 1963. Photosynthetic phosphorylation by chloroplasts isolated from rust-infected oats. *Phytopathology* 53:1376–1377.

294. Yoshioka, H., Y. Ueda and K. Chachin. 1980. Physiological studies of fruit ripening in relation to heat injury II: Effect of high temperature (40°) on the change of acid-phosphatase activity during banana fruit ripening. *J. Jpn. Soc. Food Sci. Tech.* 27:511–516.

295. Yoshioka, H., Y. Ueda and T. Iwata. 1982. Physiological studies of fruit ripening in relation to heat injury IV: Development of isoamyl acetate biosynthetic pathway in banana fruit during ripening and suppression of its development at high temperature. *J. Jpn. Soc. Food Sci. Tech.* 29:333–339.

296. Zagory, D., and M.S. Reid. 1985. Evaluation of the role of vase microorganisms in the postharvest life of cut flowers. Pp. 207–217. In: *Third International Symposium Post-Harvest Physiology of Ornamentals.* H.C.M. de Styers (ed.). Drukkerig, The Netherlands.

297. Zhang, S., and D.F. Klessig. 2001. MAPK cascades in plant defense signaling. *Trends Plant Sci.* 6:520–527.

8

MOVEMENT AND EXCHANGE OF GASES, SOLVENTS AND SOLUTES WITHIN HARVESTED PRODUCTS AND THEIR EXCHANGE BETWEEN THE PRODUCT AND ITS EXTERNAL ENVIRONMENT

Plants are composed of a diverse cross-section of organic and inorganic compounds found as solids, liquids and gases. During the growth and development of a plant or a specific organ, chemical acquisition and distribution processes of tremendous scale are operative. Molecules and ions taken up by the plant must be moved from the site of acquisition to the site of utilization. For example, pumpkin fruits* achieve a final mass of approximately 800 g (dry wt.), all of which is transported into the fruit through peduncle sieve tubes[12] having an average total diameter of only 0.05 cm². Knowing the concentration of organic material within the sieve tubes, Crafts and Lorenz[13] estimated that an average translocation velocity of 110 cm · hr⁻¹ would have to be maintained to account for the tremendous accumulation of substances within a mature fruit. Plants, therefore, have evolved in their xylem and phloem tissues elaborate systems for the movement of chemicals throughout the plant. This rapid, long distance transport in the xylem and phloem of growing plants is, however, dependent upon the movement of water; thus the ability of the plant to replenish water lost through transpiration is an essential requisite for proper functioning of the system.

Most plant products are severed from the parent plant at harvest, which generally eliminates the product's water replenishing potential, blocking the normal means for long distance movement of solutes and water throughout the product. In some cases, water can be replenished (e.g., cut flowers held in aqueous solutions) and the transport system remains partly functional for a time.

While the normal transport system is inoperative in most harvested products, the redistribution of organic and inorganic compounds continues. This is especially true in products that have a high moisture content. Transport of solutes and water is now characterized by a much slower rate of movement and generally over shorter distances. The newly dominant

---

*Cultivar 'Connecticut Field.'

forms of movement are driven by different mechanisms than those operative in xylem and phloem transport of intact plants. Redistribution of constituents within harvested products can be viewed at two levels: 1) a redistribution of compounds within the tissue or product; and 2) loss to or absorption from the external environment. In the latter case, losses normally occur as molecules in the gas phase (e.g., carbon dioxide, benzyl alcohol, esters of organic acids). While we are aware of the tremendous diversity of volatile molecules given off by plants (Chapter 4.6), the uptake of gaseous molecules may also be extensive. In addition to oxygen, which is required for respiration, a cross-section of other molecules are absorbed. In dried agronomic crops, water vapor is absorbed and in some cases uptake is extensive. Water may also enter as a liquid, for example when harvested products are washed, hydrocooled, misted on retail shelves or inadvertently exposed to rain. Without proper temperature management, water vapor may also condense on the surface of the product and be readily absorbed.

Many harvested products undergo not only extensive metabolic and compositional changes after harvest but also a significant movement and redistribution of cellular constituents within the product. Both the forces driving movement and the factors that impede movement are critical components in the redistribution process. To better understand the movement of ions and molecules, we must first look at the sources of energy that drive this movement.

1. FORCES DRIVING MOVEMENT

Energy, defined simply as the ability to do work (work = force × distance), can be separated into two general classes: kinetic and potential. **Kinetic energy** is the energy of motion. Heat energy is a form of kinetic energy. The addition of heat increases the kinetic energy level and therefore the motion of molecules. **Potential energy** is the energy of position. An example would be the energy in the chemical bonds of a sucrose molecule. For potential energy to do work, it must be released, and within plants this occurs through chemical reactions. Respiration releases potential energy, some of which ends up as one form of kinetic energy (e.g., heat) or another (e.g., mechanical). Photosynthesis converts kinetic light energy into the potential energy of the carbon bond. Plants require a continual input of free energy or else they drift energetically downward, which eventually results in the cessation of life. The ultimate source of free energy is the sun; for most postharvest products, however, additional free energy inputs are no longer possible. Therefore, the theoretical survival potential of a product is governed to a large extent by its supply of free energy, its biological requirements for the energy and the thermodynamic laws governing the expenditure of the free energy.

Not all of the total energy is available to do work. Part, the entropy factor, is lost. The remaining energy (i.e., **free energy**) can be converted to work. Free energy, therefore, in its various forms provides the force necessary to drive the movement of gases, solutes and solvent within harvested products. In many cases, differences in free energy between two locations within the product result in spontaneous movement. For example, as a rice grain desiccates, water moves from areas of high concentration or energy (the interior) to areas of low (the surface). Both the rate and direction of movement are dependent upon gradients in energy. Movement proceeds from areas of high energy (high potential) to low. Therefore, as the substance moves, it proceeds in an energetically downhill direction.

Many of the changes that occur within plants, however, must proceed in an energetically up-hill direction. These changes require an input of free energy. An example would be the ATP driven active transport of substances across the plasma membrane. Generally the substance is moving from an area of low concentration to an area of high concentration, against a gradient in energy potential; hence the need for a supplemental energy input.

Free energy, called "Gibb's free energy" after the scientist that first described it in the 1870s, is a measure of the energy within a system that can be used to do work. Free energy can be viewed at a system or a location level, where the free energy of each chemical component is added together to obtain a total. Likewise, it can also be expressed as the energy of a chemical species, such as the energy of a single type of substance (e.g., water). Since the amount of energy available is a function of the quantity of the substance (moles), free energy is usually expressed as free energy/mole—the **chemical potential.** Chemical potential therefore is a function of concentration.

Chemical potential also depends upon pressure, electrical potential and gravitational forces. The existence of a pressure gradient can cause the movement of liquids. The uptake of water into the cell due to osmosis (diffusion across a differentially permeable membrane from a region of high potential to one of lower potential) causes an increase in hydrostatic pressure within the cell. This outward force influences the chemical potential. Likewise, many substances found within the plant are not electrically neutral. The presence of a charge, whether positive (e.g., Na^+, Ca^{2+}, Mg^{2+}) or negative (NO^{3-}, Cl^-, SO_4^{2-}), influences the behavior of that substance and, therefore, its chemical potential. For charged particles, potential is expressed as **electrochemical potential** to account for charge effects. If an uneven distribution of charges occurs across a membrane, transport is influenced. Lastly, gravitational effects are operative in very tall trees; however, in harvested products, gravity has a negligible effect on chemical potential.

Many of the chemical constituents of plants can be found in more than one physical state (e.g., water is present both as vapor and a liquid). Since one state often diffuses more readily than another, net movement may be greatly influenced by the ratio between the two states. Many of the principles governing the movement of gases differ from those governing the movement of liquids. Therefore the movement of gases has been treated separately from that of solvent and solutes. Solute and water movement are discussed together, since water provides the avenue for solute transport, and the solutes in turn alter the movement of the solvent.

2. GASES: MOVEMENT AND EXCHANGE

While gases make up only a small portion of the earth's biosphere mass,* they are of paramount importance to life. Their general lack of visibility tends to minimize awareness of gases. Two of the most important gases are carbon dioxide, given off during respiration and consumed during photosynthesis, and oxygen, consumed during respiration and given off during photosynthesis. Because of their central roles in the plant's energy acquisition and utilization systems, they are critical to the overall metabolism of plants.

Water, unlike carbon dioxide and oxygen, is found as both a liquid and a gas at temperatures normally encountered during the postharvest period. Most plant products lose or gain water during storage due to the movement of gaseous water vapor into or out of the product. Alterations in the water status of the cells due to the movement of water vapor can, if of sufficient magnitude, exert a pronounced effect upon metabolism and quality.

Many other gaseous molecules are also important. Ethylene, a gaseous hormone that stimulates ripening, abscission and senescence; aromatic compounds that enhance our perception of a product; volatile insect-behavior modulating chemicals; and air pollutants can have a significant impact on the postharvest behavior and quality of a plant or plant organ.

Gaseous molecules may have beneficial or detrimental effects on the product; they may

---

*Total mass of the Earth's atmosphere is estimated to be only 5×10^{18} kg.

be produced by the plant or come from external sources; and they may function either within the product (e.g., oxygen) or externally (e.g., aromatic compounds[40]). In some cases, a specific gas may be beneficial to one product but undesirable to another. The permeation of honeydew melons in storage with volatiles from adjacent onions is decidedly detrimental.

The effect of most gases that modify plants is concentration dependent. Below a certain threshold concentration there is little or no response by the tissue. Each additional increment in concentration above this threshold level results in a progressively greater response. In some cases, the response saturates (e.g., ethylene induced leaf abscission) in a relatively narrow range and elevation of the concentration above this range does not result in an increased response.

Because of this relationship between gas concentration and the magnitude of the tissue's response, understanding what controls the concentration of a gas at the site of action is essential. The physical properties of the gas, its site of origin (external versus internal), and factors affecting the supply (e.g., rate of synthesis) and controlling movement into or out of the product all modulate the concentration at this site.

One critical characteristic of volatile compounds is the boiling point (bp). Compounds such as oxygen, carbon dioxide, and ethylene have very low boiling points (i.e. bp = -183, -78 and -169°C, respectively) and as a consequence, are always gaseous at biological temperatures. In contrast, most volatile molecules of plant origin are found in both a liquid and gaseous state [e.g., H_2O (bp 100°C), benzyl alcohol (bp 203°C)]. Since part of the movement of these compounds within the plant is by diffusion of their liquid phase, their movement and exchange differs from compounds present only as a gas. These two general classes of molecules (i.e., gases vs gas/liquid) will be discussed separately.

2.1. Properties of Gases

Gaseous molecules are in constant motion due to their kinetic energy. The velocity of this motion increases with increasing temperature and decreases with increasing size of the molecule. Carbon dioxide molecules [molecular weight (mw) = 44)] have an average velocity of approximately 1,368 km · hr^{-1} at room temperature while smaller hydrogen molecules (mw = 2.01) approach 6,437 km · hr^{-1}. This value represents an average speed, with some of each type of molecule traveling faster and others slower than their respective means. The high speed (energy) particles tend to be more reactive. For example with water molecules, the high energy particles are more likely to cause melting, evaporation or chemical reactions while the low energy particles tend to be the first to condense or freeze. Therefore the energy level of a gas (i.e., temperature) is going to affect its rate of movement into and out of harvested products.

This high speed, random movement of gases causes their molecules to expand outward, filling all the space available to them and taking the shape of whatever contains them.* This random motion also causes individual gas molecules to collide with neighboring molecules and the walls of the container. Due to the proximity of neighboring molecules and their rate of speed, one molecule may collide billions of times per second. When neighboring gas molecules do collide, typically one molecule will gain energy while the other loses it. Although the energy of each molecule is altered, the sum of the energies of the two molecules remains the

---

*The outward expansion of gases is eventually limited by gravitational forces. Fifty percent of the Earth's atmosphere is held within the first 5.5 km of the troposphere; 99% within 30 km of the surface (sea level). As the elevation increases, the density of molecules decreases.

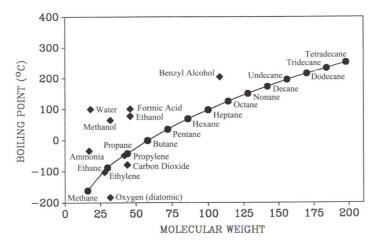

Figure 8.1. The relationship between boiling point and molecular weight of a series of saturated hydrocarbons (connected points) and other compounds of biological interest is shown. The unique physical and chemical properties of water, causes it to fall outside of the general distribution pattern.

same. Therefore, there is no net loss of energy due to collision between neighboring molecules.

Gases differ from liquids in that the distance between molecules is sufficiently great that no cohesive forces exist. Although oxygen, carbon dioxide and ethylene are not found as liquids at biological temperatures (Figure 8.1), many of the other gaseous molecules important during the postharvest period do occur in both a gaseous and liquid state. Water and essentially all of the aromatic compounds fall into this latter group.

Generally, the higher the molecular weight of a substance, the more likely it will be found predominantly as a liquid at room temperature (Figure 8.1). Larger molecules require more energy (heat) to break the cohesive forces between neighboring molecules and escape into the atmosphere. If energy is added, movement of neighboring molecules is accelerated and a greater number of the molecules near the surface of the liquid are able to escape the attractive forces of their neighbors, entering the gas phase. An example in nature of a programmed increase in energy is found in many of the *Araceae*.[37] At flowering, a part of the spathe shifts a major portion of its respiration to the alternate electron transport pathway. The alternate pathway's much lower energy trapping efficiency results in a buildup of heat. The additional heat increases the energy level of several insect attractants that are found predominantly in a liquid state, causing more of the molecules to escape the cohesive forces of the liquid and enter the atmosphere, where they can be perceived by insects that facilitate pollination.

If energy is added to a gas in the form of heat, the speed of the individual molecules increases as does their outward movement. The volume of a given mass of gas, therefore, is altered, with the change in volume being proportional to the change in temperature (Charles' Law). Likewise, by cooling a gas from temperature T_1 to temperature T_2, the volume decreases from V_1 to V_2.

$$\frac{V_1 - V_2}{T_1 - T_2} = \frac{\Delta V}{\Delta T}$$

eq. 8.1

The ratio of volume to temperature ($\Delta V/\Delta T$) remains the same for a given mass of gas regardless of changes in temperature. If the volume of the gas is restricted while the temperature increases (e.g., a sealed cold storage room in which the refrigeration system has malfunctioned), the pressure increases. The pressure can be held constant only if the volume the gas occupies increases.* For a given mass of gas held at a constant temperature, the pressure and volume of the gas are inversely proportional (Boyle's Law). The product of the volume times the pressure for a given mass of gas is the same even though the conditions (pressure or volume) change.

$$P_1 V_1 = P_2 V_2 \qquad\qquad\text{eq. 8.2}$$

The laws of Charles and Boyle can be combined into a more general gas law where volume equals temperature divided by pressure times a constant (k).

$$V = k\,(T/P) \qquad\qquad\text{eq. 8.3}$$

The constant (k) depends upon the amount and kind of gas being considered; it may be found for most gases in standard chemical tables. If the same gas is being considered under two sets of conditions, k can be ignored. Then:

$$\frac{P_1 V_1}{T_1} = \frac{P_2 V_2}{T_2} \quad \text{(for a fixed mass of gas)} \qquad\qquad\text{eq. 8.4}$$

How does this equation relate to harvested products? If the temperature (refrigerated storage), the pressure (hypobaric or low pressure storage) or the volume of gas within or around a harvested product is altered, the way the gas interacts with the product will be altered. When the gas in question has a significant effect on the product, these changes in conditions are translated to an altered response by the product. For example, decreasing the pressure within a sealed hypobaric storage system decreases the number of molecules present to interact with the product. When the gas is produced by the product (e.g., ethylene, aromatic compounds, carbon dioxide), a reduced pressure also increases the steepness of the concentration gradient from the site of origin of the gas to the exterior, accelerating the rate at which the gas molecules diffuse out of the tissue.

The gas atmosphere surrounding the surface of the earth exerts a pressure that at sea level is equal to 101 kPa (1 atmosphere). This pressure is due to the composite of all of the gases present; however, since we are often not interested in all gases, it is useful to determine the partial (part of the total) pressure exerted by one type of gas. A gas's **partial pressure** is a measure of what part of the total pressure it exerts relative to all of the gases present. An atmosphere containing 21% oxygen at sea level (101 kPa) has an oxygen partial pressure (pO_2) of 0.21 (21% of the whole). If a partial vacuum is created within a container holding the gases and the pressure is reduced to 1/10 of an atmosphere (i.e., 101 kPa $\times$ .10 = 10.1 kPa), the partial pressure of oxygen decreases proportionally.

$$\frac{0.21\ pO_2}{101\ \text{kPa}} = \frac{X\ pO_2}{10.1\ \text{kPa}} \qquad\qquad\text{eq. 8.5}$$

Therefore X = 0.021 pO_2. While nine tenths of all of the gas molecules have been removed from the container, the relative ratio of each type of gas within the container does not change, so the percent oxygen remains 21.

---

*In controlled atmosphere storage rooms this can be accomplished by using "breather" bags open to the outside that change the volume of the room to compensate for changes in barometric pressure.

2.2. Movement and Exchange of Gases

Gases move from areas of high concentration to areas of low concentration by diffusion. This movement from one point in space to another is due to the random movement of the individual molecules caused by their kinetic energy. If all of the gases within the tissue of a papaya fruit have been replaced with an inert gas such as helium and the fruit is placed in a sealed jar of nitrogen, molecules of helium begin to diffuse out of the fruit, while molecules of nitrogen begin to diffuse into it. Gradually the concentration of helium and nitrogen inside and outside of the fruit become equal, reaching what is called an **equilibrium state**. While molecules of helium continue to diffuse out of the fruit, an equal number move in the opposite direction back into the fruit. Therefore, while there is a continual exchange, there is no net movement.

If the gas is being continually formed (e.g., carbon dioxide or ethylene) or utilized (oxygen) within the tissue, a more complex situation exists. When the flow of oxygen diffusing into the fruit is equal to the rate of utilization by the tissue, the system is said to have reached a **steady state**. This differs from an equilibrium state in that there is a continual movement of the gas through the system. Since most gases of biological interest are either being produced or consumed, a true equilibrium state is not reached; however, their exchange often approaches a steady state if the tissue and its environment are not altered.

Gases may also move by **bulk flow**. Here the movement is caused by a pressure gradient, and all of the gases present move together rather than independently, as with diffusion. In nature, the bulk flow of gases in plant products is not of major importance. An example, however, can be seen in the seed dispersal mechanism that has evolved in the squirting cucumber. As the fruits ripen, the pressure within the central chamber of the fruit increases (i.e., up to 2.7 MPa).[61] Upon dehiscence, the peduncle plug separates from the fruit and the seeds are literally ejected out of the opening in the fruit wall, dispersing them over a wide area.[27] A postharvest situation where gas movement is *via* bulk flow occurs with the use of hypobaric (low pressure) storage. As the atmospheric pressure is reduced around the product, the pressure differential causes an outward flow of gases from the product to equilibrate the internal and external pressures.

2.2.1. *Conductance versus Resistance*

The rate of movement of a gas across a given space is referred to as flux density (i.e., flow and quantity). The flux density of a gas moving into or out of a harvested product can be viewed in two ways: by describing the ability or "ease" by which a gas can move through the material (conductance) or by the opposing forces, presented by the tissue, that impede movement of the gas (resistance). Resistance is the reciprocal of conductance (i.e., $c = 1/r$). It may be written:

$$\text{Flux density} = \text{Conductance} \times \text{Force} \qquad \text{eq. 8.6}$$

where force is the difference in concentration of the gas between the two sites; or:

$$\text{Flux density} = \frac{\text{Force}}{\text{Resistance}} \qquad \text{eq. 8.7}$$

Therefore, flux density is directly proportional to conductance but inversely proportional to resistance. While conductance terminology is generally more readily conceptualized, resistances are specified more commonly than conductances. Resistances are also useful when studying the movement of a gas across a series of components in sequence, for example, the diffusion of ethylene across the cytosol, plasma membrane, cell wall, intercellular channels, tissue surface and the surface boundary layer. The resistances, being in series, are additive.

2.2.2. Tissue Morphology

The interior of harvested plant products are often visualized as a dense mass of tightly packed cells. In fact, the interior contains an extensive system of intercellular gas-filled space. These spaces form an interconnecting series of channels, reaching virtually every cell, through which gases may readily diffuse. The volume of this gas-filled space varies widely between products. The cells in potato tubers are very tightly packed, with the intercellular space representing only about 1.0% of the total volume of the product. Most products are much less dense. For example, the apple has approximately 36% intercellular space.[27] Some products (e.g., pepper, muskmelon, lotus root) have a large central cavity or cavities that enhances the ability of gases to move readily throughout the tissue.

Intercellular gas volume, however, is not totally indicative of the potential for a gas to move once it is inside a product. The flux of a gas within a tissue depends upon the cross-sectional dimensions and length of the air channels, and their continuity and distribution within the tissue. In addition to the general architecture of the air channels, their condition is also extremely important. As fleshy plant products begin to senescence, membrane integrity declines and fluids may leak from the cytosol into the intercellular air spaces. Fluids found external to the plasma membrane displace an equivalent volume of gas, decreasing the effective volume of intercellular air space. This decline in volume may in turn restrict the diffusion of gases, accelerating senescence. Since gases diffuse readily in air, if the architecture and physiological condition of the intercellular air spaces are adequate, there will be little restriction (i.e., resistance) of gas movement. Therefore, only small gradients in concentration of gases over reasonable distances will occur within the tissue. However, if the diffusion of gases through the intercellular air channels is restricted, significant gradients may develop, resulting in concentrations that may contribute to the deterioration of the tissue.

Plants and their individual organs have achieved a compromise in structure relative to the diffusion of gases. While the movement of gases within the intercellular channels is in many products relatively free (i.e., low resistance), a similar unrestricted exchange between the plant and its environment would result in rapid death due to the loss of water. Early in their evolutionary path to a terrestrial existence, plants developed a waxy epidermal coating, the cuticle, over the aerial portion of the plant. With underground organs, the cuticle is replaced as the epidermal boundary layer by the periderm. In the periderm, the intercellular air channels are either absent or obstructed. Like the cuticle, the periderm seals the interior air channels, forming a boundary between the plant and its surrounding environment that greatly restricts the movement of water vapor and other gases. Therefore, unlike the intercellular air channels within most tissues, the outer surface represents a major barrier, an area of high resistance to the diffusion of gases.

Formerly it was thought that cuticle thickness was the most important parameter controlling the rate of gas diffusion. Cuticle thicknesses are known to vary widely, but in general have not been found to correlate closely with transpirational losses of water vapor. What appears to be of primary importance is the chemical composition of the cuticle and its structure and continuity.

The cuticular surface coating of aerial plant organs is not a continuous, sealed system; rather it is interspersed with localized areas of specialized cells that have a much lower resistance to gas exchange. Stomates and lenticels are the most important of these specialized gas exchange sites, however, other areas, both natural (e.g., hydathodes, stem scars) and accidental (e.g., surface punctures), may also be important.

2.2.3. Diffusion Path

For a gas to diffuse into the interior cells of a harvested product or outward from there, the individual molecules must pass along a diffusion path, from areas of high concentration to areas of low. Different sites along the pathway present varying levels of resistance to the movement of the gas molecules. To understand the net resistance to gas movement, each general step in the overall path and the level of its resistance needs to be determined. In the following example, the movement of oxygen molecules from the exterior of the product to the mitochondria of interior cells is followed (Figure 8.2). The reverse of the oxygen diffusion pathway is found for carbon dioxide molecules, originating in the mitochondria and moving outward. Since oxygen and carbon dioxide molecules have different chemical properties, the level of resistance presented by each step in the diffusion path differs.

a. Boundary layer resistance

Gas molecules moving from the exterior to the interior of a product must first pass through a boundary layer of air surrounding the product. The boundary layer is a localized area where the air is static and movement of individual molecules occurs by diffusion forces alone. Exterior to the boundary layer, air turbulence greatly facilitates movement.

   The flux of gas molecules across the boundary layer depends upon the diffusion coefficient (properties) of the gas in question, the difference in concentration (driving force) of the molecules across the boundary layer and the thickness (distance) of the layer. Boundary layers represent important barriers to the movement of many gases (e.g., water, carbon dioxide, oxygen). The thickness of the boundary layer depends upon the wind speed, size and shape of the product and surface characteristics (e.g., presence of trichomes). Air movement in storage rooms decreases the thickness of the boundary layer and thereby increases the rate at which water vapor is lost from succulent products, accelerating weight and quality losses. At the same time, it increases the rate at which oxygen molecules diffuse into the product. Storage room air movement, therefore, has both positive and negative effects. The optimum air movement rate is dependent upon the product and storage conditions.

   The boundary layer of a stored product is found adjacent to its surface; however, other boundary layers may be present and important. For example, if the product is placed inside a shipping container, the container's surface and its boundary layer represent two additional resistances that the gas molecules must overcome before approaching the product.

b. Product surface resistance

The surface of harvested products is composed of cuticle or periderm, which is generally interspersed with stomates or lenticels. Both the cuticle and periderm represent major barriers to the movement of gas and solute molecules, the presence of which allows plants to survive in an atmospheric environment without uncontrolled water loss.[32] Resistance to the diffusion of water vapor from leaves is 5 to 300 times greater for the cuticle than for open stomates. Because of this, the majority of the exchange of gases between the interior and exterior of a product happens through these specialized openings when present. Other important exchange sites are stem scars, the cut end of peduncles, surface punctures, growth cracks, and other discon-

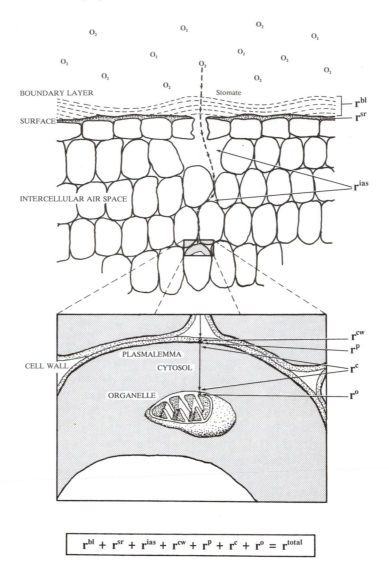

$$r^{bl} + r^{sr} + r^{ias} + r^{cw} + r^{p} + r^{c} + r^{o} = r^{total}$$

Figure 8.2. Schematic representation of the pathway of and resistances (r^{bl}—boundary layer resistance; r^{sr}—surface resistance; r^{ias}—intercellular air space resistance; r^{cw}—cell wall resistance; r^{p}—plasmalemma resistance; r^{c}—cytosol resistance; r^{o}—organelle resistance; r^{total}—total resistance) encountered by gases diffusing into or out of plant tissues. As with Ohm's Law governing the flow of electricity (resistance = voltage/current), resistances in series can be added to give the total resistance.

tinuities of the outer integument.[34] For example, most of the oxygen moves through the peduncle or peduncle scar of pepper fruits,[7] and of the gas moving through the cuticle, carbon dioxide transverses the barrier about 2 to 3 × faster than oxygen.

Unlike the lenticels, the aperture of a stomate can, under appropriate conditions, be readily altered. In response to low internal carbon dioxide, potassium is transported into the two kidney shaped guard cells of the stomata, lowering their water potential. To compensate, water moves into the guard cells, causing them to bow outward forming an elliptical pore through which gases may readily pass. Thus the resistance to gas diffusion presented by the stomates can change from extremely high (when closed) to relatively low (when open), depending upon the conditions of the product and its environment. Both decapitation (e.g., harvest of lettuce) and dark storage cause the stomates to close, greatly decreasing the exchange of gases.

While the porosity of lenticels (resistance) is not subject to the type of rapid and reversible changes seen with stomates, lenticel resistance is not necessarily static. Generally there are significant changes in resistance due to partial occlusion of the pores with time after harvest. Thus the diffusion resistance of gases into or out of products such as potato tubers tends to increase.[4] Likewise, the permeability of the cuticle is also not static but changes with response to chemical and physical effects of its surroundings.[46]

In addition to the resistance presented by individual stomata or lenticels, the size, density, distribution and condition of these surface openings is extremely important. The leaves of most dicots have a greater density of stomates on the lower surfaces, with a frequency of 40 to 300 per mm^2. When open, the combined apertures represent only 0.2 to 2% of the leaf surface but account for the bulk of the gas exchanged.

Where stomates are few or absent, exchange resistance is dominated by the cuticle (or periderm), chiefly by the physical and chemical properties of the lipids present. Solute or gas molecules must dissolve into the waxy component; thus the permeability of the cuticular waxes is proportional to the solubility and mobility of the solute in the wax barrier and inversely proportional to the path length.[14] As the solubility of the compound decreases or its size increases, permeability declines. Therefore, compounds with a low solubility in the wax component, such as water, transverse the barrier slowly, with the bulk of the diffusion as single molecules across this lipophilic barrier. Temperature alters the permeability of water through the cuticle, increasing approximately two-fold between 15 and 35°C; permeability also varies substantially among species (~2.5 orders of magnitude).[46]

c. Intercellular air space resistance

Once entering the intercellular space, gas molecules must overcome the resistances to diffusion found therein. Resistance encountered varies widely among individual types of products and their physiological condition. If the intercellular resistance is high, then a significant drop in concentration over a given distance within the tissue would be expected. These gradients in concentration are accentuated when the gas is being utilized (e.g., oxygen) or produced (e.g., carbon dioxide) by the cells along the diffusion path. Burton[9] estimates this concentration gradient for potato tubers, which have only about 1% air space. Tubers respiring at a relatively normal rate (2 mL O_2 kg^{-1} · hr^{-1}) would have a concentration gradient of only about 0.1% O_2 · cm^{-1}. In products that have much higher intercellular gas volumes (e.g., apple fruit), the gradients should be negligible. Therefore, under normal conditions, the resistance to diffusion of a gas presented by the intercellular air channels is small. However, if the intercellular air channels lack sufficient continuity due to their structure or the leakage of cellular fluids, cells only a short distance from an air space (i.e., 5 mm) may be anaerobic.

d. Cell wall resistance

Oxygen molecules diffusing from the exterior of the product to the interior must now move from the intercellular air spaces and transverse the water-filled interstices of the cell wall. This movement is accomplished by the oxygen molecules first dissolving in the water and then diffusing in the aqueous medium. Critical parameters include the rate of partitioning of oxygen into the aqueous phase, the diffusion resistance presented by the medium and the length of the path the molecules must transverse. A number of physical and chemical factors can affect each of these parameters. For example, the pH of the water held in the cell wall lattice can have a significant effect. This effect is especially true for CO_2 which can be found in solution as CO_2, H_2CO_3 or HCO_3^-, the respective concentrations being pH dependent (especially HCO_3^-). In general, cell wall resistance is considered to be quite low (i.e., ≈ 100 s $\cdot$ m^{-1}).

e. Plasmalemma resistance

The plasmalemma is the outer membrane surrounding the cell, and its properties of semiper-meability largely control the entry and exit of molecules from the cell. Since neighboring cells are connected in a continuous cell-to-cell membrane network (plasmodesmata), molecules do not necessarily need to exit through the plasmalemma to move to a neighboring cell (as will be seen, the plasmodesmata are a major diffusion pathway for volatile molecules that are found predominantly in a liquid state).

The potential for a gas in solution to diffuse through the plasmalemma depends upon the properties of the gas in question (size, polarity, etc.), the conditions of the membrane and the concentration gradient across the membrane. Molecules may move by simple diffusion across the membrane or may be actively transported to the opposite side. The resistance presented by the plasmalemma for the diffusion of dissolved gases such as carbon dioxide and oxygen is considered small to moderate (e.g., $CO_2 \approx 500$ s $\cdot$ m^{-1}).

f. Cytosol resistance

Once within the interior of the cell, the dissolved oxygen molecules must diffuse to the mito-chondria or other sites of utilization. Diffusion is not a major obstacle since the diffusion re-sistance of the cytosol is quite small (e.g., ≈ 10 s $\cdot$ m^{-1} for CO_2), although greater than in water, and the mitochondria and other organelles are found, because of the central vacuole, relatively close to the periphery of the cell.

g. Organelle resistance

Once transversing the cytosol, the oxygen molecules must pass through the outer membranes of the mitochondria (or other organelle) and subsequently the internal resistance between the interior and the site of utilization. Precise measurements of these resistances are not available; however, estimates for the diffusion of carbon dioxide into the chloroplast and across its stroma are in the range of 400–500 s $\cdot$ m^{-1}. If active transport (energy driven) or facilitated dif-fusion across the membrane occurred, the resistance would be substantially lower.

Carbon dioxide or ethylene molecules produced by the cells of a harvested product must follow the reverse of the oxygen diffusion pathway. Therefore, carbon dioxide molecules must diffuse across the organelle, its outer membranes, the cytosol, the plasmalemma, the cell wall, through the internal air channels and across the product surface and boundary layer.

2.2.4. Combined Resistances

The total or net resistance to the movement of a gas is the sum of resistances in a series (the same as Ohm's law governing resistance to the flow of electricity). The molecules move se-quentially through each resistance when they are in series. Therefore:

$$r^{total} = r^{bl} + r^{sr} + r^{ias} + r^{cw} + r^p + r^c + r^o$$

where r^{total} represents the total resistance, r^{bl} the resistance of the boundary layer, r^{sr} the surface resistance (cuticle/periderm and stomata/lenticel), r^{ias} the internal air space resistance, r^{cw} the cell wall resistance, r^p the plasmalemma resistance, r^c the cytosol resistance and r^o the resistance of the organelle. When the resistance is composed of two or more resistances in parallel (e.g., the surface resistance of the periderm + lenticel), the gas molecules can move *via* both paths. The reciprocal of the total resistance of a group of resistances in parallel is the sum of the reciprocals of the individual resistances. In this case surface resistance (r^{sr}) equals:

$$r^{sr} = \frac{r^{per}\, r^l}{r^{per} + r^l} \qquad \text{eq. 8.8}$$

where r^{per} is the resistance of the periderm and r^l that of the lenticels.

While the total resistance (r^{total}) gives us a good estimate of the ease or difficulty of getting a gas into or out of a product, it represents only part of the story. The flux of a gas is a function of force/resistance. High resistances may in some cases be compensated for by increasing the concentration gradient (driving force) or conversely, by not allowing the concentration gradient to be diminished (e.g., adequate aeration of the product).

In many cases during the postharvest period, the total resistance to diffusion of a gas is of primary interest. An estimate can be obtained by placing the product in an environment containing a known concentration of a gas, preferably one that is not used or produced by the tissue. With time the concentration of this gas within the product will equilibrate with the surrounding environment. At this time, the product is transferred to a second chamber containing air. Since the internal and external concentration of the gas (concentration gradient) and its diffusion coefficient (diffusion characteristics of the gas in question, obtained from chemical tables) are known, the rate of diffusion of the gas from the interior of the product into the surrounding air can be used to estimate the total resistance to diffusion.

3. MOVEMENT OF SOLVENT AND SOLUTES

3.1. The Aqueous Environment

Life on the planet earth occurs in an aqueous medium with water representing the single most abundant component of actively metabolizing cells. Postharvest products range widely in their water content. Very young leaves of lettuce are as much as 95% water while many dried seeds are only 5–10%. Products such as seeds that are low in water tend to display a correspondingly low level of metabolic activity. An increase in water concentration is a prerequisite for renewed activity and growth.

The physical and chemical properties of water make it a suitable chemical for a tremendous diversity of functions with the plant.

- The small molecular size, its polar nature and high dielectric constant make water an excellent solvent.
- Water functions as the transport solution, the medium in which solutes are moved throughout harvested products.
- Water represents the medium in which many of the reactions within the plant occur.
- It is a chemical reactant or product in a wide range of biochemical reactions, e.g. photosynthesis, ATP formation and respiration.
- The incompressibility of water allows it to be used for turgidity of the cells, providing support and cell elongation during growth.

- The high heat of vaporization and thermal conductivity of water make water the central temperature regulatory compound.
- Extensive hydrogen bonding between water molecules and with other polar compounds affects water's cohesive and adhesive forces that influence phloem and xylem transport, capillarity and contact angles.*
- The transport of water between cells causes a physical movement of certain plant parts (e.g., stomatal guard cells, turgor driven flower opening and closure).

While all of the aforementioned functions are essential for plant life, the first two (water's role as a solvent and as a transport medium) are of particular importance in the redistribution of compounds within harvested products and the loss of specific compounds from these products. Most chemicals will dissolve to some extent in water, and generally solubility increases with increasing temperature (exceptions would be salts, which are relatively unaffected by temperature; the solubility of calcium salts of some acids actually decreases with increasing temperature). The dissolved substance or solute can be a gas, liquid or solid.[†] When more than one solute is present in a solution, each solute generally behaves independently of the others.

Water is not the only solvent within plants. Lipids and certain other molecules also act as solvents. A number of cellular compounds are lipid soluble (e.g., chlorophyll, carotenoids). Although very slightly soluble in water, these molecules partition readily into lipids. A primary distinction between water and other solvents, however, is that water acts as a transport medium, an avenue for movement of solutes between cells, neighboring tissues and organs. Movement of lipid-soluble molecules, in contrast, tends to occur over only minute distances, e.g., within the liquid membrane component of a chloroplast. When lipid-soluble compounds need to be transported, they are generally first broken down into water-soluble precursors or subunits. Transport then occurs in an aqueous phase, with eventual reassembly upon reaching the destination.

Solutes range widely in their solubility in various solvents. For example, sugars are readily soluble in water while chlorophyll is relatively insoluble. A unit volume of solvent can dissolve only a fixed amount of solute. When the upper limit is reached, the solution is said to be saturated. When the concentration exceeds this upper limit, solid solutes will crystallize or precipitate out of the solution, while liquid solutes will form droplets of the compound.

Gases, like liquid and solid solutes, are also soluble in water, and their degree of solubility likewise varies widely. Gases such as oxygen and nitrogen are only slightly soluble in water while carbon dioxide is very soluble (approximately 100 times more soluble than oxygen). The solubility of a gas is inversely proportional to temperature. Therefore as the temperature increases, the solubility of a gas decreases. Solubility is directly proportional, however, to the partial pressure of the gas in the atmosphere over the solvent. As the concentration of carbon dioxide in the atmosphere increases, the concentration in solution increases correspondingly. The very high solubility of carbon dioxide in water is due in part to its reaction with water molecules to form a carbonate.

$$CO_2 + H_2O \rightleftharpoons H_2CO_3 \rightleftharpoons H^+ + HCO_3^-$$

---

*The oxygen atom in a water molecule is strongly electronegative, which causes it to draw electrons away from the adjacent hydrogen atoms in the molecule. This results in an asymmetric electron distribution, making water a polar molecule. As a consequence, the positively charged hydrogens are electrostatically attracted to the negatively charged oxygens of neighboring water molecules or to the negative charge on other polar molecules, leading to hydrogen bonding.
[†]Gases may also be dissolved in a gas.

The concentration of carbon dioxide in the aqueous phase is also significantly affected by pH (increasing pH → increased CO_2). While the partitioning coefficient of carbon dioxide in water is not substantially altered by pH, the equilibrium concentration of HCO_3^- is, elevating the amount of carbon dioxide that is partitioned into a given amount of water. Increasing the concentration of carbon dioxide in solution has a secondary effect; it alters the pH of the solution which, if the alteration is great enough, may alter the metabolism of the cells. The high solubility of carbon dioxide in water allows it to diffuse readily, facilitating its movement within the aqueous medium of a harvested product.

3.2. Potential Pathways for Movement in a Liquid Phase

Molecules in the liquid phase move toward areas of lower chemical potential (concentration) in harvested plant material *via* three general avenues. a) In the **apoplast**, molecules flow exterior to the protoplasts, within the cell wall and intercellular spaces. While both water and solutes are found in the apoplast, it is a especially important in the movement of water and in phloem loading of sucrose. b) The cytoplasm of neighboring cells is interconnected, forming a virtually continuous system, the **symplast**, allowing movement throughout the interior of a product without transversing the plasma membrane of each cell. Water and dissolved solutes move from cell to cell through the symplast system in response to energy gradients. For solute molecules to move from the symplast to the apoplast system, the solute must first transverse the plasma membrane of the cell; however, its semipermeability restricts the outward movement of many molecules. c) A third means is **transcellular** transport, where the molecules move across the plasma membrane of each cell in its pathway. The movement of water, for example, across the membrane is facilitated by specialized openings (**water channels** or **aquaporins**).[57] Other specialized sites (solute pores, ion channels and transporters) facilitate the transmembrane movement of other molecules.[39] Since symplast and transcellular pathways can not be separated experimentally, they collectively make up what is called "cell to cell" or "protoplasmic" flow.[58]

Water and solutes may diffuse in response to energy gradients in either the symplast, transcellular or apoplast systems. For molecules that can transverse the plasmalemma, each route may be operative; quantitatively the most important will be the path that represents the least resistance. Most of the movement of water is thought to occur within the apoplast. For molecules that do not readily transverse the plasma membrane, the symplast represents the dominant route.

Liquid solutes moving through the symplast system must transverse the plasma membrane upon reaching the surface if evaporation (loss from the product) is to occur. In some cases, specialized surface cells have ectodesmata, which allow otherwise nonpermeable molecules to move from the cytosol to the exterior. In others, the plasma membranes, due to differences in composition, may be more readily transversed by certain molecules than the plasma membrane of interior cells. Once exterior to the plasma membranes of the surface cells, volatilization of liquids or crystallization of solids may occur.

The rate at which liquid molecules can escape into the atmosphere is in part a function of how readily they or their precursors can be moved to the surface. Products with large surface areas and relatively short diffusion paths (e.g., leaves) tend to lose liquids more rapidly than bulky products. For example, spinach leaves lose water approximately 200 times faster than potato tubers under similar conditions.[26] Likewise, the physical and chemical characteristics of the product surface can have a pronounced effect on the rate of escape.

3.3. Factors Affecting the Diffusion of Solvent and Solutes

3.3.1. Forces Operative

Like gases, solvents and solutes move in the plant in response to concentration differences; however, with liquids a number of other factors are also important. Therefore, the movement of solvent and solutes is described in response to gradients in free energy or chemical potential, not just concentration, which represents only one component of free energy or chemical potential. Both the tendency for diffusion and the direction in which it occurs depend upon a gradient in chemical potential. For example, in detached fleshy products, significant differences in water concentration due to evaporation may develop, establishing a concentration or chemical potential gradient. Water moves in response to this gradient from areas of high concentration to areas of low concentration.

The presence of a solute (e.g., sugars, mineral ions) decreases the chemical potential of the solvent molecules in relation to the mole fraction present (number of solute to solvent molecules). If an area of high solute concentration is separated by a semipermeable membrane from an area of low, the solvent potential is lowest on the side of the membrane that is high in solute molecules (Figure 8.3). If the membrane is permeable to the solvent, the solvent moves across the membrane from areas of high potential (high solvent concentration) to areas of low potential (high solute concentration).

Matric effects are also important in modulating the movement of a solvent. Many of the molecules within a plant have a high affinity for solvents, especially water. Proteins, hemicellulose, pectins, starch and other molecules are hydrophilic (water-loving) and possess negative charges on their surfaces that attract the positive side of the polar water molecules. Materials that bind solvent molecules are called a matrix. This affinity to the matrix restrains the movement of the solvent molecules.

Temperature and pressure also influence the movement of liquids. Elevated temperature increases the free energy of a system, increasing movement. Diffusion will occur from points of high temperature to low. Since evaporation decreases the temperature of the surface, temperature gradients can be present within a product and affect diffusion. Likewise, increases in pressure increase the level of free energy that can also enhance solvent movement.

3.3.2. Resistance to the Movement of Solvent and Solutes

Energy from various sources provides the force required to drive the movement of solvent and solutes. The flux or movement of solute ions or molecules, however, is modulated by more than just these driving forces. Flux is equal to force divided by resistance. Both within the cell and the apoplast are various resistances that impede, to varying degrees, the free movement of ions and molecules. The two most important resistances to free diffusion are imposed by membranes and the cell wall.

Individual protoplasts within harvested products are surrounded by a plasma membrane that controls the entry and exit of ions and molecules. Various solutes display marked differences in their ability to transverse this semipermeable barrier.[43] For most solute molecules, the rate-limiting step controlling movement into or out of the cell is diffusion through this membrane. Even when a solute can diffuse readily through the membrane, the rate of diffusion is greatly restricted in comparison with its diffusion within the cytosol or extracellular water. The plasma membrane, therefore, represents a major barrier or resistance to the free movement of most molecules. Water molecules, however, while encountering some resistance, move quite readily through the membrane.

MEMBRANE TRANSPORT

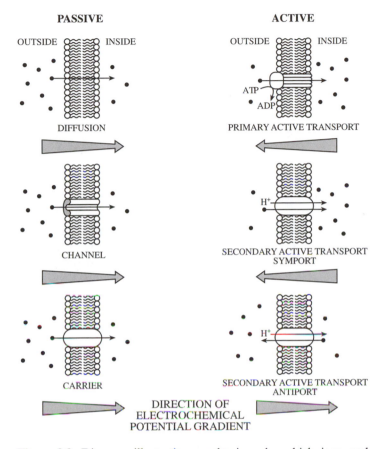

Figure 8.3. Diagram illustrating mechanisms by which ions and solutes can be moved across semipermeable membranes. Passive transport *via* simple diffusion, channels and carriers involves moving energetically downhill, from areas of high electrochemical potential to low. Active transport utilizes "pumps" to move ions and solutes energetically uphill (against the concentration gradient) and requires the input of energy such as ATP with primary active transport. Secondary active transport, a form of cotransport, is driven by coupling the passive flow of ions such as H^+ or Na^+ with the uphill flow of inorganic ions and solutes. With symporters, both move in the same direction; while with antiporters, each moves in the opposite direction.

In addition to the plasma membrane, most other subcellular organelles are surrounded by membranes. Mitochondria, for example, are enclosed within a double membrane. Dissolved oxygen molecules must diffuse from within the water of the apoplast, through the plasma membrane, across the cytosol and finally transverse the mitochondrial membrane system to participate in respiration. Both the plasma membrane and the mitochondrial membranes represent significant resistances when compared with the diffusion of oxygen within the cytosol.

The tonoplast membrane surrounding the vacuole is also of critical importance to cellular metabolism in that it controls the movement of substances into and out of the vacuole. Many of the compounds found within the vacuole would impair or terminate the normal functioning of the cell if they were allowed to enter the cytosol.

Membrane permeability is a measure of the ease with which a substance can move across a membrane, and individual substances differ in the ease at which they can do so. What imparts the semipermeable nature to membranes? Lipids represent the largest single component of membranes; thus some molecules that passively transverse the membrane must first dissolve in this lipid layer before diffusing across it. Molecules that are lipid soluble, therefore, partition (move) into the membrane much more readily than those that are more water soluble. As a molecule increases in size or polarity,* its membrane permeability decreases. Since the chemical composition varies between membranes (e.g., tonoplast vs mitochondrial), the permeability of different classes of compounds also varies. Molecules that pass readily through one membrane may be greatly restricted by another.

The ions and molecules that pass through the barrier imposed by a membrane do so by one of two general means, passive or active transport (Figure 8.3). With passive transport, molecules diffusing across the membrane move energetically downhill, driven by differences in chemical potential between the respective sides of the membrane. Movement is from the side with the higher chemical potential to the lower, with the rate of transport being in part a function of the magnitude of the difference in chemical potential across the membrane. If the molecule carries a positive or negative electrical charge, the charge becomes a part of the total potential (electrochemical potential) which drives diffusion.

In many cases, the transport of ions or molecules across a membrane goes against a chemical potential gradient or at a rate that cannot be accounted for by the laws of diffusion or energy potential. Here transport (termed active transport) takes place in an energetically uphill direction (against the chemical potential gradient) and cannot be achieved without the input of free energy (Figure 8.3).

For ions and other polar molecules, transport is facilitated by three types of membrane-transversing structures (channels, carriers and pumps, Figure 8.3),[56] each exhibiting a high degree of specificity in the solutes which are allowed to cross. Within each type are a diverse cross-section of proteins that facilitate the transport, generally of a small number of related substances.

Channels are proteins that span the membrane and function as selective pores allowing certain substances such as water and ions through.[5,33,48] Water channels are called aquaporins.[36,52] Transport is by passive diffusion that is extremely rapid when the pore is open. The pore in the channel can close in response to changing conditions or environmental signals, providing an additional level of control.

Carrier proteins, like channels, allow the movement of substances across the membrane energetically downhill.[59] Unlike channels that function *via* simple diffusion, carriers utilize facilitated diffusion, where the carrier protein binds the transported substance on one side of the membrane. Binding causes a change in the shape of the protein that shifts the substances to the other side of the membrane; upon exposure to the solution there, it is released. Carrier transport is substantially slower than movement through channels but provides a much higher degree of specificity in regard to what moves across the membrane, and it can facilitate the movement of substances that are not amendable to movement through channels.

Pumps are a form of carrier protein that requires the input of energy to drive the transport of the substance from one side of the membrane to the other. Transport, therefore, is active rather than passive, allowing the movement of substances energetically uphill, against

---

*Having a permanent electric dipole moment.

their chemical potential gradient. Pumps are extremely important in that they allow concentrating molecules. There are a cross-section of types of pumps and, within each, a variety of proteins with a narrow range of substances each can transport. The energy driving active transport can be from ATP[35] or *via* the uphill transport of one solute (or ion) coupled to the downhill transport of another[23] (Figure 8.3). The latter is a form of cotransport (called secondary active transport) which is driven by the electrochemical potential created by a proton gradient across the membrane.

Active transport is essential in plants; it allows the concentrating of certain substances and the exclusion of others. The loading of sucrose molecules from the apoplast of leaf mesophyll cells into the phloem is an example of an active transport process. If the energy input is blocked, transport ceases.

3.3.3. Factors Affecting the Rate of Diffusion

Knowing the factors that affect the chemical potential of a liquid, what then controls the rate of diffusion? The steeper the chemical potential gradient, the more rapidly diffusion will occur (assuming all other factors are held constant). Secondly, the level of permeability of a membrane to a specific compound significantly alters the rate of diffusion of the molecules. Some compounds move readily through solvents and/or membranes allowing them to respond more readily to a given gradient in chemical potential. And finally, temperature affects the rate of diffusion. High temperatures increase the average velocity of the solute molecules, increasing their rate of diffusion.

3.4. Volatilization of Solvent and Solute Molecules

The volatilization of molecules from the liquid phase occurs when the energy level of individual molecules is high enough to overcome the attractive forces of neighboring molecules. Experimentally, these forces are generally measured in pure liquids. In the plant, however, its liquid phase is exposed to a tremendous cross-section of other molecules (both solid and liquid) that can exert an influence over the liquid in question. The collective effect of these neighboring molecules modulates the general tendency of the compound in question to volatilize. The presence of solutes and solids can greatly affect the rate at which water evaporates (volatilization at a temperature below the liquid's boiling point). Adding sucrose to water increases the amount of energy required for a given amount of water to evaporate. Various liquids differ greatly in their potential for volatilization, and this difference is the basis of the separation of liquids by distillation. Likewise, temperature and pressure alter the rate of volatilization. Evaporation occurs more readily with each increment in elevation of temperature.

Volatilization, in turn, cools the liquid by the absorption of energy that occurs during the change of state (i.e., liquid → gas). The evaporation of 1 g of water from a 1000 g reservoir lowers the temperature of the remaining liquid by 0.59°C. Likewise, a boiling liquid does not increase in temperature above its boiling point, regardless of the level of energy input, due to the cooling effect of volatilization.

Evaporation also increases as the atmospheric pressure decreases. The boiling point of water decreases from 100°C at sea level to only 95°C at an elevation of one mile (1.6 km), a common altitude in many mountainous regions of the world. In hypobaric storage systems, the decreased pressure can greatly increase the loss of volatile liquids from the product. If these represent critical components of quality (e.g., volatile flavor compounds), quality can be significantly diminished.* With water, losses can be minimized by keeping the storage room atmosphere as near saturated as possible. Saturation of the air is not presently possible for most other volatile liquids, especially when some air exchange is required in the storage environment.

Evaporation differs from boiling in that evaporation is a surface phenomenon. The change in state occurs only at the surface of the liquid. With boiling, volatilization takes place within the liquid as well as at the surface. Therefore, the surface area of a liquid is an extremely important physical characteristic affecting evaporation. If other factors are held constant, evaporation increases in relation to surface area. Therefore, plant products with high surface-to-volume ratios will lose water or other volatile liquids at a faster rate than those that have a lower surface-to-volume ratio.

The net escape of liquid molecules into the atmosphere is also affected by the relative concentration of these molecules already in the gas phase. In a closed system, as the concentration in the gas phase increases, more of the randomly moving molecules return by chance to the liquid phase. When the number of molecules leaving the surface equals those returning, the atmosphere is said to be saturated (i.e., no net loss or gain in either phase).

Molecules that have escaped their liquid surface and exist in a vapor state, exert a pressure, called the **vapor pressure**. The vapor pressure is proportional at a given temperature to the net number of molecules escaping into the vapor phase. Therefore, increasing the chemical potential of the liquid (e.g., raising the temperature) increases the vapor pressure of the molecules in the vapor state. The addition of dissolved solutes, on the other hand, decreases the liquid's chemical potential, decreasing the vapor pressure.

The rate of evaporation of a liquid such as water from a plant product is proportional to the difference in the vapor pressure for water within the product and the pressure of the surrounding atmosphere. This differential is called the **vapor pressure deficit**, and as it increases, the rate of evaporation increases.

3.5. Site of Evaporation

Evaporation is a surface phenomenon, and plants have two primary surfaces from which it occurs. Surrounding each interior cell is an inner connecting series of air spaces. A portion of the liquids found exterior to the plasmalemma in the apoplast volatilize within the air space, eventually approaching an equilibrium between the liquid and gaseous phases. In the gas phase, these molecules diffuse through air channels down a concentration gradient toward the surface of the product, eventually escaping to the surrounding atmosphere. Evaporation may also occur at the surface of the product. The predominant site for water evaporation from intact plants is at this surface-exterior interface. For water, most of the evaporation actually occurs on the inner sides of stomatal guard cells and adjacent subsidiary cells. Water also evaporates from nonstomatal areas, for example, lenticels, cuts or breaks in the surface continuum.[1] The amount of cuticular and peridermal transpiration varies widely among species. Generally cuticular evaporation represents a relatively small component of the total water lost from a product.

Which of the two sites, cell surface or product surface, is the most important in the evaporation of liquids and their loss as volatiles from the plant? Based on an analogy with the movement of oxygen or carbon dioxide, one can imagine molecules of a liquid, such as water,

---

*The subsequent synthesis of volatiles by the tissue after removal from hypobaric storage may also be seriously impaired.

present in the wall matrix of interior cells, evaporating and diffusing toward the exterior of the product, where the concentration is lower. In this case, movement of the compound to the product surface is predominantly as a gas rather than a liquid. However, although some movement does occur in the vapor state in succulent products, water moves to the surface of the product predominantly in the liquid state.

Due to the presence of water in the wall matrix, the intercellular air spaces are saturated with water vapor. Saturated air not only occurs around the interior cells but surrounds the cells toward the more exterior regions of the product. The high concentration of molecules in the vapor state results in an extremely small drop in vapor state concentration (the diffusion gradient or driving force) between the interior and exterior cells. Since the rate of movement is a function of the steepness of the concentration gradient, diffusion is slow. Most of the movement of liquids between the interior and exterior of the product, therefore, occurs in the liquid state. Exceptions may occur when there is a very localized site of synthesis of a specific compound, e.g., only in the cells adjacent to the seeds of a fleshy fruit. Under such a situation, it is conceivable that significant gradients also occur in the vapor state between the site of synthesis and the exterior.

4. THE EXCHANGE OF WATER BETWEEN PRODUCT AND ENVIRONMENT

Due to the central role of water in the postharvest biology of plant products, the movement of water between the product and its environment can be of tremendous importance. Depending upon the product and the environment in which it is held, movement of water can proceed from the product to the surrounding environment, proceed from the environment into the product or, when in equilibrium, exhibit no net change. Water movement is always toward establishing an equilibrium between the product and its environment, with the individual water molecules diffusing from areas of high chemical potential toward areas of low chemical potential.

4.1. Environmental Factors Affecting Water Exchange

During storage our primary concern is to minimize the exchange of water between the product and its environment. With some products, however, prior to storage or during the initial phase of storage, conditions are created to enhance the exchange of water. For example, many grains and pulse crops are exposed to environments that are conducive for drying, i.e., the net loss of water from the product. In certain other products, it may be desirable to replace the water lost during handling and storage. At the retail level, lettuce is often briefly (10–15 min) submerged in water to rehydrate wilting leaves.

The net flux of water into or out of a detached harvested product is determined by the magnitude of the forces driving movement and the composite of resistances to movement. Environmental factors such as humidity, temperature, pressure and air movement can have a direct effect on the forces driving movement.

4.1.1. Humidity

The amount of water vapor in the air surrounding a plant product has a pronounced effect on the movement of water into or out of the tissue. Typically humidity is measured using dry and wet bulb thermometer readings from a psychrometer and a psychrometric chart (Figure 8.4);

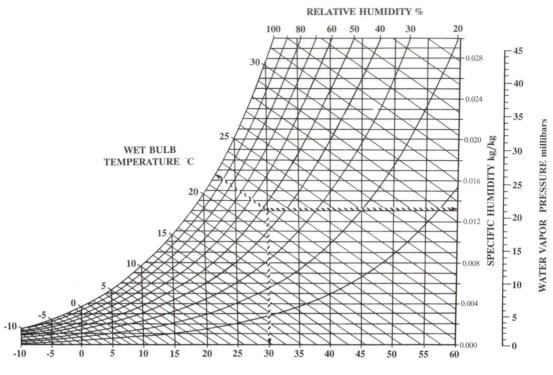

Figure 8.4. Psychrometric charts allow determination of the thermal and moisture characteristics of the atmosphere using any two parameters (e.g. wet bulb temperature, dry bulb temperature, percent relative humidity, dew point temperature, specific or absolute humidity or water vapor pressure). For example, if the dry bulb temperature is 30°C and the wet bulb temperature is 22°C, the point on the chart where the 45 degree angled wet bulb line intercepts the vertical dry bulb temperature line is the percent relative humidity (curved line) which in this case is 48%. Moving left on the horizontal line from the intercept is the dew point temperature (18°C). The absolute (specific) humidity and vapor pressure can be read by moving horizontally to the right based on a barometric pressure of 1013.25 mbar.

however, newer electronic sensors give a direct reading. The amount of water vapor present can be expressed in a number of ways: percent relative humidity, absolute humidity, vapor pressure and dew point.

The most commonly used and misused expression of the moisture content of the air is the **percent relative humidity**, a ratio of the quantity of water vapor present and the maximum amount possible at that temperature and pressure. Both temperature and pressure have a pronounced effect on the amount of water vapor the environment will hold, and seldom during the postharvest period are constant temperature conditions found. For example, there is a small but consistent fluctuation in air temperature in refrigerated storage rooms due to thermostats cycling on and off the flow of refrigerant to the coils. In unrefrigerated grain storage bins, temperature also varies with location in the bin. These differences in temperature have a significant effect upon the amount of water vapor present in the air without necessarily changing the percent relative humidity. For example, air at 40% RH, 20°C contains approximately 9 g of water vapor · kg⁻¹ of dry air while at 30°C and the same percent relative humidity, the air contains 11 g · kg⁻¹. Although the percent relative humidity is identical, the concentration of water vapor differs greatly.

A more precise measure of humidity is the **absolute humidity**.* Absolute humidity, the weight of water in a given weight of dry air (g · kg^{-1}), is independent of temperature and pressure. Therefore two storage environments at the same percent relative humidity but at different temperatures will have distinctly different absolute humidities. Water vapor pressure represents an alternate measure of the amount of water vapor in the atmosphere. Since water vapor is a gas, like oxygen and nitrogen, it makes up part of the total atmosphere. Each of the gasses in the atmosphere exerts a pressure, part of the total pressure. **Water vapor pressure** is a measure of the pressure exerted by water vapor in the atmosphere. At a given barometric pressure there is a direct relationship between the vapor pressure of water and the absolute humidity.

A useful postharvest measure is the difference in the water vapor pressure between two locations, the **water vapor pressure deficit**. Since concern is for the movement of water into or out of stored products, the difference in water vapor pressure between the interior of the product and its surrounding environment gives an estimate of how rapidly water will move between the product and its environment. The actual rate is controlled by more than just the vapor pressure deficit, however, since water exchange is also affected by the characteristics of the product itself (e.g., resistance to diffusion—waxy cuticle, amount of water present—chemical potential of the water).

In succulent plant products, the internal air spaces are considered to be saturated with water vapor.† Because of this, it was possible to develop a simple nomogram that allows determining the water vapor pressure deficit using the temperature and percent relative humidity of the environment (Figure 8.5).[62] The higher the vapor pressure deficit (i.e., the greater the gradient) the more rapid the loss of moisture from the product. Comparison of the water vapor pressure deficit between different storage conditions (temperature and/or percent relative humidity) allows ascertaining under which set of conditions water will be most readily lost from a product. For example, a product stored in a room with a vapor pressure deficit of 5.0 mbar will lose water twice as fast as the same product stored under conditions where the vapor pressure deficit is only 2.5 mbar.‡

The **dew point** is the temperature at which the air is saturated with water vapor (100% humidity). If the temperature is lowered below the dew point temperature, condensation occurs, since the air can no longer hold as much water. The dew point temperature can be determined from the air temperature (dry bulb) and the relative humidity (or wet bulb temperature) using a psychrometric chart (Figure 8.4).

Dew point is a very important and often neglected parameter in postharvest handling and storage. If a commodity is brought to a temperature in refrigerated storage below the dew point of the air outside of the storage room, water will condense on the surface of the product upon removal from storage. For some crops this can have disastrous consequences. For example, a large shipment of Chinese chestnuts was being transported from Asia in a refrigerated vessel; however, the refrigeration system malfunctioned midway in the voyage. The crew knew that live plant material would begin to heat up in a sealed, insulated room due to its respiration, so to prevent this the doors of the refrigeration rooms were opened. Due to the higher temperature and humidity of the outside air and a product temperature below the dew point, water condensed on the surface of the chestnuts. The refrigeration system was repaired within 48 hours and the rooms reclosed; however, upon reaching port, the entire shipment was a solid mass of fungal mycelium. If the operators had been cognizant

---

*Also referred to as specific humidity and humidity ratio when the two weights are in the same units (e.g., kg · kg^{-1}).
†Water-saturated internal air spaces are seldom the case in drier products such as grains and pulses.
‡Comparisons cannot, however, be made between different types of products.

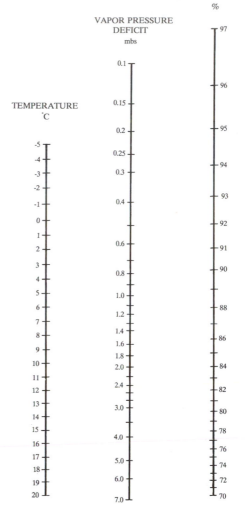

Figure 8.5. A nomogram can be used to determine the water vapor deficit between a fleshy product and its surrounding environment using the air temperature and relative humidity. A line is drawn from the air temperature (left) through the relative humidity (right) of the air (1 mbar = 0.1 kPa). The intercept on the vapor pressure deficit line gives the vapor pressure deficit between the product and its environment; the higher the vapor pressure deficit, the faster water is loss. Accurate use of the nomogram requires that the product and air to be at the same temperature and that the products internal gaseous relative humidity to be 100%. The latter is not a valid assumption with low moisture crops (e.g. grains).

of the importance of dew point, a several-hundred-thousand-dollar claim could have been avoided.

4.1.2. *Temperature*

Increasing the product temperature increases the free energy of the water molecules, which increases their movement and potential for exchange. From the heat given off during respiration, stored products normally have a slightly higher temperature than the surrounding atmosphere, which enhances water loss.

Temperature also affects the amount of moisture that can be held in the air surrounding the product. As the temperature decreases, the maximum amount of moisture that can be held by the air also decreases. Fluctuations in temperature can result in a much more rapid water loss from stored products than a constant temperature. Temperature deviations such as reaching the dew point have already been described; however, changes in temperature need not be large. For example, small changes due to the thermostat of a direct expansion refrigeration

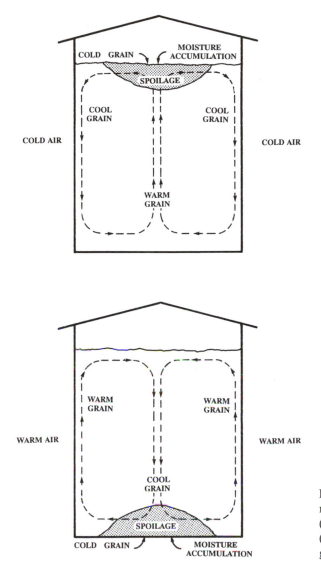

Figure 8.6. Convection currents of air causing a redistribution of moisture in stored grain when (A) the outside air is colder than the grain, and (B) when the outside air is warmer than the grain.

system cycling on and off can result in significant fluctuations in relative humidity and moisture loss from the product.[21]

In grain storage the movement of moisture from one area of a closed storage bin to another is a common problem. Although the grain moisture level may have been considered safe when the product was placed in storage, drying of one area due to a slightly elevated temperature results in the transfer of water to cooler grain, which causes wetting. Particularly noticeable in stores of greater than 70 m³, this situation can result in significant losses.[24]

When the air outside of the storage bin is cold, the grain along the walls is cooled (Figure 8.6). Air within the storage chamber moves by convection, cycling along the outer walls and returning through the center of the grain mass. Since the equilibrium between the moisture content of the grain and the air changes with temperature, warm air, which holds more moisture, rises up through the center. As the air reaches the cooler grain at the top, moisture begins to move from the air into the grain, causing an area of moisture accumulation and potential spoilage at the top. When the air outside is warm, the grain in the center is cooler and the air flow is reversed, causing water to accumulate at the base of the bin (Figure 8.6). This redistribution of moisture is a greater problem when: a) the moisture content of the grain is relatively high; b) the grain is placed in storage in warm rather than cold weather; c) there is a large difference between the temperature of the grain and the external atmosphere; and d) when tall storage structures are used.

4.1.3. Pressure

As with increasing temperature, elevated pressure increases the free energy of water molecules which in turn increases movement. However, changes in free energy due to increased pressure generally tend to be relatively insignificant in harvested products.

In contrast, reducing the pressure decreases the free energy of the water molecules. Reduced pressure can also greatly increase the concentration gradient between the liquid phase molecules of water within the tissue and the gaseous water molecules in the surrounding atmosphere. This elevation of the concentration gradient increases the net movement of water out of the tissue. The increased movement is due to the increased differential of the chemical potential of water found within the tissue (liquid phase) and that in the surrounding atmosphere exterior to the product (gas phase). The rate of evaporation is inversely proportional to pressure; for every 10% decrease in pressure there is a 10% increase in the rate of water loss.

Products placed under a partial vacuum will lose water until an equilibrium is established between the product and its environment. An equilibrium would occur in a small sealed chamber* held at a constant pressure and temperature. In practice, it is seldom this straightforward. Since postharvest products are alive, utilizing oxygen and producing carbon dioxide, fresh air must be continually introduced into the vacuum chamber. If the fresh air does not contain the correct concentration of water vapor (equilibrium concentration), a gradient will be established between the product and the environment. With succulent crops, desiccation occurs. Dryer crops (e.g., rice) either desiccate or rehydrate depending upon the direction of the gradient.

The effect of pressure on water loss from harvested products is also important at high altitudes. The cargo areas of airplanes are pressurized to an altitude of about 1500 m, where the air pressure is approximately 17.3 kPa less than at sea level. Therefore, there is approximately a 20% differential favoring accelerated water loss.

4.1.4. Air Movement

Air movement over the surface of a commodity decreases the thickness of the surface boundary layer, decreasing the boundary layer's resistance to exchange of water molecules between the product and its environment. With succulent products, this increases water loss, the increase at a specific air velocity being relative to the magnitude of the vapor pressure deficit between the product and its environment. In dry products if the water chemical potential gradient is in the opposite direction (i.e., higher in the air than the product), increasing air movement can accelerate rehydration.

The increase in rate of water loss from succulent products does not increase linearly with increasing air velocity. Rather there is a very sharp rise at relatively low air velocities, which increases at a decreasing rate with progressively higher velocities. The importance of air move-

---

*Relative to the volume of the product.

ment due to convection alone in grain stores has been described in the section of this chapter entitled *Temperature Effects.*

In a closed refrigerated storage environment, air movement also has an important influence upon the vapor pressure deficit between the product and its environment. Under refrigerated conditions, the higher the volume of air circulated per unit time over the cooling coils, the lower the differential in temperature between the return air and the delivery air. A low temperature differential decreases the amount of water removed by the refrigeration coils, decreasing the gradient in vapor pressure between the product and air. Thus, increasing the volume of air moved, up to a point, can help to maintain a high humidity within the chamber. Optimum air movement under refrigerated storage conditions is a compromise between attaining a low temperature differential between the coils and air and preventing excessive water loss due to reduction of the boundary layer around the product.

4.1.5. Light

Water loss after harvest can be significantly increased by light through its effect on stomatal aperture and/or indirectly by increasing the temperature of the product. Water losses generally increase with increasing light intensity[29] and duration. Cut roses held under constant light lose 5 times more water than those in alternating light-dark cycles of 12 hours.[11] When held in dark storage, cut roses did not display the progressive decline in water uptake typically seen with storage in the light after harvest.[15]

4.2. Plant Factors

The flux of water from a given surface area and over a fixed time interval into or out of a plant product is equal to the water vapor deficit divided by the cumulative resistances to movement:

$$\text{flux} = \frac{\text{water vapor inside} - \text{water vapor outside}}{\text{resistance to water movement}} \qquad \text{eq. 8.9}$$

Plant characteristics that decrease the water vapor pressure deficit between the interior and exterior and/or increase the resistance to water movement will reduce the rate of exchange of water between the product and its environment. The external water vapor pressure is an environmental parameter; in contrast, the internal water vapor pressure is strongly modulated by the amount of moisture present within the tissue. With succulent plant parts, the interior is generally considered to be at 100% relative humidity. However, with dryer, more durable crops (e.g., grains, pulses, nuts), this is seldom the case. While the water content, or more precisely the chemical potential of the water in the tissue, affects the internal water vapor pressure, the difference in water vapor pressures between the interior and the exterior represents the driving force.

Plant characteristics that increase the resistance to movement of water impede the rate of exchange between the product and the environment. The waxy cuticle on the surface of the product provides a major resistance to water loss, such that often only a small percentage escapes through the cuticle. The physical and chemical properties of the cuticular lipids have a dominant influence on water loss,[28] with intracuticular waxes, rather than the surface, playing a more important role.[49,50] Natural or artificial waxes[44] on the surface of the product increase the diffusion path length, increasing the resistance to flow.[6] Removal of the surface waxes can increase the permeability of water up to $1500\times$ and certain organic compounds by $9,200\times$.[51,45] Conversely, the postharvest application of surface coatings can help impede water losses in certain products.[2]

Table 8.1. Typical Surface Volume Ratios of Harvested Plant Products.*

Normal Range in Surface/Volume Ratio $cm^2 \cdot cm^{-3}$	Plant Material
50–100	Individual edible leaves (exposed surface); very small grains
10–15	Most cereal grains
5–10	Leguminous seeds, small soft fruits
2–5	Leguminous fruits; nut (except coconut); larger soft fruits (e.g. strawberry); rhubarb; shallot
0.5–1.5	Tubers; storage roots (except large yams); tap-roots (except large Swede turnips); pome, stone and citrus fruits; cucurbitous fruits (except large marrows); banana; onion
0.2–0.5	Densely packed cabbage; large Swede turnips and yams; coconut

Source: Burton.[9]

Other plant factors modulating exchange include surface trichomes, which increase boundary layer resistance. The chemical composition and structure of the product can affect how tightly water is held by the tissue. And finally, since evaporation is a surface phenomenon, the ratio of surface area to the volume of the product is of critical importance. Harvested plant products exhibit a tremendous range in surface-area-to-volume ratios (Table 8.1). Leafy crops have enormous surface areas relative to their volume and, consequently, generally loose moisture quickly when the vapor pressure deficit is conducive for water loss. Leaves exhibit surface to volume ratios of 50 to 100 $cm^2 \cdot cm^{-3}$, while large root and tuber crops[8] have ratios of only 0.2 to 0.5.

4.3. Use of Psychrometric Charts

Psychrometrics,* the measurement of heat and water vapor properties of air, is an extremely valuable tool during the postharvest period. The interrelationship among properties of the air, such as relative humidity, absolute humidity, dew point, and wet and dry bulb temperatures, are graphically illustrated in Figure 8.4. Normally, use of the chart involves initially making wet and dry bulb temperature readings.† From these two readings, the other thermal and moisture characteristics of the atmosphere can be determined using the psychrometric chart (e.g., relative humidity, dew point, vapor pressure, absolute humidity and with more elaborate charts, specific heat and other characteristics).

It is not necessary to start with just the wet and dry bulb readings; from any two parameters one can determine the remaining characteristic of the air. With any two variables it is possible to find their intercept on the chart. From that point each of the other variables can be determined. For example, if the dry bulb temperature (x-axis) is 25°C and the wet bulb temperature is 18°C (located on the curved line at the upper left hand portion of the chart and

---

*Derived from the Greek words *pyschro* for "cold" and *metron* for "measure".

†The dry bulb temperature is read directly from an ordinary glass thermometer. The wet bulb is read from a similar thermometer in which the reservoir bulb is covered with a thin layer of water (usually accomplished with a wet cotton wick) and is moved through the air at a sufficient velocity and duration until a steady temperature drop occurs due to evaporation. Evaporation cools the wet bulb; the difference between the dry and wet bulb temperatures is referred to as the wet bulb depression. The lower the moisture content in the air, the greater the evaporation and the temperature differential.

running downward at a 45° angle toward the x-axis), the relative humidity can be read on the curved line where the vertical dry bulb line and the angled wet bulb line intersect (i.e., 50%). This intercept gives a point on the chart from which the water vapor pressure and absolute (specific) humidity can also be read, i.e., by moving horizontally to the right of the intercept point.

The driving force for the movement of moisture is the difference in water vapor pressure between a product and its environment. Since water vapor pressure can be determined from many psychrometric charts or calculated from relative humidity and temperature data, it is possible to graphically express the relationship between atmospheric relative humidity and temperature conditions and the **water vapor pressure deficit** between the product and its environment. This assumes that the internal gas atmosphere of the product is saturated with water vapor, a safe assumption for succulent products, and that the product and environment are at the same temperature. The water vapor pressure deficit is read from the nomogram (Figure 8.5) by simply drawing a straight line from the temperature through the relative humidity for the environment. The water vapor pressure is read at the point of intersection and is given in millibars.

Water vapor pressure alone does not determine the rate of flux of water from a product since the resistance to water movement is not taken into consideration. Therefore, it is not possible to compare the rate of water loss between two different products (i.e., different resistances) or in many cases even among cultivars of the same product. For a given product, however, it is possible to compare two sets of environmental conditions and determine in which environment the product will lose moisture the fastest (that with the highest water vapor deficit) and how much faster (water vapor pressure of environment A ÷ the water vapor pressure of environment B).

4.4. Inhibiting the Exchange of Water Between Product and Environment

For most harvested products, once at or near their optimum moisture content, it is desirable to minimize any further change in moisture concentration during handling and storage. This includes not only succulent flowers, fruits and vegetables but also dried crops such as pulses, grains and nuts. Inhibiting the exchange of water is accomplished by minimizing the water vapor pressure deficit between the product and its environment or by increasing the resistance to exchange. Humidity and temperature are critical in minimizing the difference in water vapor pressure between product and environment.

The humidity of the surrounding environment should be maintained at a level that gives a water vapor pressure as close to that of the internal atmosphere of the product as possible. With succulent products, this generally requires very high relative humidities, 95–99%. Lower humidities may be utilized for products that exhibit a relatively high resistance to water exchange, because losses are generally at a much slower rate. With lower moisture products such as roots, tubers and corms, much lower storage humidities are required (e.g., 60–70%) or water will be taken up by the product, potentially exceeding the maximum safe concentration.

While the relationship between humidity and water exchange is relatively straightforward, temperature effects are more complex. Three thermal parameters have a pronounced effect on moisture exchange in storage: the actual temperature, the differential in temperature between product and environment and fluctuations in storage temperature. Temperature affects both the amount of water a given volume of air will hold and the level of energy of the water molecules.

Lowering the temperature decreases the maximum amount of water the air will hold; if the weight of water vapor in the air is held constant, the relative humidity will increase. The

Table 8.2. Environmental Conditions and Product Treatments Used to Minimize the Exchange of Water.

Environmental Conditions	*Product Treatments*
Lower temperature	Prevent cuts and abrasions during harvest and handling
Maintain a sufficiently high relative humidity	Rapid cooling after harvest
Minimize excessive air movement	Surface coatings (e.g. waxes)
Minimize fluctuations in air temperature	Packaging (e.g. shrink wraps)
	Decrease surface area (e.g. leaf removal for cut flowers)

water vapor pressure deficit between a product and its environment will also decrease at a given relative humidity with decreasing temperature, e.g., for apple fruit the water vapor pressure deficit at 95% relative humidity at 20°C is 1.15 mbar, while at 10°C it is only 0.6 mbar (Figure 8.5). Therefore, low temperature decreases the rate of water loss. Likewise, when the product is at a higher temperature than the environment, the gas atmosphere within a succulent product will contain more water than that of the surrounding cooler air, increasing the difference in water vapor pressure between the product and environment and the rate of water loss. As a consequence, it is desirable to cool harvested products that are to be held in refrigerated storage as quickly as possible to minimize water losses. Due to respiratory heat, however, under constant storage temperature conditions, many actively metabolizing products will remain at a slightly higher temperature than their surrounding environment. Hence there will normally be a small positive gradient favoring water loss.

Air movement is essential in refrigerated storage to remove heat generated by the stored product (see the section in this chapter entitled *Air Movement*). At the same time, air movement decreases the thickness of the boundary layer around the product, decreasing boundary layer resistance and increasing the flux of water vapor from the product. It is essential, therefore, to reach a compromise between adequate air movement for cooling and minimizing excessive water loss.

There are a number of product treatments that can be utilized to reduce water exchange after harvest (Table 8.2) Preventing cuts and abrasions to the surface of the product helps to maintain a high surface resistance to the movement of water vapor. Rapid cooling reduces the temperature differential between the product and the storage environment decreasing water loss. With some products, specialized surface treatments[2] are advantageous, e.g., waxing citrus fruit increases surface resistance, and the use of wraps and packages impedes the movement of water vapor.

4.5. Enhancing the Exchange of Water Between Product and Environment

The concentration of water in most harvested crops is of tremendous importance in that it affects the product's physical and chemical properties, its value and storage potential. Succulent products, such as lettuce or cut flowers, that lose even a relatively small percent of their total water, often decline markedly in quality and value. Dry products that are less subject to rapid quality changes with moisture concentration are generally sold by weight. Loss of water results in a decline in total weight and as a consequence, value.[11] For example, if the corn (100,000 kg @ 15.5% moisture) in a small on-the-farm storage bin is allowed to dry to 8% moisture, the total value based upon a market price of $0.11 · kg^{-1} has declined $896.74. This loss is regarded as "shrink". Shrink = [(initial moisture content - final moisture content)/(100 - final moisture content)] × mass. In the example, [(15.5 - 8)/(100 - 8)] × 100,000 kg of grain = 8152.17 kg × $0.11 · kg^{-1}. Because of this, the market price of low moisture products often depends on their moisture content.

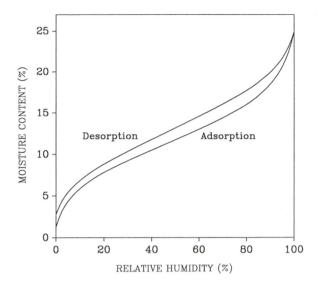

Figure 8.7. Moisture adsorption and desorption isotherms for maize starch.[18] The difference in equilibrium moisture content between wetting and drying curves is called hysteresis.

It is often desirable, therefore, to adjust the concentration of water within a product after harvest. Moisture alteration may entail accelerating the loss of water (**desorption**) or accelerating the uptake of water (**adsorption** and/or **absorption**).*

The method utilized to alter product moisture content depends upon the species in question. Both adsorption and desorption are commonly accomplished by altering the vapor pressure of the environment surrounding the product. Water is generally taken up in a vapor state; however, with some very high moisture crops (e.g., cut flowers, leaf lettuce), water may be introduced as a liquid.

Typically, the product is exposed to altered water vapor pressure conditions for a sufficient duration to reach the desired moisture concentration. Environmental conditions can be selected, so that when the product reaches an equilibrium with the environment, it is at the desired moisture concentration. Considerable time is generally required for equilibrium to be reached, with the rate of change progressively decreasing as the product approaches the point of equilibrium.

An alternative approach that accelerates exchange and decreases the length of time required to reach the desired moisture concentration is to create a very large gradient in vapor pressure between the product and its environment. The product is then either removed or the environment changed once the product reaches the desired moisture concentration, rather than when it reaches an equilibrium with the environments.

The **equilibrium moisture concentration** (i.e., the amount of moisture in the product when in equilibrium with its environment) depends not only on temperature but also on the previous moisture conditions of the material. Because of the latter, if the equilibrium moisture concentration is plotted versus the percent relative humidity (called a water sorption isotherm), two different moisture content values are obtained for each relative humidity, depending upon whether the product is adsorbing (gaining) or desorbing (losing) moisture (Figure 8.7). This

---

*Adsorbed water molecules display a molecular interaction with the adsorbing substance and are more closely bound that molecules of free water. The adsorbent and water molecules both influence the properties of the other. In contrast, absorbed water molecules are held loosely by capillary forces and exhibit the properties of free water. A combination of adsorption and absorption phenomena often occurs in biological materials.

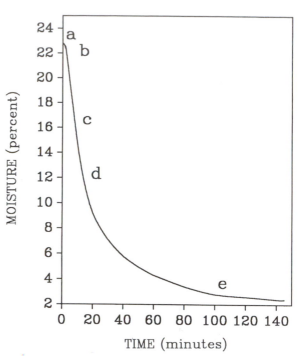

Figure 8.8. A drying rate curve illustrating three general segments: a-b initial exchange; b-c the constant rate period, and c-e, the falling rate period. The falling rate period is commonly separated into two stages; c-d and d-e, each dominated by a different process. Point c on the curve represents the critical moisture concentration.

difference between the equilibrium moisture contents at a given percent relative humidity, called **hysteresis**, can be as much as 1.5 to 2.0% moisture. While the reason for hysteresis is not fully understood, the practical implications can be significant. The difference in product moisture concentration due to adsorption versus desorption of water can, under appropriate conditions, represent the difference between the presence and absence of mold growth during storage.

4.5.1. Enhancing Water Loss

Drying has been the primary preservation process since time immemorial. For field crops, much of the drying occurs prior to harvest. Under favorable weather conditions, seeds lose water approaching an equilibrium between their moisture content and that of the surrounding environment. Due to inclement weather or other reasons, it is often necessary or desirable to harvest prior to the crop having dried sufficiently in the field for safe storage. When this occurs, some additional drying is required and may be accomplished by the use of air, *in vacuo,* direct heat, inert gas and superheated steam. Due to its low cost and convenience, air drying is the most commonly employed technique and will be focused upon here. Air drying varies from simply exposing the crop to the sun to the use of elaborate computer controlled forced air dryers.

a. Rate of water loss

Drying involves two fundamental physical processes: the transfer of heat to evaporate the water and the transfer of mass (water) within and from the product. Mass transfer occurs both as liquid diffuses to the surface and as gas diffuses outward from the product. The actual rate of water removal from a harvested product is not constant throughout the drying process but declines as the moisture content decreases (Figure 8.8). Drying curves can be separated into three general segments: initial exchange, the constant rate period, and the falling rate period. The initial rate period (a-b) is typically brief, spanning the time period required for the product to reach the temperature of the surrounding air. During this relatively brief period, some water may actually migrate toward the center of the product due to the temperature differential between the exterior (warm) and interior (cool).

The second segment of the drying curve (b-c) is characterized by a constant rate of water loss (Figure 8.8). During this period, evaporation occurs at the surface of the product, similar to evaporation from a free water surface. The constant rate period ends upon reaching the **critical moisture content**—the minimum concentration of water in the product that will sustain a flow of free water from the interior to the surface at a rate equal to the maximum rate of removal due to evaporation. Upon transversing this point (c), the drying curve enters the falling rate period, which is controlled largely by the diffusion of liquid phase moisture to the surface and its removal from the surface. The falling rate portion of the drying curve is often separated into two (occasionally more) stages. The first (c-d) is characterized by unsaturated surface drying, while the second (d-e) is controlled by the rate of diffusion of moisture to the surface. Most dried crops are harvested after passing the critical moisture content point of the drying curve.

As a product approaches its equilibrium moisture content, the net exchange of water between the product and its environment nears zero. At the equilibrium moisture content, water molecules continue to move into and out of the product; however, uptake and loss are equal. The actual moisture content of the product does not equal that of the environment; rather the chemical potential of water between the two sites is equal.

The equilibrium moisture content of a product is a function of both the moisture concentration of the air and the characteristics of the individual product. When placed in the same environment, various products exhibit different affinities for water and therefore display significantly different equilibrium moisture contents (Table 8.3). Plant species, cultivar and maturity may significantly influence the equilibrium moisture content. In addition, whether water has been adsorbed or desorbed to reach the equilibrium moisture content may also affect the precise concentration.

The rate at which a product dries is a function of the difference in the chemical potential of water within the product and that of the surrounding environment and factors that affect the exchange of moisture between the two. Important factors that alter the rate of drying are the moisture content of the air, product and air temperature, air flow rate, product surface area, the moisture content of the product, and product characteristics that enhance or retard

Table 8.3. Equilibrium Moisture Contents of Selected Grain Crops at 25°C.*

Material	*Relative Humidity (%)*									
	10	*20*	*30*	*40*	*50*	*60*	*70*	*80*	*90*	*100*
Barley	4.4	7.0	8.5	9.7	10.8	12.1	13.5	15.8	19.5	26.8
Flaxseed	3.3	4.9	5.6	6.1	6.8	7.9	9.3	11.4	15.2	21.4
Oats	4.1	6.6	8.1	9.1	10.3	11.8	13.0	14.9	18.5	24.1
Rice[†]	5.9	8.0	9.5	10.9	12.2	13.3	14.1	15.2	19.1	—
Corn[‡]	5.1	7.0	8.4	9.8	11.2	12.9	14.0	15.6	19.6	23.8
Soybeans	—	5.5	6.5	7.1	8.0	9.3	11.5	14.8	18.8	—
Wheat	5.8	7.6	9.1	10.7	11.6	13.0	14.5	16.8	20.6	—

*Source: Hall.[22]
[†]Whole grain
[‡]Shelled

water exchange (e.g., epicuticular waxes). While we have little control over most product characteristics, environmental conditions can be modified to enhance or retard the rate of drying.

Simply optimizing environmental conditions for the maximum rate of water removal from a product is seldom a satisfactory approach. Excessively rapid loss of water causes serious quality losses in many products. Seed dried too quickly develop **case hardening**, a hardening of the outer tissue. Case hardening is caused by the removal of water from the surface of the product at a more rapid rate than water at the interior can diffuse outward. The dry outer tissue tends to seal the surface of the seed, decreasing the subsequent rate of drying. When onion bulbs are partially dried prior to storage using artificial dryers, the rate of water removal must be carefully controlled. Excessively high rates of water loss can cause splitting of the outer protective surface layers, causing an undesirable loss of water later in storage.

Air temperature has a pronounced effect upon the rate of drying, with higher temperatures enhancing the rate. Air temperature may also affect the quality of the final product.[26] The temperature at which the product will undergo undesirable changes in quality (e.g., loss of germination potential, color, flavor) is the **critical temperature**. This varies widely among various products, their intended use, moisture content, type of dryer and airflow rate. For example, wheat that is to be used as seed and has a moisture content below 24% can be safely dried at 49°C; when above 24% moisture, a lower temperature (43°C) is essential to prevent quality losses.[38] Gluten quality is particularly labile when exposed to high drying temperatures, and the degree of susceptibility varies among cultivars.[60] The same grain can be dried at 60°C (>25% moisture) or 66°C (<25% moisture) if it is to be used for milling. Even higher temperatures can be used for wheat that is to be utilized as livestock feed (e.g., 82–104°C). Most commercial drying operations try to maintain the highest possible air temperature that will not impair quality because it increases the capacity of the dryers and decreases costs.

Air drying in the sun with the product placed on drying floors, roofs, court yards, roads and other suitable areas is a standard practice in many areas of the world where artificial drying is not feasible.[53] For example, in Guangdong Province of China, under favorable conditions rice can be dried from 18 to 13.5% moisture in two days.[20] This requires spreading the rice in relatively thin layers to maximize the surface area exposed (i.e., ≈2 cm deep) and stirring with rakes twice a day. Reabsorption of moisture during the night is minimized by raking the rice into piles and covering them when possible.

b. Low- versus high-temperature drying

Forced air drying depends upon the movement of air with a suitable humidity and temperature around the individual units of the product. For some crops, it is desirable to decrease the product's moisture content quickly; this is generally accomplished by elevating the temperature of the drying air. Forced air driers can be separated into two general classes: low temperature dryers and high-temperature dryers. Low-temperature dryers utilize air with little or no supplemental heating, relying largely upon suitable ambient temperature and humidity conditions. The air is passed over or around the product until a moisture equilibrium is established. The low temperature air, although requiring a greater drying time, prevents over drying of the product and potential quality losses. It also allows drying relatively deep layers of product; hence the crop can often be dried in storage, decreasing handling costs by eliminating a separate step. A disadvantage is that if the ambient conditions are undesirable, the product can gain moisture[54] unless the air flow terminated.

High-temperature driers utilize a substantially elevated air temperature to accelerate drying rate. The primary distinction between low- and high-temperature drying is that if the product were allowed to reach an equilibrium moisture content with the air in a high-temperature dryer, serious over-drying would occur. Exposure to the elevated temperature air is continued

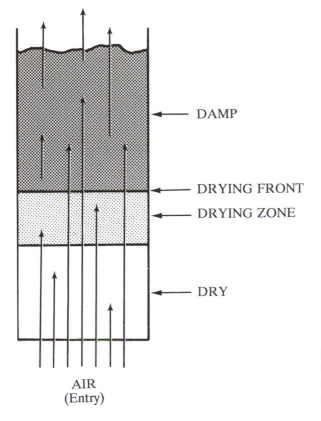

Figure 8.9. The presence of moisture zones during deep layer drying. The drying front progressively moves through the material leaving a dried product behind.

only until the product has reached the desired moisture content. The accelerated rate of drying decreases the chances of mycotoxin formation in some products.[41]

High-temperature dryers are separated into two general classes: continuous-flow and batch dryers. The former utilize a continuous movement of product through the dryer, while in the latter the product is static. Continuous-flow dryers represent the most widely used of the two classes; however, batch dryers have the advantage of simplicity.

c. Drying front and zones

In bulk drying, a relatively large volume of air, the humidity of which is lower than the equilibrium value for the product being dried, is introduced into a thick bed of the product.* The dryer air lowers the moisture content of the product in which it initially comes in contact, increasing its own moisture content. There is also a corresponding decrease in air temperature due to the energy adsorbed with the phase change of water between a liquid and a gas. As the air moves past successive units of the product, it no longer has the same properties (relative humidity and temperature) and drying potential as when first entering the bed of product. Hence the air removes less and less moisture as it proceeds. Consequently, a series of moisture zones develop within the bed (Figure 8.9). At the site of entry, the dry air removes moisture until the equilibrium moisture content is reached, resulting in a dry product. Within the dry zone there is no additional net exchange of moisture or temperature change as fresh air passes

---

*A thick layer or stratum of product.

around the product. Drying occurs in the drying zone, which moves progressively forward into the undried product. The forward edge of the drying zone is referred to as the **drying front**. During the early stages of drying, part of the moisture removed from the product near the air inlet may be reabsorbed by the cooler product near the exhaust. To prevent possible losses, product depth, air temperature and air velocity can be manipulated to minimize excessive wetting and the length of time required for the drying front to reach the moist zone.

4.5.2. *Enhancing Water Uptake*

For many harvested products it is necessary to reintroduce water when the internal concentration has dropped below the desired level. Introduction is accomplished by using moisture in either the liquid or vapor phase. The physical state of water that can be used depends upon the product, the rate of uptake required, the final concentration desired, intended use of the product and a number of other factors.

The introduction of vapor phase moisture requires considerably more time than using water in the liquid phase. The use of water vapor, however, is essential for crops that can not withstand direct wetting. Proceeding from a very dry product toward a higher moisture level, the relationship between water vapor and product moisture changes. At the low end of the curve (little water), the energy of binding between water molecules and the compounds making up the adsorbing surface is a dominant factor (Figure 8.7). The intermolecular forces involved are thought to be quite strong, accounting for the steepness of the initial segment of the curve. Chemical constituents such as starch and protein provide polar sites with which the water molecules react. The moisture deposited forms a first layer of water molecules. During the more linear portion of the isotherm, additional water molecules are being deposited upon the first layer of water molecules, forming a second layer. In this region the amount of water adsorbed is largely dependent upon the water vapor pressure of the atmosphere. In the final segment of the isotherm, the high humidity range, successive layers of water molecules are deposited. The amount of water adsorbed increases rapidly, with the vapor pressure of the air having only a moderate influence.

The rate of uptake of moisture is enhanced by a high water vapor pressure deficit between the atmosphere and product, a positive gradient favoring the movement from the atmosphere to the product. Increasing temperature and air movement also enhances the rate of uptake. For example, the effect of relative humidity and temperature on the uptake of moisture is presented in Figure 8.10. The higher the relative humidity, the greater the difference in water vapor pressure between the product and atmosphere, accelerating the rate of moisture uptake.

Liquid phase water may also be used to alter the moisture content of certain crops. Many cut flowers and foliage crops require the reintroduction of liquid phase water after harvest. When they are severed from the parent plant, water uptake ceases; however, evaporational losses continue. Unless moisture loss is inhibited (e.g., high relative humidity) or water is reintroduced, a water deficit occurs, resulting in reduced turgidity and an accelerated rate of quality loss and senescence. With the loss of turgidity, the stems of some species are no longer able to physically support the weight of the flower, resulting in a disorder called "bent neck."

When the stems of cut flowers are placed in containers of water, there is normally a gradual decline in the rate of uptake with time. This decline is in contrast to flowers that remain attached to the parent plant, where the water conductivity remains essentially constant during aging.[18] The decrease in water conductivity or increase in resistance is caused primarily by two factors: the presence of microorganisms and the introduction of air into the stem. Microorganisms found in the water reservoir enter the base of the stem, causing a direct "stem plugging" due to their physical presence and an indirect blockage due to the production of metabo-

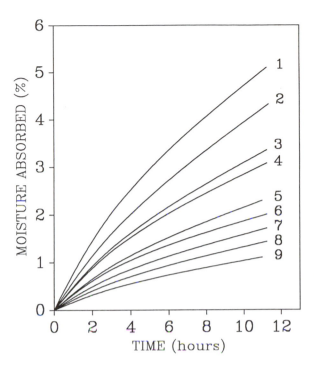

Figure 8.10. The adsorption of water vapor by corn with an initial moisture content of 9.26% (at 0 time) under various environmental conditions (1 = 27°C, 80% RH; 2 = 21°C, 80% RH; 3 = 16°C, 80% RH; 4 = 27°C, 70% RH; 5 = 21°C, 70% RH; 6 = 16°C, 70% RH; 7 = 27°C, 60% RH; 8 = 21°C, 60% RH; 9 = 16°C, 60% RH).[42]

lites by some microbes. The importance of microorganisms in increasing the resistance to water uptake has been shown through correlations between increases in their populations and decreases in water uptake,[31] the effect of chemicals that control microorganisms on water uptake, and anatomical studies.

Air entering the base of the stem during storage or shipment can disrupt the continuity of the columns of water within the stem, greatly impeding water uptake. Air may be introduced during cutting and/or dry storage or may represent the movement of dissolved gases out of the liquid phase within the stem, forming bubbles. Degassing the water by boiling or vacuum treatment[18] has been shown to prevent a decrease in conductance of microbial-free water through the stem.

A tremendous quantity of intact ornamental floral and foliage plants are marketed each year, during which time the reintroduction of water is often essential. Because they are intact, water uptake is seldom a problem; proper quality maintenance centers upon balancing the rate of water use by the plant with the reintroduction of water into the root medium. This is especially critical with bedding plants, which are generally grown in very small-volume containers. Both the volume of the root medium and its water holding capacity are critical parameters.

Increasing the volume of water available to the plant, whether by increasing the volume of the medium or the physical/chemical properties of the medium (e.g., the use of hydrophilic polymers*), potentially decreases the number of wet-dry cycles the plant goes through after removal from the production zone. Since drying represents a potential stress period when quality can be rapidly lost, decreasing the total number of cycles can decrease the number of times the plant enters a high-risk situation. Conversely, media that display too great a water holding capacity can also cause serious losses due to the occurrence of anaerobic conditions

---

*Such as polyethylene oxide.[3]

in the root zone. Selection of a medium with proper water holding and aeration properties is a production decision that can have a critical impact upon the postharvest handling and marketing period. Specific media requirements vary widely among individual species.

Cereal grains are often conditioned with water prior to milling. In wheat kernels, absorption represents a heterogeneous process.[8] Initially there is a very rapid absorption in which the pericarp is saturated by capillary imbibition.* The testa of the kernel represents the major barrier to diffusion. The embryo, although small in surface area relative to the testa, has a much lower resistance. Natural capillaries and structural flaws and cracks in the grain increase the rate of diffusion and uptake of water. The gain in product moisture decreases with increasing water temperature.

During retail marketing, water is often directly applied to a number of edible products (e.g., leaf lettuce, green onion, celery, kale, kohlrabi, parsley, watercress). Cold water sprays, in addition to decreasing product temperature and increasing the relative humidity, allow the uptake of some water by many products. Postharvest products vary widely in their ability to have water applied to their surfaces. Leafy crops which have large surface areas lose water quickly and generally benefit from water sprays. Unsprayed vegetables have been shown to display a 10–20% weight reduction during the first few days under simulated retail conditions.[9] Large surface-to-volume ratios decrease the likelihood of the occurrence of anaerobic conditions within the interior of the product.

The exposure of the surface of many products to free water, however, causes serious problems, generally seen as an increased occurrence of rots. Some species should be kept under high humid conditions (e.g., cauliflower, the floral parts of cut flowers, raspberries, mushrooms and many others) but should never come in contact with water. Dried products (e.g., grains and pulses) should be held under relatively dry atmospheric conditions and should never be wetted.

The uptake of liquid phase water by products during cooling and handling operations can, in some instances, be undesirable; it may introduce disease organisms into the product. For example, the use of water to float tomatoes out of large trucks can introduce water-borne disease organisms into the fruit through the stem scar. Uptake is greatly enhanced by the use of water with a temperature less than that of the fruit. The colder water cools the fruit floating or submerged in it, decreasing the volume of the gas atmosphere within the tissue. A partial vacuum is created that pulls water through the stem scar into the fruit until the internal and external pressures are balanced. To prevent this situation, it is recommended that the temperature of the water be approximately 1 to 2°C higher than the temperature of the product. Proper sanitation is also important (e.g., the use of chlorine).

4.6. Maximum and Minimum Acceptable Product Moisture Content

Succulent products have high moisture contents at harvest, and the loss of even a relatively small amount of water can have, in some species, a serious effect upon the physical, physiological, pathological, nutritional, economic and esthetic properties of the crop. Table 8.4 shows the maximum permissible amounts of water that can be lost from a cross-section of fruits and vegetables before becoming unmarketable. These range from a low of 3% in spinach leaves to 41% in snapbeans, indicating a wide variation among species (Figure 8.11). Deficiency symptoms typically increase with increasing water deficit (Figure 8.12). However, not all individual units within a sample display physical desiccation symptoms uniformly relative

---

*Absorption of a liquid by a solid or semisolid material *via* capillary movement.

Table 8.4. Maximum Permissible Weight Loss for a Number of Fruits and Vegetables.*

Commodity	Maximum Permissible Weight Loss (%)	Commodity	Maximum Permissible Weight Loss (%)
Apples (4 cultivars)	7.5[†]	Nectarine	21.1[†]
Asparagus	8.0	Onion	10.0
Beans, broad	6.0	Parsnip	7.0
runner	5.0	Peach (5 cvs)	16.4[†]
snap (4 cvs)	41.0[†]	Pea	5.0
Beetroot	7.0	Pear (3 cvs)	5.9[†]
Beetroot with tops	5.0	Pepper, green	7.0, 12.2[†]
Blackberries	6.0	Persimmons	13.3[†]
Broccoli	4.0	Potato	7.0
Brussels sprouts	8.0	Raspberries	6.0
Cabbage (3 cvs)	8.0, 10.9[†]	Rhubarb, forced	5.0
Carrots	8.0	Spinach	3.0
Carrots with leaves	4.0	Squash, summer	23.9[†]
Cauliflower	7.0	Sweet corn	7.0
Celery	10.0	Tomato	7.0, 6.2[†]
Cucumber	5.0	Turnip with leaves	5.0
Leek	7.0	Watercress	7.0
Lettuce (3 cultivars)	3.7		

*Source: Robinson[47] and †Hruschka.[25]

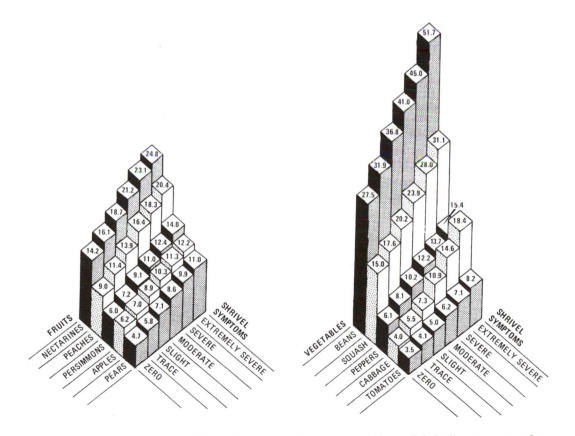

Figure 8.11. Average percent weight loss at the onset of various gradations of shriveling symptoms for selected fruits and vegetables.[25]

Figure 8.12. Shrivel symptoms in peppers ('California Wonder'). Top row, left to right—zero (6.1% weight loss), trace (8.1%), slight (10.2%); bottom row, left to right—moderate (12.2%), severe (13.7%), extremely severe (15.4%).[2]

Table 8.5. Maximum Safe Moisture Content for Stored Grain Held at 25–38°C.*

Commodity	Maximum Acceptable Moisture (%)	
	USA	*Canada*
Barley	14.5	14.8
Buckwheat	—	16.0
Corn (maize)	14.1	14.0
Oat	14.0	14.0
Pea	15.0	16.0
Rice (rough)	14.0	—
(milled)	15.0	—
Rye	16.0	14.0
Sorghum	13.0	—
Soybean	13.0	14.0
Sunflower	—	9.6
Wheat	—	14.5

*Source: Bushuk and Lee.[10]

to weight loss. For example, the actual percent weight loss of individual snap beans displaying moderate shrivel symptoms ranged from 15 to 68%.[17] Likewise, within a given lot, the onset of dehydration symptoms ranged from zero to extremely severe, indicating a tremendous range in the rate of weight loss between individual units within a sample.

Products that are stored dry (e.g., grain) have a maximum moisture concentration that will allow safe storage (Table 8.5). Exceeding this moisture level greatly increases the potential for pathogen invasion and storage rots. The actual moisture concentration at which a product can be safely stored depends upon storage temperature. Table 8.6 illustrates the relationship

Table 8.6. Estimated Maximum Storage Time for Barley at Various Seed Moisture Contents and Storage Temperatures with Respect to the Preservation of Germination.*

Storage Temperature	*Moisture Content (%)* *Maximum Storage Duration (Weeks)*								
°C	*11*	*12*	*13*	*14*	*15*	*16*	*17*	*19*	*23*
25	54	39	25	16	9	5	2.5	1	—
20	110	80	50	32	19	10	5	2	0.5
15	240	170	100	65	40	20	10	10.4	1
10	600	400	260	160	90	50	21	8.5	2
5	>1,000	1,000	600	400	200	120	50	17	4

*Source: Kreyger.[30]

between barley moisture content and storage temperature with regard to the preservation of germination. Decreasing both moisture content and storage temperature prolongs the preservation of quality.

ADDITIONAL READING

Barkla, B.J., and O. Pantoja. 1996. Physiology of ion transport across the tonoplast of higher plants. *Annu. Rev. Plant Physiol. Plant Mol. Biol.* 47:159–184.

Borghetti, M., J. Grace and A. Raschi. 1993. *Water Transport in Plants under Climatic Stress.* Cambridge University Press, Cambridge.

Boyer, J.S. 1995. *Measuring the Water Status of Plants and Soils.* Academic Press, New York.

Brett, C.T., and K.W. Waldron. 1996. *Physiology and Biochemistry of Plant Cell Walls.* Chapman & Hall, London.

Chrispeels, M.J., and C. Maurel. 1994. Aquaporins: The molecular basis of facilitated water movement through living plant cells? *Plant Physiol.* 105:9–13.

DeFelice, L.J. 1997. *Electrical Properties of Cells: Patch Clamp for Biologists.* Plenum, New York.

Gartner, B.L. (ed.). 1995. *Plant Stems: Physiology and Functional Morphology.* Academic Press, New York.

Gates, D.M., and R.B. Schmerl (eds.). 1975. *Perspectives of Biophysical Ecology.* Springer-Verlag, Berlin.

Gebhart, B. 1993. *Heat Conduction and Mass Diffusion.* McGraw-Hill, New York.

Haines, T.H. 1994. Water transport across biological membranes. *FEBS Lett.* 346:115–122.

Jackson, M.B. (ed.). 1993. *Thermodynamics of Membrane Receptors and Channels.* CRC Press, Boca Raton, FL.

Jones, H.G. 1992. *Plants and Microclimate: A Quantitative Approach to Environmental Plant Physiology.* Cambridge University Press, Cambridge.

Jones, M.N., and P. Chapman. 1995. *Micelles, Monolayers and Biomembranes.* Wiley-Liss, New York.

Kramer, P.J., and J.S. Boyer. 1995. *Water Relations of Plants and Soils.* Academic Press, New York.

Lendzian, K.J., and G. Kerstiens. 1991. Sorption and transport of gases and vapors in plant cuticles. *Rev. Environ. Contamin. Toxicol.* 121:65–128.

Leshem, Y.Y. 1992. *Plant Membranes. A Biophysical Approach to Structure, Development and Senescence.* Kluwer, Dordrecht, The Netherlands.

Linskens, H.F., and J.F. Jackson (eds.). 1996. *Plant Cell Wall Analysis.* Springer-Verlag, Berlin.

Merz, K.M., Jr., and B. Roux. 1996. *Biological Membranes: A Molecular Perspective from Computation and Experiment.* Birkhauser, Boston, MA.

Moller, I.M., and P. Brodelius (eds.). 1996. *Plant Membrane Biology.* Oxford University Press, Oxford, England.

Nobel, P.S. 1999. *Physicochemical & Environmental Plant Physiology.* Academic Press, New York.

Patrick, J.W. 1997. Phloem unloading: Sieve element unloading and post-sieve element transport. *Ann. Rev. Plant Physiol. Plant Mol. Biol.* 48:191–222.

Petty, H.R. 1993. *Molecular Biology of Membranes: Structure and Function.* Plenum, New York.

Sanders, D., and M. Tester (eds.). 1997. Ion channels. *J. Exp. Bot.* 48:353–631.

Schütz, K., and S.D. Tyerman (eds.). 1997. Water channels in *Chara corallina. J. Exp. Bot.* 48:1511–1518.

Smallwood, M., J.P. Knox and D.J. Bowles (eds.). 1996. *Membranes: Specialized Functions in Plants.* BIOS Scientific, Oxford, England.

Solomos, T. 1987. Principles of gas exchange in bulky plant tissues. *HortScience* 22:766–771.

Stein, W.D. 1990. *Channels, Carriers, and Pumps: An Introduction to Membrane Transport.* Academic Press, New York.

Steudle, E., and T. Henzler. 1995. Water channels in plants: Do basic concepts of water transport change? *J. Exp. Bot.* 46:1067–1076.

Waisel, Y., A. Eschel and U. Kafkafi. 1996. *Plant Roots: The Hidden Half.* Dekker, New York.

REFERENCES

1. Aloni, B., L. Karni, S. Moredhet, C. Yao and C. Stanghellini. 1999. Cuticular cracking in pepper fruit: II. Effect of fruit water relations and fruit expansion. *J. Hort. Res. Biotech.* 74:1–5.

2. Amarante, C., and N.H. Banks. 2001. Postharvest physiology and quality of coated fruits and vegetables. *Hort. Rev.* 26:161–238.

3. Anon. 1973. Agricultural hydrogel, concentrate 50G. *Tech. Bull.,* Union Carbide Corp., New York.

4. Banks, N.H., and S.J. Kays. 1988. Measuring internal gases and lentical resistance to gas diffusion in potato tubers. *J. Amer. Soc. Hort. Sci.* 113:577–580.

5. Barkla, B.J., and O. Pantoja. 1996. Physiology of ion transport across the tonoplast of higher plants. *Annu. Rev. Plant Physiol. Plant Mol. Biol.* 47:159–184.

6. Baur, P., H. Marzouk and J. Schönherr. 1999. Estimation of path lengths for diffusion of organic compounds through leaf cuticles. *Plant, Cell Environ.* 22:291–299.

7. Bower, J., B.D. Patterson and J.J. Jobling. 2000. Permeance to oxygen of detached *Capsicum annuum* fruits. *Aust. J. Exp. Agric.* 40:457–463.

8. Becker, H.A. 1960. On the absorption of liquid water by the wheat kernel. *Cereal Chem.* 37:309–323.

9. Burton, W.G. 1982. *Postharvest Physiology of Food Crops.* Longmans, Essex, UK.

10. Bushuk, W., and J.W. Lee. 1978. Biochemical and functional changes in cereals: Maturation, storage and germination. Pp. 1–33. In: *Postharvest Biology and Biotechnology.* H.O. Hultin and M. Milner (eds.). Food and Nutrition Press, Westport, CT.

11. Carpenter, W.J., and H.P. Rasmussen. 1973. Water uptake by cut roses (*Rosa hybrida*) in light and dark. *J. Amer. Soc. Hort. Sci.* 98:309–313.

12. Colwell, R.N. 1942. *Translocation in plants with special reference to the mechanism of phloem transport as indicated by studies on phloem exudation and on the movement of radioactive phosphorous.* Ph.D. Thesis, Univ. of Calif., Berkley, CA.

13. Crafts, A.S., and O. Lorenz. 1944. Fruit growth and food transport in cucurbits. *Plant Physiol.* 19: 131–138.

14. Crank, J., and G.S. Park. 1968. *Diffusion in Polymers.* Academic Press, London.

15. DeStigter, H.C.M. 1980. Water balance of cut and intact 'Sonia' rose plants. *Z. Pflanzenphysiol.* 99:131–140.

16. Drivel, J.W.T., and L. Duval. 1911. The shrinkage of corn in storage. *U.S. Bureau of Plt. Ind. Circ.* 81.

17. Dipman, C.W., J.L. Callahan, A.D. Michaels and S.R. Barkin. 1936. *How to Sell Fruits and Vegetables.* The Progressive Grocer, New York.

18. Durkin, D. 1979. Effect of millipore filtration, citric acid, and sucrose on peduncle water potential of cut rose flowers. *J. Amer. Soc. Hort. Sci.* 104:860–863.

19. Food and Agriculture Organization of the United Nations. 1970. Food storage manual. *Trop. Stored Prod. Cent., Min. Overseas Dev.,* Rome.

20. Food and Agriculture Organization of the United Nations. 1982. China: Postharvest grain technology. *Food and Agric. Org., United Nations, Agric. Serv. Bull.* 50, Rome.

21. Grierson, W., and W.F. Wardowski. 1975. Humidity in horticulture. *HortScience* 10:356–360.

22. Hall, C.W. 1980. *Drying and Storage of Agricultural Crops.* AVI, Westport, CT.

23. Hirshi, K.D., R.-G. Zhen, P.A. Rea and G.R. Fink. 1996. *CAX1,* an H$^+$/Ca^{2+} antiporter from *Arabidopsis. Proc. Natl. Acad. Sci.* USA 93:8782–8786.

24. Holman, L.E., and D.G. Carter. 1952. Soybean storage in farm-type bins. *Ill. Agric. Expt. Sta. Bull.* 552.

25. Hruschka, H.W. 1977. Postharvest weight loss and shrivel in five fruits and five vegetables. *USDA-ARS Mkt. Res. Rept.* 1059.

26. Inprasit, C., and A. Noomhorm. 2001. Effect of drying air temperature and grain temperature of different types of dryer and operation on rice quality. *Drying Tech.* 19:389–404.

27. Jackson, M.B., I.B. Morrow and D.J. Osborne. 1972. Abscission and dehiscence in the squirting cucumber. *Can. J. Bot.* 50:1465–1471.

28. Knoche, M., S. Peschel, M. Hinz and M.J. Bukovac. 2000. Studies on the water transport through the sweet cherry fruit surface: Characterizing conductance of the cuticular membrane using pericarp segments. *Planta* 212:127–135.

29. Kofranek, A.M., and A.H. Halevy. 1972. Conditions for opening cut chrysanthemum flower buds. *J. Amer. Soc. Hort. Sci.* 97:578–584.

30. Kreyger, J. 1964. [Investigations on drying and storage of cereals in 1963]. *Versl. Tienjarenplan Graanonderz* 10:157–164.

31. Larsen, F.E., and M. Frolich. 1969. The influence of 8-hydroxyquinoline citrate, N-dimethylaminosuccinamic acid, and sucrose on respiration and water flow in 'Red Sim' carnations in relation to flower senescence. *J. Amer. Soc. Hort. Sci.* 94:289–291.

32. Lendzian, K.J., and G. Kerstiens. 1991. Sorption and transport of gases and vapors in plant cuticles. *Rev. Environ. Contamin. Toxicol.* 121:65–128.

33. Maathuis, F.J.M., A.M. Ichida, D. Sanders and J.I. Schroeder. 1997. Roles of higher plant K$^+$ channels. *Plant Physiol.* 114:1141–1149.

34. Maguire, K.L., A. Lang, N.H. Banks, D. Hopcroft and R. Bennett. 1999. Relationship between water vapour permeance of apples and micro-croaking of the cuticle. *Posthar. Biol. Tech.* 17:89–96.

35. Martinoia, E., A. Massonneau and N. Frangne. 2000. Transport processes of solutes across the vacuolar membrane in higher plants. *Plant Cell Physiol.* 41:1175–1186.

36. Maurel, C. 1997. Aquaporins and water permeability of plant membranes. *Ann. Rev. Plant Physiol. Plant Mol. Biol.* 48:399–429.

37. Meeuse, B.J.D. 1975. Thermogenic respiration in Aroids. *Annu. Rev. Plant Physiol.* 26:117–126.

38. Nellist, M.E. 1979. Safe temperatures for drying grain. A Report to the Home Grown Cereals Authority. *N.I.A.E. Rept.* No. 29.

39. Neuhaus, H.E., and R. Wagner. 2000. Solute pores, ion channels, and metabolite transporters in the outer and inner envelope membranes of higher plant plastids. *Biochem. Biophys. Acta* 1465:307–323.

40. Nursten, H.E. 1970. Volatile compounds: The aroma of fruits. Pp. 239–268. In: *The Biochemistry of Fruits and their Products.* A.C. Hulme (ed.). Academic Press, New York.

41. Ozilgen, M., and M. Ozdemir. 2001. A review on grain and nut deterioration and design of the dryers for safe storage with special reference to Turkish hazelnuts. *Crit. Rev. Food Sci. Nutr.* 41:95–132.

42. Park, S.W., D.S. Chung and C.A. Watson. 1971. Adsorption kinetics of water vapor by yellow corn. I. Analysis of kinetic data for sound corn. *Cereal Chem.* 48:14–22.

43. Petty, H.R. 1993. *Molecular Biology of Membranes: Structure and Function.* Plenum, New York.

44. Platenius, H. 1939. Wax emulsions for vegetables. *Cornell Agri. Expt. Sta. Bull.* 723.

45. Riederer, M., and J. Schönherr. 1985. Accumulation and transport of (2,4-dichlorophenoxy) acetic acid in plant cuticles: II. Permeability of the cuticular membrane. *Ecotoxicol. Environ. Safety* 9:196–208.

46. Riederer, M., and L. Schreiber. 2001. Protecting against water loss: analysis of the barrier properties of plant cuticles. *J. Exp. Bot.* 52:2023–2032.

47. Robinson, J.E., K.M. Browne and W.G. Burton. 1975. Storage characteristics of some vegetables and soft fruits. *Ann. Appl. Biol.* 81:399–408.

48. Sanders, D., and M. Tester (eds.). 1997. Ion channels. *J. Exp. Bot.* 48:353–631.

49. Schönherr, J. 1976a. Water permeability of isolated cuticular membranes: The effect of pH and cations on diffusion, hydrodynamic permeability, and size of polar pores in the cutin matrix. *Planta* 128:113–126.

50. Schönherr, J. 1976b. Water permeability of isolated cuticular membranes: The effect of cuticular waxes on diffusion of water. *Planta* 131:159–164.

51. Schönherr, J. 1982. Resistance of plant surfaces to water loss: Transport properties of cutin, suberin and associated lipids. Pp. 153–179. In: *Encyclopedia of Plant Physiology.* Vol. 12B. *Physiological Plant Ecology.* O.L. Lange, P.S. Nobel, C.B. Osmond and H. Ziegler (eds.). Springer, Berlin.
52. Schütz, K., and S.D. Tyerman (eds.). 1997. Water channels in *Chara corallina. J. Exp. Bot.*48:1511–1518.
53. Shukla, B.D. 2001. Drying technology and equipment in India. *Drying Tech.*19:1807–1824.
54. Sinicio, R., W.E. Muir, D.S. Jayas and S. Cenkowski. 1995. Thin-layer drying and wetting of wheat. *Posthar. Biol. Tech.* 5:261–275.
55. Smith, W.H. 1938 (for 1937). Anatomy of the apple fruit. Pp. 127–133. *Rep. Fd. Invest. Bd.,* London.
56. Stein, W.D. 1990. *Channels, Carriers, and Pumps: An Introduction to Membrane Transport.* Academic Press, New York.
57. Steudle, E. 1997. Water transport across plant tissue: Role of water channels. *Biol. Cell* 89:259–273.
58. Steudle, E. 2000. Water uptake by roots: Effects of water deficit. *J. Exp. Bot.* 51:1531–1542.
59. Tanner, W., and T. Caspari. 1996. Membrane transport carriers. *Annu. Rev. Plant Physiol. Plant Mol. Biol.* 47:575–626.
60. Ugarcic-Hardi, Z., and D. Hackenberger. 2001. Influence of drying temperatures on chemical composition of certain Croatian winter wheats. *Acta Aliment.* 30:145–157.
61. von Guttenberg, H. 1926. Die Bewegungsgewebe-*Ecballium elaterium. Handb.Pflanzenanat.* 5:117–119.
62. Williams, G.D.V., and J. Brochu. 1969. Vapor pressure deficits vs. relative humidity for expressing atmospheric moisture content. *Naturaliste Can.* 96:621–636.

9

HEAT, HEAT TRANSFER AND COOLING

The flow of energy through substances is one of the underlying requirements for life. In the earth's biosphere, virtually all of the energy available for biological processes is of solar origin. Heat, a form of kinetic energy, represents one part of the total energy; upon entering a molecule, it accelerates the motion of atoms. Temperature is simply a measure of the speed of motion of the atoms. However, temperature alone does not account for the amount of heat present, only the intensity or level of heat. The amount of heat is proportional to the object's temperature multiplied by its mass.

The temperature of postharvest products is crucial, since heat affects the fluidity of membranes, the activity of enzymes, the volatility of aromatic molecules and numerous other processes. Increasing the temperature increases the rate of change during the postharvest period, and change is seldom desired. It is possible therefore to accelerate or retard change through the addition or removal of heat. The amount of heat present, in contrast to temperature alone, is extremely important when we are interested in removing heat to cool an object. Temperature alone does not tell us the heat load present (e.g., a cubic meter of air at 50°C contains only a fraction of the heat found in a cubic meter of mangos at the same temperature).

Heat flows from a warmer (higher energy) substance to a cooler (lower energy) substance. When heat is transferred, some of the fast moving atoms give up part of their energy to slower moving atoms. If you hold one end of a metal rod in a flame and the other in your hand, heat released from the chemical energy of the burning material enters the end of the rod in the fire. The speed of the metal atoms in the flame is accelerated; these in turn interact with their neighbors, causing them to speed up. This transfer of energy moves progressively up the rod until all of the atoms are vibrating rapidly. Eventually the energy is transferred to the molecules within your hand, giving the sensation of hot.

Cold is the lack of heat; objects become cold by the removal of heat. Heat always travels from a warm substance or object to a colder one. Therefore to remove heat, heat must either be adsorbed or transferred to a cooler substance.

1. MEASUREMENT OF TEMPERATURE AND UNITS OF HEAT

Temperature represents the most important single factor in postharvest quality maintenance. As a consequence, accurate measurement of temperature in harvested products is essential. Several different types of temperature measuring devices are available, each having distinct advantages and disadvantages. Glass thermometers are sealed glass tubes that have a reservoir bulb at the base containing a liquid such as mercury or alcohol. The uniform expansion of the liquid with increasing temperature elevates the level of the liquid within the tube. Therefore the liquid will rise or fall with the addition or removal of heat.

The thermometer was once calibrated in units of several different temperature scales based upon different benchmark standards. The scales used have changed with time. In 1706 Gabriel Fahrenheit, a Dutch scientist, made a number of improvements upon the original thermometer invented around 1592 by Galileo Galilei, the Italian physicist.[37] One improvement was to reference the temperature scale to the melting point of ice (32°F) and the boiling point of water (212°F). The scale gave 180 individual divisions or degrees between the freezing and boiling point of pure water. The thermometers made by Fahrenheit gained popularity because of their repeatability and quality. In 1742 Anders Celsius, a Swedish scientist, assigned the melting point of ice as 100° and the boiling point of water 0°, giving 100 equal divisions between. This was later reversed. The Celsius scale is the metric temperature scale that is now almost universally used. Since Fahrenheit and Celsius scales simply span a different number of divisions between the melting point of ice and the boiling point of water on identical thermometers, they can be readily converted mathematically, one to the other.*

The Kelvin scale was introduced in the mid-nineteenth century by the English scientist William Thomson (Lord Kelvin), who developed a universal thermodynamic scale based upon the coefficient of expansion of an ideal gas. This scale differed from the Celsius scale by making 0°K the point at which all motion of molecules should cease—absolute zero. On the Celsius scale this is -273°C (thus the melting point of ice is 273°K, boiling point of water 373°K). By utilizing the same gradations as the Celsius scale, the two are readily converted (i.e., K = C + 273). The Kelvin scale is especially applicable to chemistry studies, while the Celsius scale is the most commonly used and is the scale of choice for postharvest work.

For measuring product temperature, mercury or alcohol in glass thermometers are very inexpensive but tend to be difficult to read and have a slow response time. Other thermometers utilize metal that expands or contracts to measure changes in temperature. The most commonly used are metal dial thermometers which are easy to insert into the product and read, but are slow in response and tend to be the least accurate of the various postharvest temperature measuring devices.

Thermocouples are also utilized to measure temperature. When two dissimilar metals such as copper and nickel are welded together, upon heating, a voltage develops across the open ends of the junction. By measuring the voltage with a sensitive voltmeter calibrated against temperature, the actual temperature can be determined. Thermocouples have several advantages; they are easy to read, accurate, give fast response times, and can be used for remote sensing; however, they are more expensive. Thermistors, in contrast to thermocouples, are solid state semiconductors that allow more electrons to flow (lower resistance) through them as the temperature increases. The change in electrical flow through these metallic oxides (e.g., Mn, Co, Ni) is calibrated against temperature. Thermistors are versatile and sensitive. Another type of resistance thermometer is based upon platinum, whose resistance increases as temperature increases. These are very useful as a transfer temperature standard due to platinum's stability and inertness, though more costly (a transfer standard is a thermometer whose calibration can be traced back to a primary standard).

One final technique for measuring temperature is the use of a radiometer. Product temperature is determined by measuring the infrared irradiation given off by the substance. The amount of infrared radiation changes with product temperature. Although expensive, ra-

---

*Degrees Celsius = 5/9 (degrees F – 32); degrees Fahrenheit = 9/5 (degrees C) + 32; degrees Kelvin = degrees Celsius + 273.

diometers are fast, accurate and can measure temperature without actually contacting the product. There are also time-temperature monitors to record temperature exposure during shipping. Various types are based upon different characteristics such as chemical and enzymatic reactions.

Refrigeration systems used for initial cooling of a product after harvest, during storage, in transit, and at the retail level need to be designed with sufficient refrigeration capacity (i.e., how much heat must be removed). Excessive refrigeration capacity greatly increases costs, while an insufficient capacity results in slow or inadequate cooling of the product. The heat load is the sum of all sources of heat within or moving into the refrigerated area. In addition to product heat, heat from air infiltration, solar exposure, containers, and heat generating devices (e.g., motors, fans, lights, pumps, forklifts and people) collectively make up the total heat load. Product heat comprises the primary portion of the total heat load during initial cooling of the product. After removal from the field, product heat depends upon the temperature of the product, its cooling rate, the amount of product cooled at a given time and the specific heat of the product.

For precise heat load determinations, it is necessary to accurately determine the temperature of the product. The air temperature surrounding the product should never be used as an estimate of product temperature. In addition, the temperature is often not uniform throughout the product; this is especially true during initial cooling after harvest when there can be a significant temperature gradient from the interior (warm) to the exterior (cool). Normally the temperature of the pulp at the center of the densest part of the product is the best site to monitor. For example, the center of an apple fruit or within the stem on celery stalks or lettuce heads is the best indicator of the overall temperature status of these products. Likewise, the temperature of packaged produce should be measured on samples from the center of the package, and from center packages when palletized.

Because of the variation in temperature within a given lot of product, a measure called the mass-average temperature is used.[52] The mass-average temperature is a single value from the temperature distribution within the tissue, representing the final uniform temperature that would be reached throughout the product if moved to constant temperature conditions.

Heat was formerly measured primarily as calories (cal) or kilocalories (1000 cal or 1 kcal) and British Thermal Units (BTU). A calorie is defined as the amount of heat required to raise the temperature of 1 gm of water 1°C. Both the calorie and the BTU are strictly measures of heat and are still commonly used, although the joule (J) is the universal standard unit for energy. The use of joules is based upon the fact that mechanical work can be converted to heat, which is measured as joules or foot pounds. One joule is equal to one Newton · meter^{-1} (i.e., force × distance = work) or for our purposes, 4.187 kJ is the amount of heat required to raise the temperature of 1 kg of water 1°C. Calories, BTU's and joules can be readily interconverted.*

2. TYPES OF HEAT

2.1. Sensible Heat

When a substance is heated and the temperature rises as heat is added, the increase in heat is called sensible heat, (i.e. heat that causes a change in temperature of a substance). When heat

---

*BTUs × 1,055 = joules; BTUs × 252 = calories; calories × 4.187 = joules; calories × 0.00397 = BTUs.

Table 9.1. Heat Properties of Several Common Substances

	Specific Heat Capacity		Melting Point	Heat of Fusion		Boiling Point	Heat of Vaporization	
	$kJ \cdot kg^{-1} \cdot K^{-1}$	$cal \cdot g^{-1} \cdot K^{-1}$	°C	$kJ \cdot kg^{-1}$	$cal \cdot g^{-1}$	°C	$kJ \cdot kg^{-1}$	$cal \cdot g^{-1}$
Water	4.187	1.00	0	335	80	100	2257	540
Methanol	2.512	0.60	−97	67	16	29	272	65
Ethylene glycol	2.345	0.56	−13	188	45	198	800	191
Air	1.005	0.24						
Steam	2.010	0.48						
Ammonia	2.051	0.49	−78			−33	1314	314
Ice	2.094	0.50	0	335	80			
Copper	0.385	0.09	1080	176	42	2310		
Aluminum	0.921	0.22	658	394	94	2057		

is removed and the temperature falls, this is also sensible heat. The amount of sensible heat that is required to cause a change in temperature of a substance varies with both the kind and the amount of substance. The specific heat capacity, the amount of heat that must be added to or removed from a given mass of substance to change its temperature one degree, has been determined for numerous materials (Table 9.1). Specific heat capacity is expressed as $kJ \cdot kg^{-1} \cdot °K^{-1}$, $cal \cdot g^{-1} \cdot °C^{-1}$ or $BTU \cdot lb^{-1} \cdot °F^{-1}$. The amount of heat (S) necessary to cause a given change in temperature of a substance is determined by multiplying the mass (m) times the specific heat capacity (c), times the change in temperature (ΔT):

$$S = m \cdot c \, (\Delta T) \hspace{4cm} \text{eq. 9.1}$$

These calculations require consistent units given as:

$$kJ = (kg)(kJ \cdot kg^{-1} \cdot °K^{-1})(\Delta \, °K);$$

$$cal = (g)(cal \cdot g^{-1} \cdot °C^{-1})(\Delta \, °C);$$

$$BTU = (lb)(BTU \cdot lb^{-1} \cdot °F^{-1})(\Delta \, °F);$$

and can be readily interconverted.

2.2. Latent Heat

One of the most important effects of heat is its ability to change the physical state of pure substances. When sufficient heat is added, solids become liquids and liquids eventually become gases. For each substance at a given pressure, changes in state always occur at the same temperature. When a solid (e.g., ice) changes to a liquid (water), heat must be added; when the transition occurs in the opposite direction, water to ice, heat must be removed. Similarly, when a liquid is converted to a gas, heat must be added and conversely, removed when proceeding from a gas to a liquid. The change in heat within the substance that occurs during a change in physical state does not change temperature. This addition of heat with no change in temperature is distinct from sensible heat (heat causing a change in temperature) and is referred to as latent (i.e., hidden) heat. Latent heat is defined as heat that brings about a change of state with no change in temperature. Latent heat, illustrated in Figure 9.1, can be seen as the heat that is added between points B and C, and D and E, without causing a change in temperature. When a substance melts, the heat is called the **latent heat of melting** or when going from a liquid to a solid upon cooling, the **latent heat of fusion**. For water, 335 $kJ \cdot kg^{-1}$ (80 $cal \cdot g^{-1}$) is required.

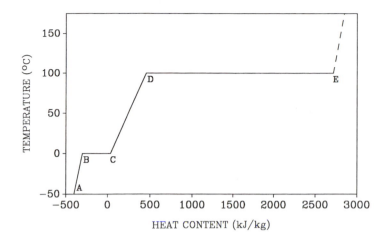

Figure 9.1. The relationship between water temperature and heat content. One kilogram of water at atmospheric pressure (101 kPa) was heated from –50°C to complete vaporization. Between points A and B the temperature of the ice increased with the addition of heat (2 kJ · kg °C × 50°C = 100 kJ). From point B to C, 335 kJ of heat were added to melt the ice (latent heat of melting) without changing its temperature. Between points C and D 420 kJ of heat were required to bring one kg of water to its boiling point (2.4 kJ · kg °C × 100°C = 420 kJ). From point D to E, 2260 kJ of heat were required to convert water to steam (latent heat of vaporization), again with no change in temperature.

The amount of heat that must be added or removed is the same in both directions for a given substance. When a liquid is converted to a gas, the latent heat is called the **latent heat of vaporization** or when cooled the **latent heat of condensation**. Again, the amount of heat that must be added or removed is identical. When water (at 100°C) is converted to steam (i.e., points D to E), 2260 kJ · kg⁻¹ (540 cal · g⁻¹) is required. When one considers the actual amounts of heat exchanged with changes in state, they are quite remarkable. For example, it takes as much heat to change 1 kg of ice to water (both at 0°C) as it does to raise the temperature of the kg of water from 0 to 80°C. Since the energy change from liquid water to gaseous is approximately 6.75 times that of ice to water, volatilization represents an excellent way to absorb heat. In fact the latent heat of vaporization is used commercially (vacuum cooling) to cool a number of harvested products. Vaporization during cooling typically results in a loss of about 2% of the water in high moisture fruits and vegetables.

The temperature at which a change in state occurs depends upon pressure. The lower the pressure, the lower the temperature at which water becomes a gas; the amount of heat required to facilitate the change of state, however, remains the same. Likewise, the greater the pressure, the higher the temperature at which the change occurs. The latent heat of vaporization of water would be of little value for cooling harvested products if the temperature of the water had to first reach 100°C. However, by lowering the pressure to 5 mm Hg (660 Pa), the pressure used for commercial vacuum cooling, water will boil at 1°C.

Since the molecular structure of each substance varies, each has different latent heat values for phase changes. Likewise, the temperature and pressure combination at which the phase change occurs differs widely between substances. For example, methanol has a latent heat of fusion of 67 kJ · kg⁻¹ (16 cal · g⁻¹) at -97°C, while water is 335 kJ · kg⁻¹ at 0°C (Table 9.1).

The cooling effect of ice comes largely from the absorption of latent heat as the ice melts at 0°C. A range of lower melting point temperatures can be obtained through the addition of salts such as NaCl or CaCl$_2$, a common practice when liquid ice* is used (e.g., for broccoli). Salt and ice mixtures can be made that reduce the melting point to -18°C, a temperature that is excessively low for direct contact with most harvested plant products. Based upon product requirements, the amount of salt added can be adjusted to give the desired melting point.

3. HEAT TRANSFER

3.1. Modes of Heat Transfer

Heat can be transferred from one substance or body to another by three processes: radiation, conduction and convection. With harvested products, heat transfer generally involves a combination of all three processes.

3.1.1. Radiation

Radiation is the transfer of heat energy by means of electromagnetic waves. These move outward from the surface of an object in all directions. Any object having a temperature above absolute zero (0°K) emits thermal radiation, the quantity and wavelength being a function of the object's temperature. The wavelengths are typically very long (i.e., longer than 4 μm for infrared). Virtually everything surrounding a harvested product can radiate heat, including water, carbon dioxide, and other molecules in the air, the sun, the soil, buildings, harvesting containers and equipment. The quantity of thermal radiation emitted by an object is proportional to the absolute temperature of the object raised to the fourth power (T^4); therefore the higher the temperature, dramatically greater is the loss of heat as thermal radiation.

3.1.2. Conduction

Conduction is the transfer of heat through direct contact or from one part of an object to another. Heat flows readily up an iron rod by conduction, iron being a very good thermal conductor. Not all substances are equally good conductors of heat; for example wood and air are poor conductors. Several very poor conductors (e.g., urethane foam and fiber-glass) are used as insulators for cold rooms and packaging houses. In general, substances that are good conductors of electricity are also good conductors of heat.

3.1.3. Convection

Convection is the transfer of heat through the movement of a fluid (i.e., a liquid or gas). When air is heated, the energy level of its molecules increases making them move faster and further apart from neighboring molecules. As the distance increases, the actual density of the warm

---

*"Liquid ice" is a slurry of ground ice and water that has sufficient fluidity to allow it to flow. The mixture is pumped into boxes of product under pressure, filling the space not occupied by the product. The water then drains, leaving the ice dispersed throughout the box.

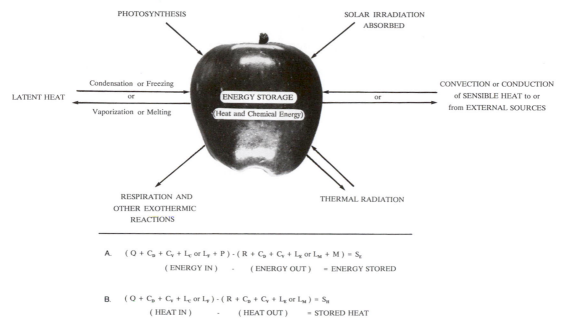

PHOTOSYNTHESIS

SOLAR IRRADIATION
ABSORBED

LATENT HEAT

Condensation or Freezing
or
Vaporization or Melting

ENERGY STORAGE
(Heat and Chemical Energy)

or

CONVECTION or CONDUCTION
of SENSIBLE HEAT to or
from EXTERNAL SOURCES

RESPIRATION AND
OTHER EXOTHERMIC
REACTIONS

THERMAL RADIATION

A. $(Q + C_D + C_V + L_C \text{ or } L_T + P) - (R + C_D + C_V + L_E \text{ or } L_M + M) = S_E$

(ENERGY IN) - (ENERGY OUT) = ENERGY STORED

B. $(Q + C_D + C_V + L_C \text{ or } L_T) - (R + C_D + C_V + L_E \text{ or } L_M) = S_H$

(HEAT IN) - (HEAT OUT) = STORED HEAT

Figure 9.2. Energy and heat balance in harvested products.

air decreases, making it lighter, causing the warmer air to rise. Heat is transferred from its source to other objects through this rising current of air. Convection also occurs in liquids, where the lighter warmer molecules rise in currents to the top.

4. SOURCES OF HEAT

Life within the earth's biosphere survives due to a continual influx of energy, virtually all of which was initially emitted by the sun. Other non-atomic sources (e.g., geothermal, cosmic, magnetic, tidal) are relatively minor contributors. Heat is but one form of the total energy within the biosphere. While energy cannot be created or destroyed (the first law of thermodynamics), it can be changed from one form to another, and much of the solar irradiation striking the earth eventually becomes heat. Re-radiation of energy back into space prevents the biosphere from progressively heating up. During the postharvest period, we are very interested in energy that is in the form of heat. Heat may enter the plant directly as absorbed irradiation or indirectly after being first absorbed by other substances in the plant's environment and then transferred to the plant. The heat moves into the plant by convection, conduction and irradiation.

5. ENERGY BALANCE

Heat load in harvested products is best understood in terms of the inputs and losses of all sources of energy (Figure 9.2). These are summarized in an energy balance equation,

$$Q + R + C_D + C_V + L + P + M + S_E = 0$$

eq. 9.2

where: Q = + energy absorbed as solar irradiation;
   R = ± energy lost or gained as thermal radiation;
   C_D = ± heat lost or gained by conduction;
   C_V = ± heat lost or gained by convection;
   L = ± heat lost or gained due to the latent heat of vaporization, condensation, melting or fusion;
   P = + energy inputs due to photosynthesis (~ 479 kJ is stored per mole of CO_2 fixed and represents a potential source of heat);
   M = - energy released by exothermic reactions within the plant (e.g., respiration, photorespiration and others); and
   S_E = ± storage of energy by the product; when positive, energy storage increases, when negative it decreases. In some instances, the energy assigned to P and M is combined with the storage term.

The terms for conduction (C_D), convection (C_V), and latent heat (L) can be either positive or negative depending upon whether there is a net gain or loss of energy. The terms for absorbed solar irradiation (Q) and photosynthesis (P) are positive, and their energy contribution ranges from relatively little to substantial for Q and zero to minor for P. Both thermal radiation (R) and exothermic metabolism (M) are negative values in the equation. The storage term, S_E, can also be positive (a net gain in energy) or negative (a net loss); however, its sign and magnitude depend upon the composite of all of the other terms in the equation.

   The equation can be simplified to:

$$(Q + C_D + C_V + [L_C \text{ or } L_F] + P) - (R + C_D + C_V + [L_E \text{ or } L_M] + M) = S_E \qquad \text{eq. 9.3}$$

or: (Energy In) - (Energy Out) = Energy Stored

6. HEAT LOAD

During the postharvest period, the heat balance of plants or plant parts is the major concern rather than total energy balance. Equation 3 can be rewritten to include just the terms that contribute significantly to the overall heat load. Energy released by exothermic reactions (M) within the plant and photosynthesis (P) can be ignored as they are rarely significant heat sources. The latent heat terms are L_C for condensation, L_F for fusion, L_E for vaporization, and L_M for melting. Their relative importance varies widely among harvested products. The storage term now becomes stored heat (S_H).

   Therefore equation 2 becomes:

$$(Q + C_D + C_V + [L_C \text{ or } L_F]) - (R + C_D + CV + [L_E \text{ or } L_M]) = S_H \qquad \text{eq. 9.4}$$

(Heat In) - (Heat Out) = Stored Heat

   Different sources of energy vary in their importance to the overall heat budget of a product, and their contribution generally changes markedly from preharvest through final utilization by the consumer. In the field, both before and after harvest, solar and thermal irradiation (Q and R) represent the primary sources of energy that contribute to the product's heat load. Solar irradiation may strike the product directly, be scattered by molecules in the atmosphere before striking the product, or be reflected from other objects in the product's environment (e.g., other plants, clouds, the soil) (Figure 9.3). Not all of the solar irradiation is absorbed by the plant; the amount absorbed depends upon the wavelength and absorption characteristics of the plant (Figure 9.4). Visible light, in the blue (400–500 nm) and red (600–700 nm) regions

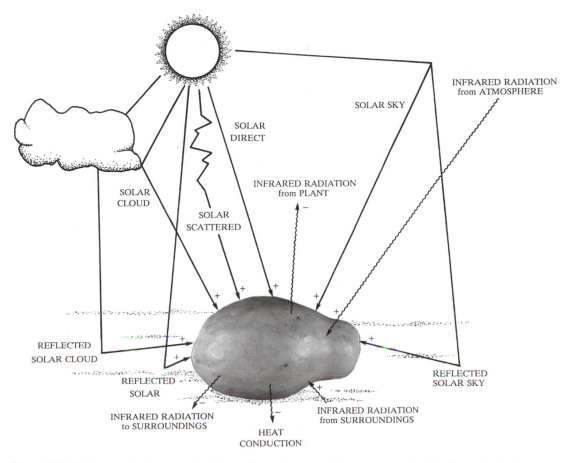

Figure 9.3. The interaction between irradiation from the sun and plant products both before and after harvest.

of the spectrum, is efficiently absorbed by green plant parts. Green (500–600 nm) and the near infrared (700–1500 nm), however, are absorbed to a much lesser extent.

Plants are very efficient absorbers of infrared or thermal irradiation. Infrared irradiation comes from the atmosphere and objects surrounding the plant. Any object with a temperature above 0°K (absolute zero) emits thermal radiation. Because of this, temperature can be measured with a radiometer and photographs can be made in complete darkness using infrared film. Aerial infrared photography is occasionally used to detect insect and disease problems in areas of large fields where stress has altered the normal level of infrared emitted by the plants. Most of the thermal irradiation from the sky is emitted from water, carbon dioxide and other molecules in the air.

Only part of the total solar irradiation striking the plant is absorbed; a significant amount is either transmitted or reflected away. Likewise not all of the absorbed irradiation is converted to heat. In chlorophyll-containing tissues of intact plants, a small part (~1%) of the total absorbed light energy is converted to chemical energy *via* the photosynthetic process. This chemical energy represents a storage form of potential heat that may eventually be liberated by respiration.

While in the sunlight, solar irradiation represents a primary source of heat for intact plants and harvested plant parts. Hence, harvesting at night or during the early morning hours and placing harvested material in the shade reduces this input of energy, thus decreasing the rate of heat build-up.

WAVELENGTH (μm)

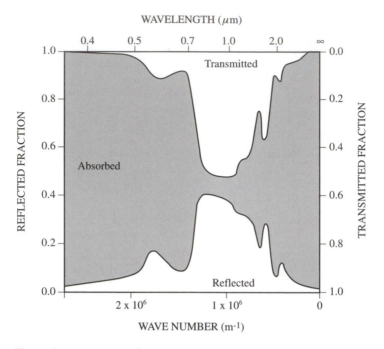

Figure 9.4. Representative fractions of the total solar irradiation at each wave length that are absorbed, transmitted and reflected by a leaf (*redrawn from Gates*[14]).

Thermal radiation of heat (R), radiation in the infrared region of the spectrum, represents an important means of dissipating heat from intact plants. It is emitted from all sides of the product, moving outward until striking another object. After harvest, plant products are consolidated into bins, boxes, trucks and other containers where each individual unit is in contact or close proximity with its neighbor. Thermal radiation emitted, now tends to largely move into the adjacent product, minimizing its impact upon decreasing the heat load of the product.

The importance of the convection and conduction term (C_V and C_D) also shifts after harvest. Convection of heat away from the plant is an important means of heat elimination in intact plants. This is especially true for leaves that have large surface areas and small volumes. During convective heat transfer, the distance between individual air or liquid molecules surrounding the product expands due to the energy transferred from the plant. This in turn decreases the density of the air, causing it to rise upward due to pressure differences, moving the thermal energy away from the plant. Both plant and environmental characteristics have a pronounced influence of the amount of convection. Convective heat exchange is proportional to the difference in absolute temperature between the product and its environment, and inversely proportional to the boundary layer resistance. Plant factors such as the shape and size of the organ in question are critical as they influence boundary layer thickness, as does environmental parameters such as wind speed.

There are two types of heat convection away from a plant part. Free convection is caused by changes in the density of the surrounding air molecules, while forced convection is caused by wind that reduces the thickness of the boundary layer and moves the heated air away from the plant. Forced convection is the dominant means of convection and increases with increasing wind speed. For example, the majority of the difference in the rate of temperature reduction between forced air and static refrigeration precooling is due to forced convection.

Table 9.2. Thermal Properties of Selected Fruits and Vegetables.*

	% H₂O	Density g · cm⁻³	Specific Heat† J · kg⁻¹ · K⁻¹	Thermal Conductivity W · m⁻¹ · K	Latent Heat J · kg⁻¹
Avocado†	94.0	1.06	3,810	0.4292	316,300
Banana	74.8	0.98	3,350	0.4811	251,200
Beet root	87.6	1.53	3,770	0.6006	293,100
Broccoli	89.9	—	3,850	0.3808	302,400
Cantaloupe	92.7	0.93	3,940	0.5711	307,000
Carrot	88.2	1.04	3,770	0.6058	293,000
Cucumber	97.0	0.95	4,103	0.5988	—
Lemon	89.3	—	3,850	1.817	295,400
Nectarines	82.9	0.99	3,770	0.5850	276,800
Peach	86.9	0.93	3,770	0.5815	288,400
Pear	—	1.00	—	0.5954	—
Peas, black-eyed	—	—	—	0.3115	246,600
Pineapple	85.3	1.01	3,680	0.5486	283,600
Turnip root	90.9	1.00	3,890	0.5625	302,400

*Source: ASHRAE,[2] Reidy,[46] Sweat,[56] Turrell and Perry[60].
†Specific heat values above freezing.

Conduction, the flow of heat between parts of a substance or neighboring substances that are in contact, can occur between any of the three physical phases (e.g., solid → solid, solid → gas, liquid → solid, gas → liquid, etc.) due to molecular and/or electronic collisions. Conduction in plants occurs in the boundary layer surrounding an organ and within the tissue. In the boundary layer, heat is transferred through the random collisions of gas molecules. The conduction of heat within the boundary layer is much slower than conduction within the plant since the thermal conductivity of air is much lower than water's, the primary constituent of most succulent plant parts (i.e., 6.0×10^{-5} cal · s⁻¹ · cm⁻¹ per degree C for air, versus 1.43×10^{-3} cal · s⁻¹ · cm⁻¹ per degree C for water).

Conduction is also responsible for the movement of heat within harvested products. This is especially important with large or dense plant parts (e.g., apple fruit, potato tubers) that can store relatively large amounts of heat. The transfer of heat through plant products is slow and moves as a wave from warm to cooler areas. Thermal waves are caused by variations in heat input that are seldom constant. In the field, heat absorption by intact plants varies markedly over a 24 hour period. Heat inputs during the postharvest period, likewise, are seldom constant. Because of this variation, heat normally moves as a wave, having both magnitude (height) and amplitude or frequency (width). Wave magnitude depends upon the temperature differential between the product and the heat source, while the frequency depends upon the duration of the heat input oscillation. The velocity of the wave through a product depends upon the product's conductivity, specific heat capacity and density, and the frequency of the thermal wave.[9] Specific thermal conductivity and heat capacity values for several postharvest products are presented in Table 9.2. Calculations of the actual heat transfer in a sample depend upon the product's thermal properties and geometry, the temperature of the heat source (or sink) and the product, and the temperature differential between the two.

After harvest the duration and magnitude of heat inputs vary widely, depending upon the source and how the product is handled. For example, heat is often transferred from the rear tires and axles of trucks upward into the load of harvested product. The frequency of the wave will depend upon a number of factors (e.g., distance traveled, number of stops).

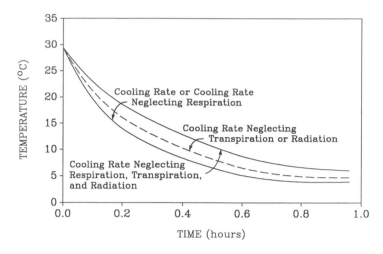

Figure 9.5. The effect of radiation, respiration and transpiration on forced air cooling of Brussels sprouts (air velocity of 40 fpm) (*redrawn from Gaffney, et al.*[13]). Metabolic heat (respiration) has a very minor effect during initial product cooling.

Conduction becomes increasingly important as the individual harvested units are consolidated, placing them in direct contact with each other. If you use the difference in thermal conductivity of air and water as a rough estimate of the difference in the heat transfer rate between two touching objects versus the object and air, the heat conducted per unit surface area is approximately 24 times more effective between touching objects than the object and air. The difference in thermal conductivity between water and air is the reason that water or hydro-cooling lowers the temperature of a harvested product faster than air cooling at equivalent temperatures.

The change in state that occurs with the evaporation of water (latent heat of vaporization) absorbs energy from the product, cooling it. Water evaporates from plants at the air-liquid interface, primarily at the cell walls of mesophyll, epidermal and guard cells. The heat removed due to evaporation can be calculated from the product of the rate of evaporation ($g \cdot cm^{-2} \cdot s^{-1}$) and latent heat of vaporization of water ($2.4 \, kJ \cdot g^{-1}$). With typical intact crop plant growing under moderate conditions,* the latent heat term in the energy balance equation (L) typically represents slightly less than 50% of the daytime heat load, the remainder being largely dissipated by radiation, convection and conduction. If transpiration is stopped, an equivalent amount of heat will be lost due to increased radiation, convection and conduction with only a 2 to 3°C rise in leaf temperature.

The rate of water loss by intact plants *via* transpiration is generally far greater than that from detached products. Stomatal closure occurs after harvest, and with the consolidation of individual plant parts in containers, evaporation is further reduced, decreasing the amount of latent heat that can be dissipated by transpiration. Lowering the atmospheric pressure decreases the temperature at which water boils, and increases the rate of evaporation (i.e., heat removed). This acceleration of evaporation, utilizing the latent heat of vaporization of water, is the thermodynamic principle involved in vacuum cooling of harvested products. Since evap-

---

*Air temperature 20°C, leaf temperature 25°C, 50% RH, net radiation 370 Wm⁻².

Table 9.3. Effect of Product Temperature on the Production of Metabolic Heat by Selected Harvested Plant Products.*

	Metabolic Heat Produced $kJ \cdot Ton^{-1} \cdot hr^{-1}$ Product Temperature				
	0°C (32°F)	5°C (40–41°F)	15°C (59–60°F)	20°C (68–70°F)	25°C (77–80°F)
Apple	24–44	53–78		145–330	179–373
Asparagus	300–640	630–1120	1236–2496	1856–2869	3965–5075
Beans, lima	112–320	208–383	1066–1328	1415–1910	
Carrot	102–218	136–281	276–572	490–1013	
Lettuce, head	63–179	141–213	339–480	543–640	780–974
Onion, dry	112–238	184–727	702–1037	824–1663	1042–2235
Orange	19–53	39–78	136–252	238–364	262–431
Radish, topped	34–102	63–141	238–451	475–616	645–945
Strawberry	131–189	175–354	756–984	1091–2089	1793–2249
Tomato, mature green		53–87	175–301	301–441	368–543

*Source: Catlin et al;[8] Haller et al;[20] Pentzer et al;[43] Scholz et al;[51] Smith;[52] Smock and Gross;[54] Tewfik and Scott.[57]

oration is a surface phenomenon, products with large surface-to-volume ratios (e.g., lettuce) are particularly suited to this type of cooling.

The condensation of water on the surface* of a harvested product releases (latent heat of condensation) the same amount of energy ($2.4 kJ \cdot g^{-1}$) as that absorbed with evaporation. Condensation has two primary effects upon the product; it adds heat and wets the surface. The importance of each is dependent upon the product in question as well as the conditions to which it is exposed.

The terms photosynthesis (P), exothermic reactions (M), and storage (S) are often grouped collectively in the energy balance equations under the storage term. The rationale for this is that photosynthesis and metabolic heat are generally very small factors in the overall heat balance of the plant. Photosynthesis represents irradiation absorbed and converted to chemical energy rather than heat while metabolic heat represents chemical energy released as heat. The actual storage term represents stored heat (S_H), with changes being seen as changes in temperature. When S_H is positive, product temperature is increasing; when negative, the temperature is decreasing.

How important is metabolic heat in the overall heat load of a harvested product? Under most situations metabolic heat has a negligible impact on the temperature of the product because of the relatively high rate of heat exchange between the product and its surrounding environment. For example, the temperature excess at the center of an apple fruit caused by metabolic heat is only 0.023°C after 2 hours of cooling.[4] With forced air cooling of Brussels sprouts (Figure 9.5), metabolic heat also represents an extremely minor component of the heat load.

Under conditions where the product has a very high rate of respiration and heat exchange is restricted (e.g., tight packing of leafy crops), metabolic heat can be significant. The production of metabolic heat decreases rapidly with a decline in product temperature (Table 9.3),

---

*The formation of dew or frost.

therefore rapid cooling is advantageous. Once cooled to the proper storage temperature, respiratory heat represents a small but significant component in the overall heat load and must be included in estimates of refrigeration requirements.

7. ENVIRONMENT AND ENVIRONMENTAL FACTORS AFFECTING HEAT TRANSFER

An intimate relationship exists between a harvested product and its environment. Thermal energy fluxes into or out of the product depending upon the relative properties of the environment and product at any point in time. The conditions of the environment have a pronounced effect upon the rate of change in product temperature. During the postharvest period, the product is normally in contact with one or more media, whether gaseous, liquid or solid, that can either add or remove heat. The specific properties of each medium in turn affect the rate and direction of heat transfer.

7.1. Gas Environments

The majority of harvested products are stored in gas atmospheres, typically air. Transfer of heat between the product and its gas environment is due to convection, conduction, absorbed irradiation and thermal radiation. Physical parameters that affect heat inputs and losses are irradiation, the temperature differential between the product and its gas environment, movement of the gas, the amount of heat the gas environment will hold (a function of volume and specific heat capacity) and the atmospheric pressure. Therefore, irradiation quality, quantity and duration; atmospheric humidity; air velocity and direction; atmospheric pressure; and water concentration interact in varying degrees of complexity to modulate heat exchange.

7.1.1. Irradiation

Irradiation from the environment surrounding the product has a major impact upon product temperature. Exposure to direct sunlight generally increases product temperature; the higher the intensity (e.g., noon versus early morning), the greater the effect. Moving harvested material out of the sun or direct sunlight, greatly reduces irradiation energy inputs. Solar irradiation also elevates the energy of other objects in the environment surrounding the product (e.g., air, plant, soil, roadways, buildings, harvesting equipment). This, in turn, can contribute to the heat load of the product by reflected, transmitted, and infrared irradiation and the convection and conduction of heat. For example, road surface temperatures as high as 61°C (air temperature of 40.6°C) are common in the summer months in many areas, and part of this heat is radiated upward into the load.[28] Surface friction of the trailer tires also generates heat that is transferred into the product held in the trailer. For example, shelled southern peas found within areas of the trailer directly over the wheels were 4.4 to 5.6°C warmer than peas found at the same depth elsewhere in the load.[26]

In intact plants, solar irradiation increases the rate of photosynthesis (P), thus part of the absorbed energy that would have otherwise been converted to heat is trapped as chemical energy. This decrease, however, is extremely small (<1%) in intact plants and is essentially nonexistent in detached and nonchlorophyllous tissues.

7.1.2. Humidity

Atmospheric humidity affects the specific heat capacity of the air, the amount of heat removed from the product due to the latent heat of vaporization (L_E or conversely given off with condensation, L_C), the amount of thermal radiation emitted from the air and the photosynthetic rate of the tissue. Increasing humidity decreases the water vapor pressure deficit between the product and its environment, decreasing the rate of evaporation. As a consequence, less heat is removed from the product. Water molecules in the air also absorb solar irradiation which is reradiated as thermal irradiation. The higher the moisture content of the air, the greater the heat capacity. Finally, humidity can alter the aperture of leaf stomata, thus the rate of photosynthesis. Depressed photosynthesis, therefore, results in a small increase in heat load.

7.1.3. Air Movement

Air movement around the product increases the rate of forced convection, accelerating heat transfer between product and environment. Relatively small increases in air velocity have significant effects. Increases in convective heat transfer make forced air cooling a much more rapid means of removing field heat than conventional room cooling.

7.1.4. Pressure

Changes in atmospheric pressure primarily affect the rate of evaporation, hence the latent heat of vaporization. There is an inverse relationship between pressure and the rate of evaporation. Vacuum cooling of harvested products is dependent upon the heat absorbed when water vaporizes and is enhanced at reduced atmospheric pressure within the cooling vessel.

7.1.5. Oxygen and Carbon Dioxide Concentration

Changes in heat balance due to an altered oxygen concentration are largely due to an effect on the rate of respiration of the tissue. While respiration is depressed at very low oxygen levels, under most situations metabolic heat is a very minor component in the overall thermal balance of a tissue.

Carbon dioxide concentration can alter the rate of photosynthesis (P) and respiration (M), but as with oxygen the effect on the overall thermal status of the product is very small. Atmospheric carbon dioxide also absorbs solar irradiation, emitting thermal irradiation, part of which may be absorbed by the plant. Harvested products being held in higher than ambient carbon dioxide concentrations (e.g., during controlled environment storage) are rarely exposed to direct solar irradiation.

7.2. Liquid Environments

While only a relatively small number of live plant products are stored in liquids, typically water, the surfaces of many are covered to varying degrees and durations with surface water. This presence of water may be intentional (e.g., during hydrocooling, transfer using water flumes, washing, and display) or unintentional (e.g., condensation upon the surface and rain). The use of ice also creates a liquid environment in that ice-product interfaces are typically quite short lived, rapidly becoming an ice-water-product interface.

The transfer of heat to or from a product into a liquid medium differs from air in that conduction and convection account for virtually all of the heat exchange. Critical properties of the exchange are: (1) the temperature differential between the product and media, (2) the velocity of the media around the product (forced convection), (3) the surface area of the product in contact with the liquid (e.g., scattered droplets versus complete coverage), (4) the thermal conductivity of the liquid, and (5) the amount of heat the liquid will hold (a function of both the volume of the liquid and the liquid's specific heat capacity). Likewise, heat inputs (e.g., irradiation, conduction) and losses (thermal radiation, conduction) from the medium affect the temperature differential between the product and the liquid. During cooling operations, heat transferred to the liquid medium is removed using some form of refrigeration or liquid exchange (e.g., the flow of cold well or river water through a cooling chamber). For most cooling rate calculations, the temperature of the liquid is considered to be constant. When heat is not removed from the liquid, the amount of heat in the liquid increases in direct proportion to the amount of heat lost by the product and the medium's surroundings (e.g., container walls).

7.3. Solid Environments

A varying percent of a harvested products' surface area may be in contact with a solid object after harvest. The solid may be adjacent product, the harvesting container, or in some cases special cooling equipment. Conduction transfers heat between the product and a solid object in contact. The rate of heat transfer is modulated by the size of the temperature gradient between the two objects, the size of the contact surface area, and the thermal properties of both objects (e.g., the rate of heat loss or absorption and the specific heat capacity). As mentioned previously, while ice is a solid, it seldom remains in direct contact with the product; rather a film of water forms at the interface as the ice melts.

Under normal conditions, solids are never in contact with the entire surface area of the product; air and/or water generally interface with the remaining area. The solid-product contact area is typically only a relatively small percentage of the total surface area, and as a consequence, product-to-solid heat transfer is seldom an efficient way to rapidly exchange heat. Exceptions would occur when the temperature differential between the two is quite high (e.g., grain on a hot drying surface). Solids, therefore, generally represent secondary sources of heat inputs or losses for harvested plant products.

8. PRODUCT FACTORS AFFECTING HEAT TRANSFER

The transfer of heat into or from a harvested product involves two processes: (1) the movement of heat within the product and (2) the absorption or loss of heat at the product's surface. Both processes are influenced by the physical and chemical characteristics of the plant material.

8.1. Internal Movement of Heat

Thermal fluxes move inward as an object gains heat and outward as the object loses heat to its environment. Transfer of heat within plant products occurs by conduction, and the rate of heat transfer is influenced by the composition and density of the object. The percent water is particularly important, since the thermal conductivity of water is far greater than that of the carbohydrate structural framework or intercellular air. Water also has a much higher specific heat capacity, therefore, much more heat must be removed to cool a saturated rice kernel than a dry one.

The relative ratio of solids and liquids to air space and their continuity is also important. Heat will move much more readily when the aqueous phase is continuous than when it must traverse intercellular air spaces.

8.2. Heat Transfer at the Surface

The transfer of heat into or out of a product is a surface phenomenon; the larger the surface-to-volume ratio, the greater the potential for transfer. Transfer at the surface is strongly modulated by the product's boundary layer, a transfer zone where the gaseous or liquid medium is in contact with the product. Generally, the thicker the boundary layer the slower the convection of heat through it. The physical characteristics of the object can have a pronounced effect upon boundary layer thickness. Objects that are relatively small and have large surface-to-volume ratios typically have thinner boundary layers than larger objects. This is due to the effect of the movement of the medium upon boundary layer thickness. The greater the movement, the thinner the layer and the faster the convective heat transfer. Structural characteristics that alter the topography of the surface (e.g., trichomes, uneven surface structure) decrease the effect of media movement on the boundary layer.

Boundary layer thickness is rarely uniform across the entire object. It is thinnest where the moving medium initially touches the object. Therefore, in addition to product shape and size, the orientation of the product relative to the flow of the medium is also important. Heat transfer will be greatest where medium first contacts the plant material.

9. HEAT REMOVAL AFTER HARVEST

Temperature change in freshly harvested plant products typically involves the removal of heat. Most plant products contain substantially more heat at harvest than is normally acceptable for subsequent handling and storage. In highly perishable fruits, vegetables and flowers, this is especially true, and it is also often the case in grain and pulse crops if their moisture content is high.

The heat contained by a product at harvest (field heat) largely comprises thermal energy from the environment surrounding the plant. To maintain the maximum storage potential of a product, it is desirable to remove this field heat as quickly as possible after harvest. For highly perishable products, the longer the removal of field heat is postponed, the shorter the time interval the product remains marketable.

Cooling after harvest can be separated into two stages: (1) the removal of field heat bringing the product temperature down to or approaching the desired storage temperature; and (2) the maintenance of that temperature through the continued removal of respiratory heat and heat moving into the storage environment (e.g., heat conducted through the walls and floor, air infiltration). For highly perishable products, the rapid removal of heat prior to storage or marketing is termed **precooling.** The existing definition for precooling is not overly rigid such that the term may be used to include cooling that occurs prior to shipment or processing.

While it is possible to remove field heat in storage, a common practice for less perishable products (e.g., nuts), the rate of heat removal is relatively slow due to the smaller refrigeration capacity of most storage rooms. The refrigeration capacity needed for rapid precooling is sub-

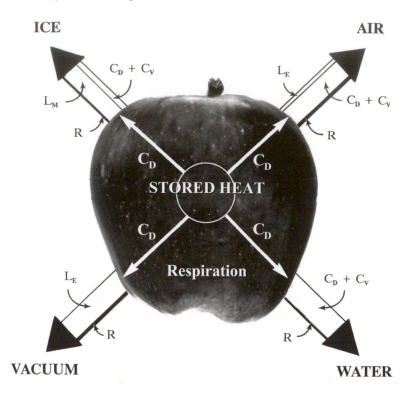

$$(\, Q \, + \, R \, + \, C_D \, + \, C_V \, + \, L_C \text{ or } L_F \,) \, - \, (\, R \, + \, C_D \, + \, C_V \, + \, L_E \text{ or } L_M \,) \, = \, S_H$$

(Minimize Inputs) - (Maximize Outputs) = Decline in Stored Energy

Figure 9.6. Methods of removal of heat from harvested plant products, R = thermal radiation, C_D = conduction, C_V = convection, L_E = latent heat of vaporization, L_M = latent heat of melting, L_C = latent heat of condensation, L_F = latent heat of fusion, S_H = stored heat, and Q = solar irradiation absorbed. The width of the area within each arrow is proportional to the contribution of each method of heat removal.

stantially greater than that needed for subsequent storage and the combination of the two is seldom an economically sound choice. As a consequence for highly perishable crops, precooling and storage are generally accomplished using different refrigeration systems.

9.1. Refrigeration

Refrigeration involves the extraction of heat from a substance, thus lowering its temperature and keeping the temperature below that of its surroundings. Since heat flows from regions of high temperature (energy) to regions of low, the removal of heat from a harvested product requires a sink for this thermal energy. Generally, a refrigeration medium is used to lower the temperature of the product. Common cooling media that come in contact with the product are air, water, and ice (Figure 9.6), each differing somewhat in the relative importance of the various means of heat transfer.

Table 9.4. Methods of Mechanical Refrigeration

Method	Description
Vapor compression	A closed system where the refrigerant cycles between liquid and vapor phases. For cooling it utilizes the heat absorbed during volatilization of the refrigerant; the gaseous refrigerant is then compressed with removal of the heat away from the cooled area. The now liquified refrigerant moves back to the low-pressure side to repeat the cycle.
Vapor absorption	A closed system where cooling is from vaporization of the refrigerant. Instead of mechanical compression of the gaseous refrigerant, as with the vapor compression technique, the gaseous refrigerant is either absorbed or reacts with a second substance [e.g. water (refrigerant) and ammonia (absorbent)], causing the pressure drop. Both the refrigerant and absorbent are then recharged, using various methods to repeat the cycle.
Air cycle	When air under pressure is allowed to expand, its temperature falls. Air removed from the cold room is compressed, with the heat removed using a water-cooled coil. The air is then allowed to expand in a cylinder against a piston (doing work), causing the air temperature to decline and be returned to the refrigerated area.
Vapor jet	This process is similar to the vapor compression system; however, the gaseous refrigerant is drawn from the evaporator and compressed using a high-pressure vapor, usually steam, that is passed through one or more nozzles (also called a thermocompressor).
Thermoelectric	The cooling effect is produced when an electrical current is passed through a junction of two dissimilar metals; one becomes cool and the other warm.

There are two basic objectives in refrigeration: first, to get the medium cold so that it will act as a heat sink, and second to keep it cold as the medium begins to absorb heat and warm up. Until the development of mechanical refrigeration, natural sources of refrigeration were relied upon almost exclusively. For example, blocks of ice cut during the winter months in the higher latitudes of the temperate zones and stored for use during the summer, cold water from streams or deep wells, and cooler air that occurs during the night and at high altitudes have been used. When air or water from streams and wells is used, heat is removed from the cooling media by replacing it with cold, new media. There is no attempt to recool the media. As ice melts, losing its cooling capacity, it is likewise replaced with new ice.

Mechanical refrigeration allows continuous removal of the heat absorbed by the media, hence minimizing the need for continuous changes and large media volumes. Heat can be removed using several refrigeration techniques: direct vaporization of a liquid, vapor compression, vapor absorption, air cycle, vapor jet and thermoelectric cooling (Table 9.4). The volatilization of liquids, absorbing energy through the latent heat of vaporization, was first described as a means of mechanical refrigeration in 1755 by William Cullen. Liquid evaporation has since become a primary means of direct cooling certain products (e.g., lettuce) in several countries.

The vapor compression technique couples vaporization and subsequent reliquefication of the refrigerant* using compression. The first patent for such a machine was issued in 1834

---

*A refrigerant differs from the cooling medium. The cooling medium is also referred to as a secondary refrigerant, in that the former is a liquid used for heat transfer in a mechanical refrigerating system that absorbs heat at a low temperature and pressure and rejects heat at a higher temperature and pressure, usually involving a change in state. The cooling medium or secondary refrigerant is a nonvolatile substance that absorbs heat from a substance or space and rejects the heat to the evaporator of the refrigeration system.[2]

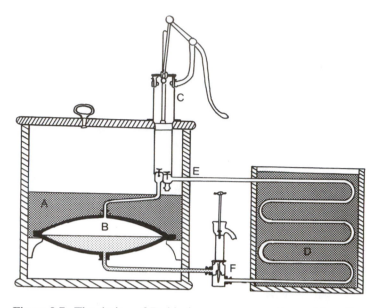

Figure 9.7. The design of Perkins's vapor compression refrigeration system patented in 1834. The refrigerant (a volatile liquid such as ether) boils in the evaporator (B) removing heat from the water held in the container (A). The volatile refrigerant is removed from the evaporator and compressed using a hand pump (C) and passes to the high pressure side *via* (E). As the refrigerant condenses on the high pressure side of the system, the heat (latent heat of condensation) moves into the water surrounding the condenser coils (D). The reliquified refrigerant then moves past a weight-loaded valve (F) that maintains the pressure differential between the two sides, and the cycle begins anew (*redrawn from Gosney*[15]).

to Jacob Perkins (Figure 9.7). The essential components are a compressor, condenser, expansion valve and evaporator (Figure 9.8). The refrigerant volatilizes in the evaporator, absorbing heat from the cooling medium surrounding the evaporator (i.e., air, water). Vaporization is caused by a drop in pressure within the closed system generated by the compressor. The gaseous refrigerant that is drawn into the compressor and compressed begins to liquefy, giving off heat (latent heat of condensation). The heat is removed using a condenser situated outside of the refrigerated area, with air or another medium passing over the coils to facilitate heat transfer. The condensed refrigerant then passes through an expansion valve to the evaporator, the low pressure side of the refrigeration system, to repeat the cycle. Therefore, heat is removed from the cooling medium by the evaporator coils utilizing the latent heat of vaporization. The heat is then transferred away from the refrigerated area and dissipated outside using the condenser coils.

Mechanical refrigeration systems are used to cool the medium that will act as a heat sink for the harvested product. In many cases, heat is removed from the cooling medium (e.g., air, water) through its continued circulation over the evaporator coils of the refrigeration unit. In other cases (e.g., ice, ice, liquid carbon dioxide or nitrogen), an expendable refrigerant is used and is metered into the cooling medium (e.g., liquid nitrogen into air, ice into water). In this case, the medium does not recycle over the evaporator coils to be recooled.

Expendable refrigerants such as ice, solid or liquid carbon dioxide, and liquid nitrogen

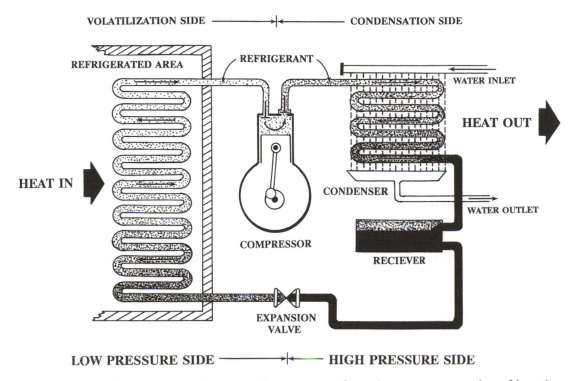

Figure 9.8. A simplified schematic of the essential components of a modern vapor compression refrigeration system. Heat is absorbed with the volatilization of the refrigerant on the low pressure side and removed with condensation of the refrigerant after being compressed on the high pressure side.

have several advantages, the primary one being that cooling is not tied to the location of the refrigeration system. Ice is often applied during field harvest of many leafy crops during hot weather. For example, turnip greens are layered with crushed ice in the beds of transport trucks for cooling during shipment to the processor. Likewise, in areas where the cost of the electricity used to run the refrigeration system varies with time of day, ice may be produced during non-peak electrical use periods and stored for use during periods when electrical costs are high.

Ice for precooling, package icing, and transit icing is typically either block ice or one of several forms of fragmentary ice. The latter are being increasingly used by produce packaging plants in that small pieces of ice are formed and are the most desirable for rapid cooling. Block ice is utilized more for field application, and is generally purchased from a supplier and transported to the field. Block ice has a much greater bulk density than fragmentary ice and requires less space. The low surface-to-volume ratio of block ice also decreases melting during transit and holding prior to use. When applied to the product, the ice is mechanically fragmented increasing the surface area and cooling rate.

The most common fragmentary ice makers produce flake, tube or plate ice. Each of the three types has certain advantages.[2] For example, flake ice is produced continuously without an intermittent thawing cycle for removal of the ice from the freezing surface, a step that is required by tube and plate ice makers. As a consequence, flake ice is colder, non-wetted on the surface and gives a maximum amount of cooling surface. For some uses, however, thicker pieces of ice may be needed, making tube or plate ice more desirable.

Fragmentary ice can be produced on a continuous or a constant number of cycles per

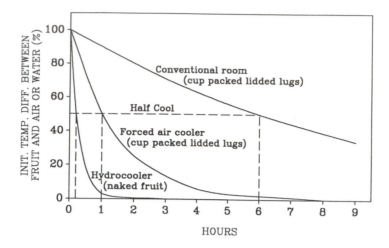

Figure 9.9. A comparison of three methods (room, forced-air and hydrocooling) for cooling peaches (*redrawn from Guillou*[19]).

hour basis, giving a relatively constant output. Ice usage, on the other hand, is typically on a batch basis. Packinghouses seldom operate 24 hours a day, and even during periods of operation, usage is seldom constant. Therefore, ice storage systems offer several advantages. First, a smaller ice-making system is required since it can be run 24 hours a day, collecting and storing the ice for use during peak periods. Secondly, ice can be produced during times of the day when electrical demand is low and power companies offer reduced rates.

Ice storage areas vary, depending upon whether short or long term storage is desired and upon the degree of automation. Ice can be removed from the storage area manually or using screw conveyors, ice rakes or other systems. Delivery systems to the site of use include screw, belt and pneumatic conveyors.

Liquid nitrogen and liquid carbon dioxide are also used to a limited extent; however, their very low temperatures (i.e., -195.8°C and -78.5°C) in contrast to ice (0°C) require much more careful control of application. Liquid nitrogen is used in some transport controlled atmosphere systems to provide both a heat sink and a source of nitrogen to maintain a low oxygen environment.

9.2. Precooling Methods

Precooling produce to remove field heat as quickly as possible after harvest is essential for slowing the rate of deterioration of highly perishable products. This generally represents a single controlled handling step after harvest and may be accomplished using several different techniques (e.g., room cooling, forced-air cooling, hydrocooling, contact icing and vacuum cooling), the appropriate choice being determined largely by type of product encountered.

9.2.1. Room Precooling

Harvested produce may be cooled by simply placing it within a refrigerated area. Typically refrigerated air is blown horizontally just below the ceiling, sweeping over and down around the

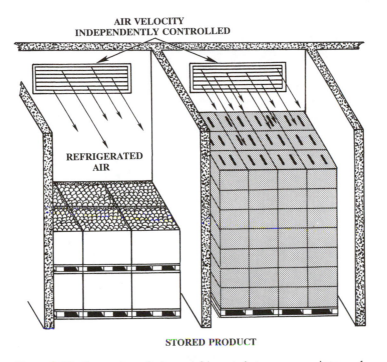

AIR VELOCITY INDEPENDENTLY CONTROLLED

REFRIGERATED AIR

STORED PRODUCT

Figure 9.10. Separation of a large refrigerated storage room into cooling bays, illustrated above, provides a way to decrease the amount of moisture lost from the product. Air circulation velocity in each bay is independent, allowing a high velocity to be used for new produce which is warm and low for product that is already cool.

containers of product below. Upon reaching the floor it moves horizontally into the return vent to be recycled. Variations of this technique (e.g., ceiling jets) are also used. Air velocities of 60 to 120 m · min^{-1} around the containers are required to minimize the length of time required for cooling. After reaching the desired product temperature, air flow is reduced to 3 to 6 m · min^{-1}, a rate sufficient to maintain product temperature while minimizing water loss. Room cooling also requires well ventilated produce containers and proper container stacking and stack spacing within the room.

The major advantage of room cooling is that the product is cooled and stored in the same place, decreasing the amount of handling required. Primary disadvantages include a relatively slow rate of cooling (Figure 9.9) and greater moisture loss from the product. The increased moisture loss is due to greater fluctuations in room temperature and a prolonged exposure to high air velocities. The first product placed in the room is subjected to high airflow rates until the last product moved in has cooled. The increased moisture loss can be largely eliminated using cooling bays within the room (Figure 9.10).

With normal air discharge, refrigerated air is blown outward from one side of the room over the containers of produce. Air velocity, however, decreases with increasing distance from the source causing produce stacked further from the fans to have less air passing over it. Ceiling jets are designed to increase the uniformity of air turbulence within the room. Air is directed downward from a number of sites in a false ceiling. Typically these discharge sites are metal or plastic nozzles, one being situated over a single pallet sized stack of produce on the floor. Air moves downward over the sides of the stack, increasing the rate of cooling. Ceiling

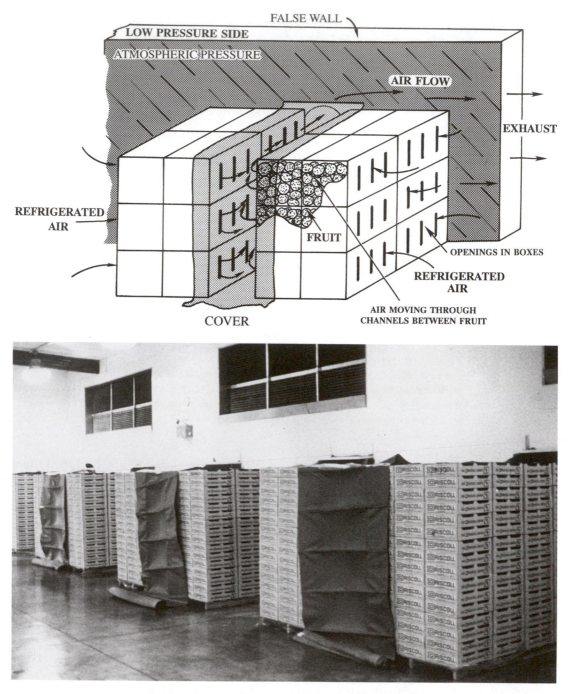

Figure 9.11. Forced air cooling utilizes a pressure drop between the room and two rows of vented containers. The space between the containers is covered with a baffle made of a heavy but flexible material. The baffle seals the top and one end. An exhaust fan placed on the opposite end (in this case mounted in the wall), pulls air from the space between the containers, creating a pressure drop. In response to the pressure differential, cold air moves through holes in the boxes, across the product and into the interior chamber from which it is exhausted. This cools the product much faster than a more static air flow system (i.e., room cooling) (*photograph courtesy of R.F. Kasmire*).

jets require correct placement of the product under the outlet, greater fan power requirements due to the increased resistance, and higher initial construction costs, while reducing the total volume that can be placed within the room.

Heat removal from containerized produce is by forced convection, conduction, radiation and a relatively small amount of evaporation. The dominant means of heat transfer is *via* forced convection due to the movement of air around the produce. Because of this, air movement into the containers is essential for rapid cooling. This is achieved using containers with holes or slots in the sides. Removal of approximately 5% of the surface area of the container sides decreases the cooling time by approximately 25%.

Large cooling rooms are often separated by internal walls to form individual bays. This allows segregating new warm produce from product that has already been partially cooled. Air circulation in each bay is independent, allowing the air velocity to be reduced in areas with cooled produce but remain high in areas with warm produce. Heat from the new produce, likewise, does not warm the cooled produce and minimizes the potential for condensation of moisture from the warm air on the already cooled product.

9.2.2. *Forced-Air Precooling*

Forced air or pressure cooling utilizes a pressure drop across opposite faces of stacks of vented produce containers to move air through the internal air spaces of the container.[20] Forced-air cooling differs from room cooling in that forced air cooling moves the air around the individual product units (e.g., individual fruits) rather than merely around the exterior of the container. This greatly accelerates the removal of heat from the product, typically reducing cooling times to 1/4 to 1/10 that of conventional room cooling (Figure 9.9). Critical components for forced air cooling are sufficient refrigeration capacity and air velocity, the use of vented containers, and utilization of a proper stacking pattern for the containers. At present there are three primary variations in how the air is moved through the containers: cold wall cooling, serpentine cooling, and the most commonly used, forced-air tunnel cooling.

With forced-air tunnel cooling, pallets or bins of product are lined up in two adjacent rows perpendicular to a single large fan, leaving an alleyway centered on the fan between the rows (Figure 9.11). The rows are generally one pallet or bin in width with the length being dependent upon the fan capacity. The alleyway between the adjacent rows is then covered with a heavy fabric cover, forming an air-return plenum. The fan draws cold air from the surrounding room through vents in the container, across the produce and into the air-return plenum or alleyway.

Cold-wall cooling utilizes a permanently constructed air plenum and fan within one or more walls of the cold room. Stacks of boxes or single pallets are placed up against the wall and cold air from the room is drawn through the containers into the return-air plenum (Figure 9.12). One advantage of this system is that the timing for cooling an individual container or stack of containers can be closely controlled, avoiding unnecessary desiccation. With tunnel cooling there may be a considerable time lag between when product at the exterior and interior of the tunnel is sufficiently cool. Cold-wall cooling is advantageous for products where desiccation is a problem (e.g., cut flowers) and where a wide range of products with differing cooling times is being handled.

Serpentine cooling is a modification of cold-wall forced-air cooling, utilizing bottom rather than side vents in the container. It is designed for produce in pallet bins, with the forklift openings at the base of the pallet being used as the air supply and return plenums (Figure 9.13), cold air is drawn from the room into a forklift opening. By blocking the back of this opening adjacent to the cold wall, the air is forced to move vertically upward and downward through the

Figure 9.12. Cold-wall forced air cooling differs from the standard forced air cooling system in that it utilizes a permanently constructed air plenum with a built-in exhaust fan. This allows the pallet or boxes to begin cooling immediately, since it is not necessary to wait for sufficient product to build a tunnel. Therefore, individual containers can be cooled for differing lengths of time, an advantageous situation when a number of different types of products need to be cooled (e.g., floral crops) (*photograph courtesy of R.F. Kasmire*).

bins. The return air plenum or forklift opening on every other vertical bin in the stack is blocked on the exterior but open at the cold wall. Therefore, air enters through one forklift opening, moves upward and downward through the bins and exits through the forklift opening at the top or bottom of the next row of bins. Serpentine cooling, requiring no space between rows of bins, is particularly desirable for large volumes of product that are handled in bulk.

Adequate container ventilation is essential for forced air cooling. Sufficient openings to permit a satisfactory volume of air flow through the container with a reasonable pressure drop are needed. Too little venting will restrict the flow of air through the containers while excessive vent openings decreases the strength of the containers. At least 5% of the outside area of the container should be open. The size, shape and arrangement of openings are similar to that described under room cooling. The use of paper wraps around individual product units, plastic bags, or liners is not conducive to forced air cooling due to the restriction of air flow.

The rate of cooling using forced air is closely related to the volume of air-flow through the container per pound of product. Air-flow is a function of the fan capacity and the resistance to air-flow presented by the container and product. Resistance varies with the type of product, the size of the openings in the containers, container stacking pattern and the distance (amount of product) the air must transverse between the site of entry into the outside containers and exit from the boxes adjacent to the fan. The greater the distance, the greater the pressure drop and the longer the cooling time for a given air flow. The optimum air-flow rate and duration for proper cooling varies with the type of product being cooled.

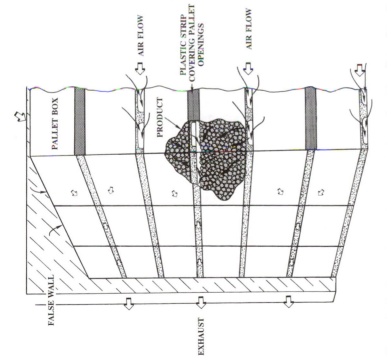

AIR FLOW

PLASTIC STRIP
COVERING PALLET
OPENINGS

PRODUCT

AIR FLOW

PALLET BOX

FALSE WALL

EXHAUST

Figure 9.13. Serpentine cooling, a variation of cold-wall cooling, is used for pallet boxes of produce. Plastic strips are placed over every other forklift opening on one end and on the opposite openings on the other. This forces the air to move either upward or downward through the pallet boxes as it moves to the fan located in the wall (*photograph courtesy of F.G. Mitchell*)

9.2.3. *Hydrocooling*

A wide range of harvested plant products can be rapidly cooled by bringing them in contact with flowing cold water. Hydrocooling is the most rapid means of cooling a wide range of succulent products,* the speed of which is largely due to the much higher heat transfer coefficient of water than air. Efficient hydrocooling has two basic requirements. (1) The water should move over the product surface contacting as much of the surface as possible. Because of this, hydrocooling is normally utilized for plant material held in bulk bins, and is seldom used after packaging. Packaging restricts water movement, greatly decreasing the cooling efficiency; in addition special water-tolerant containers are required. (2) The cooling water must be kept as cold as possible without causing chilling damage to the tissue. This is typically 0°C; however, for chill-sensitive products higher temperatures are required.

Hydrocooling is commonly utilized for stem vegetables, many leafy vegetables and some fruits (e.g., tomatoes, melons). Product requirements include: a tolerance to wetting; the product should not be susceptible to physical damage caused by the cooling water striking its surface; and the product must not be susceptible to injury by chemicals used in the water to prevent the spread of disease organisms (e.g., low levels of chlorine).

Exposure of the product to the cooling medium is by one of two methods: (1) showering the water down upon the product; or (2) submerging the product in the water. Water showers are the most commonly used in that they give excellent water movement, a prerequisite for rapid cooling. With showers, however, channeling of the water can occur if the product depth is too great. Channeling takes place due to the water following the path of least resistance, moving more rapidly through the largest openings (channels) between the individual product units. Channeling decreases the uniformity of cooling within the bin and increases the cooling time required. Proper product depth and a sufficiently high water application rate can minimize channeling.

Complete submersion of the product is an alternative to using water showers. It eliminates the problem with water channeling in that the water comes in contact with all of the product surfaces. The primary drawback to submersion is that when the product is in bulk bins, water movement is greatly restricted. Since movement of the cooling medium over the product surface is a critical component in obtaining rapid cooling, restriction can be a serious disadvantage. Restricted water movement can generally be corrected by using pumps or propellers to circulate the water around the product. A second problem with submersion is that many plant products have a density less than water and as a consequence float. To prevent this, some mechanical means of maintaining the product under water is required.

Hydrocoolers can also be separated into two general designs, conveyer versus batch, based upon whether or not the product is stationary within the cooler. The most commonly used is the conveyer hydrocooler (Figure 9.14). Here the bins of product move slowly through the water (both shower and submerged), carried by a conveyer. This allows the product to be continually placed in the cooler, minimizing down time for loading and unloading. The length of the conveyer is critical and depends upon the type and volume of product to be cooled and the amount of cooling required (initial versus final product temperature). The conveyer speed can be increased or decreased to adjust for products with different cooling rates and initial temperatures.

In a batch system the product is stationary, being loaded into the cooler and then unloaded when cooled (Figure 9.15). Batch coolers are typically easier and less expensive to construct and generally lend themselves to better insulation as more than 50% of the heat load may be from sources other than the product.[58]

---

*A possible exception would be vacuum cooling of certain leafy products.

Figure 9.14. Conveyer hydrocoolers move containers through a chamber in which ice water is showered down upon the product. The water is then recooled and recirculated through the system. Conveyer speed can be adjusted to increase or decrease the length of time the product is in the cooler (*photograph courtesy of R.F. Kasmire*).

Figure 9.15. Batch hydrocoolers differ from conveyer hydrocoolers in that the product is stationary. As with conveyer coolers, ice water cascades down upon the product until it is adequately cooled. While less expensive to build and operate, batch coolers are less efficient since they cannot be used while product is being loaded or unloaded (*photograph courtesy of R.F. Kasmire*).

Heat absorbed by the cooling water must be removed for efficient cooling. This can be achieved by recooling the water using mechanically refrigerated cooling coils or ice bunkers, or when available, by continually introducing cold, new water into the cooler. In some areas of the world, mountain streams and deep wells offer sufficiently cold water for use in cooling. The amount of refrigeration capacity needed for continuous cooling is presented in Table 9.5. The greater the product mass to be cooled, the greater the refrigeration needed; requirements are expressed as the amount of refrigeration per unit weight of produce. Refrigeration capacity requirements also increase as the amount of heat that must be removed increases and with heat leakage into the cooling medium from the surrounding environment. Typically only about 50% of the total refrigeration is used to absorb product heat, the remaining heat is from the environment. With proper insulation, the percent absorbed from the product can be increased to around 80%.

Refrigeration requirements are still commonly presented as "tons of refrigeration," the amount of cooling produced when one ton (2000 pounds) of ice (0°C) melts during a 24 hour period. The metric system does not have a comparable unit of measure; however, one ton converts to approximately 3.54 kJ · sec⁻¹ or 3.54 kW.*

---

*The energy absorbed as ice melts is equal to the latent heat of melting (336 kJ · kg⁻¹) times the weight of ice (1 ton = 907 kg), or 305,659 kJ · ton⁻¹. Dividing this by the time component (24 hr) in seconds (60 sec × 60 min · hr⁻¹ × 24 hr · day⁻¹ = 86,400) gives 1 ton of refrigeration equal to 3.54 kJ · sec⁻¹ or 3.54 kW (1 kW = 1 kJ · sec⁻¹).

Table 9.5. Refrigeration Capacity Needed for Continuous Precooling.*[†]

Change in Product Temperature		% of Total Refrigeration Used to Absorb Heat from the Product					
°C	°F	50%		65%		80%	
		Tons[‡]	kW	Tons	kW	Tons	kW
40	70	20.0	70.7	15.4	54.6	12.7	45.0
33	60	17.1	60.4	13.2	46.9	10.9	38.6
28	50	14.2	50.1	11.1	39.2	9.1	32.1
22	40	11.3	39.8	8.9	31.5	7.3	25.7
17	30	8.4	29.6	6.7	23.8	5.4	19.3
11	20	5.4	19.3	4.5	16.1	3.6	12.9

*Source: After Guillou.[19]

[†]The fraction of the refrigeration needed to cool the product will depend on the amount of insulation and heat leakage into the system.

[‡]Tons or kW of refrigeration per 1,000 kg of product per hour.

9.2.4. Icing

Ice has been used to cool harvested produce since pre-Roman times. In the postharvest period it is utilized for temporary cooling during transport from the field (e.g., leafy greens), for package icing during shipment to retail outlets, and in displays of produce at the retail level. The most widely utilized technique, package icing, represents a relatively fast cooling method that can be used for a number of products that are tolerant to contact with water and ice; this excludes most chill sensitive products. A number of the root and stem vegetables are iced, as are some flower type vegetables (e.g., broccoli), green onions, Brussels sprouts and others.

Icing is a relatively simple operation and in some cases is accomplished in the field. Truck loads of leafy greens for processing are often given alternating layers of ice during hot weather as loading proceeds. Cooling effectiveness increases with increasing contact between the ice and the product. Therefore techniques that facilitate contact (e.g., small pieces and use of liquid ice) hasten product cooling.

The primary disadvantage of icing products, for which ice can be safely used, is that the weight of the ice substantially increases the shipping weight. For relatively warm produce (i.e., 35°C), the additional weight may equal 35 to 40% of the product weight.

Several forms of fragmented ice are used—crushed, flake, snow, and liquid.[11] Body icing, using slurry mixes of ice and water (i.e., liquid ice), has increased substantially in popularity in recent years. This technique is especially useful for products that are non-uniform in size or configuration (e.g., broccoli). Liquid ice gives a much greater degree of initial contact between the product and the ice, and it can be applied after the boxes have been palletized.

Body icing involves pumping an ice-water slurry, from an agitated storage tank, through the openings in the sides of the boxes of dry packed product (Figure 9.16). Slurries range in the water to ice ratio from 1:1 to 1:4 and often contain a small quantity of salt to lower the melting point. The liquid nature of the slurry allows the ice to move throughout the box filling all of the void volume of the container, reaching all the crevices and holes around the individual units of the product. After removal from the body icing machine, the water drains, leaving a relatively solid mass of crushed ice in which the product is embedded. The principal advantage of liquid icing is the much greater contact between the ice and product. When the boxes are palletized prior to application, proper orientation of the openings is required for an unre-

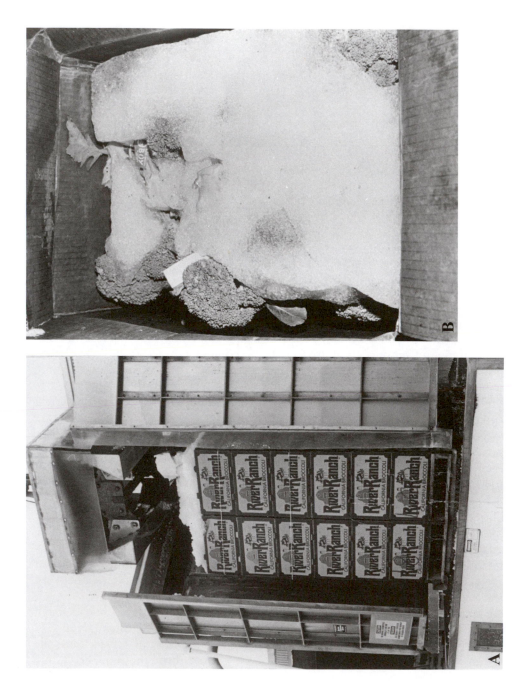

Figure 9.16. Liquid icing (also referred to as body icing) is a form of package icing that utilizes a slurry of ground ice and water which is pumped (A) into palletized boxes. Moving through the ventilation holes in the containers as a liquid, liquid icing has the advantage over older means of ice application, in that it moves around all of the individual units of produce, maximizing contact. The water present drains from the slurry upon filling, leaving a relatively solid mass of crushed ice within the container. (B). Illustrates the condition of a box of body iced broccoli upon reaching the retail market.

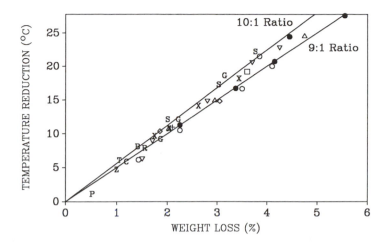

Figure 9.17. The relationship between product temperature reduction and the percent weight loss of selected vegetables when vacuum cooling is used. The products tested fell into one of two general classes based upon the ratio of the temperature drop to % moisture loss (i.e., 10:1 and 9:1; °F:%H$_2$O). Most products exhibit a 5 to 6°C (9 to 10°F) temperature drop with each percent weight loss of water △—artichoke; G—asparagus; S—Brussels sprouts; C—cabbage; R—carrot; O—cauliflower; X—celery; ▽—corn; ●—lettuce; +—mushrooms; □—green onions; ◇—peas; P—potatoes; T—potatoes, skinned; B—snap beans; Z—zucchini squash) (*redrawn from Barger*[6]).

stricted flow of ice throughout the load. The boxes used for iced products are waterproofed, which increases their cost and makes recycling more difficult.

Ice may also be applied manually using rakes or shovels, or with automatic mechanical package icers. Regardless of the technique of application, special water-tolerant boxes are required.

9.2.5. *Vacuum Cooling*

Vacuum cooling is a very rapid and uniform method of cooling used extensively in some production areas for certain crops having large surface to volume ratios. Vacuum cooling involves decreasing the pressure around the product to a level that lowers the boiling point of water to 0°C (i.e., 4.6 mm Hg). The conversion of liquid water to a gas absorbs heat, i.e. the latent heat of vaporization (2260 kJ · kg^{-1} of water). Since evaporation is a surface phenomenon, products with large surface to volume ratios are the most effectively cooled (e.g., leafy crops such as lettuce, cabbage, Brussels sprouts). Several crops with lower surface to volume ratios (e.g., mushrooms) are also vacuum cooled; however, a greater cooling time is generally required.

Since vacuum cooling requires the volatilization of water, some water is lost from the product. Generally around 1% moisture is lost for every 5 to 6°C (9 to 10°F) drop in temperature (Figure 9.17). For most products adequate cooling represents about a 3% loss of water. Losses can be reduced, however, by spraying the surface of the product with water. This, however, requires water tolerant boxes.

The packaging material must not significantly retard the escape of water vapor since evaporation and, consequently, cooling would be greatly impeded. Exposed heads of lettuce, therefore, cool faster than those held in boxes. When sealed in polyethylene bags, no evapora-

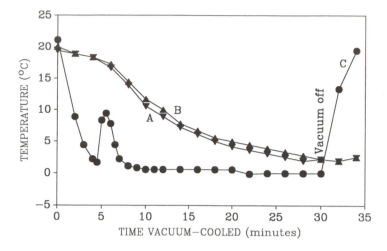

Figure 9.18. Changes in lettuce leaf temperature (A), butt (B) temperature and the wet bulb temperature (C) within the vacuum chamber with time during vacuum cooling. The sharp rise in wet bulb temperature at approximately 5 to 6 minutes is due to the rapid volatilization of water causing a transient water saturation of the chamber environment (*redrawn from Barger*[5]).

tive cooling occurs. Perforated polyethylene bags or loosely sealed polystyrene shrink wraps do allow adequate cooling.

During cooling, product temperature is routinely monitored using a temperature probe inserted into a representative sample. Cooling is relatively uniform throughout pallet loads of product; however, individual tissues will vary in their actual temperature. For example, with head lettuce the base of the stem is typically several degrees warmer than the leaves, due to differences in surface to volume ratios and the amount of water lost. Initial product temperature has little influence on the final product temperature, although it does affect the length of time required for cooling to a specific temperature. When individual product units with different initial temperatures are cooled together, the warmer product will have lost more moisture than the cooler when the final temperature is reached.

During cooling the wet bulb temperature within the vacuum chamber decreases rapidly as the initial vacuum is established (Figure 9.18). Four to 6 minutes after the start of evacuation, however, the wet bulb temperature rises dramatically (e.g., 8.3°C).* The sharp rise in wet bulb temperature is called the flash point and occurs when the product begins to lose moisture rapidly and the tank momentarily becomes saturated with water vapor. After traversing the flash point, product temperature begins to decline rapidly, the rate tending to decrease as the actual product temperature declines.

Commercial vacuum cooling operations commonly utilize a pressure of 4.6 mm Hg, the point at which water will boil at 0° (Figure 9.19). Under these conditions lettuce at typical field temperatures is cooled to 1°C in approximately 20–25 minutes. Slightly lower pressures (e.g., 4.0 mm Hg, -1.7°C boiling point) can be used to achieve a lower final temperature or a faster cooling rate (16 to 18 minutes for lettuce) with little risk of freezing. When higher final temperatures are required, the length of the cooling cycle is shortened rather than selecting a higher pressure.

---

*The precise length of time will depend on the rate at which the vacuum is established.

Figure 9.19. Products such as lettuce are vacuum cooled by rolling pallet loads into the cooler, sealing the chamber and reducing the pressure to 4.6 mm Hg for a sufficient duration to give adequate cooling. The product is then rolled out of the opposite end of the chamber (as additional product is being placed in) and moved to conventional cold storage rooms.

The principal disadvantages of vacuum cooling are the high initial equipment cost and the need for skilled operators. Nevertheless, in areas where large volumes of amenable crops are produced over extended periods, vacuum cooling is used extensively.

9.3. Selection of Precooling Method

A number of factors collectively dictate the precooling method to be used. These include product considerations, packinghouse size and operating procedures, and market demand. The specific requirements of many products exclude some precooling methods. For example, floral crops are never hydrocooled due to the damage that would be sustained by their typically delicate floral structure. For other crops (e.g., strawberry), free water on the surface greatly increases the risk of disease. Some products (e.g., squash and potatoes Figure 9.20) are not cooled rapidly enough or sufficiently using vacuum cooling. When more than one precooling method meets necessary product requirements, the economics of cooling becomes an increasingly major consideration. For example, air cooling of palleted boxes of oranges costs 40% more per box than hydrocooling.[8] Likewise, factors such as personal preference, convenience and equipment availability may also enter into the decision.

9.4. Cooling Rate

For highly perishable products it is desirable to remove the field heat as quickly and economically as possible after harvest, since these products may deteriorate significantly during slow

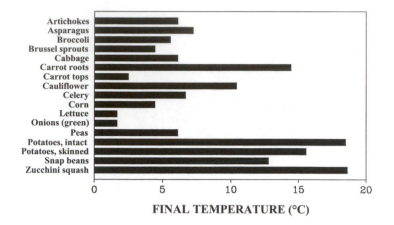

Figure 9.20. A comparison of temperature reduction using vacuum cooling for various types of vegetables under similar conditions [i.e., similar initial product temperature (20 to 22°C), minimum pressure (4.0 to 4.6 mm Hg), condenser temperature (–1.7 to 0°C), and cooling duration (25 to 30 min)] (*redrawn from Barger*[6]).

cooling. A very rapid temperature reduction does not damage the tissue unless the lower temperature limit of the product is exceeded. The actual rate of temperature decrease varies with the cooling method and conditions, the product in question and other considerations (e.g., packaging). Typically, hydrocooling and vacuum cooling (leafy products) are the most rapid means of cooling. The rate of temperature decline with forced air cooling can also be quite rapid if a large volume of air is used per unit volume of product. Icing and room cooling are typically slower, requiring substantially longer interval before adequate cooling is achieved.

For efficient management of a precooling operation, it is desirable to know how long a particular product must be precooled to reach a specific temperature. Knowing the amount of time required for proper precooling gives more control over the flow of produce through the packinghouse for marketing or storage. For a specific precooling technique, the rate of cooling is determined by the temperature of the product, the temperature of the cooling medium, and characteristics of the product. For example, a muskmelon fruit requires a longer cooling time than a pepper fruit when at the same initial temperature and exposed to the same precooling conditions. Cooling rate equations for specific products can be generated from experimental temperature-time response data or from theoretical relationships for specific products or classes within a product type (e.g., size, packaging method). One parameter for estimating the length of time a product needs to be precooled is referred to as the half-cooling time. The theoretical concepts governing precooling of harvested plant products were developed in the 1950's;[47,58] however, they were not widely used by the industry until the introduction of simplified charts or nomographs.[55]

Half-cooling time is the amount of time required to reduce the temperature difference between the product and its surroundings (cooling medium) by one-half. If the product temperature is 34°C and the coolant temperature is 0°C, the length of time required to reduce the product temperature 17°C is the half-cooling time. Since the difference in temperature between the product and cooling media is the critical factor, half-cooling time is independent of the initial temperature of the product and remains constant during cooling. Nomographs have been developed for most products that take into account all possible product-cooling medium temperature combinations. In addition, other factors such as cooling method, fruit size, crating, type of container, and trimming need to be considered in that each can have a pronounced

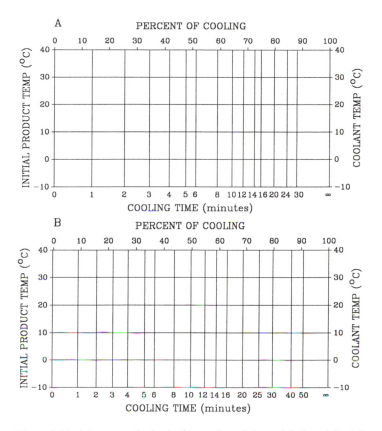

Figure 9.21. Nomographs for hydrocooling globe artichokes, (size 36) when completely exposed (A) and in crates with the lids off and the paper liner open at the top (B) (*redrawn from Stewart and Covey*[55]).

effect on the half-cooling time. For example, a separate nomograph is used for exposed artichokes versus artichokes in crates with the lids open (Figures 9.21 A and B). Half-cooling times are not applicable to vacuum cooling since heat removal is due to the latent heat of vaporization rather than the coolant.

For most products it is necessary to decrease the temperature to less than one-half the difference between the product and the coolant, the temperature achieved by leaving the product in the coolant for a single half-cooling time. As a consequence, it is necessary to utilize several half-cooling times in succession. For example, the temperature drop from an initial product temperature of 30°C and a coolant temperature of 0°C is 15°C for one half-cooling time. If the product remains in the coolant for a second half-cooling time (x minutes) after reaching 15°C, (i.e., a second half cooling period or cycle), the product temperature now drops from 15 to 7.5°C. A third cycle would decrease the temperature to 3.75°C. While the actual rate of cooling is fastest when the temperature differential between the product and coolant is greatest, the half cooling time remains constant (Figure 9.22).

Nomographs allow us to bypass calculations since the cooling time required can be read directly on the x-axis if the desired final product temperature is known. This is illustrated in Figure 9.21A, a nomograph for fully exposed globe artichokes. The nomograph is read by placing a straight edge on the initial commodity temperature (left vertical-axis) and the coolant temperature (right vertical-axis). Where the line intersects the desired product temperature (horizontal lines between the left and right vertical-axis), the time is read directly be-

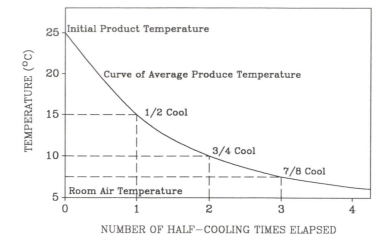

Figure 9.22. Cooling curve showing the drop in temperature from the initial product temperature (25°C) through one (1/2 cool), two (3/4 cool) and three (7/8 cool) half-cooling times. The end of the third half-cooling period represents the seven eighths cooling time. The actual rate of temperature drop and time required to achieve seven eighths cooling varies with product and cooling conditions.

low on the bottom horizontal-axis. For example, completely exposed artichokes (Figure 9.21A) with an initial temperature of 26.7°C (80°F) cooled with 4.4°C (40°F) water can be reduced to 10°C (50°F) in 16 minutes.

Having a separate nomograph for every type and size of product, packaging system, cooling method, and other possible variables would entail a tremendous number of graphs. However, if the half-cooling time is known for each product under a given set of conditions (Table 9.6), then a single universal nomograph (Figure 9.23) can be used. The universal nomograph is read as one would the nomograph for an individual product (i.e., product temperature, coolant temperature and desired product temperature) with the exception that the time scale is given in the number of half-cooling periods or cycles rather than minutes. To convert the half-cooling periods to time, the half-cooling time for the commodity is multiplied times the number of half-cooling periods. Therefore crated broccoli with a half-cooling time (HCT) of 5.8 minutes (Table 9.6) is cooled from 26.7°C (80°F) to 10°C (50°F) in 4.4°C (40°F) water in 2 half-cooling periods (HCP) (Figure 9.21) or 11.6 minutes (2 HCP x 5.8 min. HCT = 11.6 min.).

Seven-eighths cooling time is the length of time the product must be in the coolant to reduce its temperature through seven-eighths of the initial difference between the product and coolant (Figure 9.22). This equals three half-cooling times [i.e., for each half-cooling time the product temperature drops by one-half, therefore 1/2 → 1/4 → 1/8 or a 7/8 reduction (8/8 - 7/8 = 1/8) in the initial temperature difference]. The seven-eighths cooling time for a given product and cooling method is, like the half-cooling time, the same regardless of the initial product and coolant temperature differential. Typically the seven-eighths cooling time is a more practical means of expressing cooling time in the commercial trade, since half-cooling time is somewhat ambiguous (i.e., two half-cooling times do not reduce the temperature of the product to the temperature of the coolant).

Another method for determining the cooling rate of a product and therefore the length of time the product must be cooled is the cooling coefficient. A **cooling coefficient** denotes the

Table 9.6. Half-Cooling Times for Selected Hydrocooled Commodities.

Crop	Treatment	Half-Cooling Time (min)	Crop	Treatment	Half-Cooling Time (min)
Globe artichoke	Exposed	12.8	Sweet corn in husk	Exposed	20
	Crafted	15.5		Crated	28
Asparagus	Exposed	1.1	Peas in pod	Exposed	1.9
	Crafted	2.2		Basket	2.8
Broccoli	Exposed	5.0	Potatoes	Exposed	11
	Crafted	5.8		Jumble-stack	11
Brussels sprouts	Exposed	4.4	Radishes	Exposed, bunched	1.1
	Crafted	4.8		Crated, bunched	1.9
Cabbage	Exposed	69		Exposed, topped	1.6
	Crafted	81		Crafted, topped	2.2
Carrots	Exposed	3.2	Tomatoes	Exposed	10
	Crafted	4.4		Jumble-stack	11
Cauliflower	Exposed	7.2	Cantaloupes	45 size	11
Celery	Exposed	5.8		36 size	20
	Crated	9.1		27 size	20

Source: After Stewart and Covey[55]

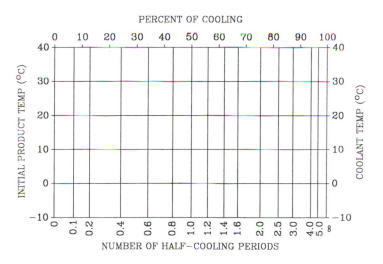

Figure 9.23. General nomograph for cooling harvested products. The nomograph is read like one for an individual product (e.g., Figure 9.21); however, the reading from the x-axis is in half-cooling periods rather than minutes. The length of time required to cool a product to a specific temperature is determined by multiplying the number of half-cooling periods by the half-cooling time for the particular product (e.g., Table 9.6).

change in product temperature per unit change of cooling time for each degree difference between the product and the coolant. The cooling coefficient (C_c) is equal to C_{hr}, the change in product temperature per unit change in cooling time (°C · hr⁻¹), divided by the average temperature differential between the product and the coolant (T_p - T_c), or

$$C_c = \frac{C_{hr}}{(T_p - T_c)} \quad \text{or} \quad C_{hr} = C_c\,(T_p - T_c) \qquad \text{eq. 9.5}$$

Therefore, the cooling coefficient can be used to calculate the length of time required to reduce the product temperature to a desired level.

9.5. Refrigeration During Storage, Transport, Sales and Consumer Holding

Refrigeration is used not only during precooling and storage but also during transit, retail sales, and by the consumer. Storage refrigeration is accomplished primarily using mechanical refrigeration in industrialized areas. While mechanical refrigeration is highly visible in the literature on storage, being the most common means utilized in research, the actual percent of the earth's population that has access to mechanical refrigeration is relatively low. Most countries that are not highly industrialized utilize mechanical refrigeration only in major metropolitan areas and for high value products, using other means or no refrigeration in smaller outlying towns, villages and rural districts. In some cases, ice purchased daily represents the primary means of refrigeration of perishable products. For others, caves and cold water from streams are the only means of refrigeration.

Harvested plant products are moved in volume from sites of production to sites of utilization by truck, railroad, ships, and air freight, each of which may be refrigerated.[27] Trucks are designed for short or long hauls, and for local delivery, or for operation within a specific temperature range. Transit refrigeration is essential for highly perishable crops during warm weather in that air and road temperatures may be extreme (e.g., 41 and 61°C, respectively[28]). In recent years, intermodal transport has become increasingly common. Truck trailers are moved part of the journey by railroad and containerized products are routinely handled using several modes in tandem (e.g., truck–rail–truck, truck–ship–truck). Refrigerated intermodal containers allow transferring the product from one carrier mode to another without direct handling of the product. Advantages include continuous refrigeration of the product during transit, reduced damage due to handling, and lower losses.

Refrigeration at the retail level involves the use of display refrigerators for merchandising the produce and walk-in coolers used for the storage of produce not in the sales area. Display refrigeration systems are designed for an attractive presentation of the product, ease of access for self-service, and sufficient refrigeration for short-term protection of the product. Generally, highly perishable produce should be sold within 24 to 48 hours from the time they are placed in the display area. Typically display cases are run at 1.5 to 7.5°C.

Product delivered to the store in refrigerated trucks should be quickly placed in a refrigerated storage area. Separate storage areas are required for plant versus animal products and larger retail outlets typically have at least two refrigerated storage areas for plant products, one at approximately 2°C and a second at 12 to 16°C for chilling sensitive products such as bananas. For smaller outlets there is only one room at about 2°C. As a general rule, the volume of the storage rooms should be equivalent to the capacity of display area.

Home refrigerators are used for the storage of produce at the consumer level. Their use tends to be concentrated in the more industrialized areas of the world. Likewise, home refrigeration is generally less important in the more northern and southern regions of the temperate zones due to lower ambient temperatures during much of the year. The storage volume of typical home refrigerators range from 28–56 liter upward, with typical units being 425–565 L and the temperature range from 3 to 7°C.

10. HEAT INPUT AFTER HARVEST

While much of this chapter has focused upon the removal of heat from harvested products, there are occasions when many perishable products must be protected against low ambient temperatures or require specialized heat treatments. Therefore, heat may occasionally need to be added to elevate the product temperature.

10.1. Addition of Heat to Maintain a Minimum Safe Product Temperature

The addition of heat to maintain a safe storage temperature is especially important in mountainous areas and during the winter months in the temperate zones as one moves progressively further away from the equator. In these areas, most cold storage rooms are situated in a temperature controlled building. Therefore, there is a controlled and relatively constant ambient temperature surrounding at least part of the storage area. Most mechanical refrigeration systems now have the capacity for some heating. During cold weather, critical times for exposure to low temperatures extremes typically occur during transfer from one site to another (e.g., movement from wholesale storage to retail stores, or movement from the retail store to the purchaser's home). Most trucks, trailers, railroad cars and refrigerated containers utilized in these areas have the potential for supplemental heating and can maintain any desired temperature between 0 and 21°C.

Heating is accomplished using: (1) reverse cycle, or hot gas operation of mechanical refrigeration units; (2) fuel-burning heaters that utilize alcohol, kerosene, butane, propane or charcoal; (3) electrical resistance heaters powered by generators run from any of a number of sources (e.g., wheel driven generators on railroad cars). The reverse cycle and fuel-burning heaters are the most common. When electrical heaters are utilized, some of the evaporator coil tubes contain tubular electrical heating elements which are used for both heating and defrosting the evaporator coils. Reverse cycle or hot gas systems are the most common in modern refrigeration systems. In some of the older mechanical refrigerated systems and when other forms of refrigeration are used (e.g., ice bunker-railroad car), portable thermostatically controlled fuel burning heaters are employed.

Portable heaters have been used in railroad cars for over 50 years, with earlier heaters utilizing charcoal as a fuel. Newer heaters are alcohol fueled since alcohol provides adequate heat production and complete combustion (charcoal heaters release carbon monoxide, ethylene and other gases into the storage area due to incomplete combustion). Alcohol heaters at their lowest setting (i.e., the pilot burning) release about $176\ W \cdot hr^{-1}$ and about $1{,}760\ W \cdot hr^{-1}$ at their maximum with methanol as the fuel. Most systems now utilize a 1:1 ratio of methanol and isopropanol, which gives a greater heat yield and heat potential per tank of fuel. Newer high output alcohol heaters can generate as much as $3{,}810\ W \cdot hr^{-1}$ using this fuel mixture.

10.2. Specialized Heat Treatments

The intentional elevation of product temperature after harvest is used for a number of purposes and crops.[17,31] Excluded from this topic are temperature increases associated with drying (Chapter 8). Short-term exposure to high temperature is used for direct pathogen and insect control, indirect control of pathogens through the healing of surface wounds (invasion sites) prior to storage, inhibition of water loss, decreased chilling injury and in some instances to facilitate sprouting, flowering or other responses.[33]

The temperature to which the product is raised and the duration of exposure are critical parameters in thermal treatments.[18,42] With insects and diseases, successful treatment results in significant damage or death to the target organism with little or no reduction in product quality. Therefore the target organism must be more susceptible than the product to the imposed temperature treatment.[39,40] Due to variation among types of products, the response to a given treatment can vary substantially. Therefore, no single treatment or system works equally well for all applications; individual treatments must be tailored to commodity requirements.

Introduction of heat may be *via* several means, the selection of which is dictated largely by crop requirements and treatment objectives: (1) elevated storage room temperatures; (2) high temperature forced air; (3) hot water dips; and (4) vapor heat.[1] In some instances (i.e., inhibition of certain pathogens and insects), brushing[12,44] or high pressure sprays significantly enhance control. Elevated temperature in conjunction with controlled gas atmospheres has also been demonstrated to facilitate control of certain insect pests.[64] Thermal treatments generally have very close tolerances that if exceeded damage the commodity, reducing its quality and life expectancy. Advantages of thermal treatments, however, include being non-polluting and relatively inexpensive, and the absence of toxic residues on the product.

10.2.1. Curing

Curing involves exposure to conditions that facilitate wound healing or the development of a protective surface layer in certain products. After harvest the product is given a short-term high temperature treatment. For example, freshly harvested sweetpotatoes are exposed to 29°C (80–90 %RH) for 5 to 7 days.[34,62] This stimulates the formation of wound periderm at sites on the roots where the surface is broken, decreasing the incidence of soft rot. In the tropics, due to higher ambient temperatures, the addition of supplemental heat is seldom required for adequate curing. Many bulbs and corms are also cured (i.e., gladiolus corms—10 days at 27 to 29°C; hyacinth bulbs—several weeks at 25 to 27°C; Dutch iris bulbs—10 to 15 days at 32°C; narcissus bulbs—4 days at 30°C; and tulip bulbs 1 week at 26°C).[22]

10.2.2. Inhibition of Chilling Injury

A relatively wide range of fruits and vegetables have reduced chilling injury symptoms after short exposures to a high temperature treatment. The chilling symptoms inhibited vary with commodity (e.g., scald in apple;[32] surface pitting in avocado,[65] citrus,[7] cucumber,[35] brown stain in cactus pear[50]). For example, grapefruit temperature is elevated to 27 to 29°C for 48 hours[11,24] or held at 21°C for 7 days[23] after harvest, prior to reducing the product temperature to the normal storage level (i.e., 10 to 16°C). Treatment decreases the incidence of chilling injury during subsequent storage.

10.2.3. Insect Disinfestations

Postharvest thermal treatments are required to disinfest a number of economically important fruits, vegetables, nuts, flowers, and ornamentals of insect pests. Disinfestation is especially important when the product is transported into areas where the pest does not occur (e.g., various tephritid fruit flies, mango seed weevil) or the importing country has a "zero tolerance" for insects. While there are many quarantined insects and arthropods, fruit flies are especially important in that they attack a very wide range of crops.[33] For example, the Caribbean fruit fly (*Anastrepha suspense*) can be controlled in mango[38] and several other fruits by exposure to air temperatures of 51.5°C for 125 minutes. In addition to fruit flies, other insects of importance in which thermal treatments have been tested include coddling moth,[39] leafroller,[63] New Zealand flower thrip,[36] two-spotted spider mite[10] and others. Failure to control the spread of insect pests can result in extensive product losses as well as expensive quarantine and eradication procedures.

10.2.4. Pathogen Control

Brief exposure for some crops to high temperatures is done to control pathogens. Both water and air are used depending upon the product in question. In some instances, very brief exposure to steam (100°C for 3 sec) is used.[1] For example, brief exposure of mango fruit to hot water (55°C) helps to control *Colletotrichum gloesporioides* and *Botryodiplodia theobromal.*[48] Likewise, exposure of gladiolus corms to hot water (56°C) decreases the incidence of *Curvularia trifolii* f. sp. *gladioli.*[29] Hot air is used for *Penicillium expansum* control on apple (38°C for 4 days + 4% $CaCl_2$).[49]

Heat treatments have been the physical method of choice for disease control. A 20 minute, 49°C water-bath dip has been recommended and used widely for effective control of some postharvest diseases of papaya. Similar heat treatments have been tested for other fruits with varying degrees of success.[9]

10.2.5. Conditioning Treatments

Thermal treatments have also been demonstrated to be useful for a diverse array of other problems. For example, pre-storage or pre-marketing high temperature treatments have been shown to increase firmness in apple,[30] inhibit asparagus gravitropism,[41] inhibit yellowing in kale[61] and inhibit sprouting in potato.[45] In some instances, thermal treatments may also be imposed after storage to simulate flowering of certain crops. Exposure of gladiolus corms and lily bulbs to hot water accelerates sprouting.[25]

ADDITIONAL READING

ASHRAE. 1998. *Refrigeration Handbook.* American Society of Heating, Refrigeration, Air-Conditioning Engineers, Atlanta, GA.
Basra, A.S. 2001. *Crop Responses and Adaptations to Temperature Stress.* Food Products Press, New York.
Dennis, D.T. 1987. *The Biochemistry of Energy Utilization in Plants.* Blackie, Glasgow, Scotland.
Fitter, A. 2002. *Environmental Physiology of Plants.* Academic Press, San Diego, CA.
Gates, D.M., and R.B. Schmerl (eds.). 1975. *Perspectives of Biophysical Ecology.* Springer-Verlag, Berlin.
Gosney, W.B. 1982. *Principles of Refrigeration.* Cambridge University Press, Cambridge.
Hanan, J.J. 1984. *Plant Environmental Measurement.* Bookmaker Guide, Longmont, CO.
Heap, R., M. Kierstan and G. Ford. 1998. *Food Transportation.* Blackie, London.
Jones, H.G. 1992. *Plants and Microclimate: A Quantitative Approach to Environmental Plant Physiology.* Cambridge University Press, Cambridge.
Mitchell, F.G., R. Guillou and R.A. Parsens. 1972. Commercial cooling of fruits and vegetables. *Univ. Calif. Ext. Manual* 43.
Nobel, P.S. 1999. *Physiochemical & Environmental Plant Physiology.* Academic Press, San Diego, CA.
Stoecker, W.F. 1998. *Industrial Refrigeration Handbook.* McGraw-Hill, New York.
Thompson, J.F., P.E. Brecht, R.T. Hinsch, and A.A. Kader. 2000. Marine container transport of chilled perishable produce. *Univ. Calif. Div. Ag. Nat. Res. Publ.* 21595, Oakland, CA.
Thompson, J.F., F.G. Mitchell, T.R. Rumsey, R.F. Kasmire, and C.H. Crisosto. 1998. Commercial cooling of fruits, vegetables, and flowers. *Univ. Calif. Div. Ag. Nat. Res. Publ.* 21567, Oakland, CA.
Thompson, J.F. 1999. Cold-storage systems. Pp. 339–361. In: *CIGR Handbook of Agricultural Engineering.* vol. 4. F.W. Bakker-Arkema (ed.). Amer. Soc. Agr. Eng., St. Joseph, MI.
Woodward, F.I. 1987. *Climate and Plant Distribution.* Cambridge University Press, Cambridge.

REFERENCES

1. Afek, U., J. Orenstein and E. Nuriel. 1999. Steam treatment to prevent carrot decay during storage. *Crop Protection* 18:639–642.

2. ASHRAE. 1967. *Handbook of Fundamentals.* Amer. Soc. Heating, Refrigeration and Air Conditioning Engineers. New York.
3. ASHRAE. 1983–1986. *Handbook of Fundamentals.* Amer. Soc. Heating, Refrigeration and Air Conditioning Engineers. New York.
4. Awberry, J.H. 1927. The flow of heat in a body generating heat. *Phil. Mag.* 4:629–638.
5. Barger, W.R. 1961. Factors affecting temperature reduction and weight-loss in vacuum-cooled lettuce. *USDA Mkt. Res. Rept.* 469.
6. Barger, W.R. 1963. Vacuum precooling. A comparison of cooling of different vegetables. *USDA Mkt. Res. Rept.* 600.
7. Ben-Yehoshua, S., B. Shapiro and R. Moran. 1987. Individual seal-packaging enables the use of curing at high temperatures to reduce decay and heat injury of citrus fruits. *HortScience* 22:777–783.
8. Catlin, P.B., F.G. Mitchell and A.S. Greathead. 1959. Studies on strawberry quality. *Calif. Agric.* 13(2):11, 16.
9. Couey, H.M. 1989. Heat treatment for control of postharvest disease and insect pests of fruit. *HortScience* 24:198–202.
10. Dentener, P.R., S.E. Lewthwaite, J.H. Maindonald and P.G. Connolly. 1998. Mortality of twospotted spider mite (Acari: Tetranychidae) after exposure to ethanol at elevated temperatures. *J. Econ. Entomol.* 91:767–772.
11. Eddy, D.E. 1965. Manufacture, storage handling and uses of fragmentary ice. *Amer. Soc. Heat, Refrig. Air Cond. Eng. J.* 7:66.
12. Fallik, E., S. Grinberg, S. Alkalai, O. Yekutiel, A. Weisblum, R. Regev, H. Beres and E. Bar-Lev. 1999. A unique rapid hot water treatment to improve storage quality of sweet pepper. *Postharv. Biol. Tech.* 15:25–32.
13. Gaffney, J.J., C.D. Baird and K.V. Chau. 1985. Methods for calculating heat and mass transfer in fruits and vegetables individually and in bulk. *Trans. Amer. Soc. Heat. Refrig. Air Eng.* 91:333–352.
14. Gates, D.M. 1965. Energy, plants, and ecology. *Ecology* 46:1–13.
15. Gosney, W.B. 1982. *Principles of Refrigeration.* Cambridge University Press, New York.
16. Grierson, W. 1974. Chilling injury in tropical and subtropical fruit. V. Effect of harvest date, degreening, delayed storage and peel color on chilling injury of grapefruit. *Proc. Trop. Reg. Amer. Soc. Hort. Sci.* 18:66–72.
17. Grondeau, C., and R. Samson. 1994. A review of thermopathy to free plant materials from pathogens, especially seeds from bacteria. *Crit. Rev. Plant Sci.* 13:57–75.
18. Guillou, R. 1960. Coolers for fruits and vegetables. *Calif. Agr. Expt. Sta. Bull.* 773.
19. Guillou, R. 1963. Pressure cooling for fruits and vegetables. *ASHRAE J.* 5(11):45–49.
20. Haller, M.H., P.L. Harding, J.M. Lutz and D.H. Rose. 1932. The respiration of some fruits in relation to temperature. *Proc. Amer. Soc. Hort. Sci.* 28:583–589.
21. Hallman, G.J. 2000. Factors affecting quarantine heat treatment efficacy. *Postharv. Biol. Tech.* 21:95–101.
22. Hardenburg, R.E., A.E. Watada and C.Y. Wang. 1986. The commercial storage of fruits, vegetables and florist and nursery stock. *USDA-ARS Agr. Hbk.* 66.
23. Hatton, T.T., and R.H. Cubbedge. 1982. Conditioning Florida grapefruit to reduce chilling injury during low-temperature storage. *J. Amer. Soc. Hort. Sci.* 107:57–60.
24. Hawkins, L.A., and W.R. Barger. 1926. Cold storage of Florida grapefruit. *Proc. Trop. Reg. Amer. Soc. Hort. Sci.* 18:66–72.
25. Hosoki, T. 1984. Effect of hot water treatment on respiration, endogenous ethanol and ethylene production from gladiolus corms and Easter lily bulbs. *HortScience* 19:700–701.
26. Hurst, W. 1986. Unpublished data. University of Georgia, Athens, GA.
27. International Institute of Refrigeration. 1995. Guide to refrigerated transport.*Internat. Inst. Refrig.,* Paris
28. Kasmire, R.F., and R.T. Hinsch. 1962. Factors affecting transit temperatures in truck shipments of fresh produce. *Univ. of Calif. Perish. Handling Trans.* Suppl. 1, 10 pp.
29. Kelling, K., and R.-M. Niebisch. 1985. Nachiveis der Curvularia—Krankheit an Gladiolenpflanzgut. *Gartenbau* 32(7):218–219.
30. Klein, J.D., and S. Lurie. 1992. Prestorage heating of apple fruit for enhanced postharvest quality: interaction of time and temperature. *HortScience* 27:326–328.

31. Lurie, S. 1998. Postharvest heat treatments of horticultural crops. *Hort. Rev.* 22:91–121.
32. Lurie, S., J.D. Klein and R. Ben-Arie. 1990. Postharvest heat treatment as a possible means of reducing superficial scald of apples. *J. Hort. Sci.* 65:503–509.
33. Lurie, S., and J.D. Klein. 2003. Temperature preconditioning. In: *USDA Hbk.* 66.
34. Lutz, J.M. 1943. Factors influencing the relative humidity of the air immediately surrounding sweet potatoes during curing. *Proc. Amer. Soc. Hort. Sci.* 43:255–258
35. McCollum, G., H. Doostdar, R. Mayer and R. McDonald. 1995. Immersion of cucumber in heated water alters chilling-induced physiological changes. *Postharv. Biol. Tech.* 6:55–64.
36. McLaren, G.F., R.M. McDonald, J.A. Fraser, R.R. Marshall, K.J. Rose and A.J. Ford. 1997. Disinfestation of New Zealand flower thrips from stonefruit using hot water. *Acta Hort.* 464:524.
37. Middleton, W.E.K. 1966. *A History of the Thermometer and Its Use in Meteorology.* Johns Hopkins Press, Baltimore, MD.
38. Miller, W.R., and R.E. McDonald. 1991. Quality of stored 'Marsh' and Ruby Red' grapefruit after high-temperature, forced-air treatment. *Am. Soc. Hort. Sci.* 26(9):1188–1191.
39. Neven, L.G. 2000. Physiological responses of insects to heat. *Postharv. Biol. Tech.* 21:103–111.
40. Paull, R.E., and J.W. Armstrong. 1994. Insect pests and fresh horticultural products: Treatments and responses. Pp. 1–36. In: *Insect Pests and Fresh Horticultural Products: Treatments and Responses.* R.E. Paull and J.W. Armstrong (eds.). CAB Intern., Wallingford, England.
41. Paull, R.E., and N.J. Chen. 1999. Heat treatment prevents postharvest geotropic curvature of asparagus spears (*Asparagus officinallis* L.). *Postharv. Biol. Tech.* 16:37–41.
42. Paull, R.E., and N.J. Chen. 2000. Heat treatments and fruit ripening. *Postharv. Biol. Tech.* 21:21–37.
43. Pentzer, W.T., R.L. Perry and G.C. Hanna. 1936. Precooling and shipping California asparagus. *Calif. Agr. Expt. Sta. Bull.* 600.
44. Prusky, D., Y. Fuchs, I. Kobiler, I. Roth, A. Weksler, Y. Shalom, F. Fallik, G. Zauberman, E. Pesis, M. Akerman, O. Ykutiely, A. Weisblum, R. Regev and L. Artes. 1999. Effect of hot water brushing, prochloraz treatment and waxing on incidence of black spot decay caused by *Alternaria alternata* in mango fruits. *Postharv. Biol. Tech.* 15:165–174.
45. Ranganna, B., G.S.V. Raghavan and A.C. Kushalappa. 1998. Hot water dipping to enhance storability of potatoes. *Postharv. Biol. Tech.* 13:215–223.
46. Reidy, G.A. 1968. Values for thermal properties of foods gathered from the literature. M.S. Thesis, Mich. St. Univ., East Lansing, MI.
47. Sainsbury, G.F. 1951. Improved fruit cooling methods. *Refrig. Eng.* 59:464–469, 506, 508–509.
48. Sampaio, V.R. 1983. Controle em pos-colheita das podridoes da manga Bourbon, conservada em camara fria. *Anais da Escola Superior de Agricultura "Luiz de Queiroz"* 40(1):519–526.
49. Sams, C.E., W.S. Conway, J.A. Abbott, R.J. Lewis and N. Ben-Shalom. 1993. Firmness and decay of apples following postharvest pressure infiltration of calcium and heat treatment. *J. Amer. Soc. Hort. Sci.* 118:623–627.
50. Schirra, M., G. Barbera, S. D'Aquino, T. La Mantia and R.E. McDonald. 1996. Hot dips and high-temperature conditioning to improve shelf quality of late-crop cactus pear fruit. *Trop. Sci.* 36:159–165.
51. Scholz, E.W., H.B. Johnson and W.R. Buford. 1963. Heat-evolution rates of some Texas-grown fruits and vegetables. *J. Rio Grande Valley Hort. Soc.* 17:170–175.
52. Smith, R.E., and A.H. Bennett. 1965. Mass-average temperature of fruits and vegetables during transient cooling. *Trans. Amer. Soc. Agri. Eng.* 8:249–252.
53. Smith, W.H. 1957. The production of carbon dioxide and metabolic heat by horticultural produce. *Modern Refrig.* 60:493–496.
54. Smock, R.M., and C.R. Gross. 1950. Studies on respiration of apples. *N.Y. (Cornell) Agr. Expt. Sta. Man.* 297.
55. Stewart, J.K., and M.H. Covey. 1963. Hydrocooling vegetables: a practical guide to predicting final temperatures and cooling times. *USDA Mkt. Res. Rept.* 637.
56. Sweat, V.E. 1974. Experimental values of thermal conductivity of selected fruits and vegetables. *J. Food Sci.* 39:1080–1083.
57. Tewfik, S., and L.E. Scott. 1954. Respiration of vegetables as affected by postharvest treatment. *J. Agr. Food Chem.* 2:415–417.
58. Thevenot, R. 1955. Precooling. *Proc. Ninth Intern. Cong. Refrig.* (Paris), 0.10:0051–0071.

59. Toussaint, W.D., T.T. Hatlow and G. Abshier. 1955. Hydrocooling peaches in the North Carolina sandhills. *North Carol. St. Agr. Expt. Sta. Infor. Ser.* 39.

60. Turrell, F.M., and R.L. Perry. 1957. Specific heat and heat conductivity of citrus fruit. *Proc. Amer. Soc. Hort. Sci.* 70:261–265.

61. Wang, C.Y. 1998. Heat treatment affects postharvest quality of kale and collard, but not of Brussels sprouts. *HortScience* 33:881–883.

62. Weimer, J.R., and L.L. Harter. 1921. Wound cork formation in the sweet potato. *J. Agr. Res.* 21:637–647.

63. Whiting, D.C., L.E. Jamieson, K.J. Spooner and M. Lay-Yee. 1999. Combination of high-temperature controlled atmosphere and cold storage as a quarantine treatment against *Ctenopseustis obliquana* and *Epiphyas postvittana* on 'Royal Gala' apples. *Postharv. Biol. Tech.* 16:119–126.

64. Whiting, D.C., G.M. O'Connor, J. van den Huevel and J.H. Maindonald. 1995. Comparative mortalities of six tortricid (Lepidoptera) species to two high-temperature controlled atmospheres and air. *J. Econ. Entomol.* 88:1365–1370.

65. Woolf, A.B., and M. Lay-Yee. 1997. Pretreatments at 38°C of 'Hass' avocado confer thermotolerance to 50°C hot water treatments. *HortScience* 32:705–708.

APPENDIX I. PHENOTYPES OF TRANSFORMED PLANTS REGISTERED IN THE UNITED STATES (JUNE, 2002)

2,4-D tolerant
ACC oxidase level decreased
Adventious root formation
  increased
Agrobacterium resistant
Alkaloids reduced
Altered amino acid composition
Altered lignin biosynthesis
Altered maturing
Altered morphology
Altered plant development
Alternaria daucii resistant
Alternaria resistant
Alternaria solani resistant
Aluminum tolerant
Ammonium assimilation
  increased
AMV resistant
Animal feed quality improved
Anthocyanin produced in seed
Anthocyanin sequestration
  supressed
Anthracnose resistant
Anthracnose susceptible
Antibiotic produced
Antibody produced
Antioxidant enzyme increased
Antiprotease producing
Aphid resistant
Apple scab resistant
Aspergillus resistant
Auxin metabolism and increased
  tuber solids

β-1,4-endoglucanase
Bacterial leaf blight resistant
Bacterial resistant
Bacterial soft rot resistant
Bacterial speck resistant
Bacterial wilt resistant
Bacteriocins suppressed
BCTV resistant
Black shank resistant
Blackspot bruise resistant
BLRV resistant
Blue mold resistant
BNYVV resistant
Botrytis cinerea resistant
Botrytis resistant
BPMV resistant
Bromoxynil tolerant
Brown spot resistant
Bruising reduced
Burkholderia glumae
BYDV resistant
BYMV resistant
Caffeine levels reduced
Calmodulin level altered
Calonectria resistant
CaMV resistant
Capable of growth on defined
  synthetic media
Carbohydrate level increased
Carbohydrate metabolism altered
Carotenoid content altered
Carotenoid metabolism altered
Catalase level reduced

CBI
Cell wall altered
Cercospora resistant
Chestnut blight resistant
Chloroacetanilide tolerant
Chlorophyll increased
Chlorsulfuron tolerant
Citrus canker resistant
Clavibacter resistant
Closterovirus resistant
CLRV resistant
CMV resistant
Cold intolerant
Cold tolerant
Coleopteran resistant
Colletotrichum resistant
Color altered
Color pigment restored
Color sectors in seeds
Colorado potato beetle resistant
Colored sectors in leaves
Common rust susceptible
Constitutive expression of
  glutamine synthetase
Corn earworm resistant
Corynebacterium sepedonicum
  resistant
Cottonwood leaf beetle resistant
Cre recombinase produced
Criconnemella resistant
Crown gall resistant
Crown rot resistant
Cucumovirus resistant

Cutting rootability increased
Cyanamide tolerance
Cylindrosporium resistant
CyMV resistant
Dalapon tolerant
Delayed softening
Development altered
Digestibility improved
Disease resistant general
Disulfides reduced in endospenn
DNA synthesis altered
Dollar spot resistant
Downy mildew resistant
Drought tolerant
Dry matter content increased
Dwarfed
Ear mold resistant
Environmental stress reduced
Epidermal cells increased on juvenile leaves
Erucic acid altered
Erwinia carotovora resistant
Erwinia resistant
Ethylene metabolism altered
Ethylene synthesis reduced
European Corn Borer resistant
Expression optimization
Extended flower life
Eyespot resistant
Fall armyworm resistant
Fatty acid level altered
Fatty acid metabolism altered
Feed properties altered
Female sterile
Fenthion susceptible
Fertility altered
Fiber quality altered
Fiber strength altered
Fire blight resistant
Flavor enhancer
Flower and fruit abscission reduced
Flower and fruit set altered
Flower color altered
Flowering altered
Flowering time altered
Frogeye leaf spot resistance
Fruit firmness increased
Fruit invertase level decreased
Fruit pectin esterase level decreased
Fruit polygalacturonase level decreased
Fruit ripening altered

Fruit ripening delayed
Fruit rot resistant
Fruit solids increased
Fruit sugar profile altered
Fruit sweetness increased
Fumonisin degradation
Fungal postharvest resistant
Fungal resistant
Fungal resistant general
Fungal susceptibility
Fusarium ear rot susceptible
Fusarium resistant
Geminivirus resistant
Gene expression altered
Gene tagging
Glucuronidase expressing
Glyphosate tolerant
Gray lead spot resistant
Gray leaf spot resistant
Gray leaf spot susceptible
Growth rate altered
Growth rate increased
Growth rate reduced
Halogenated hydrocarbons metabolized
Heat stable glucanase produced
Heat tolerant
Heavy metal bioremediation
Heavy metals sequestered
Helminthosporium resistant
Herbicide tolerance
Hordothionin produced
Hygromycin tolerant
Imidazole tolerant
Imidazolinone tolerant
Improved fruit quality
Increased phosphorus
Increased protein levels
Increased stalk strength
Increased transformation frequency
Inducible DNA modification
Industrial enzyme produced
Industrial enzymes produced
Insect predator resistance
Insect resistant general
Insect susceptible
Insecticidal protein
Iron levels increased
Isoxaflutole resistant
Isoxazole tolerant
Kanamycin resistant
Larger fruit
Leafblight resistant

Leaf senescence delayed
Leaf spot resistant
Leafhopper resistant
Lepidopteran resistant
Lesser cornstalk borer resistant
Lignin levels decreased
Lipase expressed in seeds
LMV resistant
Loss of systemic resistance
Lysine level increased
Male sterile
Male sterile nuclear
Male sterile reversible
Marssonina resistant
MCDV resistant
MCMV resistant
MDMV resistant
MDMV-B resistant
Mealybug wilt virus resistant
Melamtsora resistant
Melanin produced in cotton fibers
Meloidogyne resistant
Metabolism altered
Methionine level increased
Methotrexate resistant
Mexican Rice Borer resistant
Mildew resistant
Modified growth characteristics
Mutator transposon suppressed
Mycotoxin deficient
Mycotoxin degradation
Mycotoxin production inhibited
Mycotoxin restored
Nepovirus resistant
Nicotine levels reduced
Nitrogen metabolism altered
Non-lesion forming mutant
Norther leaf blight susceptible
Northern com leaf blight resistant
Novel protein produced
Nucleocapsid protein produced
Nutritional quality altered
Oblique banded leafroller resistant
Oil profile altered
Oil quality altered
Oleic acid content altered in seed
Oomycete resistant
Oxidative stress tolerant
Parthenocarpy
Pathogenesis related proteins level increased

Pectin esterase level reduced
PEMV resistant
Pernospora resistant
Peroxidase levels increased
PeSV resistant
Pharmaceutical proteins
  produced
Phoma resistant
Phosphinothricin resistant
Phosphinothricin tolerant
Photosynthesis enchanced
Phratora leaf beetle resistant
Phytate reduced
Phytoene synthase activity
  increased
Phytophthora resistant
Pigment composition altered
Pigment metabolism altered
PLRV resistant
Pollen visual marker
Polyamine metabolism altered
Polygalacturonase level reduced
Polymer produced
Polyphenol oxidase levels altered
Potyvirus resistant
Powdery mildew resistant
PPV resistant
Pratylenchus vulnus I resistant
Processing characteristics altered
Prolonged shelf-life
Protein altered
Protein levels increased
Protein lysine level increased
Protein quality altered
Proteinase inhibitors level
  constitutive
Protoporphyrinogen oxidase
  inhibitor tolerant
PRSV resistant
PRV resistant
PSbMV resistant
Pseudomonas syringae resistant
PSRV resistant
PStV resistant
PVA resistant
PVX resistant
PVY resistant
PVY susceptible
Pyricularia oryzae
Pythium resistant
RBDV resistant
Recombinase produced
Rhizoctonia resistant

Rhizoctonia solani resistant
Ring rot resistance
Root-knot nematode resistant
Rubber yield increased
Rust resistant
Salicylic acid level reduced
Salt tolerance increased
SbMV resistant
Sclerotinia resistant
SCMV resistant
SCYLV resistant
Secondary metabolite increased
Seed color altered
Seed composition altered
Seed methionine storage
  increased
Seed number increased
Seed quality altered
Seed set reduced
Seed size increased
Seed storage protein
Seed weight increased
Selectable marker
Senescence altered
Septoria resistant
Shorter stems
Smut resistant
SMV resistant
Sod web worm resistant
Soft rot fungal resistant
Soft rot resistant
Solids increased
Solids soluble increased
Southern rust susceptible
Southern corn leaf blight
  resistant
Southern leaf blight susecptible
Southwestern corn borer
  resistant
Spectromycin resistant
SPFMV resistant
Sphaeropsis fruit rot resistant
SqMV resistant
SrMV resistant
Stanol increased
Starch level increased
Starch metabolism altered
Starch reduced
Sterile
Steroidal glycoalkaloids reduced
Sterols increased
Sterols modified
Stewart's wilt susceptible

Storage protein
Storage protein altered
Streptomyces scabies resistant
Stress tolerant
Sugar alcohol levels increased
Sugar cane borer resistant
Sugar content altered
Sulfometuron tolerant
Sulfonylurea susceptible
Sulfonylurea tolerant
Syringomycin deficient
Tetracycline binding protein
  produced
TEV resistant
Thelaviopsis resistant
Thermostable protein produced
TMV resistant
Tobamovirus resistant
ToMoV resistant
ToMV resistant
ToRSV resistant
Transformation frequency
  increased
Transposon activator
Transposon elements inserted
Transposon inserted
Transposon movement supressed
Trifolitoxin producing
Trifolitoxin resistant
TRV resistant
Tryptophan level increased
TSWV resistant
Tuber solids increased
TVMV resistant
TYLCV resistant
Tyrosine level increased
Venturia resistant
Verticillium dahliae resistant
Verticillium resistant
Virulence reduced
Visual marker
Visual marker inactive
Vivipary increased
Western corn rootworm resistant
White mold resistant
WMV2 resistant
WSMV resistant
Xanthomonas campestris
  diffenbachiae resistant
Xanthomonas campestris resistant
Xanthomonas oryzae resistant
Yield increased
ZYMV resistant

APPENDIX II.
SENSITIVITY OF FLOWER
PETALS TO ETHYLENE*

Species	Sensitivity[†]	Symptoms[‡]	Species	Sensitivity[†]	Symptoms[‡]
Abelia schumanii	4	a	*Asphodelus albus*	0	w
Acanthus hungaricus	4	a	*Aster novi-belgii*	0	w
Acanthus spinosus	4	a	*Baldellia ranunculoides*	4	w
Achillea filipendula	0	w	*Bergenia cordifolia*	0	w
Aconitum napellus	3	a	*Bloomeria aurantiaca*	3	w
Aeschynanthus sp.	4	a	*Borago officinalis*	4	a
Agastachefoeniculum	4	a	*Brassica napus*	4	a
Alisma parviflora	4	w	*Brodiaea californica*	0	w
Alliaria petiolata	4a		*Brunnera macrophylla*	4	a
Allium caeruleum	1	w	*Buglossoides*	4	a
Allium cernuum	0	w	*purpurocaerulea*		
Allium sphaerocephalon	0	w	*Calceolaria* sp.	4	a
Alstromeria pelegrina	2–3	wa	*Calluna vulgaris*	0	w
(cvs. Carmen, Marina,			*Camassia leichtlinii*	0	w
Orchid)			*Camassia quamash*	0	w
Althaea officinalis	3	w	*Campanula garganica*	4	w
Anagallis arvensis	3	a	*Campanula glomerata*	3	w
Andromeda sp.	3	a	*Campanula pyramidalis*	4	w
Anemone (hybrid)	4	a	*Canna hybrid*	0	w
(cv. Elegans)			(3 cultivars)		
Anethum graveolens	0	w	*Cardamine pratensis*	4	a
Anigozanthos spp.	0	w	*Carpathea pomeridiana*	4	w
(3 species)			*Cattleya* (hybrid)	2–3	w
Antirrhinum majus	3	a	*Centaurea cyanus*	0–1	w
Arabis caucasia	4	a	*Centranthus ruber*	3	a
Armeria maritima	3	w	(cv. Albus)		
Armeria pseudoarmeria	4	w	*Centranthus ruber*	3	a
Asclepias tuberosa	2	w	(cv. Coccineus)		
Asperula tinctoria	3	a	*Cephalaria alpina*	2–3	wa
Asphodeline lutea	0	w	*Cephalaria gigantea*	2	wa

*After: Woltering, E.J. and W.G. van Doorn. 1988. *J. Exp. Bot.* 39:1605–1616; van Doorn, W.G. 2000. *Ann. Bot.* 87:447–456.

[†]Sensitivity: 0 = insensitive; 1 = low; 2 = intermediate; 3 = high; 4 = very high.

[‡]Petal symptoms: w = wilting; a = abscission; wa = wilting and abscission; w/wa = wilting, sometimes wilting and abscission; c = coloration; y = yellowing.

Species	Sensitivity[†]	Symptoms[‡]	Species	Sensitivity[†]	Symptoms[‡]
Ceratostigma plumbaginoides	4	w	*Dianthus caryophyllus* (standard) (cvs. Lena, Le Reve Salmon Sim, Nora Barlo, Orange Triumph, Scania)	4	w
Chasmanthe aethiopica	0	w			
Cheiranthus sp.	4	a			
Chelidonium majus	4	a			
Chelone barbatus	3	a	*Dicentra formosa*	2	wa
Chelone obliqua	1–2	a	*Dicentra* (hybrid)	2	wa
Chrysanthemum maximum	0	w	*Dorotheanthus bellidiformis*	4	w
Chrysanthemum morifolium (cvs. Horim, Spider, Westland)	0	w	*Dracocephalum nutans*	4	a
			Echeveria setosa	0	w
			Echium plantagineum	4	a
Chrysanthemum parthenium	0	w	*Edraianthus graminifolius*	3	w
			Eremurus (hybrid)	0	w
Chrysanthemum segetum	0	w	*Erica gracilis*	0	w
Claytonia sp.	3	w	*Erica hiemalis*	0	w
Colchicum autumnale	0	w	*Erica tetralix*	0	w
Colchicum speciosum	0	w	*Erigeron* (hybrid)	0	w
Columnea krakatau	4	a	*Erysimum cuspidatum*	3	a
Columnea nesse	4	a	*Erythronium americanum*	3	wa
Commelina sp.	4	w	*Eschscholzia* sp.	4	a
Convolvulus arvensis	4	w	*Eucomis bicolor*	0	w
Conium maculatum	0	w	*Euphorbia fulgens* (cvs. Albatros, Algevo, Oranje)	1	w
Convallaria majalis	0	w			
Corydalis sp.	2	wa			
Crassula falcata	0	w	*Exacum affine*	0	w
Crocosmia × *crocosmiiflora*	1–2	wa	*Forsythia* × *intermedia*	4	w
			Freesia (hybrids) (cvs. Aurora, Ballerina, Royal Blue)	0	w
Crocus chrysanthus	0	w			
Crossandra sp.	4	a			
Cyclamen (hybrid)	4	a	*Fumaria* sp.	2	wa
Cymbidium (hybrid) (cvs. Greenland King Authur, Salvator)	4	cyw	*Galanthus nivalis*	0	w
			Galium aparine	3	a
			Galtonia candicans	0	w
Cymbidium (hybrid) (cvs. Showgirl Malibu, Showgirl Stardust, Sir Lancelot, Evening Star)	4	cw	*Galtonia* sp.	0	w
			Gaultheria shallon	3	a
			Gentiana dahurica	0	w
			Gentiana kochiana	0	w
			Gentiana sino-ornata	0	w
Cyrtanthus purpureus	2	w	*Geranium gracile*	4	a
Dahlia (hybrid)	1	w	*Geranium nodosum*	4	a
Delospermum cooperi	3	w	*Geranium sanguineum*	4	a
Delospermum lyndenburgensis	4	w	*Gerbera jamesonii* (cvs. Agnes, Beatrix, Veronica)	0–1	w
Delphinium ajacis	4	a			
Dendrobium phalaenopsis	3	w	*Geum* (hybrid)	4	a
Deutzia scabra 'Macropetala'	4	a	*Gladiolus* (hybrid)	0–1	w
			Gloriosa superba (cv. Rothschildiana)	0	w
Deutzia schneideriana	4	a			
Dianthus barbarus		4 w	*Gratiola officinalis*	3	a
Dianthus caryophallus (spray) (cvs. Mini Star, Silvery Pink)	4	w	*Gypsophila paniculata*	4	w
			Helianthus annuus	0	w
			Helipterum manglesii	0	w

Species	Sensitivity[†]	Symptoms[‡]	Species	Sensitivity[†]	Symptoms[‡]
Helipterum roseum	0	w	*Limonium latifolium*	3	w
Hemerocallis (cvs. Black Prince, Invictus, Mabel Fuller, Mrs. J. Tigert)	0	w	*Lindolofia stylosa*	4	a
			Liriope koreana	0	w
			Lithops dorothea	3	w
			Lobelia cardinalis	3	w
Hemerocallis lilio-asphodelus	0	w	*Lobelia siphylitica*	2–3	w
			Lonicera × *heckrottii* (cv. Goldflame)	4	a
Hippeastrum × *ackermannii*	3	w	*Lunaria rediviva*	4	a
Hosta lancifolia	0	w	*Lychnis chalcedonica*	4	w
Hosta latifolia	0	w	*Lycopersicon esculentum*	4	wa
Hosta tardiana	0	w	*Lysimachia ciliata*	3	a
Hosta undulata	0	w	*Lysimachia clethroides*	4	a
Hyacinthus orientalis	1–2	w	*Lysimachia punctata*	4	a
Hyacinthoides non-scipta	0	w	*Malva alcea*	4	w
Incarvillea delavayi	4	a	*Malva sylvestris*	4	w
Iochroma (hybrid)	3	wa	*Matthiola incana*	2	w
Ipomoea alba	4	w	*Mentha suaveolens*	4	a
Iris (hybrid) (cvs. Ideal, Prof. Blaauw, Symphony, Witte van Vliet)	0–1	w	*Mertensia paniculata*	4	a
			Mesembryanthemum productus	3	w
Iris germanica	0	w	*Monopsis* sp.	3	w
Iris halophyta	0	w	*Muscari armeniacum*	0–1	w
Iris sibirica	0	w	*Narcissus pseudonarcissus* (cvs. Carlton, Dutch Master, Golden Harvest)	0	w
Ixia flexuosa	0	w			
Ixora (hybrid)	3	a			
Jasmium officinale	4	a	*Nerine mansellii*	0	w
Kalanchoë blossfeldiana	2	w	*Nerine sarniensis*	0	w
Kalmia latifolia	4	a	*Nicotiana tabacum*	4	wa
Kohleria (hybrid) (cv. Eriantha)	4	a	*Nerine bowdenii*	0	w
			Nierembergia sp.	3	wa
Kniphofia (hybrid)	1	w	*Nigella damascena*	4	a
Kirengeshoma palmata	4	a	*Nothoscordum aureum*	0	w
Lachenalia sp.	0	w	*Omphalodes verna*	4	a
Laurentia fluviatilis	3	w	*Ornithogalum thyrsoides*	0	w
Lavatera maritima	4	w	*Ornithoglossum parviflorum*	0	w
Leicesteria formosa	4	a			
Leucothoe axillaris	3	a	*Papaver rhoeas*	4	a
Leucothoe walterii	3	a	*Paphiopedilum* (hybrid)	2–3	w
Lewisia cotyledon	4	w	*Patrinia gibbosa*	3	a
Liatris spicata	0	w	*Penstemon cobaea*		4 a
Ligustrum ovalifolium	4	a	*Penstemon heterophyllus*	4	a
Lilium (hybrid)			*Penstemon serrulatus*	3	a
(cv. Brunello–Oriental hybrid)	1	w	*Pentas lanceolata*		4 a
(cv. Montenegro–Oriental)	0	w	*Petunia* hybrid	4	wa
			Phalaenopsis (hybrid)	3	w
(cv. Star Gazer–Aseatic hybrid)	2	w	*Phlox paniculata*	3–4	wa
			Phygelius sp.	3	a
(cv. Woodruff Memory–Aseatic hybrid)	0–1	w	*Physostegia virginiana*	3	wa
			Phyteuma scheuchzeri	2	w
			Pieris japonica	0	w
Lilium martagon	3	wa	*Plumbago auriculata*	4	w

Species	Sensitivity[†]	Symptoms[‡]	Species	Sensitivity[†]	Symptoms[‡]
Polemonium foliosissimum	2	wa	*Sisyrinchium californicum*	4	w
			Sisyrinchium laevigatum	4	w
Polianthes tuberosa	0	w	*Solanum dulcamara*	3	w
Polygonatum odoratum	0	w	*Solidago* (hybrid)	0	w
Portulaca grandiflora	4	w	*Streptocarpus* (hybrid)	4	a
Portulaca umbraticola	4	w	*Succisella inflexa*	3	w
Potentilla (cv. Gibson Scarlet)	3	a	*Symphytum cordatum*	4	a
			Symphytum grandiflorum	4	a
Potentilla grandiflora	4	a	*Symphytum ottomanum*	4	a
Primula denticulata	2–3	wa	*Syringa vulgaris*	4	a
Primula rosea (cv. Grandiflora)	2–3	wa	*Thunbergia alata*	4	a
			Tiarella cordifolia	0	w
Primula vialii	2	w	*Tigridia pavonia*	0	w
Pulmonaria officinalis	4	a	*Torenia* (hybrid)	4	a
Quamoclit coccinea	4	w	*Trachelium caeruleum*	3	w
Rhododendron brachycarpum	4	a	*Tradescantia* (hybrid)	4	w
			Tricyrtis latifolia	0	w
Rhododendron (hybrid) (several cvs)	4	a	*Triteleia laxa*	0	w
			Tritonia crocata	0	w
Ribes aureum	3	a	*Tulbaghia violacea*	0	w
Rosa (hybrid) (cvs. Amsterdam, Sonia)	3	a	*Tulipa gesneriana*		
Rosa (hybrid) (cvs. Betty, Director Riggers, Fanal, Friedrich Heyer)	4	a	(cvs. Ad Rem, Gander's Rhapsody, Rosario)	0	a
			(cvs. Atilla, Barcelona, Bastogne, Blenda, Monte Carlo, Negrita, Prominence, Recreado, Yokohama)	1	w
Rubia tinctorum	3	a			
Rudbeckia (hybrid)	0	w			
Sabatia sp.	0	w			
Sagittaria lancifolia	4	w	(cvs. Golden Apeldoorn, White Dream)	1–2	w
Saintpaulia confusa	4	a			
Saintpaulia tongwensis	4	a			
Salvia (cv. Mainacht)	4	a	(cv. Lucky Strike)	2	w
Salvia × *superba*	4	a	(cvs. Alba, Leen van der Mark, Lustige Witwe)	1	w/wa
Sambucus nigra	4	a			
Sandersonia aurantiaca	0	w			
Sansevieria sp.	0	w	(cv. Pink Impression)	2	w/wa
Satureja vulgaris	4	a	*Vaccinium macrocarpon*	4	a
Saxifraga apiculata (cv. Gregor Mendel)	0	wa	*Valeriana officinalis*	4	a
			Veronica longifolia (cv. Blauriesin)	4	a
Saponaria (hybrid)	4	w			
Saxifraga × *arendsii* (cv. Schneeteppich)	0	wa	*Veronica orchidea*	4	a
			Veronica spicata (cv. Alba)	4	a
Saxifraga litacina	0	wa	*Viburnum henryi*	4	a
Scabiosa caucasica	2–3	w	*Viburnum lobophyllum*	4	a
Sedum (hybrid)	0	w	*Weigela florida*	0	w
Sedum spectabile	0	w	*Weigela* (cv. Gustave Mallet)	0	w
Sedum spurium	1	w			
Sempervivum sp.	0	w	*Zebrina pendula*	3	w
Sinningia cardinalis	4	a	*Zephyranthes candida*	0	w
Sisyrinchium angustifolium	4	w	*Zinnia elegans*	0	w

APPENDIX III.
APPROXIMATE RESPIRATORY AND ETHYLENE PRODUCTION RATES FOR A CROSS-SECTION OF FRUITS AND VEGETABLES*

Commodity	Respiratory Rate ($mg\ CO_2 \cdot kg^{-1} \cdot hr^{-1}$) Temperature						Ethylene Production Rate ($\mu L\ C_2H_4 \cdot kg^{-1} \cdot h^{-1}$)
	0°C	5°C	10°C	15°C	20°C	25°C	
Apple							
Fall	3	6	9	15	20	—†	varies greatly
Summer	5	8	17	25	31	—	varies greatly
Apricot	6	—	16	—	40	—	<0.1 (0°C)
Arazá (ripe)	—	—	601	—	1283	—	—
Artichoke	30	43	71	110	193	—	<0.1
Asian Pear	5	—	—	—	25	—	varies greatly
Asparagus	60	105	215	235	270	—	2.6 (20°C)
Atemoya	—	—	119	168	250	—	200 (20°C)
Avocado	—	35	105	—	190	—	>100 (ripe; 20°C)
Banana (ripe)	—	—	80	140**	280	—	5.0 (15°C)
Basil	36	—	71	—	167	—	very low***
Beans							
Snapbean (f)‡	20	34	58	92	130	—	<0.05 (5°C)
Long (f)‡	40	46	92	202	220	—	<0.05 (5°C)
Beets	5	11	18	31	60	—	<0.1 (0°C)
Blackberry	19	36	62	75	115	—	varies; 0.1 to 2.0
Blueberry	6	11	29	48	70	101	varies; 0.5 to 10.0
Bok Choy	6	11	20	39	56	—	<0.2
Breadfruit	—	—	—	329	—	480	1.2
Broccoli	21	34	81	170	300	—	<0.1 (20°C)
Brussels Sprouts	40	70	147	200	276	—	<0.25 (7.5°C)
Cabbage	5	11	18	28	42	62	<0.1 (20°C)
Carambola	—	15	22	27	65	—	<3.0 (20°C)
Carrot (topped)	15	20	31	40	25	—	<0.1 (20°C)
Cassava	—	—	—	—	—	40	1.7 (25°C)

Commodity	Respiratory Rate (mg $CO_2 \cdot kg^{-1} \cdot hr^{-1}$) Temperature						Ethylene Production Rate ($\mu L\ C_2H_4 \cdot kg^{-1} \cdot h^{-1}$)
	0°C	5°C	10°C	15°C	20°C	25°C	
Cauliflower	17	21	34	46	79	92	<1.0 (20°C)
Celeriac	7	13	23	35	45	—	<0.1 (20°C)
Celery	15	20	31	40	71	—	<0.1 (20°C)
Cherimoya	—	—	119	182	300	—	200 (20°C)
Cherry, Sweet	8	22	28	46	65	—	<0.1 (0°C)
Chervil	12	—	80	—	170	—	very low
Chicory	3	6	13	21	37	—	<0.1 (0°C)
Chinese Cabbage	10	12	18	26	39	—	<0.1 (20°C)
Chinese Chive	54	—	99	—	432	—	very low
Chive	22	—	110	—	540	—	very low
Coconut	—	—	—	—	—	50	very low
Coriander	22	30	—	—	—	—	very low
Cranberry	4	5	8	—	16	—	0.6 (5°C)
Cucumber	—	—	26	29	31	37	0.6 (20°C)
Currant, Black	16	28	42	96	142	—	—
Dill	22	—	103	324	—	—	<0.1 (20°C)
Dragon Fruit	—	—	—	—	105	—	<0.1
Durian	—	—	—	—	265††	—	40 (ripe)
Eggplant							
American	—	—	—	69‡‡	—	—	0.4 (12.5°C)
Japanese	—	—	—	131‡‡	—	—	0.4 (12.5°C)
White Egg	—	—	—	113‡‡	—	—	0.4 (12.5°C)
Endive/Escarole	45	52	73	100	133	200	very low
Fennel	19§§	—	—	—	32	—	4.3 (20°C)
Fig	6	13	21	—	50	—	0.6 (0°C)
Garlic							
Bulbs	8	16	24	22	20	—	very low
Fresh Peeled	24	35	85	—	—	—	very low
Ginger	—	—	—	—	6**	—	very low
Ginseng	6	—	15	33	—	95	very low
Gooseberry	7	12	23	52	81	—	—
Grape, American	3	5	8	16	33	39	<0.1 (20°C)
Grape, Muscadine	10§§	13	—	—	51	—	<0.1 (20°C)
Grape, Table	3	7	13	—	27	—	<0.1 (20°C)
Grapefruit	—	—	—	<10	—	—	<0.1 (20°C)
Guava	—	—	34	—	74	—	10 (20°C)
Honey Dew Melon	—	8	14	24	30	33	very low
Horseradish	8	14	25	32	40	—	<1.0
Jerusalem Artichoke	10	12	19	50	—	—	—
Jicama	6	11	14	—	6	—	very low
Kiwifruit (ripe)	3	6	12	—	19	—	75
Kohlrabi	10	16	31	46	—	—	<0.1 (20°C)
Leek	15	25	60	96	110	115	<0.1
Lemon	—	—	11	19	24	—	<0.1 (20°C)
Lettuce							
Head	12	17	31	39	56	82	very low
Leaf	23	30	39	63	101	147	very low
Lime	—	—	<10	—	—	—	<0.1 (20°C)

Commodity	Respiratory Rate ($mg\ CO_2 \cdot kg^{-1} \cdot hr^{-1}$) Temperature						Ethylene Production Rate ($\mu L\ C_2H_4 \cdot kg^{-1} \cdot h^{-1}$)
	0°C	5°C	10°C	15°C	20°C	25°C	
Litchi	—	13	24	—	60	102	very low
Longan	—	7	21	—	42	—	very low
Longkong	—	—	45†††	—	—	—	4.0
Loquat	11‡‡‡	12	31	—	80	—	very low
Luffa	14	27	36	63	79	—	<0.1 (20°C)
Mamey Apple	—	—	—	—	—	35	400.0 (27°C)
Mandarin (Tangerine)	—	6	8	16	25	—	<0.1 (20°C)
Mango	—	16	35	58	113	—	1.5 (20°C)
Mangosteen	—	—	—	—	—	21	0.03
Marjoram	28	—	68	—	—	—	very low
Mint	20	—	76	—	252	—	very low
Mushroom	35	70	97	—	264	—	<0.1 (20°C)
Nectarine (ripe)	5	—	20	—	87	—	5.0 (0°C)
Netted Melon	6	10	15	37	55	67	55.0
Nopalitos	—	18	40	56	74	—	very low
Okra	21‡‡	40	91	146	261	345	0.5
Olive	—	15	28	—	60	—	<0.5 (20°C)
Onion	3	5	7	7	8	—	<0.1 (20°C)
Orange	4	6	8	18	28	—	<0.1 (20°C)
Oregano	22	—	101	—	176	—	very low
Papaya (ripe)	—	5	—	19	80	—	8.0
Parsley	30	60	114	150	199	274	very low
Parsnip	12	13	22	37	—	—	<0.1 (20°C)
Passion Fruit	—	44	59	141	262	—	280.0 (20°C)
Pea							
Garden	38	64	86	175	271	313	<0.1 (20°C)
Edible Pod	39	64	89	176	273	—	<0.1 (20°C)
Peach (ripe)	5	—	20	—	87	—	5.0 (0°C)
Pepper	—	7	12	27	34	—	<0.2 (20°C)
Persimmon	6	—	—	—	22	—	<0.5 (20°C)
Pineapple	—	2	6	13	24	—	<1.0 (20°C)
Plum (ripe)	3	—	10	—	20	—	<5.0 (0°C)
Pomegranate	—	6	12	—	24	—	<0.1 (10°C)
Potato (cured)	—	12	16	17	22	—	<0.1 (20°C)
Prickly Pear	—	—	—	—	32	—	0.2 (20°C)
Radicchio	8	13§§§	23****	—	—	45	0.3 (6°C)
Radish							
Topped	16	20	34	74	130	172	very low
Bunched with tops	6	10	16	32	51	75	very low
Rambutan (mature)	—	—	—	—	—	70	very low
Raspberry	17§§	23	35	42	125	—	≤ 12.0 (20°C)
Rhubarb	11	15	25	40	49	—	—
Rutabaga	5	10	14	26	37	—	<0.1 (20°C)
Sage	36	—	103	—	157	—	very low
Salad Greens							
Rocket Salad	42	113	—	—	—	—	very low
Lamb's Lettuce	12	67****	81	—	139	—	very low
Salsify	25	43	49	—	193	—	very low

Commodity	Respiratory Rate (mg $CO_2 \cdot kg^{-1} \cdot hr^{-1}$) Temperature						Ethylene Production Rate ($\mu L\ C_2H_4 \cdot kg^{-1} \cdot h^{-1}$)
	0°C	5°C	10°C	15°C	20°C	25°C	
Sapodilla	—	—	—	—	—	16	3.7 (20°C)
Sapote	—	—	—	—	—	—	>100 (20°C)
Southern Pea							
Intact (f)‡	24§§	25	—	—	148	—	—
Shelled (s)‡	29§§	—	—	—	126	—	—
Spinach	21	45	110	179	230	—	very low
Sprouts (mung bean)	23	42	96	—	—	—	<0.1 (10°C)
Squash, Summer	25	32	67	153	164	—	<1.0 (20°C)
Squash, Winter	—	—	99‡‡	—	—	—	very low
Star Apple	—	—	—	—	38	—	0.1 (20°C)
Strawberry	16	—	75	—	150	—	<0.1 (20°C)
Sweet Corn	41	63	105	159	261	359	very low
Swiss Chard	19§§	—	—	—	29	—	0.14 (20°C)
Tamarillo	—	—	—	—	27	—	<0.1
Tarragon	40	—	99‡‡	—	234	—	very low
Thyme	38	—	82	—	203	—	very low
Tomatillo (mature green)	—	13	16	—	32	—	10.0 (20°C)
Tomato	—	—	15	22	35	43	10.0 (20°C)
Truffles	28	35	45	—	—	—	very low
Turnip	8	10	16	23	25	—	very low
Water Chestnut	10	25	42	79	114	—	—
Water Convolvulus	—	—	—	—	—	100	<2.0
Watercress	22	50	110	175	322	377	<1.0 (20°C)
Watermelon	—	4	8	—	21	—	<1.0 (20°C)
Wax Apple	—	—	5	—	10	—	very little

*Data from: USDA Handbook 66 (a compilation of work by many scientists, for specific references, see the Hbk.) and other sources; †— = data not available; ‡f = intact fruit (pod with seed); §1 day after harvest; **at 13°C; ††at 22°C; ‡‡at 12.5°C; §§at 2°C; ***very low is considered to be <0.05 $\mu L \cdot kg^{-1} \cdot h^{-1}$; †††at 9°C; ‡‡‡at 1°C; §§§at 6°C; **** at 7.5°C.

APPENDIX IV.
LOW TEMPERATURE
KILLING POINT OF YOUNG
AND MATURE ROOTS OF
SELECTED WOODY
ORNAMENTALS*

| | Killing Temperature | |
Commodity	Young Roots[†] (°C)	50% of Root System[†] (°C)
Acer palmatum cv. Atropurpureum		-10.0
Buxus sempervirens	-3	-9.4
Cornus florida	-6	-6.7
Cotoneaster adpressa praecox		-12.2
Cotoneaster dammeri	-5	
Cotoneaster dammeri cv. Skogsholmen	-7	
Cotoneaster horizontalis		-9.4
Cotoneaster microphyllus	-4	
Cryptomeria japonica		-8.9
Cytisus × praecox		-9.4
Daphne cneorum		-6.7
Euonymus alata cv. Compacta	-7	
Euonymus fortunei cv. Argenteopmarginatus		-9.4
cv. Colorata		-15.0
cv. Carrieri		-9.4
Euonymus kiautschovica (*E. patens*)	-6	
Hedera helix cv. Baltica	-9.4	
Hypericum spp.	-5	
Ilex cornuta cv. Dazzler	-4	
cv. Helleri.	-5	
cv. Hetzi		-6.7
cv. Stokes		-6.7
cv. Convexa		-6.7
Ilex cv. Nellie Stevens	-5	
cv. San Jose	-6	
Ilex glabra		-9.4
Ilex opaca	-5	-6.7
Ilex × meserveae cv. Blue Boy	-5	
Juniperus conferta	-11	
Juniperus horizontalis cv. Douglasii		-17.8
cv. Plumosa	-11	-17.8
Juniperus squamata cv. Meyeri	-11	
Kalmia latifolia	-9	
Koelreuteria paniculata	-9	
Leucothoe fontanesiana	-7	-15.0
Magnolia stellata	-6	-5.0
Magnolia × soulangeana		-5.0
Mahonia bealei	-4	
Pachysandra terminalis	-9.4	
Picea glauca		-23.3
Picea omorika		-23.3
Pieris floribunda		-15.0
Pieris japonica	-9	-12.2
cv. Compacta	-9.4	
Potentilla fruticosa		-23.3
Pyracantha coccinea		-7.8
cv. Lalandei	-4	
Rhododendron carolinianum	-11	-17.8
Rhododendron catawbiense cv. Roseum Elegans		
Rhododendron catawbiense		-17.8
Rhododendron Exbury Hybrid	-8	
Rhododendron cv. Gibralter		-12.2
cv. Hino Crimson	-7	
cv. Hinodegiri.(azalea)		-12.2
cv. Purple Gem	-9	
cv. P.I.M. Hybrid		-23.3
Rhododendron prunifolium.	-7	
Rhododendron schlippenbachii	-9	
Stephanandra incisa cv. Crispa	-8	
Taxus × media cv. Hicksii.		
cv. Nigra.		-8
Viburnum carlesii Hemsl.		-12.2
Viburnum plicatum var. *tomentosum*	-7	-9.4
Vinca minor		-9.4

*Data from: Havis, J.R. 1976. *HortScience* 11:385–386; Studer, E.J., P.L. Stepoukus, G.L. Good and S.C. Wiest. 1978. *HortScience* 13:172–174.
[†] Measurements made on artificially acclimatized plants.

APPENDIX V.
RECOMMENDED STORAGE TEMPERATURE AND RELATIVE HUMIDITY, CHILLING SENSITIVITY, APPROXIMATE STORAGE LIFE, HIGHEST FREEZING POINT, WATER CONTENT AND SPECIFIC HEAT OF FRUITS*

Commodity	Temperature (°C)	Chill Sensitive[†]	Relative Humidity (%)	Approximate Storage Life	Highest Freezing Point[‡] (°C)	Water Content (%)	Specific Heat (Btu/lb/°F)[§]
Apple CA storage**	-1–4		90–95	3–12 mo[††]	-1.5	84.1	0.87
Air storage	-1–4		90–95	2–4 mo[††]	-1.5	84.1	0.87
Apricot	-0.5–0		90–95	1–2 mo	-1.0	85.4	0.88
Asian Pear	0		90	3–5 mo	-1.5	—	—
Atemoya	10–13	+	90–95	2–4 wk	—	—	—
Arazá	12–13	+	90–95	2 wk	—	—	—
Avocado	5–13[††,‡‡]	+	85–95	2–8 wk	-0.3	76.0	0.81
Bananas, green	13–14	+	90–95	2–4 wk	-0.7	75.7	0.81
Berries							
Blackberry	-0.5–0		90	2–14 d	-0.7	84.8	0.88
Blueberry	-0.5–0		90	2 wk	-1.2	83.2	0.86
Cranberry	2–4		90–95	2–4 mo	-0.8	87.4	0.90
Currant, Black	-0.5–0		90–95	1.5 wk	-1.0	84.7	0.88
Current, Red	0–1		90–95	2.5 wk	—	—	—
Dewberry	-0.5–0		90–95	2–3 d	-1.2	84.5	0.88
Elderberry	-0.5–1		90–95	1–2 wk	—	79.8	0.84
Gooseberry	0–1		90–95	3 wk	-1.0	88.9	0.91
Loganberry	-0.5–0		90–95	2–3 d	-1.2	83.0	0.86
Raspberry	-0.5–0		90–95	2–3 d	-1.0	82.5	0.86
Strawberry	0		90–95	5–7 d	-0.7	89.9	0.92
Breadfruit	12–14	+	90–95	3 wk	—	—	—
Carambola	1–4		90–95	3–5 wk	—	90.4	0.92
Cherimoya	10–13	+	90–95	2–3 wk	—	—	—
Cherries, sour	0		90–95	3–7 d	-1.7	83.7	0.87

Commodity	Temperature (°C)	Chill Sensitive[†]	Relative Humidity (%)	Approximate Storage Life	Highest Freezing Point[‡] (°C)	Water Content (%)	Specific Heat (Btu/lb/°F)[§]
Cherries, sweet	-1--0.5		90–95	2–3 wk	-1.8	80.4	0.84
Coconut	0–1.5		80–85	1–2 mo	-0.9	46.7	0.58
Dates	-18–0[‡‡]		70–75	6–12 mo[‡‡]	-15.7	22.5	0.38
Dragon fruit	10	+	90	2 wk	—	—	—
Durian	15	+	85–95	6 wk	—	—	—
Fig, fresh	-0.5–0		85–90	7–10 d	-2.4	78.0	0.82
Grapefruit	12–15	+	85–90	6–8 wk	-1.0	89.1	0.91
Grape, table	-1–0	+	90–95	1–6 mo[‡‡]	-2.1	81.6	0.85
Grape, Muscadine	-0.5–0		90	1–4 wk	-1.2	81.9	0.86
Guava	8–10	+	90	2–3 wk	—	83.0	0.86
Kiwifruit	0		90–95	4–5 mo	-1.6	82.0	0.86
Lemon	7–12[§§]	+	85–90	1–6 mo[§§]	-1.4	87.4	0.90
Litchi	2–5	+	90–95	3–5 wk	—	—	—
Lime	10	+	95	6–8 wk	-1.6	89.3	0.91
Longan	4–7	+	90–95	2–3 wk	—	—	—
Longkong	11–13	+	85–90	2 wk	—	—	—
Loquat	0		90	3–4 wk	—	86.5	0.89
Lychee	1.5		90–95	3–5 wk	—	81.9	0.86
Mango	10–13[§§]	+	85–90	2–4 wk	-0.9	81.7	0.85
Mangosteen	12–14	+	85–90	3 wk	—	—	—
Mardarin(Tangerine)	4		90–95	2–4 wk	-1.0	87.3	0.90
Nectarine	-1–0		90–95	2–4 wk	-0.9	81.8	0.85
Olive, fresh	5–7.5[‡‡]	+	85–90	4–6 wk[‡‡]	-1.4	80.0	0.84
Orange, Calif./Ariz.	3–8	+	90–95	3–8 wk	-1.2	85.5	0.88
Orange, Fla./Texas	1–2	+	90–95	8–10 wk	-0.7	86.4	0.89
Papaya	7–13	+	90–95	2–4 wk[***]	-0.9	88.7	0.91
Passion fruit	7–9	+	90–95	3–5 wk	—	75.1	0.80
Peach	-1–0		90–95	2–4 wk	-0.9	89.1	0.91
Pear	-1		90–95	2–7 mo[‡‡]	-1.5	83.2	0.87
Persimmon, Jpn.	0		90–95	3–4 mo	-2.1	78.2	0.83
Pineapple	7–12[§§]	+	85–95	2–4 wk[‡‡]	-1.1	85.3	0.88
Plums and Prune	-1–0		90–95	2–5 wk	-0.8	86.6	0.89
Pomegranate	5–6	+	90–98	2–3 mo	-3.0	82.3	0.86
Prickly Pear	5–8	+	90–95	2–5 wk	—	—	—
Quince	-0.5–0		90	2–3 mo	-2.0	83.8	0.87
Rambutan	8–15	+	90–95	2 wk	—	—	—
Sapodilla	12–16	+	85–90	2–3 wk	—	—	—
Tamarind	20	+	90–95	7 wk	—	—	—
Wax Apple	10–14	+	90–95	10–14 d	—	—	—

[*]Data from *USDA Handbook* 66 in which leading authorities have provided an overview of the literature and storage recommendations for individual crops. For additional information, see Handbook 66.

[†]Commodities that sustain injury after a period of exposure to temperatures above freezing but below a critical threshold temperature.

[‡]Data from Whiteman, T.M. 1957. *USDA Mkt. Res. Rpt.* 196.

[§]Specific heat in Btu/lb/°F was calculated as: $S = 0.008 \times$ (% water in food) + 0.20. In metric units of kJ/kg/°C, $S = 0.0335 \times$ (% water in food) + 0.8374. From Siebel, J.E. 1892. *Ice Refrig.* 2:256–257.

[**]Controlled atmosphere storage.

[††]Varies with cultivar.

[‡‡]See text for maturity and cultivar differences.

[§§]See *USDA Hbk.* 66 for details.

[***]Varies with maturity.

APPENDIX VI. RECOMMENDED STORAGE TEMPERATURE AND RELATIVE HUMIDITY, CHILLING SENSITIVITY, APPROXIMATE STORAGE LIFE, HIGHEST FREEZING POINT, WATER CONTENT AND SPECIFIC HEAT OF VEGETABLES*

Commodity	Temperature (°C)	Chill Sensitive[†]	Relative Humidity (%)	Approximate Storage Life	Highest Freezing Point[‡] (°C)	Water Content (%)	Specific Heat (Btu/lb/°F)[§]
Artichoke, globe	0		>95	2 wk	-1.1	83.7	0.87
Artichoke, Jerusalem	-0.5–0		90–95	6–12 mo	-2.2	79.8	0.84
Asparagus	0–2		95–98	2–3 wk	-0.6	93.0	0.94
Basil	12	+	95–100	2 wk	—	—	—
Beans, green or snap	5–7**	+	>95	8–12 d	-0.7	88.9	0.91
Beans, lima	3–5**	+	95	5–7 d	-0.6	66.5	0.73
Bean sprouts	0		95–100	7–9 d	—	88.8	0.91
Beets, bunched	0		98–100	10–14 d	-0.4	—	—
Beets, topped	0		98–100	8–10 mo	-0.9	87.6	0.90
Bitter gourd	10–12	+	90–95	2 wk	—	—	—
Bok choy	0–2		>95	3 wk	—	—	—
Broccoli	0		98–100	2–3 wk	-0.6	89.9	0.92
Brussels sprouts	0		95–100	3–5 wk	-0.8	84.9	0.88
Cabbage, early	0		98–100	3–6 wk	-0.9	92.4	0.94
Cabbage, late	0		98–100	5–6 mo	-0.9	92.4	0.94
Cactus "leaves"	5	+	95–99	3 wk	—	—	—
Carrot, bunched	0		95–100	8–10 d	—	—	—
Carrot, mature	0		98–100	7–9 mo	-1.4	88.2	0.91

Commodity	Temperature (°C)	Chill Sensitive[†]	Relative Humidity (%)	Approximate Storage Life	Highest Freezing Point[‡] (°C)	Water Content (%)	Specific Heat (Btu/lb/°F)[§]
Carrot, immature	0		98–100	2–3 wk	-1.4	88.2	0.91
Cassava	0–5	+	85–90	1 mo	—	—	—
Cauliflower	0		95–98	3 wk	-0.8	91.7	0.93
Celeriac	0–2		97–98	6–8 mo	-0.9	88.4	0.91
Celery	0		>95	5–7 mo	-0.5	93.7	0.95
Chervil	0		95–100	1 wk	—	—	—
Chicory, witloof	0		95–100	2–4 wk	-0.5	5.1	0.96
Chinese cabbage	0		95–100	2–3 mo	—	95.0	0.96
Cilantro	0–1		95–100	2 wk	—	—	—
Collard	0		95–100	10–14 d	-0.8	86.9	0.90
Coriander	0		95–100	2 wk	—	—	—
Corn Salad	0–2		95–100	10–14 d	—	—	—
Cucumber	10–13	+	95	10–14 d	-0.5	96.1	0.97
Dandelion	0–2		95–100	10–14 d	—	—	—
Dill	0		95–100	2 wk	—	—	—
Eggplant	10–12	+	90–95	>2 wk	—	—	—
Endive and Escarole	0		95–100	2–3 wk	—	—	—
Fennel	0		90–95	2 wk	—	—	—
Garlic	0		60–70	6–9 mo	-0.8	61.3	0.69
Ginger	12–14	+	85	2–3 mo	—	87.0	0.90
Ginseng	0		>95	2 mo	—	—	—
Horseradish	-2–-1		98–100	8–12 mo	-1.8	74.6	0.80
Jicama	13–15	+	80–90	2–4 mo	—	—	—
Kale	0		95–100	2–3 wk	-0.5	86.8	0.89
Kohlrabi	0		98–100	2–3 mo	-1.0	90.3	0.92
Leek	0		95–100	2–3 mo	-0.7	85.4	0.88
Lettuce	0		98–100	2–3 wk	-0.2	94.8	0.96
Luffa gourd	10–12	+	90–95	2 wk	—	—	—
Melons							
Canary	10	+	90–95	3 wk	—	—	—
Cantaloupe	2–7	+	95	10–14 d	-1.2	92.0	0.94
Casaba	10	+	90–95	3 wk	-1.0	92.7	0.94
Crenshaw	10	+	90–95	2 wk	-1.0	92.7	0.94
Honey Dew	7	+	90–95	3 wk	-0.9	92.6	0.94
Persian	7	+	90–95	2 wk	-0.8	92.7	0.94
Watermelon	10–15	+	90	2–3 wk	-0.4	92.6	0.94
Miners Lettuce	0–2		95–100	10–14 d	—	—	—
Mizuna	0–2		95–100	10–14 d	—	—	—
Mushroom	0–1		95	3–4 wk	-0.9	91.1	0.93
Okra	7–10	+	90–95	7–14 d	-1.8	89.8	0.92
Onion, green	0		95–100	3–4 wk	-0.9	89.4	0.91
Onion, dry	0		65–75	6–9 mo	-0.8	87.5	0.90
Onion, sweet—CA[††]	1–2		65–75	5–6 mo			
sweet—air	1–2		65–75	3–4 mo			
Parsley	0		95–100	1–2 mo	-1.1	85.1	0.88
Parsnip	0–1		98	4–6 mo	-0.9	78.6	0.83
Pea, green	0		95–98	1–2 wk	-0.6	74.3	0.79
Pea stems/leaves	0		95–100	1 wk	—	—	—
Pea, southern	4–5	+	95	6–8 d	—	66.8	0.73

Commodity	Temperature (°C)	Chill Sensitive[†]	Relative Humidity (%)	Approximate Storage Life	Highest Freezing Point[‡] (°C)	Water Content (%)	Specific Heat (Btu/lb/°F)[§]
Pepper, sweet, chilli	7	+	90–95	2–3 wk	-0.7	92.4	0.94
Pepino	5		90–95	21 d	—	—	—
Plantain	12–13	+	90–95	1–5 wk	—	—	—
Potato, early crop	9–12**		90–95	**	-0.6	81.2	0.85
Potato, late crop	**		95–98	5–12	-0.6	77.8	0.82
Pumpkin	10–13	+	50–70	2–3 mo	-0.8	90.5	0.92
Purslane	0–2		95–100	10–14 d	—	—	—
Radicccho	3–5		90–95	20–30 d	—	—	—
Radish,							
spring—topped	0		95–100	3–4 wk	-0.7	94.5	0.96
spring—bunched	0		95–100	1–2 wk	—	—	—
Radish, winter	0		95–100	2–4 mo	—	—	—
Rhubarb	0		95–100	2–4 wk	-0.9	4.9	0.96
Rocket Salad	0–2		95–100	7–10 d	—	—	—
Rutabaga	0		98–100	4–6 mo	-1.1	89.1	0.91
Salsify	0		95–98	3–4 mo	-1.1	79.1	0.83
Savory	0		95–100	3 wk	—	—	—
Sorrel, French	0–2		95–100	10–14 d	—	—	—
Southern pea, inshell	4–5		95–100	6–8 d	—	66.8	0.73
Southern pea, shelled	4–5		95–100	2–3 d	—	—	—
Spinach	0		95–100	10–14 d	-0.3	92.7	0.94
Sprouts							
Alfalfa	0		95–100	7 d	—	—	—
Bean (kidney, pinto, navy)	0		95–100	5–10 d	—	—	—
Broccoli	0		95–100	5–10 d	—	—	—
Buckwheat	0		95–100	5–10 d	—	—	—
Clover	0		95–100	5–10 d	—	—	—
Garbanzo	0		95–100	5–10 d	—	—	—
Green pea	0		95–100	5–10 d	—	—	—
Lentil	0		95–100	5–10 d	—	—	—
Mung bean	0		95–100	8–9 d	—	—	—
Mustard	0		95–100	5–10 d	—	—	—
Onion	0		95–100	5–10 d	—	—	—
Radish	0		95–100	5–7 d	—	—	—
Soybean	0		95–100	5–10 d	—	—	—
Sunflower	0		95–100	5–10 d	—	—	—
Watercress	0		95–100	5–10 d	—	—	—
Winter cress	0		95–100	5–10 d	—	—	—
Squash, summer	5–10	+	95	1–2 wk	-0.5	94.0	0.95
Squash, winter	10–13	+	50–70[††]	1–6 mo	-0.8	85.1	0.88
Sweet corn	0		95–98	2 wk	-0.6	73.9	0.79
Sweetpotato	14	+	90	4–12 mo	-1.3	68.5	0.75
Swiss Chard	0		95–98	1–2 wk	—	91.1	0.93
Tamarillo	3		90–95	5–9 mo	—	—	—
Tomatillo	5–10	+	80–90	2–3 wk	—	—	—
Taro (Dasheen)	7–10	+	85–90	4–5 mo	—	73.0	0.78
Tomato, mature—green	19–20	+	90–95	1–3 wk	-0.6	93.0	0.94
Tomato, firm—ripe	7	+	90–95	4–7 d	-0.5	94.1	0.95

Commodity	Temperature (°C)	Chill Sensitive[†]	Relative Humidity (%)	Approximate Storage Life	Highest Freezing Point[‡] (°C)	Water Content (%)	Specific Heat (Btu/lb/°F)[§]
Truffle	0		90–95	20–30 d	—	—	—
Turnip	0		95	4–5 mo	-1.0	91.5	0.93
Turnip greens	0		95–100	10–14 d	-0.2	90.3	0.92
Turnip–rooted parsley	0		95–100	4–6 mo	—	—	—
Water chestnut	0–2		98–100	1–2 mo	—	78.3	0.83
Water convolvulus	12–14	+	90–95	10–12 d	—	—	—
Watercress	0		95–100	2–3 wk	-0.3	93.3	0.95
Yam	16	+	70–80	6–7 mo	—	73.5	0.79

*Data from *USDA Handbook* 66 in which leading authorities have provided an overview of the literature and storage recommendations for individual crops. For additional information, see Hbk 66.

[†]Commodities that sustain injury after a period of exposure to temperatures above freezing but below a critical threshold temperature.

[‡]Data from Whiteman, T.M. 1957. *USDA Mkt. Res. Rpt.* 196.

[§]Specific heat in Btu/lb/°F was calculated as: $S = 0.008 \times$ (% water in food) $+ 0.20$. In metric units of kJ/kg/°C, $S = 0.0335 \times$ (% water in food) $+ 0.8374$. From Siebel, J.E. 1892. *Ice Refrig.* 2:256–257.

**See *USDA Hbk.* 66 for additional details.

[††]Varies with cultivar.

[‡‡]Controlled atmosphere storage.

APPENDIX VII. RECOMMENDED STORAGE TEMPERATURE AND RELATIVE HUMIDITY, APPROXIMATE STORAGE LONGEVITY AND HIGHEST FREEZING POINT FOR CUT FLOWERS, FLORIST GREENS, BULBS, CORMS, RHIZOMES, TUBERS AND ROOTS*

Commodity	Storage Temperature (°C)	Relative Humidity (%)	Approximate Storage Period[†]	Highest Freezing Point[‡] (°C)
Cut Flowers				
Acacia spp.	4	90–95	3–4 wk	-3.5
Achillea filipendulina	1–2	90–95	7–12 d	—
Achillea millefolium	1–2	90–95	3–4 d	—
Achillea ptarmica	1–2	90–95	5–8	—
Aconitum spp.	7	90–95	7–10	—
Agastache foeniculum	1–2	90–95	6–10	—
Ageratum houstonianum	1–2	90–95	7–10	—
Agrostemma githago	4	90–95	5–7	—
Allium sphaerocephalon	0–2	90–95	4 wk	—
Allium giganteum	3–6	90–95	2 wk	—
Alstroemeria spp.	0–1	90–95	1 wk	—
Amaranthus caudatus	2–5	90–95	7–10 d	—
Ammi majus	3–4	90–95	5–8 d	—
Anemone coronaria	3–8	90–95	1–2 d	—
Anemone spp.	0–1	90–95	1 wk	-2.1
Anthurium spp.[§]	12.5–20	90–95	1–2 wk	—
Antirrhinum majus	0–2	90–95	1–2 wk	—
Artemisia annua	1–2	90–95	5–7 d	—
Asclepias tuberosa	4–7	90–95	8–10 d	—
Aster ericoides	0–1	90–95	2–3	—

Commodity	Storage Temperature (°C)	Relative Humidity (%)	Approximate Storage Period[†]	Highest Freezing Point[‡] (°C)
Aster novae-angliae	4	90–95	5 d	—
Aster novi-belgii	4	90–95	5 d	—
Aster, China	0–1	90–95	1–3 wk	-0.9
Aster, Yellow	0–1	90–95	—	—
Astilbe × arendsii	1–4	90–95	2–4 d	—
Astrantia major	1–2	90–95	5–7 d	—
Baptisia australis	1–2	90–95	7–10	—
Bird-of-Paradise	6–7	85–90	1–3	—
Bouvardia	0–2	90–95	1 wk	—
Buddleia davidii	3–4	90–95	1–2 d	—
Buddleia spp.	4	90–95	1–2 d	—
Calendula	4	90–95	3–6 d	—
Calla	0–1	90–95	1 wk	—
Callicarpa spp.	0–2	90–95	2–4 d	—
Callistephus chinensis	1–2	90–95	5–7 d	—
Camellia**	7	90–95	3–6 d	-0.7
Campanula spp.	2	90–95	7–10 d	—
Candytuft	4	90–95	3 d	—
Carnation buds	0–1	90–95	4–5 wk	-0.7
Carnation, miniature	0–1	90–95	2 wk	—
Carnation[††]	0–1	90–95	2–4 wk	-0.7
Carthamus tinctorius	2–4	90–95	7 d	—
Caryopteris × clandonensis	1–2	90–95	10 d	—
Caryopteris incana	1–4	90–95	3–5 d	—
Celosia argentea	2–5	90–95	1–2 wk	—
Centaurea americana	2–5	90–95	2–3 d	—
Centaurea cyanus	2–5	90–95	2–3 d	—
Centaurea macrocephala	3–4	90–95	1–2 wk	—
Centaurea moschata	2–5	90–95	2–3 d	—
Centranthus ruber	4	90–95	3–5 d	—
Chrysanthemum	0–1	90–95	2–4 wk	-0.8
Cirsium japonicum	3–5	90–95	1–2 d	—
Clarkia spp.	3–5	90–95	5–7 d	—
Clarkia	4	90–95	3 d	—
Columbine	4	90–95	2 d	-0.5
Consolida spp.	3–5	90–95	1–2 d	—
Coreopsis spp.	2–5	90–95	1–2 d	—
Coreopsis	4	90–95	3–4 d	—
Cornflower	4	90–95	3 d	-0.6
Cosmos bipinnatus	2–4	90–95	3–4 d	—
Cosmos spp.	4	90–95	3–4 d	—
Crocosmia hybrids	1–3	90–95	4 d	—
Crocus	0.5–2	90–95	1–2 wk	—
Dahlia hybrids	3–4	80	1–2 d	—
Dahlia	4	90–95	3–5 d	—
Daisy, English	4	90–95	3 d	—
Daisy, Marguerite	0–1	90–95	1 wk	—
Daisy, Shasta	4	90–95	7–8 d	-1.1
Delphinium hybrids	3–6	90–95	1–2 d	—
Delphinium	0–1	90–95	1–2 d	-1.6

Commodity	Storage Temperature (°C)	Relative Humidity (%)	Approximate Storage Period[†]	Highest Freezing Point[‡] (°C)
Dianthus barbatus	4	90–95	7–10 d	—
Digitalis purpurea	1–2	90–95	7 d	—
Echinacea purpurea	4	90–95	7–10 d	—
Echinops bannaticus	4	90–95	7–10 d	—
Emilia javanica	2	90–95	3–6 d	—
Eremurus spp.	2	90–95	2–3 wk	—
Eryngium planum	4	90–95	7–10 d	—
Eucharis**	7–10	90–95	7–10 d	—
Euphorbia marginata	6–13	90–95	7–10 d	—
Eustoma grandiflorum	1–2	90–95	1–2 wk	—
Feverfew	4	90–95	3 d	-0.6
Forget-me-not	4	90–95	1–2 d	—
Foxglove	4	90–95	1–2 d	—
Freesia × hybrida	1–2	95	2–3 d	—
Freesia	0–0.5	90–95	10–14 d	—
Gaillardia	4	90–95	3 d	—
Gardenia	0–1	90–95	2 wk	-0.6
Gerbera	0–1	90–95	1 wk	—
Ginger	12.5–15	90–95	4–7 d	—
Gladiolus spp.	3–5	90–95	6–8 d	—
Gladiolus[††]	0–1	90–95	5–8 d	-0.3
Gloriosa	4–7	90–95	4–7d	—
Godetia	10	90–95	1 wk	—
Gomphrena globosa	3–4	90–95	1–2 d	—
Goniolimon tataricum	3–5	90–95	1–2 wk	—
Gypsophila paniculata	0–1	90	1–3 wk	—
Gypsophila paniculata	1–2	90–95	2–3 wk	—
Heather	4	90–95	1–3 wk	-1.8
Helianthus annuus	3–5	90–95	1 wk	—
Helichrysum bracteatum	3–5	90–95	1 wk	—
Heliconia	10–12.5	90–95	10 d	—
Helleborus orientalis	3–5	90–95	1 wk	—
Hyacinth[††]	0–0.5	90–95	2 wk	-0.3
Hydrangea macrophylla	1–2	90–95	1–2 wk	—
Ilex spp.	0	90–95	1–3 wk	—
Iris xiphium	1	90–95	5–10 d	—
Iris, bulbous	-0.5–0	90–95	1–2 wk	-0.8
Laceflower	4	90–95	3 d	—
Lathyrus odoratus	2	90–95	2–3 d	—
Liatris spicata	0–2	90–95	5 d	—
Liatris	0–1	90–95	2 wk	—
Lilac, forced	4	90–95	4–6 d	—
Lilium hybrids	1	90–95	4 wk	—
Lily	0–1	90–95	2–4 wk	-0.5
Lily-of-the-Valley[††]	0–1	90–95	2–3 wk	—
Limonium perezii	2–3	90–95	4–5 d	—
Limonium sinuatum	3–5	90–95	2 wk	—
Lisianthus	0–1	90–95	—	—
Lupine	4	90–95	3 d	—
Lysimachia clethroides	3–5	90–95	10–12 d	—

Commodity	Storage Temperature (°C)	Relative Humidity (%)	Approximate Storage Period[†]	Highest Freezing Point[‡] (°C)
Marigolds	4	90–95	1–2 wk	—
Matthiola incana	3–5	90–95	1 wk	—
Mignonette	4	90–95	3–5 d	—
Narcissus spp.	0–1	90–95	8 d	—
Narcissus, diffodils[††]	0–1	90	1–2 wk	-0.1
Nigella damascena	3–5	90–95	2–3 d	—
Orchid				
Cattleya[§,**]	7–10	90–95	2 wk	-0.3
Cymbidium	0–4	90–95	2 wk	—
Vanda	13	90–95	5 d	—
Ornithogalum arabicum	4	90–95	4–6 wk	—
Ornithogalum	4	90–95	4–6 wk	—
Paeonia hybrids	0–2	75–80	4 wk	—
Peony, tight bud	0–1	90–95	2–6 wk	-1.1
Phlox paniculata	3	90–95	1–3 d	—
Phlox spp.	4	90–95	1–3 d	—
Physalis alkekengi	3–5	90–95	1 wk	—
Physostegia virginiana	4	90–95	1–2 wk	—
Poinsettia	10–15	90–95	4–7 d	-1.1
Polianthes tuberosa	0–5	90–95	5–10 d	—
Poppy	4	90–95	3–5 d	—
Primrose	4	90–95	1–2 d	—
Protea	0–1	90–95	7–10 d	—
Ranunculus asiaticus	1–2	90–95	1 wk	—
Ranunculus	0–5	90–95	7–10 d	-1.7
Rose, dry pack[††]	0–1	90–95	2 wk	-0.5
Rose[††]	0–1	90–95	4–5 d	-0.5
Salvia leucantha	2–4	90–95	3–4 d	—
Scabiosa caucasica	3–5	90–95	1 wk	—
Snapdragon	0–1	90–95	1–2 wk	-0.9
Snowdrop	4	90–95	2–4 d	—
Solidago hybrids	3–5	90–95	5 d	—
Squill	0–0.5	90–95	2 wk	—
Statice[††]	0–1	90–95	3–4 wk	—
Stephanotis[**]	4	90–95	1 wk	—
Stevia	4	90–95	3 d	—
Stock	0–1	90–95	3–5 d	-0.4
Strawflower[††]	2–4	90–95	3–4 wk	—
Sunflower	0–1	90–95	2 wk	—
Sweet pea	-0.5–0	90–95	2 wk	-0.9
Sweet-William	7	90–95	3–4 d	—
Thalictrum spp.	3–5	90–95	1 wk	—
Trachelium caeruleum	4	90–95	1–2 d	—
Triteleia spp.	3–5	90–95	4 d	—
Tuberose	0	90–95	—	—
Tulip (bulbs attached)[††]	0–1	85	2–7 wk	—
Veronica spp.	3–5	90–95	1 wk	—
Violet	1–5	90–95	3–7 d	-1.8
Waxflower	0–1	90–95	2 wk	—
Zantedeschia aethiopica	3	90–95	1 wk	—

Commodity	Storage Temperature (°C)	Relative Humidity (%)	Approximate Storage Period[†]	Highest Freezing Point[‡] (°C)
Zinnia elegans	2–3	90–95	5 d	—
Zinnia	4	90–95	5–7	—
Florist Greens and Decorative Foliage				
Adiantum (Maidenhair)	0–4	90–95	—	—
Asparagus (Plumosa)[‡‡]	2–4	90–95	2–3 wk	-3.3
Asparagus (Sprengeri)[‡‡]	2–4	90–95	2–3 wk	—
Buxus (boxwood)	2–4	90–95	—	—
Camellia	4	90–95	—	—
Cedar[‡‡]	0	90–95	—	—
Chamaedorea	12.5	90–95	1–2 wk	—
Cordyline	7–10	90–95	2–3 wk	—
Croton	2–4	90–95	—	—
Dieffenbachia[‡‡]	13	90–95	—	—
Dracaena	2–4	90–95	—	-1.6
Dagger & wood ferns	0	90–95	2–3 mo	-1.7
Eucalyptus	0–1	90–95	1–3 wk	-1.8
Fir	0–1	90–95	—	—
Galax[‡‡]	0	90–95	—	—
Hedera	2–4	90–95	2–3 wk	-1.2
Holly[††,‡‡]	0–1	90–95	3–5 wk	-2.8
Juniper	0	90–95	1–2 mo	—
Leatherleaf (Baker fern)	1–6	90–95	1–2 mo	—
Leucothoe, drooping	2–4	90–95	—	—
Magnolia	2–4	90–95	2–4 wk	-2.8
Mistletoe[††]	0	90–95	3–4 wk	-3.9
Mountain laurel	0	90–95	2–4 wk	-2.5
Myrtus (myrtle)	2–4	90–95	—	—
Palm	7	90–95	—	—
Philodendron	2–4	90–95	—	—
Pine	0–1	90–95	—	—
Pittosporum	2–4	90–95	2–3 wk	—
Podocarpus	7	90–95	—	-2.3
Pothos	2–4	90–95	—	—
Rhododendron	0	90–95	2–4 wk	-2.5
Salal (Lemon leaf)[‡‡]	-0.5–0	90–95	2–3 wk	-2.9
Scotch-broom	4	90–95	2–3 wk	—
Smilax, southern[‡‡]	4	90–95	—	—
Spruce	0–1	90–95	—	—
Staghorn fern	13	90–95	—	—
Vaccinium (huckleberry)[‡‡]	0	90–95	1–4 wk	-3.0
Woodwardia fern	0–4	90–95	—	—
Bulbs, Corms, Rhizomes, Tubers and Roots[§§]				
Achimenes	10–15	Prevent drying	—	—
Acidanthera	20	Dry, vented	—	—
Alliums	20–23	Dry, vented	—	—
Allium giganteum	25–28	Dry, vented	—	—
Alstroemeria	1–3	Prevent drying	—	—
Alstroemeria belladonna	13–23	Prevent drying	—	—
Anemone blanda	9–17	Dry, vented	—	—
Anemone coronaria (summer)	15–25	Dry, vented	3–4 mo	—

Commodity	Storage Temperature (°C)	Relative Humidity (%)	Approximate Storage Period[†]	Highest Freezing Point[‡] (°C)
Anemone coronaria (winter)	10–13	Dry, vented	3–4 mo	—
Anemone fulgens	9–17	Dry, vented	—	—
Anigozanthos	2–20	Prevent drying	—	—
Begonia, tuberous	2–5	Prevent drying	3–5 mo	-0.5
Bletilla orchid	2–4	—	—	—
Brodiaea laxa	20–25	—	—	—
Caladium	23–25	Dry, vented	—	-1.3
Camassia	17–20	Prevent drying	—	—
Canna	5–10	Prevent drying	—	—
Chionodoxa	20	Prevent drying	—	—
Clivia	13	Prevent drying	—	—
Colchicum	17–23	Prevent drying	—	—
Convallaria	-2	Keep frozen-in	12 mo	—
Crocosmia	2–5	Prevent drying	—	—
Crocus	17–20	Dry, vented	2–3 mo	—
Cyclamen	9	Prevent drying	—	—
Cypella herbertii	4–10	—	—	—
Dahlia	5–10	Prevent drying	5 mo	-1.8
Endymion	20	Prevent drying	—	—
Eranthis	5	Prevent drying	—	—
Eremurus	5–7	Prevent drying	—	—
Erythronium	5–9	Prevent drying	—	—
Eucharis	20	Prevent drying	—	—
Eucomis	13–20	Dry, vented	—	—
Freesia	30	Dry, vented	3–4 mo	—
Fritillaria imperialis	23–25	Prevent drying	—	—
Fritillaria meleagris	9–13	Prevent drying	—	—
Galanthus	17	Prevent drying	—	—
Galtonia	17–20	Prevent drying	—	—
Gladiolus[††]	2–10	Dry, vented	5–8 mo	-2.1
Gloriosa	10–18	Prevent drying	3–4 mo	—
Gloxinia	5–9	Prevent drying	5–7 mo	-0.8
Haemanthus	10–15	Dry, vented	—	—
Hemerocallis	7–10	Prevent drying	1 mo	—
Hippeastrum	2–13	Prevent drying	5 mo	-0.6
Hyacinthus[††]	17–20	Dry, vented	2–5 mo	-1.5
Hymenocallis	7–10	Prevent drying	—	—
Iris, Dutch hybrids[††]	17	Dry, vented	4–12 mo	—
Iris, English hybrids	0–5	Prevent drying	—	—
Iris, German hybrids	20–23	Dry, vented	—	—
Ixia	20–25	Dry (65–70%), vented	—	—
Ixiolirion	20	Dry, vented	—	—
Liatris	-2–2	Prevent drying	—	—
Lilium longiflorum[††]	2–7	Prevent drying	1–10 mo	-1.7
Lilium hybrids and species	-2–2	Prevent drying	—	—
Lycoris	13–17	Dry, vented	—	—
Montbretia	2–5	Prevent drying	—	—
Muscari	20	Dry, vented	2–4 mo	—
Narcissus (hardy cvs.)	17	Dry, vented	2–4 mo	-1.3
Narcissus (Paperwhites)	2–30	Dry, vented	2–4 mo	—

Commodity	Storage Temperature (°C)	Relative Humidity (%)	Approximate Storage Period[†]	Highest Freezing Point[‡] (°C)
Nerine	5–9	Prevent drying	—	—
Ornithogalum dubium	9–30	Dry, vented	—	—
Ornithogalum nutans	20	Prevent drying	—	—
Ornithogalum thyrsoides	23–25	Prevent drying	—	—
Ornithogalum umbellatum	20	Prevent drying	—	—
Oxalis adenophylla	17–20	Prevent drying	—	—
Oxalis deppei	2–5	Dry, vented	—	—
Peony	0–2		5 mo	—
Persica meleagris	2–5	Prevent drying	—	—
Polianthes	20	Dry, vented	—	—
Primula	7–10	Dry, vented	—	—
Puschkinia	20–23	Dry, vented	—	—
Ranunculus (summer)	17–20	Dry, vented	—	—
Ranunculus (winter)	10–13	Dry, vented	—	—
Scadoxus	20–23	Dry, vented	—	—
Scilla siberica	20–23	Dry, vented	—	—
Sparaxis	25	Dry, vented	—	—
Tigridia	2–5	In closed boxes	—	—
Trillium	0–2	—	—	—
Triteleia laxa	17–20	Dry, vented	—	—
Tulipa[††]	17	Dry, vented	2–6 mo	-2.4
Watsonia	4–7	—	—	—
Zantedeschia	7–10	Dry, vented	—	-2.5
Zephyranthes	17–20	Dry, vented	—	—

*Data from *USDA Handbook* 66 in which leading authorities have provided an overview of the literature and storage recommendations for individual crops (for additional information, see Handbook 66) and Armitage, A.M. and J.M. Laushman. 2003. *Speciality Cut Flowers.* Timber Press, Portland, OR.

[†]Approximate storage period = time interval that should give adequate life expectancy after removal from storage.

[‡]Data from Whiteman, T.M. 1957. *USDA Mkt. Res. Rpt.* 196.

[§]Stems of orchids and some antheriums should be placed in water vials.

**Not placed in water for handling and storage.

[††]See *USDA Hbk.* 66 for details.

[‡‡]Usually held in moisture-retentive shipping cases.

[§§]Typically relative humidity conditions in the 70–90% range.

APPENDIX VIII.
SELECTED BOOKS ON THE POSTHARVEST BIOLOGY AND TECHNOLOGY OF PERISHABLE PLANT PRODUCTS*

Alzamora, S.M., M.S. Tapia and A. López-Malo. 2000. *Minimally Processed Fruits and Vegetables: Fundamental Aspects and Applications.* Aspen Publishers, Gaithersburg, MD.

Anderson, J.A., and A.W. Alcock. 1954. *Storage of Cereal Grains and Their Products.* American Association of Cereal Chemists, St. Paul, MN.

Anonymous 1793. *Plan d'établissement public qui maintiendra toujours le prix du bled dans la balance de 20 à 26 liv. le setier, c'est-à-dire, le pain à 8, 9 et 10 sols au plus les 4 liv. dans toute la République, sans mettre aucune imposition ni sols additionnels : présenté par deux citoyens de Paris, section de la Réunion.* Impr. de Roblot, Paris.

Anonymous. 1989. *Guide to Food Transport: Fruits and Vegetables.* Mercantila Press, Copenhagen, Denmark.

Appert, J. (translated by P. Skinner). 1987. *The Storage of Food Grains and Seeds.* [*Le stockage des produits viumes et semenciers.*]. Macmillan, Basingstoke, England.

Arthey, V.D. 1972. *Quality of Horticultural Products.* Halstead Press, New York.

Ashimogo, G. 1995. *Peasant Grain Storage and Marketing in Tanzania: A Case Study of Maize in Sumbawanga District.* Köster, Berlin.

Bainer, H.M. 1910–1920. *Storage of Fruits and Vegetables.* Industrial Dept., Atchison, Topeka & Santa Fe Railway Chicago, IL.

Bala, B.K. 1997. *Drying and Storage of Cereal Grains.* Science Pub., Enfield, NH.

Barkai-Golan, R. 2001. *Postharvest Diseases of Fruits and Vegetables: Development and Control.* Elsevier, Amsterdam.

Bartz, J.A., and J.K. Brecht (eds.). 2003. *Postharvest Physiology and Pathology of Vegetables.* Dekker, New York.

Bellingham, C. 1604. *The Fruiterers Secrets.* R. Bradock, London.

Bishop, C., R. Pringle and A.K. Thompson. 2002. *Potatoes Postharvest.* CAB Intern., Wallingford, England.

Bishop, C.F.H., and W.F. Maunder. 1980. *Potato Mechanisation and Storage.* Farming Press, Ipswich, England.

Blokhin, P.V. 1981. *Aerozheloba dlia Transportirovaniia Zerna.* Kolos, Moskva.

Bondoux, P. 1994. *Enfermedades de Conservación de Frutos de Pepita, Manzanas, y Peras.* Ediciones Mundi-Prensa, Madrid.

---

*Not included are university/government publications and symposium proceedings. When multiple editions have been published, the most recent is cited.

Boumans, G. 1985. *Grain Handling and Storage.* Elsevier, Amsterdam.

Brassard, M., and J. Richardson. 1989. *Worldwide Selection of Exotic Produce: Selection, Preparation, Storage.* Héritage, Saint-Lambert, Québec.

Brooker, D.B., F.W. Bakker-Arkema and C.W. Hall. 1992. *Drying and Storage of Grains and Oilseeds.* Van Nostrand Reinhold, New York.

Bubel, M., and N. Bubel. 1979. *Root Cellaring: The Simple No-processing Way to Store Fruits and Vegetables.* Rodale Press, Emmaus, PA.

Buchanan, B.B., W. Gruissem and R.L. Jones. 2000. *Biochemistry & Molecular Biology of Plants.* American Society of Plant Physiologists, Rockville, MD.

Burton, W.G. 1967. *The Potato: A Survey of Its History and of Factors Influencing Its Yield, Nutritive Value, Quality and Storage.* H. Veenman & Zonen, Wageningen, The Netherlands.

Burton, W.G. 1982. *Postharvest Physiology of Food Crops.* Longman, Essex, England.

Calderon, M., and R. Barkai-Golan. 1990. *Food Preservation by Modified Atmospheres.* CRC Press, Boca Raton, FL.

Chekravarty, A., A.S. Mujumdar, G.S. Vijaya Raghavan and H.S. Ramaswamy. 2003. *Handbook of Postharvest Technology: Cereals, Fruits, Vegetables, Tea, and Spices.* Dekker, New York.

Chioffi, N., G. Mead and L.M. Thompson. 1991. *Keeping the Harvest: Preserving Your Fruits, Vegetables & Herbs.* Storey Communications, Pownal, VT.

Christensen, C.M., and H.H. Kaufmann. 1969. *Grain Storage: The Role of Fungi in Quality Loss.* University of Minnesota Press, Minneapolis, MN.

Christensen, C.M., and R.A. Meronuck. 1986. *Quality Maintenance in Stored Grains and Seeds.* University of Minnesota Press, Minneapolis, MN.

Christensen, C.M. 1982. *Storage of Cereal Grains and Their Products.* American Association of Cereal Chemists, St. Paul, MN.

Cotton, R.T. 1950. *Insect Pests of Stored Grain and Grain Products: Identification, Habits, and Methods of Control.* Burgess Pub., Minneapolis, MN.

Cotton, R.T. 1963. *Pests of Stored Grain and Grain Products.* Burgess Pub., Minneapolis, MN.

Credland, P.F., D.M. Armitage, C.H. Bell, P.M. Cogan and E. Highley. 2002. *Advances in Stored Product Protection.* CAB Intern., Cambridge, MA.

Dasgupta, M.K., and N.C. Mandal. 1989. *Postharvest Pathology of Perishables.* Oxford & IBH Pub. Co., New Delhi.

Debney, H.G. 1980. *Handling and Storage Practices for Fresh Fruit and Vegetables: Product Manual.* Australian United Fresh Fruit and Vegetable Association, South Yarra, Victoria, Australia.

Déchalotte, J.F. 1829. *Traité Sur les Substances et Projet d'un Approvisionnement de Réserve en Grains Pour Toute la France, Sans Qu'il en Coute Rien au Trésor.* Huzard, Paris.

Dejean, J.F.A. (Ste-Fare Bontemps, Chevalier). 1824. *Economie Publique Résumé de Toutes les Expériences Faites pour Constater la Bonté du Procédé Proposé par M. le Comte Dejean, Pour la Conservation Illimitée des Grains et Farines.* Bachelier, Paris.

Dennis, C. 1983. *Post-harvest Pathology of Fruits and Vegetables.* Academic Press, New York.

Dinglinger, G.F. 1768. *Die beste Art, Korn-Magazine und Frucht-Böden anzulegen auf welchen das Getrayde niemahls, weder vom weissen noch schwarzen Wurm, angestecket werden kan: eine preis-schrift.* Im verlag der Richterschen buchhandlung, Hannover, Germany.

Donskova, S.V. 1981. *Ekonomika Khraneniia i Pererabotki Zerna/Pod Redaktsiei.* Kolos, Moskva.

Dris, R., and R. Niskanen. 2001. *Crop Management and Postharvest Handling of Horticultural Products.* Science Pub., Enfield, NH.

Farm Electric Centre. 1975. *Vegetable Storage : A Guide to the Practical Design of Installations.* National Agricultural Centre, Kenilworth, England.

Farm Electric Centre. 1994. *Energy Efficient Cooling & Storage for Fruit and Vegetables.* Stoneleigh, England.

Faust, B.C. 1824. *Kornvereine, Kornhäuser, Kornpapiere in jeder ansehnlichen Stadt des Deutschen Vaterlandes ein Schreiben an den Herrn Baumeister Geinitz zu Altenburg.* Bückeburg, Hanover, Germany.

Fidler, J.C., B.G. Wilkinson, K.L. Edney and R.O. Sharples. 1973. *The Biology of Apple and Pear Storage.* Commonwealth Agri. Bur., East Malling, England.

Filippov, A.N. 1972. *Zadachnik po Kursu Organizatsii i Planirovaniia Proizvodstva na Predpriiatiiakh Khraneniia i Pererabotki Zerna. [Book of Problems for a Course on Organization and Planning Grain Storage and Processing Plants].* Kolos, Moskva.

Franz, F.C. 1805. *Staatswirthschaftliche Abhandlung über ältere und neuere Magazin- und Versorgungsanstalten in ökonomisch-physikalischer und historisch-politischer Hinsicht nach dem gegenwärtigen Zeitbedürfnis.* Bey Gottfried Adolph Grau, Hof, Germany.

Friend, J., and M.J.C. Rhodes (eds.). 1981. *Recent Advances in the Biochemistry of Fruits and Vegetables.* Academic Press, New York.

Gall, L. 1825. *Was könnte helfen? Immerwährende Getraidelagerung, um jeder Noth des Mangels und des Ueberflusses auf immer zu begegnen, und Credit-Scheine, durch die Getraidevorräthe verbürgt, um der Alleinherrschaft des Geldes ein Ende zu machen.* Bei F.A. Gall, Trier, Germany.

Gardner, B.L. 1979. *Optimal Stockpiling of Grain.* Lexington Books, Lexington, MA.

Gast, R.H. 1929. *Western Vegetables a Handbook and Buying Guide for the Receiver of California and Arizona Vegetable Specialities.* Western Growers Protective Association, Los Angeles, CA.

Goodenough, F.W., and R.K. Atkins (eds.). 1981. *Quality in Stored and Processed Vegetables and Fruit.* Academic Press, London.

Golob, P., G. Farrell and J.E. Orchard (eds.). 2002. *Crop Post-Harvest: Science and Technology.* Blackwell, Oxford, England.

Haard, N.F., and D.K. Salunkhe. 1975. *Postharvest Biology and Handling of Fruits and Vegetables.* AVI, Westport, CT.

Hall, C.W. 1980. *Drying and Storage of Agricultural Crops.* AVI, Westport, CT.

Harris, K.L., and C.J. Lindblad. 1978. *Postharvest Grain Loss Assessment Methods: A Manual of methods for the Evaluation of Postharvest Losses.* American Association of Cereal Chemists, St. Paul, MN.

Harris, T.G., and J. Minor. 1979. *Grain Handling and Storage.* T.G. Harris and J. Minor, (s.l.).

He, S. 1987. *You Zhi Gan Ju Zai Pei, Chu Cang he Jia Gong.* [*Cultivation, Storage and Processing of Improved Citrus Fruits*]. Zhejiaing ke xue ji shu chu ban she: Zhejiang Sheng Xing hua shu dian fa xing, Hangzhou Shi, China.

Herrero Alvaro, A. 1982. *Enfermedades y Fisiopatias de Peras y Manzanas en Conservacion Frigorifica.* [*Diseases and Physiology of Peas and Apples in Refrigeration*]. Dilagro, Lerida, Spain.

Hills, L.D., and E.H. Heywood. 1946. *Rapid Tomato Ripening.* Fabr and Faber, London.

Hoogland, M., P. Holen and B. Oranje. 2000. *Granaries.* Agromisa, Wageningen, The Netherlands.

Hulme, A.C. (ed.). 1971. *The Biochemistry of Fruits and Their Products.* 2 vol. Academic Press, New York.

Huls, M.E. 1986. *Grain Storage Buildings: A Bibliography.* Vance Bibliographies, Monticello, IL.

Hultin, H.O., and M. Milner. 1978. *Postharvest Biology and Biotechnology.* Food Nutrition Press, Westport, CT.

James, W.O. 1953. *Plant Respiration.* Clarendon Press, Oxford, England.

Johnson, T. 1696. *A General Proposal for the Building of Granaries.* (n.s.), London.

Jones, D.B. 1943. *The Effect of Storage of Grains on Their Nutritive Value.* National Research Council, Washington, D.C.

Katerere, M., and D.P. Giga. 1990. *Grain Storage Losses in Zimbabwe.* Enda, Dakar, Senegal.

Kays, S.J. 1991. *Postharvest Physiology and Handling of Perishable Plant Products.* Van Nostrand Reinhold, New York.

Kays, S.J., and R.E. Paull. 2004. *Postharvest Biology.* Exon Press, Athens, GA.

Khusro, A.M. 1973. *Buffer Stocks and Storage of Foodgrains in India.* Tata McGraw-Hill, New Delhi.

Knee, M. (ed.) 2002. *Fruit Quality and Its Biological Basis.* CRC Press, Boca Raton, FL.

Kostychev, S. (translated by C.J. Lyon).1927. *Kostychev's Plant Respiration.* Blakiston's Son, Philadelphia, PA.

Lagerlöf, S. 1719. *Walmente tanckar om spannemåhls skiötzel, sparande, och uppläggande i kornhus och magaziner så att den i många åhr kan bewaras och spisning finnas i landet när krig, misswext, och dyr tijd infaller.* Tryckt hos Joh. L. Horrn, Stockholm, Sweden.

Lasteyrie, C. de. 1825. *Des Fosses a Conserver les Grains et de la Manière de les Construire.* (n.s., F. Didot), Paris.

Lawton, R., D. Goddard and N. Marshall. 2001. *Controlled Atmosphere Storage and the Transport of Fruit and Vegetables.* Cambridge Refrigeration Technology, Cambridge.

Lenehan, J.J. 1986. *Grain Drying and Storage: Principles of Drying and Storing Combinable Crops.* An Faras Talúntais, Dublin.

Lieberman, M. (ed.). 1983. *Postharvest Physiology and Crop Protection.* Plenum Press, New York.

Loewer, O.J., T.C. Bridges and R.A. Bucklin. 1994. *On-farm Drying and Storage Systems.* American Society of Agricultural Engineers, St. Joseph, MI.

Maarse, H. (ed.). 1991. *Volatile Compounds in Foods and Beverages.* Dekker, New York.

Marble, L.M. 1921. *Specialized Storage of Fruits and Vegetables: A Preliminary Statement Regarding the Life Conditions of Fruits and Vegetables in Storage, Together With a Statement Regarding the Marble Laboratory, Inc.* Marble Laboratory Inc., Canton, PA.

Marolles, 1830. *Des Moyens de Procurer des Secours à la Classe Indigente dans les Années de Disette.* Imprimerie de F. Didot Frères, Paris.

McLean, K.A. 1989. *Drying and Storing Combinable Crops.* Farming Press, Ipswich, England.

Metlitskii, L.V., E.G. Sal'Kova, N.L. Volkind, V.I. Bondarev and V.Ya. Yanyuk (translated by A.K. Dhote). 1983. *Controlled Atmosphere Storage of Fruits.* [*Khranenie Plodov v Reguliruemoi Gazovoi Srede*]. Amerind, New Delhi.

Midwest Plan Service. 1988. *Grain Drying, Handling, and Storage Handbook.* Midwest Plan Service, Ames, IA.

Miller, J.D., and H.L. Trenholm. 1994. *Mycotoxins in Grain: Compounds Other Than Aflatoxin.* Eagan Press, St. Paul, MN.

Misra, R.P. 1988. *Community Storage of Foodgrains: An O.R. Approach.* Harman Pub., New Delhi.

Mitra, S.K. 1997. *Postharvest Physiology and Storage of Tropical and Subtropical Fruits.* CAB Intern., Wallingford, England.

Mohsenin, N.N. 1986. *Physical Properties of Plant and Animal Materials.* Gordon and Breach Scientific, New York.

Mörner, A. 1810. *Undersökning om Oväldigheten af de få kallade Oväldiga Tankar angående Herr Stats-Sekreteraren Järtas åtgard I frågan om Riksdagsmannen Oxelbergs Memorial rörande Spannemåls-Magazins-Directionen.* Tryckt hos Olof Grahn, Stockholm, Sweden.

Multon, J.L. (translated by D. Marsh). 1988. *Preservation and Storage of Grains, Seeds and Their By-products: Cereals, Oilseeds, Pulses, and Animal Feed.* [*Conservation et Stockage des Grains et Graines et Produits Dérivés*]. Lavoisier Pub., New York.

Munton & Fison, Ltd. 1959. *Insect Pests in Stored Cereals.* Stowmarket, England.

Nosan Gyoson Bunka Kyokai. 1976. *Kyabetsu, Hakusai, Horenso ta: Kiso Seiri to Oyo Gijutsu.* [*Cabbage, Chinese Cabbage, spinach and Others*]. 1976. Nosan Gyoson Bunka Kyokai, Tokyo.

Osterloh, A. 1996. *Lagerung von Obst und Südfrüchten.* [*Storage of Fruits and Tropical Fruits*]. E. Ulmer, Stuttgart, Germany.

Oxley, T.A. 1948. *The Scientific Principles of Grain Storage.* Northern Pub., Liverpool, England.

Pantastico, E.B. 1979. *Postharvest Losses of Fruits and Vegetables in Developing Countries – An Action Program.* Postharvest Teaching and Research Center, Los Banos, Philippines.

Pantastico, E.B. 1979. *Fisiología de la Postrecolección, Manejo y Utilización de Frutas y Hortalizas Tropicales y Subtropicales.* Compañia Edit., Continental, Mexico.

Paterson, H. 1974. *Farm Based Chilled Vegetable Storage.* Electricity Council, London.

Peleg, K. 1985. *Produce Handling, Packaging and Distribution.* AVI, Westport, CT.

Pfost, D.L., and D.G. Anderson. 1978. *Smallholder Grain Storage in Kenya: Problems and Proposed Solutions.* The Institute, Manhattan, KS.

Phelps, R.H. 1979. *Storage Diseases of Fresh Fruit and Vegetables.* CARDI, St. Augustine, Trinidad.

Protsko, R.F., and V.B. Varshavskaia. 1980. *Prorastanie Korneplodov Sakharnoi Svekly i Problemy ee Khraneniia.* Nauk. Dumka, Kiev, Ukraine.

Queensland Chamber of Fruit and Vegetable Industries Co-operative. 1995. *Fruit & Vegetable Manual: An Informative and Practical Guide Including Information on the Selection, Handling, Storage and Use of a Wide Variety of Vegetables.* The Chamber, Brisbane, Queensland, Australia.

Reuning, T. 1847. *Ueber die Verhinderung des Mangels an Brodgetreide insbesondere durch öffentliche Magazinirung.* C.F. Winter, Heidelberg, Germany.

Reyaz, A. 2002. *Problem of Food Storage and Losses in India.* Mohit Publications, New Delhi.

Richardson, D.G., and M.Meheriuk (eds.). 1982. *Controlled Atmospheres for the Storage and Transport of Perishable Agricultural Commodities.* Timber Press, Beaverton, OR.

Ryall, A.L., and W.J. Lipton. 1982. *Handling, Transportation and Storage of Fruits and Vegetables.* 2 vols. AVI, Westport, CT.

Salunkhe, D.K., H.R. Bolin and N.R. Reddy. 1991. *Storage, Processing, and Nutritional Quality of Fruits and Vegetable.* CRC Press, Boca Raton, FL.

Salunkhe, D.K., and B.B. Desai. 1984. *Postharvest Biotechnology of Fruits.* 2 vols. CRC Press, Boca Raton, FL.

Salunkhe, D.K., and S.S. Kadam. 1998. *Handbook of Vegetable Science and Technology: Production, Composition, Storage, and Processing.* Dekker, New York.

Sandigliano, C. di, and V. Domenico. 1789. *Memoria Sul Modo de Agevolare il Movimento de Carri ed il Trasporto Delle Derrate. Giugno.* Presso Giammichele Briolo, Torino, Italy.

Sauer, D.B. 1992. *Storage of Cereal Grains and Their Products.* American Association of Cereal Chemists, St. Paul, MN.

Schlier, J.A. 1825. *Ueber unterirdische Getraid-Magazine verbunden mit Assecuranz-und Credit-Anstalten, oder, Wie kann der verderblichen Wohlfeilheit und der drünkenden Theurung der verschiedenen Producte und Lebensmittel, zugleich auch dem verderblichen Mangel an Geld und Credit für jetzt und allezeit am sichersten abgeholfen werden?* J. Dorbath, Bavaria, Germany.

Schmidt, P.V. 1990. *Zuckerrüben-Lagerung und Aufbereitung.* [*Storage and Processing of Sugarbeets*]. A. Bartens, Berlin.

Schultes, H.W. 1803. *Ideen über Getraide-Magazine nach ökonomisch-statistischen Ansichten samt Prüfung der dabey gemeiniglich angenommener Grundsätze, und Vorschlägen, wie durch andere damit zugleich un Verbindung zu stellende Mittel zur Verkaufs-Concurrenz, Theurung und Hungers-Noth am sichersten entfernt werden können.* C.E. Gabler, Leipzig, Germany.

Seelig, R.A., and M. Neale. 1979. *Buying, Handling and Using Fresh Vegetables.* National Restaurant Association, Chicago, IL.

Selwig, J.D. 1801. *Ideen und Erfahrungen über freien Kornhandel und Getreidemagazine zur Beherzigung und Prüfung mitgetheilt.* Gedrückt und im Verlage bey K. Reichard, Braunschweig, Germany.

Seymour, G.B., J.E. Taylor and G.A. Tucker (eds.). 1993. *Biochemistry of Fruit Ripening.* Chapman & Hall, London.

Shaw, J. 2000. *Winter Storage of Vegetables.* Lewiston-North Cache Valley Historical Board, Lewiston, UT.

Shewfelt, R.L., and S.E. Prussia (eds.). 1993. *Postharvest Handling: A Systems Approach.* Academic Press, New York.

Shirokov, E. (translated by L. Markin). 1968. *Practical Course in Storage and Processing of Fruit and Vegetables.* [*Praktikum po Khraneniyu i Pererabotke Plodov i Ovoshchei*]. Scientific Translations, Jerusalem, Israel.

Shirokov, E.P., and N.B. Saburov. 1961. *Khranenie Kapusty.* [*Storing Cabbage*]. Moskovskii Rabochii, Moskva.

Sinha, R.N., and W.E. Muir. 1973. *Grain Storage: Part of a System.* AVI, Westport, CT.

Smith, C.V., and M.C. Gough.1990. *Meteorology and Grain Storage.* World Meteorological Organization, Geneva, Switzerland

Smith, E. 1914. *Farm Storages for Fruits & Vegetables.* W.H. Cullen, London.

Snowdon, A.L. 1990. *A Color Atlas of Post-Harvest Diseases & Disorders of Fruits and Vegetables.* 2 vols. Wolfe Scientific, London.

Snowdon, A.L., and A.H.M. Ahmed. 1981. *The Storage and Transport of Fresh Fruit and Vegetables.* National Institute of Fresh Produce, London.

Steffen, S.L. 1973. *Grain Ecology: A Grain Care Manual for Growers, for Marketers, for Processors.* Reiman Associates, Milwaukee, WI.

Steffen, S.L. 1975. *Grain Ecology: Understanding Corn.* Harvestall Industries, New Hampton, IA.

Story, A. 1990. *Fresh Produce Manual: Handling & Storage Practices for Fresh Produce.* Australian United Fresh Fruit & Vegetable Association, Footscray, Victoria, Australia.

Tettinek, J. 1848. *Zur Zeit der Noth ein wohlfeiles Brod Vorschläge zur Herbeischaffung von Getreide-Vorräthen auf Rechnung und zum Nutzen der Stadtgemeinde.* Gedruckt bei F.X. Duyle, Salzburg, Austria.

Thompson, A.K. 1996. *Postharvest Technology of Fruits and Vegetables.* Blackwell Scientific, Oxford, England.

Thompson, A.K. 1998. *Controlled Atmosphere Storage of Fruits and Vegetables.* CAB Intern., Wallingford, England.

Thompson, A.K. 2003. *Fruit and Vegetables: Harvesting, Handling, and Storage.* Blackwell, Oxford, England.

Tilton, E.W., and S.O. Nelson. 1984. *Irradiation of Grain and Grain Products for Insect Control.* Council for Agricultural Science and Technology, Ames, IA.

Todd, M.L. 1980. *Psychrometrics Applied to Grain Processing: An Updated Analysis and Applications.* American Society of Agricultural Engineers, St. Joseph, MI.

Tomás-Barberán, F.A., and R.J. Robins (eds.). 1997. *Phytochemistry of Fruit and Vegetables.* Oxford Science, Oxford, England.

Vaughan, M.J. 1998. *The Complete Book of Cut Flower Care.* Timber Press, Portland, OR.

Verma, L.R., and V.K. Joshi. 2000. *Postharvest Technology of Fruits and Vegetables: Handling, Processing, Fermentation, and Waste Management.* Indus Pub., New Delhi.

Volkind, I.L., N.N. Roslov and P.A. Mukhanov. 1983. *Modern Potato and Vegetable Storages.* [*Sovremennye Kartofele-i Ovoshchekhranilishcha*]. Amerind, New Delhi.

Weichmann, J. (ed.). 1987. *Postharvest Physiology of Vegetables.* Dekker, New York.

Weinreich, K. 1843. *Die Getreidsperren und Landes-Magazine auch als eine Veranlassung der Theuerung gedacht, nach der Geschichte und National-Ökonomie erwogen.* J. Thomann, Landshut, Germany.

Wiley, R.C. 1994. *Minimally Processed Refrigerated Fruits & Vegetables.* Chapman & Hall, New York.

Wilkin, D.R. 1990. *Integrated Pest Control Strategy for Stored Grain.* Home-Grown Cereals Authority, London.

Wills, R.B.H., D. Graham, D. Joyce and W.B. McGlasson.1998. *Postharvest: An Introduction to the Physiology and Handling of Fruit, Vegetables and Ornamentals.* Oxford Univ. Press, Oxford, England.

Wright, J. 1796. *Observations Upon the Important Object of Preserving Wheat and Other Grain from Vermin, with a Safe and Efficacious Method to Prevent the Great Depredations that are Made on Those Valuable Articles Containing also, Some Cursory Remarks on the National Advantage Arising From Our Export Corn Laws, and Proving that the Same Happy Consequences Must be Derived from the Inclosure of our Forests, Common Fields, and Waste Lands: With a Calculation Annexed, Demonstrating the Great Savings Which May Accrue From the Use of the Patent Artificial Slate-frames, on Hay Stacks.* Cooper and Graham, London.

Yu, S. 1983. *Pin Guo He li di Chan di Zhu Cang.* [*Storage of Apple and Pears in Their Producing Area*]. Nong ye chu ban she: Xin hua shu dian Beijing fa xing suo fa xing, Beijing.

Zakladnoi, G.A., and V.F. Ratanova (translated by Indira Nair). 1987. *Stored-Grain Pests and Their Control.* [*Vrediteli Khlebnykh Zapasov i Mery Bor'by s Nimi*]. A.A. Balkema, Rotterdam, The Netherlands.

UNITS, SYMBOLS AND ANATOMICAL ABBREVIATIONS

°	degrees
°C	degrees Centigrade
°F	degrees Fahrenheit
A	adenine (nucleotide base)
ABA	abscisic acid
ACC	1-aminocyclopropane-1-carboxylic acid
ACO	1-aminocyclopropane-1-carboxylic acid oxidase
ADP	adenosine diphosphate
Ag^{2+}	silver ion
ATP	adenosine triphosphate
bp	boiling point
bp	base pairs
Bt	*Bacillus thuringiensis*
BTU	British thermal unit
c	specific heat capacity (chapter 9)
c	conductance
C	crystal (anatomical symbol, chapter 2)
C	cytosine (nucleotide base)
$C_6H_{12}O_6$	hexose sugar
$CaCl_2$	calcium chloride
cal	calorie
C_c	cooling coefficient
C_D	conduction
cDNA	complementary DNA
Ch	chromosome (s)
$-CH_3$	methyl group
-CHO	aldehyde group
C_{hr}	change in temperature per unit change in cooling time ($°C \cdot hr^{-1}$)
CIC	character impact compound
$cm \cdot hr^{-1}$	centimeters per hour
$cm \cdot sec^{-1}$	centimeters per second
$cm^2 \cdot cm^{-3}$	centimeters square per centimeter cubed
CN	hydrogen cyanide
CO	carbon monoxide
CO_2	carbon dioxide
-COOH	carboxyl group
Cs	cristae
C_V	convection
cv	cultivar
CW	cell wall
CW1	primary cell wall
CW2	secondary cell wall
Da	Daltons
DGDG	digalactosyl diglyceride
DNA	deoxyribonucleic acid
E	epidermal cell
ER	endoplasmic reticulum
ETS	electron transport system
FAD	flavin adenine dinucleotide
$FADH_2$	reduced flavin adenine dinucleotide
FAO	Food and Agriculture Organization of the United Nations
FMN	flavin mononucleotide
$FMNH_2$	reduced flavin mononucleotide
fpm	feet per minute
Fru	fructose
ft	feet
G (Gr)	granum (anatomical symbol, chapter 2)
G	guanine (nucleotide base)
g	gram
$g \cdot cm^{-2} \cdot s^{-1}$	grams per centimeter square per second
$g \cdot kg^{-1}$	grams per kilogram
GA	gibberellic acid
GB	golgi body
Glu	glucose
GM	genetically modified
H_2O	water
HCP	half-cooling period
HCT	half-cooling time

539

Hg	mercury
hv	a quantum of light energy
IAA	indoleacetic acid
IU	international units
J	Joule
K	Kelvin
k	thermal conductivity
kbp	thousand (kilo) base pairs
kDa	kiloDaltons
kg	kilogram
$kg \cdot kg^{-1}$	kilograms per kilogram
$km \cdot hr^{-1}$	kilometers per hour
kJ	kiloJoule
$kJ \cdot g^{-1}$	kiloJoules per gram
$kJ \cdot s^{-1}$	kiloJoules per second
$km \cdot hr^{-1}$	kilometers per hour
kPa	kiloPascal
kW	kiloWatts
L	liter
LB	lipid body
lb	pound
L_C	latent heat of condensation
L_E	latent heat of vaporization
L_F	latent heat of fusion
L_M	latent heat of melting
lux	luminous flux density
M	energy released by exothermic reactions within the plant
M	mitochondria (anatomical symbol, chapter 2)
m	mass
mbar	millibar
MCP	1-methylcyclopropene
Me	membrane
$mg \cdot kg^{-1} \cdot hr^{-1}$	milligrams per kilogram per hour
$mg \cdot m^{-3}$	milligrams per cubic meter
MGDG	monogalactosyl diglyceride
min	minute
Ml	middle lamella
mL	milliliter
mm	millimeter
mm Hg	millimeters of mercury
MPa	megapascal
mRNA	messenger RNA
Mt	microtubule
mw	molecular weight
N	nucleus
N_3	azide
NAD	nicotinamide adenine dinucleotide
NADH	reduced nicotinamide adenine dinucleotide
NADP	nicotinamide adenine dinucleotide phosphate

NADPH	reduced nicotinamide adenine dinucleotide phosphate
NaOCl	sodium hypochlorite
Ne	nuclear envelope
nm	nanometer
nM	nanomoles
NM	nuclear membrane
Nu	nucleolus
O	osmiophilic globule
O_2	oxygen
-OH	hydroxyl group
P	plastid (anatomical symbol, chapter 2)
P	pressure
P_S	energy inputs due to photosynthesis
PC	phosphatidyl choline
Pc	precipitated compounds
PE	phosphatidyl ethanolamine
Pe	pistil
PE	plastid envelope
PG	phosphatidyl glycerol
PGA	phosphoglycerate
PI	phosphatidyl inositol
PM	plasma membrane
pO_2	partial pressure of oxygen
PP	protoplastid
psi	pounds per square inch
pTi	tumor inducing plasmid
Q	solar irradiation
r	resistance
R	ribosome (anatomical symbol, chapter 2)
R	thermal radiation
r^{bl}	boundry layer resistance
r^c	cytosol resistance
r^{cw}	cell wall resistance
r^{ias}	internal air space resistance
RNA	ribonucleic acid
r^o	organelle resistance
r^p	plasmalemma resistance
RQ	respiratory quotient
rRNA	ribosomal RNA
r^{sr}	surface resistance
r^{total}	total resistance
S	stroma (anatomical symbol, chapter 2)
S	heat
s	second
$s \cdot m^{-1}$	seconds per meter
SAM	s-adenosyl methionine
Sb	starch body
S_E	energy stored

Se	sepal	U	uracil (nucleotide base)
S_H	heat stored	UDP	uridine diphosphate
SL	sulpholipid	Us	upper surface
SO_2	sulfur dioxide	USDA	United States Department of Agriculture
Ssc	substomatal cavity		
St	stamen	UTP	uridine triphosphate
STS	silver thiosulfate	V	vacuole (anatomical symbol, chapter 2)
T	temperature		
T	thylakoid (anatomical symbol, chapter 2)	V	volume
		Vb	vascular bundle
T	thymine (nucleotide base)	Ve	vesicle
T_c	temperature of coolant	W	Watts
TCA	tricarboxylic acid cycle	$W \cdot hr^{-1}$	Watts per hour
T-DNA	transfer DNA	$W \cdot m^{-2}$	Watts per meter square
TF	transcription factor	$\mu g \cdot g^{-1}$	micrograms per gram
T_p	temperature of product	$\mu L \cdot kg^{-1} \cdot hr^{-1}$	microliters per kilogram per hour
tRNA	transfer RNA	$\mu L \cdot L^{-1}$	microliter per liter

PLANT SPECIES INDEX

---

*Italicizing of page numbers denotes a figure.

SUBJECT INDEX

---

*Italicizing of page numbers denotes a figure or table.